Topics in Current Genetics 17

Series Editor: *Stefan Hohmann*

Topics in Current Genetics

17

Series Editor: *Stefan Hohmann*

Andrés Aguilera · Rodney Rothstein (Eds.)

Molecular Genetics of Recombination

With 70 Figures, 30 in Color; and 8 Tables

Springer

Professor Dr. ANDRÉS AGUILERA
Centro Andaluz de Biologia Molecular
y Medicina Regenerativa CABIMER
CSIC-Universidad de Sevilla
Av. Américo Vespucio s/n
41092 Sevilla
Spain
e-mail: aguilo@us.es

Professor Dr. RODNEY ROTHSTEIN
Department of Genetics & Development
Columbia University Medical Center
701 West 168th Street, HHSC 1608
New York, NY 10032-2704
USA
e-mail: rothstein@cancercenter.columbia.edu

The cover illustration depicts pseudohyphal filaments of the ascomycete *Saccharomyces cerevisiae* that enable this organism to forage for nutrients. Pseudohyphal filaments were induced here in a wild-type haploid MATa Σ1278b strain by an unknown readily diffusible factor provided by growth in confrontation with an isogenic petite yeast strain in a sealed petri dish for two weeks and photographed at 100X magnification (provided by Xuewen Pan and Joseph Heitman).

ISSN 1610-2096
ISBN-13 978-3-540-71020-2 Springer Berlin Heidelberg New York

Library of Congress Control Number: 2007921306

Springer is a part of Springer Science+Business Media
springeronline.com

Typesetting: Camera ready by editors
Production: LE-TEX Jelonek, Schmidt & Vöckler GbR, Leipzig
Cover: WMXDesign GmbH, Heidelberg

Printed on acid-free paper 39/3100/YL - 5 4 3 2 1 0

Topics in Current Genetics publishes review articles of wide interest in volumes that center around specific topics in genetics, genomics as well as cell, molecular and developmental biology. Particular emphasis is placed on the comparison of several model organisms. Volume editors are invited by the series editor for specific topics, but further suggestions for volume topics are highly welcomed.

Each volume is edited by one or several acknowledged leaders in the field, who invite authors and ensure the highest standard of content and presentation. Only solicited manuscript will be considered. All contributions are peer-reviewed. All papers are published online prior to the print version. Individual DOIs (digital object identifiers) make each article fully citable from the moment of online publication.

Subscribers to the print version of *Topics in Current Genetics* receive free access to the online version. An online-only license is also available.

Editorial office:
Topics in Current Genetics
Series Editor: Stefan Hohmann
Cell and Molecular Biology
Göteborg University
Box 462
40530 Göteborg, Sweden
FAX: +46 31 7862525
E-mail: editor@topics-current-genetics.se

Preface

Genetic recombination is an important process involved in shaping the genetic make up of progeny. Initially, genetic recombination studies focused on its relevance to sexual reproduction, including conjugation in bacteria and meiosis in eukaryotes. Increasingly, it has become evident that recombination is a DNA repair pathway crucial during DNA replication in vegetatively growing cells. It plays a critical role in preserving the integrity of the genome by mediating the repair of DNA damage, which can occur during normal cellular metabolism as a result of oxidative stress, transcription, replication fork stalling or breakdown, or after the exposure to DNA damaging agents. Until recently, much of our knowledge on the mechanisms of genetic recombination has come from studies of prokaryotic and simple eukaryotic fungal systems. However, these studies have now been significantly extended to mammals, such that a comparative picture of the general factors and mechanisms of genetic recombination is beginning to emerge. Detailed genetic and biochemical studies have led to the isolation and characterization of many of the recombination-repair proteins in *E. coli* and *S. cerevisiae*, which in turn has led to the identification of homologues in human cells. The link between recombination defects and recombination proteins in a number of tumors as well as in human hereditary syndromes makes genetic recombination a cellular process of key importance not only in basic biology but also in biomedical studies.

Since the publication of B. Lowe's (1988) and R. Kucherlapati's and G. Smith's (1989) volumes on genetic recombination, there has been a tremendous increase in the amount of research and relevant observations on genetic recombination that have greatly enhanced and altered our view of this subject. Such knowledge and information have been considered in a number of excellent reviews over the years, but the field has lacked a monograph where many of these important discoveries and views are put together in the context of modern biology. It was our goal to remedy this void and assemble this volume. In it, we have attempted to cover as many essential topics about genetic recombination as possible with the double aim of providing introductory material for the non-specialist and at the same time include the most recent and updated material on our actual knowledge of this crucial cell biological process. What are the factors necessary for and the mechanism of genetic recombination, how does recombination interconnect with other essential cellular processes such as DNA replication or meiotic chromosome segregation, how is genetic recombination controlled during the cell cycle, what is its impact on genomic integrity, and how may it impact cell division and differentiation? It is these topics, which are among the most relevant in present biological research, that we have tried to cover in this book.

The book is organized with 17 chapters written by active, qualified researchers on each topic. The chapters cover genetic recombination from both a historical and a conceptual perspective. Each chapter contains an introduction followed by an in-depth and up-to-date analysis of the current state of knowledge on its subject. In addition, the chapters were conceived and written autonomously, so that they can be read independently of the rest, even though this results in some repetitiveness of the basic concepts.

The first two chapters of the book are devoted to general recombination in Gram- and Gram+ bacteria. The chapter by B. Michel and colleagues covers not only the genes and pathways of homologous recombination in *E. coli*, but also discusses the relationship between recombination and replication. The next chapter by J. Alonso and colleagues compares the pathway and genes in Gram+ bacteria and outlines important aspects of DNA transformation in this group. The next four chapters focus on the biochemistry of genetic recombination. M. Cox and W.-D. Heyer discuss strand exchange in prokaryotes and eukaryotes, respectively. H. Klein spotlights the role of DNA helicases in mitosis, meiosis, replication and repair in eukaryotes, while M. Whitby writes about the enzymes that cleave DNA cross-strand exchanges (Holliday junctions) discovered in prokaryotes including bacteriophage and bacteria as well as those that function in eukaryotes including budding and fission yeast and mammalian cells. Next, D. Branzei and M. Foiani describe how yeast cells co-ordinate DNA replication and genetic recombination when replication forks encounter DNA damage. A. Aguilera and colleagues outline the consequences of genetic recombination that occur between sister chromatids during replication in yeast cells. The next two chapters focus on yeast model systems that underlie the mating type switch that occurs in both fission yeast (B. Arcangioli and colleagues) and in budding yeast (J. Haber), which are directed recombination events that utilize mitotic recombination proteins. Haber's chapter also includes studies on directed recombination events using meganucleases to initiate a variety of events. These chapters are followed by two that explore the cell biology of genetic recombination in yeast (M. Lisby and R. Rothstein) and mammalian cells (R. Kanaar and colleagues) demonstrating the choreography of the DNA damage response in eukaryotic cells. The surveillance of DNA damage in mammalian cells and its relationship to cancer is covered in the chapter written by M. Jasin and colleagues, which focuses on the BRCA2 tumor suppressor. N. Hunter describes the current state of affairs for meiotic recombination from the vantage point of budding yeast including a discussion of the roles of many meiotic specific genes as well as those also involved in mitotic recombination. The next two chapters focus on the biochemistry of site-specific recombination in prokaryotes (I. Grainge and D. Sherratt) and V(D)J recombination in mammals (M. Gellert). Grainge and Sherratt present an overview of the many kinds of recombinases and their roles in promoting the reaction. Gellert's chapter highlights studies on the *in vitro* biochemistry of the RAG1/2 proteins. The final chapter of the book by T. Wilson describes NHEJ throughout the animal and plant kingdoms focusing on the conservation of the enzymes and processes involved.

We are aware that in this volume many relevant aspects of genetic recombination are not covered. These include the structural biology of recombination proteins and intermediates, the physical analysis of single-molecules, recombination in phages and other model organisms such as *Drospohila, Caenorhabditis*, or *Arabidopsis*, genetic control of genome instability, systems biological studies about genetic recombination, recombination between multiple tandem arrays (rDNA), the role of histone modification in directing DNA repair processes, gene targeting and gene therapy, etc. It was impossible to assemble all of the chapters that we would have liked to have in this volume. Therefore, we had to select what, in our own view, were the most relevant topics for this volume. We hope that most will agree that many of these topics indeed are those that have contributed much to our view of this subject and importantly to modern biology in the last two decades. Perhaps in the future, we will put together an additional volume containing many of the topics that we had to omit here.

Finally, we want to thank each and every author for his/her excellent contribution. We are also indebted to the anonymous reviewers who read and made important suggestions to improve all of the chapters. We also thank Stefan Hohmann for his constant encouragement during the editing process and the staff at Springer Verlag for their continuous help and support to make this book possible.

Seville/New York, January 2007

Andrés Aguilera and Rodney Rothstein

Table of contents

List of contributors

Agarwal, Sheba
Department of Cell Biology and Genetics, Erasmus MC, PO Box 1738, 3000 DR Rotterdam, The Netherlands

Aguilera, Andrés
Departamento de Biología Molecular, CABIMER, CSIC-Universidad de Sevilla, Av. Américo Vespucio s/n, 41092 Sevilla, Spain
aguilo@us.es

Alonso, Juan C.
Departmento de Biotecnología Microbiana, Centro Nacional de Biotecnología, CSIC, Campus Universidad Autónoma de Madrid, Campus Universidad Autónoma de Madrid, Darwin Street 3, Cantoblanco, 28049 Madrid, Spain
jcalonso@cnb.uam.es

Arcangioli, Benoit
Department of Genomes and Genetics, Institut Pasteur, 25-28 rue du Dr Roux, Paris cedex 75724. France
barcan@pasteur.fr

Ayora, Silvia
Departmento de Biotecnología Microbiana, Centro Nacional de Biotecnología, CSIC, Campus Universidad Autónoma de Madrid, Campus Universidad Autónoma de Madrid, Darwin Street 3, Cantoblanco, 28049 Madrid, Spain

Baharoglu, Zeynep
Centre de Génétique Moléculaire du CNRS, Bâtiment 26, 1 Avenue de la Terrasse, 91198 Gif-sur Yvette Cedex, France

Branzei, Dana
FIRC Institute of Molecular Oncology Foundation, Via Adamello 16, 20139 Milan, Italy, and Dipartimento di Scienze Biomolecolari e Biotecnologie, Università degli Studi di Milano, Via Celoria 26, 20133 Milan, Italy
dana.branzei@ifom-ieo-campus.it

Carrasco, Begoña
Departmento de Biotecnología Microbiana, Centro Nacional de Biotecnología, CSIC, Campus Universidad Autónoma de Madrid, Campus Universidad Autónoma de Madrid, Darwin Street 3, Cantoblanco, 28049 Madrid, Spain

Christ, Nicole
Developmental Biology Program, Memorial Sloan-Kettering Cancer Center, 1275 York Avenue, New York, NY 10021, USA

Cortés-Ledesma, Felipe
Departamento de Biología Molecular, CABIMER, CSIC-Universidad de Sevilla, Av. Américo Vespucio s/n, 41092 Sevilla, Spain

Cox, Michael M.
Department of Biochemistry, University of Wisconsin-Madison, 433 Babcock Drive, Madison, Wisconsin, USA
cox@biochem.wisc.edu

Essers, Jeroen
Department of Cell Biology and Genetics, Erasmus MC, PO Box 1738, 3000 DR Rotterdam, The Netherlands, and Department of Radiation Oncology, Erasmus MC, PO Box 1738, 3000 DR Rotterdam, The Netherlands
j.essers@erasmusmc.nl

Foiani, Marco
FIRC Institute of Molecular Oncology Foundation, Via Adamello 16, 20139 Milan, Italy, and Dipartimento di Scienze Biomolecolari e Biotecnologie, Università degli Studi di Milano, Via Celoria 26, 20133 Milan, Italy
marco.foiani@ifom-ieo-campus.it

Gellert, Martin
Laboratory of Molecular Biology, National Institute of Diabetes and Digestive and Kidney Diseases, National Institutes of Health, Building 5, Room 241, 9000 Rockville Pike, Bethesda, MD 20892-0540, USA
gellert@helix.nih.gov

Grainge, Ian
Department of Biochemistry, University of Oxford, South Parks Road, Oxford. OX1 3QU, United Kingdom

Haber, James E.
Rosenstiel Center and Department of Biology, Brandeis University, Waltham, MA 02454-9110 USA
haber@brandeis.edu

Heyer, Wolf-Dietrich
Section of Microbiology and Section of Molecular and Cellular Biology, Center for Genetics and Development, University of California, Davis, Davis, CA 95616-8665, USA
wdheyer@ucdavis.edu

Holmes, Allyson
Department of Genomes and Genetics, Institut Pasteur, 25-28 rue du Dr Roux, Paris cedex 75724. France

Hunter, Neil
Sections of Microbiology and Molecular & Cellular Biology, University of California Davis, Davis, CA 95616, USA
nhunter@ucdavis.edu

Jasin, Maria
Developmental Biology Program, Memorial Sloan-Kettering Cancer Center, 1275 York Avenue, New York, NY 10021, USA
m-jasin@ski.mskcc.org

Kanaar, Roland
Department of Cell Biology and Genetics, Erasmus MC, PO Box 1738, 3000 DR Rotterdam, The Netherlands, and Department of Radiation Oncology, Erasmus MC, PO Box 1738, 3000 DR Rotterdam, The Netherlands

Klein, Hannah L
Department of Biochemistry, New York University School of Medicine, 550 First Avenue, New York, NY 10016, USA
hannah.klein@med.nyu.edu

Lisby, Michael
Department of Genetics, Institute of Molecular Biology and Physiology, University of Copenhagen, Øster Farimagsgade 2A, DK-1353 Copenhagen K, Denmark
mlisby@my.molbio.ku.dk

Lestini, Roxane
Centre de Génétique Moléculaire du CNRS, Bâtiment 26, 1 Avenue de la Terrasse, 91198 Gif-sur Yvette Cedex, France

Michel, Bénédicte
Centre de Génétique Moléculaire du CNRS, Bâtiment 26, 1 Avenue de la Terrasse, 91198 Gif-sur Yvette Cedex, France
benedicte.michel@cgm.cnrs-gif.fr

Moynahan, Mary Ellen
Department of Medicine, Memorial Sloan-Kettering Cancer Center, 1275 York Avenue, New York, NY 10021, USA.

Prado Félix
Departamento de Biología Molecular, CABIMER, CSIC-Universidad de Sevilla, Av. Américo Vespucio s/n, 41092 Sevilla, Spain

Roseaulin, Laura
Department of Genomes and Genetics, Institut Pasteur, 25-28 rue du Dr Roux, Paris cedex 75724. France

Rothstein, Rodney
Department of Genetics & Development, Columbia University Medical Center, 701 West 168th Street, New York, NY 10032, USA
rothstein@cancercenter.columbia.edu

Sanchez, Humberto
Departmento de Biotecnología Microbiana, Centro Nacional de Biotecnología, CSIC, Campus Universidad Autónoma de Madrid, Campus Universidad Autónoma de Madrid, Darwin Street 3, Cantoblanco, 28049 Madrid, Spain

Sherratt, David J
Department of Biochemistry, University of Oxford, South Parks Road, Oxford. OX1 3QU, United Kingdom
sherratt@bioch.ox.ac.uk

Whitby, Matthew C.
Department of Biochemistry, University of Oxford, South Parks Road, Oxford OX1 3QU, UK
matthew.whitby@bioch.ox.ac.uk

Wilson, Thomas E.
Department of Pathology, University of Michigan Medical School, Ann Arbor MI, Rm. 2065 BSRB, 109 Zina Pitcher Place, Ann Arbor, MI 48109-2200, USA
wilsonte@umich.edu

Genetics of recombination in the model bacterium *Escherichia coli*

Bénédicte Michel, Zeynep Baharoglu, and Roxane Lestini

Abstract

Homologous recombination in bacteria was originally discovered as a process that not only exchanges genetic material but also provides one of the major pathways of DNA-damage repair. Homologous exchanges and DNA repair illustrate the dual role of recombination which acts both to promote genetic diversity and to conserve genomic integrity. In this review, we will first describe the genetics of enzymes that act at different steps of the homologous recombination process in *Escherichia coli*, with an emphasis on the most recent results. We will then describe recent advances in our understanding of the role of homologous recombination during DNA repair. Recombination enzymes act on DNA at single- or double-strand interruptions generated as a result of nucleotide lesions or replication impairment. Although generally they can and often do promote genetic exchange, some recombination enzymes also fulfill various non-recombinogenic important functions, such as the signaling of DNA damage and the remodeling of arrested replication forks.

1 Introduction

Homologous recombination uses sequence homology to promote DNA exchange, resulting in new combinations of sequences. Bacteria and their phages have historically been an important source of understanding of the molecular basis of the reaction. Bacterial genetics allowed the identification of genes involved in homologous recombination, which led to the definition of pathways. The two main pathways turned out to characterize the two main substrates for homologous recombination: DNA double strand ends (RecBCD-dependent recombination pathway) and DNA single-strand gaps (RecF-dependent recombination pathway). In parallel, the bacterial enzymes were purified and their enzymatic activities defined three key steps of the reaction: initiation, strand invasion, and resolution. Microorganisms are nowadays the tool of choice for understanding the physiological roles of homologous recombination, the action of homologous recombination proteins in processes other than genetic exchange, and the links between recombination and other biological processes.

Several important reviews have been published in the few past years, which address different aspects of homologous recombination in bacteria, such as the exhaustive description of the genetics and biochemistry of homologous recombination in *Escherichia coli* by A. Kuzminov (1999) and "Genetic Recombination" by David Leach (Blackwell Science 1996). The biochemical properties of the enzymes that participate in homologous recombination are described in chapters by Kowalczykoswki and Cox, in this book. In addition, entire issues of different journals have been dedicated to various aspects of homologous recombination, as the special issues in TiBS (Vol 25 N°4, 2000), and PNAS (Vol 98 N°15, 2001). Most of our knowledge about bacterial recombination comes from studies of the model bacterium *E. coli*. Homologues to the main *E. coli* recombination genes can be found in the genome of most enterobacteria (Rocha et al. 2005) and the characterization of some of these genes has been performed, as for example in *Neisseria gonorrhoeae* (Kline et al. 2003; Kline and Seifert 2005b, 2005a) and *Helicobacter pylori* (Fischer and Haas 2004; Kang et al. 2004; Pinto et al. 2005). We will describe here recombination in *E. coli*.

2 Genes and pathways

2.1 The key steps of the homologous recombination reaction

2.1.1 Homology search and pairing

Homologous recombination in *E. coli* operates by two pathways that act on two different kinds of substrates (Fig. 1). In both pathways the exchange of homologous sequences is catalyzed by the central recombination enzyme RecA. The active form of RecA is a filament of proteins bound to single-strand DNA (ssDNA). RecA in the RecA filament catalyzes homology search, strand-exchange, and branch migration of the resulting strand-exchange point (Singleton and Xiao 2001; Xiao and Singleton 2002; Cox et al. 2005), for review see (Cox 2003; Bell 2005). *In vivo* the formation of the RecA filament is impeded by the presence of the single-strand DNA binding protein (SSB) (Madiraju et al. 1988; Eggler et al. 2003). RecA is an abundant protein and direct localization of wild type and mutant RecA-GFP fusion proteins in *E. coli* cells confirmed the role of RecA during DNA repair and suggested the presence of RecA molecules in storage structures (Renzette et al. 2005).

2.1.2. Preparing the RecA filament. RecBCD-dependent recombination

The two classical recombination pathways are defined by the DNA substrate and the pre-synaptic enzymes that facilitate RecA binding to ssDNA. The "RecBCD" pre-synaptic enzyme recognizes a double strand DNA (dsDNA) end. This heterotrimeric complex has several activities that enable it to unwind dsDNA while degrading the resulting ssDNA. Upon encountering a specific octameric sequence

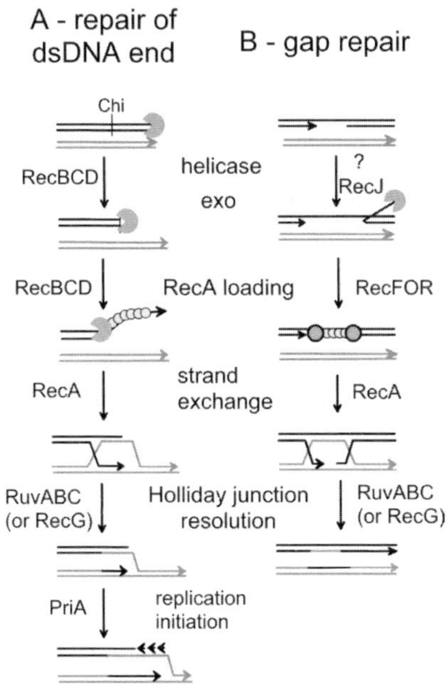

Fig. 1. Models for the mode of action of the main homologous recombination proteins in *E. coli* at dsDNA ends or ssDNA gaps. At dsDNA ends, RecBCD degrades DNA until it encounters a Chi site, its helicase-nuclease activity is then modified to produce a 3' ended ssDNA to which it loads RecA. At gaps, the 5' ssDNA exonuclease RecJ enlarges the single-strand DNA region (possibly with the help of various helicases as no specific helicase is required for gap repair); RecF, RecO, RecR promote RecA binding to SSB-coated DNA. The synaptic step (homology search and strand exchange) is always performed by RecA and results in the formation of a Holliday junction (X structure). The post-synaptic steps are the migration and the resolution of Holliday junctions. Migration can be performed by RuvAB or by RecG and resolution is made by RuvABC. In addition, RecBCD-mediated recombination is always coupled with PriA-dependent replication restart. The black and grey lines are the two recombining molecules and arrowheads indicate the 3' ends of the DNA strands. Indented circles: nucleases (RecBCD or RecJ), small circles: RecA.

named Chi, the 5' to 3' DNA degradation activity of RecBCD is increased while the 3' to 5' DNA degradation activity is abolished and RecA is loaded onto the 3' ended ssDNA, forming a RecA filament (Fig. 1A). *RecB* and *recC* null mutants are totally deficient for homologous recombination at dsDNA ends whereas a *recD* mutant is only deficient for DNA degradation and Chi recognition. The molecular basis of these phenotypes have now been defined in detail (Jockovich and Myers 2001; Dillingham et al. 2003; Taylor and Smith 2003; Singleton et al. 2004; Handa et al. 2005; Spies et al. 2005).

2.1.3 Preparing the RecA filament. RecFOR-dependent recombination

The pre-synaptic RecF, RecO, and RecR recombination proteins (called RecFOR) allow RecA to bypass the SSB barrier, as exemplified by the existence of RecA mutant proteins that have an increased affinity for ssDNA *in vitro* and bypass the need for the RecFOR pre-synaptic proteins *in vivo* (Kowalczykowski 1991; Wang et al. 1993). RecFOR are required for the repair of ssDNA gaps (Fig. 1B). Current models agree on the formation of two complexes, RecFR and RecOR. RecOR is generally thought to be responsible for rendering SSB-coated ssDNA accessible to RecA. RecFR targets dsDNA or dsDNA-ssDNA junctions and is responsible for the targeting of RecA to the ssDNA region of gaps (Shan et al. 1997; Webb et al. 1997; Bork et al. 2001; Morimatsu and Kowalczykowski 2003). Recent X-ray analysis of RecO and RecR from *Deinococcus radiodurans* confirmed the existence of a RecO-RecR complex and provided insight into the DNA-binding mode of these two proteins, revealing that RecR molecules form a ring structure that can encircle both dsDNA and ssDNA (Lee et al. 2004; Leiros et al. 2005). The 5' to 3' ssDNA exonuclease RecJ is also required for most RecFOR-dependent recombination reactions. RecJ is likely to act by enlarging gaps; notably, a direct interaction *in vivo* between RecJ and SSB has been observed in a high throughput interactome study (Butland et al. 2005). After RecJ- RecFOR-assisted formation of a RecA filament, either the central ssDNA part of the gap, or a displaced ssDNA 3' end invades the homologous sequence (Fig. 1B).

2.1.4 Resolution of recombination intermediates

Strand invasion results in the formation of a four-arm dsDNA intermediate called a Holliday junction (HJ). The main enzyme responsible for the resolution of HJs in bacteria is the RuvABC complex (Fig. 1). *RuvA* and *ruvB* genes are part of the SOS regulon, a set of genes induced by DNA damaging agents; RuvAB form a complex able to catalyze HJ branch migration. The *ruvC* gene is next to the *ruvAB* operon but is expressed independently and is not SOS-inducible. RuvC binds RuvAB to form a RuvABC complex able to resolve HJs by nicking two opposite strands (reviewed in West 1997). The interrupted strands are joined by ligase, resulting in strand exchange. The outcome of the recombination reaction depends on the pair of strands cleaved by RuvC in the HJ, which is determined by the direction of RuvAB branch migration (van Gool et al. 1999; Cromie and Leach 2000; Michel et al. 2000).

The RecG helicase is required in *ruvABC* mutants for Hfr conjugation and P1 transduction (which are dsDNA-end recombination events) and the *recG* mutation is highly synergistic with *ruvABC* mutations in UV irradiated cells (Lloyd and Buckman 1991). The genetic properties of the *recG* mutant, combined with the capacity of the RecG helicase to catalyse branch migration of HJs and the conversion of three strands junctions into HJs, led to the proposal that RecG acts as an alternative resolvase in *E. coli* (Whitby et al. 1994; Whitby and Lloyd 1995b). However, RecG is not associated with any known activity able to cleave DNA, suggesting that resolution occurs when HJs are moved to a pre-existing nick or gap. In addition to its participation in homologous recombination, RecG unwinds R-loops *in vitro* and *in vivo* (Vincent et al. 1996), and targets D-loops and fork structures (McGlynn et al. 1997; McGlynn and Lloyd 1999). The molecular mechanism of action of the protein, deduced from abundant biochemical studies and the determination of its crystal structure, is compatible with the unwinding of different kinds of three- and four-arms DNA molecules (Singleton et al. 2001), but this information did not ascribe a precise physiological role(s) for RecG *in vivo*.

2.1.5 Double-strand break (DSB) repair requires replication re-initiation

The last function essential for the formation of a viable recombinant during DSB repair, PriA, was identified by reverse genetics (reviewed in Marians 2000; Sandler and Marians 2000) (Fig. 1A). PriA catalyzes, with the help of proteins named PriB, PriC, and DnaT, the initiation of replication from D-loop or fork structures (Liu and Marians 1999). PriA recognizes the D-loop formed by strand invasion during RecBCD-catalyzed DSB repair and promotes the loading of the replicative helicase, DnaB, which, in turn, promotes the assembly of a functional replisome and initiation of replication from the recombination intermediate (Xu and Marians 2003). *PriA* mutants are deficient for DSB repair and highly sensitive to UV irradiation. They were not isolated during screening for recombination-deficient mutants presumably because *priA* mutants are very slow growing and rapidly acquire suppressor mutations (Sandler et al. 1996). Recently, the restriction enzyme I-Sce1 was used to further characterize the DSB repair pathway. As in homologous recombination during Hfr conjugation, the repair of DSBs made by I-Sce1-cleavage requires RecBCD, RecA, either RuvABC or RecG, and PriA (Meddows et al. 2004).

2.1.6 Homologous recombination in the bacterial chromosome can lead to dimerisation and occurs by RecF and RecBCD pathways

Intra- or inter-molecular recombination in circular molecules can lead to dimer formation. Dimers are resolved at a specific site *dif*, located opposite to the origin on the chromosome, by the specific recombinases, XerC and XerD. Dimer resolution is coupled to cellular division as it also requires the action of the septum-associated translocase FtsK (reviewed in Sherratt 2003; Lesterlin et al. 2004). When assayed by the extent of dimer formation or quantification of recombination between closely-spaced long tandem-repeats, spontaneous recombination in the bacterial chromosomes occurs by either the RecF or the RecBCD pathway and is significantly decreased only when both pathways are inactivated (Galitski and Roth 1997; Steiner and Kuempel 1998). When gene conversion between two non-tandem sequences was measured, recombination was found to depend only on the RecBCD pathway (Zieg and Kushner 1977; Nowosielska et al. 2004).

2.2 Alternative pathways of DSB repair

2.2.1 The RecF-pathways of DSB repair

The first assay used to detect and quantify homologous recombination was Hfr conjugation. Because this process relies mainly on recombination initiated at dsDNA ends, *recA* and *recBC* were the first recombination genes discovered. Mutations that suppress the defect of *recBC* mutants in Hfr recombination were isolated and named *sbc* for suppressors of *recBC*. Two suppressor genes were identified, named *sbcA* and *sbcB*. The *sbcA* mutation activates a cryptic prophage-encoded recombination process that requires RecE and RecT proteins (Noirot et al. 2003, and ref therein). *SbcB*, a gene also identified independently under the name of *xonA*, encodes a 3' to 5' exonuclease. *RecBC sbcB* mutants were later found to lack another nuclease, the SbcCD complex (Gibson et al. 1992). It is generally believed that the inactivation of these nucleases allows recombination to proceed by a RecBCD-independent pathway by preventing the degradation of the ssDNA recombination substrate with a 3' end. The genes that are required for Hfr recombination in a *recBC sbcB sbcCD* background defined what is called the "RecF pathway" of homologous recombination (reviewed in Clark and Sandler 1994). These genes are *recQ, recJ*, *recF*, *recO*, *recR*, *recN*, and *ruvABC* (Fig. 2). dsDNA end recombination *via* the "RecF-pathway" combines the activity of the RecQ 5' to 3' helicase and the RecJ 5' to 3' exonuclease to provide a 3' ended ssDNA on which RecFOR facilitates the formation of a RecA filament. RecN is a coiled-coil protein proposed to bind DNA ends and bring them in close proximity, which could be required because dsDNA end recombination is less efficient when it is catalyzed by the combined action of RecQ, RecJ and RecFOR proteins than by the highly specialized and efficient RecBCD complex. Confirming a role for RecN in DSB repair, RecN was recently shown to be required for the repair of I-SceI DSBs specifically when the number of DSBs is higher than one per chromosome (Meddows et al. 2005).

2.2.2 Exchange of pre-synaptic functions

In *recBC scbB sbcCD* mutants, the helicase, exonuclease and RecA loading activities of RecBCD can be replaced by RecQ, RecJ, and RecFOR, respectively (Fig. 2). Certain proteins of the RecFOR-pathway were later shown to also collaborate with partly inactivated RecBCD complexes to promote DSB repair in SbcB^{+} SbcCD^{+} cells. The observation that *recJ* inactivation renders the *recD* mutant highly sensitive to UV irradiation suggested that RecJ supplies the exonuclease activity required for RecBC action in UV-irradiated *recD* mutant cells (Lloyd et al. 1988; Lovett et al. 1988). However, *recJ recD* double mutants are capable of DSB repair (Lloyd et al. 1988; Ivancic-Bace et al. 2005), suggesting that the role of RecJ in UV-irradiated *recD* mutants is more complex. The *recB*^{D1080A} mutant is capable of homologous recombination although it produces an enzyme that does not have nuclease activity and is unable to load RecA protein onto ssDNA. In an elegant genetic study, K. Brcic-Kostic and co-workers showed that RecFOR and

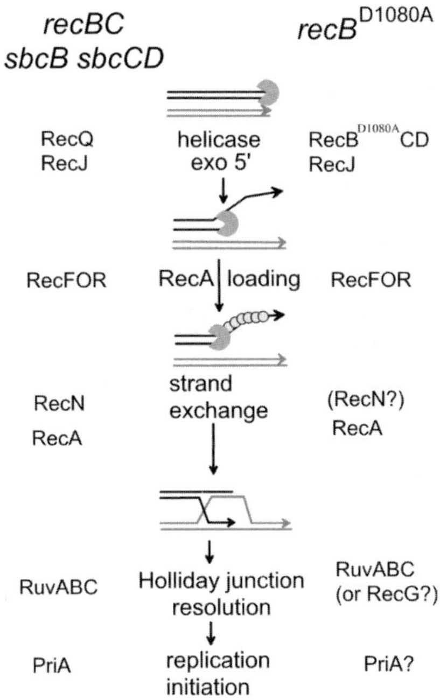

Fig. 2. Models for DSB repair by alternative pathways. In *recBC sbcB sbcCD* mutants, RecQ helicase and RecJ exonuclease provide a 3' ssDNA end on which RecFOR loads RecA. RecN is required for the formation of recombinants, presumably to facilitate intermolecular interactions. Resolution requires RuvABC and completion of the recombination reaction requires PriA (Kogoma 1997; Bidnenko et al. 1999). In the *recB*^{D1080A} mutant, the helicase function of RecB^{D1080} CD acts in concert with RecJ and RecFOR for the formation of a RecA filament. Requirements for RecN, RuvABC, RecG and PriA were not tested. Symbols are as in Fig. 1.

RecJ are required for recombination in this mutant, indicating that the RecB^{D1080A}/RecC/RecD complex uses the exonuclease activity of RecJ and the RecA-loading activity of RecFOR to initiate dsDNA end recombination (Fig. 2) (Amundsen and Smith 2003; Ivancic-Bace et al. 2003, 2005).

2.3 Homologous recombination in plasmids

Plasmid recombination was shown to require RecF, RecO and RecJ (as the plasmids used lacked Chi sites, they were committed to RecFOR-initiated recombination) (Cohen and Laban 1983; Kolodner et al. 1985). *RecQ* was not tested; *recN* inactivation had no effect in agreement with the role of this protein in joining DNA ends. Inactivation of either or both *ruvAB* and *recG* also had no effect (Kolodner et al. 1985; Lloyd and Buckman 1991; Lovett et al. 2002). RecFOR-dependent UV-repair was also observed to be unaffected by *ruvAB, ruvC* or *recG* mutations (see below 3.1.1).

2.4 Ligase and polymerase I

Ligase is presumably required for the final closure of recombination products. In a ligase *ts* mutant, recombination was increased (due to an increased level of recombination substrate) or decreased (due to a defect in recombination completion) depending on the conditions (Zieg et al. 1978). Similarly, recombination was found to be increased or decreased in different *polA* mutants, presumably because certain *polA* mutations cause the occurrence of DNA gaps or breaks, increasing the level of recombination substrate, while others prevent a step essential for the completion of the recombination reaction (Konrad 1977; Zieg et al. 1978; Nowosielska et al. 2004).

2.5 Proteins that antagonize homologous recombination

Homologous recombination is counteracted in *E. coli* by UvrD, a DNA helicase essential for nucleotide excision repair and mismatch repair (reviewed in Tuteja and Tuteja 2004). Consequently, *uvrD* mutants exhibit an increased level of homologous recombination in all assays tested (Zieg et al. 1978; Bierne et al. 1997b; Petranovic et al. 2001). The anti-recombination action of UvrD *in vivo* correlates with a capacity to undo a recombination intermediate and to remove RecA from ssDNA *in vitro* (Morel et al. 1993; Veaute et al. 2005). UvrD also counteracts a non-recombinogenic binding of RecA to blocked replication forks (Flores et al. 2005).

The *E. coli recX* gene is expressed downstream of *recA* in the same operon. *RecX* is widely conserved and was first recognized as an anti-recombination protein in *Pseudomonas aeruginosa* (Sano 1993). More recently, the *E. coli* RecX protein was shown to antagonize *in vitro* the extension of RecA filaments (Stohl et al. 2003; Drees et al. 2004). The *in vitro* anti-recombination activity of RecX is counteracted by RecF, which may be one of the reasons why the inactivation of the *E. coli recX* gene only weakly affects recombination *in vivo* (Pages et al. 2003; Lusetti et al. 2006).

Finally, homeologous recombination, which corresponds to strand-exchange between diverged sequences, is antagonized by the mismatch repair proteins and facilitated by the induction of the SOS response (Matic et al. 1995).

3 The repair of DNA lesions

Among the *E. coli* recombination mutants, the *recA* mutant is one of the most sensitive to DNA damaging agents, which indicates a major role for RecA in the repair of DNA lesions. This role corresponds to at least three well-characterized activities of the RecA filament: initiation of homologous recombination, induction of the SOS response and promotion of lesion bypass by an SOS-induced translesion polymerase. Cytological techniques also revealed that RecA binding is tightly controlled *in vivo* as SOS induction in response to UV irradiation is an oscillatory process in individual cells (Friedman et al. 2005). In addition, UV-irradiation causes a transient decrease in the rate of DNA synthesis (as measured by incorporation of labeled thymidine) and RecA is required for the recovery of a normal rate of DNA synthesis after irradiation, a process called "replisome reactivation" (Witkin 1991). Clearly, two processes associated with recombination proteins are involved in DNA repair: homologous recombination and repair without genetic exchange. These processes are intertwined and deciphering the roles of recombination proteins in lesion repair is still in progress.

3.1 RecFOR- dependent DNA repair

RecFOR mutants are sensitive to intra-strand crosslinking agents (e.g. UV), inter-stand crosslinking agents (e.g. mitomycin C) and DNA damaging agents that cause base modifications (e.g. methylating agents) (Keller et al. 2001; Bhattacharya and Beck 2002). The best studied of all these DNA damaging agents is UV-radiation and, like RecA, RecFOR participates in all known UV-repair processes: post-replicative gap repair, SOS induction, UV-induced mutagenesis, and replisome reactivation (Wang and Smith 1983; Thoms and Wackernagel 1987; Whitby and Lloyd 1995a; Courcelle et al. 1997; Liu et al. 1998). Nevertheless, *recFOR* mutants are much less sensitive to UV-irradiation than *recA* mutants. The survival of *recFOR* mutants to UV-irradiation is strongly dependent on RecBCD and may result from the conversion of post-replicative gaps into DSBs, which can then be repaired by RecBCD (Wang and Smith 1986).

3.1.1 RecFOR-dependent gap repair

Post-replicative gaps were first shown to accumulate in *uvrA* mutants after UV irradiation (Rupp and Howard-Flanders 1968). In cells proficient for recombination, the low molecular weight DNA fragments synthesized after UV irradiation are joined into high molecular weight chains. The first hint that RecF was also involved in gap repair was the accumulation of post-replicative ssDNA fragments in UV-irradiated chromosomes of *recF* mutants (Wang and Smith 1983). *RecR* and *recO* mutants were later shown to exhibit the same phenotype (Tseng et al. 1994). In contrast, *RecN* is not required for daughter-strand gap repair in UV-treated cells (Wang and Smith 1988; Tseng et al. 1994). Similarly, UV-irradiated *recJ* and *recQ* single mutants do not accumulate chromosomal single-stranded fragments (Tseng et al. 1994), suggesting that daughter-strand gaps due to UV-irradiation are accessible to RecFOR without a requirement for RecQ and RecJ. The high level of survival of the *recQ* mutant after UV-irradiation and lack of synergy between *recQ* and *recBC* mutations confirm that RecQ is not essential for gap repair in

UV-irradiated cells. However, in a UV-irradiated *recB recJ* double mutant, both the survival after UV-irradiation and gap repair are strongly affected, which intriguingly suggests that gap repair could rely on RecBC in a *recJ* single mutant (Wang and Smith 1988).

The observation of DNA strand-exchange following UV irradiation suggests that gaps are repaired by a recombinational process involving the formation and the resolution of HJs, at least in wild type cells capable of HJ resolution (Fig. 4, steps E) (Rupp and Howard-Flanders 1968; reviewed in Rupp 1996). However, inactivation of *ruvA* or *ruvC* or *recG* did not affect the repair of daughter-strand gaps (Tseng et al. 1994). This result suggests that in the absence of a resolvase another mechanism that does not involve the resolution of HJs takes place. HJ resolution is not needed in models such as the one proposed in Fig. 4F, the 3' DNA end blocked at a lesion switches to the sister chromatid, uses the intact strand as a template for the synthesis of a DNA sequence complementary to the lesion (F1) and, because the strand invasion intermediate is not be stabilized by RuvAB, this newly synthesized DNA is then displaced (by RecG or another helicase) and returns to its original template (F2). Reactions in which a 3' end uses the sister chromatid to copy a missing information before returning to its original template were proposed to occur at blocked forks (strand-switching model, Higgins et al. 1976) and at DSBs (synthesis-dependent strand annealing model, SDSA, reviewed in Paques and Haber 1999).

3.1.2 Other RecFOR-dependent repair processes in UV irradiated cells

RecFOR mutants show delayed SOS induction, in agreement with the idea that the role of RecFOR is to promote the binding of RecA to gaps (Thoms and Wackernagel 1987; Whitby and Lloyd 1995a). The defect in the formation of RecA filaments when RecFOR are absent is probably also the reason for some decrease in UV-induced mutagenesis in *recFOR* mutants (Liu et al. 1998; Maul and Sutton 2005). Finally, one of the major defects of *recFOR* mutants after UV irradiation is a total cessation of replisome reactivation, which is as dramatic as in *recA* mutants. This role of RecFOR is described below in Section 4.2.1.

3.2 RecBC-dependent recombinational repair

In addition to being sensitive to crosslinking agents and base modification agents, *recBC* mutants and *recN* mutants are highly sensitive to all agents that cause the formation of DNA double-strand breaks such as gamma irradiation, bleomycin or nitric oxide (Sargentini and Smith 1986; Knezevic-Vukcevic and Simic 1991; Schapiro et al. 2003). In UV irradiated cells, DSBs can result from the simultaneous repair of closely-spaced lesions, leading to nucleotide excision repair (NER)-dependent homologous recombination (reviewed in Smith 2004). In NER deficient strains, DSBs result from the accumulation of ssDNA gaps. The observation that *recBC* mutants are UV^S while *recN* mutants are not, and that the *recN* mutation is synergistic with *recF* and not with *recBC* (Lloyd et al. 1988), suggests that RecN might participate in RecBC-catalyzed repair under conditions where DSBs accumulate. How HJs are resolved during recombinational repair of UV-induced DSBs can be deduced from the study of *recFOR* mutants, in which recombinational repair relies solely on RecBCD. *RecFOR* mutations are not synergistic with *recG* inactivation whereas they are highly synergistic with *ruvABC* mutations (our unpublished observations), suggesting that RuvABC is the main resolvase involved.

4 Recombination and replication

The existence of a direct link between replication and recombination was first suggested in the early 1990 by the observation that blocking replication stimulates homologous and illegitimate recombination (Bierne and Michel 1994; Horiuchi et al. 1994). This observation led to the proposal that replication inactivation creates recombination substrates. Conversely, genetic studies of the *priA* mutant and biochemical characterization of the PriA protein hinted that recombination substrates were used for replication re-initiation (Kogoma et al. 1996; Sandler et al. 1996). These independent sets of observations were the foundations of a new field of research on the interplay between replication and recombination (reviewed in Marians 2000; Michel 2000).

This field is complex for several reasons. First, replication impairment can stimulate chromosome rearrangements by a variety of different mechanisms, including RecA-dependent and RecA-independent events. Second, under several conditions of defective chromosome replication, both blocked forks (due to arrest of the leading-strand polymerase) and gaps (due to arrest of the lagging-strand polymerase) co-exist. Blocked forks and gaps can be used as recombination substrates and it is sometimes difficult to ascribe an experimental observation to the action of a recombination protein at forks *versus* gaps. Finally, in addition to true recombination events, replication inactivation also triggers actions of recombination proteins that do not lead to genetic exchanges. A well-known example is the induction of the SOS response by RecA. Consequently, a requirement for a recombination protein in a replication mutant does not necessarily mean the occurrence of a recombination reaction.

4.1 Replication inactivation induces RecA-independent recombination

In the beginning of the 1980s, the first reports appeared that recombination between short-homologous sequences was RecA-independent (Foster et al. 1981; Albertini et al. 1982). A purely replicative process, called replication slippage, was proposed to account for these RecA-independent rearrangements (Fig. 3A; deletions and amplifications can be produced by replication slippage (Albertini et al. 1983). In support of the replication slippage model, the occurrence of

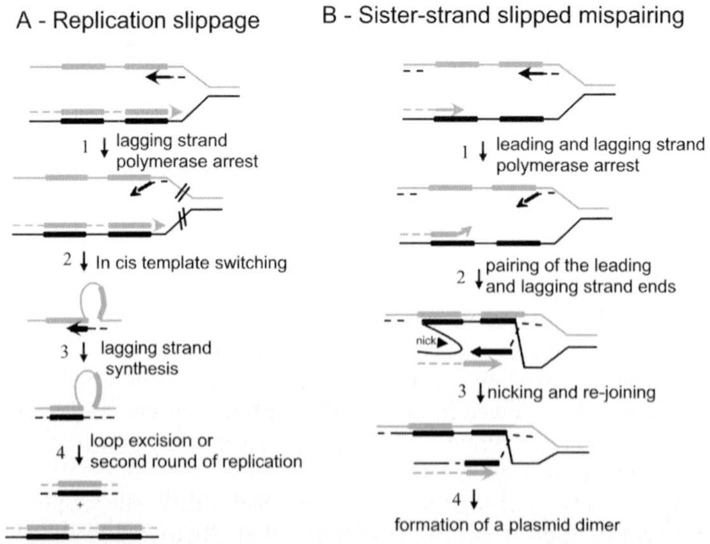

Fig. 3. Models for RecA-independent recombination between tandemly repeated sequences. A- Replication slippage : step 1, the lagging strand polymerase dissociates; step 2, the 3' end of the lagging strand separates from its template and anneals with the homologous repeat; step 3, DNA synthesis resumes from this annealed substrate, producing a sequence deleted of one of the repeats and of the intervening sequence; step 4, excision of the looped-out sequence on the lagging-strand template or a following round of replication produces a deleted dsDNA molecule. B – Sister-strand slipped mispairing (Saveson and Lovett 1997a): step 1, nascent strands are displaced and will mispair with each other; step 2, with pairing of the parental strands, a recombinational intermediate is produced; step 3 and 4, nicking of one of the parental strands and resolution of the Holliday junction causes sister-chromatid exchange producing, in a circular molecule, a dimeric deletion product. To account for the observation that RecA-independent recombination is not affected by the absence of RuvABC, RecG or both, in plasmids as in chromosome (Lovett et al. 1993; Bierne et al. 1997c), it was proposed that replication through the Holliday junction (Saveson and Lovett 1997b), or the action of an unknown resolvase activity (Lovett et al. 2002), can lead to the dimeric deleted product (step 4). Full lines are template strands, dashed lines are newly-synthesized strands, black and grey boxes are the tandemly repeated sequences. Arrowheads are the 3' ends of growing strands.

rearrangements without strand-exchange was demonstrated directly with the use of labeled-DNA molecules (D'Alencon et al. 1994). It was found that RecA-independent rearrangements could also occur between longer sequences, provided that these sequences were close enough (Lovett and Feschenko 1996; Bierne et al. 1997c), reviewed in (Lovett 2004). Recombination between short-homologous sequences and RecA-independent recombination between long-homologous sequences are stimulated by the presence of palindromes between the sequences and by mutations that affect the replication machinery, leading to the notion RecA-independent rearrangements occur upon replication impairment. Several laboratories proposed that replication slippage was taking place mainly during lagging-strand synthesis, because secondary structures may form more readily in the single-strand regions present on the lagging-strand template. Furthermore, *in vitro* the lagging-strand polymerase is more weakly associated with the replication machinery than the leading-strand polymerase (McHenry 2003), and, the weaker stability of the lagging-strand replicative polymerase may be the reason why *in vivo* it can be replaced by an alternative, error prone polymerase (Pol II and Pol IV) upon replication impairment (Banach-Orlowska et al. 2005).

An alternative model to replication slippage was proposed to account for RecA-independent recombination, based on the observation that these events were associated with a high rate of dimerisation of the recombining molecule when plasmids were used as substrates (Lovett et al. 1993). Plasmid dimers were proposed to form by a slippage reaction occurring between two opposite strands at the replication fork (Fig. 3B; reviewed in Lovett 2004). Moreover, due to the presence of two replication origins on dimers, they have a selective advantage over plasmid monomers (Mazin et al. 1996).

Interestingly, mutants have been isolated in which RecA-independent recombination is decreased (Foti et al. 2005; Goldfless et al. 2006) and these are affected for functions involved in multi-protein complex stability and check-points (*obgE* and *dnaK)*. These mutations were suggested to affect replication slippage as well as RecA-independent sister-chromatid exchange. The influence of *obgE* and *dnaK* genes on replication-catalyzed rearrangements suggests the association of these proteins with the replication machinery, possibly to control replisome stability during replication progression.

Replication arrest also stimulates RecA-independent recombination between short homologous or non-homologous sequences by break and join illegitimate recombination processes (Bierne et al. 1997a). The full replication machinery has to be inactivated in order to produce a DNA end susceptible to be joined to a non-related sequence. These observations strengthen the idea that sites where replication is inactivated or slowed down are fragile.

4.2 Recombination proteins participate in the resetting of replication forks

4.2.1 Role of recombination proteins in the replication of UV irradiated cells: replisome reactivation

After UV irradiation, the rate of DNA replication (as measured by the incorporation of tritiated thymidine) decreases for about 20 minutes. The recovery of a normal replication rate (replisome reactivation) implies replication restart from blocked forks and not from *oriC*, the chromosome origin, since it is independent of DnaA for at least one round of replication (Khidir et al. 1985), and is delayed in the absence of PriA (Rangarajan et al. 2002). Replisome reactivation is abolished in several recombination mutants: *recA, recF, recO, recR*, and in several

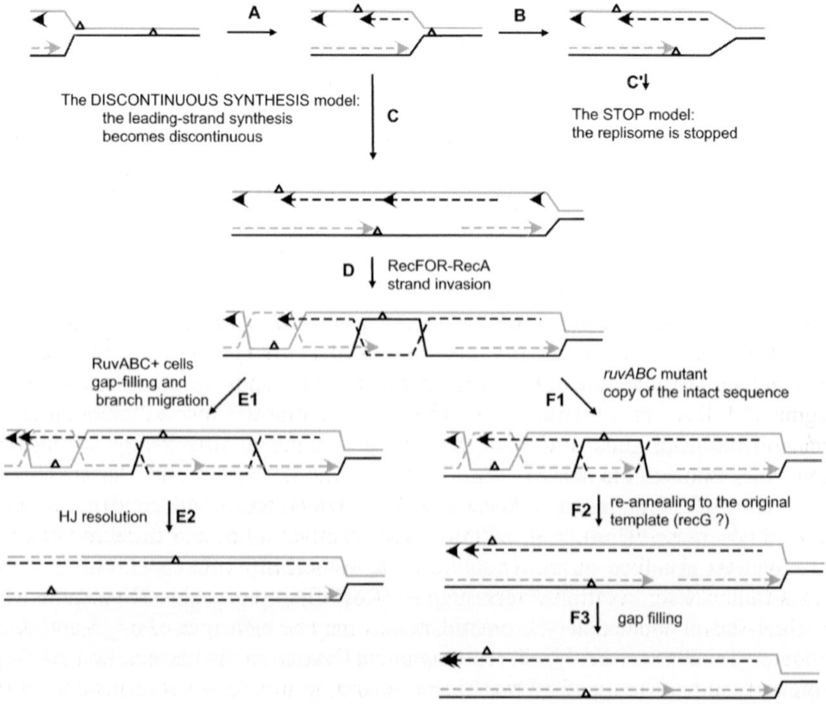

Fig. 4. Model for replication resumption after UV irradiation. A replication fork progressing from left to right encounters one-strand DNA lesions, such as UV lesions. We assume that both strands are equally susceptible to lesions and in the example drawn, the first lesion encountered is on the lagging-strand template and the second one on the leading-strand template. (A) the lagging strand polymerase is arrested, leaving a lagging-strand gap, progression of the replication fork is unaffected; (B) the leading strand is arrested; (C') recent models propose that replication is completely blocked by a leading strand arrest (Courcelle et al. 1997; McGlynn and Lloyd 2002), but these models imply the formation of less than one gap per chromosome on average (only when the first encountered lesion is on the lagging strand template); (C) based on daughter-strand gap formation, previous models proposed a re-priming of DNA synthesis on both strands (reviewed in (Kuzminov 1999); (D) daughter-strand gaps produced by discontinuous DNA synthesis on both strands are substrates for RecFOR-initiated recombination; here the RecA-bound gaps invade the complementary strand, displacing the newly synthesized strand that can be used as a template; (E1-E2) in RuvABC+ cells, resolution of the HJ leads to strand exchange; (F1-F3) in *ruvABC* mutants, after copying the intact sequence, the invading strands become unpaired and switch back to their original position. In both cases, the DNA lesions are made double-stranded and thus accessible to NER. Symbols are as in Fig. 3. Triangles are DNA lesions.

replication mutants *priA, polB* (Pol II), *umuCD* (Pol V) (Khidhir et al. 1985; Courcelle et al. 1997; Rangarajan et al. 2002; Chow and Courcelle 2004). It is not affected by a *recB* mutation, indicating that it does not involve the formation of a DNA double strand end (Khidhir et al. 1985), or by *ruvAB, recG* or the simultaneous inactivation of *ruvAB* and *recG*, suggesting that it does not require the resolution of a Holliday junction (Donaldson et al. 2004). In spite of 20 years of research, the role of recombination proteins in replisome reactivation is still unclear, mainly because the mode of residual DNA synthesis that occurs at low rate during the first 20 minutes after irradiation is unknown. The contribution of DnaA to this residual DNA synthesis has been suggested, implying a full blockage of ongoing forks and *de novo* DNA synthesis only from *oriC* (Khidir et al. 1985; Fig. 4, the stop model). Alternatively, the accumulation of gaps after UV irradiation has lead to the proposal of models in which replication continues beyond the lesions on both strands in a discontinuous fashion (Fig. 4; reviewed in Rupp 1996; Wang 2005). Replication fork block and residual synthesis of short fragments from *oriC* as well as a switch to a discontinuous mode of DNA synthesis on both strands could account for the slow rate of replication and the synthesis of chromosomes as fragments following UV-irradiation. The recent demonstration that priming could occur on the leading as well as on the lagging strand *in vitro* brings strong biochemical support to the discontinuous synthesis model (Heller and Marians 2006). A similar model has been proposed in eukaryotes (Lopes et al. 2006). The SOS-induced polymerases required for replisome reactivation in *E. coli* have not been tested yet for daughter-strand gap repair; nevertheless, as the recombination enzymes required for replisome reactivation (RecFOR, RecA) are also required for daughter-strand gap repair, the two processes may be tightly linked (Fig. 4; Kuzminov et al. 1999). The delay in replisome reactivation in the absence of PriA suggests that the re-loading of a DnaB helicase and, in turn, a novel replisome, is required for the recovery of a continuous mode of leading-strand synthesis.

4.2.2 Role of recombination proteins after the encounter of a single strand interruption: the replication fork collapse model

In an important review, Andrei Kuzminov collected data supporting the idea that when a replisome encounters a single-strand interruption, it runs off the DNA end, copying the interrupted strand to the end and being completely disassembled (Fig. 5) (Kuzminov 1995). The reaction was called collapse, a term that was unfortunately later used for a variety of different reactions involving replication fork arrest, with or without DNA breakage and/or replisome disassembly. Additional experimental support for the run-off of replication forks encountering a DNA interruption was provided: replication runs off when it encounters a nick made by a specific nicking protein (Kuzminov 2001), or when it encounters the fork of the previous replication round arrested at a physiological site of replication arrest (Bidnenko et al. 2002). Replication fork run-off was also reported in DNA repair mutants in which strand-interruptions accumulate due to the incomplete repair of a damaged nucleotide (Bradshaw and Kuzminov 2003; Kouzminova and Kuzminov 2004, 2006). The dsDNA end created by the run-off of a replication fork is repaired by classical dsDNA end homologous recombination, as it requires RecBCD, RecA, RuvABC, and PriA. The occurrence of replication run-off can be considered as one of the “raison d’être” for PriA-dependent re-initiation of replication from recombination intermediates.

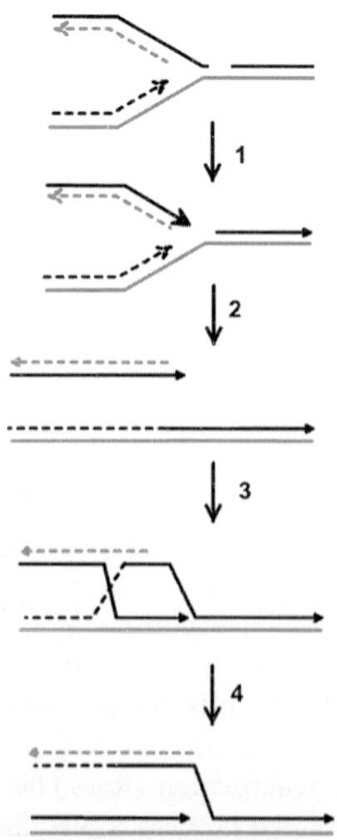

Fig. 5. Model for the run-off of a replication fork encountering a single-strand interruption. (replication fork collapse model (Kuzminov 1995). Step 1, the replication fork encounters a nick in the lagging strand template; step 2, the replisome dissociates from the DNA, leaving behind a dsDNA end; step 3, RecBCD-promoted recombinational repair of the dsDNA end creates a recombination intermediate with a D-loop (shown in Fig; 1A); step 4, the Holliday junction is resolved and replication restarts from the D-loop with the use of PriA protein. Symbols are as in Fig. 3.

4.2.3 Role of recombination proteins in replication mutants: the replication fork reversal model

The observation that replication arrest was a major cause of DNA rearrangements, prompted studies aimed at understanding the role of recombination proteins in replication mutants. It appeared that this role was more than resetting broken or run-off replication forks, as certain replication mutants required RecBC for viability, but did not require RecA. Genetic studies, combined with the direct observation of RuvABC-dependent chromosome breakage in these replication mutants, led to the proposal of the replication fork reversal model (RFR, Fig. 6; Seigneur et al. 1998). This model proposes the formation at blocked forks of a dsDNA end by annealing of leading and lagging strand ends without *bona fide* DNA breakage. This dsDNA end can only be processed by RecBCD, which either recombines or degrades it. In a reversed fork, the dsDNA end is adjacent to a Holliday junction, and, in the absence of RecBCD, resolution of this Holliday junction by RuvABC leads to fork breakage (reviewed in Michel et al. 2001, 2004).

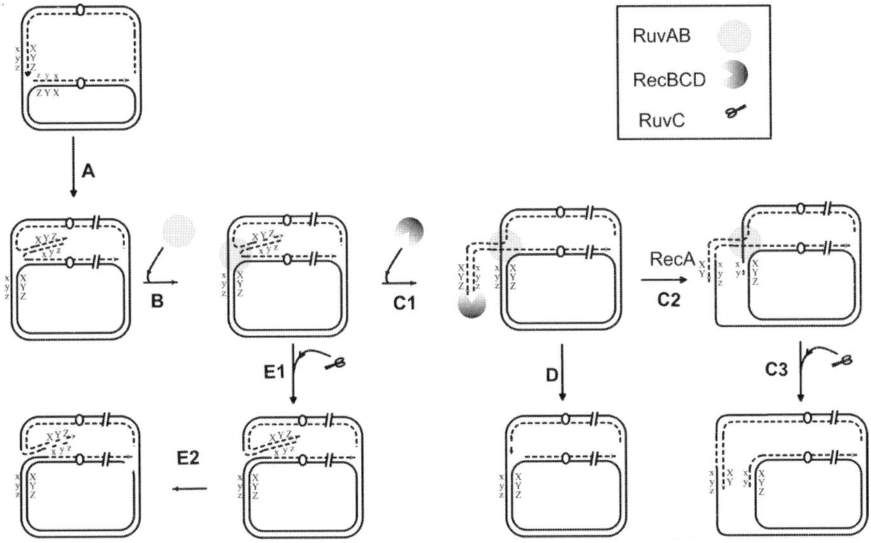

Fig. 6. Replication fork reversal model. (adapted from Seigneur et al. 1998). In step A the replication fork is blocked. The two newly synthesized strands anneal forming a Holliday junction which is stabilized by RuvAB binding (step B). In RecBCD$^+$ cells, RecBCD binds to the double strand tail (C1) and either initiates a genetic exchange mediated by RecA (C2), (C3), or degrades the tail up to the RuvAB-bound HJ and displaces the RuvAB complex (D). Both pathways allow PriA-mediated replication restart. In *recBC* mutants, RuvABC-mediated resolution of the RuvAB-bound Holliday junction leads to fork breakage (E1). A similar reaction at both forks results in the formation of a linear chromosomes (E2). Continuous and discontinuous lines represent the template and the newly synthesized strands of the chromosome respectively, the arrows indicate the 3'end of the growing strands.

Replication fork reversal (RFR) has been observed in several replication mutants, including those impaired for a replicative helicase (Seigneur et al. 1998), polymerase III (Flores et al. 2001; Grompone et al. 2002), or for replication restart (*priA*, Grompone et al. 2004b). RFR was also observed in an *nrdA101* mutant, affected for the ribonucleotide reductase, or when the enzyme is inactivated by HU treatment (E. Guarino and E. Guzman, personal communication). It does not occur in mutants impaired for replication initiation (Seigneur et al. 1998), in mutants impaired for replicative topoisomerases (Grompone et al. 2003, 2004a), and at physiological replication arrest sites (Ter, Bidnenko et al. 2002), which implies that forks arrested by different means differ in their structure or by the presence of different sets of remaining associated replication proteins.

In vitro, RecA and RecG were shown to be able to reverse DNA molecules that mimic fork structures (McGlynn and Lloyd 2000; Robu et al. 2001, 2004). Based on measures of survival of recombination mutants after UV irradiation, the RecG protein was proposed to revert forks arrested at a UV lesion (reviewed in McGlynn and Lloyd. 2002). However, fork-blockage after UV irradiation is difficult to reconcile with the possible discontinuous synthesis of chromosomes on both strands (see 4.2.1) and the proposed role of RecG in replication restart is at odds with the requirement for RecFOR, RecA but not RecG for replisome reactivation (Donaldson et al. 2004).

Studies on the mechanism of the replication fork reversal reaction in various replication mutants showed that it is catalyzed by different pathways depending on the cause of replication arrest. Replication fork reversal requires RecA in the replicative helicase mutant *dnaB,* but not in any other replication mutant (Seigneur et al. 2000; Grompone et al. 2002). In two polymerase III mutants (*dnaE, holD*) and in the *rep* helicase mutant, the conversion of the fork into a HJ requires RuvAB (Baharoglu et al. 2006). This observation suggests that RuvAB plays two different roles in *E. coli*: resolution of HJs made by homologous recombination and conversion of inactivated replication forks into HJs. RuvAB is required for replication fork reversal only in certain cases of replication arrest, which reinforces the notion that the disassembly of inactivated replication forks and hence their accessibility to recombination/repair/replication restart proteins, may depend on the cause of arrest.

The role of RuvABC proteins in the remodeling of inactivated replication forks illustrates how the understanding of the functional interactions between recombination, DNA repair and chromosome replication machines in bacteria is essential for our comprehension of genetic stability.

Acknowledgments

The authors are very grateful to Dr. Martin Marinus for very helpful reading of the manuscript. Research in the author's laboratory is supported by the ACI "Microbiologie" FNS 2003, by the ACI "BCMS 2004" and by INRA.

References

Albertini AM, Hofer M, Calos MP, Miller JH (1982) On the formation of spontaneous deletions: the importance of short sequence homologies in the generation of large deletions. Cell 29:319-328

Albertini AM, Hofer M, Calos MP, Tlsty TD, Miller JH (1983) Analysis of spontaneous deletions and gene amplification in the lac region of *Escherichia coli.* Cold Spring Harb Symp Quant Biol 47 Pt 2:841-850

Amundsen SK, Smith GR (2003) Interchangeable parts of the *Escherichia coli* recombination machinery. Cell 112:741-744

Baharoglu Z, Petranovic M, Flores MJ, Michel B (2006) RuvAB is essential for replication forks reversal in certain replication mutants. EMBO J 25:596-603

Banach-Orlowska M, Fijalkowska IJ, Schaaper RM, Jonczyk P (2005) DNA polymerase II as a fidelity factor in chromosomal DNA synthesis in *Escherichia coli.* Mol Microbiol 58:61-70

Bell CE (2005) Structure and mechanism of *Escherichia coli* RecA ATPase. Mol Microbiol 58:358-366

Bhattacharya R, Beck DJ (2002) Survival and SOS induction in cisplatin-treated *Escherichia coli* deficient in Pol II, RecBCD and RecFOR functions. DNA Repair (Amst) 1:955-966

Bidnenko V, Ehrlich SD, Michel B (2002) Replication fork collapse at replication terminator sequences. EMBO J 21:3898-3907

Bidnenko V, Seigneur M, Penel-Colin M, Bouton MF, Dusko Ehrlich S, Michel B (1999) sbcB sbcC null mutations allow RecF-mediated repair of arrested replication forks in rep recBC mutants. Mol Microbiol 33:846-857

Bierne H, Ehrlich SD, Michel B (1997a) Deletions at stalled replication forks occur by two different pathways. EMBO J 16:3332-3340

Bierne H, Michel B (1994) When replication forks stop. Mol Microbiol 13:17-23

Bierne H, Seigneur M, Ehrlich SD, Michel B (1997b) uvrD mutations enhance tandem repeat deletion in the *Escherichia coli* chromosome via SOS induction of the RecF recombination pathway. Mol Microbiol 26:557-567

Bierne H, Vilette D, Ehrlich SD, Michel B (1997c) Isolation of a dnaE mutation which enhances RecA-independent homologous recombination in the *Escherichia coli* chromosome. Mol Microbiol 24:1225-1234

Bork JM, Cox MM, Inman RB (2001) The RecOR proteins modulate RecA protein function at 5 ' ends of single-stranded DNA. EMBO J 20:7313-7322

Bradshaw JS, Kuzminov A (2003) RdgB acts to avoid chromosome fragmentation in *Escherichia coli*. Mol Microbiol 48:1711-1725

Butland G, Peregrin-Alvarez JM, Li J, Yang W, Yang X, Canadien V, Starostine A, Richards D, Beattie B, Krogan N, Davey M, Parkinson J, Greenblatt J, Emili A (2005) Interaction network containing conserved and essential protein complexes in *Escherichia coli*. Nature 433:531-537

Chow KH, Courcelle J (2004) RecO acts with RecF and RecR to protect and maintain replication forks blocked by UV-induced DNA damage in *Escherichia coli*. J Biol Chem 279:3492-3496

Clark AJ, Sandler SJ (1994) Homologous genetic recombination: the pieces begin to fall into place. Crit Rev Microbiol 20:125-142

Cohen A, Laban A (1983) Plasmidic recombination in *Escherichia coli* K-12: the role of recF gene function. Mol Gen Genet 189:471-474

Courcelle J, CarswellCrumpton C, Hanawalt PC (1997) recF and recR are required for the resumption of replication at DNA replication forks in *Escherichia coli*. Proc Natl Acad Sci USA 94:3714-3719

Cox JM, Tsodikov OV, Cox MM (2005) Organized unidirectional waves of ATP hydrolysis within a RecA filament. PLoS Biol 3:e52

Cox MM (2003) The bacterial RecA protein as a motor protein. Annu Rev Microbiol 57:551-577

Cromie GA, Leach DRF (2000) Control of crossing over. Mol Cell 6:815-826

Dalencon E, Petranovic M, Michel B, Noirot P, Aucouturier A, Uzest M, Ehrlich SD (1994) Copy-choice illegitimate DNA recombination revisited. EMBO J 13:2725-2734

Dillingham MS, Spies M, Kowalczykowski SC (2003) RecBCD enzyme is a bipolar DNA helicase. Nature 423:893-897

Donaldson JR, Courcelle CT, Courcelle J (2004) RuvAB and RecG are not essential for the recovery of DNA synthesis following UV-induced DNA damage in *Escherichia coli*. Genetics 166:1631-1640

Drees JC, Lusetti SL, Chitteni-Pattu S, Inman RB, Cox MM (2004) A RecA filament capping mechanism for RecX protein. Mol Cell 15:789-798

Eggler AL, Lusetti SL, Cox MM (2003) The C terminus of the *Escherichia coli* RecA protein modulates the DNA binding competition with single-stranded DNA-binding protein. J Biol Chem 278:16389-16396

Fischer W, Haas R (2004) The RecA protein of *Helicobacter pylori* requires a posttranslational modification for full activity. J Bacteriol 186:777-784

Flores MJ, Bierne H, Ehrlich SD, Michel B (2001) Impairment of lagging strand synthesis triggers the formation of a RuvABC substrate at replication forks. EMBO J 20:619-629

Flores MJ, Sanchez N, Michel B (2005) A fork-clearing role for UvrD. Mol Microbiol 57:1664-1675

Foster TJ, Lundblad V, Hanley-Way S, Halling SM, Kleckner N (1981) Three Tn10-associated excision events: relationship to transposition and role of direct and inverted repeats. Cell 23:215-227

Foti JJ, Schienda J, Sutera VA Jr, Lovett ST (2005) A bacterial G protein-mediated response to replication arrest. Mol Cell 17:549-560

Friedman N, Vardi S, Ronen M, Alon U, Stavans J (2005) Precise temporal modulation in the response of the SOS DNA repair network in individual bacteria. PLoS Biol 3:e238

Galitski T, Roth JR (1997) Pathways for homologous recombination between chromosomal direct repeats in *Salmonella typhimurium*. Genetics 146:751-767

Gibson FP, Leach DRF, Lloyd RG (1992) Identification of sbcD mutations as cosuppressors of recBC that allow propagation of DNA palindromes in *Escherichia-Coli* K-12. J Bacteriol 174:1222-1228

Grompone G, Bidnenko V, Ehrlich SD, Michel B (2004a) PriA is essential for viability of the *Escherichia coli* topoisomerase IV parE10(Ts) mutant. J Bacteriol 186:1197-1199

Grompone G, Ehrlich D, Michel B (2004b) Cells defective for replication restart undergo replication fork reversal. EMBO Rep 5:607-612

Grompone G, Ehrlich SD, Michel B (2003) Replication restart in gyrB *Escherichia coli* mutants. Mol Microbiol 48:845-854

Grompone G, Seigneur M, Ehrlich SD, Michel B (2002) Replication fork reversal in DNA polymerase III mutants of *Escherichia coli*: a role for the beta clamp. Mol Microbiol 44:1331-1339

Handa N, Bianco PR, Baskin RJ, Kowalczykowski SC (2005) Direct visualization of RecBCD movement reveals cotranslocation of the RecD motor after chi recognition. Mol Cell 17:745-750

Heller RC, Marians KJ (2006) Replication fork reactivation downstream of a blocked nascent leading strand. Nature 439:557-562

Higgins NP, Kato K, Strauss B (1976) A model for replication repair in mammalian cells. J Mol Biol 101:417-425

Horiuchi T, Fujimura Y, Nishitani H, Kobayashi T, Hidaka M (1994) DNA replication fork blocked at the Ter site may be an entrance for the RecBCD enzyme into duplex DNA. J Bacteriol 176:4656-4663

Ivancic-Bace I, Peharec P, Moslavac S, Skrobot N, Salaj-Smic E, Brcic-Kostic K (2003) RecFOR function is required for DNA repair and recombination in a RecA loading-deficient recB mutant of *Escherichia coli*. Genetics 163:485-494

Ivancic-Bace I, Salaj-Smic E, Brcic-Kostic K (2005) Effects of recJ, recQ, and recFOR mutations on recombination in nuclease-deficient recB recD double mutants of *Escherichia coli*. J Bacteriol 187:1350-1356

Jockovich ME, Myers RS (2001) Nuclease activity is essential for RecBCD recombination in *Escherichia coli*. Mol Microbiol 41:949-962

Kang J, Tavakoli D, Tschumi A, Aras RA, Blaser MJ (2004) Effect of host species on recG phenotypes in Helicobacter pylori and *Escherichia coli*. J Bacteriol 186:7704-7713

Keller KL, OverbeckCarrick TL, Beck DJ (2001) Survival and induction of SOS in *Escherichia coli* treated with cisplatin, UV-irradiation, or mitomycin C are dependent on the function of the RecBC and RecFOR pathways of homologous recombination. Mutat Res DNA Repair 486:21-29

Khidhir MA, Casaregola S, Holland IB (1985) Mechanism of transient inhibition of DNA synthesis in ultraviolet- irradiated *E. coli*: inhibition is independent of recA whilst recovery requires RecA protein itself and an additional, inducible SOS function. Mol Gen Genet 199:133-140

Kline KA, Sechman EV, Skaar EP, Seifert HS (2003) Recombination, repair and replication in the pathogenic Neisseriae: the 3 R's of molecular genetics of two human-specific bacterial pathogens. Mol Microbiol 50:3-13

Kline KA, Seifert HS (2005a) Mutation of the priA gene of *Neisseria gonorrhoeae* affects DNA transformation and DNA repair. J Bacteriol 187:5347-5355

Kline KA, Seifert HS (2005b) Role of the Rep helicase gene in homologous recombination in Neisseria gonorrhoeae. J Bacteriol 187:2903-2907

Knezevic-Vukcevic J, Simic D (1991) RecBC promoted repair of bleomycin damage in *Escherichia coli*. Biochimie 73:497-500

Kogoma T (1997) Stable DNA replication: interplay between DNA replication, homologous recombination, and transcription. Microbiol Mol Biol Rev 61:212-238

Kogoma T, Cadwell GW, Barnard KG, Asai T (1996) The DNA replication priming protein, PriA, is required for homologous recombination and double-strand break repair. J Bacteriol 178:1258-1264

Kolodner R, Fishel RA, Howard M (1985) Genetic recombination of bacterial plasmid DNA: effect of RecF pathway mutations on plasmid recombination in *Escherichia coli*. J Bacteriol 163:1060-1066

Konrad EB (1977) Method for the isolation of *Escherichia coli* mutants with enhanced recombination between chromosomal duplications. J Bacteriol 130:167-172

Kouzminova EA, Kuzminov A (2004) Chromosomal fragmentation in dUTPase-deficient mutants of *Escherichia coli* and its recombinational repair. Mol Microbiol 51:1279-1295

Kouzminova EA, Kuzminov A (2006) Fragmentation of replicating chromosomes triggered by uracil in DNA. J Mol Biol 355:20-33

Kowalczykowski SC (1991) Biochemistry of genetic recombination - energetics and mechanism of DNA strand exchange. Ann Rev Biophysics Biophysic Chem 20:539-575

Kuzminov A (1995) Collapse and repair of replication forks in *Escherichia coli*. Mol Microbiol 16:373-384

Kuzminov A (1999) Recombinational repair of DNA damage in *Escherichia coli* and bacteriophage lambda. Microbiol Mol Biol Rev 63:751-813

Kuzminov A (2001) Single-strand interruptions in replicating chromosomes cause double-strand breaks. Proc Natl Acad Sci USA 98:8241-8246

Lee BI, Kim KH, Park SJ, Eom SH, Song HK, Suh SW (2004) Ring-shaped architecture of RecR: implications for its role in homologous recombinational DNA repair. EMBO J 23:2029-2038

Leiros I, Timmins J, Hall DR, McSweeney S (2005) Crystal structure and DNA-binding analysis of RecO from *Deinococcus radiodurans*. EMBO J 24:906-918

Lesterlin C, Barre FX, Cornet F (2004) Genetic recombination and the cell cycle: what we have learned from chromosome dimers. Mol Microbiol 54:1151-1160

Liu J, Marians KJ (1999) PriA-directed assembly of a primosome on D loop DNA. J Biol Chem 274:25033-25041

Liu YH, Cheng AJ, Wang TCV (1998) Involvement of recF, recO, and recR genes in UV-radiation mutagenesis of *Escherichia coli*. J Bacteriol 180:1766-1770

Lloyd RG, Buckman C (1991) Genetic analysis of the recG locus of *Escherichia coli*-K-12 and of itsrole in recombination and DNA repair. J Bacteriol 173:1004-1011

Lloyd RG, Porton MC, Buckman C (1988) Effect of recF, recJ, recN, recO and ruv mutations on ultraviolet survival and genetic recombination in a recD strain of *Escherichia coli* K12. Mol Gen Genet 212:317-324

Lopes M, Foiani M, Sogo JM (2006) Multiple mechanisms control chromosome integrity after replication fork uncoupling and restart at irreparable UV lesions. Mol Cell 21:15-27

Lovett ST (2004) Encoded errors: mutations and rearrangements mediated by misalignment at repetitive DNA sequences. Mol Microbiol 52:1243-1253

Lovett ST, Drapkin PT, Sutera VA Jr, Gluckman-Peskind TJ (1993) A sister-strand exchange mechanism for recA-independent deletion of repeated DNA sequences in *Escherichia coli*. Genetics 135:631-642

Lovett ST, Feschenko VV (1996) Stabilization of diverged tandem repeats by mismatch repair: evidence for deletion formation via a misaligned replication intermediate. Proc Natl Acad Sci USA 93:7120-7124

Lovett ST, Hurley RL, Sutera VA Jr, Aubuchon RH, Lebedeva MA (2002) Crossing over between regions of limited homology in *Escherichia coli*. RecA-dependent and RecA-independent pathways. Genetics 160:851-859

Lovett ST, Luisi-DeLuca C, Kolodner RD (1988) The genetic dependence of recombination in recD mutants of *Escherichia coli*. Genetics 120:37-45

Lusetti SL, Hobbs MD, Stohl EA, Chitteni-Pattu S, Inman RB, Seifert HS, Cox MM (2006) The RecF protein antagonizes RecX function via direct interaction. Mol Cell 21:41-50

Madiraju MV, Templin A, Clark AJ (1988) Properties of a mutant recA-encoded protein reveal a possible role for *Escherichia coli* recF-encoded protein in genetic recombination. Proc Natl Acad Sci USA 85:6592-6596

Marians KJ (2000) PriA-directed replication fork restart in *Escherichia coli*. Trends Biochem Sci 25:185-189

Matic I, Rayssiguier C, Radman M (1995) Interspecies gene exchange in bacteria: The role of SOS and mismatch repair systems in evolution of species. Cell 80:507-515

Maul RW, Sutton MD (2005) Roles of the *Escherichia coli* RecA protein and the global SOS response in effecting DNA polymerase selection *in vivo*. J Bacteriol 187:7607-7618

Mazin AV, Timchenko TV, Saparbaev MK, Mazina OM (1996) Dimerization of plasmid DNA accelerates selection for antibiotic resistance. Mol Microbiol 20:101-108

McGlynn P, AlDeib AA, Liu J, Marians KJ, Lloyd RG (1997) The DNA replication protein PriA and the recombination protein RecG bind D-loops. J Mol Biol 270:212-221

McGlynn P, Lloyd RG (1999) RecG helicase activity at three- and four-strand DNA structures. Nuc Ac Res 27:3049-3056

McGlynn P, Lloyd RG (2000) Modulation of RNA polymerase by (P)ppGpp reveals a RecG-dependent mechanism for replication fork progression. Cell 101:35-45

McGlynn P, Lloyd RG (2002) Recombinational repair and restart of damaged replication forks. Nat Rev Mol Cell Biol 3:859-870

McHenry CS (2003) Chromosomal replicases as asymmetric dimers: studies of subunit arrangement and functional consequences. Mol Microbiol 49:1157-1165

Meddows TR, Savory AP, Grove JI, Moore T, Lloyd RG (2005) RecN protein and transcription factor DksA combine to promote faithful recombinational repair of DNA double-strand breaks. Mol Microbiol 57:97-110

Meddows TR, Savory AP, Lloyd RG (2004) RecG helicase promotes DNA double-strand break repair. Mol Microbiol 52:119-132

Michel B (2000) Replication fork arrest and DNA recombination. Trends Biochem Sci 25:173-178

Michel B, Flores MJ, Viguera E, Grompone G, Seigneur M, Bidnenko V (2001) Rescue of arrested replication forks by homologous recombination. Proc Natl Acad Sci USA 98:8181-8188

Michel B, Grompone G, Flores MJ, Bidnenko V (2004) Multiple pathways process stalled replication forks. Proc Natl Acad Sci USA 101:12783-12788

Michel B, Recchia GD, PenelColin M, Ehrlich SD, Sherratt DJ (2000) Resolution of Holliday junctions by RuvABC prevents dimer formation in rep mutants and UV-irradiated cells. Mol Microbiol 37:180-191

Morel P, Hejna JA, Ehrlich SD, Cassuto E (1993) Antipairing and strand transferase activities of *E. coli* helicase-II (UvrD). Nuc Ac Res 21:3205-3209

Morimatsu K, Kowalczykowski SC (2003) RecFOR proteins load RecA protein onto gapped DNA to accelerate DNA strand exchange: a universal step of recombinational repair. Mol Cell 11:1337-1347

Noirot P, Gupta RC, Radding CM, Kolodner RD (2003) Hallmarks of homology recognition by RecA-like recombinases are exhibited by the unrelated *Escherichia coli* RecT protein. EMBO J 22:324-334

Nowosielska A, Calmann MA, Zdraveski Z, Essigmann JM, Marinus MG (2004) Spontaneous and cisplatin-induced recombination in *Escherichia coli*. DNA Repair (Amst) 3:719-728

Pages V, Koffel-Schwartz N, Fuchs RP (2003) recX, a new SOS gene that is co-transcribed with the recA gene in *Escherichia coli*. DNA Repair (Amst) 2:273-284

Paques F, Haber JE (1999) Multiple pathways of recombination induced by double-strand breaks in *Saccharomyces cerevisiae*. Microbiol Mol Biol Rev 63:349-404

Petranovic M, Zahradka K, Zahradka D, Petranovic D, Nagy B, SalajSmic E, Petranovic D (2001) Genetic evidence that the elevated levels of *Escherichia coli* helicase II antagonize recombinational DNA repair. Biochimie 83:1041-1047

Pinto AV, Mathieu A, Marsin S, Veaute X, Ielpi L, Labigne A, Radicella JP (2005) Suppression of homologous and homeologous recombination by the bacterial MutS2 protein. Mol Cell 17:113-120

Rangarajan S, Woodgate R, Goodman MF (2002) Replication restart in UV-irradiated *Escherichia coli* involving pols II, III, V, PriA, RecA and RecFOR proteins. Mol Microbiol 43:617-628

Renzette N, Gumlaw N, Nordman JT, Krieger M, Yeh SP, Long E, Centore R, Boonsombat R, Sandler SJ (2005) Localization of RecA in *Escherichia coli* K-12 using RecA-GFP. Mol Microbiol 57:1074-1085

Robu ME, Inman RB, Cox MM (2001) RecA protein promotes the regression of stalled replication forks *in vitro*. Proc Natl Acad Sci USA 98:8211-8218

Robu ME, Inman RB, Cox MM (2004) Situational repair of replication forks: Roles of RecG and RecA proteins. J Biol Chem 12:10973-10981

Rocha EP, Cornet E, Michel B (2005) Comparative and evolutionary analysis of the bacterial homologous recombination systems. PLoS Genet 1:e15

Rupp (1996) DNA repair mechanisms. In: W.D. R (ed). American Society of Microbiology Press, Washington DC, pp 2277-2294

Rupp WD, Howard-Flanders P (1968) Discontinuities in the DNA synthesized in an excision-defective strain of *Escherichia coli* following ultraviolet irradiation. J Mol Biol 31:291-304

Sandler SJ, Marians KJ (2000) Role of PriA in replication fork reactivation in *Escherichia coli*. J Bacteriol 182:9-13

Sandler SJ, Samra HS, Clark AJ (1996) Differential suppression of priA2::kan phenotypes in *Escherichia coli* K-12 by mutations in priA, lexA, and dnaC. Genetics 143:5-13

Sano Y (1993) Role of the recA-related gene adjacent to the recA gene in Pseudomonas aeruginosa. J Bacteriol 175:2451-2454

Sargentini NJ, Smith KC (1986) Quantitation of the involvement of the recA, recB, recC, recF, recJ, recN, lexA, radA, radB, uvrD, and umuC genes in the repair of X-ray- induced DNA double-strand breaks in *Escherichia coli*. Radiat Res 107:58-72

Saveson CJ, Lovett ST (1997a) Enhanced deletion formation by aberrant DNA replication in *Escherichia coli*. Genetics 146:457-470

Saveson CJ, Lovett ST (1997b) Enhanced deletion formation by aberrant DNA replication in *Escherichia coli*. Genetics 146:457-470

Schapiro JM, Libby SJ, Fang FC (2003) Inhibition of bacterial DNA replication by zinc mobilization during nitrosative stress. Proc Natl Acad Sci USA 100:8496-8501

Seigneur M, Bidnenko V, Ehrlich SD, Michel B (1998) RuvAB acts at arrested replication forks. Cell 95:419-430

Seigneur M, Ehrlich SD, Michel B (2000) RuvABC-dependent double-strand breaks in dnaBts mutants require RecA. Mol Microbiol 38:565-574

Shan Q, Bork JM, Webb BL, Inman RB, Cox MM (1997) RecA protein filaments: End-dependent dissociation from ssDNA and stabilization by RecO and RecR proteins. J Mol Biol 265:519-540

Sherratt DJ (2003) Bacterial chromosome dynamics. Science 301:780-785

Singleton MR, Dillingham MS, Gaudier M, Kowalczykowski SC, Wigley DB (2004) Crystal structure of RecBCD enzyme reveals a machine for processing DNA breaks. Nature 432:187-193

Singleton MR, Scaife S, Wigley DB (2001) Structural analysis of DNA replication fork reversal by RecG. Cell 107:79-89

Singleton SF, Xiao J (2001) The stretched DNA geometry of recombination and repair nucleoprotein filaments. Biopolymers 61:145-158

Smith KC (2004) Recombinational DNA repair: the ignored repair systems. Bioessays 26:1322-1326

Spies M, Dillingham MS, Kowalczykowski SC (2005) Translocation by the RecB motor is an absolute requirement for {chi}-recognition and RecA protein loading by RecBCD enzyme. J Biol Chem 280:37078-37087

Steiner WW, Kuempel PL (1998) Sister chromatid exchange frequencies in *Escherichia coli* analyzed by recombination at the dif resolvase site. J Bacteriol 180:6269-6275

Stohl EA, Brockman JP, Burkle KL, Morimatsu K, Kowalczykowski SC, Seifert HS (2003) *Escherichia coli* RecX inhibits RecA recombinase and coprotease activities *in vitro* and *in vivo*. J Biol Chem 278:2278-2285

Taylor AF, Smith GR (2003) RecBCD enzyme is a DNA helicase with fast and slow motors of opposite polarity. Nature 423:889-893

Thoms B, Wackernagel W (1987) Regulatory role of recF in the SOS response of *Escherichia coli*: impaired induction of SOS genes by UV irradiation and nalidixic acid in a recF mutant. J Bacteriol 169:1731-1736

Tseng YC, Hung JL, Wang TCV (1994) Involvement of RecF pathway recombination genes in postreplication repair in UV-irradiated *Escherichia coli* cells. Mutat Res 315:1-9

Tuteja N, Tuteja R (2004) Unraveling DNA helicases. Motif, structure, mechanism and function. Eur J Biochem 271:1849-1863

van Gool AJ, Hajibagheri NM, Stasiak A, West SC (1999) Assembly of the *Escherichia coli* RuvABC resolvasome directs the orientation of Holliday junction resolution. Genes Dev 13:1861-1870

Veaute X, Delmas S, Selva M, Jeusset J, Le Cam E, Matic I, Fabre F, Petit MA (2005) UvrD helicase, unlike Rep helicase, dismantles RecA nucleoprotein filaments in *Escherichia coli*. EMBO J 24:180-189

Vincent SD, Mahdi AA, Lloyd RG (1996) The RecG branch migration protein *Escherichia coli* dissociates R-loops. J Mol Biol 264:713-721

Wang TC (2005) Discontinuous or semi-discontinuous DNA replication in *Escherichia coli*? Bioessays 27:633-636

Wang TC, Smith KC (1983) Mechanisms for recF-dependent and recB-dependent pathways of postreplication repair in UV-irradiated *Escherichia coli* uvrB. J Bacteriol 156:1093-1098

Wang TC, Smith KC (1986) Postreplicational formation and repair of DNA double-strand breaks in UV-irradiated *Escherichia coli* uvrB cells. Mutat Res 165:39-44

Wang TC, Smith KC (1988) Different effects of recJ and recN mutations on the postreplication repair of UV-damaged DNA in *Escherichia coli* K-12. J Bacteriol 170:2555-2559

Wang TCV, Chang HY, Hung JL (1993) Cosuppression of recF-mutation, recR-mutation and recO-mutation by mutant recA alleles in *Escherichia coli* cells. Mutat Res 294:157-166

Webb BL, Cox MM, Inman RB (1997) Recombinational DNA repair: The RecF and RecR proteins limit the extension of RecA filaments beyond single-strand DNA gaps. Cell 91:347-356

West SC (1997) Processing of recombination intermediates by the RuvABC proteins. Ann Rev Genet 31:213-244

Whitby MC, Lloyd RG (1995a) Altered SOS induction associated with mutations in recF, recO and recR. Mol Gen Genet 246:174-179

Whitby MC, Lloyd RG (1995b) Branch migration of three-strand recombination intermediates by RecG, a possible pathway for securing exchanges initiated by 3'-tailed duplex DNA. EMBO J 14:3302-3310

Whitby MC, Vincent SD, Lloyd RG (1994) Branch migration of Holliday junctions: Identification of RecG protein as a junction specific: DNA helicase. EMBO J 13:5220-5228

Witkin EM (1991) RecA protein in the SOS response - milestones and mysteries. Biochimie 73:133-141

Xiao J, Singleton SF (2002) Elucidating a key intermediate in homologous DNA strand exchange: structural characterization of the RecA-triple-stranded DNA complex using fluorescence resonance energy transfer. J Mol Biol 320:529-558

Xu L, Marians KJ (2003) PriA mediates DNA replication pathway choice at recombination intermediates. Mol Cell 11:817-826

Zieg J, Kushner SR (1977) Analysis of genetic recombination between two partially deleted lactose operons of *Escherichia coli* K-12. J Bacteriol 131:123-132

Zieg J, Maples VF, Kushner SR (1978) Recombinant levels of *Escherichia coli* K-12 mutants deficient in various replication, recombination, or repair genes. J Bacteriol 134:958-966

Baharoglu, Zeynep
Centre de Génétique Moléculaire du CNRS, Bâtiment 26, 1 Avenue de la Terrasse, 91198 Gif-sur Yvette Cedex, France

Lestini, Roxane
Centre de Génétique Moléculaire du CNRS, Bâtiment 26, 1 Avenue de la Terrasse, 91198 Gif-sur Yvette Cedex, France

Michel, Bénédicte
Centre de Génétique Moléculaire du CNRS, Bâtiment 26, 1 Avenue de la Terrasse, 91198 Gif-sur Yvette Cedex, France
benedicte.michel@cgm.cnrs-gif.fr

Homologous recombination in low dC + dG Gram-positive bacteria

Humberto Sanchez, Begoña Carrasco, Silvia Ayora, and Juan C. Alonso

Abstract

Homologous recombination is a process involved in the maintenance of chromosome integrity, in shaping the evolution of pathogens, in the resistance to antibiotic treatment, and profoundly affecting evolution. In low dC + dG Gram-positive bacteria genetic recombination of a non-replicative homologous DNA, which enters into the cell via transduction or conjugation, proceeds mainly by the double-strand break repair machinery, and this process can be limited by the host restriction system. However, transformation is not limited by sequence divergence (up to 17%), with the activity of restriction endonucleases, or the mismatch repair system only marginally affecting it. The transforming non-replicative homologous DNA, which has a single-stranded nature, is paired with recipient DNA by RecA protein, with the help of redundant accessory proteins, and the intermediates are resolved by a D-loop resolvase. If the transforming DNA does not share homology with the recipient (plasmid establishment), DNA replication and certain recombination proteins other than RecA are required.

1 Proteins required for recombinational repair

The faithful replication and maintenance of the genome(s) are of primary importance for living organisms. However, free radicals generated during essential metabolic processes, or exposure to exogenous agents may damage the DNA. DNA damage, if unrepaired or misrepaired, can induce permanent changes that may lead to severely impairment of cellular functions. In response to this threat, cells have developed a variety of mechanisms, being recombination an essential and ultimate resource for the re-establishment of replication and faithful chromosomal segregation. In many bacteria, one major pathway involving homologous recombination (HR) and a minor one involving non-homologous end joining (NHEJ) have been described.

Escherichia coli, a Gram-negative bacterium, is the model that has became a paradigm to study HR. However, the Gram-positive bacteria with low dG + dC content in their DNA form a homogeneous group, which is evolutionary separated by more than 1.5 billion years from the Gram-negative bacteria, a time divergence larger than the one between plants and animals. This group of bacteria (also termed Firmicutes) appears to be the earliest-branching bacterial phylum

(Ciccarelli et al. 2006). It is typically divided into the Clostridia class, which is anaerobic, and the Bacilli class, which are obligate or facultative aerobes, and the Mollicutes class, which lack the cell wall and are obligate parasites or endosymbionts. Many of these bacteria are relevant to health (*Clostridium*, *Listeria*, *Staphylococcus*, *Streptococcus*, etc.) as well as to biotechnological purposes (*Bacillus*, *Lactococcus*, *Lactobacillus*, etc). DNA repair and genetic recombination in the low dG + dC Gram-positive bacteria is best understood in *Bacillus subtilis*. Genetic screens with *B. subtilis* mutants, which show decreased survival in response to DNA-damaging agents and/or altered rates of recombination during natural transformation, have led to the identification of at least 35 gene products involved in HR (Fernandez et al. 2000; Sanchez et al. 2005, 2006; Tables 1 and 2). There are at least two gene products [*ykoV* (also termed *ku*) and *ykoU* (*ligD*)] involved in NHEJ present in *B. subtilis*, whereas in other Firmicutes, only the Ku homolog is present, or even both are absent (Weller et al. 2002; Bowater and Doherty 2006; Table 2).

In *B. subtilis*, some genes other than *recA* (formerly termed *recE*), which is central to all pathways of homologous recombination, have been placed into seven different epistatic groups, four of them involved in the early steps of the recombination reaction (α, β, δ, and ζ groups), two involved in the late steps of recombination (the ε and η group), and one with an unkown role (the γ group). The genes classified into the epistatic groups are: α [*recF*, *recL*, *recO* and *recR* (formerly *recM*)], β [*addA* and *addB* (also termed *rexAB* in some species of the Bacilli class, order Lactobacillales)], γ [*recP* and *recH*], δ [*recN*], ε [*ruvA* (formerly *recB2*), *ruvB*, *recV* (formerly *recD*) and *recU* (formerly *recG*, *prfA*)], ζ [*recJ*, *recQ* and *recS*] and η [*recG* (also termed *mmsA* in *Streptococcus pneumoniae*)] (Fernandez et al. 2000; Sanchez et al. 2005; Table 1). In Tables 1 and 2, *B. subtilis* was selected as the prototype of the Bacilli class, *Clostridium perfringens* as Clostridia class, and *Mycoplasama mycoides* was selected as the prototype of the Mollicutes class. Many of the genes classified into the seven epistatic groups are conserved into the Firmicutes, but the different end-processing avenues, namely AddAB, RecJ-RecQ, and RecJ-RecS (Sanchez et al. 2006), are absent in the Mollicutes class (Petit 2005; Rocha et al. 2005), and within the Bacilli class, order Lactobacillales, the RecQ and RecS proteins showed a lower level of similarity than expected from the evolutionary distances. Furthermore, from the genes involved in the late steps of the recombination reaction, the RecU Holliday junction (HJ) resolvase (Ayora et al. 2004) is not observed within species classified in the Clostridia class, and RecG in the Mollicutes class (Petit 2005; Rocha et al. 2005).

In addition to these genes products, there are a number of unclassified functions (Table 2) that are involved in several recombination processes as: in RecA modulation (RecX, HelD, PcrA, SsbA), in processing of branched intermediates [Sms (also termed RadA), SubA, Mfd], in re-assembly of the replicative helicase (PriA, DnaB and DnaD), in recombinational repair [RecD (also termed YrrC), RecK

Table 1. Designation of classified *B. subtilis* recombination genes and their *bona fide* or putative orthologs in representative bacteria

Bsu[a]	*Cpe*[b]	*Mmy*[b]	*Mtu*[b]	*Dra*[b]	*Ccr*[b] (α)	*Bpe*[b] (β)	*Eco*[b] (γ)	*Dvu*[b] (δ)	*Hpy*[b] (ε)	*Tpa*[b]	*Npc*[b]	**Role of gene product**
recA	*recA*	*recA*	*recA*	*recA*	*recA*	*recA*	*recA*[e]	*recA*	*recA*	*recA*	*recA*	Strand exchange & ATPase
α *recO*	*recO*	*recO*	*recO*	No	*recO*	*recO*	*recO*[e]	*recO*	No	*recO*	*recO*	binds and anneals ssDNA
α *recR*	*recR*	*recR*	*recR*	*recR*	*recR*	*recR*	*recR*[e]	*recR*	*recR*	*recR*	*recR*	binds ss & dsDNA
α *recF*	*recF*	*recF/N*[c]	*recF*	*recF*	*recF*	No	*recF*[e]	No	No	*recF*	*recF*	binds ss & dsDNA
α *recL*	?	?	?	?	?	?	?	?	?	?	?	unknown
β *addAB*	*addAB*	*recD* ?	*RecBCD*[d]	*recD* ?	*addAB*	*addAB*	*recBCD*	*addA* ?	*addB* ?	*addA* ?	No	Exonuclease, DNA helicase
γ *recH*	?	?	?	?	?	?	?	?	?	?	?	unknown
γ *recP*	?	?	?	?	?	?	?	?	?	?	?	unknown
δ *recN*	*recN*	*recN/F*[c]	*recN*	*recN*	*recN*[d]	*recN*	*recN*[e]	*recN*	*No*	*recN*	*recN*	ATPase & ssDNA binding
ε *ruvAB*	*ruvAB*	*ruvAB*	*ruvAB*	*ruvAB*	*ruvAB*	*ruvAB*	*ruvAB*[e]	*ruvAB*	*ruvAB*	*ruvAB*	*ruvAB*	HJ binding & processing
ε *recV*	?	?	?	?	?	?	?	?	?	?	?	unknown
ε *recU*	No	*recU*	*ruvC*	*ruvC*	*ruvC*	*ruvC*	*ruvC*[e]	*ruvC*	*ruvC*	*ruvC*	*ruvC*	HJ cleavage
ζ *recS*	*recS*	No	No	No	No	No	No	No	No	No	No	putative DNA helicase
ζ *recQ*	*recQ*	No	No	*recQ*	*recQ*	*recQ*	*recQ*[e]	No	No	No	No	DNA helicase
ζ *recJ*	*recJ*	No	No	*recJ*	*recJ*	*recJ*	*recJ*[e]	*recJ*	*recJ*	*recJ*	*recJ*	ssDNA exonuclease
η *recG*	*recG*	*No*	*recG*	*recG*	*recG*	*recG*	*recG*[e]	*No ?*	*recG*	*recG*	*recG*	HJ processing

[a]The *B. subtilis* epistatic groups are indicated by Greek letter. [b]Orthologous genes in representative bacteria. [c]Shares homology with both *rec*F and *recN* genes. [d]Missing in certain genera of the class. [e]Missing in obligatote parasites of γ-proteobacteria. Bacteria representatives of Firmicutes (Bacilli: *Bsu, B. subtilis*; Clostridia: *Cpe, C. perfringens*; Mollicutes: *Mmy, Mycoplasma mycoides*); Actinobacteria: *Mtu, Mycobacterium tuberculosis*; Deinococci: *Dra, Deinococcus radiodurans*; proteobacteria (α-proteobacteria: *Ccr, Caulobacter crescentus*; β-proteobacteria: *Bpe, Bordetella pertusis*; γ-proteobacteria: *Eco, Escherichia coli*;δ-proteobacteria: *Dvu, Desulfovibrio vulgaris*; ε-proteobacteria: *Hpy, Helicobacter pylori*), spirochetes: *Tpa, Treponema pallidum*; cyanobacteria: *Npc*, Nostoc PCC7120. No, indicates the probable absence of the gene; ?, denotes that the sequence of the gene product is unknown.

(formerly RecB3), SbcC, SbcD, SbcC2 (YhaN), SbcD2 (YhaO)], in the maintenance of separated chromosomes with the proper structure and supercoiling levels (Smc-ScpA-ScpB complex, Hbsu, LrpC, Topo I, Topo II, Topo III and Topo IV), and in sister chromosome separation [RipX-CodV working in concert at *dif*] (Fernandez et al. 1997; 2000; Champoux 2001; Sherratt 2003; our unpublished results; Lopez et al. 2005; Lopez-Torrejon et al. 2006; Table 2). From this group of genes, there is a subgroup which corresponds to genes that are essential for cell proliferation (e.g. the *dnaB, dnaD, pcrA, ssbA, topA, gyrA, gyrB, parC, parE* and *hbs* genes), another subgroup whose defect renders cells temperature sensitive for growth (e.g. the *smc, scpA* or *scpB* genes), a defect in the third subgroup (e.g. *priA, recX, helD, sms, subA, mfd, lrpC, recD, recK, sbcC, sbcD*, and *ripX* genes) renders cells sensitive to some DNA damaging agents, and finally the remaining group only shows a defect in certain recombination deficient strains (e.g. *ligD, ku, topB, mutS2, sbcC2* and *sbcD2* genes). Within some species of the Bacilli class, order Bacillales (e.g. *Stpahylococcus aureus*) the *smc* and *mutS2* genes seem to be essential (Forsyth et al. 2002) and in the order Lactobacillales (e.g. in *S. pneumoniae*) the *ruvA, ruvB* and *recU* genes, classified within the ε epistatic group, seem to be essential (Thanassi et al. 2002). Unlike in *B. subtilis* and in *S. aureus* cells (Petit and Ehrlich 2002), the conserved *pcrA* gene is not essential in *S. pneumoniae* (Forsyth et al. 2002; Thanassi et al. 2002). Naturally transformable bacteria (*B. subtilis* and *S. pneumoniae* among others) usually have two Ssb proteins: one essential (SsbA), which is required for DNA replication and repair during vegetative growth (Forsyth et al. 2002; Thanassi et al. 2002), whereas the other (SsbB or YwpH), which is not essential, is required for optimal natural transformation (Lindner et al. 2004). Recently, it was shown that SsbA and SsbB are tyrosine phosphorylated proteins, and that in response to DNA damage, the degree of phosphorylation of both proteins was reduced. The physiological role of SsbA, SsbB, and *E. coli* Ssb phosphorylation in the regulation of DNA metabolism remains to be addressed (Mijakovic et al. 2006).

The *B. subtilis* RecA, RecF, RecG, RecJ, RecN, RecO, RecR, RuvAB, PriA, and SsbA proteins have homologous counterparts with the same designation in selected representatives of actinobacteria, ε-, δ-, α-, β-, and γ-proteobacteria, spirochetes and cyanobacteria (see Rocha et al. 2005; Tables 1 and 2) and their role will be discussed below. The AddAB, Hbsu, PcrA, Smc-ScpAB, and RipX-CodV proteins, which are also broadly distributed among bacteria, and RecU, which is only present in Firmicutes and archaea, have functional analogues with different names in γ-proteobacteria: the RecBCD, HU, UvrD/Rep, MukFEB, XerCD, and RuvC proteins, respectively). There are many functions that do not have a counterpart in γ-proteobacteria [as the RecS, RecD (YrrC), MutS2 (YshD), SbcC2 (YhaN), SbcD2 (YhaO), DnaB and DnaD functions)] and vice versa, γ-proteobacteria functions that do not have a counterpart in low dG + dC Gram-positive bacteria (namely SbcB, PriB, PriC, DnaT) (Table 2). Furthermore, the corresponding gene products of several *B. subtilis* recombination mutants (namely, *recH, recK, recL, recP* and *recV*) have not yet been identified (Fernandez et al. 2000; Tables 1 and 2).

Table 2. Designation of *B. subtilis* unclassified genes required for recombinational repair and their bona fide or putative orthologs in representative bacteria

Bsu[a]	*Cpe*[b]	*Mmy*[b]	*Mtu*[b]	*Dra*[b]	*Ccr*[b] (α)	*Bpe*[b] (β)	*Eco*[b] (γ)	*Dvu*[b] (δ)	*Hpy*[b] (ε)	*Tpa*[b]	*Npc*[b]	**Role of gene product**
smc-scpAB[c]	*smc-scpAB*	?	*smc-scpAB*	*smc-scpAB*	*smc-scpAB*	*smc-scpAB*	*mukFEB*[b]	No	No	*smc-scpAB*	*smc-scpAB*	Segregation, DNA repair
recX	*recX*	No	*recX*	No	No	*recX*	*RecX*[d]	No	No	*recX*	No	RecA modulator
recK	?	?	?	?	?	?	?	?	?	?	?	unknown
recD	*recD*	*recD*	*No*	*recD*	No	No	No	No	No	No	No	DNA helicase
pcrA[c]	*pcrA*	*pcrA*	*pcrA*	*pcrA*	*pcrA*	*pcrA*	*rep/uvrD*	*pcrA*	*pcrA*	*pcrA*	*pcrA*	DNA helicase
priA	*priA*	No	*priA*	No	*priA*	*priA*	*priA*	*priA*	*priA*	*priA*	*priA*	DNA helicase
No	No	No	No	No	No	No	priC[d]	No	No	No	No	DNA helicase
dnaB[c]	No	No	No	No	No	No	No	No	No	No	No	Primosome assembly
dnaD[c]	*dnaD*	No	No	No	No	No	No	No	No	No	No	Primosome assembly
sms	*sms*	No	*sms*	*sms*	*sms*	*sms*	*radA*	No	*sms*	*sms*	No	HJ processing
subA	*subA*	No	No	No	No	No	No	No	No	No	No	HJ processing
yrrK[c]	*yrrK*	*YrrK*	*yrrK*	*yrrK*	*yrrK*	*yrrK*	*yqgF*	*yrrK*	*yrrK*	*No*	*yrrK*	Putative HJ resolvase
mfd	*mfd*	No	*mfd*	*mfd*	*mfd*	*mfd*	*mfd*	*mfd*	*mfd*	*mfd*	*mfd*	Transcription repair factor
mutS2	*mutS2*	No	No	*mutS2*	No	No	No	No	*mutS2*	*mutS2*	*mutS2*	Processing intermediates
ripX-codV	*ripX-codV*	No	*xerCD*	*xerCD*	*xerD*	*xerD*	*xerCD*	*codV*	*xerCD*	*xprB-codV*	*xerD*	Site-specific recombinase
No	No	No	No	No	No	*sbcB*	*sbcB*	No	No	No	No	Exonuclease I
sbcCD	*sbcCD*	No	No	*sbcCD*	No	No	*sbcCD*[d]	No	No	*sbcCD*	*sbcCD*	Binds & degrades DNA
sbcC2D2	*sbcC2D2*	No	No	No	No	No	No	No	No	No	No	To be characterised
ykoV (ku)	No	No	ku	No	*ku*	*ku*	No	No	No	No	No	DNA end binding
ykoU (ligD)	No	No	*lig*	No	*lig*	No	No	No	No	No	No	Multifunctional protein[e]

[a]The designation of the genes and their former and/or alternative name denoted between parentheses. [b]Ortholog genes in other bacteria. [c]Denotes that the gene product is essential. [d]Missing in obligatote parasites of γ-proteobacteria. [e]The gene product has DNA ligase and polymerase activity. The abbreviation of the representative bacteria are the same as in Table 1. No, indicates the probable absence of the gene and ? denotes that the sequence of the gene product is unknown.

The study of different suppressors has uncovered an unknown but common role for many recombination functions as for the HelD (also termed helicase IV) and PcrA proteins. It was shown that the requirement of PcrA is suppressed by inactivation of the *recF, recL, recO* or *recR* genes, and that the absence of *helD* partially suppressed the requirement for the *recF, recL, recO*, or *recR* gene products (Carrasco et al. 2001; Petit and Ehrlich 2002). It was also shown that the Sms, SubA and Mfd proteins play an unknown but active role in the stabilization and/or processing of branched DNA molecules (Carrasco et al. 2004b). The absence of Sms, SubA, or Mfd partially suppressed the DNA repair and segregation defect of *ruvAB* or *recG* cells (Carrasco et al. 2004b, our unpublished results). The Sms, Mfd and PcrA proteins are broadly distributed functions (Table 2), HelD is less conserved and SubA is restricted to low dC + dG Gram-positive bacteria.

The main role of DNA topoisomerases (Topo I, Topo II, and Topo IV) is to modulate the supercoiling levels of chromosomes and thereby to coordinate the main cellular DNA transactions (Champoux 2001). Topo IV is also responsible for untangling catenanes and knots (Espeli and Marians 2004). In contrast to the other three enzymes, Topo III, which is not essential, might be specifically involved in disentangling recombination intermediates, as an alternative to the RuvAB-RecV-RecU action (Carrasco et al. 2004b; Lopez et al. 2005). The Smc, ScpA (member of the kleisins family), and ScpB proteins form the ternary Smc-ScpA-ScpB complex, which is involved in DNA repair (Dervyn et al. 2004), and in the condensation and segregation of the bacterial chromosome immediately after replication (Mascarenhas et al. 2002; Lehmann 2005). The Smc-ScpA-ScpB complex or its analogue in γ-proteobacteria (MukFEB) is well conserved among bacteria (Table 2). In low dC + dG Gram-positive bacteria there are two SbcCD-like complexes (SbcCD and SbcC2D2), but no information is available about the biochemical activities associated with them. LrpC was shown to bridge two separate DNA strands, promote joint molecule formation and to bind specifically to Holliday junctions (HJ) (Lopez-Torrejon et al. 2006). LrpC belongs to the Lrp/AsnC family of proteins, which is highly conserved in bacteria and archaea (Thaw et al. 2006), but the role of other members of this family in recombination remains to be tested.

In *E. coli* cells, the replication apparatus can be reloaded, at stalled or collapsed forks, by a PriA- or by a PriC-dependent restart pathway (Heller and Marians 2005; Heller and Marians 2006), but the presence of the PriC PriB and DnaT assembly factors is not obvious in low dC + dG Gram-positive bacteria (Petit 2005). In *B. subtilis*, the DNA remodelling DnaD protein recruits the D-loop bound PriA to the membrane-associated DnaB, what leads to the loading of the replicative helicase (termed DnaC in *B. subtilis*) with the help of the helicase loader DnaI onto single-stranded DNA (ssDNA) (Bruand et al. 2005; Zhang et al. 2005). DnaB and DnaD are essential functions that also play an active role in DNA replication initiation at *ori*C, suggesting again a fundamental difference in DNA metabolism between the Gram-negative and the low dC + dG Gram-positive bacteria. The DnaB protein is not observed within species classified in the Clostridia class and PriA, DnaB and DnaD are absent in the Mollicutes class (Petit 2005).

In mammalian cells, the repair of DNA double strand breaks (DSBs) by the NHEJ pathway is critical for genome stability. Until recently, it was assumed that this DSB repair pathway was restricted to the eukarya. However, a functionally homologous prokaryotic NHEJ repair apparatus has now been identified and characterised (Weller et al. 2002; Wilson et al. 2003; Bowater and Doherty 2006). *B. subtilis* has a single homolog of the mammalian heterodimer Ku70/Ku80, the Ku (also termed YkoV) protein, and a protein with limited homology to Ligase IV or LigD (termed YkoU) (Aravind and Koonin 2001; Weller and Doherty 2001; Bowater and Doherty 2006; Tables 2 and 3). The *ku* (*ykoV*) and *ligD* (*ykoU*) genes, which encode *bona fide* proteins of the NHEJ complex, are also present in certain actinobacteria, γ-, β-, α- and δ-proteobacteria (Aravind and Koonin 2001; Doherty et al. 2001; Weller and Doherty 2001; Table 2). Genetic and cytological studies have revealed that the *B. subtilis* Ku (YkoV) protein is associated with the nucleoid during spore germination, and that the *ku* and *ku ligD* mutant strains are significantly more sensitive to dry heat (a procedure known to cause DSBs) than the wild type spores (Wang et al. 2006). Null mutations in *ku*, *ligD* (Weller et al. 2002), *sbcC* and *ku sbcC* show a mild sensitization, during both vegetative growth and stationary phase, to ionizing radiation and to mitomycin C (MMC) addition (Mascarenhas et al. 2006). Furthermore, *ku* and *sbcC* are epistatic one to another, (Weller and Doherty 2001; Mascarenhas et al. 2006). In contrast, the *addAB* and *recN* mutations render cells sensitive and very sensitive to ionizing radiation and MMC, respectively, and the *ku* and *sbcC* mutations, markedly increase the sensitivity of *recN* or *addAB* cells to ionizing radiation and to MMC addition (Mascarenhas et al. 2006; Table 3). These experiments support the conclusion that HR is the major pathway for repairing DSBs, and that NHEJ serves as a backup system for DSB repair in growing or in stationary cells. They also suggest that SbcC plays a role in NHEJ. However, *sbcC* is also epistatic with *recA* in *B. subtilis*, and both SbcC and Sbc2(YhaN) proteins interact with the DnaG (primase) protein (Noirot-Gros et al. 2002), and cytological studies revealed that SbcC colocalises with the replication factory (Mascarenhas et al. 2006), indicating that SbcC does not solely act in NHEJ. The putative role of the *B. subtilis* SbcC2 (YhaN) and SbcD2 (YhaO) proteins in NHEJ remains to be unravelled.

Some genes involved in recombinational repair are regulated by the SOS response, whose regulator, the LexA (also termed DinR) protein, is found in many bacterial species. In *B. subtilis*, about 33 genes with LexA binding sites exhibit RecA-dependent induction by both MMC and UV radiation, and among them, the DNA repair genes *recA*, *ruvAB*, *pcrA*, *uvrBA*, *uvrC*, *polY*1 (also termed *dinB*), *polY*2 (*yqjW* or *umuC*) and *parCE* (Au et al. 2005), but *recN* induction in response to DNA damage (our unpublished results) does not seem to depend on LexA (see Au et al. 2005). Alternatively, as it has been shown in *S. pneumoniae,* the cellular response to DNA damage or to general stress conditions is coordinated by the competence regulatory cascade (see below) (Prudhomme et al. 2006).

Some integrated bacterial viruses (prophages) also provide functions that might contribute to the repair of DSBs in the host strain. The prototype is the SKIN prophage, which is present in the genome of some *B. subtilis* strains. It encodes proteins with a significant degree of identity with RecE (a 5′- to 3′exonuclease), with

Table 3. Proteins involved in NHEJ in various systems

Activity	*B. subtilis*	*S. cerevisiae*	*H. sapiens*
End binding	Ku (YkoV)	Hdf1/Hdf2	Ku70/ku80
Protein kinase	?	?	DNA-PK_{cs}
Multifunctional[a]	Lig (YkoU)	?	?
DNA ligase	?	Dnl4/Lif1	LigIV/XRCC4
End bridging	SbcCD	MRX	MRN
DNA polymerase	?	Pol4	Pol μ or Pol λ
5′ flap endonuclease	?	Rad27	FEN1

[a]DNA ligase and polymerase activity.

RecT (an ATP-independent recombinase) (Clark and Sandler 1994), and with RusA (a HJ resolvase) which are encoded in *E. coli* in different cryptic prophages (namely *recE* and *recT* by Rac and *rusA* by DLP12). Deletion of the SKIN prophage shows no DNA repair or segregation phenotype (Carrasco et al. 2004a). Finally, some virus-encoded 5′- to 3′exonuclease and ATP-independent recombinase activities (e.g. SPP1 bacteriophage) are usually required for the generation of concatemeric viral DNA, and play an active role in the generation of transducing particles (Viret et al. 1991; Ayora et al. 2002; Martinez-Jimenez et al. 2005).

2 Recombination avenues

During normal growth the replication fork may be stalled at DNA lesions, at proteins trapped at specific sites, or it might collapse at nicks or DSBs in the DNA template, and their restoration relies on replication reactivated by HR. At the stalled fork single-strand gaps (SSGs) accumulate. Daughter strand gap or SSG repair is poorly characterised in low dG + dC Gram-positive bacteria. A DSB arises if a replication fork is collapsed at a single-stranded (ss) DNA nick in the template DNA. In *B. subtilis* cells, more than 99.9 % of the DSBs created by the site-specific HO endonuclease may be repaired by HR, and the rest (see Dempsey and Dubnau 1989) are supposed to be repaired by NHEJ. To understand the cellular response to DSBs in *B. subtilis*, we subdivide its repair response into six general steps: (a) damage recognition, (b) end-processing, (c) DSB "*coordination*", (d) RecA loading, (e) homologous pairing and strand exchange, and (f) branch migration and resolution (Table 4). Genetic and molecular analysis of both eukaryotes and prokaryotes, together with the identification of the proteins that promote the predicted steps of the DSB repair model, validate the general applicability of it (Table 4).

Table 4. Proteins involved in DSB repair in various systems

Activity	**Bacteria**		**Archaea**	**Human**
	E. coli	***B. subtilis***		
Damage recognition	RecN ?	RecN	Mre11/Rad50	MRN[a]
End processing	RecBCD	AddAB	?	?
	RecQ-RecJ[b]	RecQ(S)[c]/RecJ		
DSB coordination	SbcCD ?	RecN ?	Mre11/Rad50	MRN
	MukB ?	SMC/ScpAB	?	SMC1/3/Scc1/3
	Ssb	SsbA	SSB/RPA	RPA
Recombinase load-	RecBCD	AddAB?	?	
ing	RecFOR	RecOR(L)F		Rad52
Recombinase	RecA	RecA	RadA	Rad51
protein				
Mediator proteins	RecOR	RecOR (L?)	?	Rad52
	RecF	RecF	RadB	Rad51BCDXrcc2/3
	?	RecL ?	Rad54	Rad54
	?	?	?	BRCA2
Recombinase modu-	DinI	?	?	?
lators	RecX	RecX ?	?	?
	UvrD	PcrA	?	?
	?	RecU	?	Rad51C/Xrcc3 ?
Replication re-start	PriAB/DnaTBC	PriA/DnaDBIC	?	?
	PriC/Rep/DnaBC	?	?	?
Branch migration	RuvAB	RuvAB/RecV	?	?
	RecG	RecG	?	?
	?	RecS/RecQ ?	Hjm	RecQ family ?
	?	?		Rad54 ?
HJ resolution	RuvC	RecU	Hjc/Hje	Rad51C/Xrcc3 ?
protein(s)	RecQ-Topo III	RecQ-Topo III ?	?	BLM/Topo IIIα
	?	?	XPF	Mus81/Eme1

[a]Mre11, Rad50 and Nbs1 in humans, and Mre11, Rad50 and Xrs2 in yeast. [b]In the *recBC sbcB sbcC* background RecQ and RecJ are involved in end processing. [c]Two RecQ-like helicases (RecS and RecQ) in concert with RecJ process DNA ends in an otherwise wild type *B. subtilis* strain.

2.1 DNA damage recognition

Cytological studies have revealed that upon induction of random or site-specific DSBs, which can be accomplished by the expression of the HO-endonuclease, the nucleoids fuse. 15 to 30 minutes after induction of DSBs, RecN localises as a discrete focus on the nucleoids in a majority of the cells, forming mainly a discrete repair centre (RC) (Kidane et al. 2004). Under this condition, growth resumes ~ 180 minutes after induction of the DSBs. Other HR proteins assemble at the RC in a discrete temporal order, being RecN the first protein recruited (Kidane et al. 2004). RecN forms one discrete focus or RC in all tested mutant strains (Kidane et al. 2004; Sanchez and Alonso 2005; Sanchez et al. 2005, 2006). Recently it has been shown that ~ 2 % of *add*AB Δ*recJ*, ~ 35% of Δ*recA* and ~ 5% of Δ*recU* mutant cells contain RecN-induced foci under normal growth conditions, what suggests that these mutants accumulate unrepaired DSBs (Kidane et al. 2004) or

SSGs (our unpublished results). These results together suggest that the RecN protein, which directs the recruitment of proteins at DSBs, is the main player in step (a) (see Section 2). RecN is highly ubiquitous among Bacteria and Firmicutes but absent in the Mollicutes class. RecN initiates assembly at DNA ends, by specifically binding to the 3′-ends, and organises the assembly of a single network of protein-protein interactions at a RC (Kidane et al. 2004; Sanchez et al. 2005; Sanchez et al. 2006). However, in the absence of RecN few RCs were still observed, what suggests that in this null mutant strain an unknown factor may organise the RCs (Kidane et al. 2004).

In eukaryotes, the Mre11-Rad50- Nbs1 (Xrs2) complex binds directly to the DNA ends, and appears to be the earliest sensor of a DSB (Lisby and Rothstein 2004). Interestingly, both Rad50 and RecN are SMC-like proteins, and may serve similar functions in the detection of DSBs.

2.2 DNA end-processing

Genetic and biochemical studies revealed that nucleolytic resection of the DNA ends, to generate a 3′-terminated ssDNA molecule, can be performed by AddAB or RecJ in concert with RecQ and/or RecS. The AddAB enzyme, which comprises one set of helicase motifs and two distinct nuclease activities, recognises a short Chi sequence, and degrades one of the two DNA strands to generate the a recombinogenic 3′-tailed duplex molecule (Chedin and Kowalczykowski 2002; Chedin et al. 2006). Alternatively, the putative RecJ 5′to 3′ ssDNA exonuclease may generate a 3′-terminated ssDNA molecule by degrading the 5′-terminated ssDNA upon the action of any of the two putative RecQ-like (RecQ or RecS) 3' to 5' DNA helicases (Sanchez et al. 2006). This is consistent with the observation that AddAB and RecJ are the "*major*" nucleases that process the dsDNA ends to generate a 3′-tailed duplex DNA to which RecA will bind (Sanchez et al. 2006), and with the observation that *addA*5 *addB*72 Δ*recQ* and *addA*5 *addB*72 Δ*recS* cells are less sensitive than *addA*5 *addB*72 Δ*recJ* cells to different DNA damaging agents (Sanchez et al. 2006). The fine mechanisms of end-processing via RecQ-RecJ or RecS-RecJ is poorly understood. The RecS protein contains the DExH-box helicase motif and part of the RecQ conserved C-terminal (RQC) domain, whereas the RecQ protein contains the DExH, RQC and the RNase D C-terminal (HRDC) domains. The *recQ* gene is broadly distributed, and the *recS* gene is mainly restricted to low dC + dG Gram-positive bacteria (Fernandez et al. 1998; Table 4).

An evolutionary study of the bacterial end-processing machineries revealed that in the 115 deposited non-redundant sequenced genomes, the RecJ-RecQ-like (RecQ and/or RecS) avenue, might be more widely spread (~ 81%) than the AddAB/RecBCD nuclease/helicase (~ 48%) avenue. Only five among the 115 genomes contain an AddAB/ RecBCD enzyme and lack the RecJ-RecQ-like nuclease-helicase system. In only those species that are obligate intracellular parasites or obligate endosymbionts (~ 15% of total non-redundant sequenced genomes), both the RecQ(S)-RecJ and AddAB/RecBCD functions are missing (Sanchez et al. 2006). In some species within the Mollicutes class the RecD (YrrC) helicase alone or in concert with an exonuclease domain of another protein (e.g. DNA Pol I) might process the ends.

2.3 DSB coordination

In the absence of the two end-processing nucleases, namely in *addA*5 *addB*72 Δ*recJ* cells, multiple RecN-promoted repair foci accumulate within the cells, and RecA cannot form threads or filaments (Kidane and Graumann 2005a; Sanchez et al. 2006). We proposed that immediately after DNA damage RecN binds to the ssDNA tail of the duplex molecule forming repair foci, and after end-processing RecN facilitates the tethering of these DNA ends together to form mainly one discrete focus or RC (Kidane et al. 2004; Kidane and Graumann 2005a). Indeed, in the presence of ATP RecN specifically binds to the 3′-OH end of the ssDNA tail, and promotes a large nucleoprotein assembly (Sanchez and Alonso 2005; Sanchez et al. 2006).

After induction of DSBs SbcC forms discrete foci and they mostly co-localise with the stationary chromosomal replicase complex, whereas RecN forms a RC at any discrete location on the nucleoids (Mascarenhas et al. 2006). It is likely, therefore, that SbcCD alone or in concert with Ku, which are epistatic, may modulate a minor DNA repair avenue unlinked to RecN-promoted RCs (Mascarenhas et al. 2006). This is consistent with the observation that RecN and SbcC are not epistatic and mutations in HR genes (*addAB*, *recJ* or *recN*) do not increase the frequency of NHEJ (Mascarenhas et al. 2006). It remains to be established if there is any coordination between the HR and NHEJ pathways. In eukaryotic cells upon end-processing, checkpoint activation and Mre11 dissociation from the DSB site, the commitment to NHEJ or HR takes place (Lisby and Rothstein 2005).

2.4 RecA loading, homologous pairing and strand exchange

Upon resection of the DNA ends, the SSGs are covered by SsbA protein. *In vivo* and *in vitro* evidence shows that the SsbA or SsbB proteins bind ssDNA without affecting RecN-promoted protein-DNA networks (Kidane and Graumann 2005b; Sanchez and Alonso 2005). Both RecN and SsbA (or SsbB) proteins must be displaced from the ssDNA in order to be coated with RecA. Little is known concerning RecA loading onto ssDNA and its regulation in *B. subtilis* cells [step (d) in Section 2; Table 4]. Cytological studies reveal that *B. subtilis* RecA forms highly dynamic threads (filamentous) structures across the nucleoids (Kidane and Graumann 2005a). *In vitro* studies reveal that the formation of a RecA nucleoprotein filament allows strand invasion of the 3′-ssDNA broken end into its homologous undamaged partner, forming a 3-strand intermediate (D-loop) (Carrasco et al. 2005). However, it is unknown whether AddAB can directly load RecA onto ssDNA, as it is the case with its *E. coli* RecBCD counterpart, although genetic studies suggest that it can (Chedin and Kowalczykowski 2002).

Cytological studies revealed that RecN may direct to a RC RecO and RecR, perhaps in concert with RecL. *In vitro* studies reveal that RecO reverses the negative effect of SsbA on the ATPase activity of RecA and may promote displacement of SsbA from ssDNA (C. Manfredi, unpublished results). Concomitantly with RecA promoting strand exchange, RecF, which binds ssDNA or dsDNA (Ayora and Alonso 1997), co-localised with the RecN-RecO-RecA multiprotein complex (Kidane et al. 2004). This temporal order is consistent with the observation that RecF failed to form any foci in *recO* cells and RecF foci accumulate later than RecA loading (Kidane et al. 2004; Sanchez and Alonso 2005). The role of the RecFLOR complex in RecA loading and SsbA displacement is consistent with the fact that: i) certain RecA mutants or the over-expression of a bacteriophage Ssb protein partially suppresses the RecFLOR defect (Alonso and Luder 1991); ii) the absence of RecA modulators as HelD or PcrA partially suppress the RecFLOR defect (Carrasco et al. 2001; Petit and Ehrlich 2002), and iii) mutants lacking *rec*F, *rec*L, *rec*O or *rec*R exhibit a delayed and reduced induction of SOS-regulated genes (Gassel and Alonso 1989).

In eukaryotic cells DNA damage recognition, resection of the DNA ends, checkpoint activation, and NHEJ and HR commitment can proceed at any stage of the cell cycle, whereas Rad51 loading is restricted to S and G_2 phase. Here, Rad52 and Rad55-Rad57, in yeast, or Rad52 in concert with the Rad51 paralogs, in humans, ensure the loading of multiple Rad51 molecules onto ssDNA (Sung et al. 2003; Krogh and Symington 2004; Table 4).

2.5 Branch migration and resolution

Once a D-loop is formed, by extension into the homologous region of the strand exchange reaction a HJ is formed. The RuvAB translocase alone, or in concert with the uncharacterised RecV protein [RuvAB-(RecV)] may recruit RecU to the HJ to process it. Alternatively, in the absence of RuvAB-(RecV), RecU alone, or with the help of the branch migration translocase RecG, is loaded at the HJ (Sanchez et al. 2005). RecU bound to a HJ modulates RecA-promoted strand exchange (Carrasco et al. 2005) and catalyses the cleavage of opposite arms of the junction at certain preferred sequences (Ayora et al. 2004; McGregor et al. 2005). The RecU HJ-resolvase is not observed in some species of the Clostridia class and some species of the Mollicutes class, suggesting that resolution might be provided in these bacteria by the ubiquitous YrrK protein, which has been predicted to be an "HJ-resolvase" (Table 2). The *B. subtilis* YrrK protein, which has an RNase H fold, was unable to cleave mobile or static HJs (S. Ayora, unpublished results).

Random resolution of HJs is expected to yield equal numbers of crossover (CO) and non-crossover (NCO) products, but generally bacteria have circular genomes and resolution towards COs generates dimeric chromosomes (Sherratt 2003). The branch migration translocases may help RecU to catalyse cleavage of the HJs towards NCOs, because in the absence of any of the branch migration translocases (*ruvAB* or *recG*) accumulation of anucleated cells and linked chromosomes were observed (Carrasco et al. 2004b). This is consistent with the observations that i) a null *subA* mutant allele suppresses the repair and segregation defect of Δ*ruvAB* or Δ*recG* cells (Carrasco et al. 2002, 2004b); ii) Δ*ripX* and Δ*recG* are not epistatic (our unpublished results); and iii) the defect on chromosomal segregation is synergistic in the synthetically lethal Δ*recU* Δ*smc* strain, and additive in Δ*recU* Δ*sp0J* cells (Pedersen and Setlow 2000).

Cytological studies have revealed that RecU, which fails to form any foci in *ruvAB* cells, forms a single, discrete focus on the nucleoid upon induction of DSBs and co-localises with RecN at a RC (Sanchez and Alonso 2005). These data indicate that repair of DSBs is a sequential processes.

Biochemical evidences suggested that RecU and the human Rad51 paralogs might share some features in common: i) both RecU and the Rad51BCD-XRCC2 complex bind specifically to HJs and help Rad51 RecA and, respectively, to initiate DNA strand exchange (Lio et al. 2003; Ayora et al. 2004; Carrasco et al. 2005); ii) both RecU and Rad51C show synergy in their binding affinity with RecA and Rad51, respectively (Kurumizaka et al. 2001; Sigurdsson et al. 2001; Carrasco et al. 2005); and iii) both RecU and the Rad51C-XRCC3 complex are involved in the processing and resolution of HJs (Ayora et al. 2004; Liu et al. 2004; Carrasco et al. 2005). The suggested role of human Rad51 paralogs on the positive and negative modulation of the Rad51 function remains to be documented.

3 Horizontal gene transfer

The discovery of the link between the *S. pneumoniae transforming principle*, which was early described by F. Griffith in 1928, and naked DNA by Avery, MacLeod, and McCarty in 1944, the description of *E. coli* conjugation by Lederberg and Tatum in 1946, and viral general transduction in *Salmonella typhimurium* by Zinder and Lederberg in 1951 gave rise to the first insights into the mechanics of genetic recombination, and how bacteria can acquire new genetic information by horizontal gene transfer (HGT). HGT is a powerful mechanism to provide templates for DNA repair, to increase fitness of a bacterial population or to supply nutrients for bacterial proliferation. The analysis of complete genomes has suggested that HGT events, by any of these three general mechanisms of genetic recombination, are relevant mechanisms for genome plasticity, and responsible for the rapid spread of antibiotic resistance (Jain et al. 1999).

3.1 Transport and uptake of dsDNA or ssDNA

During the infectious cycle newly replicated bacteriophage genomes are packaged into the preformed procapsids. At this stage, non-viral double-stranded (ds) DNA (general transduction) or amplified non-viral plasmid dsDNA (restricted transduction) is also encapsidated by the viral packaging machinery via recombination-dependent replication, albeit with a low frequency (in $< 0.001\%$ of infected cells) (Deichelbohrer et al. 1985). If homology between the host and the viral DNA is present, the amount of non-viral dsDNA encapsidated goes up to 0.5 % of total viral particles (high frequency transfer) (Alonso et al. 1986). The transducing dsDNA, which has a non-infectious nature, is then injected into a new host (Viret et al. 1991; Canchaya et al. 2003). However, in Gram-negative bacteria, the acquisition of plasmid DNA by a new host can be also done by an infectious viral particle (Canchaya et al. 2003).

With the exception of high dC + dG Gram-positive bacteria, which seem to be characterised by the ability to transfer dsDNA, conjugation and transformation in the other bacteria mediate the transport of ssDNA across one or more membranes. Transfer of ssDNA from one bacterium to another through the conjugation apparatus of self-transmissible and mobilisable plasmids, conjugative transposons, and integrative conjugative elements takes place with specific polarity. There are several types of conjugative mechanisms, being the greatest distinctions between those of Gram-positive and the ones from Gram-negative bacteria (Grohmann et al. 2003; Lawley et al. 2003; Chen et al. 2005). The mechanisms of conjugative gene transfer of conjugative plasmids, transposons, or integrative elements were recently reviewed and will not be further discussed here (Grohmann et al. 2003; Lawley et al. 2003; Chen et al. 2005; Frost et al. 2005; Thomas and Nielsen 2005).

Genetic transformation also involves the ability of a natural competent cell to take up exogenous DNA from the environment (Chen and Dubnau 2004; Claverys et al. 2006). With the exception of *Helicobacter pylori*, which uses a conjugation-like machinery, Gram-positive and Gram-negative bacteria use related proteins to import the linear ssDNA (Chen and Dubnau 2004). The major distinction between the competence machineries between these microorganisms comes from the differences in their membranes. The mechanisms of DNA uptake during bacterial transformation were recently reviewed and will not be further discussed here (Chen and Dubnau 2004; Chen et al. 2005).

4 Fate of the incoming DNA

The assimilation of the linear ssDNA transferred by conjugation, which is rapidly converted into dsDNA after transfer, or the assimilation of linear transduced dsDNA injected by a defective bacteriophage into a recipient strain follows the DSB repair process. It is likely that RecN protects the 3′-OH end of the transducing or conjugative non-replicative homologous DNA upon entry, whereas the 5′-end is processed via the AddAB, RecJ-Q/S avenues (Section 2.2). RecA is then loaded, by different mediators, onto the 3′-ssDNA region, and filaments. RecA catalyses strand invasion of the 3′-ssDNA into the intact homologous recipient duplex and forms a D-loop intermediate, and by extension of the strand exchange reaction, a HJ (Section 2.5). Finally, RecG or RuvAB alone or in concert with RecV promote branch migration (Sanchez et al. 2005). Then, RecU resolves the HJs (Carrasco et al. 2004a). Alternatively, RecQ and Topo III, in concert, may catalyse resolution of the HJ. This is consistent with the observation that chromosomal transduction is blocked by mutations in recombination functions as in *rec*A, *addAB recF*, *addAB recH*, or *recF recH* cells (Fernandez et al. 2000).

The establishment of transduced concatemeric linear self-replicative dsDNA (plasmid transduction) or unit-length conjugative plasmid DNA takes place in a similar way as the circularisation of viral molecules, which use for circularisation the terminal redundancy present in the concatemeric head-to-tail plasmid molecules, or the small duplicated region of the transferred origin. In short, nucleolytic resection of the DNA ends to generate a 3-terminated ssDNA can be performed by AddAB or by RecJ-Q/S, and any of the different strand annealing proteins (e.g. RecO, RecU or LrpC) may promote the re-circularisation of the redundant molecule. This is consistent with the observation that plasmid transduction or plasmid conjugation is independent of RecA. The requirement of RecO, RecU, and RecS proteins for plasmid establishment was observed (Alonso et al. 1992; Fernandez et al. 2000). The oligomeric state of the re-circularised molecule could be resolved either by the site-specific resolution system, provided that a resolution site is present in the plasmid molecule, or via HR (Ceglowski et al. 1993). In the case of mobile genetic elements, however, site-specific recombination promotes circularization of the incoming DNA molecule (Frost et al. 2005; Thomas and Nielsen 2005). The β serine recombinase is the prototype of a group of site-specific recombinases encoded by large conjugative plasmids of low dC + dG Gram-positive bacteria. The β recombinase (Alonso et al. 1996) differs from serine site-specific recombinases of Gram-negative bacteria (Tn*3* and γδ resolvases or Hin and Gin DNA invertases). β recombinase promotes the formation of an elaborated synaptic complex (Canosa et al. 2003), but lacks the "topological" filter, because it catalyses resolution between two directly oriented recombination sites (*six* sites) and both resolution and DNA inversion between two inversely oriented *six* sites (Canosa et al. 1998). The β synaptic complex formation requires an architectural, bacterial HBsu or eukaryotic HMGB, protein that bends the intervening DNA segment to allows direct contact between the β dimers (Alonso et al. 1995; Canosa et al. 2003). In spite of the differences the β recombinase and Hbsu form a synaptic complex topologically similar to that of Tn*3* and γδ resolvases (Canosa et al. 2003).

Upon entry into the cells, the DNA fragments are integrated in a RecA-dependent process into the resident chromosome, without extensive DNA replication [< 300-nucleotides (nt)], when a sufficiently large region of mutual homology is available (100-nt) (Fernandez et al. 2000). Except *rec*A mutants that are reduced more than four orders of magnitude in chromosomal transformation, a defect on any function classified with the α, β, γ, δ, ε, ζ, or η epistatic groups (Table 1), when present in an otherwise Rec^{+} strain, does not alter more than threefold the frequency of chromosomal transformation (Fernandez et al. 2000). It is likely that redundant avenues are functional in *B. subtilis*, because double mutants as *recO addAB*, *recO recH*, or *addAB recH* cells are blocked for both chromosomal and plasmid transformation (Fernandez et al. 2000).

During the uptake process, one of the two strands of the transforming dsDNA is taken up into the competent cell, with simultaneous hydrolysis of the complementary strand. DNA uptake in *S. pneumoniae* proceeds in 3′ to 5′ direction (Mejean and Claverys 1988) and this may hold true for other natural competent cells. The RecA and SsbB proteins co-localise with ComGA at the cell pole (Kidane and Graumann 2005b), but very little is known about the role of SsbA proteins during genetic transformation (Lindner et al. 2004). The incoming ssDNA is bound by RecA, and RecN protein (Kidane and Graumann 2005b). RecN, which oscillates from pole to pole and becomes static at one pole when transforming DNA is added (Kidane and Graumann 2005b), protects the 3′-OH end of the incoming ssDNA, and may enhance SsbB displacement and RecA loading onto the incoming ssDNA (Kidane and Graumann 2005b; Sanchez and Alonso 2005). RecA forms threads extending from the pole to the centrally located nucleoid (Kidane and Graumann 2005b; Sanchez and Alonso 2005). In the absence of homology with the recipient, the DNA persists in the cell for ~ 30 minutes and is then degraded by cellular nucleases and used as a nutrient (Lacks 1988).

Comparison of well documented transformation systems in Firmicutes, those of *B. subtilis* (representative of the Bacillales class), and of *S. pneumoniae* (representative of the Lactobabillales class) showed that the two species independently evolved competence regulatory circuits, adapted to their own lifestyle (*B. subtilis* is a soil bacterium, and *S. pneumoniae* is a human pathogen) (Martin et al. 2006). Competence induction, which is inhibited in stationary phase, develops in ~ 90 % of *S. pneumoniae* cells in an exponentially growing culture, and it triggers killing of the non-competent siblings cells. By contrast, competence develops at the onset of stationary phase in ~ 10 % of *B. subtilis* cells, and synthesis of secondary metabolits blocks proliferation of sibling cells (Chen et al. 2005; Claverys et al. 2006). The expression of the *smf* (also termed *dprA*), *ssbA*, *ssbB*, and *recA* genes is induced, together with competence specific genes, but expression of end-processing (namely *addA*, *addB*, *recJ*, *recS*, or *recQ*), or RecA-modulators genes (*recF*, *recO*, or *recR*) are not induced upon competence induction (Berka et al. 2002; Ogura et al. 2002). This different induction of recombination functions correlates well with the single stranded nature of the incoming DNA.

If the incoming ssDNA shares homology, a mismatch-free segment of ~ 50-nt or more at both ends of the donor strand, with recipient DNA is rapidly and efficiently integrated (e.g. chromosomal transformation) (Majewski and Cohan 1999). The RecA nucleoprotein filament invades a homologous region on the bacterial genome, forming a D-loop intermediate (Fernandez et al. 2000). The assimilated taken up ssDNA forms a linearly synapsed segment, which cannot be extended to a HJ due to its single-stranded nature (Fig. 1A). The RecU protein might bind to the D-loop and enhance RecA-promoted strand invasion, but inhibits RecA-promoted strand exchange (Carrasco et al. 2005). The RecG or the RuvAB-(RecV) translocases could extend the heteroduplex region at a lower ATP cost than RecA (Hedayati et al. 2002). Simultaneous excision of the recipient and the donor DNA requires the intervention of a putative D-loop resolvase, with the subsequent elimination of the displaced recipient strand. Finally, complete restoration of the duplex "transformed" form requires gap filling by polymerase I and joining of the donor marker by DNA ligase (Ceglowski et al. 1990; Fernandez et al. 2000). Which enzyme is responsible for the cleavage of the D-loop intermediate is

Fig. 1 (overleaf). Recombination of chromosomal (A, B, C) and plasmid DNA (D, E). In A, Chromosomal transformation mediated by linear synapsis. The black lines represent recipient chromosomal DNA, and the grey line homologous donor DNA. The arrowhead on the grey DNA denotes the 3′-end and the pointing triangles indicate the simultaneous incision by a putative D-loop resolvase. Degradation of the displaced recipient DNA, and ligation of donor and recipient DNA would generate the recombinant product. In B, transformation of homologous DNA flanking a heterologous segment (resistance marker) by a double CO event. The donor DNA (grey line at the left hand-side) pairs with recipient chromosomal DNA in a RecA-dependent manner, the filled dotted line denotes the heterologous marker and the empty dotted line vector DNA. In a second step, pairing takes place on the right-hand side, with displacement of the heterologus DNA segment. The putative D-loop resolvase cleaves the strands, and the recombinant product is generated. In C, transformation of a homologous segment flanked by heterologous DNA with autonomous replication potential. The complementary strand is synthesised, annealing of complementary strands renders a circular intermediate that integrates by a single CO event. In D, establishment of a monomeric plasmid DNA, which shares homology with the recipient. RecA promotes strand invasion on the homologous region of bacterial chromosome (or on resident plasmid) to form a D-loop intermediate. Synthesis of the complementary strand, and elongation of the coming strand using recipient DNA as a template, followed by recombination between the terminal redundancy, renders a monomeric plasmid molecule. In E, entering of multimeric ssDNA and reconstitution of a monomeric plasmid molecule. After the failure of RecA in the search for homology, replication of the complementary strand takes place to generate a linear dsDNA molecule with terminal redundancy. In a RecO-, RecU- and RecS-dependent manner the plasmid circularises. The ssDNA tail is degraded and the ends ligated. In a second step ~ 95% of the plasmid molecules monomerise.

unknown. The involvement of the essential and highly conserved among bacteria YrrK protein, rather than the RecU or RusA resolvases, in the resolution of the D-loop has not yet been demonstrated, (Aravind et al. 2000; Table 2). It has been observed that: i) the RecU HJ resolvase fails to cleave D-loop structures, and ii) in the absence of both *rec*U and *rus*A HJ resolvases, chromosomal transformation is not affected more than fourfold (Ayora et al. 2004; Carrasco et al. 2005).

The presence of heterologous DNA, provided that homology is present at both DNA ends, does not drastically interfere with transformation, and is broadly used for the generation of gene disruptions with subsequent replacement of the recipient chromosomal sequence (ectopic integration of a marker, Figs. 1B and 1C). As revealed in Fig. 1B, transforming DNA containing a non-homologous genetic marker (filled dotted line), but linearised within a homologous region > 200-nucleotides (grey line), is assimilated via a RecA-dependent linear synapsis. The extension of the heteroduplex might lead *to transient pairing* between the heterologous donor and recipient sequences. Then RecA promotes the invasion of the other adjacent homologous segment and searches for homology, looping out the non-homologous sequence, and recombination by a second and independent

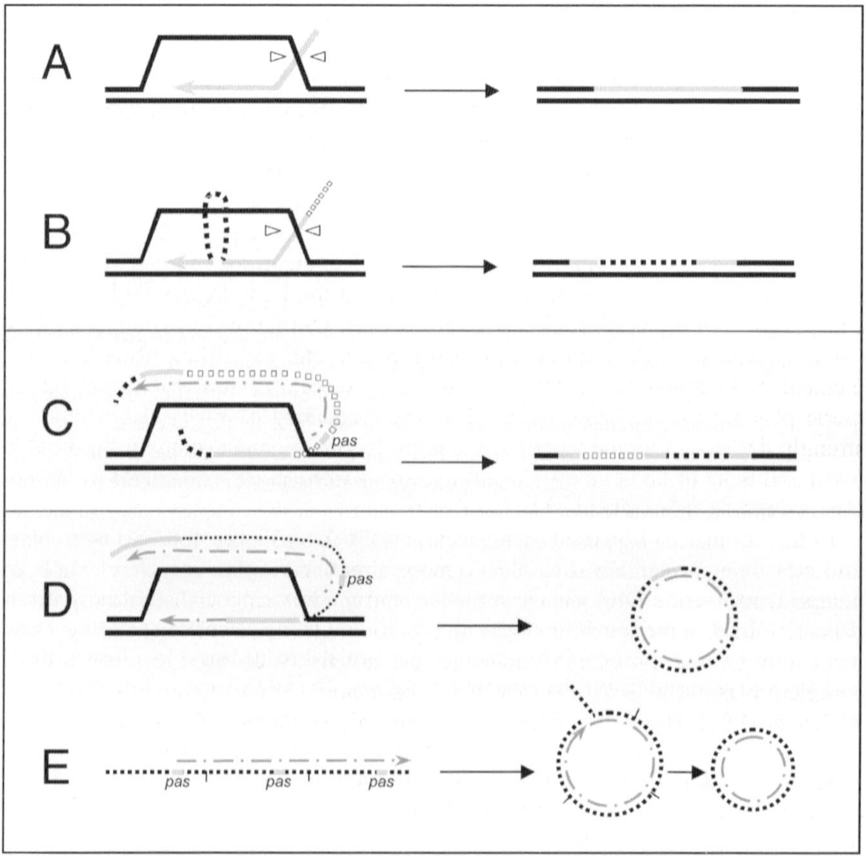

crossover event leads to full integration of the DNA marker flanked by homologous DNA region without deletions (Itaya 1995, Fig. 1B). The generated D-loop intermediate is then processed, as described for normal chromosomal transformation (see Fig. 1A) without integration of the vector DNA (Fig. 1B, open dotted lines). This type of insertion event takes place with a low frequency (~ 1% of total recombination events), but integration of a second marker without gene disruption event takes place with very low frequency (~ 0.01 % of total recombination events) (Prudhomme et al. 2002, our unpublished results). If during the take up process the nick is outside the homologous region (e.g. in the non-homologous genetic marker, filled dotted lines in Fig. 1C) integration of the donor DNA into the recipient is via a replication-dependent "*circular*" intermediate. Synthesis of the complementary strand at any primosome assemble site (*pas*) will render a dsDNA molecule (Fig. 1C) that can be circularised (see below). Since the circular molecule has no functional replication origin, it remains associated with the chromosome until RecA promoted strand invasion allows the integration of the entire molecule, containing a duplication of the homologous segment and insertion of the vector DNA (Lacks 1988, open dotted lines). This single CO event (Fig. 1C) is ~

25-fold more efficient than replacement by the double CO (reviewed in Lacks 1988).

Plasmid transformation can follow two different avenues, depending if it shows or not homology with the recipient. If there is homology with the recipient DNA (> 100-nt), a monomeric plasmid molecule is established in a RecA-dependent manner (Fig. 1D), by a process similar to the one described for the early steps of chromosomal transformation (Fig. 1A). Integration of a segment of chromosomal DNA into the plasmid will generate internal redundancy required for recircularisation by *de novo* DNA synthesis (see Fig. 1D). The incoming monomeric plasmid ssDNA, which has the vector portion intact, and the homologous region at both ends, is covered by RecA as described for chromosomal transformation (Fig. 1A). Complementary DNA synthesis at any *pas* region and gap filling using the complementary recipient strand DNA, followed by circularisation will render monomeric plasmid establishment (Fig. 1D). The frequency of such establishment is strongly dependent on the length of the homologous segment (ranging from 0.2 to 6 kb) and is increased over 1000-fold in *S. pneumoniae* and ~ 100-fold in *B. subtilis* competent cells (reviewed in Lacks 1988).

In the absence of homology with recipient DNA (< 50-nt), an oligomeric plasmid molecule is established 1000-fold more effective than monomers by a RecA-independent avenue, and such event is an intramolecular recombination process (Fig. 1E). In *B. subtilis* cells a single oligomeric molecule is sufficient for replicon reconstitution and monomeric molecules are not freely diffused to reconstitute a complete plasmid as it was shown for *S. pneumoniae* competent cells (reviewed by Lacks 1988). However, when a donor plasmid contains a directly repeated sequence, monomeric forms can be transferred to *B. subtilis* competent cells (Michel et al. 1982). Here, most of the plasmid established consists of a portion of the original plasmid, and only ~ 10% of the transformed cells bear the entire donor plasmid (Michel et al. 1982).

Transformation of oligomeric plasmid DNA, in the absence of homology with recipient DNA does not require RecA, but requires both DNA replication and certain recombination functions other than RecA. Since the incoming oligomeric DNA has an ssDNA nature, RecN may protect the 3′-OH end and RecA polymerase on it at the entry place, and then searches for homology. If RecA fails to find a homologous region, it should disassemble of the ssDNA, by an unknown mechanism, and the replication machinery at any *pas* region (or at the lagging strand origin in rolling circle replicating plasmids) initiates the synthesis the complementary strand (Fig. 1E). Pairing of one of the incoming ssDNA ends with the unwound newly replicated strand results in the circularisation of the plasmid molecule, which is then ligated. This is consistent with the observation that plasmid transformation is not affected in a *rec*A strain, is increased in *rec*N cells, but it is reduced 25- to 100-fold in the absence of the pairing RecO (epistatic group α) and RecU (ε) proteins, or the putative RecS (ζ) helicase (Fernandez et al. 2000). Many other recombination proteins are required for plasmid establishment, but due to the redundancy of the recombination apparatus, they can be only scored as double mutants (e.g. *addAB recO*, *addAB recF*, *addAB recH* are blocked in plasmid transformation) and their role remains unknown (Fernandez et al. 2000).

5 Barriers for HGT

In low dC + dG Gram-positive bacteria restriction endonucleases can operate as barriers for transducing or conjugating dsDNA, however the DNA entering through natural transformation, because of its single stranded nature, is not a barrier for interspecies recombination (Humbert et al. 1995; Rossolillo and Albertini 2001). Here, hemimethylated DNA, which is generated during synthesis of the complementary strand of the transforming molecule, by the action of the modification enzyme is resistant to the restriction endonucleases and escapes degradation (Lacks 1988). Indeed, DNA segments with sequence divergence from 4% to 17% from the recipient fail to prevent recombination during transformation; therefore transformational recombination provides a way to generate genetic diversity in these microorganisms (Humbert et al. 1995; Rossolillo and Albertini 2001, our unpublished results). Furthermore, structures reminiscent to mosaic *pbp* genes, found in clinical penicillin resistant *S. pneumoniae* strains, could be generated under laboratory conditions (Humbert et al. 1995). It is likely that the mismatch repair machinery of low dC + dG Gram-positive bacteria has not evolved to cope with excess of mismatches and to abort interspecies recombination.

Acknowledgements

This work was supported by grants BMC2006-01062 from Dirección General de Investigación (DGI) to JCA and BMC2006-02907 from DGI to SA. We are grateful to Peter Graumann for the communication of unpublished results. We wish to offer our apologies to all researchers whose important work is not cited due to space limitations.

References

Alonso JC, Ayora S, Canosa I, Weise F, Rojo F (1996) Site-specific recombination in Gram-positive theta-replicating plasmids. FEMS Microbiol Lett 142:1-10

Alonso JC, Luder G (1991) Characterization of *recF* suppressors in *Bacillus subtilis*. Biochimie 73:277-280

Alonso JC, Luder G, Trautner TA (1986) Requirements for the formation of plasmid-transducing particles of *Bacillus subtilis* bacteriophage SPP1. EMBO J 5:3723-3728

Alonso JC, Luder G, Trautner TA (1992) Intramolecular homologous recombination in *Bacillus subtilis* 168. Mol Gen Genet 236:60-64

Alonso JC, Weise F, Rojo F (1995) The *Bacillus subtilis* histone-like protein Hbsu is required for DNA resolution and DNA inversion mediated by the beta recombinase of plasmid pSM19035. J Biol Chem 270:2938-2945

Aravind L, Koonin EV (2001) Prokaryotic homologs of the eukaryotic DNA-end-binding protein Ku, novel domains in the Ku protein and prediction of a prokaryotic double-strand break repair system. Genome Res 11:1365-1374

Aravind L, Makarova KS, Koonin EV (2000) Survey and summary: Holliday junction resolvases and related nucleases: identification of new families, phyletic distribution and evolutionary trajectories. Nucleic Acids Res 28:3417-3432

Au N, Kuester-Schoeck E, Mandava V, Bothwell LE, Canny SP, Chachu K, Colavito SA, Fuller SN, Groban ES, Hensley LA, O'Brien TC, Shah A, Tierney JT, Tomm LL, O'Gara TM, Goranov AI, Grossman AD, Lovett, CM (2005) Genetic composition of the *Bacillus subtilis* SOS system. J Bacteriol 187:7655-7666

Ayora S, Alonso JC (1997) Purification and characterization of the RecF protein from *Bacillus subtilis* 168. Nucleic Acids Res 25:2766-2772

Ayora S, Carrasco B, Doncel E, Lurz R, Alonso JC (2004) *Bacillus subtilis* RecU protein cleaves Holliday junctions and anneals single-stranded DNA. Proc Natl Acad Sci USA 101:452-457

Ayora S, Missich R, Mesa P, Lurz R, Yang S, Egelman, EH, Alonso JC (2002) Homologous-pairing activity of the *Bacillus subtilis* bacteriophage SPP1 replication protein G35P. J Biol Chem 277:35969-35979

Berka RM, Hahn J, Albano M, Draskovic I, Persuh, M, Cui X, Sloma A, Widner W, Dubnau D (2002) Microarray analysis of the *Bacillus subtilis* K-state: genome-wide expression changes dependent on ComK. Mol Microbiol 43:1331-1345

Bowater R, Doherty AJ (2006) Making ends meet: repairing breaks in bacterial DNA by non-homologous end-joining. PLoS Genet 2:e8

Bruand C Velten M, McGovern S, Marsin S, Serena C, Ehrlich SD, Polard P (2005) Functional interplay between the *Bacillus subtilis* DnaD and DnaB proteins essential for initiation and re-initiation of DNA replication. Mol Microbiol 55:1138-1150

Canchaya C, Fournous G, Chibani-Chennoufi S, Dillmann ML, Brussow H (2003) Phage as agents of lateral gene transfer. Curr Opin Microbiol 6:417-424

Canosa I, Lopez G, Rojo F, Boocock MR, Alonso JC (2003) Synapsis and strand exchange in the resolution and DNA inversion reactions catalysed by the β recombinase. Nucleic Acids Res 31:1038-1044

Canosa I, Lurz R, Rojo F, Alonso JC (1998) β Recombinase catalyzes inversion and resolution between two inversely oriented *six* sites on a supercoiled DNA substrate and only inversion on relaxed or linear substrates. J Biol Chem 273:13886-13891

Carrasco B, Ayora S, Lurz R, Alonso JC (2005) *Bacillus subtilis* RecU Holliday-junction resolvase modulates RecA activities. Nucleic Acids Res 33:3942-3952

Carrasco B, Cozar MC, Lurz R, Alonso JC, Ayora S (2004a) Genetic recombination in *Bacillus subtilis* 168: contribution of Holliday junction processing functions in chromosome segregation. J Bacteriol 186:5557-5566

Carrasco B, Cozar MC, Lurz R, Alonso JC, Ayora S (2004b) Genetic recombination in *Bacillus subtilis* 168: Contribution of Holliday junction-processing functions in chromosome segregation. J Bacteriol 186:5557-5566

Carrasco B, Fernandez S, Asai K, Ogasawara N, Alonso JC (2002) Effect of the *recU* suppressors *sms* and *subA* on DNA repair and homologous recombination in *Bacillus subtilis*. Mol Genet Genomics 266:899-906

Carrasco B, Fernandez S, Petit MA, Alonso JC (2001) Genetic recombination in *Bacillus subtilis* 168: effect of Δ*helD* on DNA repair and homologous recombination. J Bacteriol 183:5772-5777

Ceglowski P, Boitsov A, Karamyan N, Chai S, Alonso JC (1993) Characterization of the effectors required for stable inheritance of *Streptococcus pyogenes* pSM19035-derived plasmids in *Bacillus subtilis*. Mol Gen Genet 241:579-585

Ceglowski P, Luder G, Alonso JC (1990) Genetic analysis of *recE* activities in *Bacillus subtilis*. Mol Gen Genet 222:441-445

Champoux JJ (2001) DNA topoisomerases: structure, function, and mechanism. Annu Rev Biochem 70:369-413

Chedin F, Handa N, Dillingham MS, Kowalczykowski SC (2006) The AddAB helicase/nuclease forms a stable omplex with its cognate χ sequence during translocation. J Biol Chem 281:18610-18617

Chedin F, Kowalczykowski SC (2002) A novel family of regulated helicases/nucleases from Gram-positive bacteria: insights into the initiation of DNA recombination. Mol Microbiol 43:823-834

Chen I, Christie PJ, Dubnau D (2005) The ins and outs of DNA transfer in bacteria. Science 310:1456-1460

Chen I, Dubnau D (2004) DNA uptake during bacterial transformation. Nat Rev Microbiol 2:241-249

Ciccarelli FD, Doerks T, von Mering C, Creevey CJ, Snel B, Bork P (2006) Toward automatic reconstruction of a highly resolved tree of life. Science 311:1283-1287

Clark AJ, Sandler SJ (1994) Homologous genetic recombination: the pieces begin to fall into place. Crit Rev Microbiol 20:125-142

Claverys JP, Prudhomme M, Martin B (2006) Induction of competence regulons as a general response to stress in Gram-Positive bacteria. Annu Rev Microbiol, in press

Deichelbohrer I, Alonso JC, Luder G, Trautner TA (1985) Plasmid transduction by *Bacillus subtilis* bacteriophage SPP1: effects of DNA homology between plasmid and bacteriophage. J Bacteriol 162:1238-1243

Dempsey LA, Dubnau DA (1989) Identification of plasmid and *Bacillus subtilis* chromosomal recombination sites used for pE194 integration. J Bacteriol 171:2856-2865

Dervyn E, Noirot-Gros MF, Mervelet P, McGovern S, Ehrlich SD, Polard P, Noirot P (2004) The bacterial condensin/cohesin-like protein complex acts in DNA repair and regulation of gene expression. Mol Microbiol 51:1629-1640

Doherty AJ, Jackson SP, Weller GR (2001) Identification of bacterial homologues of the Ku DNA repair proteins. FEBS Lett 500:186-188

Espeli O, Marians KJ (2004) Untangling intracellular DNA topology. Mol Microbiol 52:925-931

Fernandez S, Ayora S, Alonso JC (2000) *Bacillus subtilis* homologous recombination: genes and products. Res Microbiol 151:481-486

Fernandez S, Rojo F, Alonso JC (1997) The *Bacillus subtilis* chromatin-associated protein Hbsu is involved in DNA repair and recombination. Mol Microbiol 23:1169-1179

Fernandez S, Sorokin A, Alonso JC (1998) Genetic recombination in *Bacillus subtilis* 168: effects of *recU* and *recS* mutations on DNA repair and homologous recombination. J Bacteriol 180:3405-3409

Forsyth RA, Haselbeck RJ, Ohlsen KL, Yamamoto RT, Xu H, Trawick JD, Wall D, Wang L, Brown-Driver V, Froelich JM, C KG, King P, McCarthy M, Malone C, Misiner B, Robbins D, Tan Z, Zhu Zy ZY, Carr G, Mosca DA, Zamudio C, Foulkes JG, Zyskind JW (2002) A genome-wide strategy for the identification of essential genes in *Staphylococcus aureus*. Mol Microbiol 43:1387-1400

Frost LS, Leplae R, Summers AO, Toussaint A (2005) Mobile genetic elements: the agents of open source evolution. Nat Rev Microbiol 3:722-732

Gassel M, Alonso JC (1989) Expression of the *recE* gene during induction of the SOS response in *Bacillus subtilis* recombination-deficient strains. Mol Microbiol 3:1269-1276

Grohmann E, Muth G, Espinosa M (2003) Conjugative plasmid transfer in Gram-positive bacteria. Microbiol Mol Biol Rev 67:277-301

Hedayati MA, Steffen SE, Bryant FR (2002) Effect of the *Streptococcus pneumoniae* MmsA protein on the RecA protein-promoted three-strand exchange reaction. Implications for the mechanism of transformational recombination. J Biol Chem 277:24863-24869

Heller RC, Marians KJ (2005) The disposition of nascent strands at stalled replication forks dictates the pathway of replisome loading during restart. Mol Cell 17:733-743

Heller RC, Marians KJ (2006) Replication fork reactivation downstream of a blocked nascent leading strand. Nature 439:557-562

Humbert O, Prudhomme M, Hakenbeck R, Dowson CG, Claverys JP (1995) Homeologous recombination and mismatch repair during transformation in *Streptococcus pneumoniae*: saturation of the Hex mismatch repair system. Proc Natl Acad Sci USA 92:9052-9056

Itaya M (1995) Toward a bacterial genome technology: integration of the *Escherichia coli* prophage lambda genome into the *Bacillus subtilis* 168 chromosome. Mol Gen Genet 248:9-16

Jain R, Rivera MC, Lake JA (1999) Horizontal gene transfer among genomes: the complexity hypothesis. Proc Natl Acad Sci USA 96:3801-3806

Kidane D, Graumann PL (2005a) Dynamic formation of RecA filaments at DNA double strand break repair centers in live cells. J Cell Biol 170:357-366

Kidane D, Graumann PL (2005b) Intracellular protein and DNA dynamics in competent *Bacillus subtilis* cells. Cell 122:73-84

Kidane D, Sanchez H, Alonso JC, Graumann PL (2004) Visualization of DNA double-strand break repair in live bacteria reveals dynamic recruitment of *Bacillus subtilis* RecF, RecO and RecN proteins to distinct sites on the nucleoids. Mol Microbiol 52:1627-1639

Krogh BO, Symington LS (2004) Recombination proteins in yeast. Annu Rev Genet 38:233-271

Kurumizaka H, Ikawa S, Nakada M, Eda K, Kagawa W, Takata M, Takeda S, Yokoyama S, Shibata T (2001) Homologous-pairing activity of the human DNA-repair proteins Xrcc3·Rad51C. Proc Natl Acad Sci USA 98:5538-5543

Lacks SA (1988) Mechanisms of genetic recombination in Gram-positive bacteria. American Society for Microbiology, Washington, D.C. pp 43-86

Lawley TD, Klimke WA, Gubbins MJ, Frost LS (2003) F factor conjugation is a true type IV secretion system. FEMS Microbiol Lett 224:1-15

Lehmann AR (2005) The role of SMC proteins in the responses to DNA damage. DNA Repair (Amst) 4:309-314

Lindner C, Nijland R, van Hartskamp M, Bron S, Hamoen LW, Kuipers OP (2004) Differential expression of two paralogous genes of *Bacillus subtilis* encoding single-stranded DNA binding protein. J Bacteriol 186:1097-1105

Lio YC, Mazin AV, Kowalczykowski SC, Chen DJ (2003) Complex formation by the human Rad51B and Rad51C DNA repair proteins and their activities *in vitro*. J Biol Chem 278:2469-2478

Lisby M, Rothstein R (2004) DNA damage checkpoint and repair centers. Curr Opin Cell Biol 16:328-334

Lisby M, Rothstein R (2005) Localization of checkpoint and repair proteins in eukaryotes. Biochimie 87:579-589

Liu Y, Masson JY, Shah R, O'Regan P, West SC (2004) RAD51C is required for Holliday junction processing in mammalian cells. Science 303:243-246

Lopez CR, Yang S, Deibler RW, Ray SA, Pennington JM, Digate RJ, Hastings PJ, Rosenberg, SM, Zechiedrich EL (2005) A role for topoisomerase III in a recombination pathway alternative to RuvABC. Mol Microbiol 58:80-101

Lopez-Torrejon G, Martinez-Jimenez MI, Ayora S (2006) Role of LrpC from *Bacillus subtilis* in DNA transactions during DNA repair and recombination. Nucleic Acids Res 34:120-129

Majewski J, Cohan FM (1999) DNA sequence similarity requirements for interspecific recombination in *Bacillus*. Genetics 153:1525-1533

Martin B, Quentin Y, Fichant G, Claverys JP (2006) Independent evolution of competence regulatory cascades in streptococci? Trends Microbiol 8:339-345

Martinez-Jimenez MI, Alonso JC, Ayora S (2005) *Bacillus subtilis* bacteriophage SPP1-encoded gene *34.1* product is a recombination-dependent DNA replication protein. J Mol Biol 351:1007-1019

Mascarenhas J, Sanchez H, Tadesse S, Kidane D, Krishnamurthy M, Alonso JC, Graumann PL (2006) *Bacillus subtilis* SbcC protein plays an important role in DNA inter-strand cross-link repair. BMC Mol Biol 7:20

Mascarenhas J, Soppa J, Strunnikov AV, Graumann PL (2002) Cell cycle-dependent localization of two novel prokaryotic chromosome segregation and condensation proteins in *Bacillus subtilis* that interact with SMC protein. EMBO J 21:3108-3118

McGregor N, Ayora S, Sedelnikova S, Carrasco B, Alonso JC, Thaw P, Rafferty J (2005) The structure of *Bacillus subtilis* RecU Holliday junction resolvase and its role in substrate selection and sequence-specific cleavage. Structure 13:1341-1351

Mejean V, Claverys JP (1988) Polarity of DNA entry in transformation of *Streptococcus pneumoniae.* Mol Gen Genet 213:444-448

Michel B, Niaudet B, Ehrlich SD (1982) Intramolecular recombination during plasmid transformation of *Bacillus subtilis* competent cells. EMBO J 1:1565-1571

Mijakovic I, Petranovic D, Macek B, Cepo T, Mann M, Davies J, Jensen PR, Vujaklija D (2006) Bacterial single-stranded DNA-binding proteins are phosphorylated on tyrosine. Nucleic Acids Res 34:1588-1596

Noirot-Gros MF, Dervyn E, Wu LJ, Mervelet P, Errington J, Ehrlich SD, Noirot P (2002) An expanded view of bacterial DNA replication. Proc Natl Acad Sci USA 99:8342-8347

Ogura M, Yamaguchi H, Kobayashi K, Ogasawara N, Fujita Y, Tanaka T (2002) Whole-genome analysis of genes regulated by the *Bacillus subtilis* competence transcription factor ComK. J Bacteriol 184:2344-2351

Pedersen LB, Setlow P (2000) Penicillin-binding protein-related factor A is required for proper chromosome segregation in *Bacillus subtilis*. J Bacteriol 182:1650-1658

Petit MA (2005) Mechanisms of homologous recombination in bacteria Cambridge University press, Cambridge, pp 3-32

Petit MA, Ehrlich D (2002) Essential bacterial helicases that counteract the toxicity of recombination proteins. EMBO J 21:3137-3147

Prudhomme M, Attaiech L, Sanchez G, Martin B, Claverys JP (2006) Antibiotic stress induces genetic transformability in the human pathogen *Streptococcus pneumoniae*. Science 313:89-92

Prudhomme M, Libante V, Claverys JP (2002) Homologous recombination at the border: insertion-deletions and the trapping of foreign DNA in *Streptococcus pneumoniae*. Proc Natl Acad Sci USA 99:2100-2105

Rocha EP, Cornet E, Michel B (2005) Comparative and evolutionary analysis of the bacterial homologous recombination systems. PLoS Genet 1:e15

Rossolillo P, Albertini AM (2001) Functional analysis of the *Bacillus subtilis yshD* gene, a *mutS* paralogue. Mol Gen Genet 264:809-818

Sanchez H, Alonso JC (2005) *Bacillus subtilis* RecN binds and protects 3'-single-stranded DNA extensions in the presence of ATP. Nucleic Acids Res 33:2343-2350

Sanchez H, Kidane D, Castillo Cozar M, Graumann PL, Alonso JC (2006) Recruitment of *Bacillus subtilis* RecN to DNA double-strand breaks in the absence of DNA end processing. J Bacteriol 188:353-360

Sanchez H, Kidane D, Reed P, Curtis FA, Cozar MC, Graumann PL, Sharples GJ, Alonso JC (2005) The RuvAB Branch Migration Translocase and RecU Holliday Junction Resolvase Are Required for Double-Stranded DNA Break Repair in *Bacillus subtilis*. Genetics 171:873-883

Sherratt DJ (2003) Bacterial chromosome dynamics. Science 301:780-785

Sigurdsson S, Van Komen S, Bussen W, Schild D, Albala JS, Sung P (2001) Mediator function of the human Rad51B-Rad51C complex in Rad51/RPA-catalyzed DNA strand exchange. Genes Dev 15:3308-3318

Sung P, Krejci L, Van Komen S, Sehorn MG (2003) Rad51 recombinase and recombination mediators. J Biol Chem 278:42729-42732

Thanassi JA, Hartman-Neumann SL, Dougherty TJ, Dougherty BA, Pucci MJ (2002) Identification of 113 conserved essential genes using a high-throughput gene disruption system in *Streptococcus pneumoniae*. Nucleic Acids Res 30:3152-3162

Thaw P, Sedelnikova S. E, Muranova T, Wiese S, Ayora S, Alonso JC, Brinkman AB, Akerboom J, van der Oost J, Rafferty JB (2006) Structural insight into gene transcriptional regulation and effector binding by the Lrp/AsnC family. Nucleic Acids Res 34:1439-1449

Thomas CM, Nielsen KM (2005) Mechanisms of, and barriers to, horizontal gene transfer between bacteria. Nat Rev Microbiol 3:711-721

Viret JF, Bravo A, Alonso JC (1991) Recombination-dependent concatemeric plasmid replication. Microbiol Rev 55:675-683

Wang ST, Setlow B, Conlon EM, Lyon JL, Imamura D, Sato T, Setlow P, Losick R, Eichenberger P (2006) The forespore line of gene expression in *Bacillus subtilis*. J Mol Biol 358:16-37

Weller GR, Doherty AJ (2001) A family of DNA repair ligases in bacteria? FEBS Lett 505:340-342

Weller GR, Kysela B, Roy R, Tonkin LM, Scanlan E, Della M, Devine SK, Day JP, Wilkinson A, d'Adda di Fagagna F, Devine KM, Bowater RP, Jeggo PA, Jackson SP, Doherty AJ (2002) Identification of a DNA nonhomologous end-joining complex in bacteria. Science 297:1686-1689

Wilson TE, Topper LM, Palmbos PL (2003) Non-homologous end-joining: bacteria join the chromosome breakdance. Trends Biochem Sci 28:62-66

Zhang W, Carneiro MJ, Turner IJ, Allen S, Roberts CJ, Soultanas P (2005) The *Bacillus subtilis* DnaD and DnaB proteins exhibit different DNA remodelling activities. J Mol Biol 351:66-75

Alonso, Juan C.
Departmento de Biotecnología Microbiana, Centro Nacional de Biotecnología, CSIC, Campus Universidad Autónoma de Madrid, Campus Universidad Autónoma de Madrid, Darwin Street 3, Cantoblanco, 28049 Madrid, Spain
jcalonso@cnb.uam.es

Ayora, Silvia
Departmento de Biotecnología Microbiana, Centro Nacional de Biotecnología, CSIC, Campus Universidad Autónoma de Madrid, Campus Universidad Autónoma de Madrid, Darwin Street 3, Cantoblanco, 28049 Madrid, Spain

Carrasco, Begoña
Departmento de Biotecnología Microbiana, Centro Nacional de Biotecnología, CSIC, Campus Universidad Autónoma de Madrid, Campus Universidad Autónoma de Madrid, Darwin Street 3, Cantoblanco, 28049 Madrid, Spain

Sanchez, Humberto
Departmento de Biotecnología Microbiana, Centro Nacional de Biotecnología, CSIC, Campus Universidad Autónoma de Madrid, Campus Universidad Autónoma de Madrid, Darwin Street 3, Cantoblanco, 28049 Madrid, Spain

The bacterial RecA protein: structure, function, and regulation

Michael M. Cox

Abstract

The bacterial RecA protein is the prototypical recombinase, promoting the central steps of DNA pairing and strand exchange in genetic recombination and recombinational DNA repair. RecA homologs are present in virtually all organisms from bacteria to humans. RecA is a multifunctional protein. As a recombinase, the protein binds to DNA in the form of a helical filament, and exhibits a DNA-dependent ATPase activity. As a nucleoprotein filament, RecA promotes a series of easily monitored DNA strand exchange reactions *in vitro*. In addition to its role as a recombinase, the *E. coli* RecA protein is also a key component of the regulatory system that controls the induction of the SOS response, and it plays a direct role in the UV mutagenesis promoted by DNA polymerase V. RecA protein is subject to multiple layers of regulation. RecA is autoregulated by its own C-terminus. Many other proteins, including the RecF, RecO, RecR, DinI, RecX, RdgC, PsiB, and SSB proteins, have either a demonstrated or probable role in modulating where and when RecA-mediated recombination events occur.

1 The role of recombination in DNA metabolism

In bacteria, the major function of homologous genetic recombination is the recombinational DNA repair of replication forks that have stalled or collapsed at the site of an encounter with DNA damage (Kuzminov 1999, Cox et al. 2000, Kowalczykowski 2000, Cox 2001b, Cox 2002). Replication fork demise occurs often even under normal growth conditions in bacteria (Cox et al. 2000, Cox 2001b, Cox 2002). Estimates of the frequency of recombinational DNA repair under normal growth conditions vary. Studies with a variety of *rec* mutants suggest that at least 10-20% of all replication forks originating at the bacterial origin of replication are halted by DNA damage and must undergo recombinational DNA repair (Zavitz and Marians 1992, Kuzminov 1996, Cox 1998, Kuzminov 1999, Cox et al. 2000, Cox 2001a, Cox 2001b, Cox 2002, Michel et al. 2004, Friedberg 2005, Heller and Marians 2005, Kreuzer 2005). Following repair, additional systems act to restart replication (Marians 2000b, Marians 2000a) and deal with the dimeric chromosomes sometimes produced by recombination (Barre et al. 2001).

There are at least two major pathways for recombinational DNA repair of stalled replication forks. If the replication fork encounters an unrepaired DNA lesion, a gap in the DNA is generated. Repair can occur via regression (backward movement) of the stalled fork to create a distinctive Holliday junction sometimes dubbed a "chicken foot" (Postow et al. 2001). The chicken foot can be processed in several different ways (Dillingham and Kowalczykowski 2001, Michel et al. 2001, Cox 2002, McGlynn and Lloyd 2002, McGlynn 2004, Michel et al. 2004, Friedberg 2005, Heller and Marians 2005, Kreuzer 2005). If the replication fork encounters a break in one strand (as might be the case where a lesion was undergoing repair), a double strand break is generated. In this situation, the classic RecBCD pathway predominates, processing the broken end and promoting strand invasion to reconstitute a fork structure (Dillingham and Kowalczykowski 2001, Michel et al. 2001, Cox 2002, McGlynn and Lloyd 2002, McGlynn 2004, Michel et al. 2004, Friedberg 2005, Heller and Marians 2005, Kreuzer 2005). There are multiple variants of these repair pathways and recombination enzymes can be considered an assemblage of activities that can be adapted to the damage situation.

DNA damage may also be bypassed by a replication fork, leaving the lesion behind in a DNA gap. This would bring into play a form of postreplication DNA gap repair (Rupp and Howard-Flanders 1968, Smith 2004). Recent *in vitro* work demonstrated that de novo priming of both leading and lagging strand synthesis downstream of non-coding lesions in the template DNA enables replication forks to re-initiate and continue replication with limited hindrance, leaving gaps in the nascent strands to be filled in by recombination (Heller and Marians 2006). This suggests that postreplication gap repair might comprise a substantial fraction of the fork-related recombinational repair in bacterial cells. However, the *in vitro* model system employed may not adequately mimic the sensitivity of forks *in vivo* to DNA damage. Heavy DNA damage, from UV or other sources, halts DNA replication (when the experiment is controlled to prevent DnaA-dependent replication initiation from *oriC*) and induces the SOS response (Setlow et al. 1963, Courcelle and Hanawalt 2003, Courcelle et al. 2004). Cellular replication forks can be halted by at least some types of damage, and a complete picture of the situation *in vivo* is not yet available.

2 The RecA protein of *Escherichia coli*

2.1 Overview

The RecA protein of *E. coli* (M_r 37,842; 352 amino acids) is a multifunctional protein, with roles in the induction of the SOS response to DNA damage, SOS mutagenesis, and general recombination processes such as recombinational DNA repair (Brendel et al. 1997, Roca and Cox 1997, Cox 1998, Cox 2001a, Cox 2001b, Lusetti and Cox 2002). In the context of SOS induction, filaments of RecA protein formed on DNA facililtate an autocatalytic cleavage of the LexA repressor (Little 1991). In SOS mutagenesis, RecA is required to stimulate DNA polymerase V in its lesion bypass function (Pham et al. 2001, Pham et al. 2002, Schlacher et al. 2005, Schlacher et al. 2006).

This chapter focuses only on the role of RecA protein in recombination and recombinational DNA repair. RecA protein promotes a DNA strand exchange reaction *in vitro* that mimics the postulated central steps in homologous genetic recombination *in vivo*. The reaction is stimulated by the single-stranded DNA binding protein of *E. coli* (SSB). RecA is also a DNA-dependent ATPase, and some aspects of the DNA strand exchange reaction require ATP hydrolysis. The RecA protein is found in virtually all bacteria, with certain endosymbionts being the only apparent exceptions (Moran and Baumann 2000, Tamas et al. 2002). An activity of this type has been a part of bacterial physiology for over 1.5 billion years (Roca and Cox 1990, Brendel et al. 1997, Roca and Cox 1997). Structural and functional homologs of RecA have been found in bacteriophage (the T4 UvsX protein; Jiang et al. 1993), in archaeae (RadA; Sandler et al. 1996b, Seitz et al. 1998, Seitz et al. 2001), and in eukaryotic cells (Rad51 and Dmc1; Shinohara et al. 1992, Ogawa et al. 1993, Bishop 1994, Sung 1994, Gupta et al. 1997, Baumann and West 1998, New et al. 1998, Passy et al. 1999, Gupta et al. 2001, Sehorn et al. 2004) For a description of the important eukaryotic homolog Rad51, see the chapter by Heyer in this volume.

2.2 Structure

The bacterial RecA protein is a highly conserved polypeptide chain. Primary sequence alignments of the RecA proteins from many scores of bacterial species have been published (Karlin and Brocchieri 1996, Brendel et al. 1997, Roca and Cox 1997). The EcRecA protein consists of 352 amino acid residues (Mr 37,842). With the EcRecA as a reference, the percent of identical amino acid residues in bacterial homologs range from 49% for *Mycoplasma pulmonis* to 100% for *Shigella flexneri*.

The number of reported x-ray crystal structures of bacterial RecA proteins has been increasing, including four from *E. coli* (Story and Steitz 1992, Story et al. 1992, Xing and Bell 2004b, Xing and Bell 2004a), two from *M. tuberculosis* (Datta et al. 2000, Datta et al. 2003a), one from *Deinococcus radiodurans* (Rajan and Bell 2004), and two from *Mycobacterium smegmatis* (Datta et al. 2003b, Krishna et al. 2006). These include several complexes with nucleotide cofactors, and a range of collapsed and extended filament forms. Structural information has been enriched by the solution of the structures of RecA homologs from archaeans and eukaryotes (Conway et al. 2004, Wu et al. 2004, Qian et al. 2005, Wu et al. 2005). Notably, none of the structures includes bound DNA, leaving open many critical questions about how RecA and its homologs interact with DNA. A 24 monomer filament based on the EcRecA structure of Story and Steitz (Story et al. 1992), along with a ribbon representation of a RecA monomer, are presented in Fig. 1. In the crystals, monomers tend to pack so as to form a right-handed helical filament with six monomers/turn (Fig. 1). The filament revealed in the earliest

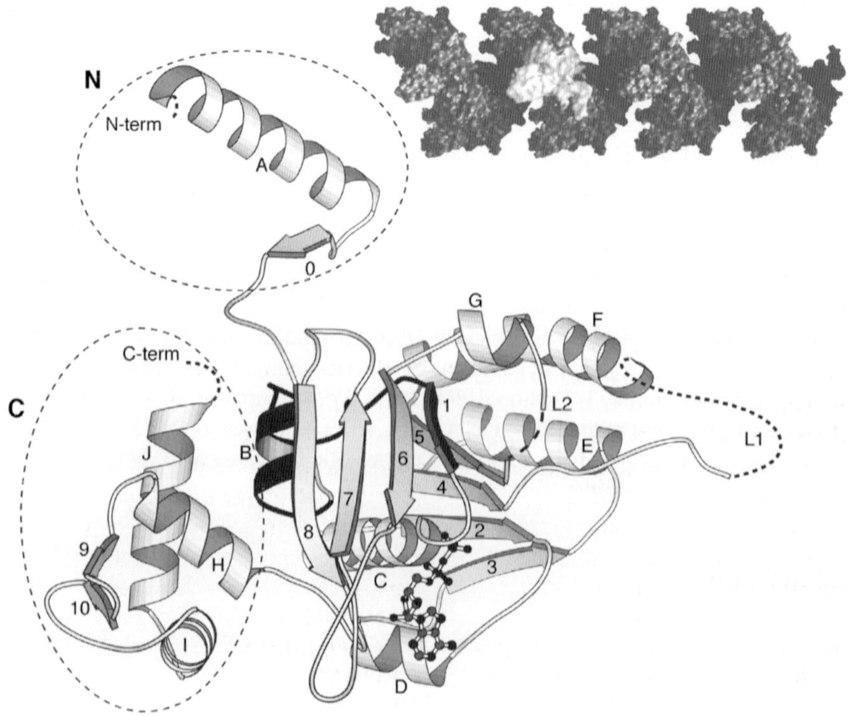

Fig. 1. Structure of RecA protein. A monomer is shown in this ribbon diagram. The domain labeled N and denoted by a dashed line circle is the N-terminal domain. The C-terminal domain is similarly outlined and denoted with a C. The remainder of the protein is the core domain, and a bound ADP molecule (in ball and stick representation) is also evident. The structure is based on that reported by Story and Steitz, 1992. A 24 monomer segment of a RecA filament based on the same structure is shown in the inset at upper right, with one of the subunits colored white.

crystal structures were not as extended as the active filaments visualized by electron microscopy, and likely represented an inactive conformation. EM image reconstructions have provided a glimpse of the active filament (Egelman 1993, Egelman and Stasiak 1993, Yu et al. 2001). The structural information garnered from electron microscopy has suggested a revised monomer-monomer interface in the active filament (Egelman 1993, Egelman and Stasiak 1993, Yu et al. 2001), one that is consistent with recent published structures of RecA homologs (Conway et al. 2004, Wu et al. 2004).

The RecA structures have revealed a central core domain and two smaller domains at the amino (N) and carboxyl (C) termini (Fig. 1). The core domain of the RecA protein (residues 34-269) is the part most highly conserved among bacterial species, and is also structurally homologous to several proteins to which it bears very little to no sequence similarity. The structural units of hexameric helicases are RecA-like domains (Bird et al. 1998, Egelman 2000). Furthermore, 120 α-carbon atoms of the core domain of RecA can be spatially aligned with the mitochondrial F_1-ATPase and the cobalamin nucleotide loop assembly protein CobU, with an RMSD of less than 2 Å (Story et al. 1993, Thompson et al. 1998). The core domain region of residues 47-74 is especially well conserved with 14 invariant residues. The *E. coli* sequence GPESSGKT matches the consensus Walker A box (Walker et al. 1982)(also referred to as the P-loop) amino acid consensus sequence (G/A)XXXXGK(T/S) found in a number of NTP-binding proteins. The RecA K72R mutant protein is ATPase deficient while retaining nucleotide binding and DNA pairing function (Rehrauer and Kowalczykowski 1993, Shan et al. 1996).

Within the RecA structure, there are two regions implicated in DNA binding, consisting of residues 151-176 and 190-227. These loops, disordered in most of the available structures, are commonly referred to as L1 (residues 157-164) and L2 (residues 195-209). Both of these regions are well-conserved among bacterial RecA proteins, although the conservation does not extend to the archaeal and eukaryotic homologues. The loop L1 residues are ordered in one MtRecA structure and, are oriented into the groove (Datta et al. 2000). Detailed mutagenesis of loops L1 and L2 have been carried out by the Knight (Nastri and Knight 1994) and Camerini-Otero (Hortnagel et al. 1999) groups, respectively. DNA cross-linking studies support a role for these loops in DNA binding (Malkov and Camerini-Otero 1995, Wang and Adzuma 1996). Residues outside of these regions, such as Tyr103 (Morimatsu and Horii 1995), Lys183 (Morimatsu and Horii 1995, Rehrauer and Kowalczykowski 1996), and in the region of residues 233-243 (Rehrauer and Kowalczykowski 1996) have also been shown to cross-link to DNA. Many details of the RecA-DNA interaction remain to be elucidated.

The C-terminal domain (residues 270-352) exhibits the least amount of sequence conservation. This domain is positioned distal to the filament axis in the polymer structure (the dark lobes in Fig. 1). Egelman and co-workers have observed C-terminal domain movement relative to the core domain that may be responsible for (or diagnostic of) the "active" or "inactive" state of the RecA filament (Yu et al. 2001). The last 25 residues of the RecA protein are disordered in most of the crystal structures, with one of the likely numerous conformations being visualized only in a recent structure of *M. smegmatis* RecA protein (Krishna et al. 2006). This region includes a high concentration of negatively charged residues, a characteristic shared by many but not all bacterial RecA sequences (Roca and Cox 1997). Other ssDNA binding proteins such as SSB of *E. coli* (Williams et al. 1983) and the gene 32 protein of phage T4 (Lonberg et al. 1981) also have highly negatively charged C-terminal regions that modulate DNA affinity and protein-protein interactions (Lonberg et al. 1981, Williams et al. 1983, Benedict and Kowalczykowski 1988, Tateishi et al. 1992, Genschel et al. 2000, Lusetti et al. 2001, Witte et al. 2003, Cadman and McGlynn 2004). The negatively charged C-terminus of RecA autoregulates all aspects of RecA function, as detailed in section 3.

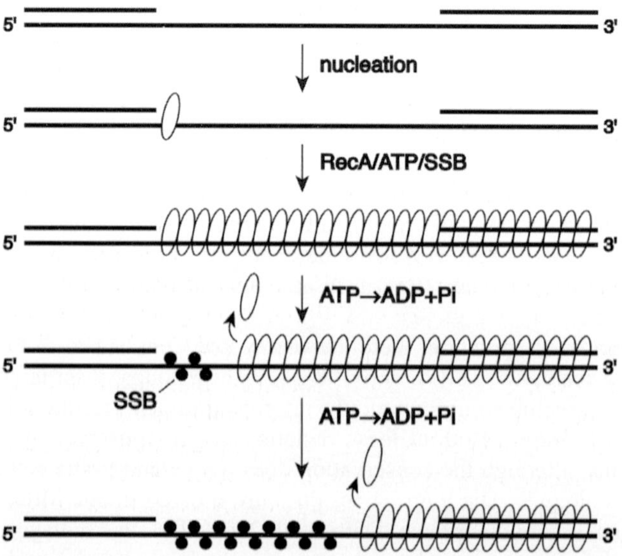

Fig. 2. A multistep pathway for the binding of ssDNA is shown. RecA protein first nucleates on the DNA, a step that may involve more than the one subunit depicted. The filament is then extended 5' to 3' when ATP is present, and the extension can continue into any contiguous duplex DNA. ATP hydrolysis leads to disassembly at the opposite end. Vacated ssDNA is bound by SSB.

2.3 Binding to DNA

RecA protein filaments both assemble and disassemble in a largely unidirectional (5' to 3') and end-dependent manner, with monomers added at one end and subtracted from the other (Fig. 2) (Register and Griffith 1985, Lindsley and Cox 1990b, Shan et al. 1997, Arenson et al. 1999, Bork et al. 2001b). This is true on both ssDNA and dsDNA. The unidirectionality has been inferred in experiments in which ATP has been actively hydrolyzed at some point in the protocol. Filaments formed in the presence of ATPγS are very stable, and the polarity of assembly with this non-hydrolyzed (or very weakly hydrolyzed) analogue is now being examined in several laboratories using single-molecule protocols. Without ATP hydrolysis, there is no reason, in principle, for the assembly process to favor one or the other end of a RecA filament. The monomer-monomer interfaces should be identical at both ends.

The DNA within a RecA filament is extended about 1.5-1.6X and underwound by about 40% (Stasiak and Di Capua 1982, Pugh et al. 1989). One RecA monomer binds to 3 nucleotides or base pairs of DNA. The helical filament thus has 18 bp and 6 monomers per right-handed turn (6.2 monomers per turn as seen in the electron microscope (Yu et al. 2001)). NMR studies have revealed a new conformation of DNA within the extended RecA filament, one in which the 2' methylene group of each nucleotide stacks on the adjacent base (Shibata et al. 2001). The assembly of a RecA filament is limited by a slow nucleation step (Kowalczykowski et al. 1987, Pugh and Cox 1987, Pugh and Cox 1988, Lavery and Kowalczykowski 1990). The subsequent extension of the filament is relatively rapid, although good rate constants are not yet available.

RecA filament disassembly can occur in two ways. First, if ATP is not regenerated and ADP levels are allowed to build up, there is a rapid dissociation of the RecA filament from DNA when the ADP/ATP ratio nears 1.0 (Cox et al. 1983, Lee and Cox 1990, Ellouze et al. 1999). A more ordered, demonstrably end-dependent, and largely unidirectional disassembly of filaments is seen when ATP hydrolysis occurs in the presence of a regeneration system and with ATP concentrations sufficient to operate at V_{max} (Lindsley and Cox 1990b, Shan et al. 1997, Arenson et al. 1999, Cox et al. 2005). For RecA filaments bound to ssDNA, the rate of end-dependent (5' to 3') filament disassembly is 60 – 70 monomers min^{-1} (Arenson et al. 1999) On dsDNA, the rate of disassembly increases to approximately 120 monomers min^{-1} at neutral pH (Cox et al. 2005). The different rates reflect different filament states operative on ssDNA vs dsDNA, as described in the next section.

2.4 ATP hydrolysis and RecA filament states

RecA protein is a DNA-dependent ATPase, with a K_m for ATP that varies with conditions and cofactors but is often on the order of 50-100 μM. At ATP saturation, the k_{cat} is about 30 min^{-1} on ssDNA, and 20 min^{-1} on dsDNA (Lusetti and Cox 2002, Cox 2003, Cox et al. 2005). Also hydrolyzed efficiently is dATP, with measured k_{cat} values slightly higher than ATP (Menetski and Kowalczykowski 1989). Whereas the rates of ATP/dATP hydrolysis observed with RecA protein are not particularly robust, they are up to two orders of magnitude higher than the rates observed for the archaeal and eukaryotic homologues of RecA.

ATP is hydrolyzed throughout a RecA filament, with all filament subunits participating. There is no increase in ATP hydrolytic rates for RecA monomers near a filament end (Brenner et al. 1987). In general, this means that only a very small fraction of the ATP hydrolytic events that occur in a contiguous RecA filament – those that occur on the disassembly end – are coupled to RecA dissociation from the DNA. In RecA filaments that are bound to circular single-stranded DNAs, such as ϕX174 viral DNA, there are typically 2-3 breaks where RecA dissociation and replacement are occurring, based on challenge experiments where the RecAK72 R is available in excess to replace any dissociated monomers (Shan and Cox 1996, Lusetti et al. 2004b). The RecA protein monomers at all other locations in these filaments are hydrolyzing ATP at 2-3 per second without dissociating.

It is clear that ATP hydrolysis is important for RecA protein function *in vivo*. Two mutant RecA proteins that bind but do not hydrolyze ATP have been described, K72R (Rehrauer and Kowalczykowski 1993, Shan et al. 1996) and E96D (Campbell and Davis 1999). In both cases, the altered RecA protein is functional

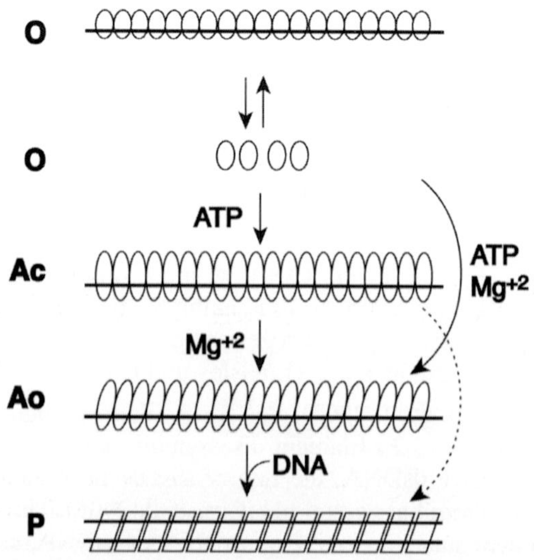

Fig. 3. RecA has at least four different filament states. In the absence of DNA or nucleotide cofactor, or in the presence of ADP, the RecA forms collapsed filaments designated O. When bound to ssDNA, in the presence of ATP or its weakly hydrolyzed analog ATPγS, RecA forms an extended filament in a state designated A. With minimal Mg ion present, the Ac form is present, with a limited capacity to promote DNA strand exchange. Addition of free Mg ion converts this into the Ao state, with an enhanced capacity to promote DNA strand exchange. Addition of a second DNA strand converts the filament to the P state, a filament state characterized by higher levels of cooperativity, lower levels of ATP hydrolysis, and more rapid rates of filament disassembly.

for many key activities, forming nucleoprotein filaments and promoting DNA pairing and facilitating the autocatalytic cleavage of LexA protein. In both cases, some aspects of DNA strand exchange function are compromised by the mutation, as described in detail later. In both cases, the mutant gene delivers a phenotype equivalent to a complete *recA* deletion or null mutation (Konola et al. 1994). The situation is quite different for the eukaryotic homolog Rad51. The K191R mutation in the yeast Rad51 protein, which corresponds to the K72R mutation of *E. coli* RecA, yields an altered Rad51 that is functional for recombinase functions both *in vitro* or *in vivo* (Sung and Stratton 1996), although overexpression of the altered protein is needed for full biological function (Shinohara et al. 1992, Sung and Stratton 1996). The situation is similar for the human Rad51 protein (Morrison et al. 1999).

The different rates of ATP hydrolysis observed when the *E. coli* RecA protein is bound to ssDNA and dsDNA again reflect different filament states. Evidence has accumulated for at least four distinct filament states (Fig 3) (Shan et al. 1996, Cox 2003, Haruta et al. 2003). In the absence of DNA, or when ADP levels are high, a state designated O is observed. This is a collapsed and inactive filament

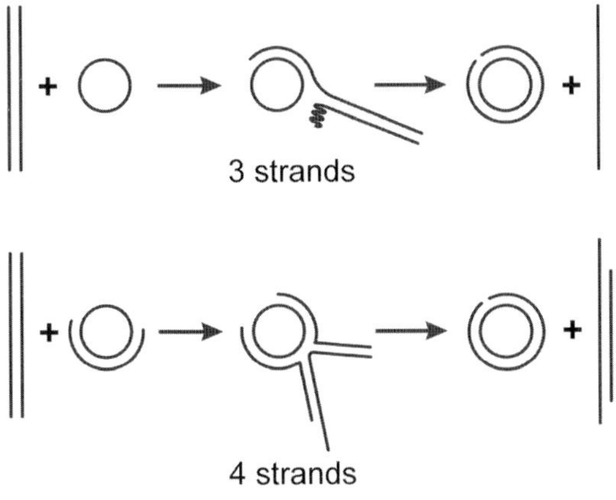

Fig. 4. Model DNA strand exchange reactions promoted by the RecA protein. The form of the DNA substrates is chosen for convenience in these widely-used reactions. RecA protein forms a filament on the ssDNA, or the duplex DNA with a gap. DNA strand exchange begins when the second DNA molecule (the linear duplex) is aligned with homologous sequences in the first, and initiates a strand exchange. The strand exchange proceeds unidirectionally around the DNA circle until it is completed.

state. When ssDNA and ATP is added, the O state must disassemble (Lee and Cox 1990, Yu and Egelman 1992) and RecA then re-assembles a filament on the ssDNA in a state designated A. There are two forms of state A that depend on the level of Mg ion present (Shan et al. 1996, Lusetti et al. 2003a). At low Mg ion (little or none in excess relative to the ATP present), the Ac state is observed, characterized by a limited capacity to promote DNA strand exchange. When 6-8 mM Mg ion is added in excess to the ATP present, the Ao state is observed, characterized by a more robust capacity to promote DNA strand exchange with a wide array of DNA substrates. When a second DNA strand is added, as when RecA is bound to dsDNA or is promoting DNA strand exchange, the RecA protein converts to a state designated P (Haruta et al. 2003). The P state has the highest capacity for the initiation of DNA strand exchange. The different forms of the RecA filament are here described as states rather than conformations, since it is assumed that there are multiple conformations of RecA accessed during the ATP hydrolytic cycles taking place within each of the defined A and P states.

2.5 DNA strand exchange is a multi-step process

The most common model reactions used for in vitro studies of RecA protein-mediated DNA strand exchange are outlined in Fig. 4. DNA substrates are generally derived from bacteriophage DNA and the reaction can involve either 3 or 4 strands. This reaction has been well-studied (Kowalczykowski et al. 1994, Kowalczykowski and Eggleston 1994, Roca and Cox 1997, Cox 1999, Lusetti and Cox 2002, Cox 2003). A RecA helical filament first forms on the ssDNA (or the gapped duplex DNA in B). The nucleoprotein filament then binds to a second homologous duplex DNA and aligns it with the bound ssDNA. A strand switch occurs to form a region of hybrid duplex. This process is propagated unidirectionally, 5' to 3' relative to the ssDNA within the original nucleoprotein filament, until strand exchange is completed (Cox and Lehman 1981, Kahn et al. 1981, West et al. 1981). RecA protein hydrolyzes ATP during this reaction with a monomer k_{cat} of about 20 min^{-1}, characteristic of the P state.

DNA pairing occurs within the filament, as originally proposed by Paul Howard-Flanders (Howard-Flanders et al. 1984) and confirmed many times. However, the 4-stranded DNA pairing intermediate suggested by Howard-Flanders has never been observed. Physical studies have generally demonstrated that no more than three DNA strands can be readily accommodated within the interior helical groove of a RecA filament (Müller et al. 1990, Takahashi et al. 1991, Wittung et al. 1994, Cox 1995, Kubista et al. 1996, Roca and Cox 1997). DNA pairing in an efficient 4-strand exchange reaction is always initiated within the single-strand gap; i.e. productive 4-strand exchanges must be initiated as 3-strand reactions (Conley and West 1990, Lindsley and Cox 1990a, Chow et al. 1992, Shan and Cox 1998).

The problem of DNA pairing inside a RecA filament is thus reduced to an interaction between a RecA-bound ssDNA and a homologous duplex. In principle, the duplex could approach the ssDNA via either its major or minor grooves. A minor groove-first pathway for DNA pairing is currently favored by the evidence (Kumar and Muniyappa 1992, Baliga et al. 1995, Frank-Kamenetskii and Mirkin 1995, Podyminogin et al. 1995, Podyminogin et al. 1996, Zhou and Adzuma 1997, Gupta et al. 1999, Rice et al. 2000, Xiao and Singleton 2002). In this scheme, homologous alignment involves standard Watson-Crick base pairing. As the duplex is bound, it becomes extended and underwound such that its bases would be free to flip and "sample" the bound ssDNA for complementarity (Gupta et al. 1999). Studies by Radding and colleagues indicate that the base flipping occurs mainly at A:T base pairs (Gupta et al. 1999). The RecA filament appears to stabilize the products of DNA strand exchange, using binding energy to promote the strand switch (Adzuma 1992). The fundamental DNA pairing reaction has been subjected to kinetic analysis, using simplified systems employing short oligonucleotides for DNA substrates (Yancey-Wrona and Camerini-Otero 1995, Bazemore et al. 1997). These studies have revealed that the reaction proceeds minimally in three to four steps, with a rapid second order DNA alignment followed by several slower first order process that likely involve the completion of strand exchange, perhaps some conformational changes, and removal of the displaced strand (Yancey-Wrona and Camerini-Otero 1995, Bazemore et al. 1997, Gumbs and Shaner 1998, Folta-Stogniew et al. 2004, Xiao et al. 2006).

The 1.5 to 1.6 fold extension of the DNA effected when an active RecA protein filament forms on it may play a direct role in the subsequent homology search leading to DNA pairing, as laid out in a simple but elegant model proposed by Bruinsma and colleagues (Klapstein et al. 2004). If two homologous DNAs have the same rise per base pair, placing them side by side in alignment will mean they will be in alignment at every position. Moving one of the DNAs relative to the other by one nucleotide or base pair puts every nucleotide or base pair *out* of alignment. If, instead, one DNA is extended and the other is not, only a subset of nucleotides or base pairs in one may be in alignment with the other, but moving one DNA relative to the other now places different nucleotides or base pairs in alignment. In effect, there are many potential productive alignments every time the two DNAs come together, instead of just one (Klapstein et al. 2004). The productive alignment could be the nucleation site for pairing, extending to create a larger aligned region by spooling the second DNA into the RecA nucleoprotein filament.

The single-strand DNA binding protein of *E. coli* (SSB) facilitates DNA strand exchange, binding to the displaced single-strand product and facilitating its release from the filament (Lavery and Kowalczykowski 1992).

2.6 The role of ATP hydrolysis in DNA strand exchange

ATP hydrolysis is not required to promote the fundamental process of DNA strand exchange within the filament. RecA protein can promote limited DNA strand exchange (typically yielding up to1-2 kbp of hybrid DNA) under conditions in which ATP is not hydrolyzed. These conditions include the use of ATP analogues which are bound but not hydrolyzed by wild type RecA protein (ATPγS and an ADP•AlF_4^- complex) (Menetski et al. 1990, Rosselli and Stasiak 1990, Kim et al. 1992a, Kim et al. 1992b, Kowalczykowski and Krupp 1995), as well as the use of the mutant RecA K72R, which binds but does not hydrolyze dATP (Rehrauer and Kowalczykowski 1993, Shan et al. 1996). However, when ATP is not hydrolyzed, DNA strand exchange generally halts long before the reaction reaches completion with the bacteriophage DNAs commonly employed (Menetski et al. 1990, Jain et al. 1994, Shan et al. 1996). ATP hydrolysis allows the reaction to go to completion, renders the DNA strand exchange reaction unidirectional (Jain et al. 1994, Shan et al. 1996) and allows it to bypass substantial DNA structural barriers (Rosselli and Stasiak 1991, Kim et al. 1992a, Shan et al. 1996). These structural barriers can include a heterologous insertion of 100 bp or more in the duplex DNA substrate. In addition, ATP hydrolysis is required for any DNA strand exchange involving two duplex DNAs (the 4-strand reaction, Fig. 4) (Kim et al. 1992b, Shan et al. 1996), and for RecA-mediated regression of a model replication fork (Robu et al. 2001, Robu et al. 2004). These results imply that ATP hydrolysis is directly coupled to the later stages of DNA strand exchange.

Notably, when DNA strand exchange is initiated, RecA protein filaments are rapidly converted from the A state to the P state. The addition of the homologous dsDNA to a reaction mixture containing RecA nucleoprotein complexes bound to ssDNA and hydrolyzing ATP leads to a 30% drop in the rate of ATP hydrolysis within 2 min (Schutte and Cox 1987). The new rate, characteristic of the P state, is subsequently maintained throughout the strand exchange reaction and beyond, given sufficient ATP regeneration (Schutte and Cox 1987). The decline in rate is dependent on homology between the two DNAs. If a duplex DNA with half the homology is added to the reaction, the drop is half of that observed with full length dsDNA. If a longer DNA is added that is partially homologous to the ssDNA, the drop is proportional to the length of homology that exists. In effect, the RecA nucleoprotein filament senses all of the available homology at a very early stage of DNA strand exchange, even though the exchange itself may not be completed for another 15 – 20 min (Schutte and Cox 1987).

NTPases can generally be classified according to one of three biological functions: motor proteins, molecular timing devices, or recycling functions (Alberts and Miake-Lye 1992). The ATPase activity of RecA is often portrayed as a recycling function (Alberts and Miake-Lye 1992, West 1992, Kowalczykowski et al. 1994, Kowalczykowski and Eggleston 1994), causing the dissociation of RecA monomers from the filament after DNA strand exchange has occurred. ATP hydrolysis is indeed coupled to the end-dependent disassembly of RecA filaments (Arenson et al. 1999, Cox et al. 2005). However, ATP hydrolysis is also coupled to DNA strand exchange, in such a way as to allow it to overcome significant barriers and proceed at predictable rates (Kim et al. 1992b, Kim et al. 1992a, Jain et al. 1994, Bedale and Cox 1996, Shan et al. 1996, MacFarland et al. 1997, Shan and Cox 1998, Cox et al. 2005).

The role of ATP hydrolysis in RecA-mediated DNA strand exchange has been a subject of controversy for over 20 years. Since RecA filaments can remain on the heteroduplex product of typical DNA strand exchange reactions with SSB binding to the displaced single strand (Lindsley and Cox 1990b, Ullsperger and Cox 1995, Lusetti and Cox 2002), the early notion that net RecA filament disassembly, and/or filament reassembly on the displaced single strand, played a mechanistic role in the movement of the DNA branch (Howard-Flanders et al. 1984, Konforti and Davis 1992, Morel et al. 1994) has been largely discredited. There are two additional models that attempt to explain the coupling between ATP hydrolysis and DNA strand exchange. The first is the RecA redistribution model (Menetski et al. 1990, Rehrauer and Kowalczykowski 1993, Kowalczykowski and Krupp 1995). This model begins with discontinuities in the RecA filament, where DNA strand exchange halts when ATP is not hydrolyzed. ATP hydrolysis serves to recycle RecA protein so as to fill in the discontinuities. The filament may stay largely intact and bound to the hybrid DNA duplex at the end of a reaction as consistent with observation.The second model envisions a RecA-facilitated DNA rotation, coupled to ATP hydrolysis (Cox 1994, Shan et al. 1996, Roca and Cox 1997, Cox 2003, Cox et al. 2005). This model depicts the RecA ATPase as a motor activity with RecA recycling being a secondary function. ATP hydrolysis is organized throughout the filament and coupled to DNA rotation so as to effect branch movement during strand exchange or fork regression. ATP hydrolysis throughout the filament is thus coupled to DNA strand exchange, and not simply the ATP hydrolysis that occurs at a filament end or discontinuity. Tests of both models have been reviewed (Cox 2003) and are continuing (Cox et al. 2005).

3 Regulation of RecA function

Recombination plays a critical role in DNA repair and genome maintenance. However, it is equally critical to regulate where, when, and how recombination takes place. When unregulated, recombination can lead to genome instability and carcinogenesis. Meiotic recombination in eukaryotes is under tight regulation, ensuring the proper spacing and complete chromosomal coverage of the recombination events that are needed for proper chromosome segregation (Thompson and Schild 1999, Cohen and Pollard 2001, de Massy 2003, Hillers and Villeneuve 2003). Defects in many recombination functions result not only in DNA repair defects, but also in more general genomic instability. Much of this is associated with stalled replication forks (Chakraverty and Hickson 1999, Myung et al. 2001, van Gent et al. 2001, Venkitaraman 2001, Bjergbaek et al. 2002, Kolodner et al. 2002, Myung and Kolodner 2002, Osborn et al. 2002, Thompson and Schild 2002). Regulation determines which pathway is used to correct a double strand break in DNA in eukaryotes (Haber 2000, Lieber et al. 2003, Slupphaug et al. 2003, Aylon and Kupiec 2005, Jeggo and Lobrich 2005, Yurchenko et al. 2006). Humans with mutations conferring a hyperrec phenotype have an increased risk of cancer (Bishop and Schiestl 2003). Mitotic recombination is regulated at least in part by the mismatch repair system in eukaryotes (Datta et al. 1996, Chen and Jinks-Robertson 1998). In many cases, the operative regulatory mechanisms are not well defined.

The pattern is not limited to eukaryotes. Mutations in some bacterial recombination proteins reduce homologous recombination and associated repair processes, but can produce large increases in illegitimate recombination (Lovett and Sutera 1995, Hanada et al. 1997, Hanada et al. 2000, Lovett et al. 2002). Much of this genomic instability is again associated with stalled replication forks (Bierne and Michel 1994, Hanada et al. 1997, Hyrien 2000).

The activity of RecA protein, and presumably all related recombinases, is regulated on at least three levels. First, *recA* gene expression is controlled within the SOS regulon (Foster 2005, Friedberg et al. 2005, Kreuzer 2005). Second, RecA protein is subject to autoregulation. Its activities are suppressed, to degrees that vary with conditions, by the C-terminus (and perhaps other parts) of the protein. Third, the activity of RecA protein is modulated by a growing array of other proteins. The second and third modes of regulation are addressed here.

3.1 Autoregulation by the RecA C-terminus

The C-terminal 25 amino acid residues of RecA protein represent only a small part of the C-terminal domain. This will be referred to here as the C-terminus (as opposed to the entire domain). Over half of these terminal 25 residues have side chains that are either negatively charged (seven of the last seventeen are Glu or Asp residues) or contain hydroxyl groups (six Ser or Thr residues). Positively charged amino acid side chains are absent. Sequence conservation in this part of the protein is quite limited even when comparisons are limited to other bacterial

RecA proteins. The major feature of the primary structure, found in most but not all bacterial RecA sequences, is the preponderance of negatively charged residues in this region. A few RecA proteins, notably from *Bacteroides* and *Mycoplasma* species, lack this protein segment altogether (Roca and Cox 1997). In a few other species, particularly *Streptomyces*, the C-terminus is lengthened and exhibits a preponderance of positively charged residues (Roca and Cox 1997).

In the first two decades of RecA research, several C-terminal deletion mutants of the *E. coli* RecA protein were characterized. Ogawa and colleagues described RecAΔC25 (RecA5327) (Tateishi et al. 1992), and Kowalczykowski and colleagues described a RecA mutant in which a fragment of the protein, approximately the C-terminal 15% of the of the RecA polypeptide, had been spontaneously proteolyzed during storage (Benedict and Kowalczykowski 1988). These altered proteins exhibited a faster nucleation of filament formation on dsDNA, reducing the long lag in dsDNA-dependent ATP hydrolysis observed with wild-type RecA (Pugh and Cox 1987, Pugh and Cox 1988). Both C-terminal deletion mutants were shown to be proficient in the key RecA protein reaction of DNA pairing. Shorter C-terminal deletions of RecA protein were also constructed and characterized. A 17 residue C-terminal deletion mutant does not affect UV resistance, induction of the SOS response, or Weigle reactivation (Larminat and Defais 1989). There is a small effect on conjugational recombination only when the wild-type and mutant proteins are both present *in vivo* (Larminat and Defais 1989). Removal of about 18 residues from the C-terminus produces a substantial conformational difference in RecA filaments bound to dsDNA as seen in electron micrographs (Yu and Egelman 1991).

A more complete picture of the function of the C-terminus was revealed in the study of a set of C-terminal deletions involving the removal of 6, 13, 17, or 25 amino acid residues (Eggler et al. 2003, Lusetti et al. 2003a, Lusetti et al. 2003b). This work revealed that the C-terminal 17 amino acid residues of RecA protein (a region that includes all 7 of the negatively charged residues) act more broadly as a kind of autoregulatory flap. Removal of 17 C-terminal amino acid residues (RecAΔC17) enhances a wide range of RecA activities beyond binding to dsDNA. The deletion mutant no longer requires free Mg^{2+} ion for optimal strand exchange activity, indicating that access to the Ao state is regulated to some extent by the C-terminus (Lusetti et al. 2003a). The pH-rate profile for the DNA strand exchange is shifted sharply upwards in in the C-terminal deletion mutants (Lusetti et al. 2003b). Whereas bound SSB protein represents a barrier to the nucleation of wild type RecA protein, RecAΔC17 rapidly displaces SSB on single-stranded DNA even without the assistance of a mediator protein (e.g., the RecOR proteins described later). This indicates an intrinsic capacity of RecA to displace SSB that is modulated by the C-terminus. It also suggests that RecOR may not act by displacing SSB and creating a nucleation site for RecA, but instead may interact with the RecA C-terminus and facilitate the intrinsic process of SSB displacement by RecA. The non-recombination functions of RecA are also enhanced by the C-terminal deletions. The LexA protein is cleaved more rapidly when interacting with RecAΔC17 bound to duplex DNA (S. Lusetti and M. Cox, unpublished results). RecAΔC17 is also more effective in stimulating the activity of DNA polymerase V (Pham et al. 2002, Schlacher et al. 2005, Schlacher et al. 2006). Thus, the C-terminal peptide appears to modulate virtually every RecA function. As such, it is a logical interaction point for other proteins that modulate RecA function.

3.2 Proteins that modulate RecA function

Classically, the RecF, RecO, and RecR proteins (often abbreviated RecFOR) have been highlighted as functions necessary to load RecA protein onto SSB-coated DNA at single-strand gaps (Umezu et al. 1993, Umezu and Kolodner 1994, Sandler 2001). The RecBCD helicase/nuclease has a RecA loading function on the single-strand segments it creates at DNA ends (Anderson and Kowalczykowski 1997, Churchill et al. 1999, Arnold and Kowalczykowski 2000, Spies et al. 2005, Spies and Kowalczykowski 2006). These functions seemed sufficient to target RecA filaments to the locations requiring them, yet recent work has shown that RecA regulation is much more complex. The RecF protein may have multiple functions (Sandler 1996, Rangarajan et al. 2002). Several additional proteins play important roles. We now turn to a description of the activities of these proteins. For a description of RecBCD, see the chapter by S. Kowalczykowski, this volume.

3.3 The single-strand DNA binding protein (SSB)

SSB plays a complex role in RecA reactions. RecA filament nucleation is inhibited, and under some conditions blocked entirely, if SSB is allowed to coat the DNA prior to RecA addition (Kowalczykowski et al. 1987, Lavery and Kowalczykowski 1990, Umezu and Kolodner 1994, Shan et al. 1997, Bork et al. 2001a). This inhibition of binding nucleation is overcome in the bacterial cell by the mediator proteins, RecO and RecR (Umezu and Kolodner 1994, Shan et al. 1997, Bork et al. 2001a). However, Wt EcRecA protein does not bind well to secondary structure in ssDNA, and addition of SSB after RecA protein disrupts the secondary structure and allows RecA to form a contiguous filament on the DNA (Kowalczykowski and Krupp 1987). As already mentioned above, SSB also facilitates DNA strand exchange by binding to the displaced DNA strand.

3.4 The RecFOR proteins

Mediator proteins are as ubiquitous as recombinases, ensuring the targeted assembly of recombinase filaments. In *E. coli*, the RecF, RecO, and RecR proteins function in this capacity, and perhaps have other functions as well.

3.4.1 The RecF protein

The *recF* gene was discovered (Horii and Clark 1973) as a UV sensitive, recombination-deficient mutant in a *recBC sbcBC* background. Although it has never been implicated in replication, the gene is contained in an operon that also includes the *dnaA*, *dnaN*, and *gyrB* genes. The sequenced *recF* gene encodes a 357 amino acid polypeptide (40.5 kDa). The RecF protein has been purified and characterized *in vitro* (Griffin and Kolodner 1990, Madiraju and Clark 1991, Madiraju and Clark 1992, Umezu et al. 1993, Umezu and Kolodner 1994, Webb et al. 1995, Webb et al. 1997, Webb et al. 1999). It binds to ssDNA with an apparent stoichiometry of 1 RecF monomer per 15 nucleotides (Madiraju and Clark 1991). In the presence of ATP, the RecF protein also binds to dsDNA (Madiraju and Clark 1992, Webb et al. 1995). The protein contains a consensus nucleotide-binding fold (Walker A box). The protein binds ATP, and has a weak dsDNA-dependent ATPase activity (k_{cat} about 1.0 min^{-1}) (Webb et al. 1995, Webb et al. 1999). ATP hydrolysis leads to RecF dissociation from DNA (Webb et al. 1999).

3.4.2 RecO protein

The *recO* gene (Kolodner et al. 1985) is situated in an operon with the *rnc* gene, which encodes ribonuclease III, and the *era* gene, which encodes a GTP-binding protein with sequence similarities to the yeast RAS proteins (Ahnn et al. 1986). The sequenced *recO* gene encodes a protein with 242 amino acids (26 kDa) and includes a Walker A box (Morrison et al. 1989, Takiff et al. 1989). The purified protein binds to both ssDNA and dsDNA and behaves as a monomer in solution (Umezu et al. 1993, Luisi-DeLuca and Kolodner 1994, Umezu and Kolodner 1994, Luisi-DeLuca 1995). It promotes an ATP-independent renaturation of complementary DNA strands (Luisi-DeLuca and Kolodner 1994). Binding or hydrolysis of ATP has not been reported. The protein forms a functional complex with the RecR protein, as described below.

The structure of the *D. radiodurans* RecO protein has been determined (Makharashvili et al. 2004). The protein has three structural domains, including an N-terminal domain which features an OB-fold, a novel α-helical domain, and an unusual zinc-binding domain. Sequence alignments indicate that this structural pattern is found in other bacterial RecO proteins.

3.4.3 RecR protein

The *recR* gene (Mahdi and Lloyd 1989a, Mahdi and Lloyd 1989b) is cotranscribed with the *dnaX* gene and shares an operon with a small open reading frame of unknown function called orf-12. The *recR* gene encodes a 201 amino acid protein (22 kDa). The sequence includes two putative DNA-binding motifs (helix-turn-helix and zinc finger) (Alonso et al. 1993). The purified protein has been examined *in vitro*, where it was studied in concert with the RecF and RecO proteins (Umezu et al. 1993, Umezu and Kolodner 1994, Webb et al. 1995, Shan et al. 1997, Webb et al. 1997). There is no indication that the *E. coli* RecR protein alone binds directly to DNA, although the RecR proteins from *D. radiodurans* (Lee et al. 2004) and *B. subtilis* (Alonso et al. 1993) do. The *E. coli* RecR protein is a dimer in solution (Umezu and Kolodner 1994).

The structure of the *D. radiodurans* RecR protein (44% identity with EcRecR) has been determined (Lee et al. 2004). The protein crystallizes as a tetrameric ring with a central hole large enough to accommodate a molecule of dsDNA (Lee et al. 2004). The mechanistic implications of this structure have not yet been explored.

Of the RecFOR proteins, RecR is the most common protein in bacterial genomes (Rocha et al. 2005). RecF is the least common.

3.4.4 Interaction of RecF, O, and R proteins

Several lines of evidence indicate that these 3 proteins function at the same stage of recombination, and tie them to a role in displacing SSB and modulating RecA filament assembly. The phenotypes of mutations in the 3 genes are very similar, defining them as an epistatic group (Smith 1989, Clark and Sandler 1994). Mutations in all three genes are suppressed by *recA*441 (E38K, I298V), *recA730* (E38K), and *recA*803 (V37M) mutations (Wang et al. 1993). *In vitro*, the same RecA441 (previously tif) and RecA803 proteins exhibit an enhanced capacity to displace SSB and bind ssDNA (Lavery and Kowalczykowski 1988, Madiraju et al. 1992). In addition, a gene in bacteriophage λ called *ninB* or *orf* (described further below) has been identified which can replace *recF*, *recO*, and *recR* functions in lambda recombination (Sawitzke and Stahl 1992, Sawitzke and Stahl 1994). *In vivo*, mutant bacteria missing any of the *recFOR* functions exhibit a delayed activation of the SOS response that might reflect slow formation of the RecA filaments required to facilitate LexA cleavage (Madiraju et al. 1988, Whitby and Lloyd 1995). *E. coli* strains in which SSB is overexpressed exhibit a *recFOR*-like phenotype (Moreau 1988), again suggesting that these proteins function together in overcoming the barrier to RecA filament nucleation represented by SSB.

A more detailed examination of the literature, however, shows that the roles of these proteins are not always confluent. In particular, RecF protein appears to have a distinct role that may not always intersect with that of RecO and RecR. In general, the distinctions show up in genetic studies where the underlying molecular mechanisms are poorly understood. RecF activity can be toxic to the cell at least in some contexts. The effect of RecO loss is moderated in *recOrecR* or *recOrecF* strains, suggesting that RecF and RecR are doing something deleterious to the cell in the absence of RecO. In a strain lacking the function of PriA protein (a helicase that plays a key role in restart of replication forks that stall or collapse away from the replication origin (Marians 2000b, Marians 2000a)), the additional loss of RecO is about 10 times more deleterious than the loss of either RecF or RecR (Grompone et al. 2004). In the presence of a different *priA* mutation, *recF* mutants are more deleterious than *recO* or *recR* mutants (Sandler et al. 1996a). The apparent discrepancy may be explained by the extra steps taken in the former study (Grompone et al. 2004) to avoid the appearance of suppressors in the very sick *recOpriA* strains. Both studies, however, draw a clear distinction between the effects of *recF* and *recO* mutations in the *priA* background. The RecF protein, but not

RecO or RecR, is needed for the *in vivo* function of DNA polymerase V and mutagenic TLS (Rangarajan et al. 2002). This work suggests that RecF may work with RecOR in some processes and independently in others. A number of bacterial species with sequenced genomes possess homologues of the *recF* and *recR* genes, but no *recO* gene (Sandler 2001). In *Bacillus subtilis* (which has all three genes), RecF protein recruitment to repair foci is preceded by the appearance of RecO protein (and by RecA protein) by several minutes (Kidane et al. 2004). Overexpression of RecF protein in *E. coli* reduces SOS induction, UV resistance, and viability at 42ºC (Sandler and Clark 1993). The overexpression of the RecOR proteins suppresses many of the deleterious effects of either RecF overexpression (Sandler 1994) or a *recF* null mutation (Sandler and Clark 1994). These varied results suggest that the current pictures of RecFOR and RecF function require expansion.

The functional distinction between RecF and RecO proteins is also quite evident *in vitro* (Umezu et al. 1993, Umezu and Kolodner 1994, Webb et al. 1995, Shan et al. 1997, Webb et al. 1997). RecR protein forms alternative complexes with RecF and RecO protein (Webb et al. 1995, Shan et al. 1997, Webb et al. 1997, Bork et al. 2001a, Morimatsu and Kowalczykowski 2003). RecF protein generally interferes with RecOR function (Webb et al. 1995, Shan et al. 1997, Webb et al. 1997, Bork et al. 2001a, Morimatsu and Kowalczykowski 2003).

The RecOR complex stimulates RecA protein binding to ssDNA coated with SSB, in a process that is not further stimulated by RecF protein under most conditions (Umezu et al. 1993, Umezu and Kolodner 1994, Shan et al. 1997, Bork et al. 2001a). RecO and RecR proteins remain associated with the RecA filament after it is formed (Umezu and Kolodner 1994, Shan et al. 1997). In addition to stimulating nucleation of RecA filament formation on SSB-coated ssDNA, the RecOR complex prevents a net end-dependent dissociation from linear ssDNA (Shan et al. 1997), although it does not suppress RecA dissociation altogether. It is likely that the presence of RecOR leads to rapid nucleation that leads to a rapid replacement of any RecA that dissociates from ssDNA. The RecO and RecR proteins are not active independently in these processes.

There is no evidence yet reported for an interaction between the RecO and RecF proteins. With one exception (Morimatsu and Kowalczykowski 2003), there are no reports of a stimulatory effect of RecF on any RecA activity. However, the presence of RecO and RecR appears to nullify a strong inhibitory effect of RecF on RecA-mediated reactions *in vitro* (Umezu et al. 1993).

The RecF and RecR proteins form a complex in an ATP and DNA-dependent fashion (Webb et al. 1995). The RecFR complex binds primarily to dsDNA, and the complex is stable enough to halt RecA filament extension (Webb et al. 1997). The RecR protein stimulates the RecF ATPase, but reduces the rate of RecF transfer from one DNA to another (Webb et al. 1995, Webb et al. 1999).

Optimized *in vitro* reconstitution of several steps of one major pathway for recombination-dependent replication restart requires the presence of the RecOR proteins (Xu and Marians 2003). RecF protein reduces the stimulation provided by RecOR (Xu and Marians 2003). The effects of RecOR on RecF inhibition of RecA *in vitro* and RecF overexpression *in vivo* (described above) may reflect a similar RecO antagonism of RecFR function.

It is always possible that the failure to detect the formation or activity of a RecFOR complex could reflect a failure to find the right reaction conditions. One recent study has provided evidence that RecF, O, and R can act together to facilitate RecA protein filament formation on SSB-coated DNA gaps (Morimatsu and Kowalczykowski 2003). The stimulatory effect of RecF is observed, however, only in the presence of levels of SSB that are in 6-8 fold excess of that required to saturate the available ssDNA. A useful overview model is that RecOR is necessary and sufficient to load RecA protein onto SSB-coated ssDNA, and that RecF plays a role in targeting this process to the ends of ssDNA gaps as proposed by Morimatsu et al. (Morimatsu and Kowalczykowski 2003). More substantiation of this model is needed. RecF protein has an additional function in antagonizing the function of RecX protein, as described below. This links the RecFOR proteins into a wider network of RecA regulation.

3.5 The DinI and RecX proteins

These two proteins are related in the sense that they have opposing activities, each antagonizing the function of the other.

3.5.1 The RecX protein

The RecX protein (19 kDa) is encoded by a widespread bacterial gene often found just downstream or even overlapping the *recA* gene (Sano 1993, De Mot et al. 1994, Papavinasasundaram et al. 1997, Vierling et al. 2000, Yang et al. 2001). In a few cases, the gene is found in another region of the chromosome (Stohl and Seifert 2001). In *E. coli*, the *recX* gene is just downstream of the recA gene, and expressed from the *recA* promoter via a 5-10% transcriptional readthrough of a hairpin sequence separating the two genes (Pages et al. 2003). In some bacterial species, RecX protein is necessary to overcome deleterious effects of overexpression of RecA protein, implying that RecX is a negative modulator of RecA expression or function (Sano 1993, Papavinasasundaram et al. 1998, Vierling et al. 2000, Sukchawalit et al. 2001). Deletion of the gene in *E. coli* produces no clear phenotype (Pages et al. 2003), although overexpression of the *recX* gene can reduce the induction of the SOS response (Stohl et al. 2003). When purified, both the *Mycobacterium* RecX (Venkatesh et al. 2002) and the *E. coli* RecX protein (Stohl et al. 2003) inhibit the ATPase and strand exchange activities of RecA protein *in vitro*. The RecX protein binds deep within the major helical groove of an AMPPNP-stabilized RecA filament (VanLoock et al. 2003).

Purified RecX blocks the extension of RecA filaments during assembly, almost certainly by capping the filament (Drees et al. 2004a). When RecA filaments have been formed on circular ssDNAs, there is generally no net dissociation and ATP hydrolysis proceeds at a constant steady state. There are generally breaks in the filaments where dissociation at a disassembly end can occur, but the resulting ends are quickly filled in by growth of the trailing filament assembly end. When RecX protein is added at relatively low concentrations (about one RecX per 20-100 bound RecA monomers), a net disassembly of the RecA filaments occurs that takes 10-15 min to complete. Whereas RecX blocks RecA filament assembly, RecA filament disassembly proceeds unabated. The RecA C-terminus plays a significant role in the RecX-RecA interaction (Drees et al. 2004b). Mutations in the RecA C-terminus moderate the interaction (Drees et al. 2004b).

The RecF protein physically interacts with the RecX protein and protects RecA from the inhibitory effects of RecX (Lusetti et al. 2006). *In vitro*, efficient RecA filament formation onto single-stranded DNA binding protein (SSB)-coated circular single-stranded DNA (ssDNA) in the presence of RecX occurs only when all of the RecFOR proteins are present. The RecOR proteins promote RecA filament nucleation onto SSB-coated single-stranded DNA. When RecX is present, substantial RecA filament extension (after RecOR-mediated nucleation) does not occur unless RecF protein is also present (Lusetti et al. 2006). *In vivo*, RecF protein counters a RecX-mediated inhibition of plasmid recombination (Lusetti et al. 2006). Thus, a significant positive contribution of RecF to RecA filament assembly is to antagonize the effects of the negative modulator RecX, specifically during the extension phase of RecA filament assembly.

3.5.2 The DinI protein

DinI is a small (81 amino acids) protein that is induced very early in the SOS response (Kenyon and Walker 1980, Yasuda et al. 1996, Yasuda et al. 1998). Overexpression of the DinI protein in *E. coli* results in UV sensitivity and inhibits the induction of the SOS response (Yasuda et al. 1998). An early proposal suggested that DinI plays a role in bringing the SOS response to an end (Yasuda et al. 1998, Voloshin et al. 2001, Yasuda et al. 2001), but recent work calls this hypothesis into question (Lusetti et al. 2004b). DinI has been purified by several different research groups. DinI inhibits the RecA-mediated cleavage of the UmuD protein (Yasuda et al. 2001). Little effect on RecA filaments was noted in this work, and LexA cleavage was not affected to the same extent. DinI can disrupt RecA filaments (Voloshin et al. 2001), but only when very large excesses (> 20 fold) of DinI are present (Lusetti et al. 2004b). The structure of the DinI protein has been solved by NMR (Ramirez et al. 2000). The C-terminal 17 amino acid residues of DinI features six negatively charged residues, arranged much like they are in the 17 C-terminal amino acid residues of the RecA protein. Another NMR study suggested that the DinI protein binds to the core domain of RecA (Yoshimasu et al. 2003).

Rather than an inhibitor, DinI is actually a potent stabilizer of RecA protein filaments. At DinI concentrations more closely stoichiometric with RecA, DinI strongly stabilizes RecA filaments. Filament disassembly is almost completely suppressed. The effect can be seen dramatically in the electron microscope (Lusetti et al. 2004b). Further, most DNA strand exchange is not blocked by the DinI protein. In one instance (with duplex DNAs cut with restriction enzymes that leave a 5' single strand extension), the initiation of strand exchange is slowed by DinI, but DinI has no effect on the reaction once it is initiated.

The interaction between DinI and RecA is modulated by the RecA C-terminus. Removal of the C-terminal 17 amino acid residues of RecA strongly enhances the interaction between the two proteins (Lusetti et al. 2004b). This is consistent with the C-terminus of RecA being a target for RecA modulators.

This work has led to a new hypothesis for DinI action in which DinI is a selective modulator. During SOS, DinI is induced with somewhat faster kinetics than RecA (Voloshin et al. 2001), consistent with an activator function; its concentration declines late in SOS. The only RecA function reliably suppressed by DinI is the cleavage of UmuD protein, a subunit of DNA polymerase V that is activated by a RecA-mediated autocatalytic cleavage (similar to what is seen with the LexA protein). Thus, the presence of DinI early in SOS could suppress the activation of DNA polymerase V while leaving most other RecA activities intact. This would have the effect of delaying the onset of the mutagenic phase of the SOS response. Thus, DinI may regulate the temporal course of the SOS response, allowing non-mutagenic DNA repair processes to proceed early and delaying the onset of mutagenic translesion DNA polymerase activity.

The role of DinI is distinct from that of Rec(F)OR. DinI does not affect the loading of RecA onto SSB-coated ssDNA (Lusetti et al. 2004b). Instead, DinI stabilizes RecA filaments after they are formed.

The DinI and RecX proteins each antagonize the function of the other. DinI protein stabilizes the RecA filament, and RecX destabilizes it (Lusetti et al. 2004a). With sufficient DinI protein present, a challenge with low concentrations of RecX protein has no effect on RecA filaments. (Lusetti et al. 2004a). If DinI protein is added after the RecX protein challenge, the filaments recover. High concentrations of RecX (nearly stoichiometric with RecA protein) do displace the DinI protein and destabilize the RecA filament. In addition to its capping function, the RecX protein appears to compete with DinI for binding sites in the RecA filament groove. This idea meshes well with the binding of RecX within the RecA filament groove as seen in the EM studies of Egelman and colleagues (VanLoock et al. 2003).

3.6 The PsiB and RdgC proteins

These proteins are implicated in the modulation of RecA filament formation and/or function, but have been characterized minimally *in vitro*.

3.6.1 The PsiB protein

The PsiB protein is encoded by a number of conjugative plasmids (Delver and Belogurov 1997, Sarno et al. 2002). These proteins are generally small, ~140 amino acid residues giving a molecular mass of 15-16 kDa. PsiB inhibits the induction of SOS when conjugation is taking place, presumably by interacting with the RecA protein (Bagdasarian et al. 1986). The *psiB* gene is transferred early in conjugation and transcribed transiently (Bagdasarian et al. 1986). None of the known PsiB proteins have been studied *in vitro*.

When bacterial cells are starved, there is a transient induction of the SOS response that leads to genome-wide hypermutation. This is called adaptive mutation, and it requires RecA function both directly and for the SOS induction (McKenzie et al. 2000, McKenzie et al. 2001). In this context, DinI protein has little effect on the SOS response, but PsiB is a potent inhibitor (in cells with an F plasmid) (McKenzie et al. 2000). These results again suggest that PsiB is an inhibitor of RecA function (and reinforce the hypothesis that DinI is not an inhibitor. Also of interest, the IncN plasmid pKM101 (broad host range) encodes both a PsiB protein and a RecX protein (Delver and Belogurov 1997).

3.6.2 The RdgC protein

RdgC is a fairly abundant DNA binding protein that appears to affect the function of RecA and RecFOR. The *rdgC* gene is located near *sbcC* and *sbcD* on the *E. coli* chromosome, encoding a protein of 34 kDa (Ryder et al. 1996). A deletion of the *rdgC* gene by itself has little obvious effect. However, the same deletion in a *recBCsbcBC* background is viable only if the RecA and RecF proteins are functional (hence, recombination-dependent growth or rdg) (Ryder et al. 1996). The RdgC protein also appears to be important in a *priA* background. The poor viability of *priA* mutants is suppressed by certain mutations in *dnaC* such as *dnaC*212 These allow the DnaC protein to circumvent PriA in the pathway by which the DnaB helicase is loaded onto a repaired fork structure during replication restart (Sandler et al. 1999, Sandler and Marians 2000). Introducing an *rdgC* deletion into a *priAdnaC212* background confers a slow growth phenotype (Moore et al. 2003). Suppressors arise rapidly in these strains, and they are quite informative. First, the slow growth is suppressed by mutations that eliminate RecF, RecO or RecR function (Moore et al. 2003). This suggests that the slow growth is caused by inappropriate loading of RecA protein. It also suggests that the RdgC protein has a role in preventing this inappropriate loading or function of RecA protein. The slow growth is also suppressed by certain mutations in the *ssb* gene (R97C and Δ115-144)(Moore et al. 2003) that could define interaction points for Rec(F)OR. The RdgC protein is present at about 1000 copies per cell (compared to about 800-1700 tetramers of SSB) (Moore et al. 2003), and *in vitro* the protein binds to both ssDNA and dsDNA (Moore et al. 2003).

The *Escherichia coli* RdgC protein is a potential negative regulator of RecA function. RdgC inhibits RecA protein-promoted DNA strand exchange, ATPase activity, and RecA-dependent LexA cleavage *in vitro* (Drees et al. 2006). There is no apparent interaction between RdgC protein and RecA, and RdgC inhibition appears to involve a simple competition for DNA binding sites, especially on duplex DNA. The capacity of RecA to compete with RdgC is improved by the DinI protein. When RdgC protein is bound to the homologous duplex DNA, DNA strand exchange catalyzed by RecA nucleoprotein filaments formed on single-stranded DNA is inhibited (Drees et al. 2006). RdgC protein exists in solution as a mixture of oligomeric states in equilibrium, most likely as monomers, dimers, and tetramers. In the electron microscope, the RdgC protein coats duplex DNA (Drees et al. 2006). It is not clear whether RdgC is a dedicated regulator of recombination. If the observed inhibition of RecA function *in vitro* is physiologically relevant, there must exist some interaction between RdgC and another recombination or replication protein that would serve to target RdgC to sites of recombinational activity.

3.7 The UvrD helicase

In every organism, there appear to be helicases that antagonize recombination functions. This is perhaps best characterized in yeast. The yeast Srs2 helicase negatively modulates recombination (Aguilera and Klein 1988, Aboussekhra et al. 1989). Additional work suggests that Srs2 actively removes recombination complexes or structures *in vivo* (Kaytor et al. 1995, Milne et al. 1995, Schild 1995, Chanet et al. 1996). The yeast Sgsl protein is a helicase from the RecQ family (Gangloff et al. 1994). A combination of *srs2* and *sgs1* null mutants results in a near-lethal slow growth phenotype (Lee et al. 1999, Klein 2001, Fabre et al. 2002, Mankouri et al. 2002). Mutations in a number of recombination functions, including Rad51, Rad52, Rad55, and Rad57, alleviate the defect in the *sgs1 srs2* double mutant (Klein 2001, Fabre et al. 2002). The work indicates that Sgs1 and Srs2 are involved in the removal of toxic recombination intermediates, and can partially substitute for each other. *In vitro*, the Srs2 helicase will disrupt Rad51 protein filaments and interfere with their DNA strand exchange activity (Krejci et al. 2003, Veaute et al. 2003).

In bacteria, parallels are evident but less developed. The RuvA and B proteins displace RecA filaments *in vitro* (Adams et al. 1994), but the genetics provides only limited support for this role *in vivo*. Null mutants of the *uvrD* gene have phenotypes implicating it in recombinational DNA repair (Mendonca et al. 1995). UvrD protein is induced as part of the SOS response, and has been demonstrated to displace RecA filaments.

The Ec UvrD protein (or *E. coli* DNA helicase II) is an 82 kDa helicase protein, whose gene is located at about 84 minutes on the *E. coli* chromosome. UvrD is homologous to the somewhat smaller *E. coli* Rep helicase, and can even form heterodimers with Rep (Wong et al. 1993). Deletion of both the *rep* and *uvrD* genes in *E. coli* is lethal (Washburn and Kushner 1991). UvrD protein is a 3' to 5' helicase (Matson 1986) (as is Rep and the yeast Srs2 helicase) and unwinds duplex DNA best when there is a 3' single strand extension upon which to bind and initiate. However, UvrD also exhibits significant unwinding activity even when initiating the reaction at a nick (Runyon et al. 1990) or blunt end, and this capability could be important for repair systems designed to address strand breaks. In addition, UvrD will unwind RNA-DNA hybrids in a reaction more robust than the unwinding of DNA (Matson 1989), perhaps suggesting a role in replication fork repair on the lagging strand. The UvrD helicase functions as a dimer (Ali et al. 1999, Maluf et al. 2003), although the protein binds well to single-stranded DNA as a monomer (Mechanic et al. 1999, Velankar et al. 1999, Maluf et al. 2003).

As Rep helicase has been shown to remove proteins from the DNA (Yancey-Wrona and Matson 1992), it has been proposed that Rep might serve to remove proteins in the path of the replication fork (DNA synthesis is slowed 50% in cells lacking Rep function (Lane and Denhardt 1975)). However, UvrD is uniquely able to displace RecA protein from ssDNA *in vitro* (Veaute et al. 2005). Rep helicase has a much abbreviated capacity to displace RecA in side by side assays (Veaute et al. 2005).

The genetic studies of the *uvrD* gene generally bears out a role in eliminating RecA and perhaps other recombination proteins from sites where recombinational DNA repair is underway. UvrD plays a role in many aspects of DNA metabolism, although its detailed molecular function remains somewhat enigmatic. This helicase is involved in both methyl-directed mismatch repair (Lahue et al. 1989, Modrich 1989) and the DNA excision repair mediated by the UvrABC excinuclease (Kumura et al. 1985). A role in chromosomal replication is suggested by the constitutive induction of the SOS response that is observed in many *uvrD* mutant cells (Ossanna and Mount 1989, George et al. 1994), presumably because replication forks are stalling in these strains. Cells lacking UvrD function have a defect in recombinational DNA repair (Howard-Flanders and Bardwell 1981, Lloyd 1983, Mendonca et al. 1993). At the same time, *uvrD*$^{-}$ strains have a hyperrecombination phenotype, with large increases in illegitimate recombination (Washburn and Kushner 1991, Lovett and Sutera 1995). Certain alterations of UvrD can suppress the phenotypes of *ruvB* and *recJ* (Lovett and Sutera 1995) mutations, further suggesting a complex involvement in recombination processes. UvrD has a demonstrated anti-recombinase function *in vivo* that may involve the destabilization of recombination intermediates, the complexes that form them, or both (Morel et al. 1993, Lovett and Sutera 1995, Petranovic et al. 2001). In strains lacking the RecBCD pathway, *recQ* and *uvrD* null mutations are synthetically lethal (Mendonca et al. 1995). A fork-clearing role has been proposed for UvrD protein, based on the suppression of the lethality of *uvrD* mutants in DNA polymerase III ts backgrounds by mutations in the *recA*, *recFOR*, *recJ*, and *recQ* genes (Flores et al. 2005).

4 Regulation summary

Bacterial genetic recombination and recombinational DNA repair is clearly under the regulation of an elaborate network of positive and negative effectors. RecA protein appears to be the principle target of regulation. The RecFOR proteins promote RecA filament formation in various ways. The DinI protein stabilizes RecA filaments and alters their function by inhibiting UmuD cleavage. RecX protein blocks RecA filament extension. RdgC protein inhibits RecA by blocking access to duplex DNA. PsiB may be a RecA inhibitor. The UvrD helicase dismantles RecA filaments on DNA. The RecF and DinI proteins antagonize the activities of RecX. The biochemical functions of all of these proteins require further elucidation. It seems likely that we do not yet have a complete picture of the regulatory network.

References

Aboussekhra A, Chanet R, Zgaga Z, Cassier-Chauvat C, Heude M, Fabre F (1989) RADH, a gene of *Saccharomyces cerevisiae* encoding a putative DNA helicase involved in DNA repair. Characteristics of radH mutants and sequence of the gene. Nucl Acids Res 17:7211-7219

Adams DE, Tsaneva IR, West SC (1994) Dissociation of RecA filaments from duplex DNA by the RuvA and RuvB DNA repair proteins. Proc Natl Acad Sci USA 91:9901-9905

Adzuma K (1992) Stable synapsis of homologous DNA molecules mediated by the *Escherichia coli* RecA protein involves local exchange of DNA strands. Genes Dev 6:1679-1694

Aguilera A, Klein HL (1988) Genetic control of intrachromosomal recombination in *Saccharomyces cerevisiae*. I. Isolation and genetic characterization of hyper-recombination mutations. Genetics 119:779-90

Ahnn J, March PE, Takiff HE, Inouye M (1986) A GTP-binding protein of *Escherichia coli* has homology to yeast RAS proteins. Proc Natl Acad Sci USA 83:8849-8853

Alberts B, Miake-Lye R (1992) Unscrambling the puzzle of biological machines: The importance of the details. Cell 68:415-420

Ali JA, Maluf NK, Lohman TM (1999) An oligomeric form of *E. coli* UvrD is required for optimal helicase activity. J Mol Biol 293:815-834

Alonso JC, Stiege AC, Dobrinski B, Lurz R (1993) Purification and properties of the RecR protein from *Bacillus subtilis* 168. J Biol Chem 268:1424-1429

Anderson DG, Kowalczykowski SC (1997) The translocating RecBCD enzyme stimulates recombination by directing RecA protein onto ssDNA in a chi-regulated manner. Cell 90:77-86

Arenson TA, Tsodikov OV, Cox MM (1999) Quantitative analysis of the kinetics of end-dependent disassembly of RecA filaments from ssDNA. J Mol Biol 288:391-401

Arnold DA, Kowalczykowski SC (2000) Facilitated loading of RecA protein is essential to recombination by RecBCD enzyme. J Biol Chem 275:12261-12265

Aylon Y, Kupiec M (2005) Cell cycle-dependent regulation of double-strand break repair - A role for the CDK. Cell Cycle 4:259-261

Bagdasarian M, Bailone A, Bagdasarian MM, Manning PA, Lurz R, Timmis KN, Devoret R (1986) An inhibitor of SOS induction, specified by a plasmid locus in *Escherichia coli*. Proc Natl Acad Sci USA 83:5723-5726

Baliga R, Singleton JW, Dervan PB (1995) RecA.oligonucleotide filaments bind in the minor groove of double-stranded DNA. Proc Natl Acad Sci USA 92:10393-10397

Barre FX, Soballe B, Michel B, Aroyo M, Robertson M, Sherratt D (2001) Circles: The replication-recombination-chromosome segregation connection. Proc Natl Acad Sci USA 98:8189-8195

Baumann P, West SC (1998) Role of the human Rad51 protein in homologous recombination and double-stranded break repair. Trends Biochem Sci 23:247-251

Bazemore LR, Takahashi M, Radding CM (1997) Kinetic analysis of pairing and strand exchange catalyzed by RecA. Detection by fluorescence energy transfer. J Biol Chem 272:14672-14682

Bedale WA, Cox M (1996) Evidence for the coupling of ATP hydrolysis to the final (extension) phase of RecA protein-mediated DNA strand exchange. J Biol Chem 271:5725-5732

Benedict RC, Kowalczykowski SC (1988) Increase of the DNA strand assimilation activity of RecA protein by removal of the C terminus and structure-function studies of the resulting protein fragment. J Biol Chem 263:15513-15520

Bierne H, Michel B (1994) When replication forks stop. Mol Microbiol 13:17-23

Bird LE, Subramanya HS, Wigley DB (1998) Helicases: a unifying structural theme? Curr Opin Struct Biol 8:14-18

Bishop AJR, Schiestl RH (2003) Role of homologous recombination in carcinogenesis. Exp Mol Pathol 74:94-105

Bishop DK (1994) RecA homologs Dmc1 and Rad51 interact to form multiple nuclear complexes prior to meiotic chromosome synapsis. Cell 79:1081-1092

Bjergbaek L, Cobb JA, Gasser SM (2002) RecQ helicases and genome stability: Lessons from model organisms and human disease. Swiss Med Wkly 132:433-442

Bork JM, Cox MM, Inman RB (2001a) The RecOR proteins modulate RecA protein function at 5' ends of single-stranded DNA. EMBO J 20:7313-7322

Bork JM, Cox MM, Inman RB (2001b) RecA protein filaments disassemble in the 5' to 3' direction on single-stranded DNA. J Biol Chem 276:45740-45743

Brendel V, Brocchieri L, Sandler SJ, Clark AJ, Karlin S (1997) Evolutionary comparisons of RecA-like proteins across all major kingdoms of living organisms. J Mol Evol 44:528-541

Brenner SL, Mitchell RS, Morrical SW, Neuendorf SK, Schutte BC, Cox MM (1987) RecA protein-promoted ATP hydrolysis occurs throughout RecA nucleoprotein filaments. J Biol Chem 262:4011-4016

Cadman CJ, McGlynn P (2004) PriA helicase and SSB interact physically and functionally. Nucl Acids Res 32:6378-6387

Campbell MJ, Davis RW (1999) On the *in vivo* function of the RecA ATPase. J Mol Biol 286:437-445

Chakraverty RK, Hickson ID (1999) Defending genome integrity during DNA replication: a proposed role for RecQ family helicases. Bioessays 21:286-94

Chanet R, Heude M, Adjiri A, Maloisel L, Fabre F (1996) Semidominant mutations in the yeast Rad51 protein and their relationships with the Srs2 helicase. Mol Cell Biol 16:4782-4789

Chen WL, Jinks-Robertson S (1998) Mismatch repair proteins regulate heteroduplex formation during mitotic recombination in yeast. Mol Cell Biol 18:6525-6537

Chow SA, Chiu SK, Wong BC (1992) RecA protein-promoted homologous pairing and strand exchange between intact and partially single-stranded duplex DNA. J Mol Biol 223:79-93

Churchill JJ, Anderson DG, Kowalczykowski SC (1999) The RecBC enzyme loads RecA protein onto ssDNA asymmetrically and independently of chi, resulting in constitutive recombination activation. Genes Dev 13:901-911

Clark AJ, Sandler SJ (1994) Homologous genetic recombination: the pieces begin to fall into place. Crit Rev Microbiol 20:125-142

Cohen PE, Pollard JW (2001) Regulation of meiotic recombination and prophase I progression in mammals. Bioessays 23:996-1009
Conley EC, West SC (1990) Underwinding of DNA associated with duplex-duplex pairing by RecA protein. J Biol Chem 265:10156-10163
Conway AB, Lynch TW, Zhang Y, Fortin GS, Fung CW, Symington LS, Rice PA (2004) Crystal structure of a Rad51 filament. Nat Struct Biol 11:791-796
Courcelle J, Hanawalt PC (2003) RecA-dependent recovery of arrested DNA replication forks. Ann Rev Genetics 37:611-646
Courcelle J, Belle JJ, Courcelle CT (2004) When replication travels on damaged templates: bumps and blocks in the road. Res Microbiol 155:231-237
Cox JM, Tsodikov OV, Cox MM (2005) Organized unidirectional waves of ATP hydrolysis within a RecA filament. PLOS Biology 3:231-243
Cox MM, Lehman IR (1981) Directionality and polarity in RecA protein-promoted branch migration. Proc Natl Acad Sci USA 78:6018-6022
Cox MM, Soltis DA, Lehman IR, DeBrosse C, Benkovic SJ (1983) ADP-mediated dissociation of stable complexes of RecA protein and single-stranded DNA. J Biol Chem 258:2586-2592
Cox MM (1994) Why does RecA protein hydrolyze ATP. Trends Biochem Sci 19:217-222
Cox MM (1995) Alignment of three (but not four) DNA strands in a RecA protein filament. J Biol Chem 270:26021-26024
Cox MM (1998) A broadening view of recombinational DNA repair in bacteria. Genes Cells 3:65-78
Cox MM (1999) Recombinational DNA repair in bacteria and the RecA protein. Prog Nucl Acids Mol Biol 63:310-366
Cox MM, Goodman MF, Kreuzer KN, Sherratt DJ, Sandler SJ, Marians KJ (2000) The importance of repairing stalled replication forks. Nature 404:37-41
Cox MM (2001a) Recombinational DNA repair of damaged replication forks in *Escherichia coli*: questions. Ann Rev Genetics 35:53-82
Cox MM (2001b) Historical overview: Searching for replication help in all of the rec places. Proc Natl Acad Sci USA 98:8173-8180
Cox MM (2002) The nonmutagenic repair of broken replication forks via recombination. Mut Res 510:107-120
Cox MM (2003) The bacterial RecA protein as a motor protein. Ann Rev Microbiol 57:551-577
Datta A, Adjiri A, New L, Crouse GF, Jinksrobertson S (1996) Mitotic crossovers between diverged sequences are regulated by mismatch repair proteins in *Saccharomyces cerevisiae*. Mol Cell Biol 16:1085-1093
Datta S, Prabu MM, Vaze MB, Ganesh N, Chandra NR, Muniyappa K, Vijayan M (2000) Crystal structures of *Mycobacterium tuberculosis* RecA and its complex with ADP-AlF4: implications for decreased ATPase activity and molecular aggregation. Nucl Acid Res 28:4964-4973
Datta S, Ganesh N, Chandra NR, Muniyappa K, Vijayan M (2003a) Structural studies on MtRecA-nucleotide complexes: insights into DNA and nucleotide binding and the structural signature of NTP recognition. Proteins 50:474-485
Datta S, Krishna R, Ganesh N, Chandra NR, Muniyappa K, Vijayan M (2003b) Crystal structures of *Mycobacterium smegmatis* RecA and its nucleotide complexes. J Bacteriol 185:4280-4284
de Massy B (2003) Distribution of meiotic recombination sites. Trends Genet 19:514-522

De Mot R, Schoofs G, Vanderleyden J (1994) A putative regulatory gene downstream of recA is conserved in gram-negative and gram-positive bacteria. Nucl Acids Res 22:1313-1314

Delver EP, Belogurov AA (1997) Organization of the leading region of incn plasmid pkm101 (r46) - a regulon controlled by cup sequence elements. J Mol Biol 271:13-30

Dillingham MS, Kowalczykowski SC (2001) A step backward in advancing DNA replication: rescue of stalled replication forks by RecG. Mol Cell 8:734-736

Drees JC, Lusetti SL, Chitteni-Pattu S, Inman RB, Cox MM (2004a) A RecA filament capping mechanism for RecX protein. Mol Cell 15:789-798

Drees JC, Lusetti SL, Cox MM (2004b) Inhibition of RecA protein by the *Escherichia coli* RecX protein - Modulation by the RecA C terminus and filament functional state. J Biol Chem 279:52991-52997

Drees JC, Chitteni-Pattu S, McCaslin DR, Inman RB, Cox MM (2006) Inhibition of RecA protein function by the RdgC protein from *Escherichia coli*. J Biol Chem 281:4708-4717

Egelman E (2000) A common structural core in proteins active in DNA recombination and replication. Trends Biochem Sci 25:180-181

Egelman EH (1993) What do x-ray crystallographic and electron microscopic structural studies of the RecA protein tell us about recombination? Curr Opin Struct Biol 3:189-197

Egelman EH, Stasiak A (1993) Electron microscopy of RecA-DNA complexes: two different states, their functional significance and relation to the solved crystal structure. Micron 24:309-324

Eggler AL, Lusetti SL, Cox MM (2003) The C terminus of the *Escherichia coli* RecA protein modulates the DNA binding competition with single-stranded DNA-binding protein. J Biol Chem 278:16389-16396

Ellouze C, Selmane T, Kim HK, Tuite E, Norden B, Mortensen K, Takahashi M (1999) Difference between active and inactive nucleotide cofactors in the effect on the DNA binding and the helical structure of RecA filament - Dissociation of RecA-DNA complex by inactive nucleotides. Eur J Biochem 262:88-94

Fabre F, Chan A, Heyer WD, Gangloff S (2002) Alternate pathways involving Sgs1/Top3, Mus81/ Mms4, and Srs2 prevent formation of toxic recombination intermediates from single-stranded gaps created by DNA replication. [erratum appears in Proc Natl Acad Sci U S A. 2003 Feb 4;100(3):1462]. Proc Natl Acad Sci USA 99:16887-16892

Flores MJ, Sanchez N, Michel B (2005) A fork-clearing role for UvrD. Mol Microbiol 57:1664-1675

Folta-Stogniew E, O'Malley S, Gupta R, Anderson KS, Radding CM (2004) Exchange of DNA base pairs that coincides with recognition of homology promoted by *E. coli* RecA protein. Mol Cell 15:965-975

Foster PL (2005) Stress responses and genetic variation in bacteria. Mut Res 569:3-11

Frank-Kamenetskii MD, Mirkin SM (1995) Triplex DNA structures. Ann Rev Biochem 64:65-95

Friedberg EC (2005) Suffering in silence: The tolerance of DNA damage. Nature Reviews Mol Cell Biol 6:943-953

Friedberg EC, Walker GC, Siede W, Wood RD, Schultz RA, Ellenberger T (2006) DNA Repair and Mutagenesis, 2nd edition, ASM Press Washington DC

Gangloff S, McDonald JP, Bendixen C, Arthur L, Rothstein R (1994) The yeast type I topoisomerase Top3 interacts with Sgs1, a DNA helicase homolog: a potential eukaryotic reverse gyrase. Mol Cell Biol 14:8391-8398

Genschel J, Curth U, Urbanke C (2000) Interaction of *E. coli* single-stranded DNA binding protein (SSB) with exonuclease I. The carboxy-terminus of SSB is the recognition site for the nuclease. Biol Chem 381:183-192

George JW, Brosh RM Jr, Matson SW (1994) A dominant negative allele of the *Escherichia coli* uvrD gene encoding DNA helicase II. A biochemical and genetic characterization. J Mol Biol 235:424-435

Griffin TJ, Kolodner RD (1990) Purification and preliminary characterization of the *Escherichia coli* K-12 recF protein. J Bacteriol 172:6291-6299

Grompone G, Sanchez N, Ehrlich SD, Michel B (2004) Requirement for RecFOR-mediated recombination in priA mutant. Mol Microbiol 52:551-562

Gumbs OH, Shaner SL (1998) Three mechanistic steps detected by FRET after presynaptic filament formation in homologous recombination. ATP hydrolysis required for release of oligonucleotide heteroduplex product from RecA. Biochemistry 37:11692-11706

Gupta RC, Bazemore LR, Golub EI, Radding CM (1997) Activities of human recombination protein Rad51. Proc Natl Acad Sci USA 94:463-468

Gupta RC, Folta-Stogniew E, O'Malley S, Takahashi M, Radding CM (1999) Rapid exchange of A:T base pairs is essential for recognition of DNA homology by human Rad51 recombination protein. Mol Cell 4:705-714

Gupta RC, Golub E, Bi B, Radding CM (2001) The synaptic activity of HsDmc1, a human recombination protein specific to meiosis. Proc Natl Acad Sci USA 98:8433-8439

Haber JE (2000) Partners and pathways - repairing a double-strand break. Trends Genet 16:259-264

Hanada K, Ukita T, Kohno Y, Saito K, Kato J, Ikeda H (1997) RecQ DNA helicase is a suppressor of illegitimate recombination in *Escherichia coli*. Proc Natl Acad Sci USA 94:3860-3865

Hanada K, Iwasaki M, Ihashi S, Ikeda H (2000) UvrA and UvrB suppress illegitimate recombination: Synergistic action with RecQ helicase. Proc Natl Acad Sci USA 97:5989-5994

Haruta N, Yu XN, Yang SX, Egelman EH, Cox MM (2003) A DNA pairing-enhanced conformation of bacterial RecA proteins. J Biol Chem 278:52710-52723

Heller RC, Marians KJ (2005) The disposition of nascent strands at stalled replication forks dictates the pathway of replisome loading during restart. Mol Cell 17:733-743

Heller RC, Marians KJ (2006) Replication fork reactivation downstream of a blocked nascent leading strand. Nature 439:557-562

Hillers KJ, Villeneuve AM (2003) Chromosome-wide control of meiotic crossing over in *C. elegans*. Curr Biol 13:1641-1647

Horii Z, Clark AJ (1973) Genetic analysis of the RecF pathway to genetic recombination in *Escherichia coli* K12: isolation and characterization of mutants. J Mol Biol 80:327-344

Hortnagel K, Voloshin ON, Kinal HH, Ma N, Schaffer-Judge C, Camerini-Otero RD (1999) Saturation mutagenesis of the *E. coli* RecA loop L2 homologous DNA pairing region reveals residues essential for recombination and recombinational repair. J Mol Biol 286:1097-1106

Howard-Flanders P, Bardwell E (1981) Effects of recB21, recF143, and uvrD152 on recombination in lambda bacteriophage-prophage and Hfr by F- crosses. J Bacteriol 148:739-43

Howard-Flanders P, West SC, Stasiak A (1984) Role of RecA protein spiral filaments in genetic recombination. Nature 309:215-219

Hyrien O (2000) Mechanisms and consequences of replication fork arrest. Biochimie 82:5-17

Jain SK, Cox MM, Inman RB (1994) On the role of ATP hydrolysis in RecA protein-mediated DNA strand exchange III. Unidirectional branch migration and extensive hybrid DNA formation. J Biol Chem 269:20653-20661

Jeggo PA, Lobrich M (2005) Artemis links ATM to double strand break rejoining. Cell Cycle 4:359-362

Jiang H, Giedroc D, Kodadek T (1993) The role of protein-protein interactions in the assembly of the presynaptic filament for T4 homologous recombination. J Biol Chem 268:7904-7911

Kahn R, Cunningham RP, Das Gupta C, Radding CM (1981) Polarity of heteroduplex formation promoted by *Escherichia coli* RecA protein. Proc Natl Acad Sci USA 78:4786-4790

Karlin S, Brocchieri L (1996) Evolutionary conservation of RecA genes in relation to protein structure and function. J Bacteriol 178:1881-1894

Kaytor MD, Nguyen M, Livingston DM (1995) The complexity of the interaction between RAD52 and SRS2. Genetics 140:1441-1442

Kenyon CJ, Walker GC (1980) DNA-damaging agents stimulate gene expression at specific loci in *Escherichia coli*. Proc Natl Acad Sci USA 77:2819-2823

Kidane D, Sanchez H, Alonso JC, Graumann PL (2004) Visualization of DNA double-strand break repair in live bacteria reveals dynamic recruitment of *Bacillus subtilis* RecF, RecO and RecN proteins to distinct sites on the nucleoids. Mol Microbiol 52:1627-1639

Kim JI, Cox MM, Inman RB (1992a) On the role of ATP hydrolysis in RecA protein-mediated DNA strand exchange. I. Bypassing a short heterologous insert in one DNA substrate. J Biol Chem 267:16438-16443

Kim JI, Cox MM, Inman RB (1992b) On the role of ATP hydrolysis in RecA protein-mediated DNA strand exchange. II. Four-strand exchanges. J Biol Chem 267:16444-16449

Klapstein K, Chou T, Bruinsma R (2004) Physics of RecA-mediated homologous recognition. Biophys J 87:1466-1477

Klein HL (2001) Mutations in recombinational repair and in checkpoint control genes suppress the lethal combination of srs2Delta with other DNA repair genes in *Saccharomyces cerevisiae*. Genetics 157:557-565

Kolodner R, Fishel RA, Howard M (1985) Genetic recombination of bacterial plasmid DNA: effect of RecF pathway mutations on plasmid recombination in *Escherichia coli*. J Bacteriol 163:1060-1066

Kolodner RD, Putnam CD, Myung K (2002) Maintenance of genome stability in *Saccharomyces cerevisiae*. Science 297:552-557

Konforti BB, Davis RW (1992) ATP hydrolysis and the displaced strand are two factors that determine the polarity of RecA-promoted DNA strand exchange. J Mol Biol 227:38-53

Konola JT, Logan KM, Knight KL (1994) Functional characterization of residues in the P-loop motif of the RecA protein ATP binding site. J Mol Biol 237:20-34

Kowalczykowski SC, Clow J, Somani R, Varghese A (1987) Effects of the *Escherichia coli* SSB protein on the binding of *Escherichia coli* RecA protein to single-stranded DNA.

Demonstration of competitive binding and the lack of a specific protein-protein interaction. J Mol Biol 193:81-95

Kowalczykowski SC, Krupp RA (1987) Effects of *Escherichia coli* SSB protein on the single-stranded DNA-dependent ATPase activity of *Escherichia coli* RecA protein. Evidence that SSB protein facilitates the binding of RecA protein to regions of secondary structure within single-stranded DNA. J Mol Biol 193:97-113

Kowalczykowski SC, Dixon DA, Eggleston AK, Lauder SD, Rehrauer WM (1994) Biochemistry of homologous recombination in *Escherichia coli.* Microbiol Rev 58:401-65

Kowalczykowski SC, Eggleston AK (1994) Homologous pairing and DNA strand-exchange proteins. Annu Rev Biochem 63:991-1043

Kowalczykowski SC, Krupp RA (1995) DNA-strand exchange promoted by RecA protein in the absence of ATP: implications for the mechanism of energy transduction in protein-promoted nucleic acid transactions. Proc Natl Acad Sci USA 92:3478-3482

Kowalczykowski SC (2000) Initiation of genetic recombination and recombination-dependent replication. Trends Biochem Sci 25:156-165

Krejci L, Van Komen S, Li Y, Villemain J, Reddy MS, Klein H, Ellenberger T, Sung P (2003) DNA helicase Srs2 disrupts the Rad51 presynaptic filament. Nature 423:305-309

Kreuzer KN (2005) Interplay between DNA replication and recombination in prokaryotes. Ann Rev Microbiol 59:43-67

Krishna R, Manjunath GP, Kumar P, Surolia A, Chandra NR, Muniyappa K, Vijayan M (2006) Crystallographic identification of an ordered C-terminal domain and a second nucleotide-binding site in RecA: new insights into allostery. Nucl Acids Res 34:2186-2195

Kubista M, Simonson T, Sjöback R, Widlund H, Johansson A 1996 Towards an understanding of the mechanism of DNA strand exchange promoted by RecA protein, p. 49-59. In Sarma RH and MH Sarma (ed.), Biological structure and function: Proceedings of the ninth Conversation, The State University of New York. Adenine press, New York

Kumar KA, Muniyappa K (1992) Use of structure-directed DNA ligands to probe the binding of recA protein to narrow and wide grooves of DNA and on its ability to promote homologous pairing. J Biol Chem 267:24824-24832

Kumura K, Sekiguchi M, Steinum AL, Seeberg E (1985) Stimulation of the UvrABC enzyme-catalyzed repair reactions by the UvrD protein (DNA helicase II). Nucl Acids Res 13:1483-1492

Kuzminov A (1996) Recombinational repair of DNA damage. R. G. Landes Company, Georgetown, Texas

Kuzminov A (1999) Recombinational repair of DNA damage in *Escherichia coli* and bacteriophage lambda. Microbiol Mol Biol Rev 63:751-813

Lahue RS, Au KG, Modrich P (1989) DNA mismatch correction in a defined system. Science 245:160-164

Lane HE, Denhardt DT (1975) The rep mutation. IV. Slower movement of replication forks in *Escherichia coli* rep strains. J Mol Biol 97:99-112

Larminat F, Defais M (1989) Modulation of the SOS response by truncated RecA proteins. Mol Gen Genet 216:106-112

Lavery PE, Kowalczykowski SC (1988) Biochemical basis of the temperature-inducible constitutive protease activity of the RecA441 protein of *Escherichia coli*. J Mol Biol 203:861-874

Lavery PE, Kowalczykowski SC (1990) Properties of recA441 protein-catalyzed DNA strand exchange can be attributed to an enhanced ability to compete with SSB protein. J Biol Chem 265:4004-4010

Lavery PE, Kowalczykowski SC (1992) A postsynaptic role for single-stranded DNA-binding protein in recA protein-promoted DNA strand exchange. J Biol Chem 267:9315-9320

Lee BI, Kim KH, Park SJ, Eom SH, Song HK, Suh SW (2004) Ring-shaped architecture of RecR: implications for its role in homologous recombinational DNA repair. EMBO J 23:2029-2038

Lee JW, Cox MM (1990) Inhibition of RecA protein-promoted ATP hydrolysis. II. Longitudinal assembly and disassembly of RecA protein filaments mediated by ATP and ADP. Biochemistry 29:7677-7683

Lee SK, Johnson RE, Yu SL, Prakash L, Prakash S (1999) Requirement of yeast SGS1 and SRS2 genes for replication and transcription. Science 286:2339-2342

Lieber MR, Ma YM, Pannicke U, Schwarz K (2003) Mechanism and regulation of human non-homologous DNA end-joining. Nat Rev Mol Cell Biol 4:712-720

Lindsley JE, Cox MM (1990a) On RecA protein-mediated homologous alignment of 2 DNA molecules - 3 strands *versus* 4 strands. J Biol Chem 265:10164-10171

Lindsley JE, Cox MM (1990b) Assembly and disassembly of RecA protein filaments occurs at opposite filament ends: relationship to DNA strand exchange. J Biol Chem 265:9043-9054

Little JW (1991) Mechanism of specific LexA cleavage - autodigestion and the role of RecA coprotease. Biochimie 73:411-422

Lloyd RG (1983) lexA dependent recombination in uvrD strains of *Escherichia coli*. Mol Gen Genet 189:157-161

Lonberg N, Kowalczykowski S, Paul L, von Hippel P (1981) Interactions of Bacteriophage T4-coded gene 32 protein with nucleic acids. III. Binding properties of two specific proteolytic digestion products of the protein (G32P*I and G32P*III). JMB 145:123-138

Lovett ST, Sutera VA (1995) Suppression of recJ exonuclease mutants of *Escherichia coli* by alterations in DNA helicases II (UvrD) and IV (HelD). Genetics 140:27-45

Lovett ST, Hurley RL, Sutera VA, Aubuchon RH, Lebedeva MA (2002) Crossing over between regions of limited homology in *Escherichia coli*: RecA-dependent and RecA-independent pathways. Genetics 160:851-859

Luisi-DeLuca C, Kolodner R (1994) Purification and characterization of the *Escherichia coli* RecO protein. Renaturation of complementary single-stranded DNA molecules catalyzed by the RecO protein. J Mol Biol 236:124-138

Luisi-DeLuca C (1995) Homologous pairing of single-stranded DNA and superhelical double-stranded DNA catalyzed by RecO protein from *Escherichia coli*. J Bacteriol 177:566-572

Lusetti SL, Inman RB, Cox MM (2001) Short C-terminal deletions of the RecA protein: I. Effects on duplex DNA binding. J Biol Chem submitted

Lusetti SL, Cox MM (2002) The bacterial RecA protein and the recombinational DNA repair of stalled replication forks. Ann Rev Biochem 71:71-100

Lusetti SL, Shaw JJ, Cox MM (2003a) Magnesium ion-dependent activation of the RecA protein involves the C terminus. J Biol Chem 278:16381-16388
Lusetti SL, Wood EA, Fleming CD, Modica MJ, Korth J, Abbott L, Dwyer DW, Roca AI, Inman RB, Cox MM (2003b) C-terminal deletions of the *Escherichia coli* RecA protein - Characterization of *in vivo* and *in vitro* effects. J Biol Chem 278:16372-16380
Lusetti SL, Drees JC, Stohl EA, Seifert HS, Cox MM (2004a) The DinI and RecX proteins are competing modulators of RecA function. J Biol Chem 279:55073-55079
Lusetti SL, Voloshin ON, Inman RB, Camerini-Otero RD, Cox MM (2004b) The DinI protein stabilizes RecA protein filaments. J Biol Chem 279:30037-30046
Lusetti SL, Hobbs MD, Stohl EA, Chitteni-Pattu S, Inman RB, Seifert HS, Cox MM (2006) The RecF protein antagonizes RecX function via direct interaction. Mol Cell 21:41-50
MacFarland KJ, Shan Q, Inman RB, Cox MM (1997) RecA as a motor protein. Testing models for the role of ATP hydrolysis in DNA strand exchange. J Biol Chem 272:17675-17685
Madiraju MV, Templin A, Clark AJ (1988) Properties of a mutant *recA*-encoded protein reveal a possible role for *Escherichia coli recF*-encoded protein in genetic recombination. Proc Natl Acad Sci USA 85:6592-6596
Madiraju MV, Clark AJ (1992) Evidence for ATP binding and double-stranded DNA binding by *Escherichia coli* RecF protein. J Bacteriol 174:7705-7710
Madiraju MV, Lavery PE, Kowalczykowski SC, Clark AJ (1992) Enzymatic properties of the RecA803 protein, a partial suppressor of recF mutations. Biochemistry 31:10529-10535
Madiraju MVVS, Clark AJ (1991) Effect of RecF protein on reactions catalyzed by RecA protein. Nucleic Acids Res 19:6295-6300
Mahdi AA, Lloyd RG (1989a) The recR locus of *Escherichia coli* K-12: molecular cloning, DNA sequencing and identification of the gene product. Nucl Acids Res 17:6781-6794
Mahdi AA, Lloyd RG (1989b) Identification of the recR locus of *Escherichia coli* K-12 and analysis of its role in recombination DNA repair. Mol Gen Genet 216:503-510
Makharashvili N, Koroleva O, Bera S, Grandgenett DP, Korolev S (2004) A novel structure of DNA repair protein RecO from *Deinococcus radiodurans*. Structure 12:1881-1889
Malkov VA, Camerini-Otero RD (1995) Photocross-links between single-stranded DNA and *Escherichia coli* RecA protein map to loops L1 (amino acid residues 157-164) and L2 (amino acid residues 195-209). J Biol Chem 270:30230-30233
Maluf NK, Fischer CJ, Lohman TM (2003) A dimer of *Escherichia coli* UvrD is the active form of the helicase *in vitro*. J Mol Biol 325:913-935
Mankouri HW, Craig TJ, Morgan A (2002) SGS1 is a multicopy suppressor of srs2: functional overlap between DNA helicases. Nucl Acids Res 30:1103-1113
Marians KJ (2000a) Replication and recombination intersect. Curr Opin Genet Dev 10:151-156
Marians KJ (2000b) PriA-directed replication fork restart in *Escherichia coli*. Trends Biochem Sci 25:185-189
Matson SW (1986) *Escherichia coli* helicase II (urvD gene product) translocates unidirectionally in a 3' to 5' direction. J Biol Chem 261:10169-10175
Matson SW (1989) *Escherichia coli* DNA helicase II (uvrD gene product) catalyzes the unwinding of DNA.RNA hybrids *in vitro*. Proc Natl Acad Sci USA 86:4430-4434
McGlynn P, Lloyd RG (2002) Genome stability and the processing of damaged replication forks by RecG. Trends Genet 18:413-419

McGlynn P (2004) Links between DNA replication and recombination in prokaryotes. Curr Opin Gen Develop 14:107-112

McKenzie GJ, Harris RS, Lee PL, Rosenberg SM (2000) The SOS response regulates adaptive mutation. Proc Natl Acad Sci USA 97:6646-6651

McKenzie GJ, Lee PL, Lombardo MJ, Hastings PJ, Rosenberg SM (2001) SOS mutator DNA polymerase IV functions in adaptive mutation and not adaptive amplification. Mol Cell 7:571-579

Mechanic LE, Hall MC, Matson SW (1999) *Escherichia coli* DNA helicase II is active as a monomer. J Biol Chem 274:12488-12498

Mendonca VM, Kaiser-Rogers K, Matson SW (1993) Double helicase II (uvrD)-helicase IV (helD) deletion mutants are defective in the recombination pathways of *Escherichia coli*. J Bacteriol 175:4641-4651

Mendonca VM, Klepin HD, Matson SW (1995) DNA helicases in recombination and repair: construction of a delta uvrD delta helD delta recQ mutant deficient in recombination and repair. J Bacteriol 177:1326-1335

Menetski JP, Kowalczykowski SC (1989) Enhancement of *Escherichia coli* RecA protein enzymatic function by dATP. Biochemistry 28:5871-5881

Menetski JP, Bear DG, Kowalczykowski SC (1990) Stable DNA heteroduplex formation catalyzed by the *Escherichia coli* RecA protein in the absence of ATP hydrolysis. Proc Natl Acad Sci USA 87:21-25

Michel B, Flores MJ, Viguera E, Grompone G, Seigneur M, Bidnenko V (2001) Rescue of arrested replication forks by homologous recombination. Proc Natl Acad Sci USA 98:8181-8188

Michel B, Grompone G, Flores MJ, Bidnenko V (2004) Multiple pathways process stalled replication forks. Proc Natl Acad Sci USA 101:12783-12788

Milne GT, Ho T, Weaver DT (1995) Modulation of *Saccharomyces cerevisiae* DNA double-strand break repair by SRS2 and RAD51 Genetics. 139:1189-1199

Modrich P (1989) Methyl-directed DNA mismatch correction. J Biol Chem 264:6597-6600

Moore T, McGlynn P, Ngo HP, Sharples GJ, Lloyd RG (2003) The RdgC protein of *Escherichia coli* binds DNA and counters a toxic effect of RecFOR in strains lacking the replication restart protein PriA. EMBO J 22:735-745

Moran NA, Baumann P (2000) Bacterial endosymbionts in animals. Curr Opin Microbiology 3:270-275

Moreau PL (1988) Overproduction of single-stranded-DNA-binding protein specifically inhibits recombination of UV-irradiated bacteriophage DNA in *Escherichia coli*. J Bacteriol 170:2493-2500

Morel P, Hejna JA, Ehrlich SD, Cassuto E (1993) Antipairing and strand transferase activities of *E. coli* helicase II (UvrD). Nucleic Acids Res 21:3205-3209

Morel P, Stasiak A, Ehrlich SD, Cassuto E (1994) Effect of length and location of heterologous sequences on RecA-mediated strand exchange. J Biol Chem 269:19830-19835

Morimatsu K, Horii T (1995) The DNA-binding site of the RecA protein. Photochemical cross-linking of Tyr103 to single-stranded DNA. Eur J Biochem 228:772-778

Morimatsu K, Kowalczykowski SC (2003) RecFOR proteins load RecA protein onto gapped DNA to accelerate DNA strand exchange: A universal step of recombinational repair. Mol Cell 11:1337-1347

Morrison C, Shinohara A, Sonoda E, Yamaguchi-Iwai Y, Takata M, Weichselbaum RR, Takeda S (1999) The essential functions of human Rad51 are independent of ATP hydrolysis. Mol Cell Biol 19:6891-6897

Morrison PT, Lovett ST, Gilson LE, Kolodner R (1989) Molecular analysis of the *Escherichia coli* recO gene. J Bacteriol 171:3641-3649

Müller B, Koller T, Stasiak A (1990) Characterization of the DNA binding activity of stable RecA-DNA complexes: interaction between the two DNA binding sites within RecA helical filaments. J Mol Biol 212:97-112

Myung K, Datta A, Chen C, Kolodner RD (2001) SGS1, the *Saccharomyces cerevisiae* homologue of BLM and WRN, suppresses genome instability and homeologous recombination. Nature Genet 27:113-116

Myung K, Kolodner RD (2002) Suppression of genome instability by redundant S-phase checkpoint pathways in *Saccharomyces cerevisiae*. Proc Natl Acad Sci USA 99:4500-4507

Nastri HG, Knight KL (1994) Identification of residues in the L1 region of the RecA protein, which are important to recombination or coprotease activities. J Biol Chem 269:26311-26322

New JH, Sugiyama T, Zaitseva E, Kowalczykowski SC (1998) Rad52 protein stimulates DNA strand exchange by Rad51 and replication protein A. Nature 391:337-338

Ogawa T, Yu X, Shinohara A, Egelman EH (1993) Similarity of the yeast RAD51 filament to the bacterial RecA filament. Science 259:1896-1899

Osborn AJ, Elledge SJ, Zou L (2002) Checking on the fork: the DNA-replication stress-response pathway. Trends Cell Biol 12:509-516

Ossanna N, Mount DW (1989) Mutations in uvrD induce the SOS response in *Escherichia coli*. J Bacteriol 171:303-307

Pages V, Koffel-Schwartz N, Fuchs RP (2003) recX, a new SOS gene that is co-transcribed with the recA gene in *Escherichia coli*. DNA Repair 2:273-284

Papavinasasundaram KG, Movahedzadeh F, Keer JT, Stoker NG, Colston MJ, Davis EO (1997) Mycobacterial recA is cotranscribed with a potential regulatory gene called recX. Mol Microbiol 24:141-153

Papavinasasundaram KG, Colston MJ, Davis EO (1998) Construction and complementation of a recA deletion mutant of *Mycobacterium smegmatis* reveals that the intein in *Mycobacterium tuberculosis* recA does not affect RecA function. Mol Microbiol 30:525-534

Passy SI, Yu X, Li ZF, Radding CM, Masson JY, West SC, Egelman EH (1999) Human Dmc1 protein binds DNA as an octameric ring. Proc Natl Acad Sci USA 96:10684-10688

Petranovic M, Zahradka K, Zahradka D, Petranovic D, Nagy B, Salaj-Smic E (2001) Genetic evidence that the elevated levels of *Escherichia coli* helicase II antagonize recombinational DNA repair. Biochimie 83:1041-1047

Pham P, Bertram JG, O'Donnell M, Woodgate R, Goodman MF (2001) A model for SOS-lesion-targeted mutations in *Escherichia coli*. Nature 409:366-370

Pham P, Seitz EM, Saveliev S, Shen X, Woodgate R, Cox MM, Goodman MF (2002) Two distinct modes of RecA action are required for DNA polymerase V-catalyzed translesion synthesis. Proc Natl Acad Sci USA 99:11061-11066

Podyminogin MA, Meyer RB, Gamper HB (1995) Sequence-specific covalent modification of DNA by cross-linking oligonucleotides. Catalysis by RecA and implication for the mechanism of synaptic joint formation. Biochemistry 34:13098-13108

Podyminogin MA, Meyer RB, Gamper HB (1996) RecA-catalyzed, sequence-specific alkylation of DNA by crosslinking oligonucleotides. Effects of length and nonhomologous base substitution. Biochemistry 35:7267-7274

Postow L, Ullsperger C, Keller RW, Bustamante C, Vologodskii AV, Cozzarelli NR (2001) Positive torsional strain causes the formation of a four-way junction at replication forks. J Biol Chem 267:2790-2796

Pugh BF, Cox MM (1987) Stable binding of RecA protein to duplex DNA. Unraveling a paradox. J Biol Chem 262:1326-1336

Pugh BF, Cox MM (1988) General mechanism for RecA protein binding to duplex DNA. J Mol Biol 203:479-493

Pugh BF, Schutte BC, Cox MM (1989) Extent of duplex DNA underwinding induced by RecA protein binding in the presence of ATP. J Mol Biol 205:487-492

Qian XG, Wu Y, He YJ, Luo Y (2005) Crystal structure of *Methanococcus voltae* RadA in complex with ADP: Hydrolysis-induced conformational change. Biochemistry 44:13753-13761

Rajan R, Bell CE (2004) Crystal structure of RecA from *Deinococcus radiodurans*: Insights into the structural basis of extreme radioresistance. J Mol Biol 344:951-963

Ramirez BE, Voloshin ON, Camerini-Otero RD, Bax A (2000) Solution structure of DinI provides insight into its mode of RecA inactivation. Protein Sci 9:2161-2169

Rangarajan S, Woodgate R, Goodman MF (2002) Replication restart in UV-irradiated *Escherichia coli* involving pols II, III, V, PriA, RecA and RecFOR proteins. Mol Microbiol 43:617-628

Register JC, III, Griffith J (1985) The direction of RecA protein assembly onto single strand DNA is the same as the direction of strand assimilation during strand exchange. J Biol Chem 260:12308-12312

Rehrauer WM, Kowalczykowski SC (1993) Alteration of the nucleoside triphosphate (NTP) catalytic domain within *Escherichia coli* recA protein attenuates NTP hydrolysis but not joint molecule formation. J Biol Chem 268:1292-1297

Rehrauer WM, Kowalczykowski SC (1996) The DNA binding site(s) of the *Escherichia coli* RecA protein. J Biol Chem 271:11996-2002

Rice KP, Chaput JC, Cox MM, Switzer C (2000) RecA protein promotes strand exchange with DNA substrates containing isoguanine and 5-methyl isocytosine. Biochemistry 39:10177-10188

Robu ME, Inman RB, Cox MM (2001) RecA protein promotes the regression of stalled replication forks *in vitro*. Proc Natl Acad Sci USA 98:8211-8218

Robu ME, Inman RB, Cox MM (2004) Situational repair of replication forks - Roles of RecG and RecA proteins. J Biol Chem 279:10973-10981

Roca AI, Cox MM (1990) The RecA protein: structure and function. CRC Crit Rev Biochem Mol Biol 25:415-456

Roca AI, Cox MM (1997) RecA protein: structure, function, and role in recombinational DNA repair. Prog Nuc Acid Res Mol Biol 56:129-223

Rocha EPC, Cornet E, Michel B (2005) Comparative and evolutionary analysis of the bacterial homologous recombination systems. PLOS Genetics 1:e15

Rosselli W, Stasiak A (1990) Energetics of RecA-mediated recombination reactions - without ATP hydrolysis RecA can mediate polar strand exchange but is unable to recycle. J Mol Biol 216:335-352

Rosselli W, Stasiak A (1991) The ATPase activity of RecA is needed to push the DNA strand exchange through heterologous regions. EMBO J 10:4391-4396

Runyon GT, Bear DG, Lohman TM (1990) *Escherichia coli* helicase II (UvrD) protein initiates DNA unwinding at nicks and blunt ends. Proc Natl Acad Sci USA 87:6383-6387

Rupp WD, Howard-Flanders P (1968) Discontinuities in the DNA synthesized in an excision-defective strain of *Escherichia coli* following ultraviolet irradiation. J Mol Biol 31:291-304

Ryder L, Sharples GJ, Lloyd RG (1996) Recombination-dependent growth in exonuclease-depleted recBC sbcBC strains of *Escherichia coli* K-12. Genetics 143:1101-1114

Sandler SJ, Clark AJ (1993) Use of high and low level overexpression plasmids to test mutant alleles of the recF gene of *Escherichia coli* K-12 for partial activity. Genetics 135:643-654

Sandler SJ (1994) Studies on the mechanism of reduction of UV-inducible sulAp expression by recF overexpression in *Escherichia coli* K-12. Mol Gen Genet 245:741-749

Sandler SJ, Clark AJ (1994) RecOR suppression of recF mutant phenotypes in *Escherichia coli* K-12. J Bacteriol 176:3661-3672

Sandler SJ (1996) Overlapping functions for *recF* and *priA* in cell viability and UV-inducible SOS expression are distinguished by *dnaC809* in *Escherichia coli* K-12. Mol Microbiol 19:871-880

Sandler SJ, Samra HS, Clark AJ (1996a) Differential suppression of priA2::kan phenotypes in *Escherichia coli* K-12 by mutations in priA, lexA, and dnaC. Genetics 143:5-13

Sandler SJ, Satin LH, Samra HS, Clark AJ (1996b) recA-like genes from three archaean species with putative protein products similar to Rad51 and Dmc1 proteins of the yeast *Saccharomyces cerevisiae*. Nucl Acids Res 24:2125-2132

Sandler SJ, Marians KJ, Zavitz KH, Coutu J, Parent MA, Clark AJ (1999) dnaC mutations suppress defects in DNA replication- and recombination-associated functions in priB and priC double mutants in *Escherichia coli* K-12. Mol Microbiol 34:91-101

Sandler SJ, Marians KJ (2000) Role of PriA in replication fork reactivation in *Escherichia coli*. J Bacteriol 182:9-13

Sandler SJ (2001) RecFOR protein. Encyclopedia of Life Sciences Nature Publishing Group: online

Sano Y (1993) Role of the recA-related gene adjacent to the recA gene in *Pseudomonas aeruginosa* J Bacteriol 175:2451-2454

Sarno R, McGillivary G, Sherratt DJ, Actis LA, Tolmasky ME (2002) Complete nucleotide sequence of *Klebsiella pneumoniae* multiresistance plasmid pJHCMW1. Antimicrobial Agents & Chemotherapy 46:3422-3427

Sawitzke JA, Stahl FW (1992) Phage lambda has an analog of *Escherichia coli* recO, recR and recF genes. Genetics 130:7-16

Sawitzke JA, Stahl FW (1994) The phage lambda orf gene encodes a trans-acting factor that suppresses *Escherichia coli* recO, recR, and recF mutations for recombination of lambda but not of *E. coli*. J Bacteriol 176:6730-6737

Schild D (1995) Suppression of a new allele of the yeast RAD52 gene by overexpression of RAD51, mutations in srs2 and ccr4, or mating-type heterozygosity. Genetics 140:115-127

Schlacher K, Leslie K, Wyman C, Woodgate R, Cox MM, Goodman MF (2005) DNA polymerase V and RecA protein, a minimal mutasome. Mol Cell 17:561-572

Schlacher K, Pham P, Cox MM, Goodman MF (2006) Roles of DNA polymerase V and RecA protein in SOS damage-induced mutation. Chem Rev 106:406-419

Schutte BC, Cox MM (1987) Homology-dependent changes in adenosine 5'-triphosphate hydrolysis during RecA protein promoted DNA strand exchange: evidence for long paranemic complexes. Biochemistry 26:5616-5625

Sehorn MG, Sigurdsson S, Bussen W, Unger VM, Sung P (2004) Human meiotic recombinase Dmc1 promotes ATP-dependent homologous DNA strand exchange. Nature 429:433-437

Seitz EM, Brockman JP, Sandler SJ, Clark AJ, Kowalczykowski SC (1998) RadA protein is an archaeal RecA protein homolog that catalyzes DNA strand exchange. Genes Dev 12:1248-1253

Seitz EM, Haseltine CA, Kowalczykowski SC (2001) DNA Recombination and repair in the Archaea. In "Archaea: Ancient Microbes, Extreme Environments, and the Origin of Life", Blum P, ed, Advances in Applied Microbiology 50:101-169

Setlow RB, Swenson PA, Carrier WL (1963) Thymine dimers and inhibition of DNA synthesis by ultraviolet irradiation of cells. Science 142:1464-1466

Shan Q, Cox MM (1996) RecA protein dynamics in the interior of RecA nucleoprotein filaments. J Mol Biol 257:756-774

Shan Q, Cox MM, Inman RB (1996) DNA strand exchange promoted by RecA K72R. Two reaction phases with different Mg2+ requirements. J Biol Chem 271:5712-5724

Shan Q, Bork JM, Webb BL, Inman RB, Cox MM (1997) RecA protein filaments: end-dependent dissociation from ssDNA and stabilization by RecO and RecR proteins. J Mol Biol 265:519-540

Shan Q, Cox MM (1998) On the mechanism of RecA-mediated repair of double-strand breaks: no role for four-strand DNA pairing intermediates. Mol Cell 1:309-317

Shibata T, Nishinaka T, Mikawa T, Aihara H, Kurumizaka H, Yokoyama S, Ito Y (2001) Homologous genetic recombination as an intrinsic dynamic property of a DNA structure induced by RecA/Rad51-family proteins: A possible advantage of DNA over RNA as genomic material. Proc Natl Acad Sci USA 98:8425-8432

Shinohara A, Ogawa H, Ogawa T (1992) Rad51 protein involved in repair and recombination in S. cerevisiae is a RecA-like protein. Cell 69:457-470

Slupphaug G, Kavli B, Krokan HE (2003) The interacting pathways for prevention and repair of oxidative DNA damage. Mut Res 531:231-251

Smith GR (1989) Homologous recombination in prokaryotes: enzymes and controlling sites. Genome 31:520-527

Smith KC (2004) Recombinational DNA repair: the ignored repair systems. Bioessays 26:1322-1326

Spies M, Dillingham MS, Kowalczykowski SC (2005) Translocation by the RecB motor is an absolute requirement for chi-recognition and RecA protein loading by RecBCD enzyme. J Biol Chem 280:37078-37087

Spies M, Kowalczykowski SC (2006) The RecA binding locus of RecBCD is a general domain for recruitment of DNA strand exchange proteins. Mol Cell 21:573-580

Stasiak A, Di Capua E (1982) The helicity of DNA in complexes with RecA protein. Nature (London) 299:185-186

Stohl EA, Seifert HS (2001) The recX gene potentiates homologous recombination in *Neisseria gonorrhoeae*. Mol Microbiol 40:1301-1310

Stohl EA, Brockman JP, Burkle KL, Morimatsu K, Kowalczykowski SC, Siefert HS (2003) *Escherichia coli* RecX inhibits RecA recombinase and coprotease activities *in vitro* and *in vivo*. J Biol Chem 278:2278-2285

Story RM, Steitz TA (1992) Structure of the RecA Protein-ADP complex. Nature 355:374-376

Story RM, Weber IT, Steitz TA (1992) The structure of the *E. coli* RecA protein monomer and polymer. Nature 355:318-325

Story RM, Bishop DK, Kleckner N, Steitz TA (1993) Structural relationship of bacterial RecA proteins to recombination proteins from bacteriophage T4 and yeast. Science 259:1892-1896

Sukchawalit R, Vattanaviboon P, Utamapongchai S, Vaughn G, Mongkolsuk S (2001) Characterization of *Xanthomonas oryzae* pv. oryzae recX, a gene that is required for high-level expression of recA. FEMS Microbiology Letters 205:83-89

Sung P (1994) Catalysis of ATP-dependent homologous DNA pairing and strand exchange by yeast RAD51 protein. Science 265:1241-1243

Sung P, Stratton SA (1996) Yeast Rad51 recombinase mediates polar DNA strand exchange in the absence of ATP hydrolysis. J Biol Chem 271:27983-27986

Takahashi M, Kubista M, Nordén B (1991) Co-ordination of multiple DNA molecules in RecA fiber evidenced by linear dichroism spectroscopy. Biochimie 73:219-226

Takiff HE, Chen SM, Court DL (1989) Genetic analysis of the rnc operon of *Escherichia coli*. J Bacteriol 171:2581-90

Tamas I, Klasson L, Canback B, Naslund AK, Eriksson AS, Wernegreen JJ, Sandstrom JP, Moran NA, Andersson SGE (2002) 50 million years of genomic stasis in endosymbiotic bacteria. Science 296:2376-2379

Tateishi S, Horii T, Ogawa T, Ogawa H (1992) C-terminal truncated *Escherichia coli* RecA protein RecA5327 has enhanced binding affinities to single- and double-stranded DNAs. J Mol Biol 223:115-129

Thompson LH, Schild D (1999) The contribution of homologous recombination in preserving genome integrity in mammalian cells. Biochimie 81:87-105

Thompson LH, Schild D (2002) Recombinational DNA repair and human disease. Mut Res 509:49-78

Thompson TB, Thomas MG, Escalante-Semerena JC, Rayment IR (1998) Three-dimensional structure of adenosylcobinamide kinase/adenosylcobinamide phoshate guanylyltransferase from *Salmonella typhimurium* determined to 2.3 resolution. Biochemistry 37:686-695

Ullsperger CJ, Cox MM (1995) Quantitative RecA protein binding to the hybrid duplex product of DNA strand exchange. Biochemistry 34:10859-10866

Umezu K, Chi NW, Kolodner RD (1993) Biochemical interaction of the *Escherichia coli* RecF, RecO, and RecR proteins with RecA protein and single-stranded DNA binding protein. Proc Natl Acad Sci USA 90:3875-3879

Umezu K, Kolodner RD (1994) Protein interactions in genetic recombination in *Escherichia coli*. Interactions involving RecO and RecR overcome the inhibition of RecA by single-stranded DNA-binding protein. J Biol Chem 269:30005-30013

van Gent DC, Hoeijmakers JHJ, Kanaar R (2001) Chromosomal stability and the DNA double-stranded break connection. Nat Rev Genet 2:196-206

VanLoock MS, Yu X, Yang S, Galkin VE, Huang H, Rajan SS, Anderson WF, Stohl EA, Seifert HS, Egelman EH (2003) Complexes of RecA with LexA and RecX differentiate between active and Inactive RecA nucleoprotein filaments. J Mol Biol 333:345-354

Veaute X, Jeusset J, Soustelle C, Kowalczykowski SC, Le Cam E, Fabre F (2003) The Srs2 helicase prevents recombination by disrupting Rad51 nucleoprotein filaments. Nature 423:309-312

Veaute X, Delmas P, Selva M, Jeusset J, Le Cam E, Matic I, Fabre F, Petit MA (2005) UvrD helicase, unlike Rep helicase, dismantles RecA nucleoprotein filaments in *Escherichia coli*. EMBO J 24:180-189

Velankar SS, Soultanas P, Dillingham MS, Subramanya HS, Wigley DB (1999) Crystal structures of complexes of PcrA DNA helicase with a DNA substrate indicate an inchworm mechanism. Cell 97:75-84

Venkatesh R, Ganesh N, Guhan N, Reddy MS, Chandrasekhar T, Muniyappa K (2002) RecX protein abrogates ATP hydrolysis and strand exchange promoted by RecA: Insights into negative regulation of homolgous recombination. Proc Natl Acad Sci USA 99:12091-12096

Venkitaraman AR (2001) Chromosome stability, DNA recombination and the BRCA2 tumour suppressor. Curr Opin Cell Biology 13:338-343

Vierling S, Weber T, Wohlleben W, Muth G (2000) Transcriptional and mutational analyses of the *Streptomyces lividans* recX gene and its interference with RecA activity. J Bacteriol 182:4005-4011

Voloshin ON, Ramirez BE, Bax A, Camerini-Otero RD (2001) A model for the abrogation of the SOS response by an SOS protein: a negatively charged helix in DinI mimics DNA in its interaction with RecA. Genes Dev 15:415-427

Walker JE, Saraste M, Runswick MJ, Gay NJ (1982) Distantly related sequences in the α- and β-subunits of ATP synthase, myosin, kinases and other ATP-requiring enzymes and a common nucleotide binding fold. EMBO J 1:945-951

Wang TC, Chang HY, Hung JL (1993) Cosuppression of recF, recR and recO mutations by mutant recA alleles in *Escherichia coli* cells. Mutat Res 294:157-166

Wang Y, Adzuma K (1996) Differential proximity probing of two DNA binding sites in the *Escherichia coli* RecA protein using photo-cross-linking methods. Biochemistry 35:3563-3571

Washburn BK, Kushner SR (1991) Construction and analysis of deletions in the structural gene (uvrD) for DNA helicase II of *Escherichia coli*. J Bacteriol 173:2569-2575

Webb BL, Cox MM, Inman RB (1995) An interaction between the *Escherichia coli* RecF and RecR proteins dependent on ATP and double-stranded DNA. J Biol Chem 270:31397-31404

Webb BL, Cox MM, Inman RB (1997) Recombinational DNA repair - the RecF and RecR proteins limit the extension of RecA filaments beyond single-strand DNA gaps. Cell 91:347-356

Webb BL, Cox MM, Inman RB (1999) ATP hydrolysis and DNA binding by the *Escherichia coli* RecF protein. J Biol Chem 274:15367-15374

West SC, Cassuto E, Howard-Flanders P (1981) Heteroduplex formation by RecA protein: polarity of strand exchanges. Proc Natl Acad Sci USA 78:6149-6153

West SC (1992) Enzymes and molecular mechanisms of genetic recombination. Annu Rev Biochem 61:603-640

Whitby MC, Lloyd RG (1995) Altered SOS induction associated with mutations in recF, recO and recR. Molec Gen Genet 246:174-179

Williams KR, Spicer EK, LoPresti MB, Guggenheimer RA, Chase JA (1983) Limited proteolysis studies on the *Escherichia coli* single-stranded DNA binding protein: evidence for a functionally homologous domain in both the *Escherichia coli* and T4 DNA binding proteins. J Biol Chem 258:3346-3355

Witte G, Urbanke C, Curth U (2003) DNA polymerase III chi subunit ties single-stranded DNA binding protein to the bacterial replication machinery. Nucl Acids Res 31:4434-4440

Wittung P, Nordén B, Kim SK, Takahashi M (1994) Interactions between DNA molecules bound to RecA filament. Effects of base complementarity. J Biol Chem 269:5799-5803

Wong I, Amaratunga M, Lohman TM (1993) Heterodimer formation between *Escherichia coli* Rep and UvrD proteins. J Biol Chem 268:20386-20391

Wu Y, He Y, Moya IA, Qian XG, Luo Y (2004) Crystal structure of archaeal recombinase RadA: A snapshot of its extended conformation. Mol Cell 15:423-435

Wu Y, Qian XG, He YJ, Moya IA, Luo Y (2005) Crystal structure of an ATPase-active form of rad51 homolog from *Methanococcus voltae* - Insights into potassium dependence. J Biol Chem 280:722-728

Xiao J, Singleton SF (2002) Elucidating a key intermediate in homologous DNA strand exchange: Structural characterization of the RecA-triple-stranded DNA complex using fluorescence resonance energy transfer. J Mol Biol 320:529-558

Xiao J, Lee AM, Singleton SF (2006) Construction and evaluation of a kinetic scheme for RecA-mediated DNA strand exchange. Biopolymers 81:473-496

Xing X, Bell CE (2004a) Crystal structures of *Escherichia coli* RecA in a compressed helical filament. J Mol Biol 342:1471-1485

Xing X, Bell CE (2004b) Crystal structures of *Escherichia coli* RecA in complex with MgADP and MnAMP-PNP. Biochemistry 43:16142-16152

Xu LW, Marians KJ (2003) PriA mediates DNA replication pathway choice at recombination intermediates. Mol Cell 11:817-826

Yancey-Wrona JE, Matson SW (1992) Bound Lac repressor protein differentially inhibits the unwinding reactions catalyzed by DNA helicases. Nucl Acids Res 20:6713-6721

Yancey-Wrona JE, Camerini-Otero RD (1995) The search for DNA homology does not limit stable homologous pairing promoted by RecA protein. Curr Biol 5:1149-1158

Yang MK, Chou ME, Yang YC (2001) Molecular characterization and expression of the recX gene of *Xanthomonas campestris* pv. citri. Current Microbiology 42:257-263

Yasuda T, Nagata T, Ohmori H (1996) Multicopy suppressors of the cold-sensitive phenotype of the pcsA68 (dinD68) mutation in *Escherichia coli.* J Bacteriol 178:3854-3859

Yasuda T, Morimatsu K, Horii T, Nagata T, Ohmori H (1998) Inhibition of *Escherichia coli* RecA coprotease activities by DinI. EMBO J 17:3207-3216

Yasuda T, Morimatsu K, Kato R, Usukura J, Takahashi M, Ohmori H (2001) Physical interactions between DinI and RecA nucleoprotein filament for the regulation of SOS mutagenesis. EMBO J 20:1192-202

Yoshimasu M, Aihara H, Ito Y, Rajesh S, Ishibe S, Mikawa T, Yokoyama S, Shibata T (2003) An NMR study on the interaction of *Escherichia coli* DinI with RecA-ssDNA complexes. Nucl Acids Res 31:1735-1743

Yu X, Egelman EH (1991) Removal of the RecA C-terminus results in a conformational change in the RecA-DNA filament. J Struct Biol 106:243-254

Yu X, Egelman EH (1992) Structural data suggest that the active and inactive forms of the RecA filament are not simply interconvertible. J Mol Biol 227:334-346

Yu X, Jacobs SA, West SC, Ogawa T, Egelman EH (2001) Domain structure and dynamics in the helical filaments formed by RecA and Rad51 on DNA. Proc Natl Acad Sci USA 98:8419-8424

Yurchenko V, Xue Z, Sadofsky MJ (2006) SUMO modification of human XRCC4 regulates its localization and function in DNA double-strand break repair. Mol Cell Biol 26:1786-1794

Zavitz KH, Marians KJ (1992) ATPase-deficient mutants of the *Escherichia coli* DNA replication protein PriA are capable of catalyzing the assembly of active primosomes. J Biol Chem 267:6933-6940

Zhou X, Adzuma K (1997) DNA strand exchange mediated by the *Escherichia coli* RecA protein initiates in the minor groove of double-stranded DNA. Biochemistry 36:4650-4661

Cox, Michael M.

Department of Biochemistry, University of Wisconsin-Madison, 433 Babcock Drive, Madison, Wisconsin, USA

cox@biochem.wisc.edu

Biochemistry of eukaryotic homologous recombination

Wolf-Dietrich Heyer

Abstract

The biochemistry of eukaryotic homologous recombination caught fire with the discovery that Rad51 is the eukaryotic homolog of the bacterial RecA and T4 UvsX proteins; and this field is still hot. The core reaction of homologous recombination, homology search and DNA strand invasion, along with the proteins catalyzing it, are conserved throughout evolution in principle. However, the increased complexity of eukaryotic genomes and the diversity of eukaryotic cell biology pose additional challenges to the recombination machinery. It is not surprising that this increase in complexity coincided with the evolution of new recombination proteins and novel support pathways, as well as changes in the properties of those eukaryotic recombination proteins that are evidently conserved in evolution. In humans, defects in homologous recombination lead to increased cancer predisposition, underlining the importance of this pathway for genomic stability and tumor suppression. This review will focus on the mechanisms of homologous recombination in eukaryotes as elucidated by the biochemical analysis of yeast and human proteins.

1 Introduction

Homologous recombination (HR) is a ubiquitous cellular pathway that performs template-dependent, high-fidelity repair of complex DNA damage caused by endogenous and exogenous sources including DNA double-stranded breaks (DSBs), DNA gaps, and inter-strand crosslinks. Historically, the interest in HR has been sparked by its essential role during meiosis where recombination is initiated by a specific DSB delivered by the Spo11 transesterase. Moreover, HR is essential in preserving genome stability through its role in the recovery of stalled or collapsed replication forks, as well as its function in telomere maintenance. This review will discuss the mechanisms of HR and its potential regulation as elucidated by the biochemical analysis of recombination proteins from the yeast *Saccharomyces cerevisiae* and humans. The discussion will concentrate on the core proteins (and their homologs) defined by the *RAD52* epistasis group and on a number of context-specific factors that are involved in HR in some sub-pathways. The reader is

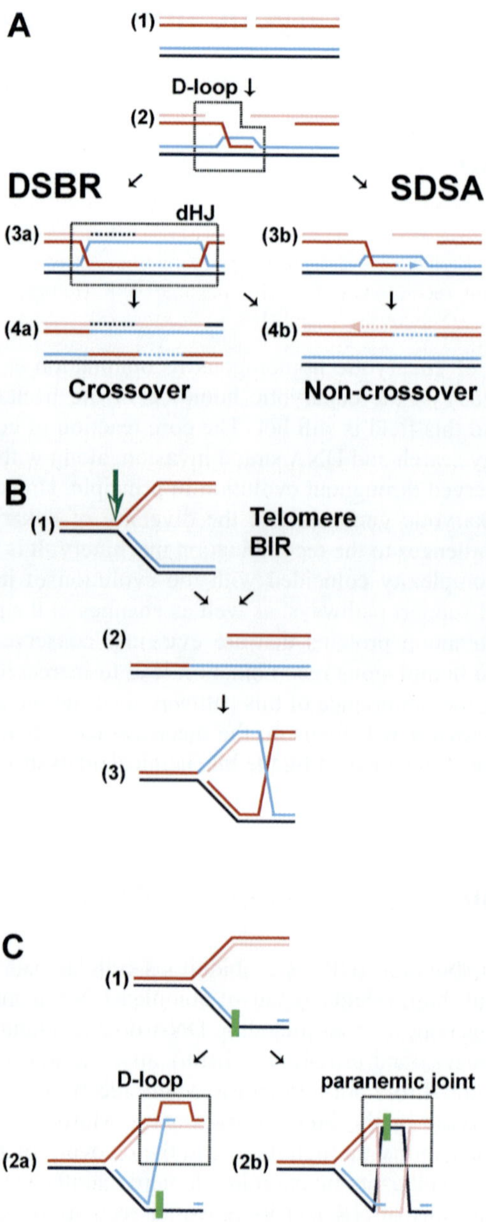

referred to excellent earlier reviews of recombination models and proteins in eukaryotes (Paques and Haber 1999; Symington 2002; Sung et al. 2003; West 2003; Krogh and Symington 2004; Wyman et al. 2004), as well as to reviews of the paradigmatic *Escherichia coli* and phage T4 recombination proteins (Kowalczykowski et al. 1994; Beernink and Morrical 1999), in addition to the

Fig. 1 (overleaf). Pathways of homologous recombination. A. Repair of a frank DSB by double-strand break repair (DSBR) and synthesis-dependent strand annealing (SDSA). After Rad51-mediated D-loop formation, the DSBR and SDSA pathways split. In DSBR, a double Holliday junction (dHJ) is generated that can be resolved by a HJ-specific endonuclease into crossover and non-crossover products or dissolved by the action of BLM-TopoIIIα leading to non-crossover products only. In SDSA, the invading strand reanneals after DNA repair synthesis with the second DSB end without generating a dHJ intermediate, leading to non-crossover products only. B. Repair of a one-sided DSB. Cleavage of a stalled replication fork yields a one-sided DSB, a situation that is similar to break-induced replication (BIR) and recombination at chromosome ends (telomeres). Rad51-mediated DNA strand invasion (D-loop) can establish a replication fork with a single Holliday junction. C. Bypass of DNA damage blocking the lagging strand of a replication fork. Rad51-mediated DNA strand invasion using the blocked 3' end leads to D-loop formation and may proceed either by a DSBR-type pathway (involving a dHJ) or an SDSA-type pathway (without dHJ). This pathway requires a 5'-3' DNA helicase that peels the blocked strand off the template. Alternatively, Rad51 assembles a filament on the template strand to form a paranemic joint (no free end available), allowing the blocked end to use the new sister strand as a template, giving rise to a nicked dHJ. After DNA synthesis either a DSBR-type pathway (involving a dHJ) or an SDSA-type pathway (without dHJ) may ensue. Paranemic joint formation likely requires an additional factor(s) to stabilize the joint.

archaeal recombination proteins (Seitz et al. 2001). The discussion will be limited to recombination in somatic (vegetative) cells as this volume provides a contribution dedicated to meiotic recombination (Hunter). Particular emphasis is given to the assembly and function of the central recombination catalyst, the Rad51-ssDNA presynaptic filament. The article devoted entirely to the late stages of HR (Whitby) will elaborate this topic in much greater detail than attempted here.

2 Homologous recombination in different contexts

HR functions in different contexts with some variations on the types of substrates and intermediates encountered by the recombination machinery. The primary realm of discussing HR has been in the repair of a frank (i.e. two-ended) DSB induced either by ionizing radiation (IR) or an endonuclease (Fig. 1A). Three subpathways were proposed to lead to repair of a frank DSB: double-strand break repair (DSBR), synthesis-dependent strand annealing (SDSA), and single-strand annealing (SSA). The enzymatic and mechanistic requirements for these pathways differ. DSBR and SDSA require DNA strand invasion mediated by the Rad51 filament (and the corresponding co-factors needed for filament assembly and function). On the contrary, SSA is Rad51-independent and does not involve strand invasion but rather reannealing of RPA-coated ssDNA. This rather specialized pathway requires direct repeat sequences flanking the breaks and leads to deletion of one repeat and of the intervening DNA. SSA becomes quite relevant with model substrates containing repeated DNA sequences (for review see Paques and Haber 1999; Krogh and Symington 2004), but is not discussed further here. The

Table 1. Homologous recombination proteins in *E. coli, S. cerevisiae* and humans

	E. coli	*S. cerevisiae*	Human
Initiation	RecBCD	-	-
	SbcCD	Mre11-Rad50-Xrs2	Mre11-Rad50-Nbs1
	RecQ	Sgs1	RecQL, RecQ4, RecQ5, BLM, WRN
	RecJ	ExoI	ExoI
	UvrD	Srs2	Fbh1?
Homologous pairing and DNA strand exchange	RecA	Rad51, Dmc1	Rad51, Dmc1
	SSB	RPA	RPA
	RecF(R)	Rad55-Rad57	Xrcc3-Rad51C
		Shu1-Psy3-Shu2-Csm2	Xrcc2-Rad51D-Sws1
			Rad51B
	RecO(R)	Rad52	Rad52
	-	Rad59	-
	-	-	Brca2-Dss1
	-	Rad54, Rdh54/Tid1	Rad54
			Rad54B
DNA Heteroduplex Extension	RuvAB	Rad54	Rad54
	RecG	-	-
	RecQ	Sgs1	RecQL, RecQ4, RecQ5, BLM, WRN
Resolution/ Dissolution	RuvC	-	Resolvase A
	RecQ-TopoIII	Sgs1-TopIII-Rmi1	BLM-TopIII α-BLAP75
	-	Mus81-Mms4	Mus81-Mms4/Eme1

DSBR and SDSA pathways diverge after D-loop formation, requiring different sets of activities in postsynapsis that function on different types of junctions (Fig. 1A). The repair of a one-sided DSB, formed during break-induced replication (BIR), telomere maintenance or recovery of a broken replication fork brings variations on this theme (Fig. 1B). This pathway likely involves a single Holliday junction instead of the double Holliday junction during DSBR. Gap repair after replication fork blockage poses yet a different substrate for Rad51 filament assembly and likely involves different junction types (Fig. 1C). The different pathways leading to D-loop formation and the processing of the D-loop formed by Rad51 and its cofactors are likely reflected in the differential requirement for specific protein factors, termed here context-specific factors.

3 Biochemistry of recombination proteins

The seminal discovery that Rad51 represents the eukaryotic homolog of the bacterial homologous pairing and DNA strand exchange protein RecA represented a breakthrough in the biochemistry of eukaryotic recombination (Aboussekhra et al.

1992; Shinohara et al. 1992). The demonstration that Rad51 forms a filament as well as functions in ways that are highly similar to RecA allowed for application of the paradigms established with bacterial RecA and T4 UvsX to eukaryotes (Ogawa et al. 1993; Sung 1994). Table 1 lists HR proteins in the budding yeast *S. cerevisiae* and humans. The bacterial proteins are listed for comparison and their biochemistry is elaborated in another dedicated contribution in this volume (Cox). Figure 2 illustrates a number of commonly used *in vitro* recombination assays and a brief discussion of their properties.

HR can be conceptually divided into three stages: presynapsis, synapsis, and postsynapsis (see Fig. 3). In presynapsis, the DNA lesion is processed, if necessary, to form a Rad51-ATP-ssDNA filament, which is also known as the presynaptic filament. Synapsis is defined by homology search and DNA strand exchange, leading to the D-loop intermediate by DNA strand invasion, which is the hallmark of a Rad51-dependent recombination reaction. All ensuing steps constitute postsynapsis including the release of Rad51 from the heteroduplex product DNA, mismatch repair (MMR), DNA synthesis, and processing of the various junction intermediates. Our understanding of events during postsynapsis is poor, and the potential pathways display significant variety after the initial D-loop is formed (Fig. 1). We would like to use the term context-specific factors for those HR proteins that lead to and from the core reaction of Rad51 filament formation and function starting with various substrates (frank or one-sided DSBs, gaps) and process the D-loop intermediate into different pathways of resolution/dissolution or annealing. The requirement for such factors will depend on the context in which the recombination core reaction occurs, and will vary with the specific biochemical or genetic assays utilized. The discussion will focus on the proteins from *S. cerevisiae* but will introduce human proteins where they are unique, or when their function diverged or differs from that of their yeast counterparts.

3.1 Structure of the presynaptic Rad51 filament

S. cerevisiae Rad51 is a ~43 kDa (400 amino acids) protein and shares a core ATPase domain with its homologs, RecA, UvsX, and the archaeal RadA. This domain includes the Walker A and B boxes with structural similarity to the ATPase domains of DNA motor proteins and that of the F1-ATPase. The ATP-bound form of Rad51 undergoes a conformational change necessary for DNA binding. The binding cooperativity leads to filament formation with a stoichiometry of one Rad51 protomer per three-four nucleotides. Rad51 forms a right-handed filament with a helical pitch of 130Å, as determined by crystal structure analysis of a Rad51 fragment lacking the N-terminal 79 amino acids (Conway et al. 2004). The lacking N-terminal amino acids almost exactly correspond to the budding yeast-specific N-terminal extension of Rad51 (Shinohara et al. 1993). Although the crystals were grown in the presence of DNA, the DNA was not visible in the crystal, and the ATP was replaced, probably by a sulfate. Yet, the filament likely represents the DNA-bound form of Rad51. This is consistent with electron microscopic (EM) studies of Rad51/RecA filaments, suggesting flexible filaments with a pitch that varies with the particular protein and the bound nucleotide cofactor (Ogawa et al. 1993; Yu et al. 2001). The crystal structure of the Rad51 filament likely represents the extended, active form, whereas the crystal structure of the RecA filament has a much shorter pitch (83Å) (Story et al. 1992), which is comparable to the pitch determined for inactive filaments by EM (Yu et al. 2001). The Rad51 crystal structure revealed that the nucleotide-binding pocket of one protomer is in direct contact with the ATPase domain of the neighboring subunit, providing a structural basis for coordinated ATPase activity in the Rad51 filament. The conserved N-terminal domain of Rad51, which contains a DNA binding site, was also found in contact with the ATPase domain of the next protomer. This arrangement possibly provides a basis for cooperative DNA binding and for linking the ATPase cycle to DNA binding (Galkin et al. 2006). The crystallographic analysis revealed an asymmetry in the filament due to alternating conformations of the nucleotide-binding pocket. Conway et al. (2004) noted the similarity of this arrangement to hexameric helicases where only a subset of the ATPase sites is active at any given time (Singleton et al. 2000). The crystal structure of the Rad51-ssDNA filament represents a significant advance and provides a sound basis for understanding the interaction of the accessory proteins as well as their roles in the assembly, function, and turnover of the Rad51 filament.

3.1.1 Rad51 versus RecA - cousins not brothers

While RecA and Rad51 are homologous proteins that form relatively equivalent structures, the biochemical and structural analysis has revealed interesting differences between the two proteins. These differences are the result of different evolutionary constraints and selection. Understanding the basis for these differences will enable us to elucidate the functional environment of the Rad51 filament. The ATPase core is structurally conserved between all RecA homologs with recognizable sequence similarity (Conway et al. 2004); however, the N- and C-terminal extensions vary extensively between these proteins (Shinohara et al. 1993). The equivalent of the C-terminal DNA binding site of RecA resides at the N-terminus in Rad51 (Aihara et al. 1999), but the significance of this is not understood.

In RecA, the ATPase cycle is tightly coupled to DNA binding, such that DNA binding requires ATP binding and ATP-hydrolysis triggers the release of DNA (see Cox, this volume; Bianco et al. 1998). RecA has significant preference in binding ssDNA, due to a kinetic barrier that restricts binding to dsDNA. The high cooperativity leads to filament formation in the 5'-3' direction with the addition of ATP-bound subunits on the growing end and the much slower loss of protomers after ATP hydrolysis at the initiating end. On ssDNA, RecA hydrolyzes about 30 ATP min^{-1}, leading to a dynamic situation of filament assembly and disassembly. After DNA strand exchange, RecA is bound to the product heteroduplex DNA and ATP hydrolysis (20 ATP min^{-1}) results in dissociation of RecA from the dsDNA. This frees up the 3'-OH end of the invading strand, allowing access by DNA polymerase to extend the D-loop formed by RecA (Xu and Marians 2002).

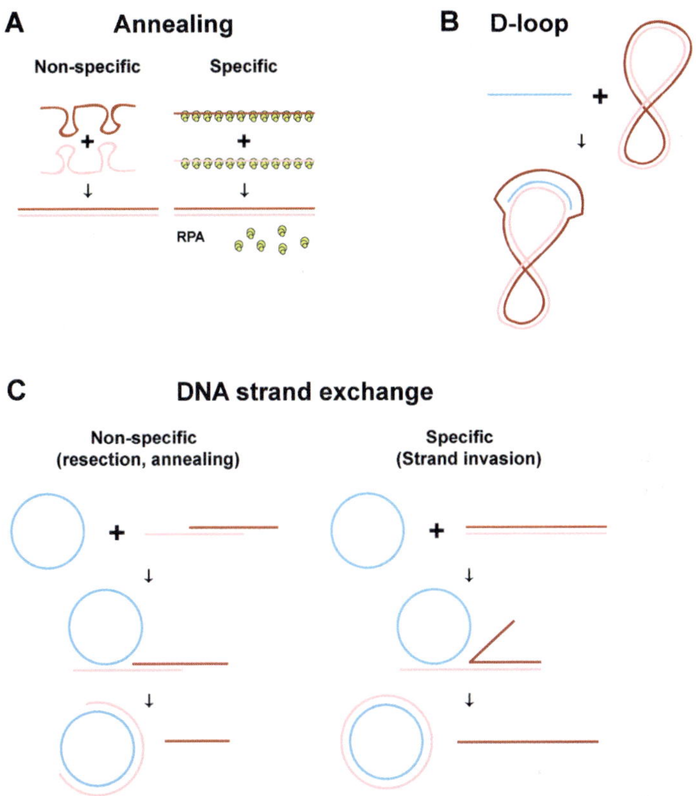

Fig. 2. Model reactions to study homologous recombination *in vitro*. A. DNA annealing. Annealing of protein-free DNA is a relatively non-specific reaction that can occur when proteins aggregate DNA non-specifically. Annealing of RPA-coated ssDNA instead is a highly specific reaction, catalyzed by the UvsY, RecO, and Rad52 proteins. B. D-loop reaction with either linear ssDNA (shown) or linear tailed DNA (not shown) and a supercoiled dsDNA substrate. Note that the RPA requirement for this reaction depends on the length and secondary-structure potential of the ssDNA. Low level, non-specific apparent D-loop formation can occur in particular when the supercoiled dsDNA substrate has been prepared by procedures involving DNA denaturation. C. Three strand DNA strand exchange between circular ssDNA and linear dsDNA. A well-known potential artifact with this assay is exonucleoytic resection followed by DNA strand annealing that will lead to the formation of intermediates and products that resemble DNA strand exchange intermediates. D-loop and DNA strand exchange reactions typically follow strict order of addition protocols: ssDNA + Rad51 → +RPA → +dsDNA. Inhibition of *in vitro* recombination by early RPA addition is overcome by the mediator proteins. Inhibition by early addition of dsDNA can be overcome by the Rad54 motor protein. Release of the product DNA is achieved in all reactions by treatment with detergent and proteinase, thus side-stepping a requirement for turnover by the proteins.

Rad51's biochemical properties differ from those observed in RecA in several ways, but the significance of these differences is poorly understood. First, unlike RecA, Rad51 has only a slight preference for binding ssDNA and can readily bind dsDNA (Zaitseva et al. 1999). It is likely that co-factors, probably one or several of the mediator proteins discussed below, target Rad51 to form filaments on ssDNA. Second, Rad51 binds DNA with significantly less cooperativity than RecA, forming only short filaments at limiting protein concentrations (Kiianitsa et al. 2002). Third, RecA has a clear polarity in filament formation as well as in subsequent DNA strand exchange. RecA polymerizes in a 5' to 3' direction, ensuring coverage of the 3' end to make it more invasive. The issue of polarity with Rad51 appears somewhat unresolved. Some results argue in favor of a DNA strand exchange polarity (and hence filament formation polarity) of yeast and human Rad51 opposite to that of RecA (Sung and Robberson 1995; Baumann and West 1997; Solinger and Heyer 2001). Another study using an N-terminal truncation of yeast Rad51 was shown to lack polarity (Namsaraev and Berg 2000). Fourth, yeast Rad51 exhibits significantly lower ATPase activity than RecA (0.7 ATP min^{-1} on ssDNA, 40 x less than RecA; 0.1 ATP min^{-1} on dsDNA, 200 x less than RecA) (Sung 1994), and similar values were obtained with the human protein (Tombline and Fishel 2002). The lower ATPase activity of Rad51 will significantly curtail the dynamics of the filament. This may lead to less biased polar growth and possibly allow growth in both directions depending on the substrates and factors that nucleate the filament (see below). The lower cooperativity of the Rad51-ssDNA binding may create the problem of multiple nucleation events on a given ssDNA tail. Frequent independent nucleations resulting initially in short filaments will unlikely lead to long contiguous filaments, because with a binding site size of three nucleotides there is a two in three probability to be out of register. Dynamic rearrangement of the mini-filaments would be required to form a single Rad51-ssDNA filament. Compared to RecA filaments, Rad51 filaments on duplex DNA are significantly more stable. Rad51 remains bound to DNA even in the ADP-bound form posing a problem for Rad51 turnover after DNA strand exchange (Zaitseva et al. 1999; Tombline et al. 2002; Bugreev and Mazin 2004). These biochemical properties of Rad51 have to be viewed in the context of the eukaryotic cofactors of Rad51. The specific eukaryotic cofactors may function in addition to nucleating the filament in modulating cooperativity of Rad51 DNA binding, Rad51 filament dynamics on ssDNA to form functional presynaptic filaments and dissociation of Rad51 from dsDNA after DNA strand exchange.

3.1.2 Human Rad51 versus yeast Rad51 - brothers not twins

Besides the fundamental homology between RecA and Rad51, the above discussion focused on the differences that selection imposed on the two proteins. The yeast and human Rad51 proteins exhibit significantly more sequence homology with each other (57% or 66% identity, depending on direction of comparison) than with RecA (26% and 29% identity, respectively) (Shinohara et al. 1992, 1993). However, yeast and human Rad51 proteins also exhibit significant differences. The yeast Rad51 protein is 61 amino acids longer (400 versus 339) than its human counterpart. This difference accounts for the budding yeast specific N-terminal extension of about 75 amino acids, which is the site of species-specific interaction of yeast Rad51 with the Rad52 protein (Donovan et al. 1994). Unfortunately, this segment is lacking in the Rad51 crystal structure (Conway et al. 2004). A major difference between the yeast and human Rad51 proteins was noted immediately in the much lower efficiency of the human protein in DNA strand exchange reactions (Baumann et al. 1996). The efficiency of the *in vitro* recombination activities of the human Rad51 protein can be significantly enhanced by curtailing its binding to dsDNA by addition of 100 mM ammonium sulfate (Sigurdsson et al. 2001a) and inhibiting its ATPase through the addition of calcium ions (Bugreev and Mazin 2004). These reaction conditions produce filaments of more regular structure with decreased dynamics as visualized by atomic force microscopy (Ristic et al. 2005). Interestingly, both protocols were reported not to stimulate the yeast Rad51 protein. Moreover, the strict nucleotide cofactor requirement for DNA binding by RecA and yeast Rad51 at neutral pH is relaxed with human Rad51 (Bianco et al. 1998; Zaitseva et al. 1999; Chi et al. 2006). These structural and biochemical changes suggest that the yeast and human Rad51 proteins will have somewhat different dynamic properties when bound to DNA and understanding these differences may provide a key to understanding the differences in the complexity and function of the cofactors for both proteins (Table 1). While the dynamics of Rad51 filament assembly and filament function is amenable to biochemical analysis, *in vivo* it is difficult to distinguish between functional and non-functional Rad51 assemblies by cytology or chromatin immunoprecipitation (ChIP).

3.2 Presynapsis: different pathways leading to Rad51 filament formation and the function of distinct mediator proteins

3.2.1 Different pathways to generate ssDNA for Rad51 filament formation

Context-specific factors during presynapsis are likely to be required in the processing of various types of DNA damage to single-stranded DNA amenable for Rad51 filament formation. In the context of a frank or one-sided DSB, nucleases or helicase/nuclease combinations are needed to generate a single-stranded tail. In the case of IR-induced breaks, the Mre11-Rad50-Xrs2 complex is needed to process ends with non-standard chemistry. However, with clean nuclease-induced breaks (HO, EndoSceI), the requirement for this complex appears diminished and other nucleases, including Exo1, can function. Furthermore, in *E. coli* RecQ/J function in end-processing; a similar role of the eukaryotic homologs Sgs1/Exo1 has not been clearly demonstrated.

Gap repair on the lagging strand of stalled replication forks (Fig. 1C) may not require further resection, but poses specific topological challenges. In the case of Rad51 filament formation on the uninterrupted ssDNA molecule, DNA strand invasion would lead to a paranemic joint that may require specific stabilization. It is also possible that gap repair involves a helicase that dissociates the stalled 3'-end from the template to create a substrate for Rad51 filament formation on the interrupted strand. This scenario requires a 5'-3' DNA helicase translocating on the uninterrupted strand that is not stalled by the DNA damage that led to the initial stall of the replicative polymerase. The 5'-3' requirement rules out the RecQ family and the Srs2 helicase, which display the opposite polarity (3' to 5').

3.2.2 The problem of forming a filament - learning from actin and tubulin

The RecA/Rad51/RadA proteins are among the few proteins in nature that acquired the propensity to form helical filaments, and their structural similarity to actin filaments has been pointed out before (Egelman 2003). The dynamic instability of the Rad51 filament, the role of the nucleotide cofactor cycle, and the roles of cofactors can be productively compared to other filament forming proteins, e.g. actin and tubulin, where these processes are understood in significant detail (reviewed in Moritz and Agard 2001; Pollard and Borisy 2003; Zigmond 2004). This comparison reveals a number of potentially general characteristics that may help in understanding the complexity and function of the cofactors required for the Rad51 filament (Table 1, see Fig. 3).

Nucleation (the binding of the first subunit) is the rate-limiting step in forming a filament. The filaments are polar with two distinctive ends, displaying dynamic instability correlated to the nucleotide cofactor cycle. The faster growing end adds triphosphate-bound protomers and the slower growing end accumulates the diphosphate-bound protomer. While the filament can grow in both directions, the difference in growth speed leads to treadmilling. Proteins can cap either end of the filament, leading to strongly biased polar filament growth. At high concentration, the protomers can nucleate filament formation independently, but *in vivo* and under limiting *in vitro* conditions, nucleation requires specific mechanisms, providing the basis for regulated filament assembly.

In actin and tubulin filaments, central nucleation factors, the Arp2/3 complex and the γ-tubulin complexes respectively, contain paralogs of the filament protomer and additional subunits to initiate filament formation. These paralogs are incapable of forming an extended polar filament. The Arp2/3 complex, as well as the γ–tubulin complexes, anchor the minus end of the protomer preventing minus end growth and depolymerization, thus leading to plus end-directed filament growth. Arp2/3 is a major regulatory target for actin filament formation. A similar role can be envisioned for the Rad51 paralogs and Rad51 mimics (Brca2), as discussed below. An alternative mode of nucleating the actin filament is catalyzed by formins, a family of proteins with no recognizable homology to actin. These proteins bind to the plus-end of the actin filament and remain processively associated with the growing end. This possibility has not yet been considered for Rad51 cofactors. The continued association of cofactors with the filament could lead to a function also downstream of filament nucleation (Rad51C-Xrcc3, see Section 3.4). Proteins binding monomeric protomers can bias filament growth and dynamics. Profilin catalyzes nucleotide exchange from ADP-G-actin to the ATP-G-actin form. In addition, by binding to the plus end of the protomer, profilin targets the

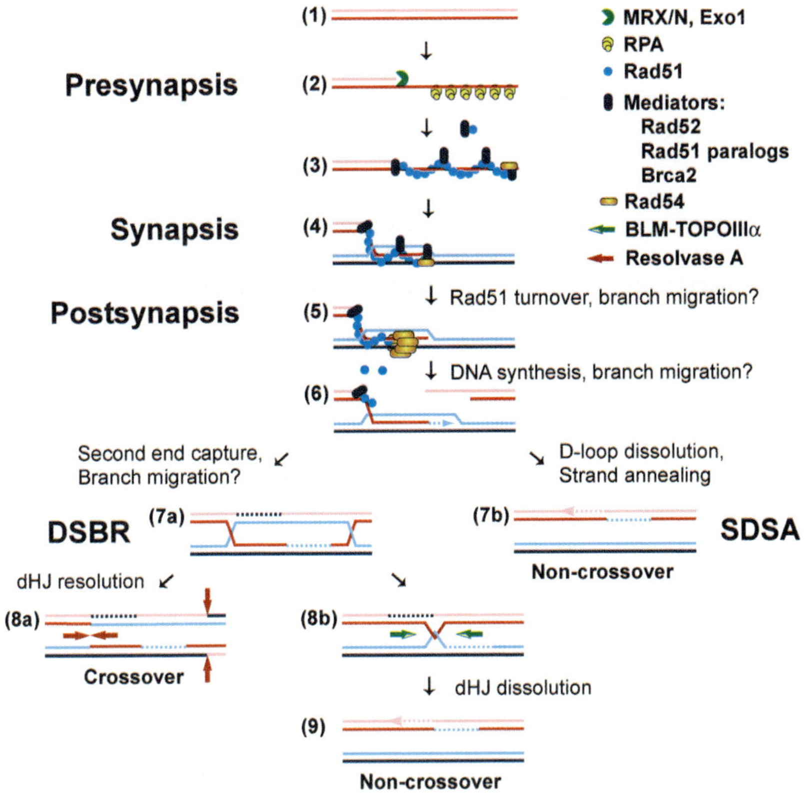

Fig. 3. Mechanistic stages of homologous recombination. HR can be conceptually divided into three stages. First, during presynapsis the ends are processed and the Rad51 filament is assembled. The potential functions of cofactors in Rad51 filament assembly are indicated as anchoring the non-growing end, binding to the growing end or binding the filament laterally. In addition, cofactors may control the Rad51 protomer pool. Second, in synapsis the Rad51 filament undergoes homology search and DNA strand invasion, likely in conjunction with Rad54 protein. Third, postsynapsis comprises all ensuing steps including branch migration, Rad51 turnover, and DNA synthesis, common to the DSBR and SDSA pathways, as well as the subpathway-specific functions of dHJ resolution by a putative resolvase (Resolvase A) and dHJ dissolution by BLM-TOPOIIIα. The SDSA sub-pathway requires a protein, likely a DNA helicase, to dissolve the D-loop and likely employs Rad52 in reannealing the broken ends. The representation of the proteins is for illustration purposes only and does not imply specific stoichiometries or a specific oligomeric assembly status.

minus end of the protomer to the plus end of the growing filament. The importance of regulating the nucleotide cycle of Rad51 has recently been appreciated (roles of Rad51D-Xrcc2 and Ca^{++}, see below), but it is not known if any cofactor delivers Rad51 protomers to the growing end of the filament. Filament stability is also regulated by lateral binding of proteins that stabilize or destabilize actin filaments. Tropomyosin is one such factor that stabilizes actin filaments by lateral binding, and a conceptually similar role has been identified for Rad54 in presynapsis (see below). Actin filament dynamics are tightly linked to the nucleotide cycle. ATP hydrolysis is required for dissociation of the protomer from the filament, but the cofactor cofilin is important for efficient dissociation of ADP-actin from the filament. An active role in Rad51 filament turnover has been proposed for Rad54 protein (see Section 3.4).

Detailed understanding of the actin and tubulin systems provides guidance for identifying the exact mechanisms of Rad51 co-factors. In difference to actin and tubulin, Rad51 (and RecA) form a filament on an underlying lattice, DNA, with defined physical properties, which dictates the register of the filament and confines the filament structure to some degree. Moreover, the DNA provides a lattice for motor proteins (dsDNA: Rad54, Tid1/Rdh54 discussed in Section 3.4; ssDNA: Srs2 discussed in Chapter 4; see also Table 1) that affect Rad51 filament stability.

3.2.3 The problem of forming a filament on DNA - learning from Escherichia coli

Bacterial RecA protein forms a highly cooperative filament and faces the same nucleation problem as Rad51. The two major HR pathways in *E. coli* employ two different strategies to nucleate the RecA filament. In the RecBCD pathway, filament nucleation is coupled to DSB processing and RecA is loaded by the resecting RecBCD helicase/nuclease (Spies and Kowalczykowski 2006). The C-terminal domain of the RecB subunit interacts with a conserved element of the RecA fold, mimicking the initiating protomer. In the RecF pathway, the RecFOR complex targets filament nucleation to the dsDNA-ssDNA transition, forcing filament growth toward the 3' end of the ssDNA tail and limiting growth towards the duplex DNA (Morimatsu and Kowalczykowski 2003). The RecO protein facilitates SSB displacement, and RecO is able to reanneal ssDNA coated with SSB, the prokaryotic ssDNA binding protein (Fig. 2) (Kantake et al. 2002; Morimatsu and Kowalczykowski 2003). While the RecOR complex alone can nucleate RecA filament formation on SSB-coated ssDNA, RecF protein targets the nucleation to a dsDNA-ssDNA junction. RecF lacks obvious sequence homology with RecA and no structural information is available to suggest whether it is structurally related to RecA. In the RecF pathway, nucleation competes with ongoing resection by RecJ (or other nucleases), and it is unclear how this competition is balanced. The absence of a RecBCD homolog in eukaryotes and the similarity between the RecO and Rad52 proteins (Table 1) suggests that the RecF pathway may provide a paradigm for Rad51 filament nucleation anchoring the minus end of the filament to a dsDNA-ssDNA transition.

3.2.4 The problem of forming a Rad51 filament on RPA-coated single-stranded DNA

Forming a Rad51 filament on ssDNA faces two distinct but related mechanistic challenges. First, like all filament forming proteins, Rad51 has to overcome the rate-limiting step of nucleation (binding of the first protomer), which determines the time and place of filament formation. Second, Rad51 faces a more specific challenge of having to replace RPA, the eukaryotic ssDNA binding protein, from ssDNA. RPA binds to ssDNA with higher affinity and specificity than Rad51. These challenges are similar to the bacterial RecA protein, but the biochemical properties of Rad51 discussed above (little preference for ssDNA binding, low filament dynamics, lower cooperativity) create additional problems.

3.2.5 Proteins functioning in Rad51-ssDNA filament assembly and stability

RPA: RPA is a hetero-trimeric complex with subunits of about 70, 30, and 14 kDa, containing six ssDNA binding sites or OB-folds, of which four actively bind to ssDNA in a sequential fashion (for review see Wold 1997; Bochkarev and Bochkareva 2004). RPA is an abundant cellular protein with the highest known affinity to ssDNA of any cellular protein, so that any ssDNA generated during replication, repair, or recombination can be expected to be an ssDNA-RPA complex *in vivo*. RPA engages in a host of protein interactions to function in these contexts. In recombination, RPA serves to counteract secondary structure in ssDNA. This is particularly important for Rad51-mediated reactions, because Rad51 readily binds to dsDNA that forms as secondary structure in ssDNA. This would interfere with forming a functional presynaptic filament on ssDNA. *In vitro* yeast and human Rad51 are strongly stimulated by RPA in reactions employing long ssDNA substrates that have potential to form secondary structures, whereas reactions using ssDNA substrates devoid of secondary structure can largely dispense with RPA (Sung 1994; Sugiyama et al. 1997; Sigurdsson et al. 2001a). Another feature of RPA function in the DNA strand exchange reaction is that it may bind to the displaced strand, preventing the reverse reaction during DNA strand exchange (Fig. 2). It is unclear whether this property is relevant *in vivo* or whether RPA would have access to the displaced strand in a D-loop (Fig. 2). Lastly, reannealing of RPA coated ssDNA by Rad52 protein is a highly specific *in vitro* reaction that distinguishes Rad52 from many other proteins that non-specifically reanneal protein-free DNA (Fig. 2) (Sugiyama et al. 1998). This reaction lies at the heart of the SSA pathway that is not discussed here, but is reviewed elsewhere (Krogh and Symington 2004). Moreover, this mechanism is likely to be relevant for the capture of the second end in the formation of the double Holliday junction and in the annealing step of SDSA (Fig. 1).

While RPA is required for efficient DNA strand exchange by Rad51, the stimulation is only observed using a strict order of addition protocol, in which RPA is added to the ssDNA after Rad51 (Sung 1994; Sugiyama et al. 1997). This allows Rad51 to nucleate on ssDNA, and the binding cooperativity of Rad51 displaces RPA. Simultaneous addition of both proteins to ssDNA or pre-incubation of ssDNA with RPA strongly inhibits Rad51-mediated DNA strand exchange. Since both proteins are expected to be present simultaneously *in vivo*, this poses a problem and calls for catalysts of this reaction to displace RPA from ssDNA by Rad51.

These catalysts are called recombination mediators (Beernink and Morrical 1999; Sung et al. 2003) (Table 1).

The inhibition of Rad51-mediated DNA strand exchange reactions by the early addition of RPA provided an assay to search for mediator proteins. Two types of mediators, Rad52 (Sung 1997a; New et al. 1998; Shinohara and Ogawa 1998) and the Rad55-Rad57 heterodimer (Sung 1997b), were identified by biochemical analysis in budding yeast that appear to address the specific problems of Rad51 nucleation on RPA-coated ssDNA.

Rad52: *S. cerevisiae* Rad52 protein features an N-terminal DNA binding domain and a C-terminal Rad51 interaction domain. The 52.4 kDa budding yeast protein forms a multimeric ring-shaped structure (Shinohara et al. 1998). Three-dimensional image reconstruction of the human Rad52 ring structure revealed a heptameric ring around a central channel that can bind ssDNA on the outside face of the ring (Stasiak et al. 2000; Kagawa et al. 2002; Singleton et al. 2002). *S. cerevisiae* Rad52 specifically interacts with Rad51 and RPA (Shinohara et al. 1992; Hays et al. 1998), and is critical for the ejection of RPA from ssDNA by Rad51 (Sugiyama and Kowalczykowski 2002). The interaction of Rad52 with Rad51 was demonstrated to be critical for its mediator function using a small internal Rad52 deletion mutant specifically affecting this interaction (Krejci et al. 2002). The detailed model for the mediator function of the T4 UvsY protein, a Rad52 homolog in T4 UvsX-mediated recombination, suggests that binding of RPA covered ssDNA to the outside of the Rad52 ring kinks DNA sufficiently to favor binding of the DNA strand exchange protein (Beernink and Morrical 1999; Liu et al. 2006). The genetics and cytology are consistent with Rad52 functioning as a mediator in Rad51 filament assembly (Symington 2002; Krogh and Symington 2004). Rad51 foci, which likely represent Rad51 filaments or later pairing intermediates, do not form or very poorly form in the absence of Rad52 (Gasior et al. 1998, 2001; Lisby et al. 2004). The recombination defect in *rad52* mutants is the most extreme in budding yeast and significantly stronger than in *rad51* mutants (Symington 2002; Krogh and Symington 2004). This reflects the dual function of Rad52 as a mediator for Rad51 and in Rad51-independent SSA (see above).

Rad55-Rad57: Rad55 and Rad57 are two Rad51 paralogs in *S. cerevisiae* with 46.3 and 52.2 kDa, respectively. Both proteins share the RecA core with Rad51 but maintain different N- and C-terminal extensions (Symington 2002; Krogh and Symington 2004). Rad55 and Rad57 form a tight heterodimer, and the available biochemical and genetic evidence suggests that both proteins exclusively function as a complex (Sung 1997b). The heterodimer does not catalyze DNA strand exchange itself, but was shown to function as a mediator, like Rad52 protein, allowing DNA strand exchange when RPA was added to ssDNA at the same time as Rad51 instead of after Rad51 as in standard reactions (Sung 1997b). Unlike Rad52, Rad55-Rad57 are not known to interact with RPA physically, but Rad55 interacts directly with Rad51 (Hays et al. 1995; Johnson and Symington 1995; Sung 1997b). Based on the conceptual similarity with microtubules and actin filaments, one might speculate that the specialized paralogs Rad55-Rad57 nucleate the filament of the Rad51 protomer like γ–tubulin and the Arp2/3 complex. However, at present there is no direct mechanistic biochemical evidence to support this model. It has also been speculated that like RecFOR, Rad55-Rad57 targets nucleation to the dsDNA-ssDNA junction to force mono-directional Rad51 filament formation on ssDNA (Morimatsu and Kowalczykowski 2003). However, such substrates have not been directly tested yet with Rad55-Rad57. The use of a heterodimer (Rad55-Rad57) for nucleation may reflect that the functional unit of Rad51 in the filament appears to be a dimer as proposed from the crystal structure (Conway et al. 2004). The genetics and cytology are consistent with a role of Rad55-Rad57 in Rad51 filament formation (Symington 2002; Krogh and Symington 2004), although the requirement in the formation of DNA damage-induced Rad51 foci is less strict than for meiotic Rad51 foci (Gasior et al. 1998, 2001; Lisby et al. 2004). A particularly elegant demonstration of the presynaptic Rad55-Rad57 function was the isolation of a partially Rad55-independent Rad51 mutant, Rad51-I345T, that increased its intrinsic capacity for displacing RPA from ssDNA (Fortin and Symington 2002). In *S. cerevisiae* the general requirement of filament forming proteins for a nucleation factor and the specific requirement of Rad51 to displace RPA appear to be fulfilled by the Rad55-Rad57 and Rad52 mediators. However, it is unclear how these two mediators cooperate, and no reactions have been reconstituted containing both mediators.

Shu1-Psy1-Shu2-Csm2: Another complex containing two highly divergent Rad51 paralogs, Rdl1 and Rlp1, has been identified in the fission yeast *Schizosaccharomyces pombe* together with a third subunit Sws1 (Martin et al. 2006). The genetic and cytological characterization of the genes establish an early function in HR. Rdl1 and Rlp1 appear to be reduced versions of the human Rad51 paralogs Rad51D and Xrcc2 proteins, respectively (Table 1). Identification of this homology was complicated by the loss of the Walker A box in Rdl1 and the loss of the Walker B box in Rlp1. The authors also noted significant homology to components of the Shu1-Psy3-Shu2-Csm2 complex of budding yeast, proposing that Shu1 represents the *S. cerevisiae* Xrcc2 (Rlp1) homolog, Psy3 the Rad51D (Rdl1) homolog, and Shu2 the budding yeast Sws1 (Table1) (Martin et al. 2006). The Shu1-Psy3-Shu2-Csm2 complex was identified in *S. cerevisiae* as suppressors of the slow growth phenotype of *top3* cells and demonstrated to be factors in the *RAD52* epistasis group required for efficient HR (Shor et al. 2005). While their contribution does not appear to match the effect of the other Rad51 paralogs, Rad55 and Rad57, deletion of *SHU1* (and presumably of the other subunits as well) prolongs the half-life of MMS-induced Rad52 foci (Shor et al. 2005), suggesting this complex may be required for normal kinetics in Rad51 filament assembly. Linking the Shu1-Psy3-Shu2-Csm2 complex to Rad51 paralogs (Martin et al. 2006) provides a great impetus to initiate biochemical studies with these proteins.

In humans, the presynaptic situation is somewhat different and surprisingly more complex. In the discussion of the yeast and human Rad51 proteins it is argued that this is in part a reflection of the biochemical differences between both eukaryotic Rad51 proteins and in part a reflection of the greater complexity of human cells *vis-à-vis* pathway regulation and tissue differentiation. A major difference lies with the Rad52 protein. Rad52 is critically required for Rad51 filament formation and SSA in yeast, causing the most extreme recombination defects as a single mutant, whereas the mouse mutant causes only a very mild recombination defect (Rijkers et al. 1998). Has the role of Rad52 been usurped by other factors (Brca2, Rad51 paralogs)? Which protein is mediating reannealing of RPA-coated ssDNA during mammalian SSA?

Human Rad52: Human Rad52 was found to stimulate human Rad51-mediated DNA strand exchange. This effect was observed in the absence of RPA (Benson et al. 1998) and may involve mechanisms other than the mediator function demonstrated for yeast Rad52. In support of a mediator function speaks the genetic observation that double mutants of *XRCC3* and *RAD52* are lethal in chicken DT40 cells (Fujimori et al. 2001), suggesting a shift in balance between the individual mediators in different eukaryotes.

Human Rad51 paralogs: The number of Rad51 paralogs increased to five in vertebrates (Rad51B/Rad51L1, Rad51C/Rad51L2, Rad51D/Rad51L3, Xrcc2, Xrcc3; Table 1) (Thacker 2005). Each of the five paralogs is required for Rad51 focus formation *in vivo* in chicken DT40 cells and in mammalian cells. The individual single mutants display near identical phenotypes in chicken DT40 cells, where they were studied in parallel (Takata et al. 2000, 2001). Also the mammalian Rad51 paralog mutants display highly similar, though possibly not identical, phenotypes (Thacker 2005). The presynaptic function of the Rad51 paralogs is underscored by the significant rescue of the individual paralog mutants by Rad51 overexpression (Takata et al. 2000, 2001), recapitulating the same finding with the yeast paralogs Rad55-Rad57 (Hays et al. 1995; Johnson and Symington 1995). The initial model proposed a function of all Rad51 paralogs acting in a single complex to support Rad51 filament assembly. It is now clear that the situation is more complex. The five paralogs form different sub-assemblies with the main complexes being Rad51B-Rad51C-Rad51D-Xrcc2 and Rad51C-Xrcc3 (Masson et al. 2001b). Genetic analysis of *rad51B, rad51D,* and *xrcc3* single and double mutants in DT40 cells is consistent with both complexes having distinct functions (Yonetani et al. 2005). The biochemistry of the paralogs is hampered by their poor solubility, and the existence of further subassemblies (Rad51B-Rad51C, Rad51D-Xrcc2) complicates the interpretation. Moreover, Rad51D-Xrcc2 associates with an additional protein, Sws1. RNAi knockdown of Sws1 reduced but not eliminated spontaneous and IR-induced Rad51 focus, suggesting that Sws1 is another protein with a function in Rad51 filament formation/stabilization (Martin et al. 2006). The presence of novel subunits in Rad51 paralog complexes could explain the difficulties in expressing soluble forms of such complexes in the absence of these subunits. The purified Rad51B-Rad51C-Rad51D-Xrcc2 and Rad51C-Xrcc3 complexes preferentially bind ssDNA with no preference for tailed DNA and little to no reported dsDNA binding activity (Masson et al. 2001a, 2001b). [The potential function of the Rad51C-Xrcc3 complex in Holliday junction resolution is discussed below and more extensively in the contribution by Whitby.] It is generally believed that the Rad51 paralogs neither form filaments nor catalyze DNA strand invasion reactions, although this view has been challenged by the visualization of Rad51D-Xrcc2 and Rad51C-Xrcc3 filaments on ssDNA (Kurumizaka et al. 2001, 2002). The significance of the observed structures remains unclear. The low activity in D-loop assays (Fig. 2) identified for both complexes in these studies may be a reflection of a strand annealing activity rather than true DNA strand invasion. Such an activity may result from destabilizing duplex DNA, followed by random renaturation of single-strands, as demonstrated for the Rad51C protein (Lio et al. 2003). The Rad51B-Rad51C subassembly, but not the tetrameric complex also containing Rad51D and Xrcc3 (Masson et al. 2001b), were found to preferentially bind 3'-end tailed DNA (Lio et al. 2003). The Rad51B-Rad51C complex also partially overcame the inhibition imposed by early addition of RPA in the DNA strand exchange reaction (Fig. 2A) displaying mediator function similar to the Rad55-Rad57 complex (Sigurdsson et al. 2001b). These studies did not yet test whether Rad51B-Rad51C target filament formation to the dsDNA-ssDNA junction, as suggested by its DNA binding specificity.

Brca2: Brca2, a human breast cancer tumor suppressor protein, is also required for DNA damage-induced Rad51 focus formation *in vivo* and was identified as another mediator for Rad51 filament formation (Tarsounas et al. 2003; Pellegrini and Venkitaraman 2004). The protein contains eight BRC repeats that bind Rad51 with varying degree of affinity. In addition, Brca2 contains a unique Rad51 binding domain in the extreme C-terminus. Crystal structure analysis identified three OB-folds in the C-terminal portion of the protein, a typical ssDNA-binding motif also found in RPA, and a helix-turn-helix domain, a dsDNA-binding motif found in many transcription factors, suggesting that Brca2 might bind at an ssDNA-dsDNA transition (Yang et al. 2002). Brca2 is associated with the small 70-residue protein, Dss1, whose function remains largely mysterious. The enormous size of Brca2 with 3,418 amino acids has precluded analysis of the full-length human protein, but the analysis of Brca2 fragments and of Brca2 homologs with a smaller size than the human protein has afforded significant insights into Brca2 function. The *Ustilago maydis* Brca2 homolog Brh2 represents a diet version of human Brca2 with 1,075 amino acids containing a single BRC repeat plus the C-terminal Rad51 binding domain. Elegant biochemical experiments demonstrated that Brh2 targets Rad51 filament formation to a dsDNA-ssDNA transition and overcomes inhibition of early RPA addition (Yang et al. 2005). This is fully consistent with the analyses of human Brca2 fragments, which also support a mediator function of Brca2 for Rad51 nucleation (Yang et al. 2002; San Filippo et al. 2006). Of particular interest for the mechanism of Brca2 function is the structure of the BRC repeats that appear to mimic the Rad51 subunit interface in the filament (Yang et al. 2002), although no apparent sequence homology between the BRC repeat and the structurally corresponding Rad51 sequence can be detected. This suggests a potential mechanism for Rad51 filament nucleation in that Brca2 provides a polymerization interface for the first Rad51 protomer, similar to the Arp2/3 and γ-tubulin complexes with actin filaments and microtubules.

Why do mammalian cells need so many mediators: five Rad51 paralogs, Brca2, Sws1, and to some degree Rad52? A simple model would be to suggest that they serve as mediators on different substrates. However, the observation that all proteins are required (with the exception of Rad52 as a single mutant) for IR-induced Rad51 focus formation suggests otherwise. If Brca2 nucleates the Rad51 filament at a dsDNA-ssDNA transition and anchors the filament at the non-growing end, what do the Rad51 paralogs do? Why does human Brca2 have eight BRC repeats, and what is the function of the C-terminal Rad51 binding site? In the actin filament, two cofactors impinge on the nucleotide cycle of actin. ATP hydrolysis by actin occurs relatively fast, but dissociation of the phosphate and release of actin-ADP is slow and aided by cofilin. Instead, profilin catalyzes nucleotide exchange from ADP-G-actin to the ATP-G-actin form. Similar proteins could be expected to function in the dynamics of the Rad51 filament and its nucleotide cycle. First evidence for such a function comes from the analysis of the effect of the Rad51D-Xrcc2 complex on the ATPase cycle of Rad51 (Shim et al. 2004). Xrcc2 enhanced the Rad51 ATPase activity and the function of Rad51 filaments as assayed by DNA unwinding and DNA strand exchange, whereas the Rad51D-Xrcc2 complex exhibited less stimulation of the Rad51 ATPase activity. Such an activity would enhance the dynamic turnover of the Rad51 filament and might be critical to maintain a functional and growing Rad51 filament. This effect of Xrcc2 appears opposite to the observed stimulation of Rad51 DNA pairing activities by Ca^{++}, which was accompanied by an inhibition of the Rad51 ATPase activity that keeps the filament protomers in the active, ATP-bound state (Bugreev and Mazin 2004). It is unclear whether this is a physiological role of Ca^{++} or whether Ca^{++} replaces a protein cofactor(s). These biochemical studies are still in their infancy, but clearly suggest that protein cofactors, possibly the Rad51 paralogs, affect the nucleotide cycle of Rad51 and hence the dynamics of the Rad51 filament. Thus, it is possible that these proteins might also associate with the growing end of the filament, as found for profilin in the actin filament. This role may result in a continuous requirement for such a protein in filament formation, in addition to or instead of a requirement solely in the nucleation of the Rad51 filament. Cytologically both defects would be indistinguishable leading to the absence of detectable Rad51 foci. Physical monitoring of DNA intermediates during DSB repair may have identified a signature of such a late role. *S. cerevisiae rad57* mutants exhibit a delay in the occurrence of DNA synthesis products from the invading strand of the D-loop, consistent with a defect in DNA strand invasion (Aylon et al. 2003). The accumulation of this intermediate (D-loop + DNA synthesis; Fig. 3, Step 6) was interpreted as evidence for a late, second role of Rad57 protein in processing the extended D-loop to conversion products by the SDSA pathway (Fig. 1, Fig. 3), but this phenotype could also be a late consequence of a Rad51 filament that is less dynamic, interfering with the strand annealing step.

Rad54: Rad54 is another protein with a function in presynapsis, which appears quite different from the mediator proteins introduced before. Due to the structure of this review, the presynaptic function of Rad54 is discussed here, whereas the analysis of the roles of Rad54 protein in synapsis (see Section 3.3) and postsynapsis (see Section 3.4) is deferred to later. The extensive biochemical analyses of this protein has been more fully evaluated in other reviews (Tan et al. 2003; Heyer et al. 2006). Rad54 is a dsDNA motor protein (see below), whose ATPase activity is essential for its *in vivo* function (Clever et al. 1999). However, budding yeast Rad54 was identified to also exhibit an ATPase-independent function in presynapsis (Mazin et al. 2003; Wolner and Peterson 2005). Effects on Rad51 filament formation are difficult to distinguish from effects on filament stability using *in vivo* methods (cytology, ChIP), but biochemical experiments demonstrated that

Rad54 stabilizes Rad51 filaments rather than helping in their assembly (Solinger et al. 2002; Mazin et al. 2003). The stabilization likely occurs through inhibition of Rad51 protomer dissociation, but it is unclear whether Rad54 binds to the end of the Rad51-ssDNA filament or binds laterally. The association with the pre-synaptic filament targets Rad54 protein to the pairing site (Mazin et al. 2000a), which is likely critical in the positioning of the motor on dsDNA for its function in the later stages of HR.

In summary, the surprising complexity of pre-synapsis and Rad51 filament formation, particularly in mammalian cells, still needs to be fully explored biochemically. Formation of a functional filament is not only mechanistically complex, but its nucleation is also a likely regulatory target. Nucleation defines the transition from RPA-covered ssDNA to a Rad51 filament. While RPA-coated ssDNA may have multiple fates (NHEJ *vs.* SSA *vs.* HR for a DSB; HR *vs.* translesion synthesis [TLS] *vs.* fork regression for a gap), a Rad51 filament is committed to HR. The genetic and biochemical differences between the yeast and the human proteins suggest that different variations on the same underlying fundamental theme are at work.

3.3 Synapsis: homology search and DNA strand invasion

The ternary complex of Rad51-ATP-ssDNA, otherwise known as the presynaptic filament, has a secondary binding site for the duplex DNA used during the homology search. The DNA in the presynaptic filament is stretched to an extended state with 5.1Å per base equaling 18 nt/bp per helical repeat (Yu et al. 2001), which is believed to facilitate the homology search by a base flipping mechanism, allowing the single-strand to sample homology on the duplex DNA (Gupta et al. 1999). However, the exact mechanism of the homology search still remains to be determined. Rad51 filament formation, homology search and DNA strand exchange do not require ATP hydrolysis (Sung and Stratton 1996). In RecA protein, ATP hydrolysis serves to release the heteroduplex DNA product of the DNA strand exchange reaction and turnover of RecA, which is critical for filament dynamics (Bianco et al. 1998). Unlike RecA, Rad51 cannot catalyze a four-stranded reaction or by-pass heterology. This inability is likely a consequence of the impaired dynamics of Rad51 filaments compared to RecA, suggesting that specific cofactors (maybe a function of the mediator proteins) will be required to reconstitute such reactions *in vitro*.

Rad54: While Rad51 can perform synapsis on its own, synapsis is greatly stimulated by the dsDNA motor protein Rad54 (for review see Tan et al. 2003; Heyer et al. 2006). In budding yeast, Rad54 is an 898 amino acid (human Rad54: 748 amino acids) dsDNA motor protein/translocase and exhibits significant dsDNA-specific ATPase activity of about 1,000/min per Rad54 molecule (Petukhova et al. 1998; Swagemakers et al. 1998; Ristic et al. 2001). Rad54 translocates on dsDNA at an astounding pace of 300 bp/sec with significant processivity (Amitani et al. 2006). Although not all experiments have been performed with both the yeast and human Rad54 proteins, there is currently no reason to suggest a difference in function between both proteins. The specific physical interaction between the Rad51 and Rad54 proteins has significant functional consequences for the biochemical activities of either protein (Jiang et al. 1996; Clever et al. 1997; Golub et al. 1997). Rad51 in its DNA-bound form stimulates the ATPase and motor activity of Rad54, and *vice versa* Rad54 stimulates the pairing activities of Rad51 (Petukhova et al. 1999; Mazin et al. 2000b; Van Komen et al. 2000; Solinger et al. 2001; Kiianitsa et al. 2002; Sigurdsson et al. 2002; Mazin and Mazin 2004). In particular the D-loop reaction (Fig. 2) essentially requires the presence of Rad54 to be catalyzed by Rad51. The mechanism by which Rad54 stimulates Rad51-mediated pairing remains to be determined. It may involve sliding of the target duplex DNA during homology search, topological opening of the target duplex DNA, or the clearing of Rad51 protein bound to the target DNA. Rad54 is a member of the Snf2 family of DNA translocases, which contains paradigmatic chromatin remodeling factors. Another function of Rad54 may be its ability to remodel chromatin structure. This activity was identified in biochemical assays with nucleosomal substrates (Alexeev et al. 2003; Jaskelioff et al. 2003) and may help to overcome the inherent inhibition imposed by nucleosomes (Alexiadis and Kadonaga 2003). Rdh54/Tid1, another Snf2-like motor protein primarily involved in meiotic recombination, also stimulates Rad51-mediated DNA strand exchange in a manner that appears similar to Rad54 (Petukhova et al. 2000).

3.4 Postsynapsis: many subpathways call for context-specific factors

A multitude of processes occur after DNA strand invasion. Distinct subpathways of HR (Fig. 1, Fig. 3) require common factors for Rad51 dissociation, DNA synthesis, and branch migration from the invading 3'-end, but also distinct, context-specific factors that act on the specific intermediates generated by the discrete pathways (DSBR: second end capture, dHJ resolution/dissolution; SDSA: D-loop dissolution, reannealing; BIR: single HJ resolution; gap repair: paranemic joint processing). Little is known about these processes in eukaryotes and a dedicated chapter in this volume (Whitby) focuses on this theme; two reasons to keep the discussion brief here.

3.4.1 Rad51 turnover by Rad54

After DNA strand invasion, RecA requires ATP hydrolysis to release the resulting heteroduplex DNA to allow DNA polymerase access to the invading 3'-OH end, as shown in a reconstituted reaction (Xu and Marians 2002). Similarly, Rad51 requires turnover and its DNA binding properties suggest that its intrinsic ATPase activity is insufficient for efficient product release. The Rad54 motor protein dissociates Rad51-dsDNA filaments in a reaction that requires specific protein interaction and the dsDNA-specific ATPase activity of Rad54 (Solinger et al. 2002). The Rad54 ATPase is significantly (six-fold) stimulated by Rad51 filaments partially occupying dsDNA (Kiianitsa et al. 2002; Solinger et al. 2002), suggesting that Rad54 associates with the terminus and catalyzes processive dissociation of terminal protomers. This terminal interaction between the motor and the Rad51 filament has been visualized by EM (Kiianitsa et al. 2006), and this mechanism is consistent with the ability of Rad54 to translocate on dsDNA (Amitani et al. 2006). In addition to its ATPase-independent function in presynapsis and ATP-dependent functions during synapsis and chromatin remodeling, ATP-dependent Rad51 turnover after DNA strand invasion is the fourth role assigned to Rad54 by biochemical experiments. Data from *in vivo* assays (ChIP, cytology, genetics) are consistent with a function of Rad54 in synapsis and postsynapsis but presently these assays cannot distinguish between both possibilities. An evaluation of this evidence would exceed the frame of this review and the interested reader is referred to dedicated reviews on this subject (Tan et al. 2003; Heyer et al. 2006). It is interesting to note that *E. coli* does not have a Rad54 homolog (Table 1), possibly because RecA is self-sufficient in turnover. A role of Rad54 in Rad51 turnover helps explain some biochemical differences between RecA and Rad51 with regards to dsDNA binding and ATPase activity.

3.4.2 DNA synthesis and DNA polymerase η

DNA synthesis after DNA strand invasion is critical in restoring the genetic information compromised by a DSB and the continuity of strands in gaps. In *E. coli*, a coupled reaction between D-loop formation and extension has been reconstituted using Pol III holoenzyme, the highly processive bacterial replicative DNA polymerase (Xu and Marians 2002). Which polymerase accesses the invading 3'-end *in vivo* remains unclear. In eukaryotes this question encounters significant complexity with the identification of an entire new suite of translesion synthesis DNA polymerases (Rattray and Strathern 2003). Fractionation of human cell extracts identified that human DNA polymerase η (eta), known for its role in bypass of UV photoproducts and the UV syndrome *Xeroderma pigmentosum*, extends the invading strand in a D-loop (Fig. 2) in a reaction that could not be supported by other DNA polymerases (McIlwraith et al. 2005). This reaction may involve an interaction between Polη and Rad51 (McIlwraith et al. 2005) and its *in vivo* significance is underlined by the finding in chicken DT40 cells that Polη is required for efficient gene conversion between immunoglobulin-variable genes (Kawamoto et al. 2005). The involvement of a low fidelity polymerase in the high fidelity HR process is somewhat surprising and suggests a hand-off to another more processive, high-fidelity polymerase might occur. Genetic studies in yeast had suggested earlier the potential involvement of TLS polymerases in recombination-mediated DSB repair (Holbeck and Strathern 1997).

3.4.3 Branch migration: Rad54 and BLM

In bacteria, RuvA protein targets the RuvB motor protein to Holliday junctions. RuvB positions as hexameric rings to two opposite arms of the Holliday junction where they act as coordinated DNA pumps providing the motor force for branch migration (West 2003). Considering the comparably low ATPase and DNA strand activity of Rad51, eukaryotes would appear to require a DNA motor protein for branch migration. However, it is unclear from genetic analyses whether branch migration occurs in eukaryotes or whether the extent of heteroduplex DNA is entirely determined by the extent of the DSB resection. The Rad54 motor protein was shown to enhance branch migration in the three-stranded DNA strand exchange reaction (Fig. 2) by up to six-fold. This activity of Rad54 depended on its ATPase activity and a specific interaction with the Rad51 proteins, as branch migration in reactions catalyzed by RecA or human Rad51 was not stimulated (Solinger and Heyer 2001). The human Rad54 protein displays both specificity in DNA binding and ATPase activity for several types of junction substrates, including X and partial X-junctions, as well as forked DNA (Bugreev et al. 2006). Both yeast and human Rad54 proteins could branch migrate protein-free three-stranded and four-stranded branches in a bi-directional manner, but not branched substrates still associated with Rad51 strand exchange protein (Bugreev et al. 2006). It is unclear whether Rad54 is responsible for the ATP-dependent branch migration activity observed in human extract and partially purified fractions of resolvase A (see below) (Constantinou et al. 2002). A potential role of Rad54 in branch migration is consistent with *in vivo* data measuring conversion track length under conditions where Rad54 or an ATPase-defective Rad54 mutant protein are overexpressed (Kim et al. 2002). Branch migration of Holliday junctions formed by RecA-mediated four-stranded DNA strand exchange was demonstrated for the BLM helicase, an activity that is underpinned by the binding preference of this DNA helicase to model Holliday junctions (Karow et al. 2000). During HR, BLM is possibly targeted to junctions by its interactions with Rad51 and the Rad51 paralog, Rad51D (Wu et al. 2001; Braybrooke et al. 2003). BLM is a 3'-5' DNA helicase of the RecQ family (Table 1), and it will be interesting to learn whether the other RecQ homologs in humans display similar biochemical activities.

3.4.4 Second end capture/DNA annealing by Rad52 and Rad59

The capture of the second end during DSBR could conceivably be accomplished by a second Rad51-mediated DNA strand invasion event or by reannealing the second end to the displaced strand from the initial DNA strand invasion (Fig. 1, Fig. 3). DNA strand annealing in DSBR and SDSA (Figs. 1, 2) likely involves RPA-coated ssDNA, which can be reannealed by Rad52 protein in a highly specific reaction (Sugiyama et al. 1998). It is unclear how such an asymmetry between an invading end (Rad51 filament assembly) and non-invading end (strand annealing) can be mechanistically accomplished. Rad52-mediated reannealing in *S. cerevisiae* likely involves Rad59 (Table 1), a budding yeast specific Rad52 paralog, that enhances the Rad52 annealing reaction (Bai and Symington 1996; Davis and Symington 2001; Wu et al. 2006b).

3.4.5 Junction resolution/dissolution: the roles of BLM-TOPOIIIα, WRN, and Mus81-Mms4/Eme1

The SDSA subpathway requires dissociation of the D-loop, and this activity has been identified with two RecQ-like DNA helicases, BLM and WRN (van Brabant et al. 2000; Orren et al. 2002). BLM favored a D-loop structure with 5' invading ssDNA, a non-productive recombination intermediate, but also dissociated D-loops with 3' invading end (van Brabant et al. 2000). The possible functions of the other three human RecQ-like enzymes in such assays remain to be tested. *In vivo* analysis in yeast suggested that Srs2, another 3'-5' DNA helicase (Table 1), plays a crucial role enabling the SDSA pathway (Ira et al. 2003). However, the biochemical analysis of Srs2 does not presently support a function in dissociating D-loops (Krejci et al. 2003; Veaute et al. 2003). See Chapter 4 for a more extended discussion of Srs2.

Processing of double Holliday junctions (dHJs) lies at the heart of crossover formation in present recombination models (Fig. 1, Fig. 3). While single Holliday junctions require endonucleolytic processing, dHJs may be processed by an endonuclease, following the RuvC paradigm of bacteria (see chapter by Cox), or by a combination of a helicase and type I topoisomerase. Elegant biochemical work using an oligonucleotide-based dHJ substrate demonstrated that BLM helicase in conjunction with TOPOIIIα can process dHJs by collapsing the junctions to a hemi-catenane (Fig. 3) that is resolved by the topoisomerase activity (Wu and Hickson 2003). This activity has been termed dHJ dissolution, as opposed to resolution by an endonuclease, and leads to non-crossovers exclusively. A third subunit of the BLM-TOPOIIIα complex, BLAP75/RMI1, appears to recruit TOPOIIIα to dHJs (Raynard et al. 2006; Wu et al. 2006a). The elevated level of sister chromatid exchanges found in BLM-deficient cells can be nicely explained as a direct consequence of a failure to process dHJs to non-crossovers.

Resolvase A: RuvC provides a paradigm for the resolution of HJs and the formation of crossovers (see contribution by Cox). Eukaryotes do not contain an obvious RuvC homolog (Table 1) and the quest to identify the eukaryotic Holliday junction resolvase has been long and arduous (for reviews see Heyer et al. 2003; West 2003). Besides a mitochondrial activity, termed CCE1 (Kleff et al. 1992), the identity of the nuclear resolvase is still elusive. Biochemical fractionation of human cell extracts identified a HJ-specific endonuclease activity, termed Resolvase A, which displays striking similarity to RuvC (Liu et al. 2004). Two of the Rad51 paralogs, Rad51C and Xrcc3, are required for Resolvase A activity, but the complex in its purified form does not exhibit resolvase activity (Liu et al. 2004) suggesting these factors might be required to target the nuclease to the junction but do not constitute the nuclease function themselves. Such a late function of the Rad51 paralogs was surprising, given their suggested role in Rad51 filament formation (see above). However, these Rad51 filament co-factors might stay associated with the growing and/or nucleating end of the Rad51 filament (Fig. 3), which may position them to direct a nuclease to a junction. Identifying the nuclease of Resolvase A will be a major breakthrough, and it will be interesting to test this activity on dHJs, the likely intermediate in the DSBR pathway (Fig. 1, Fig. 3).

Mus81-Mms4/Eme1: Mus81 contains the nuclease function of the heterodimeric, DNA structure-selective endonuclease Mus81-Mms4/Eme1 (Table 1). Mus81 was first identified by its interaction with the Rad54 motor protein in *S. cerevisiae* (Interthal and Heyer 2000) and as potential substrate of the Cds1 checkpoint kinase in *S. pombe* (Boddy et al. 2000). Genetic analysis puts this enzyme squarely in the *RAD52* epistasis group, but the mutant is not sensitive to DSBs or IR damage, two classic substrates for the HR pathway. This rather confusing genetic behavior suggests that Mus81-Mms4/Eme1 is a context-specific factor required only in certain sub-pathways of HR. The chapter by Whitby will provide an in-depth analysis of this topic, and previous reviews have exhaustively discussed this subject (Haber and Heyer 2001; Heyer et al. 2003; Hollingsworth and Brill 2004). The biochemical activity of Mus81-Mms4/Eme1 (cleavage of Holliday junction versus other junction substrates including D-loops, replication forks, and 3'-flaps) appeared rather unsettled depending on the source and purity of the enzymatic preparation. A consensus has emerged to suggest that nicked HJs are the preferred *in vitro* substrate, whereas cleavage of intact HJs by the purified enzyme is rather poor (Gaillard et al. 2003; Osman et al. 2003; Fricke et al. 2005). The cleavage of junctions by Mus81-Mms4/Eme1 does not conform with the RuvC paradigm and the question remains: Can this enzyme resolve HJs? Is there a missing co-factor or post-translational modification? In what context does the nuclease function with the HR machinery and does Rad54 position the nuclease at the relevant joints? Regardless of the mechanism, Mus81-Eme1 controls the vast majority of meiotic crossovers in *S. pombe* (Smith et al. 2003), whereas it makes a more subtle contribution to crossover formation in *S. cerevisiae* (de los Santos et al. 2003).

3.4.6 Postsynaptic processing of terminal heterologies

During homologous recombination, a 3' end invades a homologous donor sequence and initiates new DNA synthesis. Using the invading DNA strand as a primer for DNA synthesis requires that non-homologous bases at the 3' end be removed. Removal of such non-homologous nucleotides can be critical when the strand interruption occurred in an area of non-homology utilizing internal homology for DNA strand invasion, as well as in strand annealing and second end capture, or in gene targeting experiments with DNA fragments that contain terminal heterology. Two pathways have been identified in *Saccharomyces cerevisiae* that process terminal heterologies on the invading strand. For heterologies greater than 30 nucleotides, the XPF family structure-selective endonuclease Rad1-Rad10 functions as 3' flap endonuclease in conjunction with the MMR proteins Msh2 and Msh3 and the Srs2 helicase (Ivanov and Haber 1995; Paques and Haber 1997). It is unclear whether the related Mus81-Mms4 endonuclease that can cleaves such substrates well *in vitro*, functions also in this capacity *in vivo*. Smaller heterologies are processed by the 3'-to-5' proofreading activity of DNA polymerase δ (Paques and Haber 1997) and possibly other DNA polymerases. The exact biochemical mechanisms of these pathways have not been analyzed yet.

4 Regulation of recombination

All fundamental DNA metabolic processes, including DNA replication and transcription, are strictly regulated, and HR is no exception. Mechanistically, HR is the pathway in which RPA-coated ssDNA is targeted for assembly of the Rad51 filament. Hence, any reaction that involves RPA needs to shield against Rad51 filament assembly, specifically DNA replication that involves large stretches of RPA-coated ssDNA on the lagging strand. In addition, chromatin modification and remodeling are required to overcome the inherently repressing character of chromatin on DNA transactions. These mechanisms will not only impact the efficiency of HR *in vivo* but may also be involved in regulating the hierarchy between the individual DSB response pathways (HR, NHEJ, translesion synthesis, fork regression, apoptosis). The mechanisms regulating HR are beginning to emerge and likely involve actively inhibiting pathways (Srs2, MMR), as well as the post-translational modification of HR proteins. Modulation of chromatin structure may provide another regulatory dimension in HR, but a discussion here would go beyond the scope of this review.

4.1 Negative regulation of HR and the roles of the Srs2 DNA helicase and MMR

The isolation of mutations that increase HR, so called hyper-rec mutations, provide genetic evidence for negative regulation of HR. Hyper-rec mutations are often identified in proteins with a normal function in DNA replication or in the recovery of stalled forks by translesion synthesis or fork regression, whose absence leads to an increased probability of lesions initiating HR. DNA damage induced by IR, UV, oxidative stress or agents that lead to fork stalling may also provide such lesions, and these agents were also found to induce recombination (Paques and Haber 1999). In these cases, recombination appears to occur as an indirect consequence of excess RPA-coated ssDNA.

A more specific mechanism of negative regulation of recombination is represented by the Srs2 DNA helicase. *SRS2* was identified genetically as a hyper-rec mutation and functions as an active anti-recombinator (Aguilera and Klein 1988; Aboussekhra et al. 1989). The Srs2 protein exhibits a 3' to 5' polarity like its homologs UvrD and Rep from bacteria (Rong and Klein 1993) and specifically dismantles the Rad51 presynaptic filament (Krejci et al. 2003; Veaute et al. 2003). When added to an ongoing DNA strand exchange reaction catalyzed by Rad51 and RPA (Fig. 2), Srs2 inhibited product formation by dissociating the Rad51 presynaptic filament, as demonstrated by EM analysis (Krejci et al. 2003; Veaute et al. 2003). However, Srs2 was unable to dissociate Rad51-made DNA joint molecules (Veaute et al. 2003). Hence, the requirement for Srs2 in the repair of a site-specific DSB by SDSA (Aylon et al. 2003; Ira et al. 2003) and the proposal that Srs2 may reverse the D-loop to allow reannealing with the second end cannot be explained by these biochemical data, suggesting that Srs2 might require other cofactors or conditions for such an activity. The biochemical data show that Rad51 presynaptic filament assembly is in a dynamic balance between nucleation/filament extension and disassembly by Srs2. Surprisingly, Srs2 lacks an obvious homolog in mammals, but an Srs2-related protein, Fbh1, is present in mammals and in *S. pombe* (Morishita et al. 2005; Osman et al. 2005). The *fbh1* mutation does suppress the requirement for mediators of Rad51 filament assembly in fission yeast, similar to the phenotype of *srs2* in *S. cerevisiae*. Furthermore, the *fbh1* mutation confers a synthetic growth defect with *srs2* or *rqh1* (encodes a RecQ homolog) mutations in *S. pombe*, which can be suppressed by a *rad51* mutation indicating the accumulation of toxic recombination intermediates in the helicase double mutants (Morishita et al. 2005; Osman et al. 2005). However, it has not yet been shown biochemically that Fbh1 can disrupt Rad51-ssDNA filaments. It is also possible that one of the mammalian RecQ-like helicases can substitute for Srs2 function, because Sgs1 overexpression can rescue an *srs2* defect in budding yeast (Mankouri et al. 2002).

The anti-recombination activity of Srs2 is recruited to replication forks by the specific interaction of Srs2 with sumolyated PCNA (Papouli et al. 2005; Pfander et al. 2005). PCNA is sumoylated on lysine 164 during S-phase as well as on lysine 127 after DNA damage. The singly and doubly sumoylated PCNA species specifically interact with Srs2, whereas unmodified PCNA does not. The Srs2 recruitment to origins of replication after hydroxy urea (HU)-induced arrest was also demonstrated *in vivo* by ChIP, and failure to recruit Srs2 led to an increase in Rad51 protein at those forks. The biochemical and genetic data of Srs2 as well as its interaction with PCNA suggests that HR is actively antagonized at replication forks, and suggests that alternative pathways of processing stalled replication forks, like translesion synthesis and fork regression, are preferred over HR. It is unclear how the Srs2 biochemical activity of Rad51 presynaptic filament dissociation relates to its other functions in DNA damage checkpoint signaling (Liberi et al. 2000), as well as in adaptation and recovery from DNA damage (Vaze et al. 2002).

Recombination, specifically the DNA strand invasion step, is also negatively regulated by the MMR system to avoid recombination between sequences that exceed a certain heterology threshold (also known as homeologous recombination) as demonstrated for *E. coli* RecA and the bacterial MMR (Worth et al. 1994). Genetic evidence for a similar mechanism in yeast meiosis has been provided (Hunter et al. 1996). The RecQ-like helicase Sgs1 functions in conjunction with the MMR pathway in suppressing homeologous recombination (Myung et al. 2001; Spell and Jinks-Robertson 2004), but the biochemistry of this pathway remains to be worked out.

4.2 Post-translational modification of HR proteins

The importance of post-translational modifications in molecular regulation of HR is evident, and the sumoylation-dependent interaction of Srs2 with PCNA illustrates how a biochemical activity can be targeted to its preferred substrate by induction of a specific protein interaction through post-translational modification. A growing number of HR proteins have been identified to be post-translationally modified, mostly by phosphorylation, often carried out by cell cycle or DNA damage checkpoint kinases (Zhou and Elledge 2000). The discussion here is limited to a few proteins (RPA, Rad51, Rad55-Rad57, Brca2) due to space constraints, and it is anticipated that these examples just scratch the surface of an enormous modification iceberg, which is waiting to be uncovered.

RPA: The eukaryotic ssDNA binding protein, RPA, works at the nexus of replication, repair, and recombination, and its middle subunit (RPA32) has been found to be phosphorylated in a cell-cycle-dependent manner and in response to DNA damage (reviewed in Wold 1997). The cell-cycle-dependent phosphorylation involves cyclin-dependent kinases, but the functional consequences remain unclear. The DNA damage-induced phosphorylation of RPA32 (and likely the large subunit RPA70) is dependent on the PI3 kinase-like kinases DNA-PK, ATM, and perhaps ATR (Mec1, Tel1 in *S. cerevisiae*), and there was considerable uncertainty about the functional consequences of DNA damage-induced RPA phosphorylation using reconstituted *in vitro* assays (Wold 1997). The confusion has been clarified to some degree by showing that DNA damage-induced RPA2 phosphorylation targets RPA to repair centers and excludes the phosphorylated RPA from active replication centers (Vassin et al. 2004). Phosphorylated RPA does associate with stalled replication forks, likely involving a phosphorylation-dependent interaction with Mre11-Rad50-Nbs1 complex (Robison et al. 2004; Vassin et al. 2004), but it is unclear whether this interaction is direct or mediated by a bridging protein. Whether RPA phosphorylation only targets the protein to a specific nuclear location or also changes some intrinsic biochemical property of RPA remains unsolved.

Human Rad51: Human Rad51 was found to be phosphorylated by the c-Abl kinase on tyrosine 54 after exposure to IR in cells transiently transfected with HsRad51 and c-Abl (Yuan et al. 1998). Rad51 phosphorylation correlated with decreased DNA binding by phosphorylated Rad51. Tyrosine 54 (tyrosine 112 in yeast Rad51) stacks against another tyrosine in the protomer interface in the crystal structure of the yeast Rad51 protein (Conway et al. 2004). Surface exposure of this site would be expected to exert a negative effect on filament formation, consistent with the observed DNA binding defect. The biological significance of this phosphorylation is unclear, and the effect is somewhat unexpected, because it is predicted to inhibit HR in response to a target lesion of this pathway. c-Abl kinase activated by IR phosphorylated human Rad51 on tyrosine 315 *in vitro* (Chen et al. 1999). The reason for this difference in the phosphorylation patterns is unclear. Tyrosine 315 phosphorylation correlated with enhanced Rad52 interaction, but the effect on Rad51 function has not been explored (Chen et al. 1999). Tyrosine 315 phosphorylation in response to DNA damage was confirmed using a phospho-tyrosine 315-specific antibody, and this residue is constitutively phosphorylated by the oncogenic BCR-ABL fusion protein found in chronic myelogenous leukemia and some acute lymphocytic leukemia patients (Slupianek et al. 2001). Biochemical analysis showed that tyrosine 315 of human Rad51 is required for DNA binding and filament formation (Takizawa et al. 2004), suggesting that phosphorylation of this residue should have a negative effect on Rad51 function. Downregulation of Rad51 in response to a target lesion of the HR pathway is unexpected, and the observed drug resistance in BCR-ABL expressing cells might be more the consequence of the increased Rad51 and Rad51 paralog expression (Slupianek et al. 2001) than an effect of Rad51 tyrosine 315 phosphorylation.

Rad55-Rad57: The Rad55-Rad57 proteins, which are critical for assembly of the Rad51 filament in yeast, are phosphorylated in response to genotoxic stress. Phosphorylation depended on the DNA damage checkpoint (Bashkirov et al. 2000). Mapping of the sites identified a complex pattern of phosphorylation sites, and analysis of three sites in Rad55 (serines 2,8,14) showed that these phosphorylation events positively regulate Rad55-Rad57 function, particularly in the recovery of stalled replication forks (Bashkirov et al. 2006). The biochemical consequences of Rad55-Rad57 phosphorylation have not been identified yet. Human Brca2 is phosphorylated on S3291 by CDK2-cyclin A during the G2/M phase of the cell cycle, which appears to negatively regulate Brca2 function in HR (Esashi et al. 2005). S3291 is part of the C-terminal Rad51 binding site of Brca2 and phosphorylation abrogates the interaction of the site with the Rad51 protein. IR counteracts S3291 phosphorylation, consistent with the expectation that Brca2 activity is required after ionizing radiation. Taking the analogy of the regulation of actin filament assembly through the Arp2/3 nucleating complex, these data suggest that also in HR the proteins controlling Rad51 filament nucleation are important regulatory targets.

These few examples should highlight that HR is not a constitutive repair pathway but is under elaborate negative and likely positive control. Cells avoid HR when unnecessary and in fact unwanted, i.e. during ongoing replication, whereas they promote HR when necessary and critical in the repair of DNA damage and in the recovery of stalled forks. These mechanisms are able to localize HR to the DNA damage site or induce general changes to the recombination machinery by a pan-nuclear response.

5 Conclusion

Homology search and DNA strand invasion are central during HR. The assembly, function, and disassembly of the Rad51 filament are critical for the central reaction. The conservation of this step in the form of the RecA/Rad51/RadA protein underlines its fundamental importance. The multitude of pathways leading to and from the central reaction in processing distinct lesions and resolving the strand invasion product in various ways are catalyzed by context-specific factors, whose requirements may vary with the lesion or postsynaptic pathway. The differences between the RecA and Rad51 cousins, as well as between the yeast and human brothers of eukaryotic Rad51 proteins, will provide a key to understanding the significant differences in the number and function of the other HR proteins that work in conjunction with the central DNA strand exchange proteins. Analogies to other filament forming proteins like actin and tubulin may elucidate the mechanistic details of these cofactors. The greater complexity of the eukaryotic genome structure and the requirement for more intricate regulation of HR in complex cells and organisms will likely be reflected in the biochemical properties of the recombination machinery and its modulation by post-translational modifications. This will provide enough fuel to keep the fire of studying recombination burning for a long while. The individual approaches to understanding homologous recombination in eukaryotes range from single-molecule biophysics to structural analysis, biochemistry, genetics, and cell biology. Their discussion is often segregated as reflected by the individual chapters in this volume. However, it is the combination of all approaches that will ultimately succeed in elucidating the mechanism and regulation of this fascinating cellular pathway.

Acknowledgements

The support of our research by the National Institutes of Health, the Susan G. Komen Breast Cancer Foundation, and University of California BIOSTAR program is gratefully acknowledged. I thank Shannon Ceballos, Tammy Doty, Kirk Ehmsen, Kristina Herzberg, Neil Hunter, Xuan Li, Jie Liu, Eugene Nadezhdin, Mike Rolfsmeier, John Roth, Erin Schwartz, William Wright, and Xiao-Ping Zhang for their critical comments on the manuscript. My sincere apologies for omitting all the references and great work that could not be discussed here due to space constraints.

References

Aboussekhra A, Chanet R, Adjiri A, Fabre F (1992) Semi-dominant suppressors of Srs2 helicase mutations of *Saccharomyces cerevisiae* map in the *RAD51* gene, whose sequence predicts a protein with similarities to procaryotic RecA protein. Mol Cell Biol 12:3224-3234

Aboussekhra A, Chanet R, Zgaga Z, Cassier Chauvat C, Heude M, Fabre F (1989) *RADH*, a gene of *Saccharomyces cerevisiae* encoding a putative DNA helicase involved in DNA repair. Characteristics of *radH* mutants and sequence of the gene. Nucl Acids Res 17:7211-7219

Aguilera A, Klein HL (1988) Genetic control of intrachromosomal recombination in *Saccharomyces cerevisiae*. I. Isolation and genetic characterization of hyper-recombination mutations. Genetics 119:779-790

Aihara H, Ito Y, Kurumizaka H, Yokoyama S, Shibata T (1999) The N-terminal domain of the human Rad51 protein binds DNA: Structure and a DNA binding surface as revealed by NMR. J Mol Biol 290:495-504

Alexeev A, Mazin A, Kowalczykowski SC (2003) Rad54 protein possesses chromatin-remodeling activity stimulated by a Rad51-ssDNA nucleoprotein filament. Nature Struct Biol 10:182-186

Alexiadis V, Kadonaga JT (2003) Strand pairing by Rad54 and Rad51 is enhanced by chromatin. Genes Dev 16:2767-2771

Amitani I, Baskin RJ, Kowalczykowski S (2006) Direct visualization of a chromatin-remodeling protein, Rad54, translocating along single-molecules of double-stranded DNA. Mol Cell 23:143-148

Aylon Y, Liefshitz B, Bitan-Banin G, Kupiec M (2003) Molecular dissection of mitotic recombination in the yeast *Saccharomyces cerevisiae*. Mol Cell Biol 23:1403-1417

Bai Y, Symington LS (1996) A Rad52 homolog is required for *RAD51*-independent mitotic recombination in *Saccharomyces cerevisiae*. Genes Dev 10:2025-2037

Bashkirov VI, Herzberg K, Rolfsmeier M, Heyer W-D (2006) Rad55 phopshorylation on serines 2,8,14 is required for efficient recombinational DNA repair and recovery of stalled replication forks. Mol Cell Biol, in press

Bashkirov VI, King JS, Bashkirova EV, Schmuckli-Maurer J, Heyer WD (2000) DNA repair protein Rad55 is a terminal substrate of the DNA damage checkpoints. Mol Cell Biol 20:4393-4404

Baumann P, Benson FE, West SC (1996) Human Rad51 protein promotes ATP-dependent homologous pairing and strand transfer reactions *in vitro*. Cell 87:757-766

Baumann P, West SC (1997) The human Rad51 protein: polarity of strand transfer and stimulation by hRP-A. EMBO J 16:5198-5206

Beernink HTH, Morrical SW (1999) RMPs: Recombination/replication mediator proteins. Trends Biochem Sci 24:385-389

Benson FE, Baumann P, West SC (1998) Synergistic actions of Rad51 and Rad52 in recombination and DNA repair. Nature 391:401-404

Bianco PR, Tracy RB, Kowalczykowski SC (1998) DNA strand exchange proteins: a biochemical and physical comparison. Front Biosci 3:570-603

Bochkarev A, Bochkareva E (2004) From RPA to BRCA2: lessons from single-stranded DNA binding by the OB-fold. Curr Opin Struct Biol 14:36-42

Boddy MN, Lopez-Girona A, Shanahan P, Interthal H, Heyer WD, Russell P (2000) Damage tolerance protein Mus81 associates with the FHA1 domain of checkpoint kinase Cds1. Mol Cell Biol 20:8758-8766

Braybrooke JP, Li JL, Wu L, Caple F, Benson FE, Hickson ID (2003) Functional interaction between the Bloom's syndrome helicase and the RAD51 paralog, RAD51L3 (RAD51D). J Biol Chem 278:48357-48366

Bugreev DV, Mazin AV (2004) Ca2+ activates human homologous recombination protein Rad51 by modulating its ATPase activity. Proc Natl Acad Sci USA101:9988-9993

Bugreev DV, Mazina OM, Mazin A (2006) Rad54 protein promotes branch migration of Holliday junctions. Nature 442:590-593

Chen G, Yuan SSF, Liu W, Xu Y, Trujillo K, Song BW, Cong F, Goff SP, Wu Y, Arlinghaus R, Baltimore D, Gasser PJ, Park MS, Sung P, Lee E (1999) Radiation-induced assembly of Rad51 and Rad52 recombination complex requires ATM and c-Abl. J Biol Chem 274:12748-12752

Chi P, Van Komen S, Sehorn MG, Sigurdsson S, Sung P (2006) Roles of ATP binding and ATP hydrolysis in human Rad51 recombinase function. DNA Repair (Amst) 5:381-391

Clever B, Interthal H, Schmuckli-Maurer J, King J, Sigrist M, Heyer WD (1997) Recombinational repair in yeast: Functional interactions between Rad51 and Rad54 proteins. EMBO J 16:2535-2544

Clever B, Schmuckli-Maurer J, Sigrist M, Glassner B, Heyer W-D (1999) Specific negative effects resulting from elevated levels of the recombinational repair protein Rad54p in *Saccharomyces cerevisiae*. Yeast 15:721-740

Constantinou A, Chen XB, McGowan CH, West SC (2002) Holliday junction resolution in human cells: two junction endonucleases with distinct substrate specificities. EMBO J 21:5577-5585

Conway AB, Lynch TW, Zhang Y, Fortin GS, Fung CW, Symington LS, Rice PA (2004) Crystal structure of a Rad51 filament. Nat Rev Struc Mol Biol 11:791-796

Davis AP, Symington LS (2001) The yeast recombinational repair protein Rad59 interacts with Rad52 and stimulates single-strand annealing. Genetics 159:515-525

de los Santos T, Hunter N, Lee C, Larkin B, Loidl J, Hollingsworth NM (2003) The Mus81/Mms4 endonuclease acts independently of double-Holliday junction resolution to promote a distinct subset of crossovers during meiosis in budding yeast. Genetics 164:81-94

Donovan JW, Milne GT, Weaver DT (1994) Homotypic and heterotypic protein associations control Rad51 function in double-strand break repair. Genes Dev 8:2552-2562

Egelman EH (2003) A tale of two polymers: New insights into helical filaments. Nat Rev Mol Cell Biol 4:621-630

Esashi F, Christ N, Gannon J, Liu YL, Hunt T, Jasin M, West SC (2005) CDK-dependent phosphorylation of BRCA2 as a regulatory mechanism for recombinational repair. Nature 434:598-604

Fortin GS, Symington LS (2002) Mutations in yeast Rad51 that partially bypass the requirement for Rad55 and Rad57 in DNA repair by increasing the stability of Rad51-DNA complexes. EMBO J 21:3160-3170

Fricke WM, Bastin-Shanower SA, Brill SJ (2005) Substrate specificity of the *Saccharomyces cerevisiae* Mus81-Mms4 endonuclease. DNA Repair 4:243-251

Fujimori A, Tachiiri S, Sonoda E, Thompson LH, Dhar PK, Hiraoka M, Takeda S, Zhang Y, Reth M, Takata M (2001) Rad52 partially substitutes for the Rad51 paralog XRCC3 in maintaining chromosomal integrity in vertebrate cells. EMBO J 20:5513-5520

Gaillard P-H, Noguchi E, Shanahan P, Russell P (2003) The endogenous Mus81-Eme1 complex resolves Holliday junctions by a nick and counternick mechanism. Mol Cell 12:747-759

Galkin VE, Wu Y, Zhang X-P, Qian X, He Y, Yu X, Heyer W-D, Luo Y, Egelman EH (2006) The RAD51/RadA N-terminal domain activates nucleoprotein filaments. Structure 14:983-992

Gasior SL, Olivares H, Ear U, Hari DM, Weichselbaum R, Bishop DK (2001) Assembly of RecA-like recombinases: Distinct roles for mediator proteins in mitosis and meiosis. Proc Natl Acad Sci USA 98:8411-8418

Gasior SL, Wong AK, Kora Y, Shinohara A, Bishop DK (1998) Rad52 associates with RPA and functions with Rad55 and Rad57 to assemble meiotic recombination complexes. Genes Dev 12:2208-2221

Golub EI, Kovalenko OV, Gupta RC, Ward DC, Radding CM (1997) Interaction of human recombination proteins Rad51 and Rad54. Nucleic Acids Res 25:4106-4110

Gupta RC, Folta-Stogniew E, O'Malley S, Takahashi M, Radding CM (1999) Rapid exchange of A : T base pairs is essential for recognition of DNA homology by human Rad51 recombination protein. Mol Cell 4:705-714

Haber JE, Heyer WD (2001) The fuss about Mus81. Cell 107:551-554

Hays SL, Firmenich AA, Berg P (1995) Complex formation in yeast double-strand break repair: Participation of Rad51, Rad52, Rad55, and Rad57 proteins. Proc Natl Acad Sci USA 92:6925-6929

Hays SL, Firmenich AA, Massey P, Banerjee R, Berg P (1998) Studies of the interaction between Rad52 protein and the yeast single-stranded DNA binding protein RPA. Mol Cell Biol 18:4400-4406

Heyer WD, Ehmsen KT, Solinger JA (2003) Holliday junctions in the eukaryotic nucleus: Resolution in sight? Trends Biochem Sci 10:548-557

Heyer WD, Li XR, Rolfsmeier M, Zhang X-P (2006) Rad54: the Swiss Army knife of homologous recombination? Nucleic Acids Res: in press

Holbeck SL, Strathern JN (1997) A role for REV3 in mutagenesis during double-strand break repair in *Saccharomyces cerevisiae*. Genetics 147:1017-1024

Hollingsworth NM, Brill SJ (2004) The Mus81 solution to resolution: generating meiotic crossovers without Holliday junctions. Genes Dev 18:117-125

Hunter N, Chambers SR, Louis EJ, Borts RH (1996) The mismatch repair system contributes to meiotic sterility in an interspecific yeast hybrid. EMBO J 15:1726-1733

Interthal H, Heyer WD (2000) *MUS81* encodes a novel Helix-hairpin-Helix protein involved in the response to UV- and methylation-induced DNA damage in *Saccharomyces cerevisiae*. Mol Gen Genet 263:812-827

Ira G, Malkova A, Liberi G, Foiani M, Haber JE (2003) Srs2 and Sgs1-Top3 suppress crossovers during double-strand break repair in yeast. Cell 115:401-411

Ivanov EL, Haber JE (1995) RAD1 and RAD10, but not other excision repair genes, are required for double-strand break-induced recombination in *Saccharomyces cerevisiae*. Mol Cell Biol 15:2245-2251

Jaskelioff M, Van Komen S, Krebs JE, Sung P, Peterson CL (2003) Rad54p is a chromatin remodeling enzyme required for heteroduplex joint formation with chromatin. J Biol Chem 278:9212-9218

Jiang H, Xie YQ, Houston P, Stemke-Hale K, Mortensen UH, Rothstein R, Kodadek T (1996) Direct association between the yeast Rad51 and Rad54 recombination proteins. J Biol Chem 271:33181-33186

Johnson RD, Symington LS (1995) Functional differences and interactions among the putative RecA homologs *RAD51*, *RAD55*, and *RAD57*. Mol Cell Biol 15:4843-4850

Kagawa W, Kurumizaka H, Ishitani R, Fukai S, Nureki O, Shibata T, Yokoyama S (2002) Crystal structure of the homologous-pairing domain from the human Rad52 recombinase in the undecameric form. Mol Cell 10:359-371

Kantake N, Madiraju M, Sugiyama T, Kowalczykowski C (2002) *Escherichia coli* RecO protein anneals ssDNA complexed with its cognate ssDNA-binding protein: A common step in genetic recombination. Proc Natl Acad Sci USA 99:15327-15332

Karow JK, Constantinou A, Li JL, West SC, Hickson ID (2000) The Bloom's syndrome gene product promotes branch migration of Holliday junctions. Proc Natl Acad Sci USA 97:6504-6508

Kawamoto T, Araki K, Sonoda E, Yamashita YM, Harada K, Kikuchi K, Masutani C, Hanaoka F, Nozaki K, Hashimoto N, Takeda S (2005) Dual roles for DNA polymerase eta in homologous DNA recombination and translesion DNA synthesis. Mol Cell 20:793-799

Kiianitsa K, Solinger JA, Heyer WD (2002) Rad54 protein exerts diverse modes of ATPase activity on duplex DNA partially and fully covered with Rad51 protein. J Biol Chem 277:46205-46215

Kiianitsa K, Solinger JA, Heyer WD (2006) Terminal association of the Rad54 protein with the Rad51-dsDNA filament. Proc Natl Acad Sci USA 103:9767-9772

Kim PM, Paffett KS, Solinger JA, Heyer WD, Nickoloff JA (2002) Spontaneous and double-strand break-induced recombination, and gene conversion tract lengths, are differentially affected by overexpression of wild-type or ATPase-defective yeast Rad54. Nucleic Acids Res 30:2727-2735

Kleff S, Kemper B, Sternglanz R (1992) Identification and characterization of yeast mutants and the gene for a cruciform cutting endonuclease. EMBO J 11:699-704

Kowalczykowski SC, Dixon DA, Eggleston AK, Lauder SD, Rehrauer WM (1994) Biochemistry of homologous recombination in *Escherichia coli*. Microbiol Rev 58:401-465

Krejci L, Song BW, Bussen W, Rothstein R, Mortensen UH, Sung P (2002) Interaction with Rad51 is indispensable for recombination mediator function of Rad52. J Biol Chem 277:40132-40141

Krejci L, Van Komen S, Li Y, Villemain J, Reddy MS, Klein H, Ellenberger T, Sung P (2003) DNA helicase Srs2 disrupts the Rad51 presynaptic filament. Nature 423:305-309

Krogh BO, Symington LS (2004) Recombination proteins in yeast. Annu Rev Genet 38:233-271

Kurumizaka H, Ikawa S, Nakada M, Eda K, Kagawa W, Takata M, Takeda S, Yokoyama S, Shibata T (2001) Homologous-pairing activity of the human DNA-repair proteins Xrcc3.Rad51C. Proc Natl Acad Sci USA 98:5538-5543

Kurumizaka H, Ikawa S, Nakada M, Enomoto R, Kagawa W, Kinebuchi T, Yamazoe M, Yokoyama S, Shibata T (2002) Homologous pairing and ring and filament structure formation activities of the human Xrcc2*Rad51D complex. J Biol Chem 277:14315-14320

Liberi G, Chiolo I, Pellicioli A, Lopes M, Plevani P, Muzi-Falconi M, Foiani M (2000) Srs2 DNA helicase is involved in checkpoint response and its regulation requires a functional Mec1-dependent pathway and Cdk1 activity. EMBO J 19:5027-5038

Lio YC, Mazin AV, Kowalczykowski SC, Chen DJ (2003) Complex formation by the human Rad51B and Rad51C DNA repair proteins and their activities *in vitro*. J Biol Chem 278:2469-2478

Lisby M, Barlow JH, Burgess RC, Rothstein R (2004) Choreography of the DNA damage response: Spatiotemporal relationships among checkpoint and repair proteins. Cell 118:699-713

Liu J, Bond JP, Morrical SW (2006) Mechanism of presynaptic filament stabilization by the bacteriophage T4 UvsY recombination mediator protein. Biochemistry 45:5493-5502

Liu YL, Masson JY, Shah R, O'Regan P, West SC (2004) RAD51C is required for Holliday junction processing in mammalian cells. Science 303:243-246

Mankouri HW, Craig TJ, Morgan A (2002) SGS1 is a multicopy suppressor of srs2: functional overlap between DNA helicases. Nucleic Acids Res 30:1103-1113

Martin V, Chahwan C, Gao H, Blais V, Wohlschlegel J, Yates JR 3rd, McGowan CH, Russell P (2006) Swsl is a conserved regulator of homologous recombination in eukaryotic cells. EMBO J 25:2564-2574

Masson JY, Stasiak AZ, Stasiak A, Benson FE, West SC (2001a) Complex formation by the human RAD51C and XRCC3 recombination repair proteins. Proc Natl Acad Sci USA 98:8440-8446

Masson JY, Tarsounas MC, Stasiak AZ, Stasiak A, Shah R, McIlwraith MJ, Benson FE, West SC (2001b) Identification and purification of two distinct complexes containing the five RAD51 paralogs. Genes Dev 15:3296-3307

Mazin AV, Alexeev AA, Kowalczykowski SC (2003) A novel function of Rad54 protein - Stabilization of the Rad51 nucleoprotein filament. J Biol Chem 278:14029-14036

Mazin AV, Bornarth CJ, Solinger JA, Heyer W-D, Kowalczykowski SC (2000a) Rad54 protein is targeted to pairing loci by the Rad51 nucleoprotein filament. Mol Cell 6:583-592

Mazin AV, Zaitseva E, Sung P, Kowalczykowski SC (2000b) Tailed duplex DNA is the preferred substrate for Rad51 protein-mediated homologous pairing. EMBO J 19:1148-1156

Mazin OM, Mazin AW (2004) Human Rad54 protein stimulates DNA strand exchange activity of hRad51 protein in the presence of Ca2+. J Biol Chem 279:52041-52051

McIlwraith MJ, Vaisman A, Liu YL, Fanning E, Woodgate R, West SC (2005) Human DNA polymerase eta promotes DNA synthesis from strand invasion intermediates of homologous recombination. Mol Cell 20:783-792

Morimatsu K, Kowalczykowski SC (2003) RecFOR proteins load RecA protein onto gapped DNA to accelerate DNA strand exchange: A universal step of recombinational repair. Mol Cell 11:1337-1347

Morishita T, Furukawa F, Sakaguchi C, Toda T, Carr AM, Iwasaki H, Shinagawa H (2005) Role of the *Schizosaccharomyces pombe* F-Box DNA helicase in processing recombination intermediates. Mol Cell Biol 25:8074-8083

Moritz M, Agard DA (2001) Gamma-tubulin complexes and microtubule nucleation. Curr Opin Struct Biol 11:174-181

Myung K, Datta A, Chen C, Kolodner RD (2001) SGS1, the *Saccharomyces cerevisiae* homologue of BLM and WRN, suppresses genome instability and homeologous recombination. Nat Genet 27:113-116

Namsaraev EA, Berg P (2000) Rad51 uses one mechanism to drive DNA strand exchange in both directions. J Biol Chem 275:3970-3976

New JH, Sugiyama T, Zaitseva E, Kowalczykowski SC (1998) Rad52 protein stimulates DNA strand exchange by Rad51 and replication protein A. Nature 391:407-410

Ogawa T, Yu X, Shinohara A, Egelman EH (1993) Similarity of the yeast RAD51 filament to the bacterial RecA filament. Science 259:1896-1899

Orren DK, Theodore S, Machwe A (2002) The Werner syndrome helicase/exonuclease (WRN) disrupts and degrades D-loops *in vitro*. Biochemistry 41:13483-13488

Osman F, Dixon J, Barr AR, Whitby MC (2005) The F-Box DNA helicase Fbh1 prevents Rhp51-dependent recombination without mediator proteins. Mol Cell Biol 25:8084-8096

Osman F, Dixon J, Doe CL, Whitby MC (2003) Generating crossovers by resolution of nicked Holliday junctions: A role of Mus81-Eme1 in meiosis. Mol Cell 12:761-774

Papouli E, Chen SH, Davies AA, Huttner D, Krejci L, Sung P, Ulrich HD (2005) Crosstalk between SUMO and ubiquitin on PCNA is mediated by recruitment of the helicase Srs2p. Mol Cell 19:123-133

Paques F, Haber JE (1997) Two pathways for removal of nonhomologous DNA ends during double-strand break repair in *Saccharomyces cerevisiae*. Mol Cell Biol 17:6765-6771

Paques F, Haber JE (1999) Multiple pathways of recombination induced by double-strand breaks in *Saccharomyces cerevisiae*. Microbiol Mol Biol Rev 63:349-404

Pellegrini L, Venkitaraman A (2004) Emerging functions of BRCA2 in DNA recombination. Trends Biochem Sci 29:310-316

Petukhova G, Stratton S, Sung P (1998) Catalysis of homologous DNA pairing by yeast Rad51 and Rad54 proteins. Nature 393:91-94

Petukhova G, Sung P, Klein H (2000) Promotion of Rad51-dependent D-loop formation by yeast recombination factor Rdh54/Tid1. Genes Dev 14:2206-2215

Petukhova G, Van Komen S, Vergano S, Klein H, Sung P (1999) Yeast Rad54 promotes Rad51-dependent homologous DNA pairing via ATP hydrolysis-driven change in DNA double helix conformation. J Biol Chem 274:29453-29462

Pfander B, Moldovan GL, Sacher M, Hoege C, Jentsch S (2005) SUMO-modified PCNA recruits Srs2 to prevent recombination during S phase. 436:428-433

Pollard TD, Borisy GG (2003) Cellular motility driven by assembly and disassembly of actin filaments. Cell 112:453-465

Rattray AJ, Strathern JN (2003) Error-prone DNA polymerases: When making a mistake is the only way to get ahead. Annu Rev Genet 37:31-66

Raynard S, Bussen W, Sung P (2006) A double Holliday junction dissolvasome comprising BLM, topoisomerase III alpha, and BLAP75. J Biol Chem 281:13861-13864

Rijkers T, VandenOuweland J, Morolli B, Rolink AG, Baarends WM, VanSloun PPH, Lohman PHM, Pastink A (1998) Targeted inactivation of mouse *RAD52* reduces homologous recombination but not resistance to ionizing radiation. Mol Cell Biol 18:6423-6429

Ristic D, Modesti M, van der Heijden T, van Noort J, Dekker C, Kanaar R, Wyman C (2005) Human Rad51 filaments on double- and single-stranded DNA: correlating regular and irregular forms with recombination function. Nucleic Acids Res 33:3292-3302

Ristic D, Wyman C, Paulusma C, Kanaar R (2001) The architecture of the human Rad54-DNA complex provides evidence for protein translocation along DNA. Proc Natl Acad Sci USA 98:8454-8460

Robison JG, Elliott J, Dixon K, Oakley GG (2004) Replication protein A and the Mre11•Rad50•Nbs1 complex co-localize and interact at sites of stalled replication forks. J Biol Chem 279:34802-34810

Rong L, Klein HL (1993) Purification and characterization of the SRS2 DNA helicase of the yeast *Saccharomyces cerevisiae*. J Biol Chem 268:1252-1259

San Filippo J, Chi P, Sehorn MG, Etchin J, Krejci L, Sung P (2006) Recombination mediator and Rad51 targeting activities of a human BRCA2 polypeptide. J Biol Chem 281:11649-11657

Seitz EM, Haseltine CA, Kowalczykowski SC (2001) DNA recombination and repair in archaea. Adv Appl Microbiol 50:101-169

Shim KS, Schmutte C, Tombline G, Heinen CD, Fishel R (2004) hXRCC2 enhances ADP/ATP processing and strand exchange by hRAD51. J Biol Chem 279:30385-30394

Shinohara A, Ogawa H, Matsuda Y, Ushio N, Ikea K, Ogawa T (1993) Cloning of human, mouse and fission yeast recombination genes homologous to RAD51 and recA. Nat Genet 4:239-243

Shinohara A, Ogawa H, Ogawa T (1992) Rad51 protein involved in repair and recombination in *S. cerevisiae* is a RecA-like protein. Cell 69:457-470

Shinohara A, Ogawa T (1998) Stimulation by Rad52 of yeast Rad51-mediated recombination. Nature 391:404-407

Shinohara A, Shinohara M, Ohta T, Matsuda S, Ogawa T (1998) Rad52 forms ring structures and co-operates with RPA in single-strand annealing. Genes Cells 3:145-156

Shor E, Weinstein J, Rothstein R (2005) Genetic screen for top3 suppressors in *Saccharomyces cerevisiae* identifies SHU1, SHU2, PSY3 and CSM2: Four genes involved in error-free DNA repair. Genetics 169:1275-1289

Sigurdsson S, Trujillo K, Song BW, Stratton S, Sung P (2001a) Basis for avid homologous DNA strand exchange by human Rad51 and RPA. J Biol Chem 276:8798-8806

Sigurdsson S, Van Komen S, Bussen W, Schild D, Albala JS, Sung P (2001b) Mediator function of the human Rad51B-Rad51C complex in Rad51/RPA-catalyzed DNA strand exchange. Genes Dev 15:3308-3318

Sigurdsson S, Van Komen S, Petukhova G, Sung P (2002) Homologous DNA pairing by human recombination factors Rad51 and Rad54. J Biol Chem 277:42790-42794

Singleton MR, Sawaya MR, Ellenbeger T, Wigley DB (2000) Crystal structure of T7 gene 4 ring helicase indicates a mechanism for sequential hydrolysis of nucleotides. Cell 101:589-600

Singleton MR, Wentzell LM, Liu YL, West SC, Wigley DB (2002) Structure of the single-strand annealing domain of human RAD52 protein. Proc Natl Acad Sci USA 99:13492-13497

Slupianek A, Schmutte C, Tombline G, Nieborowska-Skorska M, Hoser G, Nowicki MO, Pierce AJ, Fishel R, Skorski T (2001) BCR/ABL regulates mammalian RecA homologs, resulting in drug resistance. Mol Cell 8:795-806

Smith GR, Boddy MN, Shanahan P, Russell P (2003) Fission yeast Mus81•Eme1 Holliday junction resolvase is required for meiotic crossing over but not for gene conversion. Genetics 165:2289-2293

Solinger JA, Heyer W-D (2001) Rad54 protein stimulates the postsynaptic phase of Rad51 protein-mediated DNA strand exchange. Proc Natl Acad Sci USA 98:8447-8453

Solinger JA, Kiianitsa K, Heyer W-D (2002) Rad54, a Swi2/Snf2-like recombinational repair protein, disassembles Rad51:dsDNA filaments. Mol Cell 10:1175-1188

Solinger JA, Lutz G, Sugiyama T, Kowalczykowski SC, Heyer W-D (2001) Rad54 protein stimulates heteroduplex DNA formation in the synaptic phase of DNA strand exchange via specific interactions with the presynaptic Rad51 nucleoprotein filament. J Mol Biol 307:1207-1221

Spell RM, Jinks-Robertson S (2004) Examination of the roles of Sgs1 and Srs2 helicases in the enforcement of recombination fidelity in *Saccharomyces cerevisiae*. Genetics 168:1855-1865

Spies M, Kowalczykowski SC (2006) The RecA binding locus of RecBCD is a general domain for recruitment of DNA strand exchange proteins. Mol Cell 21:573-580

Stasiak AZ, Larquet E, Stasiak A, Muller S, Engel A, Van Dyck E, West SC, Egelman EH (2000) The human Rad52 protein exists as a heptameric ring. Curr Biol 10:337-340

Story RM, Weber IT, Steitz TA (1992) The structure of the *E. coli* recA protein monomer and polymer. Nature 355:318-325

Sugiyama T, Kowalczykowski SC (2002) Rad52 protein associates with replication protein A (RPA)-single-stranded DNA to accelerate Rad51-mediated displacement of RPA and presynaptic complex formation. J Biol Chem 277:31663-31672

Sugiyama T, New JH, Kowalczykowski SC (1998) DNA annealing by Rad52 Protein is stimulated by specific interaction with the complex of replication protein A and single-stranded DNA. Proc Natl Acad Sci USA 95:6049-6054

Sugiyama T, Zaitseva EM, Kowalczykowski SC (1997) A single-stranded DNA-binding protein is needed for efficient presynaptic complex formation by the *Saccharomyces cerevisiae* Rad51 protein. J Biol Chem 272:7940-7945

Sung P (1994) Catalysis of ATP-dependent homologous DNA pairing and strand exchange by yeast RAD51 protein. Science 265:1241-1243

Sung P (1997a) Function of yeast Rad52 protein as a mediator between replication protein A and the Rad51 recombinase. J Biol Chem 272:28194-28197

Sung P (1997b) Yeast Rad55 and Rad57 proteins form a heterodimer that functions with replication protein A to promote DNA strand exchange by Rad51 recombinase. Genes Dev 11:1111-1121

Sung P, Krejci L, Van Komen S, Sehorn MG (2003) Rad51 recombinase and recombination mediators. J Biol Chem 278:42729-42732

Sung P, Robberson DL (1995) DNA strand exchange mediated by a RAD51-ssDNA nucleoprotein filament with polarity opposite to that of RecA. Cell 82:453-461

Sung P, Stratton SA (1996) Yeast Rad51 recombinase mediates polar DNA strand exchange in the absence of ATP hydrolysis. J Biol Chem 271:27983-27986

Swagemakers SMA, Essers J, deWit J, Hoeijmakers JHJ, Kanaar R (1998) The human Rad54 recombinational DNA repair protein is a double-stranded DNA-dependent ATPase. J Biol Chem 273:28292-28297

Symington LS (2002) Role of *RAD52* epistasis group genes in homologous recombination and double-strand break repair. Microbiol Mol Biol Rev 66:630-670

Takata M, Sasaki MS, Sonoda E, Fukushima T, Morrison C, Albala JS, Swagemakers SMA, Kanaar R, Thompson LH, Takeda S (2000) The Rad51 paralog Rad51B promotes homologous recombinational repair. Mol Cell Biol 20:6476-6482

Takata M, Sasaki MS, Tachiiri S, Fukushima T, Sonoda E, Schild D, Thompson LH, Takeda S (2001) Chromosome instability and defective recombinational repair in knockout mutants of the five Rad51 paralogs. Mol Cell Biol 21:2858-2866

Takizawa Y, Kinebuchi T, Kagawa W, Yokoyama S, Shibata T, Kurumizaka H (2004) Mutational analyses of the human Rad51-Tyr315 residue, a site for phosphorylation in leukaemia cells. Genes Cells 9:781-790

Tan TLR, Kanaar R, Wyman C (2003) Rad54, a Jack of all trades in homologous recombination (vol 2, pg 787, 2003). DNA Repair (Amst) 2:1293

Tarsounas M, Davies D, West SC (2003) BRCA2-dependent and independent formation of RAD51 nuclear foci. Oncogene 22:1115-1123

Thacker J (2005) The RAD51 gene family, genetic instability and cancer. Cancer Lett 219:125-135

Tombline G, Fishel R (2002) Biochemical characterization of the human RAD51 protein - I. ATP hydrolysis. J Biol Chem 277:14417-14425

Tombline G, Heinen CD, Shim KS, Fishel R (2002) Biochemical characterization of the human RAD51 protein - III. - Modulation of DNA binding by adenosine nucleotides. J Biol Chem 277:14434-14442

van Brabant AJ, Ye T, Sanz M, German JL, Ellis NA, Holloman WK (2000) Binding and melting of D-loops by the Bloom syndrome helicase. Biochemistry 39:14617-14625

Van Komen S, Petukhova G, Sigurdsson S, Stratton S, Sung P (2000) Superhelicity-driven homologous DNA pairing by yeast recombination factors Rad51 and Rad54. Mol Cell 6:563-572

Vassin VM, Wold MS, Borowiec JA (2004) Replication protein A (RPA) phosphorylation prevents RPA association with replication centers. Mol Cell Biol 24:1930-1943

Vaze MB, Pellicioli A, Lee SE, Ira G, Liberi G, Arbel-Eden A, Foiani M, Haber JE (2002) Recovery from checkpoint-mediated arrest after repair of a double-strand break requires Srs2 helicase. Mol Cell 10:373-385

Veaute X, Jeusset J, Soustelle C, Kowalczykowski SC, Le Cam E, Fabre F (2003) The Srs2 helicase prevents recombination by disrupting Rad51 nucleoprotein filaments. Nature 423:309-312

West SC (2003) Molecular views of recombination proteins and their control. Nat Rev Mol Cell Biol 4:435-445

Wold MS (1997) Replication protein A: A heterotrimeric, single-stranded DNA-binding protein required for eukaryotic DNA metabolism. Annu Rev Biochem 66:61-92

Wolner B, Peterson CL (2005) ATP-dependent and ATP-independent roles for the Rad54 chromatin remodeling enzyme during recombinational repair of a DNA double strand break. J Biol Chem 280:10855-10860

Worth L, Clark S, Radman M, Modrich P (1994) Mismatch repair proteins MutS and MutL inhibit RecA-catalyzed strand transfer between diverged DNAs. Proc Natl Acad Sci USA 91:3238-3241

Wu L, Bachrati CZ, Ou JW, Xu C, Yin JH, Chang M, Wang WD, Li L, Brown GW, Hickson ID (2006) BLAP75/RMI1 promotes the BLM-dependent dissolution of homologous recombination intermediates. Proc Natl Acad Sci USA103:4068-4073

Wu L, Davies SL, Levitt NC, Hickson ID (2001) Potential role for the BLM helicase in recombinational repair via a conserved interaction with RAD51. J Biol Chem 276:19375-19381

Wu LJ, Hickson ID (2003) The Bloom's syndrome helicase suppresses crossing-over during homologous recombination. Nature 426:870-874

Wu Y, Sugiyama T, Kowalczykowski SC (2006) DNA annealing mediated by Rad52 and Rad59 proteins. J Biol Chem 281:15441-15449

Wyman C, Ristic D, Kanaar R (2004) Homologous recombination-mediated double-strand break repair. DNA Repair (Amst) 3:827-833

Xu L, Marians KJ (2002) A dynamic RecA filament permits DNA polymerase-catalyzed extension of the invading strand in recombination intermediates. J Biol Chem 277:14321-14328

Yang HJ, Jeffrey PD, Miller J, Kinnucan E, Sun YT, Thoma NH, Zheng N, Chen PL, Lee WH, Pavletich NP (2002) BRCA2 function in DNA binding and recombination from a BRCA2-DSS1-ssDNA structure. Science 297:1837-1848

Yang HJ, Li QB, Fan J, Holloman WK, Pavletich NP (2005) The BRCA2 homologue Brh2 nucleates RAD51 filament formation at a dsDNA-ssDNA junction. Nature 433:653-657

Yonetani Y, Hochegger H, Sonoda E, Shinya S, Yoshikawa H, Takeda S, Yamazoe M (2005) Differential and collaborative actions of Rad51 paralog proteins in cellular response to DNA damage. Nucleic Acids Res 33:4544-4552

Yu X, Jacobs SA, West SC, Ogawa T, Egelman EH (2001) Domain structure and dynamics in the helical filaments formed by RecA and Rad51 on DNA. Proc Natl Acad Sci USA 98:8419-8425

Yuan ZM, Huang YY, Ishiko T, Nakada S, Utsugisawa T, Kharbanda S, Wang R, Sung P, Shinohara A, Weichselbaum R, Kufe D (1998) Regulation of Rad51 function by c-Abl in response to DNA damage. J Biol Chem 273:3799-3802

Zaitseva EM, Zaitsev EN, Kowalczykowski SC (1999) The DNA binding properties of *Saccharomyces cerevisiae* Rad51 protein. J Biol Chem 274:2907-2915

Zhou BBS, Elledge SJ (2000) The DNA damage response: putting checkpoints in perspective. Nature 408:433-439

Zigmond SH (2004) Formin-induced nucleation of actin filaments. Curr Opin Cell Biol 16:99-105

Heyer, Wolf-Dietrich

Section of Microbiology and Section of Molecular and Cellular Biology, Center for Genetics and Development, University of California, Davis, Davis, CA 95616-8665, USA

wdheyer@ucdavis.edu

DNA helicases in recombination

Hannah L. Klein

Abstract

The different pathways of homologous recombination involve the recognition and pairing of homologous DNA sequences promoted by proteins that catalyze strand exchange. Other steps in recombination involve double stranded DNA unwinding for branch migration activities, and in some cases the reannealing of single DNA strands. These processes do not occur spontaneously and so require DNA helicase enzymes that unwind the DNA helices. DNA helicases also have additional and sometimes unexpected functions in homologous recombination. These include anti-recombination activities that either reverse strand pairings or destabilize the primary homologous recombination intermediate, a single stranded DNA molecule coated with Rad51 protein that is primed to promote strand exchange. This latter novel anti-recombinase activity is especially important during the repair of stalled replication forks. This review examines the roles of eukaryotic DNA helicases in promoting and antagonizing homologous recombination.

1 Recombination pathways and models

Before considering the action of DNA helicases in homologous recombination, it is useful to review homologous recombination models and consider the steps at which a DNA helicase might act. It is now clear that there are two major pathways for double strand break (DSB) repair: 1) synthesis-dependent strand annealing in which the DSB is repaired though a strand invasion with subsequent disengagement from the invaded strand and reannealing with the other side of the initial double strand break, or 2) DSB repair (DSBR) in which the second end of the DSB is captured to form an intermediate with Holliday junctions (HJs) (Fig. 1). The process begins with the processing of the DSB ends through degradation of the 5' termini to produce single stranded DNA tails with 3'OH termini. These 3' ends then engage a homologous DNA sequence through strand invasion.

The SDSA reaction results in the repair of the DSB through a gene conversion, but is not associated with exchange of the flanking sequences. The strand invasion can use the sister chromatid, a homologous chromosome, or a homologous ectopic sequence as a pairing partner. While the repair is accurate, only those events that use a homolog or ectopic template result in a genetically detectable recombination event. Strand invasion is promoted by the Rad51 nucleofilament, and the Rad54 protein, which cooperates with Rad51 to promote strand pairing and transient

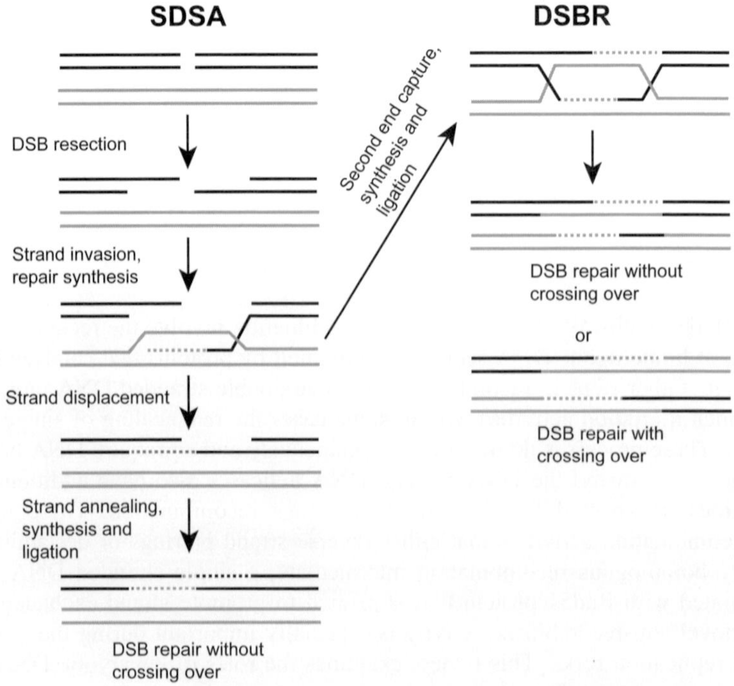

Fig. 1. DSB repair models. DSB repair is initiated by resection of a DSB to provide 3' single strand tails. Strand invasion into a homologous sequence is followed by DNA synthesis at the invading 3' end. The reaction can proceed to SDSA by strand displacement, annealing of the extended single strand end to the single stranded DNA on the other break end, followed by gap filling synthesis and ligation. The product always is noncrossover. Alternatively, after strand invasion and synthesis the second DSB end can be captured to form an intermediate with two Holliday junctions. After gap repair synthesis and ligation the structure is resolved at the Holliday junctions in a noncrossover or crossover mode.

strand separation through negative supercoiling. A DNA helicase could have a pro-recombination activity through the opening of the invaded duplex helix to stabilize the strand invasion event or an anti-recombination activity through a strand rejection activity. Ultimately, the invading strand is displaced after the 3' terminus has been extended through DNA synthesis.

The DSBR recombination event begins in the same manner, with the processing of the DSB ends to produce single strands with 3'OH termini. One end invades a homologous DNA sequence, promoted by Rad51 and Rad54 proteins. DNA synthesis extends the 3'OH terminus while forming a displaced strand called the D-loop. Once the D-loop has extended to allow annealing of the second 3'OH single strand tail, the second end becomes captured into the recombination intermediate. Subsequent ligation of the strand breaks at the captured strand results in a structure with two crossed strand junctions called Holliday junctions (HJs). Resolution of the Holliday junctions gives a noncrossover or crossover molecule, with

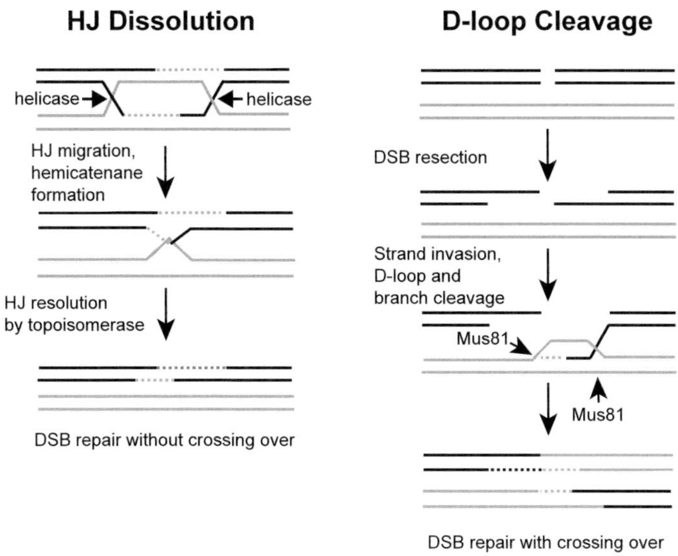

Fig. 2. Possible resolution modes of recombination intermediates. A double HJ molecule can be resolved by the action of the BLM or Sgs1 helicase to branch migrate the HJs to form a hemicatenane. This is resolved by TOPIIIα or Top3 to give noncrossover products exclusively from a double HJ intermediate. Cleavage of the D-loop intermediate at branch junctions by the Mus81/Mms4(Eme1) endonuclease, followed by gap filling synthesis and ligation results in exclusively crossover products without the formation of double HJ intermediates.

the DSB repaired by gap repair synthesis that may result in a gene conversion. Again, it is possible to propose a pro-recombination activity for DNA helicases in extending the heteroduplex to stabilize the strand invasion, in promoting branch migration, in HJ resolution through branch migration and an anti-recombination activity by strand rejection either at the initial strand invasion or in the second end capture to prevent formation of double HJs (dHJs) and crossover products.

Both of these pathways have been modified to explain specific effects of DNA helicase mutants. An alternative mechanism termed HJ dissolution involves branch migration of the HJs by a DNA helicase to form a hemicatenane structure, which can be resolved by topoisomerase III activity as noncrossover products (Ira et al. 2003; Wu and Hickson 2003; Mankouri and Hickson 2004) (Fig. 2). A second pathway that is always resolved as crossover has been proposed to account for the meiotic phenotype of the *mer3* DNA helicase mutant (Nakagawa and Ogawa 1999; Borner et al. 2004; Whitby 2005). In this model the Mer3 helicase in association with the ZMM complex stabilizes early strand invasions though the extension of heteroduplex at the strand invasion joint. Eventually a double HJ intermediate is formed through capture of the second end, and the ZMM complex at the first HJ ensures resolution only in the crossover configuration.

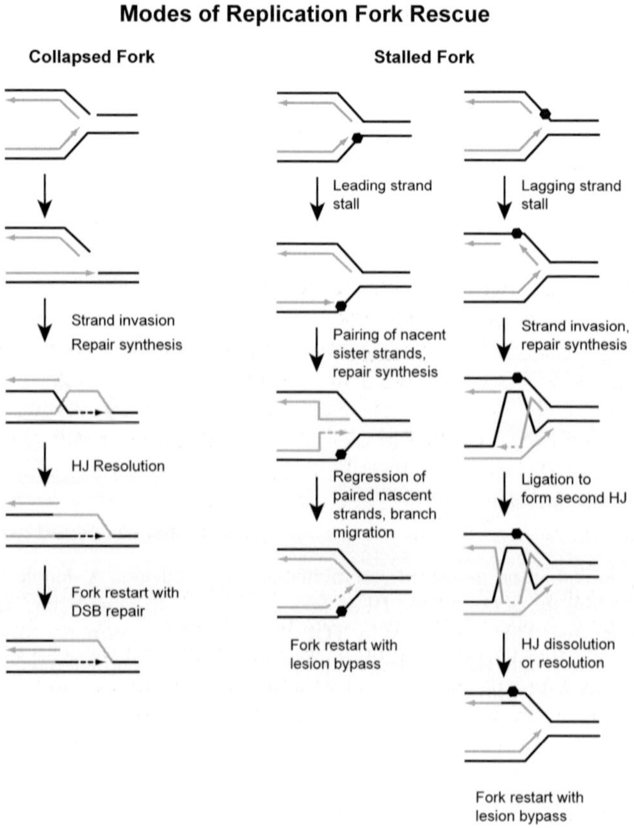

Fig. 3. Repair of a stalled or collapsed replication fork by homologous recombination. A collapsed fork occurs when a DSB forms at the fork, by replication to a single nick on the template strand or processing a stalled fork to a DSB. A one-sided invasion by the 3' end into the sister chromatid strand forms a single HJ intermediate which can be resolved to re-form the replication fork. Stalled forks occur when replication is stalled on the leading or lagging strand. Stalling on the leading strand can be overcome by fork regression, template switching and lesion bypass by repair synthesis using the nascent sister chromatid as template. Reversal of the nascent sister pairing through a helicase activity restores the replication fork. Stalling on the lagging strand can be overcome by a homologous recombination reaction that is similar to a one-sided invasion, with the 3' stalled strand invading the uninterrupted sister chromatid, forming a HJ intermediate. Capture of the nick on the lagging strand by ligation forms a double HJ intermediate, which is resolved to reform the replication fork. Resolution can occur by HJ dissolution or cleavage.

Another pathway resulting in exclusively crossover products has been proposed to occur through the action of the Mus81-Mms4/Eme1 endonuclease (Heyer et al. 2003; Osman et al. 2003; Hollingsworth and Brill 2004; Whitby 2005). This pathway involves strand invasion and a single HJ formation, but is processed by a Mus81 endonuclease that has a preference for branched DNA structures. The second end is captured through ligation to the Mus81 nicked strand. The end product is always a crossover, due to the endonuclease cleavage of the branched structures (Fig. 2).

It is also necessary to consider one-sided DSB events, which occur during DNA replication, when a replication fork collapses at a nick on the template strand, or when a stalled replication fork is processed to a DSB (Fig. 3). They may also occur when a telomere becomes uncapped and is recognized as a DSB (Lundblad and Blackburn 1993; Garvik et al. 1995; Hackett and Greider 2003). One-ended strand invasion events can rescue a collapsed replication fork through strand invasion into the intact sister chromatid, resolution of the HJ and re-establishment of the fork (Fig. 3). If a broken chromosome occurs outside the context of a replication fork, the broken end can be healed through homologous recombination using the homologous chromosome, the sister chromatid, or homologous sequences at an ectopic site. DNA synthesis may continue on both strands of the intact donor duplex through a process called break-induced replication or BIR (Malkova et al. 1996; Morrow et al. 1997; Kraus et al. 2001; Davis and Symington 2004). Similar to recombination events with two-sided DSBs, BIR may involve DNA helicases to promote stabilization of the invading strand, extension of the replication tracts, and strand rejection if homology is limited.

Lastly, homologous recombination may be used to restart replication at stalled replication forks that contain gaps on the nascent strands, but do not have DSBs (Fig. 3). A gap on either the lagging or leading strand of a replication fork can be filled in by mechanisms that involve strand invasion, branch migration and HJ resolution. Any of these steps may require DNA helicase activity. Additionally, lesion bypass mechanisms involving translesion DNA polymerases can act on some of the gapped substrates. There is probably regulation at an early step to control gap repair and replication fork restart through homologous recombination or translesion DNA polymerase synthesis.

Two additional DSB repair pathways include single strand annealing (SSA) and nonhomologous end-joining (NHEJ) (Fig. 4). Single strand annealing occurs between complementary single strand DNA sequences. The annealing process itself does not require strand invasion activities, but does entail processing of the single strand tails by the Rad1/10 endonuclease. End-joining uses microhomologies of a few nucleotides and does not require homologous recombination factors. Homologous recombination mutations may affect end-joining rates, depending on the cell cycle context of the end-joining process.

In this review, the role of DNA helicases in promoting and preventing homologous recombination will be discussed. The review will focus on eukaryotic DNA helicases, primarily the helicases of *Saccharomyces cerevisiae*. Where information is available, the recombination impact of helicases from other eukaryotic organisms will also be discussed. The bacterial helicases RecBCD, functioning early in DSB-mediated homologous recombination, RecQ which functions in the RecF pathway, and RecG which functions late in homologous recombination after

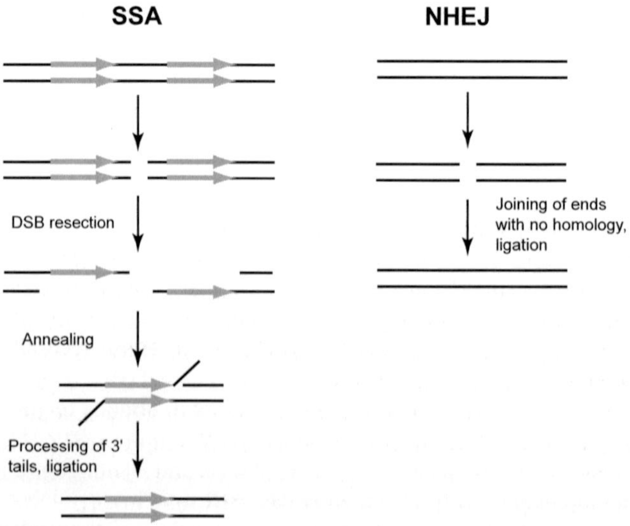

Fig. 4. Single strand annealing and nonhomologous end-joining. If a DSB occurs between directly repeated DNA sequences, the break can be repaired through DSB resection to reveal the complementary DNA sequences, annealing, trimming of the 3' tails by endonucleolytic cleavage, and ligation. The final product is a deletion between the direct repeats. End-joining occurs between DSB ends with little or no homology and involves a set or repair functions distinct from those involved in homologous recombination.

strand exchange, will not be discussed in this review. The reader is referred to the review by Stephen Kowalczykowski in this volume for a discussion of recombination in *E. coli* and the action of RecBCD. The role of helicases will be considered in the context of the various DSB repair pathways described above with a focus on general recombination. Specific types of recombination, for example, recombination at uncapped telomeres, will not be discussed and the reader is instead referred to several recent reviews (Kass-Eisler and Greider 2000; d'Adda di Fagagna et al. 2004; Harrington 2004; Maser and DePinho 2004; Blackburn 2005; Blasco 2005; Tarsounas and West 2005).

2 DNA helicases in mitotic recombination

Several helicases have been described that affect aspects of mitotic recombination. In general these helicases are related to the bacterial RecQ DNA helicase that suppresses illegitimate recombination, and to the bacterial UvrD DNA helicase, which has anti-recombination activity through the disruption of the RecA nucleoprotein filament (Veaute et al. 2005). Depending on the type of recombination assay used, a helicase mutant may exhibit either a pro-recombination or an anti-recombination activity. There are DNA helicases that have as their main activity the rescue of stalled replication forks, and only secondarily have effects on recombination. These are not considered to be helicases with an active role in homologous recombination, and will not be discussed outside of the context of their effect on homologous recombination.

2.1 Srs2

The *SRS2* gene was first described through semidominant mutants that suppressed the trimethoprim and UV sensitivities of the *rad6* and *rad18* mutants in the postreplication repair (PRR) pathway of *S. cerevisiae* (Lawrence and Christensen 1979). The gene was independently isolated several times subsequently, in screens for suppressors of *rad6* and *rad18* (Aboussekhra et al. 1989; Schiestl et al. 1990) and partially active *rad52* mutants (Kaytor et al. 1995; Milne et al. 1995; Schild 1995) and in a screen for mutants with increased mitotic recombination rates (Aguilera and Klein 1988; Rong et al. 1991). The *SRS2* gene was shown to encode a protein related to the bacterial UvrD DNA helicase (Aboussekhra et al. 1989) and later characterized as a 3'-5' DNA helicase (Rong and Klein 1993). In addition to having DNA helicase activity, the Srs2 protein can disrupt Rad51 nucleoprotein filaments (Krejci et al. 2003; Veaute et al. 2003), an activity similar to that of UvrD helicase on RecA nucleofilaments (Veaute et al. 2005). Studies on the mitotic *srs2* mutant phenotypes have shown that Srs2 has both pro- and anti-recombination roles, dependent on the nature and context of the DNA substrate and proteins that interact with Srs2.

2.1.1 Hyper-recombination phenotype

Early studies on *SRS2* showed that point mutants and the null allele were increased for mitotic gene conversion, in both intrachromosomal and interchromosomal assays (Aguilera and Klein 1988; Aboussekhra et al. 1989; Rong et al. 1991). Indeed *SRS2* was identified as the hyper-rec mutant *hpr5-1*, which was increased in gene conversion tenfold over wild type (Aguilera and Klein 1988; Rong et al. 1991). Mutants targeted to the Walker A box are completely defective in DNA helicase activity and are also hyper-rec (Van Komen et al. 2003). However, the null allele mutant does not have a strong hyper-rec phenotype (Palladino and Klein 1992; Milne et al. 1995; Liefshitz et al. 1998). Additional studies have shown that gene conversion tracts in the *srs2-101* allele, mutated in helicase domain I, are shorter than those in wild type (Rong et al. 1991). These results implicate Srs2 in HJ branch migration or stabilization of a nascent joint following strand invasion. It is possible that a mutant Srs2 protein that can bind to DNA and Rad51, but cannot translocate, acts in a dominant-negative manner to inhibit branch migration and joint stabilization, while in the absence of Srs2 another helicase can compensate for its role. Since the recombinant segregants are selected as prototrophs from heteroalleles, longer gene conversion tracts may go undetected when they do not result in a prototrophic phenotype. In fact, the *srs2Δ* null mutant did not show a decrease in gene conversion tract length (Rong et al. 1991).

2.1.2 SRS2, anti-recombination, and the post-replication repair (PRR) pathway

Diploid *srs2* mutants are as sensitive as G1 haploid mutants to DNA damage, suggesting that recombination between sister chromatids is tolerated, but recombination between homologous chromosomes is lethal (Aboussekhra et al. 1989). This combined with the reduced spore viability of the *srs2* diploid has led to the channeling model whereby *SRS2* promotes repair of lesions by the PRR pathway, and in its absence these lesions are channeled to the homologous recombination pathway (Aboussekhra et al. 1989; Schiestl et al. 1990). Consistent with this model, the suppression phenotype of *srs2* mutants is dependent on a functional homologous recombination pathway (Schiestl et al. 1990). Suppressors of the MMS sensitivity of diploid *srs2* strains were identified as semi-dominant mutations in *RAD51*, and led to the idea that Srs2 is antagonistic to Rad51 protein (Aboussekhra et al. 1992; Chanet et al. 1996). This has been borne out by biochemical studies (Krejci et al. 2003; Veaute et al. 2003). In fact, suppressors of weak *rad52* alleles turned out to be in *SRS2*, consistent with this model (Kaytor et al. 1995; Milne et al. 1995). The balance between Srs2 and Rad51 is delicate and tipping it by overexpression of Srs2 or Rad51 sensitizes cells to DNA damage and turns DNA damage into lethal events, presumably by overloading the homologous recombination repair pathway or by channeling DNA lesions to inappropriate recombination events (Kaytor et al. 1995; Milne et al. 1995; Paffett et al. 2005).

Epistasis studies of *srs2Δ* and PRR mutants in *S. cerevisiae* have placed *SRS2* in the error free subpathway. Suppression of error-free subpathway mutants by *srs2Δ* is dependent on homologous recombination (Ulrich 2001; Broomfield and Xiao 2002), and Srs2 is proposed to act as a switch between PRR and homologous recombination by either inhibiting homologous recombination, through the antagonistic action on Rad51 nucleofilaments, or through the promotion of PRR. *S. cerevisiae* Srs2 preferentially interacts with PCNA modified by sumoylation, and this is proposed to be critical for Srs2 regulation of PRR and inhibition of homologous recombination by targeting Rad51 through the PCNA interaction to stalled replication forks (Papouli et al. 2005; Pfander et al. 2005).

In contrast to *srs2* mutants in *S. cerevisiae*, *S. pombe srs2* mutants do not suppress mutants of the PRR pathway. Although the *S. pombe srs2+* gene is not involved in channeling lesions to the PRR pathway, it is involved in tolerance of DNA damage. The main function of *srs2+* is in the repair of one ended single strand breaks, such as those that would occur at collapsed replication forks (Fig. 3). A similar observation has been made with the *Neurospora crassa srs2* mutant. The *N. crassa SRS2* gene does not function in PRR, but is required for recovery from replication fork damage (Suzuki et al. 2005).

PRR mutants of *S. cerevisiae* are increased about fivefold for recombination with direct repeat substrates. The increase is dependent on *SRS2* and in the double mutants *srs2Δ* is epistatic (Liefshitz et al. 1998; Friedl et al. 2001). In this situation, *SRS2* could be acting at a later stage of HR to stabilize the nascent joint, especially when recombination occurs between short regions of homology.

2.1.3 SRS2 and DSB repair

Several studies have shown that *SRS2* is required for DSB repair. The *srs2Δ* mutant is lethal with *rad27Δ*, which is defective in the Fen1 endonuclease required for lagging strand DNA synthesis (Debrauwere et al. 2001). The spectrum of *rad27Δ* synthetic lethal mutations identified recombination repair genes involved in DSB repair, and placed an important role for this process in the repair of breaks at replication forks. *srs2Δ* mutants are unable to recover from DSBs induced by *in vivo* expression of EcoRI enzyme (Lewis et al. 1999), another indication of a failure of DSB repair. *srs2Δ* mutants show low viability after an induced DSB that elicits a checkpoint response (Vaze et al. 2002). Repair of the DSB was through a SSA mechanism (Fig. 4). Physical studies showed the formation of a recombinant product, suggesting that SSA *per se* was not affected, but the cells never returned to growth and were considered to be defective in recovery from the cell cycle arrest. This defect was partially dependent on the DNA damage checkpoint and was fully dependent on the homologous recombination factor *RAD51*. Even when the DSB could be repaired by an ectopic gene conversion event, cells were unable to recover from the arrest. These results suggest that Rad51 can be inhibitory to recovery from a DSB when Srs2 is absent or defective.

In a different study of DSB-induced ectopic recombination, *SRS2* was required for DSB repair and in its absence the DSB was hyper-resected, few recombinants were recovered and the cell viability declined rapidly (Aylon et al. 2003). Most of the surviving cells did not show a gene conversion event, suggesting that in the absence of Srs2, an alternative repair pathway is used. Whether the DSB repair events in the absence of Srs2 are associated with rearrangements was not addressed, but the reduced survival of cells that had undergone a recombination event is similar to the phenotypes of WRN and BLM deficient cells, described in sections 2.4 and 2.5.

Survivors of an induced DSB in *srs2Δ* mutants show a two to threefold excess of crossover molecules. Overexpression of Srs2 reduces the formation of crossover products. These data have led to the hypothesis that Srs2 suppresses crossover formation by promoting SDSA repair of a DSB (Fig. 1) and preventing second end capture and formation of HJs that could be resolved as crossovers (Ira et al. 2003). In support of this, overexpression of Rad51 in a *srs2Δ* mutant eliminated the noncrossover SDSA pathway. Formation of HJ molecules that can be resolved as crossovers may account for the sensitivity of *srs2Δ* cells to high copy Rad51 and the lethality with *sgs1Δ*.

Studies on DSB repair through gene conversion where there are ends of nonhomology longer than 30 nucleotides showed a requirement for *SRS2* (Paques and Haber 1997). Srs2 was proposed to act as a DNA helicase to stabilize the nascent joint between the invading single strand and the duplex homologous DNA by increasing heteroduplex formation and preventing reverse branch migration (Fig. 1). Thus Srs2 is proposed to have a role in DSB both as a helicase to promote formation of recombination intermediates, to prevent crossover formation by promoting SDSA and possibly by removing Rad51 protein from ssDNA.

2.1.4 srs2Δ genetic interactions

Srs2Δ cells grow normally and are not synthetically lethal with the recombination repair gene mutants *rad51, rad52, rad55,* and *rad57,* showing that spontaneous lesions are not DSBs (Fabre et al. 2002). However, *srs2Δ* mutants are synthetically sick or lethal with *sgs1* and *rad54* and a collection of mutants in DNA repair and sister chromatid cohesion (Gangloff et al. 2000; Klein 2001; Fabre et al. 2002; Ooi et al. 2003; Tong et al. 2004). Most of the synthetic lethal combinations are suppressed by eliminating homologous recombination, suggesting that the lethality or poor growth is due to excess recombination or accumulation of Rad51 nucleofilament on DNA gaps generated by replication stalling. DNA damage checkpoint mutations also partially rescue the synthetic phenotypes, indicating that part of the failure to grow could result from a continued checkpoint arrest and the inability of *srs2* mutants to overcome the checkpoint arrest (Klein 2001; Vaze et al. 2002). *S. pombe srs2* mutants are also lethal with *rqh1*, the *S. pombe* RecQ and Sgs1 homolog. This lethality was found not to be suppressed by loss of homologous recombination (Wang et al. 2001), but this result is probably due to an additional mutation in the *fbh1* gene in the *rhp51* strain, and subsequent studies have shown that *srs2 rqh1* lethality is suppressed by *rhp51, rhp55* and *rhp57* mutations (Doe and Whitby 2004). In this regard the *S. pombe* Srs2 seems to function similar to the *S. cerevisiae* Srs2 although *S. pombe srs2* mutants do not suppress mutants of the PRR pathway.

2.1.5 SRS2 and SGS1

Srs2 acts upstream of Sgs1 in homologous recombination, to prevent homologous recombination by acting on the Rad51 filament, while Sgs1 promotes HJ dissolution and inhibits HJ resolution as crossovers (Gangloff et al. 2000; Fabre et al. 2002). Consistent with this result, high copy Sgs1 can suppress the DNA damage sensitivity of the *srs2Δ* mutant (Mankouri et al. 2002). *srs2Δ* mutants do not accumulate X-structures at replication forks after DNA damage, while *sgs1Δ* mutants accumulate such structures, whose formation is dependent on recombination (Liberi et al. 2005). The X-structures can be prevented by overexpression of Srs2, showing that Srs2 acts at an early stage to prevent X-structure formation. Once X-structures are allowed to form, high copy Srs2 is unable to resolve the structures, but high copy Sgs1 can resolve the structures (Liberi et al. 2005). This demonstrates that Srs2 acts early to prevent homologous recombination, but is inactive on the HJ structures of the later stage of recombination. These results are consistent with the failure to observe fork pausing at the rDNA in a *srs2Δ* mutant (Torres et al. 2004b).

2.1.6 Additional roles of SRS2

Srs2 also acts during replication to ensure replication fidelity. This occurs through the DNA helicase activity of Srs2, acting on small secondary structures in the template strand DNA that can arise during replication. Srs2 blocks trinucleotide expansion and in the *srs2Δ* mutant expansions occur that are independent of homologous recombination (Bhattacharyya and Lahue 2004). Srs2 is able to unwind partial duplexes of trinucleotide repeats (Bhattacharyya and Lahue 2005). Srs2 may be recruited to the replication fork through an interaction with DNA polymerase δ to unwind hairpin structures on the template strand. This is different from its putative recruitment by SUMO-PCNA to forks stalled at lesions on the template strand (Papouli et al. 2005; Pfander et al. 2005).

Srs2 has a role in NHEJ (Fig. 4) but the effect is only a two to threefold reduction in the *srs2Δ* mutant (Hegde and Klein 2000; Wilson 2002).

2.2 Fbh1

The human hFbh1 helicase is a composite protein with a 3'-5' DNA helicase activity and a N-terminus F-box sequence that interacts with Cullin and Roc1 and is a component of SCF, which has ubiquitin ligase activity (Kim et al. 2002, 2004). A similar helicase is present in *S. pombe*, but is absent in *S. cerevisiae* (Park et al. 1997). Genetic studies on the role of the *S. pombe* Fbh1 helicase in replication and recombination show that the gene is not essential in log phase cells, but is required in stationary phase cells. The deletion is lethal combined with a *srs2Δ* or *rqh1Δ* mutation, and this lethality can be suppressed by loss of homologous recombination (Morishita et al. 2005; Osman et al. 2005). *fbh1Δ* strains are sensitive to DNA damaging agents, but have no effect on direct repeat recombination substrates (Osman et al. 2005). From genetic studies, Fbh1 is proposed to interact with the Rad22 mediator protein to control Rhp51, the Rad51 homolog, by regulating Rhp51 filament formation through both the helicase and F-box domains. Dissociation of the Rhp1 filament is proposed to block inappropriate recombination, or inappropriate Rhp51 filament formation that is inhibitory to other repair activities such as DNA repair polymerases (Osman et al. 2005). The formation of spontaneous Rad51 foci in *fbh1Δ* strains is consistent with this model.

2.3 Sgs1

The *SGS1* gene was first described in *S. cerevisiae* from a mutation that suppressed the slow growth phenotype of a *top3* mutant (Gangloff et al. 1994). The *sgs1* mutant was slightly increased for rDNA recombination and was epistatic to *top3* for rDNA recombination. Cloning of the gene revealed that it was a member of the RecQ DNA helicase family (Gangloff et al. 1994; Watt et al. 1995) and further studies confirmed the 3'-5' DNA helicase activity *in vitro* (Bennett et al. 1998). Additional phenotypes of the *sgs1Δ* mutant include increases in chromosome missegregation and recombination at non-rDNA substrates (Watt et al. 1995, 1996). An assay for spontaneous gross chromosomal rearrangements, including translocations, showed that the *sgs1Δ* mutant was increased for these events (Myung et al. 2001). The frequency of targeted gene replacement is increased in *sgs1Δ* mutants, similar to the phenotypes of BLM and WRN mutant cell lines

(Langston and Symington 2005). The *S. pombe* RecQ mutant *rqh1Δ* is sensitive to DNA damaging agents (Stewart et al. 1997). While spontaneous recombination is not increased, treatment of cells with HU gives a tenfold excess stimulation of recombination in the *rqh1Δ* mutant.

The *sgs1Δ/rqh1Δ* phenotypes suggest a role for Sgs1 in preventing recombination in response to DNA damage or in controlling the outcome of homologous recombination events (Ajima et al. 2002). Since mutant human RecQ homologs WRN and BLM result in increased aberrant recombination and sister chromatid exchange respectively, the hyper-rec phenotype of the yeast mutants is relevant to understanding how chromosome instability arises. Expression of BLM in yeast complements some of the *sgs1Δ* phenotypes (Neff et al. 1999). *sgs1Δ* mutants are also increased in illegitimate recombination and recombination between short regions of homology, and this too is suppressed by expression of the human BLM and WRN genes in yeast (Yamagata et al. 1998).

The N-terminal region of Sgs1 interacts with Top3, and the C-terminal domain contains the DNA helicase motifs. The N-terminal region complements the hyper-rec phenotype, while N-terminal truncations are hyper-rec, indicating that interaction with Top3 is essential for preventing recombination events. The helicase-dead allele of *SGS1* is unable to complement the hyper-rec phenotype, showing that the helicase activity is also essential for reducing recombination (Mullen et al. 2000; Onoda et al. 2000; Ui et al. 2005).

The Sgs1 protein has preferential binding to synthetic DNA fork structures (Bennett et al. 1998). This observation combined with genetic studies on functional overlaps between Sgs1 and Top3 have led to the hypothesis that Sgs1 acts on intermediates formed after replication forks stall (Kaliraman et al. 2001; Shor et al. 2002). In *sgs1Δ* mutants, recombination intermediate-like X-structures, dependent on Rad51, are formed at stalled replication forks (Liberi et al. 2005). The *sgs1Δ* synthetic lethal interactions are due to excess recombination as they are suppressed by *rad51Δ* mutations (Gangloff et al. 2000; Fabre et al. 2002; Bastin-Shanower et al. 2003).

Sgs1 appears to have at least two roles in homologous recombination. The first is in the resolution of recombination intermediates into a noncrossover mode via a HJ dissolution model (Fig. 2). This accounts for the mitotic hyper-rec phenotype, the finding of increased crossovers following homologous recombination repair of a DSB (Ira et al. 2003; Rockmill et al. 2003), the increase in targeted gene replacement (Langston and Symington 2005) in *sgs1Δ* mutants and the preferential binding of Sgs1 protein to forked DNA substrates. In the absence of functional Sgs1 protein, recombination intermediates are either processed to form HJs or the existing HJs are resolved as crossovers, resulting in the increase in crossovers seen in *in vivo* assays. This model is also consistent with the biochemical properties of the BLM helicase, discussed in section 2.5. A second role for Sgs1 is in rejection of heteroduplex DNA recombination intermediates containing mismatches (Sugawara et al. 2004; Goldfarb and Alani 2005). This activity functions with the mismatch repair system and seems to account for the increase in translocations in the GCR assays seen with the *sgs1Δ* mutant (Myung et al. 2001).

2.4 WRN

Werner syndrome is a human progeroid syndrome associated with defects in DNA damage repair. The WRN gene is related to RecQ (Yu et al. 1996) and encodes a 3'-5' DNA helicase (Gray et al. 1997; Shen et al. 1998) with an associated 3'-5' exonuclease activity (Huang et al. 1998). The WRN protein unwinds and degrades D-loop substrates (Orren et al. 2002), thus destroying early recombination intermediates. WRN appears to have both pro-recombination and anti-recombination phenotypes. Both outcomes seem to result from defective resolution of recombination intermediates. Human WRN-deficient cells are increased in aberrant recombination in a direct repeat reporter (Prince et al. 2001), but many of the recombinant cells have low survival, indicating a homologous recombination defect after initiation. The reduced survival of WRN-deficient cells, particularly those cells that have experienced recombination events, can be increased by expression of the bacterial HJ resolvase protein RusA or re-expression of WRN, indicating that one function of the WRN helicase is to resolve recombination intermediates that occur after DNA damage (Saintigny et al. 2002). WRN protein is a structure-specific DNA helicase that can branch migrate HJs (Constantinou et al. 2000; Mohaghegh et al. 2001), suggesting a role for WRN in resolution of HJs formed through homologous recombination at collapsed replication forks (Pichierri et al. 2001) (Fig. 3). Both the exonuclease and helicase activities of WRN are required for recombination resolution, as mutations that eliminate either activity do not reduce survival after DNA damage (Saintigny et al. 2002; Swanson et al. 2004).

Cell lines deficient in WRN do not show any increase in sister chromatid exchange, but are increased in targeted integration (Imamura et al. 2002). The increase in gene targeting is dependent on the homologous recombination factor Rad54, showing that loss of WRN can result in an increase in homologous recombination events. The increase is either due to the ability of the integration intermediate to be resolved as a crossover in the absence of WRN, or a failure to reverse the D-loop intermediate.

2.5 BLM

Bloom syndrome is a DNA damage repair defect associated with increased sister chromatid exchange and genomic instability. Similar to WRN protein, BLM functions during somatic cell growth to regulate the outcome of recombination intermediates formed at the replication fork during repair of stalled and collapsed replication forks (Wu et al. 2001; Wu and Hickson 2002; Rassool et al. 2003). The BLM gene is a member of the RecQ family (Ellis et al. 1995) and encodes a 3'-5' DNA helicase (Karow et al. 1997). The BLM protein also promotes strand annealing (Cheok et al. 2005; Machwe et al. 2005), which may function with the helicase activity for branch migration of HJs.

BLM associates with TopIIIα (Wu et al. 2000; Hu et al. 2001), similar to the interaction between bacterial RecQ and TopIII (Harmon et al. 1999), and suggests a capability to promote strand passage between DNA duplexes. BLM can branch migrate HJs and is a structure specific DNA helicase (Karow et al. 2000; Mohaghegh et al. 2001). In association with a topoisomerase activity, BLM could promote HJ dissolution (Fig. 2) to noncrossovers and such activity has been seen *in vitro* (Wu and Hickson 2003).

Another substrate of the BLM helicase is the D-loop (Fig. 1). BLM protein can bind to and dissociate D-loop structures (van Brabant et al. 2000). In addition to preventing aberrant recombination intermediates, the D-loop melting activity could promote SDSA through the disengagement of the invading strand in the D-loop (Fig. 1). The strand annealing activity of BLM might promote the reannealing step in the SDSA reaction. The *Drosophila* BLM protein, DmBlm, is required for SDSA (Adams et al. 2003), and in its absence, deletions are formed during DSB repair (McVey et al. 2004). This phenotype is reminiscent of the rearrangements seen in BLM-deficient cells during DSB repair (Gaymes et al. 2002).

Sister chromatid exchanges are suppressed by BLM either because recombination intermediates are not resolved as crossovers, or because repair intermediate never form a stable HJ intermediate, due to the D-loop dissolution and strand annealing activities. In chicken DT40 BLM knockout cells, both sister chromatid exchange and targeted integration are increased, both of which require HJ resolution as a crossover and occur through homologous recombination involving the *RAD54* gene (Wang et al. 2003).

2.6 Rad3/Rem1

The *RAD3* gene of *S. cerevisiae* encodes a DNA helicase of 5'-3' polarity (Sung et al. 1987). Rad3 functions in nucleotide excision repair and is part of the RNA polymerase II transcription machinery. Although *RAD3* is not commonly considered to be a HR gene, it is linked to homologous recombination through specific mutants, called the *rem1* alleles that increase mitotic interhomolog recombination (Golin and Esposito 1977; Malone and Hoekstra 1984; Montelone et al. 1988). The *rem1* alleles are semidominant and also affect chromosome stability and mutagenesis. Although the *rem1* mutations are located within or near the helicase domains of the Rad3 protein, the effect on DNA helicase activity is not known (Montelone and Malone 1994). However, unlike other *rad3* mutations, *rem1* alleles are lethal in combination with homologous recombination defects, suggesting that the *rem1* alleles result in the formation of DSBs (Montelone et al. 1988).

The *rem1* alleles are not the only mutants of *RAD3* that affect homologous recombination rates. Other alleles specifically stimulate DSB-promoted recombination between short homologous DNA sequences (Bailis et al. 1995). It has been proposed that the *rad3-G595R* mutation, located in the putative DNA binding domain of Rad3, stabilizes DSBs by reducing end degradation (Maines et al. 1998; Lee et al. 2000). At regions of short homology, excess DSB degradation could remove homology to the point that a stable heteroduplex might not form. A specific mutation in the SSL2 helicase of the TFIIH transcription complex also has a recombination phenotype similar to that of the *rad3-G595R* mutation (Lee et al. 2000).

2.7 Rrm3 and Pif1

Rrm3 and Pif1 are considered to be DNA helicases that help replication through protein/DNA regions that are difficult to replicate and promote genomic stability. In their absence the rDNA repeat sequence arrays are prone to rearrangements through HR mediated processes. The *RRM3* gene was first identified as a mutant that increased recombination within the rDNA array and other tandem arrays in *S. cerevisiae* (Keil and McWilliams 1993). The increased recombination occurred only within tandem repeats, not with single copy genes, and was dependent on *RAD52*. Sequence analysis of the *RRM3* gene showed that it encoded a putative DNA helicase with 30% identity to the *S. cerevisiae* Pif1 DNA helicase, and a 5'-3' DNA helicase activity was determined *in vitro* (Ivessa et al. 2002; Schmidt et al. 2002). The increased rDNA recombination in the *rrm3*Δ mutant is apparently due to the formation of rDNA circles, the result of pauses and breakage within the rDNA array during DNA replication (Ivessa et al. 2000; Ivessa et al. 2002). The Rrm3 helicase promotes replication through nonhistone protein-DNA complexes, including rDNA-protein complexes (Ivessa et al. 2003; Torres et al. 2004a). The Rrm3 helicase appears to function to prevent replication fork stalling, although its precise mechanistic role is unclear (Torres et al. 2004b). The *rrm3*Δ mutant has genetic interactions with the helicase mutants *sgs1*Δ and *srs2*Δ, suggesting that any two of these helicases are necessary to prevent the formation of unwanted or unrepairable recombination intermediates that arise due to stalled replication (Schmidt and Kolodner 2004; Torres et al. 2004b).

The *PIF1* gene of *S. cerevisiae* encodes a 5'-3' DNA helicase with a dual function and localization (Lahaye et al. 1991). In the mitochondria, the Pif1 helicase is required for repair of mitochondrial DNA damage after UV irradiation or ethidium bromide treatment. Pif1 promotes recombination between rho^+ and tandemly arranged rho^- genomes, although recombination between rho^+ genomes is normal in *pif1* mutants (Foury and Kolodynski 1983). Pif1 is also localized in the nucleus where it is required for proper telomere maturation by removing telomerase from telomeric sequences and thereby inhibiting telomerase action (Schulz and Zakian 1994; Zhou et al. 2000; Boule et al. 2005). This is another example of a DNA helicase removing a protein bound to DNA.

Pif1 also acts on the rDNA array of *S. cerevisiae*, promoting the accumulation of rDNA circles through a mechanism that requires the homologous recombination protein Rad52. In the *pif1*Δ mutant rDNA circle formation was reduced threefold (Ivessa et al. 2000).

3 DNA helicases in meiotic recombination

The DSB repair models described in Figure 1 are the major meiotic DSB-promoted recombination pathways. Meiotic recombination is essential for chromosome segregation and meiotic product viability. However, no helicase has been found that completely eliminates meiotic recombination. Nonetheless, there are several helicases that act on the meiotic recombination intermediates and affect the outcome of the meiotic recombination products.

3.1 Mer3

The *MER3* gene of *S. cerevisiae* encodes a meiosis-specific 3'-5' DNA helicase (Nakagawa and Ogawa 1999; Nakagawa et al. 2001). Meiotic DSBs are formed in the deletion mutant, but do not disappear at the time when DSBs are repaired by recombination in wild type cells. Instead the DSBs become hyper-resected and the *mer3Δ* mutant has a reduced crossover frequency. The residual crossovers show a random distribution, resulting in a high incidence of nondisjunction in the first meiotic division and reduced spore viability (Nakagawa and Ogawa 1999). Analysis of the crossover pattern shows that Mer3 has a role in crossover control through crossover interference, and in its absence crossovers are not correctly distributed (Nakagawa and Ogawa 1999). *MER3* mutants defective in an *in vitro* helicase assay have *in vivo* meiotic defects of reduced crossovers and crossover interference (Nakagawa and Kolodner 2002b). *In vitro* studies of the Mer3 DNA helicase activity show that Mer3 can unwind a HJ substrate (Nakagawa and Kolodner 2002a) and can stimulate DNA heteroduplex extension by the Rad51 recombinase (Mazina et al. 2004). The Mer3 helicase could act early in DSB-promoted recombination to unwind DSB ends for processing, or at later stages to stabilize the nascent joint molecule formed by strand invasion. Mer3 could also act in the extension of heteroduplex to promote capture of the second end. Finally, Mer3 could act on resolution of the HJ by branch migration. Defects at any of these steps would reduce crossovers. The residual crossovers that do not exhibit crossover interference in the *mer3* mutant suggest that there are *MER3*-independent pathways for crossover formation. These might be crossovers that are formed by Mus81 processing of recombination intermediates (Whitby 2005), and would be the crossovers that do not exhibit interference in *S. cerevisiae*. This pathway would be more prominent in *S. pombe*, where none of the meiotic crossovers exhibit interference and where there is a stronger dependence on *MUS81* in meiosis.

MER3 is a conserved DNA helicase. Similar to *S. cerevisiae*, the *Arabidopsis* mutant has reduced fertility, a decreased crossover frequency and the residual crossovers do not exhibit interference (Chen et al. 2005; Mercier et al. 2005).

3.2 Srs2

Several *srs2* mutants have reduced sporulation and spore viability, including Walker A box mutants that eliminate the DNA helicase activity (Krejci et al. 2004) and mutants in conserved helicase domains that have a strong mitotic hyper-recombination phenotype (Palladino and Klein 1992). These mutants show a delay in the formation of genetic recombinants in return to growth experiments and are delayed in the meiotic divisions (Palladino and Klein 1992). Genetic map distances are reduced twofold in the *srs2-101* mutant (F. Palladino and H. Klein, unpublished observations). These recombination phenotypes could reflect a role for the Srs2 helicase in heteroduplex extension for second end capture or in stabilization of the nascent joint in strand invasion. The pattern of spore inviability does not indicate a specific defect in meiotic chromosome division and nondisjunction is not increased by genetic tests in incomplete tetrads. However, expression of *SRS2* is increased in meiosis at the time of onset of meiotic recombination, suggesting a proactive role of Srs2 in meiotic recombination (Heude et al. 1995).

3.3 Sgs1

Although *sgs1* mutants are known to have reduced sporulation and spore viability, in contrast to the mitotic mutant phenotype, no increase in meiotic recombination was been found in early studies (Watt et al. 1996; Miyajima et al. 2000). In return to growth experiments recombination was decreased, suggesting that Sgs1 is needed to promote recombination in meiosis (Miyajima et al. 2000). This same group found that the meiotic defects of the *sgs1Δ* mutant could be complemented by an allele of *SGS1* that was defective for helicase activity but retained other interaction domains.

A more detailed study of the meiotic *sgs1Δ* phenotype concluded that Sgs1 regulates chromosome synapsis and meiotic crossovers (Rockmill et al. 2003). The null mutant had increased axial associations between homologous chromosomes, increased sites of chromosome synapsis, and associated with this, an increase in meiotic crossovers. However, there was no increase in meiotic gene conversion. Although the cells appeared to be proficient for meiotic recombination, the increased crossovers resulted in a checkpoint arrest at pachytene. In contrast to other studies, this group did not find a rescue of the meiotic phenotypes by a helicase-deficient allele of *SGS1*. Thus, it appears that a meiotic function of the Sgs1 helicase is to prevent some recombination intermediates from becoming crossovers at an early stage in meiosis. This could occur through heteroduplex disruption, particularly at the stage of second strand capture, to prevent HJs from being formed. Alternatively, Sgs1 prevents the occurrence of the Mus81-dependent crossover pathway (Fig. 2).

3.4 BLM

Bloom syndrome females have reduced fertility and Bloom syndrome males are infertile, suggesting an essential function of the BLM helicase in promoting meiotic recombination (German 1993). BLM protein localizes to synapsed chromosomes during zygotene, but is later dispersed in mouse spermatocytes, and colocalizes with recombination proteins, including Rad51 and the meiotic recombinase Dmc1 (Walpita et al. 1999; Moens et al. 2000, 2002). Localization of BLM to the XY pseudoautosomal region strongly suggests an active role for BLM in meiotic homologous recombination. Various roles for BLM have been proposed including the resolution of chromosome interactions without forming crossovers, to facilitating branch migration of HJs. However, since these roles are based on the *in vitro* activities of the protein or on the *in vivo* phenotypes of somatic cells, they remain speculative.

4 Replication and repair helicases

The helicases described in this section have not been directly linked to homologous recombination. Nonetheless, either through mutant phenotypes or biochemical activity these helicases impact HR and function in a recombination-related repair pathway for avoidance of DNA damage and rearrangements during replication. The RecQ-related helicases seem to promote replication fork repair in a manner that avoids crossover formation while the HEF/Mph1-related helicases promote replication fork repair in a process mediated by HR factors that also prevents crossover formation. The remaining helicases in this section are proposed to antagonize homologous recombination at an early step.

4.1 Mph1

The *MPH1* gene was discovered as a mutant with an increased spontaneous mutation rate and sensitivity to a range of DNA damaging agents, including MMS, EMS, 4NQO and camptothecin, but excluding UV and ionizing radiation (Scheller et al. 2000). The null allele mutant had a slight sporulation defect but overall spore viability was not reduced. *MPH1* encodes a 3'-5' DNA helicase (Prakash et al. 2005). The mutant is not defective in homologous recombination, but epistasis studies of the mutator phenotype have placed *MPH1* in a homologous recombination pathway that functions in error-free bypass of DNA damage (Schurer et al. 2004). The double mutant *mph1*Δ *sgs1*Δ is hyper-rec, leading to the suggestion that Sgs1 prevents recombination events in a *mph1*Δ mutant (Schurer et al. 2004). The DNA damage sensitivity spectrum of the *mph1*Δ mutant suggests that Mph1 acts during recombination-mediated replication fork rescue.

4.2 HEF/FANCM

Pyrococcus furiosus has an endonuclease related to the XPF and Mus81 endonucleases, which also contains a DNA helicase domain. This protein has been called Hef for helicase-associated endonuclease for fork-structured DNA. As the name implies, Hef works on a flap structure, with the endonuclease acting on the DNA strand at the 5' side of a nick or flap (Komori et al. 2002). The helicase domain of the protein is most closely related to the *S. cerevisiae* Mph1 DNA helicase (Komori et al. 2002; Nishino et al. 2005) and is specific for forked-structured and HJ substrates (Komori et al. 2004). The inferred polarity is 3'-5', based on the activity with fork-structured substrates (Komori et al. 2004). The Hef protein has been proposed to act on stalled replication forks through the combined endonuclease and helicase activities. It is not known if Hef promotes replication fork repair through homologous recombination, but based on the homology with Mph1, the protein could act to promote error-free bypass by dissolving recombination intermediates.

Recently a vertebrate homolog of Hef has been identified (Mosedale et al. 2005). Hef disruption in DT40 cell results in sensitivity to DNA damaging agents, including crosslinking agents, and an increase in chromosome breakage. However, Hef knockout cells are proficient in DSB repair and do not have any apparent defect in homologous recombination, although sister chromatid exchange is reported to be elevated (Mosedale et al. 2005). As the phenotypes of the DT40 *Hef-/-* cells are similar to those of Fanconi anemia (FA) mutant cell lines, the Hef protein was tested for interactions with the FA complex. Hef was found to be part of the FA complex and was identified as FANCM (Meetei et al. 2005; Mosedale et al. 2005). However, DNA helicase activity has yet to be demonstrated for the FANCM protein.

4.3 BRIP1/BACH1/FANCJ

The BRIP1 or BACH1 helicase was identified by its association with BRCA1 (Cantor et al. 2001). The helicase has a 5'-3' polarity (Cantor et al. 2004). The dominant negative allele BACH1 K52R is defective in ATP hydrolysis and is delayed in the repair of gamma irradiation-induced damage, suggesting that BACH1 is required for repair of DSBs (Cantor et al. 2001). BACH1-deficient cells are defective in the repair of DSBs and are sensitive to crosslinking agents (Litman et al. 2005). BACH1 binds to HJs but cannot unwind them (Gupta et al. 2005; Litman et al. 2005), but is reported to be able to release the strand in a D-loop and have preference for a forked duplex substrate (Gupta et al. 2005). BACH1 has been proposed to displace Rad51 filament at a late stage in homologous recombination, to aid recombination through recycling Rad51 (Litman et al. 2005; Cantor and Andreassen 2006). Such action is proposed to coordinate homologous recombination with BRCA1. Recent studies have identified BACH1 as the FANCJ protein (Levitus et al. 2005; Levran et al. 2005; Litman et al. 2005). Since Rad51 foci are induced after DNA damage in BACH1/FA-J mutants, it has been suggested that the BACH1/FANCJ-mediated repair pathway does not involve homologous recombination (Godthelp et al. 2006). It remains to be seen if BACH1/FANCJ DNA helicase can disrupt Rad51 filament from a recombination intermediate.

4.4 HEL308/MUS308

The *Drosophila mus308* mutant is sensitive to DNA crosslinking agents (Boyd et al. 1990). The protein is a composite, with DNA helicase and DNA polymerase domains (Harris et al. 1996). The human gene, called HEL308, is homologous to the DNA helicase segment of *MUS308* and encodes a 3'-5' DNA helicase (Marini and Wood 2002). A similar protein exists in *Archea*, and the *Pyrococcus furiosus* protein, called Hjm, is a helicase that can branch migrate HJs (Fujikane et al. 2005). The archeal *Methanothermobacter thermautotrophicus* protein can unwind the lagging strand at a replication fork. This may assist translesion polymerases in access at the fork for repair synthesis across a template strand lesion. The same type of activity could unwind D-loops and have anti-recombination activity. Promotion of translesion polymerase repair at a stalled fork could operate in competition with homologous recombination. Thus promotion of translesion synthesis through the helicase action of HEL308 would effectively reduce homologous recombination at stalled replication forks (Guy and Bolt 2005).

4.5 RecQ5β

RecQ5β is the largest isoform of RECQ5, a eukaryotic member of the RecQ family of 3'-5' DNA helicases (Kitao et al. 1998; Shimamoto et al. 2000). In addition to a DNA helicase activity, the protein also has a strand annealing activity (Garcia et al. 2004; Machwe et al. 2005). These dual activities have also been observed in other RecQ-related proteins (Cheok et al. 2005; Machwe et al. 2005) and are suggested to aid in fork regression or HJ migration during replication fork repair to form the hemicatenane intermediate for HJ dissolution (Fig. 2).

4.6 RecQL1

RecQL1 is another member of the RecQ family. The helicase has a 3'-5' polarity (Cui et al. 2003) and can promote strand annealing (Sharma et al. 2005). The protein has been reported to branch migrate HJs (LeRoy et al. 2005). RNAi in HeLa cells results in an increase in sister chromatid exchange, suggesting that RecQL1 helicase can prevent a crossover class of recombinants through a HJ migration or dissolution activity. Knockouts of RECQL1 and RECQL5 in DT40 cells had no DNA damage sensitive phenotype, but when these knockouts were combined with a knockout of the BLM gene, the double mutants grew slower and produced dead cells, suggesting a partial overlap in function (Wang et al. 2003).

4.7 Hmi1

HMI1 is a nuclear gene that encodes a mitochondrial DNA helicase required for maintenance of mitochondrial genomes, but appears not to be the replicative helicase for mitochondrial DNA (Sedman et al. 2000; Monroe et al. 2005). It is a 3'-5' DNA helicase homologous to UvrD and Srs2. The Hmi1 helicase is most effective on a substrate with a 3' overhang, including 3' flap structures (Kuusk et al. 2005). *hmi1*Δ mutants have no nuclear recombination phenotype (Monroe et al. 2005), but the mitochondrial genome becomes fragmented (Sedman et al. 2005). It is possible that Hmi1 functions in replication fork restart during mitochondrial DNA replication and unwinds stalled replication forks to promote sister strand recombination.

5 Conclusions

The DNA helicases that act in mitosis to promote aspects of homologous recombination or to influence recombination outcomes are not essential to recombination *per se*. Mitotic homologous recombination occurs in response to spontaneous DNA damage, which has its most deleterious effects during DNA replication. Hence, much of the recombination occurs between sister chromatids, which is a genetically silent event. However, when recombination is not correctly regulated, DNA rearrangements such as deletions and translocations and more complex events occur.

The mitotic DNA helicases fall into four major groups. The UvrD related helicases, Srs2, Fbh1 and possibly Hmi1, act in an inhibitory way by destabilizing the Rad51 nucleofilament. The same activity can act at a later stage to aid in the later steps after strand invasion. Srs2 also promotes aspects of recombination by stabilizing a nascent joint particularly when short regions of homology are involved, by promoting long regions of heteroduplex and gene conversion tracts, and by promoting the SDSA reaction. Through the promotion of SDSA, Srs2 and perhaps Fbh1 prevent second end capture and the formation of crossover products. The formation of crossovers can lead to rearrangements, but also may be deleterious for the rescue of collapsed replication forks through recombination.

The second group of DNA helicases is related to RecQ and includes Sgs1, WRN, BLM and RecQL1 and RecQ5β. These helicases function in HJ dissolution and in their absence, there can be an increase in sister chromatid exchange, DNA rearrangements, X-structures at the replication fork, and crossovers. These helicases would act at a late stage in recombination; on recombination intermediates promoted by DNA damage in mitosis and on DSB promoted meiotic recombination. In mitosis damage at the replication fork in *sgs1Δ* mutants provokes the accumulation of recombination intermediates, showing that Sgs1 acts on these intermediates to resolve them as noncrossovers. In meiosis, the recombination intermediates that involve double HJs are acted upon by Sgs1 to prevent excess crossovers.

The last two classes of helicases have some relationship to homologous recombination but their function in processing recombination intermediates has not been demonstrated. Pif1 and Rrm3 form one group. Both helicases act on protein complexes, removing them to allow proper telomere maturation or DNA replication through nonstandard chromatin domains. Their malfunction probably induces the formation of recombinogenic substrates, which accounts for the recombination phenotypes seen in the deletion mutants. The other group is formed by the helicases related to the archeal Hef helicase, which include Mph1 and the putative helicase FANCM. The link of these helicases to homologous recombination is through the finding that the *mph1*Δ mutator phenotype requires homologous recombination factors. Mph1 helicase is thought to act on intermediates that form at stalled replication forks. It may act with Mus81 endonuclease to promote resolution of collapsed fork intermediates to the reconstructed fork structure (Fig. 3) or on D-loops with Mus81 to prevent crossover products from forming in an endonuclease-assisted SDSA type of reaction.

The pro-recombinogenic and anti-recombinogenic roles of DNA helicases seem concentrated on the early steps, the Rad51 nucleofilament, and the late steps, HJ resolution, although there are examples of helicase involvement in strand rejection and branch migration. New assays for steps of the homologous recombination reactions should lead to additional roles of DNA helicases.

Acknowledgements

I thank J. Borowiec and H. Karnam for comments on the manuscript. Work in the laboratory is supported by grants from the National Institutes of Health.

References

Aboussekhra A, Chanet R, Adjiri A, Fabre F (1992) Semidominant suppressors of Srs2 helicase mutations of *Saccharomyces cerevisiae* map in the RAD51 gene, whose sequence predicts a protein with similarities to procaryotic RecA proteins. Mol Cell Biol 12:3224-3234

Aboussekhra A, Chanet R, Zgaga Z, Cassier-Chauvat C, Heude M, Fabre F (1989) RADH, a gene of *Saccharomyces cerevisiae* encoding a putative DNA helicase involved in DNA repair. Characteristics of radH mutants and sequence of the gene. Nucleic Acids Res 17:7211-7219

Adams MD, McVey M, Sekelsky JJ (2003) *Drosophila* BLM in double-strand break repair by synthesis-dependent strand annealing. Science 299:265-267

Aguilera A, Klein HL (1988) Genetic control of intrachromosomal recombination in *Saccharomyces cerevisiae*. I. Isolation and genetic characterization of hyper-recombination mutations. Genetics 119:779-790

Ajima J, Umezu K, Maki H (2002) Elevated incidence of loss of heterozygosity (LOH) in an sgs1 mutant of *Saccharomyces cerevisiae*: roles of yeast RecQ helicase in suppression of aneuploidy, interchromosomal rearrangement, and the simultaneous incidence of both events during mitotic growth. Mutat Res 504:157-172

Aylon Y, Liefshitz B, Bitan-Banin G, Kupiec M (2003) Molecular dissection of mitotic recombination in the yeast *Saccharomyces cerevisiae*. Mol Cell Biol 23:1403-1417

Bailis AM, Maines S, Negritto MT (1995) The essential helicase gene RAD3 suppresses short-sequence recombination in *Saccharomyces cerevisiae*. Mol Cell Biol 15:3998-4008

Bastin-Shanower SA, Fricke WM, Mullen JR, Brill SJ (2003) The mechanism of Mus81-Mms4 cleavage site selection distinguishes it from the homologous endonuclease Rad1-Rad10. Mol Cell Biol 23:3487-3496

Bennett RJ, Sharp JA, Wang JC (1998) Purification and characterization of the Sgs1 DNA helicase activity of *Saccharomyces cerevisiae*. J Biol Chem 273:9644-9650

Bhattacharyya S, Lahue RS (2004) *Saccharomyces cerevisiae* Srs2 DNA helicase selectively blocks expansions of trinucleotide repeats. Mol Cell Biol 24:7324-7330

Bhattacharyya S, Lahue RS (2005) Srs2 helicase of *Saccharomyces cerevisiae* selectively unwinds triplet repeat DNA. J Biol Chem 280:33311-33317

Blackburn EH (2005) Telomeres and telomerase: their mechanisms of action and the effects of altering their functions. FEBS Lett 579:859-862

Blasco MA (2005) Mice with bad ends: mouse models for the study of telomeres and telomerase in cancer and aging. Embo J 24:1095-1103

Borner GV, Kleckner N, Hunter N (2004) Crossover/noncrossover differentiation, synaptonemal complex formation, and regulatory surveillance at the leptotene/zygotene transition of meiosis. Cell 117:29-45

Boule JB, Vega LR, Zakian VA (2005) The yeast Pif1p helicase removes telomerase from telomeric DNA. Nature 438:57-61

Boyd JB, Sakaguchi K, Harris PV (1990) mus308 mutants of *Drosophila* exhibit hypersensitivity to DNA cross-linking agents and are defective in a deoxyribonuclease. Genetics 125:813-819

Broomfield S, Xiao W (2002) Suppression of genetic defects within the RAD6 pathway by srs2 is specific for error-free post-replication repair but not for damage-induced mutagenesis. Nucleic Acids Res 30:732-739

Cantor S, Drapkin R, Zhang F, Lin Y, Han J, Pamidi S, Livingston DM (2004) The BRCA1-associated protein BACH1 is a DNA helicase targeted by clinically relevant inactivating mutations. Proc Natl Acad Sci USA 101:2357-2362

Cantor SB, Andreassen PR (2006) Assessing the Link between BACH1 and BRCA1 in the FA Pathway. Cell Cycle 5:164-167

Cantor SB, Bell DW, Ganesan S, Kass EM, Drapkin R, Grossman S, Wahrer DC, Sgroi DC, Lane WS, Haber DA, Livingston DM (2001) BACH1, a novel helicase-like protein, interacts directly with BRCA1 and contributes to its DNA repair function. Cell 105:149-160

Chanet R, Heude M, Adjiri A, Maloisel L, Fabre F (1996) Semidominant mutations in the yeast Rad51 protein and their relationships with the Srs2 helicase. Mol Cell Biol 16:4782-4789

Chen C, Zhang W, Timofejeva L, Gerardin Y, Ma H (2005) The Arabidopsis ROCK-N-ROLLERS gene encodes a homolog of the yeast ATP-dependent DNA helicase MER3 and is required for normal meiotic crossover formation. Plant J 43:321-334

Cheok CF, Wu L, Garcia PL, Janscak P, Hickson ID (2005) The Bloom's syndrome helicase promotes the annealing of complementary single-stranded DNA. Nucleic Acids Res 33:3932-3941

Constantinou A, Tarsounas M, Karow JK, Brosh RM, Bohr VA, Hickson ID, West SC (2000) Werner's syndrome protein (WRN) migrates Holliday junctions and co-localizes with RPA upon replication arrest. EMBO Rep 1:80-84

Cui S, Klima R, Ochem A, Arosio D, Falaschi A, Vindigni A (2003) Characterization of the DNA-unwinding activity of human RECQ1, a helicase specifically stimulated by human replication protein A. J Biol Chem 278:1424-1432

d'Adda di Fagagna F, Teo SH, Jackson SP (2004) Functional links between telomeres and proteins of the DNA-damage response. Genes Dev 18:1781-1799

Davis AP, Symington LS (2004) RAD51-dependent break-induced replication in yeast. Mol Cell Biol 24:2344-2351

Debrauwere H, Loeillet S, Lin W, Lopes J, Nicolas A (2001) Links between replication and recombination in *Saccharomyces cerevisiae*: a hypersensitive requirement for homologous recombination in the absence of Rad27 activity. Proc Natl Acad Sci USA 98:8263-8269

Doe CL, Whitby MC (2004) The involvement of Srs2 in post-replication repair and homologous recombination in fission yeast. Nucleic Acids Res 32:1480-1491

Ellis NA, Groden J, Ye TZ, Straughen J, Lennon DJ, Ciocci S, Proytcheva M, German J (1995) The Bloom's syndrome gene product is homologous to RecQ helicases. Cell 83:655-666

Fabre F, Chan A, Heyer WD, Gangloff S (2002) Alternate pathways involving Sgs1/Top3, Mus81/ Mms4, and Srs2 prevent formation of toxic recombination intermediates from single-stranded gaps created by DNA replication. Proc Natl Acad Sci USA 99:16887-16892

Foury F, Kolodynski J (1983) pif mutation blocks recombination between mitochondrial rho+ and rho- genomes having tandemly arrayed repeat units in *Saccharomyces cerevisiae*. Proc Natl Acad Sci U S A 80:5345-5349

Friedl AA, Liefshitz B, Steinlauf R, Kupiec M (2001) Deletion of the SRS2 gene suppresses elevated recombination and DNA damage sensitivity in rad5 and rad18 mutants of *Saccharomyces cerevisiae*. Mutat Res 486:137-146

Fujikane R, Komori K, Shinagawa H, Ishino Y (2005) Identification of a novel helicase activity unwinding branched DNAs from the hyperthermophilic archaeon, *Pyrococcus furiosus*. J Biol Chem 280:12351-12358

Gangloff S, McDonald JP, Bendixen C, Arthur L, Rothstein R (1994) The yeast type I topoisomerase Top3 interacts with Sgs1, a DNA helicase homolog: a potential eukaryotic reverse gyrase. Mol Cell Biol 14:8391-8398

Gangloff S, Soustelle C, Fabre F (2000) Homologous recombination is responsible for cell death in the absence of the Sgs1 and Srs2 helicases. Nat Genet 25:192-194

Garcia PL, Liu Y, Jiricny J, West SC, Janscak P (2004) Human RECQ5beta, a protein with DNA helicase and strand-annealing activities in a single polypeptide. Embo J 23:2882-2891

Garvik B, Carson M, Hartwell L (1995) Single-stranded DNA arising at telomeres in cdc13 mutants may constitute a specific signal for the RAD9 checkpoint. Mol Cell Biol 15:6128-6138

Gaymes TJ, North PS, Brady N, Hickson ID, Mufti GJ, Rassool FV (2002) Increased error-prone non homologous DNA end-joining--a proposed mechanism of chromosomal instability in Bloom's syndrome. Oncogene 21:2525-2533

German J (1993) Bloom syndrome: a mendelian prototype of somatic mutational disease. Medicine (Baltimore) 72:393-406

Godthelp BC, Wiegant WW, Waisfisz Q, Medhurst AL, Arwert F, Joenje H, Zdzienicka MZ (2006) Inducibility of nuclear Rad51 foci after DNA damage distinguishes all Fanconi anemia complementation groups from D1/BRCA2. Mutat Res 594:39-48

Goldfarb T, Alani E (2005) Distinct roles for the *Saccharomyces cerevisiae* mismatch repair proteins in heteroduplex rejection, mismatch repair and nonhomologous tail removal. Genetics 169:563-574

Golin JE, Esposito MS (1977) Evidence for joint genic control of spontaneous mutation and genetic recombination during mitosis in *Saccharomyces*. Mol Gen Genet 150:127-135

Gray MD, Shen JC, Kamath-Loeb AS, Blank A, Sopher BL, Martin GM, Oshima J, Loeb LA (1997) The Werner syndrome protein is a DNA helicase. Nat Genet 17:100-103

Gupta R, Sharma S, Sommers JA, Jin Z, Cantor SB, Brosh RM Jr (2005) Analysis of the DNA substrate specificity of the human BACH1 helicase associated with breast cancer. J Biol Chem 280:25450-25460

Guy CP, Bolt EL (2005) Archaeal Hel308 helicase targets replication forks *in vivo* and *in vitro* and unwinds lagging strands. Nucleic Acids Res 33:3678-3690

Hackett JA, Greider CW (2003) End resection initiates genomic instability in the absence of telomerase. Mol Cell Biol 23:8450-8461

Harmon FG, DiGate RJ, Kowalczykowski SC (1999) RecQ helicase and topoisomerase III comprise a novel DNA strand passage function: a conserved mechanism for control of DNA recombination. Mol Cell 3:611-620

Harrington L (2004) Those dam-aged telomeres! Curr Opin Genet Dev 14:22-28

Harris PV, Mazina OM, Leonhardt EA, Case RB, Boyd JB, Burtis KC (1996) Molecular cloning of *Drosophila* mus308, a gene involved in DNA cross-link repair with homology to prokaryotic DNA polymerase I genes. Mol Cell Biol 16:5764-5771

Hegde V, Klein H (2000) Requirement for the SRS2 DNA helicase gene in non-homologous end joining in yeast. Nucleic Acids Res 28:2779-2783

Heude M, Chanet R, Fabre F (1995) Regulation of the *Saccharomyces cerevisiae* Srs2 helicase during the mitotic cell cycle, meiosis and after irradiation. Mol Gen Genet 248:59-68

Heyer WD, Ehmsen KT, Solinger JA (2003) Holliday junctions in the eukaryotic nucleus: resolution in sight? Trends Biochem Sci 28:548-557

Hollingsworth NM, Brill SJ (2004) The Mus81 solution to resolution: generating meiotic crossovers without Holliday junctions. Genes Dev 18:117-125

Hu P, Beresten SF, van Brabant AJ, Ye TZ, Pandolfi PP, Johnson FB, Guarente L, Ellis NA (2001) Evidence for BLM and Topoisomerase IIIalpha interaction in genomic stability. Hum Mol Genet 10:1287-1298

Huang S, Li B, Gray MD, Oshima J, Mian IS, Campisi J (1998) The premature ageing syndrome protein, WRN, is a 3'-->5' exonuclease. Nat Genet 20:114-116

Imamura O, Fujita K, Itoh C, Takeda S, Furuichi Y, Matsumoto T (2002) Werner and Bloom helicases are involved in DNA repair in a complementary fashion. Oncogene 21:954-963

Ira G, Malkova A, Liberi G, Foiani M, Haber JE (2003) Srs2 and Sgs1-Top3 suppress crossovers during double-strand break repair in yeast. Cell 115:401-411

Ivessa AS, Lenzmeier BA, Bessler JB, Goudsouzian LK, Schnakenberg SL, Zakian VA (2003) The *Saccharomyces cerevisiae* helicase Rrm3p facilitates replication past non-histone protein-DNA complexes. Mol Cell 12:1525-1536

Ivessa AS, Zhou JQ, Schulz VP, Monson EK, Zakian VA (2002) *Saccharomyces* Rrm3p, a 5' to 3' DNA helicase that promotes replication fork progression through telomeric and subtelomeric DNA. Genes Dev 16:1383-1396

Ivessa AS, Zhou JQ, Zakian VA (2000) The *Saccharomyces* Pif1p DNA helicase and the highly related Rrm3p have opposite effects on replication fork progression in ribosomal DNA. Cell 100:479-489

Kaliraman V, Mullen JR, Fricke WM, Bastin-Shanower SA, Brill SJ (2001) Functional overlap between Sgs1-Top3 and the Mms4-Mus81 endonuclease. Genes Dev 15:2730-2740

Karow JK, Chakraverty RK, Hickson ID (1997) The Bloom's syndrome gene product is a 3'-5' DNA helicase. J Biol Chem 272:30611-30614

Karow JK, Constantinou A, Li JL, West SC, Hickson ID (2000) The Bloom's syndrome gene product promotes branch migration of holliday junctions. Proc Natl Acad Sci USA 97:6504-6508

Kass-Eisler A, Greider CW (2000) Recombination in telomere-length maintenance. Trends Biochem Sci 25:200-204

Kaytor MD, Nguyen M, Livingston DM (1995) The complexity of the interaction between RAD52 and SRS2. Genetics 140:1441-1442

Keil RL, McWilliams AD (1993) A gene with specific and global effects on recombination of sequences from tandemly repeated genes in *Saccharomyces cerevisiae*. Genetics 135:711-718

Kim J, Kim JH, Lee SH, Kim DH, Kang HY, Bae SH, Pan ZQ, Seo YS (2002) The novel human DNA helicase hFBH1 is an F-box protein. J Biol Chem 277:24530-24537

Kim JH, Kim J, Kim DH, Ryu GH, Bae SH, Seo YS (2004) SCFhFBH1 can act as helicase and E3 ubiquitin ligase. Nucleic Acids Res 32:2287-2297

Kitao S, Ohsugi I, Ichikawa K, Goto M, Furuichi Y, Shimamoto A (1998) Cloning of two new human helicase genes of the RecQ family: biological significance of multiple species in higher eukaryotes. Genomics 54:443-452

Klein HL (2001) Mutations in recombinational repair and in checkpoint control genes suppress the lethal combination of srs2Delta with other DNA repair genes in *Saccharomyces cerevisiae*. Genetics 157:557-565

Komori K, Fujikane R, Shinagawa H, Ishino Y (2002) Novel endonuclease in Archaea cleaving DNA with various branched structure. Genes Genet Syst 77:227-241

Komori K, Hidaka M, Horiuchi T, Fujikane R, Shinagawa H, Ishino Y (2004) Cooperation of the N-terminal helicase and C-terminal endonuclease activities of archaeal Hef protein in processing stalled replication forks. J Biol Chem 279:53175-53185

Kraus E, Leung WY, Haber JE (2001) Break-induced replication: a review and an example in budding yeast. Proc Natl Acad Sci USA 98:8255-8262

Krejci L, Macris M, Li Y, Van Komen S, Villemain J, Ellenberger T, Klein H, Sung P (2004) Role of ATP hydrolysis in the antirecombinase function of *Saccharomyces cerevisiae* Srs2 protein. J Biol Chem 279:23193-23199

Krejci L, Van Komen S, Li Y, Villemain J, Reddy MS, Klein H, Ellenberger T, Sung P (2003) DNA helicase Srs2 disrupts the Rad51 presynaptic filament. Nature 423:305-309

Kuusk S, Sedman T, Joers P, Sedman J (2005) Hmi1p from *Saccharomyces cerevisiae* mitochondria is a structure-specific DNA helicase. J Biol Chem 280:24322-24329

Lahaye A, Stahl H, Thines-Sempoux D, Foury F (1991) PIF1: a DNA helicase in yeast mitochondria. Embo J 10:997-1007

Langston LD, Symington LS (2005) Opposing roles for DNA structure-specific proteins Rad1, Msh2, Msh3, and Sgs1 in yeast gene targeting. Embo J 24:2214-2223

Lawrence CW, Christensen RB (1979) Metabolic suppressors of trimethoprim and ultraviolet light sensitivities of *Saccharomyces cerevisiae* rad6 mutants. J Bacteriol 139:866-876

Lee BS, Bi L, Garfinkel DJ, Bailis AM (2000) Nucleotide excision repair/TFIIH helicases RAD3 and SSL2 inhibit short-sequence recombination and Ty1 retrotransposition by similar mechanisms. Mol Cell Biol 20:2436-2445

LeRoy G, Carroll R, Kyin S, Seki M, Cole MD (2005) Identification of RecQL1 as a Holliday junction processing enzyme in human cell lines. Nucleic Acids Res 33:6251-6257

Levitus M, Waisfisz Q, Godthelp BC, de Vries Y, Hussain S, Wiegant WW, Elghalbzouri-Maghrani E, Steltenpool J, Rooimans MA, Pals G, Arwert F, Mathew CG, Zdzienicka MZ, Hiom K, De Winter JP, Joenje H (2005) The DNA helicase BRIP1 is defective in Fanconi anemia complementation group J. Nat Genet 37:934-935

Levran O, Attwooll C, Henry RT, Milton KL, Neveling K, Rio P, Batish SD, Kalb R, Velleuer E, Barral S, Ott J, Petrini J, Schindler D, Hanenberg H, Auerbach AD (2005) The BRCA1-interacting helicase BRIP1 is deficient in Fanconi anemia. Nat Genet 37:931-933

Lewis LK, Westmoreland JW, Resnick MA (1999) Repair of endonuclease-induced double-strand breaks in *Saccharomyces cerevisiae*: essential role for genes associated with nonhomologous end-joining. Genetics 152:1513-1529

Liberi G, Maffioletti G, Lucca C, Chiolo I, Baryshnikova A, Cotta-Ramusino C, Lopes M, Pellicioli A, Haber JE, Foiani M (2005) Rad51-dependent DNA structures accumulate at damaged replication forks in sgs1 mutants defective in the yeast ortholog of BLM RecQ helicase. Genes Dev 19:339-350

Liefshitz B, Steinlauf R, Friedl A, Eckardt-Schupp F, Kupiec M (1998) Genetic interactions between mutants of the 'error-prone' repair group of *Saccharomyces cerevisiae* and their effect on recombination and mutagenesis. Mutat Res 407:135-145

Litman R, Peng M, Jin Z, Zhang F, Zhang J, Powell S, Andreassen PR, Cantor SB (2005) BACH1 is critical for homologous recombination and appears to be the Fanconi anemia gene product FANCJ. Cancer Cell 8:255-265

Lundblad V, Blackburn EH (1993) An alternative pathway for yeast telomere maintenance rescues est1- senescence. Cell 73:347-360

Machwe A, Xiao L, Groden J, Matson SW, Orren DK (2005) RecQ family members combine strand pairing and unwinding activities to catalyze strand exchange. J Biol Chem 280:23397-23407

Maines S, Negritto MC, Wu X, Manthey GM, Bailis AM (1998) Novel mutations in the RAD3 and SSL1 genes perturb genome stability by stimulating recombination between short repeats in *Saccharomyces cerevisiae*. Genetics 150:963-976

Malkova A, Ivanov EL, Haber JE (1996) Double-strand break repair in the absence of RAD51 in yeast: a possible role for break-induced DNA replication. Proc Natl Acad Sci USA 93:7131-7136

Malone RE, Hoekstra MF (1984) Relationships between a hyper-rec mutation (REM1) and other recombination and repair genes in yeast. Genetics 107:33-48

Mankouri HW, Craig TJ, Morgan A (2002) SGS1 is a multicopy suppressor of srs2: functional overlap between DNA helicases. Nucleic Acids Res 30:1103-1113

Mankouri HW, Hickson ID (2004) Understanding the roles of RecQ helicases in the maintenance of genome integrity and suppression of tumorigenesis. Biochem Soc Trans 32:957-958

Marini F, Wood RD (2002) A human DNA helicase homologous to the DNA cross-link sensitivity protein Mus308. J Biol Chem 277:8716-8723

Maser RS, DePinho RA (2004) Telomeres and the DNA damage response: why the fox is guarding the henhouse. DNA Repair (Amst) 3:979-988

Mazina OM, Mazin AV, Nakagawa T, Kolodner RD, Kowalczykowski SC (2004) *Saccharomyces cerevisiae* Mer3 helicase stimulates 3'-5' heteroduplex extension by Rad51; implications for crossover control in meiotic recombination. Cell 117:47-56

McVey M, Larocque JR, Adams MD, Sekelsky JJ (2004) Formation of deletions during double-strand break repair in *Drosophila* DmBlm mutants occurs after strand invasion. Proc Natl Acad Sci USA 101:15694-15699

Meetei AR, Medhurst AL, Ling C, Xue Y, Singh TR, Bier P, Steltenpool J, Stone S, Dokal I, Mathew CG, Hoatlin M, Joenje H, de Winter JP, Wang W (2005) A human ortholog of archaeal DNA repair protein Hef is defective in Fanconi anemia complementation group M. Nat Genet 37:958-963

Mercier R, Jolivet S, Vezon D, Huppe E, Chelysheva L, Giovanni M, Nogue F, Doutriaux MP, Horlow C, Grelon M, Mezard C (2005) Two meiotic crossover classes cohabit in Arabidopsis: one is dependent on MER3,whereas the other one is not. Curr Biol 15:692-701

Milne GT, Ho T, Weaver DT (1995) Modulation of *Saccharomyces cerevisiae* DNA double-strand break repair by SRS2 and RAD51. Genetics 139:1189-1199

Miyajima A, Seki M, Onoda F, Shiratori M, Odagiri N, Ohta K, Kikuchi Y, Ohno Y, Enomoto T (2000) Sgs1 helicase activity is required for mitotic but apparently not for meiotic functions. Mol Cell Biol 20:6399-6409

Moens PB, Freire R, Tarsounas M, Spyropoulos B, Jackson SP (2000) Expression and nuclear localization of BLM, a chromosome stability protein mutated in Bloom's syndrome, suggest a role in recombination during meiotic prophase. J Cell Sci 113:663-672

Moens PB, Kolas NK, Tarsounas M, Marcon E, Cohen PE, Spyropoulos B (2002) The time course and chromosomal localization of recombination-related proteins at meiosis in the mouse are compatible with models that can resolve the early DNA-DNA interactions without reciprocal recombination. J Cell Sci 115:1611-1622

Mohaghegh P, Karow JK, Brosh RM Jr, Bohr VA, Hickson ID (2001) The Bloom's and Werner's syndrome proteins are DNA structure-specific helicases. Nucleic Acids Res 29:2843-2849

Monroe DS Jr, Leitzel AK, Klein HL, Matson SW (2005) Biochemical and genetic characterization of Hmi1p, a yeast DNA helicase involved in the maintenance of mitochondrial DNA. Yeast 22:1269-1286

Montelone BA, Hoekstra MF, Malone RE (1988) Spontaneous mitotic recombination in yeast: the hyper-recombinational rem1 mutations are alleles of the RAD3 gene. Genetics 119:289-301

Montelone BA, Malone RE (1994) Analysis of the rad3-101 and rad3-102 mutations of *Saccharomyces cerevisiae*: implications for structure/function of Rad3 protein. Yeast 10:13-27

Morishita T, Furukawa F, Sakaguchi C, Toda T, Carr AM, Iwasaki H, Shinagawa H (2005) Role of the *Schizosaccharomyces pombe* F-Box DNA helicase in processing recombination intermediates. Mol Cell Biol 25:8074-8083

Morrow DM, Connelly C, Hieter P (1997) "Break copy" duplication: a model for chromosome fragment formation in *Saccharomyces cerevisiae*. Genetics 147:371-382

Mosedale G, Niedzwiedz W, Alpi A, Perrina F, Pereira-Leal JB, Johnson M, Langevin F, Pace P, Patel KJ (2005) The vertebrate Hef ortholog is a component of the Fanconi anemia tumor-suppressor pathway. Nat Struct Mol Biol 12:763-771

Mullen JR, Kaliraman V, Brill SJ (2000) Bipartite structure of the SGS1 DNA helicase in *Saccharomyces cerevisiae*. Genetics 154:1101-1114

Myung K, Datta A, Chen C, Kolodner RD (2001) SGS1, the *Saccharomyces cerevisiae* homologue of BLM and WRN, suppresses genome instability and homeologous recombination. Nat Genet 27:113-116

Nakagawa T, Flores-Rozas H, Kolodner RD (2001) The MER3 helicase involved in meiotic crossing over is stimulated by single-stranded DNA-binding proteins and unwinds DNA in the 3' to 5' direction. J Biol Chem 276:31487-31493

Nakagawa T, Kolodner RD (2002a) The MER3 DNA helicase catalyzes the unwinding of holliday junctions. J Biol Chem 277:28019-28024

Nakagawa T, Kolodner RD (2002b) Saccharomyces cerevisiae Mer3 is a DNA helicase involved in meiotic crossing over. Mol Cell Biol 22:3281-3291

Nakagawa T, Ogawa H (1999) The *Saccharomyces cerevisiae* MER3 gene, encoding a novel helicase-like protein, is required for crossover control in meiosis. Embo J 18:5714-5723

Neff NF, Ellis NA, Ye TZ, Noonan J, Huang K, Sanz M, Proytcheva M (1999) The DNA helicase activity of BLM is necessary for the correction of the genomic instability of bloom syndrome cells. Mol Biol Cell 10:665-676

Nishino T, Komori K, Tsuchiya D, Ishino Y, Morikawa K (2005) Crystal structure and functional implications of *Pyrococcus furiosus* hef helicase domain involved in branched DNA processing. Structure (Camb) 13:143-153

Onoda F, Seki M, Miyajima A, Enomoto T (2000) Elevation of sister chromatid exchange in *Saccharomyces cerevisiae* sgs1 disruptants and the relevance of the disruptants as a system to evaluate mutations in Bloom's syndrome gene. Mutat Res 459:203-209

Ooi SL, Shoemaker DD, Boeke JD (2003) DNA helicase gene interaction network defined using synthetic lethality analyzed by microarray. Nat Genet 35:277-286

Orren DK, Theodore S, Machwe A (2002) The Werner syndrome helicase/exonuclease (WRN) disrupts and degrades D-loops *in vitro*. Biochemistry 41:13483-13488

Osman F, Dixon J, Barr AR, Whitby MC (2005) The F-Box DNA helicase Fbh1 prevents Rhp51-dependent recombination without mediator proteins. Mol Cell Biol 25:8084-8096

Osman F, Dixon J, Doe CL, Whitby MC (2003) Generating crossovers by resolution of nicked Holliday junctions: a role for Mus81-Eme1 in meiosis. Mol Cell 12:761-774

Paffett KS, Clikeman JA, Palmer S, Nickoloff JA (2005) Overexpression of Rad51 inhibits double-strand break-induced homologous recombination but does not affect gene conversion tract lengths. DNA Repair (Amst) 4:687-698

Palladino F, Klein HL (1992) Analysis of mitotic and meiotic defects in *Saccharomyces cerevisiae* SRS2 DNA helicase mutants. Genetics 132:23-37

Papouli E, Chen S, Davies AA, Huttner D, Krejci L, Sung P, Ulrich HD (2005) Crosstalk between SUMO and ubiquitin on PCNA is mediated by recruitment of the helicase Srs2p. Mol Cell 19:123-133

Paques F, Haber JE (1997) Two pathways for removal of nonhomologous DNA ends during double-strand break repair in *Saccharomyces cerevisiae*. Mol Cell Biol 17:6765-6771

Park JS, Choi E, Lee SH, Lee C, Seo YS (1997) A DNA helicase from *Schizosaccharomyces pombe* stimulated by single-stranded DNA-binding protein at low ATP concentration. J Biol Chem 272:18910-18919

Pfander B, Moldovan GL, Sacher M, Hoege C, Jentsch S (2005) SUMO-modified PCNA recruits Srs2 to prevent recombination during S phase. Nature 436:428-433

Pichierri P, Franchitto A, Mosesso P, Palitti F (2001) Werner's syndrome protein is required for correct recovery after replication arrest and DNA damage induced in S-phase of cell cycle. Mol Biol Cell 12:2412-2421

Prakash R, Krejci L, Van Komen S, Anke Schurer K, Kramer W, Sung P (2005) *Saccharomyces cerevisiae* MPH1 gene, required for homologous recombination-mediated mutation avoidance, encodes a 3' to 5' DNA helicase. J Biol Chem 280:7854-7860

Prince PR, Emond MJ, Monnat RJ Jr (2001) Loss of Werner syndrome protein function promotes aberrant mitotic recombination. Genes Dev 15:933-938

Rassool FV, North PS, Mufti GJ, Hickson ID (2003) Constitutive DNA damage is linked to DNA replication abnormalities in Bloom's syndrome cells. Oncogene 22:8749-8757

Rockmill B, Fung JC, Branda SS, Roeder GS (2003) The Sgs1 helicase regulates chromosome synapsis and meiotic crossing over. Curr Biol 13:1954-1962

Rong L, Klein HL (1993) Purification and characterization of the SRS2 DNA helicase of the yeast *Saccharomyces cerevisiae*. J Biol Chem 268:1252-1259

Rong L, Palladino F, Aguilera A, Klein HL (1991) The hyper-gene conversion hpr5-1 mutation of Saccharomyces cerevisiae is an allele of the SRS2/RADH gene. Genetics 127:75-85

Saintigny Y, Makienko K, Swanson C, Emond MJ, Monnat RJ Jr (2002) Homologous recombination resolution defect in werner syndrome. Mol Cell Biol 22:6971-6978

Scheller J, Schurer A, Rudolph C, Hettwer S, Kramer W (2000) MPH1, a yeast gene encoding a DEAH protein, plays a role in protection of the genome from spontaneous and chemically induced damage. Genetics 155:1069-1081

Schiestl RH, Prakash S, Prakash L (1990) The SRS2 suppressor of rad6 mutations of *Saccharomyces cerevisiae* acts by channeling DNA lesions into the RAD52 DNA repair pathway. Genetics 124:817-831

Schild D (1995) Suppression of a new allele of the yeast RAD52 gene by overexpression of RAD51, mutations in srs2 and ccr4, or mating-type heterozygosity. Genetics 140:115-127

Schmidt KH, Derry KL, Kolodner RD (2002) *Saccharomyces cerevisiae* RRM3, a 5' to 3' DNA helicase, physically interacts with proliferating cell nuclear antigen. J Biol Chem 277:45331-45337

Schmidt KH, Kolodner RD (2004) Requirement of Rrm3 helicase for repair of spontaneous DNA lesions in cells lacking Srs2 or Sgs1 helicase. Mol Cell Biol 24:3213-3226

Schulz VP, Zakian VA (1994) The *Saccharomyces* PIF1 DNA helicase inhibits telomere elongation and de novo telomere formation. Cell 76:145-155

Schurer KA, Rudolph C, Ulrich HD, Kramer W (2004) Yeast MPH1 gene functions in an error-free DNA damage bypass pathway that requires genes from homologous recombination, but not from postreplicative repair. Genetics 166:1673-1686

Sedman T, Joers P, Kuusk S, Sedman J (2005) Helicase Hmi1 stimulates the synthesis of concatemeric mitochondrial DNA molecules in yeast *Saccharomyces cerevisiae*. Curr Genet 47:213-222

Sedman T, Kuusk S, Kivi S, Sedman J (2000) A DNA helicase required for maintenance of the functional mitochondrial genome in *Saccharomyces cerevisiae*. Mol Cell Biol 20:1816-1824

Sharma S, Sommers JA, Choudhary S, Faulkner JK, Cui S, Andreoli L, Muzzolini L, Vindigni A, Brosh RM Jr (2005) Biochemical analysis of the DNA unwinding and strand annealing activities catalyzed by human RECQ1. J Biol Chem 280:28072-28084

Shen JC, Gray MD, Oshima J, Loeb LA (1998) Characterization of Werner syndrome protein DNA helicase activity: directionality, substrate dependence and stimulation by replication protein A. Nucleic Acids Res 26:2879-2885

Shimamoto A, Nishikawa K, Kitao S, Furuichi Y (2000) Human RecQ5beta, a large isomer of RecQ5 DNA helicase, localizes in the nucleoplasm and interacts with topoisomerases 3alpha and 3beta. Nucleic Acids Res 28:1647-1655

Shor E, Gangloff S, Wagner M, Weinstein J, Price G, Rothstein R (2002) Mutations in homologous recombination genes rescue top3 slow growth in *Saccharomyces cerevisiae*. Genetics 162:647-662

Stewart E, Chapman CR, Al-Khodairy F, Carr AM, Enoch T (1997) rqh1+, a fission yeast gene related to the Bloom's and Werner's syndrome genes, is required for reversible S phase arrest. Embo J 16:2682-2692

Sugawara N, Goldfarb T, Studamire B, Alani E, Haber JE (2004) Heteroduplex rejection during single-strand annealing requires Sgs1 helicase and mismatch repair proteins Msh2 and Msh6 but not Pms1. Proc Natl Acad Sci USA 101:9315-9320

Sung P, Prakash L, Matson SW, Prakash S (1987) RAD3 protein of *Saccharomyces cerevisiae* is a DNA helicase. Proc Natl Acad Sci USA 84:8951-8955

Suzuki K, Kato A, Sakuraba Y, Inoue H (2005) Srs2 and RecQ homologs cooperate in mei-3-mediated homologous recombination repair of *Neurospora crassa*. Nucleic Acids Res 33:1848-1858

Swanson C, Saintigny Y, Emond MJ, Monnat RJ, Jr (2004) The Werner syndrome protein has separable recombination and survival functions. DNA Repair (Amst) 3:475-482

Tarsounas M, West SC (2005) Recombination at mammalian telomeres: an alternative mechanism for telomere protection and elongation. Cell Cycle 4:672-674

Tong AH, Lesage G, Bader GD, Ding H, Xu H, Xin X, Young J, Berriz GF, Brost RL, Chang M, Chen Y, Cheng X, Chua G, Friesen H, Goldberg DS, Haynes J, Humphries C, He G, Hussein S, Ke L, Krogan N, Li Z, Levinson JN, Lu H, Menard P, Munyana C, Parsons AB, Ryan O, Tonikian R, Roberts T, Sdicu AM, Shapiro J, Sheikh B, Suter B, Wong SL, Zhang LV, Zhu H, Burd CG, Munro S, Sander C, Rine J, Greenblatt J, Peter M, Bretscher A, Bell G, Roth FP, Brown GW, Andrews B, Bussey H, Boone C (2004) Global mapping of the yeast genetic interaction network. Science 303:808-813

Torres JZ, Bessler JB, Zakian VA (2004a) Local chromatin structure at the ribosomal DNA causes replication fork pausing and genome instability in the absence of the *S. cerevisiae* DNA helicase Rrm3p. Genes Dev 18:498-503

Torres JZ, Schnakenberg SL, Zakian VA (2004b) *Saccharomyces cerevisiae* Rrm3p DNA helicase promotes genome integrity by preventing replication fork stalling: viability of rrm3 cells requires the intra-S-phase checkpoint and fork restart activities. Mol Cell Biol 24:3198-3212

Ui A, Seki M, Ogiwara H, Onodera R, Fukushige S, Onoda F, Enomoto T (2005) The ability of Sgs1 to interact with DNA topoisomerase III is essential for damage-induced recombination. DNA Repair (Amst) 4:191-201

Ulrich HD (2001) The srs2 suppressor of UV sensitivity acts specifically on the RAD5- and MMS2-dependent branch of the RAD6 pathway. Nucleic Acids Res 29:3487-3494

van Brabant AJ, Ye T, Sanz M, German IJ, Ellis NA, Holloman WK (2000) Binding and melting of D-loops by the Bloom syndrome helicase. Biochemistry 39:14617-14625

Van Komen S, Reddy MS, Krejci L, Klein H, Sung P (2003) ATPase and DNA helicase activities of the Saccharomyces cerevisiae anti-recombinase Srs2. J Biol Chem 278:44331-44337

Vaze MB, Pellicioli A, Lee SE, Ira G, Liberi G, Arbel-Eden A, Foiani M, Haber JE (2002) Recovery from checkpoint-mediated arrest after repair of a double-strand break requires Srs2 helicase. Mol Cell 10:373-385

Veaute X, Delmas S, Selva M, Jeusset J, Le Cam E, Matic I, Fabre F, Petit MA (2005) UvrD helicase, unlike Rep helicase, dismantles RecA nucleoprotein filaments in *Escherichia coli*. Embo J 24:180-189

Veaute X, Jeusset J, Soustelle C, Kowalczykowski SC, Le Cam E, Fabre F (2003) The Srs2 helicase prevents recombination by disrupting Rad51 nucleoprotein filaments. Nature 423:309-312

Walpita D, Plug AW, Neff NF, German J, Ashley T (1999) Bloom's syndrome protein, BLM, colocalizes with replication protein A in meiotic prophase nuclei of mammalian spermatocytes. Proc Natl Acad Sci USA 96:5622-5627

Wang SW, Goodwin A, Hickson ID, Norbury CJ (2001) Involvement of *Schizosaccharomyces pombe* Srs2 in cellular responses to DNA damage. Nucleic Acids Res 29:2963-2972

Wang W, Seki M, Narita Y, Nakagawa T, Yoshimura A, Otsuki M, Kawabe Y, Tada S, Yagi H, Ishii Y, Enomoto T (2003) Functional relation among RecQ family helicases RecQL1, RecQL5, and BLM in cell growth and sister chromatid exchange formation. Mol Cell Biol 23:3527-3535

Watt PM, Hickson ID, Borts RH, Louis EJ (1996) SGS1, a homologue of the Bloom's and Werner's syndrome genes, is required for maintenance of genome stability in *Saccharomyces cerevisiae*. Genetics 144:935-945

Watt PM, Louis EJ, Borts RH, Hickson ID (1995) Sgs1: a eukaryotic homolog of *E. coli* RecQ that interacts with topoisomerase II *in vivo* and is required for faithful chromosome segregation. Cell 81:253-260

Whitby MC (2005) Making crossovers during meiosis. Biochem Soc Trans 33:1451-1455

Wilson TE (2002) A genomics-based screen for yeast mutants with an altered recombination/end-joining repair ratio. Genetics 162:677-688

Wu L, Davies SL, Levitt NC, Hickson ID (2001) Potential role for the BLM helicase in recombinational repair via a conserved interaction with RAD51. J Biol Chem 276:19375-19381

Wu L, Davies SL, North PS, Goulaouic H, Riou JF, Turley H, Gatter KC, Hickson ID (2000) The Bloom's syndrome gene product interacts with topoisomerase III. J Biol Chem 275:9636-9644

Wu L, Hickson ID (2002) RecQ helicases and cellular responses to DNA damage. Mutat Res 509:35-47

Wu L, Hickson ID (2003) The Bloom's syndrome helicase suppresses crossing over during homologous recombination. Nature 426:870-874

Yamagata K, Kato J, Shimamoto A, Goto M, Furuichi Y, Ikeda H (1998) Bloom's and Werner's syndrome genes suppress hyperrecombination in yeast sgs1 mutant: implication for genomic instability in human diseases. Proc Natl Acad Sci USA 95:8733-8738

Yu CE, Oshima J, Fu YH, Wijsman EM, Hisama F, Alisch R, Matthews S, Nakura J, Miki T, Ouais S, Martin GM, Mulligan J, Schellenberg GD (1996) Positional cloning of the Werner's syndrome gene. Science 272:258-262

Zhou J, Monson EK, Teng SC, Schulz VP, Zakian VA (2000) Pif1p helicase, a catalytic inhibitor of telomerase in yeast. Science 289:771-774

Klein, Hannah L
Department of Biochemistry, New York University School of Medicine, 550 First Avenue, New York, NY 10016, USA
hannah.klein@med.nyu.edu

Holliday junction resolution

Matthew C. Whitby

Abstract

The Holliday junction (HJ) is a key intermediate in homologous recombination. It can be formed both as a consequence of RecA/Rad51-catalysed strand invasion reactions, and by the reversal of impeded replication forks. HJs constitute physical connections between two DNA duplexes and therefore they have to be removed to enable DNA segregation during cell division. Nucleases, DNA helicases and topoisomerases are variously employed in different strategies of HJ removal, which in turn can influence whether a crossover or non-crossover recombinant DNA molecule is generated. This review will focus on two main types of endonuclease, the HJ resolvases and the Mus81 nucleases, which process HJs and/or related intermediates (e.g. forks, D-loops and nicked HJs). It will describe what is currently known about their mechanisms of action and biological functions.

1 A brief overview of HJ formation and processing

The classic view of how homologous recombination works is proposed in the double strand break (DSB) repair model (Fig. 1A) (Szostak et al. 1983). This describes how a 3'-ended single-strand tail, generated by the resection of a broken DNA end, invades a homologous duplex to form a displacement (D-) loop. The 3'-end then primes new DNA synthesis, which extends the D-loop enabling it to anneal to the other resected end of the break. Further DNA synthesis, and the ligation of strand discontinuities, ultimately results in the formation of a double HJ (dHJ). In the DSB repair model the dHJ is resolved by an endonuclease cleaving a pair of symmetrical strands at each junction, with the relative orientation of cleavage determining whether crossover or non-crossover recombinant DNA molecules are generated (Fig. 1A).

The symmetry of the dHJ means that its resolution should generate crossover and non-crossover recombinants with equal frequency. However, this is not always the case, as the orientation of dHJ resolution is often biased. In *Escherichia coli*, and during meiosis in budding yeast, this bias is in favour of crossovers (Cromie and Leach 2000; Allers and Lichten 2001; Borner et al. 2004). This results in the formation of chromosome dimers in *E. coli*, which would be problematic for DNA segregation without the XerCD site-specific recombinases that efficiently convert chromosome dimers back to monomers (Sherratt et al. 2004). In

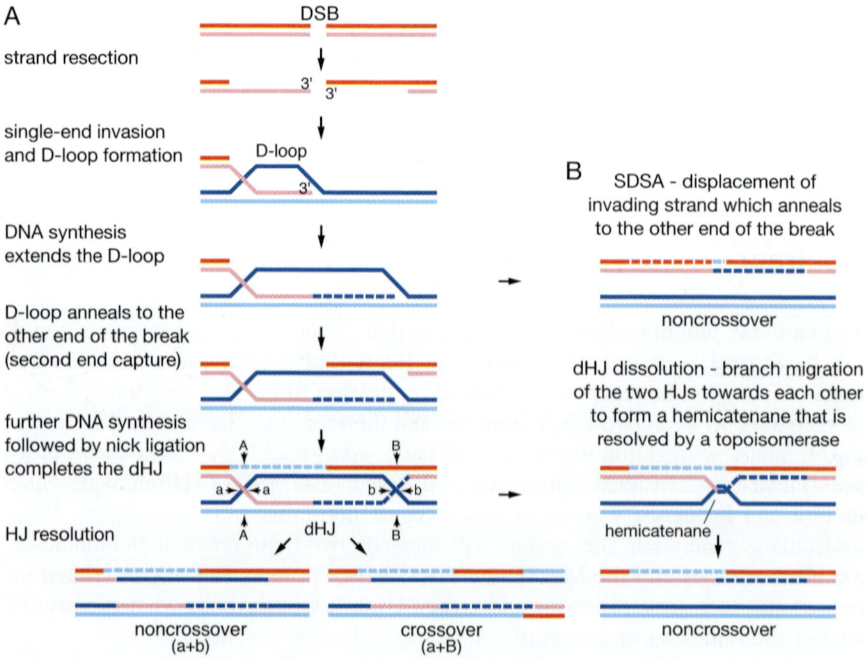

Fig. 1. Models of DSB repair by homologous recombination.

contrast the formation of crossovers during meiosis is advantageous since it promotes both genetic variability and the establishment of chiasmata - the latter being necessary for directing the correct segregation of homologous chromosomes during meiosis I (Hunter N, this volume).

In contrast to the beneficial effects of crossing over in meiosis, crossing over in vegetative cells is potentially risky. This is due in varying measure to the presence of a homologous chromosome, the abundance of repetitive DNA elements, and the failure of recombination to always act precisely between sister chromatids. Consequently, both allelic recombination and ectopic recombination occur at significant frequencies, with the risk that crossing over could result in the loss of heterozygosity and/or gross chromosome rearrangements, and subsequent disease and death. For this reason it appears that dHJs are often not made, with recombination proceeding via a mechanism such as synthesis-dependent strand annealing (SDSA) (Fig. 1B). In situations where dHJs are formed, their endonucleolytic processing seems to be curbed in favour of dissolution by a RecQ family DNA helicase acting together with topoisomerase III (Top3) (Fig. 1B) (Ira et al. 2003; Wu and Hickson 2003; Klein H, this volume).

In addition to classic DSB repair HJs may also form during the repair of one-sided DSBs (Fig. 2A). One-sided breaks are formed when a replication fork runs into a single-strand break in the DNA template (Kuzminov 2001). Strand invasion here is thought to promote reformation of the replication fork together with a single HJ (McGlynn and Lloyd 2002). It would seem that such a HJ would need to be

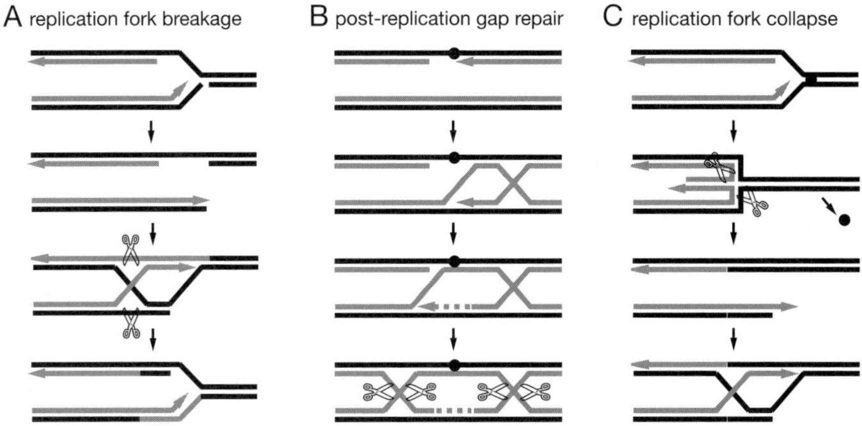

Fig. 2. Models for how HJs may be formed following replication fork breakage, blockage and collapse. Arrowheads indicate 3' ends, and solid circles indicate replication fork barriers.

removed by nucleolytic cleavage, the only alternative being to branch migrate it behind the replication fork. Here it would dissociate at strand breaks present at the sites where opposing forks merge.

Replication fork blockage can also promote HJ formation. There are a number of ways in which this may happen, and details of which mechanisms are used are still being worked out (McGlynn and Lloyd 2002). One way is for recombinases to load at single-stranded gaps exposed at blocked forks. This may be accidental or serve a purpose in protecting the fork, or in promoting template switching (Higgins et al. 1976; Courcelle et al. 1997). Alternatively, the leading or lagging strand may skip a block in one of the template strands, resulting in the formation of a daughter strand gap at which recombination could be initiated to aid repair post-replication (Fig. 2B) (Rupp et al. 1971; Lopes et al. 2006). Either of the above processes could give rise to a single or double HJ, which could be processed by nucleolytic cleavage, dissolution, or branch migration to existing strand breaks.

Impeding a replication fork may also result in the dissociation of replisome components, and the active or passive reversal of the fork (McGlynn and Lloyd 2002). Fork reversal involves the rewinding of parental strands and annealing of nascent strands to form a HJ-like structure (Fig. 2C). It has been speculated that fork reversal may provide room for lesion repair and/or a mechanism for lesion bypass by template switching (Higgins et al. 1976; Courcelle et al. 2003). However, in eukaryotes it is considered to be mainly a pathological event that is guarded against by checkpoint proteins (Lopes et al. 2001; Sogo et al. 2002). The removal of such HJs can be by branch migration, degradation of the spooled out junction arm, or nucleolytic cleavage (McGlynn and Lloyd 2002). In bacteria the cleavage of a reversed fork by a HJ resolvase generates a one-sided DSB that provokes recombination, which in turn forms a D-loop at which replisome components can reassemble (Fig. 2C) (McGlynn and Lloyd 2002). Here HJ resolution is acting to initiate recombination rather than to complete it.

2. The HJ resolvases

2.1 Structural relationships

The HJ resolvases are a major class of endonuclease that process HJs. They are typically small, homodimeric, metal ion-dependent endonucleases that bind with structure-specificity to the HJ and introduce a pair of symmetrically placed incisions in strands of opposite polarity, at or close to the junction crossover point. The products of this reaction are two nicked duplex species - the nicks displaying 5' phosphate and 3' hydroxyl termini making them directly repairable by DNA ligase.

HJ resolvases are ubiquitous being found in all domains of life. Despite this they show very little primary sequence similarity, and it was not until the X-ray crystal structures of a number of HJ resolvases had been solved that evolutionary relationships were fully appreciated (Fig. 3). We now realise that the majority of known HJ resolvases herald either from the PD...D/EXK nuclease or Integrase superfamilies. Exceptions include T4 endonuclease VII, which is a member of a heterogeneous group of nucleases that are characterised by a $\beta\beta\alpha$-Me-finger fold (Saravanan et al. 2004), and RusA that has no known relatives (Rafferty et al. 2003).

Bacterial RuvC and yeast mitochondrial Cce1 and Ydc2 are examples of HJ resolvases from the Integrase superfamily, and therefore relatives of RNase H1, HIV integrase, and Mu transposase. Both RuvC and Ydc2 exhibit similar overall folds, with each monomer consisting of a five-stranded β-sheet flanked by α helices (Ariyoshi et al. 1994; Ceschini et al. 2001). Each monomer also contains the four conserved acidic residues that form the catalytic sites in Integrase family members. The archaeal HJ resolvases Hjc and Hje, phage T7 endonuclease I, and RecU from *Bacillus subtilis* each contain the conserved catalytic domain that marks them out as members of the Nuclease superfamily, and therefore relatives of type II restriction endonucleases, λ-exonuclease, and MutH (Bond et al. 2001; Hadden et al. 2001; Nishino et al. 2001; Middleton et al. 2004; McGregor et al. 2005). Like the HJ resolvases from the Integrase superfamily, these enzymes exhibit an α/β architecture, albeit the topology of the β-sheets is quite different. Such structural similarities and differences might suggest that HJ resolvases from different superfamilies have evolved convergently (Venclovas and Siksnys 1995). However, a similarity in the metal-ion-binding loop (consisting of two parallel β-strands and one α-helix) in RuvC and Hje have lead to speculation that both Integrase and Nuclease superfamilies derive from the same ancestral divalent metal-/phosphate-binding domain (Lilley and White 2001).

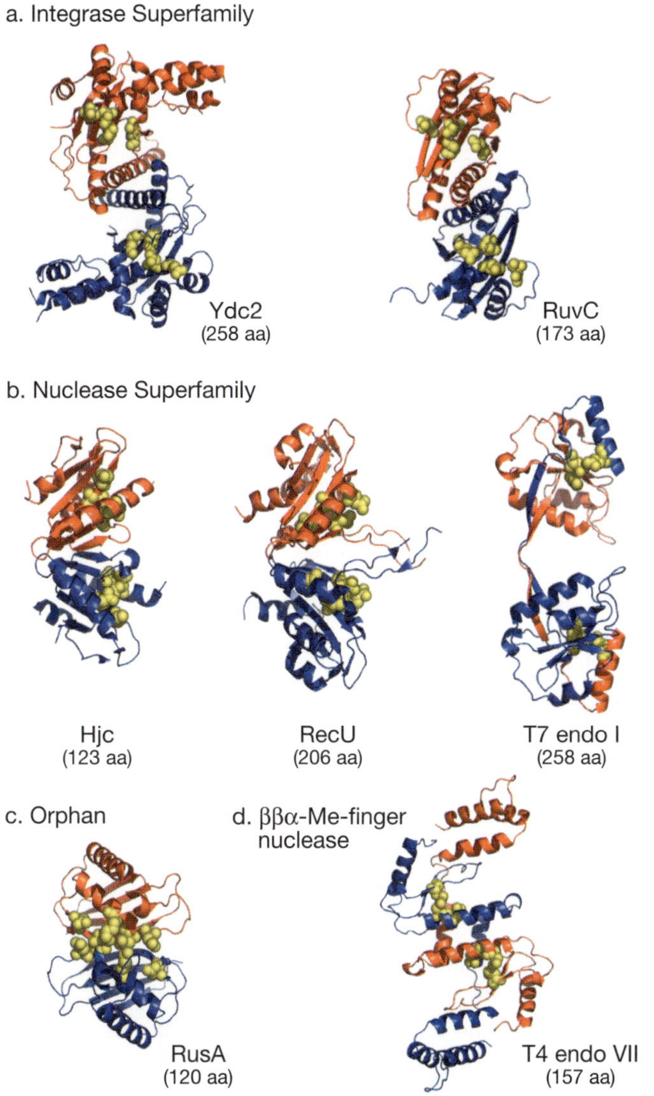

Fig. 3. Crystal structures of HJ resolvases. The structures are of *S. pombe* Ydc2 (Ceschini et al. 2001), *E. coli* RuvC (Ariyoshi et al. 1994), *Pyrococcus furiosus* Hjc (Nishino et al. 2001), *Bacillus subtilis* RecU (McGregor et al. 2005), Phage T7 endonuclease I (Hadden et al. 2001), *E. coli* RusA (Rafferty et al. 2003), and Phage T4 endonuclease VII (Raaijmakers et al. 1999). Note that crystal structures have also been obtained for Hjc and Hje from *Sulfolobus solfataricus* (Bond et al. 2001; Middleton et al. 2004). Images were generated from coordinates in the RCSB Protein Data Bank using VMD software, and are of the dimer forms of these enzymes, with individual subunits coloured in blue or red and catalytic residues in yellow. The monomer size in numbers of amino acids is given for each enzyme in parentheses.

2.2 Junction recognition and distortion

One of the main characteristics of the HJ resolvases is their ability to bind with a high degree of structure-selectivity to four-way DNA junctions. Typically a resolvase-junction complex can resist displacement by a 1000-fold excess of linear double stranded DNA of the same nucleotide sequence (e.g. Bennett et al. 1993). This does not mean that binding is necessarily exclusive to HJs, as many resolvases will bind, albeit with varying affinity, to other branched and distorted DNA structures. For some, especially the phage resolvases, this less-specific binding is accompanied by cleavage consistent with their ability to act as general debranching enzymes *in vivo* (Kemper 1997). However, for others, like RuvC and RusA, cleavage appears to be restricted to HJs (Benson and West 1994; Takahagi et al. 1994; Bolt and Lloyd 2002). It is thought that this exclusivity is due, at least in part, to sequence-specific cleavage (see Section 2.3).

To date no one has succeeded in determining the X-ray crystal structure of a HJ resolvase bound to its substrate. Therefore there is no detailed knowledge of the protein-DNA interactions that occur during HJ resolution. Nevertheless, structural determinations of HJs and resolvases in isolation, together with information from gel retardation and chemical probing experiments of protein-DNA interactions, have enabled tenable models for active resolvase-junction complexes to be proposed.

In solution HJs adopt two main conformations, and the equilibrium between these conformations depends on the prevailing concentration of divalent cations (Fig. 4A) (reviewed in Lilley 2000). In the absence of cations an unfolded structure is favoured, where the four arms of the HJ open out into an approximately four-fold symmetric square-planar configuration. Whereas in the presence of divalent cations the negative charges on the arms of the junction are neutralized, enabling them to stack pairwise. This results in the so-called stacked X configuration, which contains two continuous strands in an anti-parallel arrangement, and two discontinuous strands that exchange between the two pairs of stacked arms at the junction crossover point. It also exhibits two faces – one with major groove characteristics at the point of strand exchange, and the other with minor groove characteristics.

Generally recognition and selective binding to a HJ appears to depend on interactions over a relatively large surface area in order that the overall structure of the HJ is 'seen' by the resolvase. Each resolvase has its own favoured junction conformation that it binds to (see Fig. 4B for examples). Some prefer the open conformation, whereas others interact with a particular face of the stacked X structure. However, the ability to bind is not imposed by the conformation that the junction happens to be in, as most resolvases are capable of distorting the HJ into their own favoured conformation, which can be quite distinct from the open and stacked X forms, and often involves a degree of base-pair disruption at the junction crossover point (Bennett and West 1995b; White and Lilley 1997b, 1998; Chan et al. 1998; Declais and Lilley 2000; Kvaratskhelia et al. 2000; Declais et al. 2003; McGregor et al. 2005). One exception here is Rap from Phage λ, which does not

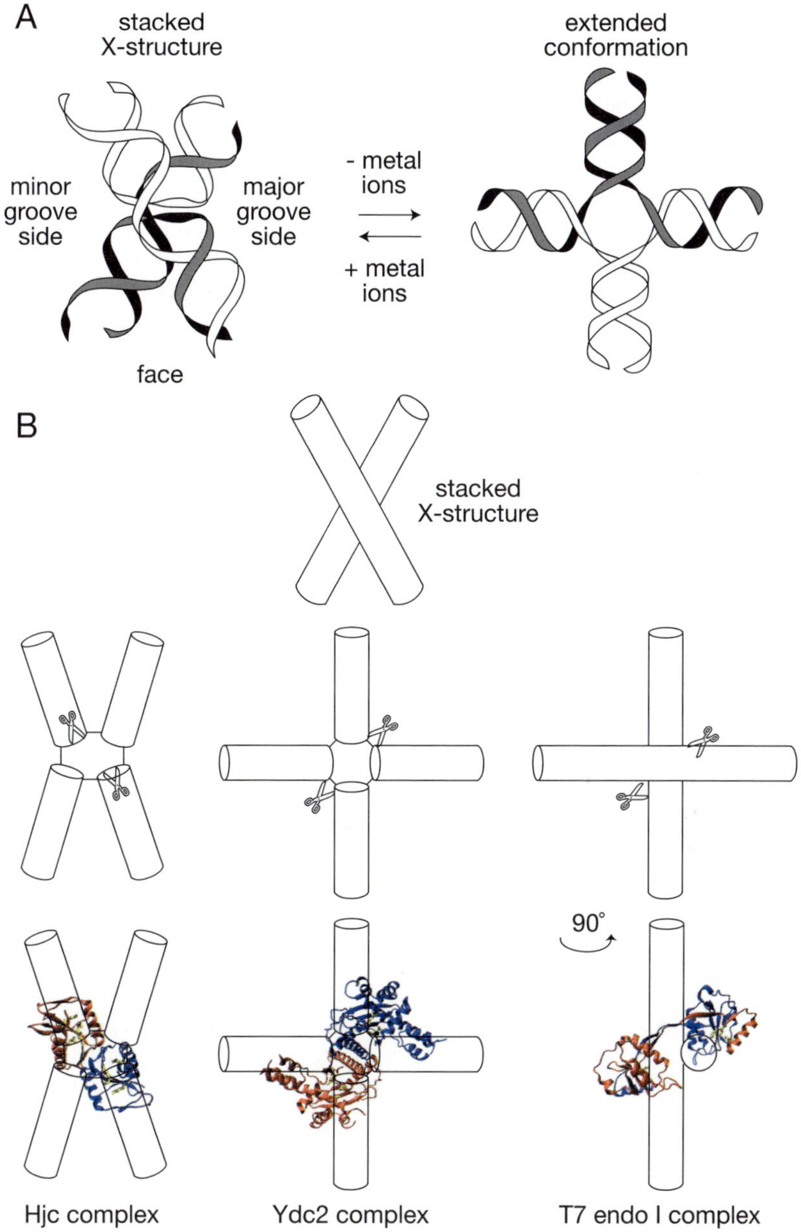

Fig. 4. The conformations of free and bound HJs. (A) Cartoons of the conformations adopted by HJs in the presence and absence of divalent cations. (B) Models of three different HJ-resolvase complexes, in each case showing how the enzyme is thought to dock with the HJ and distort its conformation away from the stacked X-structure (Bond et al. 2001; Ceschini et al. 2001; Nishino et al. 2001; Declais et al. 2003). The images of the HJ resolvases are adapted from Figure 3. Scissors indicate the position of cleavage sites.

appear to affect global HJ conformation (Sharples et al. 2004). In most cases junction distortion probably helps to orientate the scissile bonds into the active sites of the resolvase. To achieve these interactions resolvases are typically highly basic proteins (pI >8.5) with large areas of positive charge on their surfaces, which presumably enable large-scale DNA contact. Good examples of this are RuvC, Ydc2, Hjc and T4 endonuclease VII, which each exhibit an extensively flat (in the case of T4 endonuclease VII it is slightly concave), positively charged dimer face that is presumed to interact with the HJ (Ariyoshi et al. 1994; Raaijmakers et al. 1999; Bond et al. 2001; Ceschini et al. 2001; Nishino et al. 2001). Intriguingly the positive charges on the faces of Ydc2, Hjc, and T4 endonuclease VII dimers each form an 'S' shape with the two active sites being located within the elbows of the 'S'. These enzymes share no structural homology and distort the junction in different ways. Nevertheless, it would seem that an S-shape of charge provides a useful strategy for interacting with both stacked and open forms of the HJ. Another recurring feature amongst HJ resolvases (e.g. Hjc, Ydc2 and RecU) is the presence of extended or flexible loops and/or α helices at or close to the dimer interface, which in some cases form a protrusion from the surface of the protein (Bond et al. 2001; Ceschini et al. 2001; Nishino et al. 2001; McGregor et al. 2005). It is presumed that these structures interact with the centre of the HJ, and variously play roles in stabilizing disrupted base-pairs, substrate and/or sequence recognition, and communication between active sites.

2.3 Sequence-specific cleavage and the need for branch migration

For some HJ resolvases (e.g. T4 endonuclease VII and T7 endonuclease I) binding and distortion of the junction is sufficient for cleavage. However, for others (e.g. RuvC, RusA, Cce1, and Ydc2) cleavage is only performed at specific nucleotide sequences (Shah et al. 1994; Chan et al. 1997; Schofield et al. 1998; Whitby and Dixon 1997; White and Lilley 1997a; Oram et al. 1998; Fogg et al. 1999; Nishino et al. 2001). The sequence requirements are typically not that demanding, for example RuvC cleaves the consensus $5'\text{-}{}^{A}/_{T}TT\downarrow{}^{G}/_{C}\text{-}3'$ (↓ is the cleavage site) (Shah et al. 1994; Fogg et al. 1999). Nevertheless it is thought that this degree of sequence-specificity is sufficient to endow a HJ resolvase with an added level of substrate selectivity (Lilley and White 2001). This is due to the fact that HJs, unlike other branched and distorted DNAs, can relocate their crossover point to different sequences by undergoing branch migration; i.e. if the resolvase is unable to cleave a particular HJ on its initial encounter, it may be able to later after the junction has had a chance to move.

In vitro studies on the kinetics of branch migration have revealed a 1000-fold greater rate of movement in the absence of metal ions, i.e. under conditions where the junction is in an open configuration (Panyutin and Hsieh 1994; Panyutin et al. 1995). Recent single-molecule studies have further refined this view by revealing that the junction flips between open and stacked forms, but only moves (in steps that can be several base pairs) when in its open form (Karymov et al. 2005). The rate of branch migration is therefore dictated by the equilibrium between the two forms, which is in turn influenced by the prevailing concentration of divalent metal ions. Branch migration may also be affected by a HJ resolvase holding the junction in either its stacked or open conformation. One example is Ydc2, which unfolds the junction into its fully open configuration (Fig. 4B) (White and Lilley 1998). It has been suggested that, by holding the junction in its open form, Ydc2 promotes a degree of branch migration that allows short range scanning for its preferred cleavage sequence (after $5'\text{-}{}^{C}/_{T}T$) (Ceschini et al. 2001). A SAP DNA binding motif at the N-terminus of Ydc2 appears to be important for maintaining a sufficiently stable protein-junction interaction for this limited branch migration to occur (Ahn and Whitby 2003).

For some HJ resolvases (e.g. RuvC and RecU) branch migration is so important for their ability to function *in vivo* that they need to work together with a dedicated branch migration enzyme. In *E. coli* RuvA and RuvB are the branch migration enzymes that work with RuvC (reviewed in Yamada et al. 2004). The HJ is bound by two tetramers of RuvA, which hold it in an open square planar configuration and direct the assembly of hexameric rings of RuvB protein onto diametrically opposed junction arms. RuvB contains the characteristic motifs of a DNA helicase and acts as an ATP-dependent pump that drives branch migration. Efficient branch migration depends on the HJ being sandwiched between the two RuvA tetramers, which act as a stator for the process (Privezentzev et al. 2005). The precise manner in which RuvC works with RuvAB is still uncertain. It is known that the octamer of RuvA shields the HJ from RuvC, and that there must therefore be an interchange between at least one tetramer of RuvA and a dimer of RuvC (Whitby et al. 1996; Dickman et al. 2002). Gel retardation and surface plasmon resonance experiments have indicated that a tetramer of RuvA and a dimer of RuvC can sandwich a HJ *in vitro*, and co-immunoprecipitation experiments have confirmed the existence of a RuvABC complex on HJs (Whitby et al. 1996; Davies and West 1998; Dickman et al. 2002). Such data, together with observations that RuvAB can promote and direct RuvC cleavage *in vitro*, have lead to a model in which RuvABC form a functional 'resolvasome' on junction DNA, which enables RuvC to scan for cleavable sequences whilst the HJ is being branch migrated (Whitby et al. 1996; Eggleston et al. 1997; van Gool et al. 1998; Zerbib et al. 1998; van Gool et al. 1999) (Fig. 5).

2.4 The catalysis of cleavage

HJ resolvases catalyse the cleavage of the scissile phosphodiester bonds of a HJ by a hydrolysis reaction. All known HJ resolvases exhibit a cluster of aspartic and/or glutamic acid residues in each subunit, which constitute their active sites (Lilley and White 2001; Sharples 2001). Depending on the heritage of the resolvase, these critical catalytic residues are generally conserved in either the Nuclease or Integrase superfamilies (exceptions being T4 endonuclease VII and RusA). The carboxylate side chains of the catalytic acidic residues bind metal ions, which are hydrated to provide the water for hydrolysis. The general scheme of

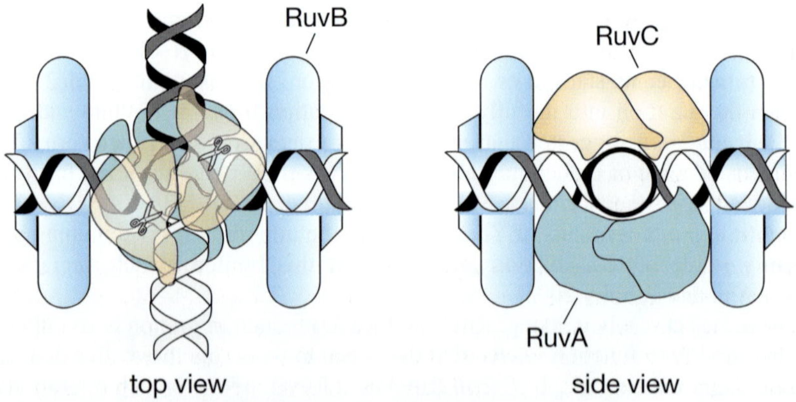

Fig. 5. Cartoon of the hypothetical RuvABC resolvasome bound to a HJ.

phosphodiester bond hydrolysis follows an S_N2-like mechanism in which the nucleophile (water) is deprotonated and attacks the hydroxide ion on the scissile phosphorous, resulting in the formation of a pentavalent intermediate. The reaction then terminates with the departure of the 3'-oxyanion. The precise catalysis of these steps has not been worked out for any HJ resolvase. However, knowledge of the catalysis performed by other Integrase and PD...D/EXK nuclease superfamily members enables extrapolations to be made (Nowotny et al. 2005; Pingoud et al. 2005). For example, we can predict that RuvC, Cce1 and Ydc2 coordinate two metal ions in each of their active sites (Nowotny et al. 2005). One metal ion would act as a base to deprotonate the water molecule, whereas both would be required to stabilize the negatively charged pentavalent transition state. The second metal ion may also coordinate a water molecule to protonate the leaving 3'-oxyanion group. HJ resolvases that are members of the Nuclease superfamily probably catalyse hydrolysis by a similar two-metal ion mechanism, possibly using the primary amine of a conserved lysine residue as a Lewis acid to help stabilize the transition state (Hadden et al. 2002; Pingoud et al. 2005). With regards to RusA, the presence of several catalytically important aspartates and a lysine residue suggests that a mechanism similar to members of the Nuclease superfamily may be used (Rafferty et al. 2003). In contrast, T4 endonuclease VII appears to use a histidine as a general base to deprotonate a water molecule, and a bound divalent metal ion to act as the Lewis acid for stabilizing the transition state (Giraud-Panis and Lilley 1996; Raaijmakers et al. 1999).

2.5 Coordination of cleavage events

HJ resolvases catalyse junction resolution by two separate cleavage events. Coordination of these events relies on the fact that resolvases function as homodimers. In some cases (e.g. T4 endonuclease VII, T7 endonuclease I and RusA) the dependence on dimer formation for activity is enforced by extensive domain swapping and/or the formation of composite active sites (Raaijmakers et al. 1999; Hadden et al. 2001; Rafferty et al. 2003). However, each active site within a resolvase dimer functions independently (Giraud-Panis and Lilley 1997; Shah et al. 1997; Birkenbihl and Kemper 1998; Fogg et al. 2000; Guan and Kumar 2005). This creates the potential problem of junction nicking, instead of productive resolution, especially if the junction branch migrates between cleavage events – a problem that would be particularly acute for resolvases like RuvC and Ydc2 that depend on branch migration to locate sequences that they can cleave. HJ resolvases promote proper junction resolution by catalysing a sufficiently fast rate of cleavage to ensure that both incisions are made within the lifetime of a single junction-resolvase complex. In the case of Hje both subunits function with equally fast kinetics (Parker and White 2005), whereas in other resolvases, like RuvC and Cce1, the second incision is made at a much faster rate than the first (Fogg and Lilley 2000; Fogg et al. 2000). The acceleration of second strand cleavage has been attributed to the nicked junction being more flexible, enabling faster docking of the scissile phosphate into its active site (Lilley and White 2001). However, at least in the case of RuvC, the nick needs to have a 5' phosphate terminus in order to stimulate second strand cleavage (Gaskell and Whitby, unpublished data). This suggests that junction flexibility is insufficient on its own to promote second strand cleavage. Possibly an interaction between the 5' phosphate and the first active site of RuvC is communicated to the second active site to promote a conformational change leading to catalysis. This kind of communication between active sites has been suggested for Ydc2 (Ceschini et al. 2001).

2.6 Directing the orientation of junction cleavage

As discussed in Section 2, the orientation of HJ resolution can be biased. Most HJ resolvases favour cleaving either the continuous or exchanging strands of a HJ (Duckett et al. 1988; Bennett and West 1995a; White and Lilley 1996, 1998; Giraud-Panis and Lilley 1998). However, it is unlikely that resolution bias is ever achieved by this mechanism of strand discrimination, since, as junction folding is influenced by the local nucleotide sequence, the designation of continuous and exchanging strands changes with branch migration (Miick et al. 1997). Resolution bias is more likely to depend on the HJ resolvase being directed by other proteins. This is true in *E. coli* where RuvAB directs the orientation of HJ cleavage by RuvC (van Gool et al. 1999). Specifically, it is the orientation of RuvB loading that directs RuvC – cleavage being on the strands that pass through the RuvB rings 3' toward the HJ. It has been proposed that RuvAB assembles on HJs in a particular orientation to avoid aborting recombination by branch migration to existing DNA ends (Cromie and Leach 2000). For the repair of a DSB this orientation would direct RuvC to resolve the dHJ into crossover products, whereas for recombination initiated from a single-strand gap non-crossover recombinants would be generated. As mentioned in Section 1 crossover recombination in *E. coli* generates chromosome dimers that are resolved into monomers prior to cell division by XerCD site-specific recombination. Studies, using the extent to which cells are dependent on XerCD for viability as a measure of the crossover frequency, have confirmed that DSB repair is biased in favour of crossovers when RuvABC is active, whereas conditions that generate single-strand gaps (e.g. replication fork stalling) give rise to non-crossovers (Cromie and Leach 2000; Michel et al. 2000).

In eukaryotes there are a number of examples where recombination is biased in favour of either a crossover or non-crossover. In most cases it appears that this bias is due to mechanisms like SDSA or dissolution of dHJs rather than directed HJ resolution (see Section 1). However, certainty about this will not be possible until we know the identity of the nuclear resolvase (see Section 2.7). Even so there is at least one situation where the putative HJ resolvase appears to be directed. This is during meiosis in budding yeast, where dHJs between homologous chromosomes are resolved predominantly, or possibly exclusively, into crossover products (Allers and Lichten 2001; Borner et al. 2004). Although the identity of the resolvase is unknown, proteins that are necessary for the crossover bias, and therefore presumed to direct the resolvase, have been characterised. These include Msh4 and Msh5, which are homologues of the bacterial mismatch repair (MMR) protein MutS (Ross-Macdonald and Roeder 1994; Hollingsworth et al. 1995). In *E. coli* a homodimer of MutS promotes MMR by binding to mismatched DNA via one of two central holes in its structure (Kunkel and Erie 2005). Interaction with a second protein MutL then activates the latent endonuclease activity of MutH. This nicks the DNA to allow the entry of a helicase and exonuclease to remove a stretch of single-stranded DNA containing the mismatch. Neither Msh4 nor Msh5 are involved in MMR, instead biochemical studies of the human homologues of these enzymes have shown that they form a heterodimeric complex that binds specifically to HJs (Snowden et al. 2004). The Msh4-Msh5 complex appears to encircle the HJ, and then, after binding ATP, can release from it to slide along the adjacent duplex DNA. This may allow further Msh4-Msh5 complexes to bind the HJ, leading to an accumulation of complexes that could direct the orientation of HJ resolution. In yeast homologues of MutL called Mlh1 and Mlh3 form a complex that functions downstream of Msh4-Msh5 to promote crossover formation (Hunter and Borts 1997; Wang et al. 1999; Santucci-Darmanin et al. 2000; Wang and Kung 2002; Argueso et al. 2004). Based on the analogy with *E. coli* MMR, it is possible that Mlh1-Mlh3 activate the putative HJ resolvase.

2.7 Searching for the elusive nuclear HJ resolvase

Despite the ubiquity of HJ resolvases, the identity of a eukaryotic nuclear resolvase has yet to be established. Evidence that such an enzyme exists has come from analysing fractionated extracts of mammalian cells for the tell-tail characteristics of specific HJ cleavage (Elborough and West 1990; Hyde et al. 1994; Constantinou et al. 2001). This approach has lead to the detection of an activity designated Resolvase A (Constantinou et al. 2002). Intriguingly, an ATP-dependent branch migration activity co-purifies with Resolvase A, suggesting that mammals may have an analogue of RuvABC (Constantinou et al. 2001). Through a combination of screening mutant cell lines and immuno-depletion experiments it has been found that full resolvase activity depends on a complex of two RAD51 paralogs, RAD51C and XRCC3 (Liu et al. 2004). Neither of these proteins has recognisable nuclease motifs, so it is unclear at present whether they are directly responsible for catalysis or act as accessory factors that facilitate or activate cleavage by another unidentified component. *In vivo* RAD51C and XRCC3 appear to have early roles in recombination, most likely acting as mediators for the assembly of RAD51 filaments at sites of DNA damage (Thacker 2005). However, there are data that suggest that they also play a late role. These include the observation that *XRCC3*$^-$ cells exhibit longer conversion tracts, which might be due to a reduction in heteroduplex stability, but could also be due to inefficient HJ resolution (Brenneman et al. 2002). Homologues of RAD51C and XRCC3 also appear to play a critical role at a late stage in meiotic recombination in *Arabidopsis* (Bleuyard and White 2004; Bleuyard et al. 2005).

Genetic approaches have also been used in the hunt for HJ processing enzymes in eukaryotic nuclei. In particular one strategy, which involves rescuing mutant phenotypes with a recombinant form of RusA, has implicated both members of the RecQ helicase family and the Mus81-Eme1 endonuclease in processing HJs (Doe et al. 2000, 2002; Boddy et al. 2001; Saintigny et al. 2002). In the case of RecQ helicases, the suppression of mutant phenotypes by RusA fits well with their proposed roles in HJ branch migration and dHJ dissolution (Constantinou et al. 2000; Karow et al. 2000; Wu and Hickson 2003). RusA suppression of *mus81* mutant phenotypes also appeared to fit with the ability of the purified enzyme to cleave HJs *in vitro*. In fact based on these data Mus81 was proposed to be the first example of a nuclear HJ resolvase (Boddy et al. 2001; Chen et al. 2001). However, this idea proved to be controversial (Haber and Heyer 2001). On one level the purists in the recombination field disagreed that Mus81 should be classed as a *bona fide* HJ resolvase because cutting was not always symmetrical, and the resolved products were therefore not directly repairable by DNA ligase (Boddy et al. 2001; Chen et al. 2001; Constantinou et al. 2002). Certainly the Mus81 activity was shown to be distinct from that of Resolvase A (Constantinou et al. 2002). At another level, analysis of substrate preferences revealed that Mus81 has relatively low levels of activity on HJs compared to various other branched DNAs, raising doubt as to whether it would be an effective HJ resolvase *in vivo* (Kaliraman et al. 2001; Constantinou et al. 2002; Doe et al. 2002; Ciccia et al. 2003; Whitby et al. 2003). It was proposed that Mus81 might be acting as a 3' flap endonuclease rather than as a HJ resolvase (de los Santos et al. 2001; Kaliraman et al. 2001; de los Santos et al. 2003). However, this did not fit well with the RusA suppression data, because RusA does not cleave 3' flaps. A third idea then emerged that suggested that Mus81 cleaves D-loops and nicked HJs (Heyer et al. 2003; Osman et al. 2003). These junctions are the natural precursors of fully ligated HJs, and therefore RusA suppression could be explained by it cleaving the HJs that form in the absence of Mus81. Although this idea has received favourable press (Hollingsworth and Brill 2004), there is still uncertainty as to how Mus81 really acts *in vivo*. What follows is a background to Mus81 together with a review of the current state of play.

3 Mus81

3.1 Mus81 is related to the XPF family of endonucleases

Mus81 was first identified in the budding yeast *Saccharomyces cerevisiae* by its physical interaction with the Rad54 recombination protein, and by its requirement for viability in the absence of the RecQ helicase Sgs1 (Interthal and Heyer 2000; Mullen et al. 2001). At the same time Mus81 was identified in the fission yeast *Schizosaccharomyces pombe* by its interaction with the forkhead-associated (FHA) domain of the checkpoint kinase Cds1 (Boddy et al. 2000). Database searches revealed potential homologues of Mus81 in other organisms including mouse and human (Chen et al. 2001). They also showed that Mus81 contains the conserved ERKX$_3$D active site motif of the XPF family of structure-specific endonucleases, which is similar to the nuclease domain in members of the PD...D/EXK nuclease superfamily (Boddy et al. 2000; Interthal and Heyer 2000; Mullen et al. 2001; Enzlin and Scharer 2002; Nishino et al. 2003) (Fig. 6). Mus81 also contains two helix-hairpin-helix (HhH) motifs, which in eukaryotic and archaeal XPFs promote dimer formation and DNA binding (de Laat et al. 1998b; Newman et al. 2005). However, unlike XPF, which contains a tandem HhH (HhH$_2$) domain near its C-terminal end, the HhH motifs in Mus81 are positioned at either end of the protein (Fig. 6). Eukaryotic XPFs function as heterodimers with a non-catalytic subunit, which is critical for the stability and activity of XPF, and may interact with other proteins to help in substrate targeting (de Laat et al. 1999). In humans this subunit is called ERCC1, and, like XPF, it contains a HhH$_2$ domain near its C-terminus, which is important for DNA binding and dimerization (Tripsianes et al. 2005; Tsodikov et al. 2005) (Fig. 6). It also contains an inactive version of the XPF catalytic domain, which acts as an additional DNA binding domain (Tsodikov et al. 2005). XPF-ERCC1 nicks duplex DNA adjacent to a 3' single-stranded flap. This activity is useful in a number of DNA repair and recombination processes, most notably nucleotide excision repair where it is necessary for removal of the lesion-containing oligonucleotide (de Laat et al. 1999).

The relatedness of Mus81 to the XPF family indicated that it too was likely to be a structure-specific endonuclease. A candidate partner for it, called Mms4, had already been identified in the same *sgs1* synthetic-lethal screen that had identified Mus81 (Mullen et al. 2001). The homologue of Mms4 in *S. pombe*, called Eme1, was later identified by its co-purification with Mus81 (Boddy et al. 2001). Mouse and human homologues were then identified by their sequence similarity to Eme1 (Abraham et al. 2003; Ciccia et al. 2003; Ogrunc and Sancar 2003; Blais et al. 2004). Interestingly, there are two Eme1-like proteins in humans designated Eme1 and Eme2, as well as two potential splice variants of Eme1 (Ciccia et al. 2003; Blais et al. 2004). The significance of this has yet to be established. Eme1 and Mms4 proteins from different organisms share only limited sequence identity with each other, and even less similarity to ERCC1. What similarity there is, is contained mainly within their C-termini, which is a region that is important for interaction with their partner protein. (Mullen et al. 2001; Ciccia et al. 2003; Fu and Xiao 2003). This region may contain a HhH domain, as well as a defective

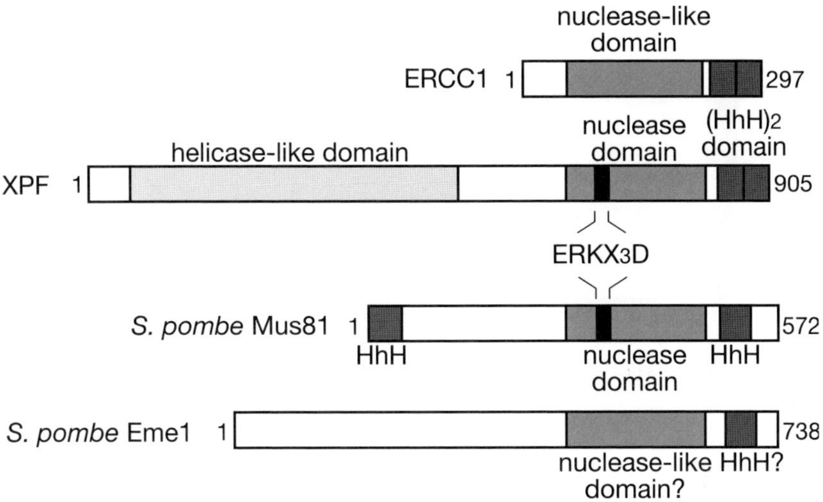

Fig. 6. A comparison of the domain structure of *S. pombe* Mus81-Eme1 and human XPF-ERCC1.

nuclease domain that could contribute to DNA binding. Based on analogy to XPF-ERCC1, it is assumed that Mus81-Eme1/Mms4 functions as a heterodimer. However, higher molecular weight complexes are seen by gel filtration, and the individual subunits can self-associate (Blais et al. 2004; Fricke et al. 2005). It is therefore possible that Mus81-Eme1/Mms4 functions as a larger complex than a heterodimer.

The analysis of recombinant Mus81-Mms4 produced in *E. coli*, affinity purified Mus81-Eme1 from *S. pombe*, and immune precipitated Mus81 from HeLa cells, each confirmed that it is a structure-specific endonuclease (Boddy et al. 2001; Chen et al. 2001; Kaliraman et al. 2001). For simplicity the Mus81-Eme1/Mms4 enzyme will be referred to as Mus81* henceforth. Like other XPF family nucleases Mus81* cleaves duplex DNA a few nucleotides 5' to a 3' flap (Bastin-Shanower et al. 2003; Whitby et al. 2003). However, whereas XPF-ERCC1 will cleave simple Y-shaped DNA, Mus81* needs a 5' terminus at the branch point in order to activate and direct cleavage (de Laat et al. 1998a; Bastin-Shanower et al. 2003).

3.2 The substrate specificity of Mus81*

A number of studies have tested Mus81*'s cleavage specificity *in vitro* (Kaliraman et al. 2001; Doe et al. 2002; Ciccia et al. 2003; Gaillard et al. 2003; Osman et al. 2003; Whitby et al. 2003; Fricke et al. 2005). The consensus view from these studies is that Mus81*'s preferred substrates are three- and four-way junctions that have an exposed 5' end at or close to the junction point (Fig. 7). Examples of this kind of junction *in vivo* are nicked HJs, D-loops, replication forks

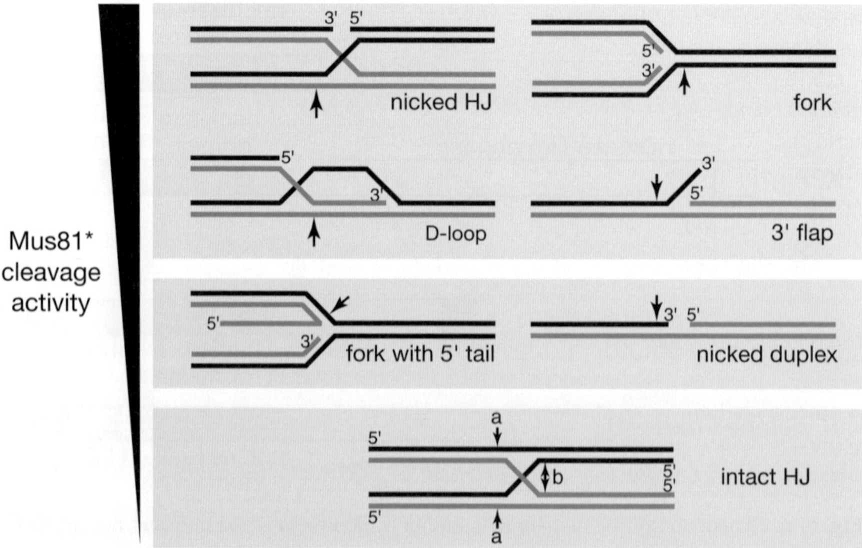

Fig. 7. The substrate specificity of Mus81*. Arrows indicate cleavage sites.

with the lagging strand at the junction point, and 3' flaps. Next in terms of preference are nicked linear duplex DNA, which is a common lesion *in vivo*, and four-way junctions with only a 3' end at the junction point, which may represent a subset of reversed replication forks. Finally, intact HJs are cleaved least well by Mus81* *in vitro* – the activity here being between 75 – 1800 times lower than on 3' flaps. These substrate preferences are conserved in both yeast and human Mus81*.

The substrate range of Mus81* shows that it has the potential to act in a number of different guises in various recombination and repair reactions *in vivo*. Some genetic data are consistent with Mus81* resolving HJs (Boddy et al. 2001; Doe et al. 2002; Odagiri et al. 2003; Osman et al. 2003; Smith et al. 2003; Blais et al. 2004). However, based on *in vitro* activity alone, a major role in HJ resolution would seem unlikely. Furthermore, the genetic data can be explained without invoking classical HJ resolution (see below). Nevertheless the idea that Mus81* might cleave intact HJs *in vivo* should not be discarded yet. It is possible that Mus81* has two modes of action – one that is activated and directed by an exposed 5' end at a junction point, and one that resolves intact HJs by a nick and counter-nick mechanism (Gaillard et al. 2003; Osman et al. 2003). Maybe the *in vitro* conditions that have been used so far only partially activate Mus81*, and for efficient HJ resolution it needs another protein or post-translational modification. The greater specific 'HJ resolvase' activity of *S. pombe* Mus81* purified from its endogenous cells versus that purified from *E. coli* has been cited as evidence for this (Gaillard et al. 2003). However, such differences between endogenous and recombinant enzyme have not been observed for either *S. cerevisiae* or human Mus81* (Ciccia et al. 2003; Fricke et al. 2005). Even if Mus81* is able to cleave HJs *in* *vivo*, it would seem that it does not have the elegant coordination of a classical HJ resolvase, as it makes both symmetrical and asymmetrical incisions producing a mixture of gapped and flapped linear duplexes that are not directly repairable by ligation (Boddy et al. 2001; Chen et al. 2001; Constantinou et al. 2002). It is possible that this seemingly haphazard behaviour is inconsequential when there are efficient gap-filling polymerases and flap endonucleases to tidy-up the resolved products. Alternatively, lack of coordination may be due to limitations of the *in vitro* conditions that have been used to study Mus81*. In summary, the jury is still out with regards to Mus81* acting as a HJ resolvase.

3.3 The role of Mus81* in meiosis

As mentioned above the formation of crossovers during meiosis in *S. cerevisiae* is achieved by the biased resolution of dHJs by an as yet unidentified HJ resolvase. The formation and correct processing of the dHJ depends on a set of meiosis-specific proteins (Zip1, Zip2, Zip3, Mer3, Msh4, and Msh5) that are collectively called the ZMM proteins (Whitby 2005; Hunter N, this volume). These proteins are also needed to enforce crossover interference, which ensures a non-clustered distribution of crossovers. However, the ZMM proteins are not required for all crossover formation – there is a back-up pathway that is dependent on Mus81*, which can achieve up to 50% of the normal levels of crossing over under certain conditions (de los Santos et al. 2003; Argueso et al. 2004; Borner et al. 2004). An appreciation of how this back-up pathway might work came mainly from studying Mus81*'s involvement in meiosis in *S. pombe* (Boddy et al. 2001; Osman et al. 2003; Smith et al. 2003). This yeast does not have a ZMM-dependent pathway for crossover formation, and instead relies solely on a Mus81-dependent pathway. Without Mus81* spore viability drops to ≤1%, and meiotic chromosome segregation fails seemingly because the homologous chromosomes remain entangled by unresolved HJs. Crossover formation amongst the few viable spores is dramatically reduced, whereas the formation of non-crossover recombinants appears unaffected. These mutant phenotypes are suppressed by the expression of RusA, providing strong evidence that Mus81* promotes crossover formation by resolving HJs or their precursors. Based on Mus81*'s substrate preferences *in vitro*, the popular view is that it cleaves the D-loop and nicked HJ that precede the dHJ (Fig. 8) (Osman et al. 2003; Hollingsworth and Brill 2004). The manner in which it cleaves these junctions means that only crossover recombinants would be formed, which is consistent with the genetic data, and provides a nice explanation for the strong crossover bias that is observed in *S. pombe* (Osman et al. 2003; Hollingsworth and Brill 2004). As discussed above, there remains the possibility that Mus81* acts on fully ligated HJs. If true then this could be a back-up to its main activity on D-loops and nicked HJs, and presumably would require additional factors to ensure that it resolved the dHJ with the appropriate bias.

It would seem that a ZMM or ZMM-like pathway, which is subject to crossover interference, is the principle pathway of crossover formation in most organisms, with *S. pombe* being a notable exception. It may also be true that, like budding

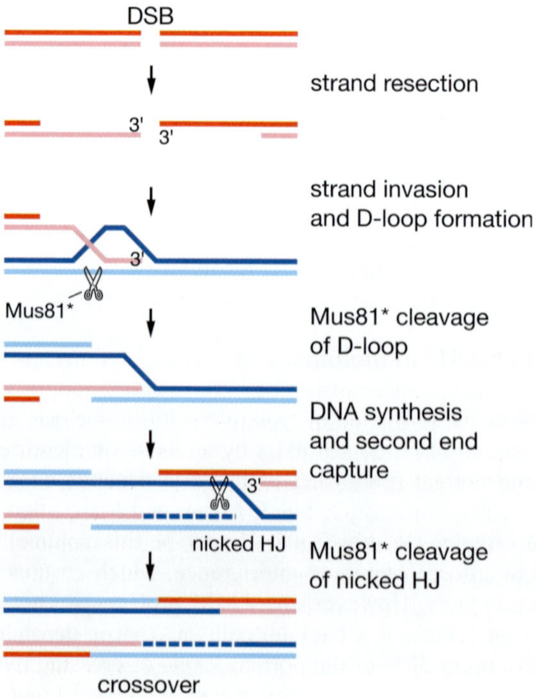

Fig. 8. Model for crossover formation by Mus81*.

yeast, many organisms (e.g. *Arabidopsis*) utilize Mus81* as a back-up pathway for crossover formation, although the worm *Caenorhabditis elegans* is at least one example where it is not used (Higgins et al. 2004; Hollingsworth and Brill 2004; Mercier et al. 2005). In mammals the detection of Mlh1-independent crossovers has been tentatively attributed to Mus81* (Guillon et al. 2005). However, unlike yeast, the Mus81* pathway of crossover formation is apparently not critical for meiosis in mammals based on the normal fertility and gametogenesis of a *MUS81* knockout mouse (McPherson et al. 2004).

3.4 Mus81 and links to cancer

Mus81* is important for the normal growth and viability of both fission yeast and mammalian cells (Boddy et al. 2000; Abraham et al. 2003; McPherson et al. 2004; Dendouga et al. 2005). So much so that mammalian *MUS81* and *EME1* mutant cells fail to proliferate in culture unless p53 is attenuated, and display an increased frequency of aneuploidy and chromosomal abnormalities, including breaks, chromosome fusions and dicentric chromosomes (Abraham et al. 2003; McPherson et al. 2004; Dendouga et al. 2005). These phenotypes indicate that Mus81* plays a caretaker role in genome maintenance, and therefore it was not surprising to learn that a *MUS81* knockout mouse is highly cancer prone (McPherson et al. 2004).

What was intriguing was that *MUS81*$^{-/+}$ heterozygote mice were as cancer prone as their *MUS81*$^{-/-}$ homozygote littermates, indicating that *MUS81* is a haploinsufficient tumour suppressor gene. However, a more recent publication has shed doubt on the validity of these observations by reporting that an independently made *MUS81* knockout mouse exhibits no cancer proneness (Dendouga et al. 2005). Further investigations are required to reconcile the differences between these studies.

3.5 Mus81 and DSB repair in vegetative cells

What exactly Mus81*'s critical roles are in genome maintenance is unclear. We can however discount a number of potential functions where it is evident that its presence is unnecessary. Based on the lack of γ-ray sensitivity of *mus81*/*eme1* mutants, one of these is DSB repair, which in some ways is surprising given the critical role that it can play in this process during meiosis (Boddy et al. 2000; Interthal and Heyer 2000; Abraham et al. 2003; McPherson et al. 2004; Dendouga et al. 2005). Mus81* is also dispensable for gene targeting in mammalian cells and intragenic recombination in budding yeast (Interthal and Heyer 2000; Abraham et al. 2003; Odagiri et al. 2003; McPherson et al. 2004). The apparent lack of involvement of Mus81* in DSB-induced recombination is consistent with the low frequency of crossing-over that is generally associated with mitotic recombination. However, in some cases these low-levels of crossing over can be partly attributed to Mus81*. For example, in *S. pombe* 10% of recombinants in a plasmid gap repair assay are crossovers, and half of these are dependent on *mus81* (Osman et al. 2003; Sun and Whitby, unpublished data). As most of the recombinants are likely to derive from a strand invasion event (i.e. a D-loop will be formed), it would be interesting to know what prevents Mus81* from generating a greater proportion of crossovers. One possibility is that competition with proteins that promote repair by pathways such as SDSA limit Mus81*'s action.

Although Mus81* appears to be largely dispensable for the repair of two-sided DSBs in mitotic cells, it does seem to be needed when a one-sided DSB is formed by replication fork breakage. The evidence for this comes mainly from the striking sensitivity to the topoisomerase I (Top1) poison camptothecin (CPT) that yeast *mus81*/*eme1*/*mms4* mutants exhibit (Doe et al. 2002; Liu et al. 2002; Vance and Wilson 2002; Bastin-Shanower et al. 2003). In contrast *MUS81* mutant mouse embryonic fibroblasts are relatively insensitive to CPT, suggesting that Mus81* may be redundant for the repair of broken replication forks in mammals (Dendouga et al. 2005). CPT inhibits Top1 at the religation stage of its reaction cycle causing an accumulation of single-strand breaks with Top1 covalently attached to the 3' terminus (Liu et al. 2000). It is thought that replication forks run-off at these breaks to create one-sided DSBs, which in yeast are repaired by *RAD52*-dependent strand invasion. The hypersensitivity of yeast *mus81*/*eme1*/*mms4* mutants to CPT is suppressed by the expression of RusA, suggesting that Mus81*'s role is to process an intermediate that either is or will become a HJ (Doe et al. 2002; Bastin-Shanower et al. 2003). For example, the repair of a broken fork by strand invasion would involve a D-loop that Mus81* could cleave. Mus81* may also utilize its 3' flap endonuclease activity to remove a section of DNA containing the covalently bound Top1, providing an alternative to processing by the tyrosyl DNA phosphodiesterase Tdp1 or Rad1-Rad10 (Liu et al. 2002). Epistasis analysis in fission yeast suggests that Mus81* can promote fork repair via a pathway that is dependent on Rad22 (the *S. pombe* homologue of Rad52) but independent of Rad51 (Doe et al. 2004). These data do not discount Mus81* from cleaving some Rad51-dependent D-loops, but do raise the possibility that it acts on DNA junctions made solely by Rad22. Rad52/Rad22 can perform a limited strand invasion reaction *in vitro*, so it is possible that some D-loops may be formed without Rad51 *in vivo*.

3.6 Mus81 and stalled replication forks

Similar to the HJ resolvases in bacteria, Mus81* has been implicated in processing DNA junctions that form when replication forks are blocked. The evidence for this comes in part from the hypersensitivity of yeast *mus81*/*eme1*/*mms4* mutants to genotoxins such as ultra violet (UV) light and the alkylating agent methyl methanesulfonate (MMS), which are known to induce lesions that perturb DNA replication (Boddy et al. 2000; Interthal and Heyer 2000; Doe et al. 2002). Furthermore, in human cells Mus81* localizes to regions of UV-induced damage in S-phase cells (Gao et al. 2003). As with CPT the UV and MMS hypersensitivities of the yeast mutants are suppressed by RusA, again suggesting that Mus81* is needed to process a DNA junction that either is a HJ or will become a HJ (Doe et al. 2002; Bastin-Shanower et al. 2003; Odagiri et al. 2003).

Mus81*'s involvement with processing junctions that arise when forks are perturbed is also indicated by the synthetic interactions that yeast *mus81*/*eme1*/*mms4* mutants exhibit with a network of genes that are involved in aspects of DNA replication. These interactions include the reduction in the restrictive temperatures for thermosensitive alleles of Polα and Polδ in *S. pombe*, and the synthetic poor growth or lethality with mutants of Elg1, which forms part of an alternative Replication Factor C complex in *S. cerevisiae*, and Swi1 and Swi3 that are components of the so-called fork protection complex in *S. pombe* (Boddy et al. 2000; Bellaoui et al. 2003; Noguchi et al. 2003, 2004). In each case the impairment of replication fork progression and/or stability may lead to fork reversal and/or single-strand gaps at which recombination could act. Mus81*'s involvement with processing the DNA junctions that would be formed from such transactions is indicated by two-dimensional gel electrophoresis analysis of the rDNA in *S. pombe*, which shows an accumulation of X-shaped DNA molecules in both *pol1-1* ((at the restrictive temperature) (*pol1-1* is a temperature sensitive allele of the catalytic subunit of Polα)) and *swi1* mutant backgrounds in the absence of *mus81* (Gaillard et al. 2003; Noguchi et al. 2004). Intriguingly, the very poor growth of a *mus81 swi1* double mutant, together with its accumulation of X-shaped DNA, is suppressed by deleting *rad22* but not by deleting either *rad51* or *rad54* (Noguchi et al. 2004). This is another example where Mus81 seems to be working on DNA junctions that are formed without Rad51. Further examples include Mus81's involvement in the response to MMS-induced damage, and in dealing with problems that arise in a *rnh202* mutant in *S. cerevisiae*, which is believed to be impaired for Okazaki fragment maturation (Ii and Brill 2005). However, once again it is important to emphasise that these data do not exclude Mus81* from working on a subset of Rad51-dependent junctions as suggested by its physical interaction with Rad54 in *S. cerevisiae* (Interthal and Heyer 2000). Furthermore, Mus81* is not the sole enzyme that prevents the accumulation of DNA junctions at stalled forks. In yeast it overlaps in this activity both with the RecQ helicases (together with Top3 and Rmi1), and with the Smc5-Smc6 complex (Boddy et al. 2000; Mullen et al. 2001; Fabre et al. 2002; Torres-Rosell et al. 2005; Pebernard et al. 2006).

There are three main types of model that have emerged to explain Mus81*'s involvement in events that stem from replication fork stalling: 1) it cleaves the stalled fork to promote replication restart similar to the HJ resolvases in bacteria (Kaliraman et al. 2001; Constantinou et al. 2002; Doe et al. 2002; Whitby et al. 2003); 2) it processes the D-loops and/or HJs that stem from events initiated at single-strand gaps or from the extruded arm of a reversed fork (Doe et al. 2002; Whitby et al. 2003; Noguchi et al. 2004); and 3) it cleaves 3' flaps that may form from the over-replication of DNA during SDSA initiated at single-strand gaps (Fabre et al. 2002; Ii et al. 2005). With regards to the first model, the fact that $Mus81^+$ cells are viable in the absence of known pathways for DSB repair suggests that cleaving stalled forks is not common during normal growth (Fabre et al. 2002). In fact this is a potentially dangerous activity that can result in increased deletions and genome rearrangements. In fission yeast this danger is avoided, when replication forks are stalled by hydroxyurea-mediated dNTP depletion, by the phosphorylation of Mus81 at a T-X-X-F motif. This phosphorylated form of Mus81 is bound by the FHA domain of Cds1 causing Mus81 to delocalize from chromatin (Kai et al. 2005). Distinguishing between models 2 and 3 is harder, since both are reconcilable with much of the available data. For example, the failure of Mus81* to cleave a 3' flap could result in it being used to initiate strand invasion leading to the formation of a HJ, thereby explaining the RusA suppression data (Fabre et al. 2002; Ii et al. 2005). Time will tell which, if either, of these models is correct.

3.7 Mus81 and inter-strand cross-link repair

Perhaps one of the more critical roles of Mus81* is in inter-strand cross-link (ICL) repair. This is seen in mouse *$MUS81^{-/-}$* cells and animals, which, although not particularly hypersensitive to CPT and UV, are hypersensitive to ICL-inducing agents such as mitomycin-C (MMC) (Abraham et al. 2003; McPherson et al. 2004; Dendouga et al. 2005). ICLs are total barriers to DNA replication, and necessitate the formation of a DSB together with recombination functions for their repair. It has been suggested that Mus81* might be the enzyme that makes the DSB – possibly cleaving replication forks that are blocked by ICLs (Niedernhofer et al. 2004). However, based on the kinetics of MMC-induced γ-H2AX foci accumulation, and the detection of MMC-induced DSBs by pulsed-field gel electrophoresis, it would appear that mouse *MUS81*$^{-/-}$ cells are not defective in making the DSB that initiates ICL repair (Dendouga et al. 2005). What Mus81*'s function is in ICL repair therefore remains unclear. In humans ICL repair depends on the Fanconi anemia proteins. Faconi anemia is a disease caused by defects in any one of at least 12 genes, which is characterized at the cellular level by hypersensitivity to MMC, and at the whole organism level by genomic instability and cancer predisposition (Kennedy and D'Andrea 2005). One of the Fanconi anemia core complex proteins is FANCM, which has an XPF-like nuclease domain as well as a helicase domain (Meetei et al. 2005). FANCM is related to the archaeal Hef protein, which, like Mus81*, favours cleaving forks (Komori et al. 2002; Roberts and White 2005). Intriguingly, the nuclease domain in FANCM is largely dispensable for resistance to ICLs (Meetei et al. 2005). Perhaps this is because of an overlap in activity with Mus81*.

4 Future perspectives

A great deal is known about the HJ resolvases, however two major issues remain to be resolved. First, structural determinations of HJ resolvases bound to junction DNA are needed to gain real insight into the mechanisms of substrate recognition and cleavage. Second, we are still awaiting the identification of a eukaryotic nuclear HJ resolvase. It will be interesting to see how well this is conserved, and whether some eukaryotes have forsaken the use of a classical HJ resolvase in favour of the Mus81* solution to resolution. With regards to Mus81* much remains to be discovered. For example, the structure of Mus81-Eme1 and how it binds and cleaves DNA junctions needs to be worked out. Progress here has already been made by the determination of two archaeal XPF-DNA structures (Newman et al. 2005; Nishino et al. 2005). Despite differences in domain architecture, and it acting as a homodimer, archaeal XPF exhibits similar substrate specificities as Mus81* (Roberts and White 2005). Knowledge of its structure therefore provides a useful guide to understanding Mus81. Further studies are required to determine whether other XPF family endonucleases fulfil similar roles to Mus81* in some organisms. For example, the XPF homologue MEI-9 generates meiotic crossovers in *Drosophila* (Yildiz et al. 2002) – does it do this by resolving dHJs or by cleaving D-loops and/or nicked HJs? Finally, what are Mus81*'s critical roles in vegetative cells and how is it controlled in different organisms? The extensive repertoire of junctions that Mus81* can cleave suggests that it may have multifarious roles. Identifying its true substrate range *in vivo* should narrow down the options, and hopefully resolve the argument as to whether Mus81* should be classed as a HJ resolvase.

Acknowledgements

I would like to thank Fikret Osman and Alexander Lorenz for critical reading of the manuscript. Work in my laboratory is funded by a Senior Research Fellowship in Basic Biomedical Research from the Wellcome Trust.

References

Abraham J, Lemmers B, Hande MP, Moynahan ME, Chahwan C, Ciccia A, Essers J, Hanada K, Chahwan R, Khaw AK, McPherson P, Shehabeldin A, Laister R, Arrowsmith C, Kanaar R, West SC, Jasin M, Hakem R (2003) Eme1 is involved in DNA damage processing and maintenance of genomic stability in mammalian cells. EMBO J 22:6137-6147

Ahn J, Whitby MC (2003) The role of the SAP motif in promoting Holliday junction binding and resolution by SpCCE1. J Biol Chem 278:29121-29129

Allers T, Lichten M (2001) Differential timing and control of noncrossover and crossover recombination during meiosis. Cell 106:47-57

Argueso JL, Wanat J, Gemici Z, Alani E (2004) Competing crossover pathways act during meiosis in *Saccharomyces cerevisiae*. Genetics 168:1805-1816

Ariyoshi M, Vassylyev DG, Iwasaki H, Nakamura H, Shinagawa H, Morikawa K (1994) Atomic structure of the RuvC resolvase: a Holliday junction-specific endonuclease from *E. coli*. Cell 78:1063-1072

Bastin-Shanower SA, Fricke WM, Mullen JR, Brill SJ (2003) The mechanism of Mus81-Mms4 cleavage site selection distinguishes it from the homologous endonuclease Rad1-Rad10. Mol Cell Biol 23:3487-3496

Bellaoui M, Chang M, Ou J, Xu H, Boone C, Brown GW (2003) Elg1 forms an alternative RFC complex important for DNA replication and genome integrity. EMBO J 22:4304-4313

Bennett RJ, Dunderdale HJ, West SC (1993) Resolution of Holliday junctions by RuvC resolvase: cleavage specificity and DNA distortion. Cell 74:1021-1031

Bennett RJ, West SC (1995a) RuvC protein resolves Holliday junctions via cleavage of the continuous (noncrossover) strands. Proc Natl Acad Sci USA 92:5635-5639

Bennett RJ, West SC (1995b) Structural analysis of the RuvC-Holliday junction complex reveals an unfolded junction. J Mol Biol 252:213-226

Benson FE, West SC (1994) Substrate specificity of the *Escherichia coli* RuvC protein. Resolution of three- and four-stranded recombination intermediates. J Biol Chem 269:5195-5201

Birkenbihl RP, Kemper B (1998) Endonuclease VII has two DNA-binding sites each composed from one N- and one C-terminus provided by different subunits of the protein dimer. EMBO J 17:4527-4534

Blais V, Gao H, Elwell CA, Boddy MN, Gaillard PH, Russell P, McGowan CH (2004) RNA interference inhibition of Mus81 reduces mitotic recombination in human cells. Mol Biol Cell 15:552-562

Bleuyard JY, Gallego ME, Savigny F, White CI (2005) Differing requirements for the *Arabidopsis* Rad51 paralogs in meiosis and DNA repair. Plant J 41:533-545

Bleuyard JY, White CI (2004) The *Arabidopsis* homologue of Xrcc3 plays an essential role in meiosis. EMBO J 23:439-449

Boddy MN, Gaillard P-HL, McDonald WH, Shanahan P, Yates JR 3rd, Russell P (2001) Mus81-Eme1 are essential components of a Holliday junction resolvase. Cell 107:537-548

Boddy MN, Lopez-Girona A, Shanahan P, Interthal H, Heyer WD, Russell P (2000) Damage tolerance protein Mus81 associates with the FHA1 domain of checkpoint kinase Cds1. Mol Cell Biol 20:8758-8766

Bolt EL, Lloyd RG (2002) Substrate specificity of RusA resolvase reveals the DNA structures targeted by RuvAB and RecG *in vivo*. Mol Cell 10:187-198

Bond CS, Kvaratskhelia M, Richard D, White MF, Hunter WN (2001) Structure of Hjc, a Holliday junction resolvase, from *Sulfolobus solfataricus*. Proc Natl Acad Sci USA 98:5509-5514

Borner GV, Kleckner N, Hunter N (2004) Crossover/noncrossover differentiation, synaptonemal complex formation, and regulatory surveillance at the leptotene/zygotene transition of meiosis. Cell 117:29-45

Brenneman MA, Wagener BM, Miller CA, Allen C, Nickoloff JA (2002) XRCC3 controls the fidelity of homologous recombination: roles for XRCC3 in late stages of recombination. Mol Cell 10:387-395

Ceschini S, Keeley A, McAlister MS, Oram M, Phelan J, Pearl LH, Tsaneva IR, Barrett TE (2001) Crystal structure of the fission yeast mitochondrial Holliday junction resolvase Ydc2. EMBO J 20:6601-6611

Chan SN, Harris L, Bolt EL, Whitby MC, Lloyd RG (1997) Sequence specificity and biochemical characterization of the RusA Holliday junction resolvase of *Escherichia coli*. J Biol Chem 272:14873-14882

Chan SN, Vincent SD, Lloyd RG (1998) Recognition and manipulation of branched DNA by the RusA Holliday junction resolvase of *Escherichia coli*. Nucleic Acids Res 26:1560-1566

Chen XB, Melchionna R, Denis CM, Gaillard PH, Blasina A, Van de Weyer I, Boddy MN, Russell P, Vialard J, McGowan CH (2001) Human Mus81-associated endonuclease cleaves Holliday junctions *in vitro*. Mol Cell 8:1117-1127

Ciccia A, Constantinou A, West SC (2003) Identification and characterization of the human Mus81-Eme1 endonuclease. J Biol Chem 278:25172-25178

Constantinou A, Chen XB, McGowan CH, West SC (2002) Holliday junction resolution in human cells: two junction endonucleases with distinct substrate specificities. EMBO J 21:5577-5585

Constantinou A, Davies AA, West SC (2001) Branch migration and Holliday junction resolution catalyzed by activities from mammalian cells. Cell 104:259-268

Constantinou A, Tarsounas M, Karow JK, Brosh RM, Bohr VA, Hickson ID, West SC (2000) Werner's syndrome protein (WRN) migrates Holliday junctions and co-localizes with RPA upon replication arrest. EMBO Rep 1:80-84

Courcelle J, Carswell-Crumpton C, Hanawalt PC (1997) *recF* and *recR* are required for the resumption of replication at DNA replication forks in *Escherichia coli*. Proc Natl Acad Sci USA 94:3714-3719

Courcelle J, Donaldson JR, Chow KH, Courcelle CT (2003) DNA damage-induced replication fork regression and processing in *Escherichia coli*. Science 299:1064-1067

Cromie GA, Leach DR (2000) Control of crossing over. Mol Cell 6:815-826

Davies AA, West SC (1998) Formation of RuvABC-Holliday junction complexes *in vitro*. Curr Biol 8:725-727

de Laat WL, Appeldoorn E, Jaspers NG, Hoeijmakers JH (1998a) DNA structural elements required for ERCC1-XPF endonuclease activity. J Biol Chem 273:7835-7842

de Laat WL, Jaspers NG, Hoeijmakers JH (1999) Molecular mechanism of nucleotide excision repair. Genes Dev 13:768-785

de Laat WL, Sijbers AM, Odijk H, Jaspers NG, Hoeijmakers JH (1998b) Mapping of interaction domains between human repair proteins ERCC1 and XPF. Nucleic Acids Res 26:4146-4152

de los Santos T, Hunter N, Lee C, Larkin B, Loidl J, Hollingsworth NM (2003) The Mus81/Mms4 endonuclease acts independently of double-Holliday junction resolution to promote a distinct subset of crossovers during meiosis in budding yeast. Genetics 164:81-94

de los Santos T, Loidl J, Larkin B, Hollingsworth NM (2001) A role for *MMS4* in the processing of recombination intermediates during meiosis in *Saccharomyces cerevisiae*. Genetics 159:1511-1525

Declais AC, Fogg JM, Freeman AD, Coste F, Hadden JM, Phillips SE, Lilley DM (2003) The complex between a four-way DNA junction and T7 endonuclease I. EMBO J 22:1398-1409

Declais AC, Lilley DM (2000) Extensive central disruption of a four-way junction on binding CCE1 resolving enzyme. J Mol Biol 296:421-433

Dendouga N, Gao H, Moechars D, Janicot M, Vialard J, McGowan CH, Heyer WD, Ehmsen KT, Solinger JA (2005) Disruption of murine Mus81 increases genomic instability and DNA damage sensitivity but does not promote tumorigenesis. Mol Cell Biol 25:7569-7579

Dickman MJ, Ingleston SM, Sedelnikova SE, Rafferty JB, Lloyd RG, Grasby JA, Hornby DP (2002) The RuvABC resolvasome. Eur J Biochem 269:5492-5501

Doe CL, Ahn JS, Dixon J, Whitby MC (2002) Mus81-Eme1 and Rqh1 involvement in processing stalled and collapsed replication forks. J Biol Chem 277:32753-32759

Doe CL, Dixon J, Osman F, Whitby MC (2000) Partial suppression of the fission yeast *rqh1*(-) phenotype by expression of a bacterial Holliday junction resolvase. EMBO J 19:2751-2762

Doe CL, Osman F, Dixon J, Whitby MC (2004) DNA repair by a Rad22-Mus81-dependent pathway that is independent of Rhp51. Nucleic Acids Res 32:5570-5581

Duckett DR, Murchie AI, Diekmann S, von Kitzing E, Kemper B, Lilley DM (1988) The structure of the Holliday junction, and its resolution. Cell 55:79-89

Eggleston AK, Mitchell AH, West SC (1997) *In vitro* reconstitution of the late steps of genetic recombination in *E. coli*. Cell 89:607-617

Elborough KM, West SC (1990) Resolution of synthetic Holliday junctions in DNA by an endonuclease activity from calf thymus. EMBO J 9:2931-2936

Enzlin JH, Scharer OD (2002) The active site of the DNA repair endonuclease XPF-ERCC1 forms a highly conserved nuclease motif. EMBO J 21:2045-2053

Fabre F, Chan A, Heyer WD, Gangloff S (2002) Alternate pathways involving Sgs1/Top3, Mus81/ Mms4, and Srs2 prevent formation of toxic recombination intermediates from single-stranded gaps created by DNA replication. Proc Natl Acad Sci USA 99:16887-16892

Fogg JM, Lilley DM (2000) Ensuring productive resolution by the junction-resolving enzyme RuvC: large enhancement of the second-strand cleavage rate. Biochemistry 39:16125-16134

Fogg JM, Schofield MJ, Declais AC, Lilley DM (2000) Yeast resolving enzyme CCE1 makes sequential cleavages in DNA junctions within the lifetime of the complex. Biochemistry 39:4082-4089

Fogg JM, Schofield MJ, White MF, Lilley DM (1999) Sequence and functional-group specificity for cleavage of DNA junctions by RuvC of *Escherichia coli*. Biochemistry 38:11349-11358

Fricke WM, Bastin-Shanower SA, Brill SJ (2005) Substrate specificity of the *Saccharomyces cerevisiae* Mus81-Mms4 endonuclease. DNA Repair (Amst) 4:243-251

Fu Y, Xiao W (2003) Functional domains required for the *Saccharomyces cerevisiae* Mus81-Mms4 endonuclease complex formation and nuclear localization. DNA Repair (Amst) 2:1435-1447

Gaillard PH, Noguchi E, Shanahan P, Russell P (2003) The endogenous Mus81-Eme1 complex resolves Holliday junctions by a nick and counternick mechanism. Mol Cell 12:747-759

Gao H, Chen XB, McGowan CH (2003) Mus81 endonuclease localizes to nucleoli and to regions of DNA damage in human S-phase cells. Mol Biol Cell 14:4826-4834

Giraud-Panis MJ, Lilley DM (1996) T4 endonuclease VII. Importance of a histidine-aspartate cluster within the zinc-binding domain. J Biol Chem 271:33148-33155

Giraud-Panis MJ, Lilley DM (1997) Near-simultaneous DNA cleavage by the subunits of the junction- resolving enzyme T4 endonuclease VII. EMBO J 16:2528-2534

Giraud-Panis MJ, Lilley DM (1998) Structural recognition and distortion by the DNA junction-resolving enzyme RusA. J Mol Biol 278:117-133

Guan C, Kumar S (2005) A single catalytic domain of the junction-resolving enzyme T7 endonuclease I is a non-specific nicking endonuclease. Nucleic Acids Res 33:6225-6234

Guillon H, Baudat F, Grey C, Liskay RM, de Massy B (2005) Crossover and noncrossover pathways in mouse meiosis. Mol Cell 20:563-573

Haber JE, Heyer WD (2001) The fuss about Mus81. Cell 107:551-554

Hadden JM, Convery MA, Declais AC, Lilley DM, Phillips SE (2001) Crystal structure of the Holliday junction resolving enzyme T7 endonuclease I. Nat Struct Biol 8:62-67

Hadden JM, Declais AC, Phillips SE, Lilley DM (2002) Metal ions bound at the active site of the junction-resolving enzyme T7 endonuclease I. EMBO J 21:3505-3515

Heyer WD, Ehmsen KT, Solinger JA (2003) Holliday junctions in the eukaryotic nucleus: resolution in sight? Trends Biochem Sci 28:548-557

Higgins JD, Armstrong SJ, Franklin FC, Jones GH (2004) The *Arabidopsis* MutS homolog AtMSH4 functions at an early step in recombination: evidence for two classes of recombination in *Arabidopsis*. Genes Dev 18:2557-2570

Higgins NP, Kato K, Strauss B (1976) A model for replication repair in mammalian cells. J Mol Biol 101:417-425

Hollingsworth NM, Brill SJ (2004) The Mus81 solution to resolution: generating meiotic crossovers without Holliday junctions. Genes Dev 18:117-125

Hollingsworth NM, Ponte L, Halsey C (1995) MSH5, a novel MutS homolog, facilitates meiotic reciprocal recombination between homologs in *Saccharomyces cerevisiae* but not mismatch repair. Genes Dev 9:1728-1739

Hunter N, Borts RH (1997) Mlh1 is unique among mismatch repair proteins in its ability to promote crossing-over during meiosis. Genes Dev 11:1573-1582

Hyde H, Davies AA, Benson FE, West SC (1994) Resolution of recombination intermediates by a mammalian activity functionally analogous to *Escherichia coli* RuvC resolvase. J Biol Chem 269:5202-5209

Ii M, Brill SJ (2005) Roles of SGS1, MUS81, and RAD51 in the repair of lagging-strand replication defects in *Saccharomyces cerevisiae.* Curr Genet 48:213-225

Interthal H, Heyer WD (2000) MUS81 encodes a novel helix-hairpin-helix protein involved in the response to UV- and methylation-induced DNA damage in *Saccharomyces cerevisiae*. Mol Gen Genet 263:812-827

Ira G, Malkova A, Liberi G, Foiani M, Haber JE (2003) Srs2 and Sgs1-Top3 suppress crossovers during double-strand break repair in yeast. Cell 115:401-411

Kai M, Boddy MN, Russell P, Wang TS (2005) Replication checkpoint kinase Cds1 regulates Mus81 to preserve genome integrity during replication stress. Genes Dev 19:919-932

Kaliraman V, Mullen JR, Fricke WM, Bastin-Shanower SA, Brill SJ (2001) Functional overlap between Sgs1-Top3 and the Mms4-Mus81 endonuclease. Genes Dev 15:2730-2740

Karow JK, Constantinou A, Li JL, West SC, Hickson ID (2000) The Bloom's syndrome gene product promotes branch migration of Holliday junctions. Proc Natl Acad Sci USA 97:6504-6508

Karymov M, Daniel D, Sankey OF, Lyubchenko YL (2005) Holliday junction dynamics and branch migration: single-molecule analysis. Proc Natl Acad Sci USA 102:8186-8191

Kemper B (1997) Branched DNA resolving enzymes. In: Nickoloff JA, Hoekstra MF (eds) DNA Damage and Repair. Vol 1. DNA Repair in Prokaryotes and Lower Eukaryotes. Humana Press Inc., Totowa, NJ, pp 179-204

Kennedy RD, D'Andrea AD (2005) The Fanconi Anemia/BRCA pathway: new faces in the crowd. Genes Dev 19:2925-2940

Komori K, Fujikane R, Shinagawa H, Ishino Y (2002) Novel endonuclease in Archaea cleaving DNA with various branched structure. Genes Genet Syst 77:227-241

Kunkel TA, Erie DA (2005) DNA mismatch repair. Annu Rev Biochem 74:681-710

Kuzminov A (2001) Single-strand interruptions in replicating chromosomes cause double-strand breaks. Proc Natl Acad Sci USA 98:8241-8246

Kvaratskhelia M, Wardleworth BN, Norman DG, White MF (2000) A conserved nuclease domain in the archaeal Holliday junction resolving enzyme Hjc. J Biol Chem 275:25540-25546

Lilley DM (2000) Structures of helical junctions in nucleic acids. Q Rev Biophys 33:109-159

Lilley DM, White MF (2001) The junction-resolving enzymes. Nat Rev Mol Cell Biol 2:433-443

Liu C, Pouliot JJ, Nash HA (2002) Repair of topoisomerase I covalent complexes in the absence of the tyrosyl-DNA phosphodiesterase Tdp1. Proc Natl Acad Sci USA 99:14970-14975

Liu LF, Desai SD, Li TK, Mao Y, Sun M, Sim SP (2000) Mechanism of action of camptothecin. Ann N Y Acad Sci 922:1-10

Liu Y, Masson JY, Shah R, O'Regan P, West SC (2004) RAD51C is required for Holliday junction processing in mammalian cells. Science 303:243-246

Lopes M, Cotta-Ramusino C, Pellicioli A, Liberi G, Plevani P, Muzi-Falconi M, Newlon CS, Foiani M (2001) The DNA replication checkpoint response stabilizes stalled replication forks. Nature 412:557-561

Lopes M, Foiani M, Sogo JM (2006) Multiple mechanisms control chromosome integrity after replication fork uncoupling and restart at irreparable UV lesions. Mol Cell 21:15-27

McGlynn P, Lloyd RG (2002) Recombinational repair and restart of damaged replication forks. Nat Rev Mol Cell Biol 3:859-870

McGregor N, Ayora S, Sedelnikova S, Carrasco B, Alonso JC, Thaw P, Rafferty J (2005) The structure of *Bacillus subtilis* RecU Holliday junction resolvase and its role in substrate selection and sequence-specific cleavage. Structure 13:1341-1351

McPherson JP, Lemmers B, Chahwan R, Pamidi A, Migon E, Matysiak-Zablocki E, Moynahan ME, Essers J, Hanada K, Poonepalli A, Sanchez-Sweatman O, Khokha R, Kanaar R, Jasin M, Hande MP, Hakem R (2004) Involvement of mammalian Mus81 in genome integrity and tumor suppression. Science 304:1822-1826

Meetei AR, Medhurst AL, Ling C, Xue Y, Singh TR, Bier P, Steltenpool J, Stone S, Dokal I, Mathew CG, Hoatlin M, Joenje H, de Winter JP, Wang W (2005) A human ortholog of archaeal DNA repair protein Hef is defective in Fanconi anemia complementation group M. Nat Genet 37:958-963

Mercier R, Jolivet S, Vezon D, Huppe E, Chelysheva L, Giovanni M, Nogue F, Doutriaux MP, Horlow C, Grelon M, Mezard C (2005) Two meiotic crossover classes cohabit in *Arabidopsis*: one is dependent on *MER3*, whereas the other one is not. Curr Biol 15:692-701

Michel B, Recchia GD, Penel-Colin M, Ehrlich SD, Sherratt DJ (2000) Resolution of Holliday junctions by RuvABC prevents dimer formation in rep mutants and UV-irradiated cells. Mol Microbiol 37:180-191

Middleton CL, Parker JL, Richard DJ, White MF, Bond CS (2004) Substrate recognition and catalysis by the Holliday junction resolving enzyme Hje. Nucleic Acids Res 32:5442-5451

Miick SM, Fee RS, Millar DP, Chazin WJ (1997) Crossover isomer bias is the primary sequence-dependent property of immobilized Holliday junctions. Proc Natl Acad Sci USA 94:9080-9084

Mullen JR, Kaliraman V, Ibrahim SS, Brill SJ (2001) Requirement for three novel protein complexes in the absence of the Sgs1 DNA helicase in *Saccharomyces cerevisiae*. Genetics 157:103-118

Newman M, Murray-Rust J, Lally J, Rudolf J, Fadden A, Knowles PP, White MF, McDonald NQ (2005) Structure of an XPF endonuclease with and without DNA suggests a model for substrate recognition. EMBO J 24:895-905

Niedernhofer LJ, Odijk H, Budzowska M, van Drunen E, Maas A, Theil AF, de Wit J, Jaspers NG, Beverloo HB, Hoeijmakers JH, Kanaar R (2004) The structure-specific endonuclease Ercc1-Xpf is required to resolve DNA interstrand cross-link-induced double-strand breaks. Mol Cell Biol 24:5776-5787

Nishino T, Komori K, Ishino Y, Morikawa K (2003) X-ray and biochemical anatomy of an archaeal XPF/Rad1/Mus81 family nuclease: similarity between its endonuclease domain and restriction enzymes. Structure 11:445-457

Nishino T, Komori K, Ishino Y, Morikawa K (2005) Structural and functional analyses of an archaeal XPF/Rad1/Mus81 nuclease: asymmetric DNA binding and cleavage mechanisms. Structure (Camb) 13:1183-1192

Nishino T, Komori K, Tsuchiya D, Ishino Y, Morikawa K (2001) Crystal structure of the archaeal Holliday junction resolvase Hjc and implications for DNA recognition. Structure 9:197-204

Noguchi E, Noguchi C, Du LL, Russell P (2003) Swi1 prevents replication fork collapse and controls checkpoint kinase Cds1. Mol Cell Biol 23:7861-7874

Noguchi E, Noguchi C, McDonald WH, Yates JR 3rd, Russell P (2004) Swi1 and Swi3 are components of a replication fork protection complex in fission yeast. Mol Cell Biol 24:8342-8355

Nowotny M, Gaidamakov SA, Crouch RJ, Yang W (2005) Crystal structures of RNase H bound to an RNA/DNA hybrid: substrate specificity and metal-dependent catalysis. Cell 121:1005-1016

Odagiri N, Seki M, Onoda F, Yoshimura A, Watanabe S, Enomoto T (2003) Budding yeast *mms4* is epistatic with *rad52* and the function of Mms4 can be replaced by a bacterial Holliday junction resolvase. DNA Repair (Amst) 2:347-358

Ogrunc M, Sancar A (2003) Identification and characterization of human MUS81-MMS4 structure-specific endonuclease. J Biol Chem 278:21715-21720

Oram M, Keeley A, Tsaneva I (1998) Holliday junction resolvase in *Schizosaccharomyces pombe* has identical endonuclease activity to the CCE1 homologue YDC2. Nucleic Acids Res 26:594-601

Osman F, Dixon J, Doe CL, Whitby MC (2003) Generating crossovers by resolution of nicked Holliday junctions: a role for Mus81-Eme1 in meiosis. Mol Cell 12:761-774

Panyutin IG, Biswas I, Hsieh P (1995) A pivotal role for the structure of the Holliday junction in DNA branch migration. EMBO J 14:1819-1826

Panyutin IG, Hsieh P (1994) The kinetics of spontaneous DNA branch migration. Proc Natl Acad Sci USA 91:2021-2025

Parker JL, White MF (2005) The endonuclease Hje catalyses rapid, multiple turnover resolution of Holliday junctions. J Mol Biol 350:1-6

Pebernard S, Wohlschlegel J, McDonald WH, Yates JR 3rd, Boddy MN (2006) The nse5-nse6 dimer mediates DNA repair roles of the smc5-smc6 complex. Mol Cell Biol 26:1617-1630

Pingoud A, Fuxreiter M, Pingoud V, Wende W (2005) Type II restriction endonucleases: structure and mechanism. Cell Mol Life Sci 62:685-707

Privezentzev CV, Keeley A, Sigala B, Tsaneva IR (2005) The role of RuvA octamerization for RuvAB function *in vitro* and *in vivo*. J Biol Chem 280:3365-3375

Raaijmakers H, Vix O, Toro I, Golz S, Kemper B, Suck D (1999) X-ray structure of T4 endonuclease VII: a DNA junction resolvase with a novel fold and unusual domain-swapped dimer architecture. EMBO J 18:1447-1458

Rafferty JB, Bolt EL, Muranova TA, Sedelnikova SE, Leonard P, Pasquo A, Baker PJ, Rice DW, Sharples GJ, Lloyd RG (2003) The structure of *Escherichia coli* RusA endonuclease reveals a new Holliday junction DNA binding fold. Structure 11:1557-1567

Roberts JA, White MF (2005) An archaeal endonuclease displays key properties of both eukaryal XPF-ERCC1 and Mus81. J Biol Chem 280:5924-5928

Ross-Macdonald P, Roeder GS (1994) Mutation of a meiosis-specific MutS homolog decreases crossing over but not mismatch correction. Cell 79:1069-1080

Rupp WD, Wilde CE 3rd, Reno DL, Howard-Flanders P (1971) Exchanges between DNA strands in ultraviolet-irradiated *Escherichia coli*. J Mol Biol 61:25-44

Saintigny Y, Makienko K, Swanson C, Emond MJ, Monnat RJ Jr (2002) Homologous recombination resolution defect in werner syndrome. Mol Cell Biol 22:6971-6978

Santucci-Darmanin S, Walpita D, Lespinasse F, Desnuelle C, Ashley T, Paquis-Flucklinger V (2000) MSH4 acts in conjunction with MLH1 during mammalian meiosis. FASEB J 14:1539-1547

Saravanan M, Bujnicki JM, Cymerman IA, Rao DN, Nagaraja V (2004) Type II restriction endonuclease R.KpnI is a member of the HNH nuclease superfamily. Nucleic Acids Res 32:6129-6135

Schofield MJ, Lilley DM, White MF (1998) Dissection of the sequence specificity of the Holliday junction endonuclease CCE1. Biochemistry 37:7733-7740

Shah R, Bennett RJ, West SC (1994) Genetic recombination in *E. coli*: RuvC protein cleaves Holliday junctions at resolution hotspots *in vitro*. Cell 79:853-864

Shah R, Cosstick R, West SC (1997) The RuvC protein dimer resolves Holliday junctions by a dual incision mechanism that involves base-specific contacts. EMBO J 16:1464-1472

Sharples GJ (2001) The X philes: structure-specific endonucleases that resolve Holliday junctions. Mol Microbiol 39:823-834

Sharples GJ, Curtis FA, McGlynn P, Bolt EL (2004) Holliday junction binding and resolution by the Rap structure-specific endonuclease of phage lambda. J Mol Biol 340:739-751

Sherratt DJ, Soballe B, Barre FX, Filipe S, Lau I, Massey T, Yates J (2004) Recombination and chromosome segregation. Philos Trans R Soc Lond B Biol Sci 359:61-69

Smith GR, Boddy MN, Shanahan P, Russell P (2003) Fission yeast Mus81.Eme1 Holliday junction resolvase is required for meiotic crossing over but not for gene conversion. Genetics 165:2289-2293

Snowden T, Acharya S, Butz C, Berardini M, Fishel R (2004) hMSH4-hMSH5 recognizes Holliday junctions and forms a meiosis-specific sliding clamp that embraces homologous chromosomes. Mol Cell 15:437-451

Sogo JM, Lopes M, Foiani M (2002) Fork reversal and ssDNA accumulation at stalled replication forks owing to checkpoint defects. Science 297:599-602

Szostak JW, Orr-Weaver TL, Rothstein RJ, Stahl FW (1983) The double-strand-break repair model for recombination. Cell 33:25-35

Takahagi M, Iwasaki H, Shinagawa H (1994) Structural requirements of substrate DNA for binding to and cleavage by RuvC, a Holliday junction resolvase. J Biol Chem 269:15132-15139

Thacker J (2005) The RAD51 gene family, genetic instability and cancer. Cancer Lett 219:125-135

Torres-Rosell J, Machin F, Farmer S, Jarmuz A, Eydmann T, Dalgaard JZ, Aragon L (2005) SMC5 and SMC6 genes are required for the segregation of repetitive chromosome regions. Nat Cell Biol 7:412-419

Tripsianes K, Folkers G, Ab E, Das D, Odijk H, Jaspers NG, Hoeijmakers JH, Kaptein R, Boelens R (2005) The structure of the human ERCC1/XPF interaction domains reveals a complementary role for the two proteins in nucleotide excision repair. Structure 13:1849-1858

Tsodikov OV, Enzlin JH, Scharer OD, Ellenberger T (2005) Crystal structure and DNA binding functions of ERCC1, a subunit of the DNA structure-specific endonuclease XPF-ERCC1. Proc Natl Acad Sci USA 102:11236-11241

van Gool AJ, Hajibagheri NM, Stasiak A, West SC (1999) Assembly of the *Escherichia coli* RuvABC resolvasome directs the orientation of Holliday junction resolution. Genes Dev 13:1861-1870 van Gool AJ, Shah R, Mezard C, West SC (1998) Functional interactions between the Holliday junction resolvase and the branch migration motor of *Escherichia coli*. EMBO J 17:1838-1845

Vance JR, Wilson TE (2002) Yeast Tdp1 and Rad1-Rad10 function as redundant pathways for repairing Top1 replicative damage. Proc Natl Acad Sci USA 99:13669-13674

Venclovas C, Siksnys V (1995) Different enzymes with similar structures involved in Mg(2+)-mediated polynucleotidyl transfer. Nat Struct Biol 2:838-841

Wang TF, Kleckner N, Hunter N (1999) Functional specificity of MutL homologs in yeast: evidence for three Mlh1-based heterocomplexes with distinct roles during meiosis in recombination and mismatch correction. Proc Natl Acad Sci USA 96:13914-13919

Wang TF, Kung WM (2002) Supercomplex formation between Mlh1-Mlh3 and Sgs1-Top3 heterocomplexes in meiotic yeast cells. Biochem Biophys Res Commun 296:949-953

Whitby MC (2005) Making crossovers during meiosis. Biochem Soc Trans 33:1451-1455

Whitby MC, Bolt EL, Chan SN, Lloyd RG (1996) Interactions between RuvA and RuvC at Holliday junctions: inhibition of junction cleavage and formation of a RuvA-RuvC-DNA complex. J Mol Biol 264:878-890

Whitby MC, Dixon J (1997) A new Holliday junction resolving enzyme from *Schizosaccharomyces pombe* that is homologous to CCE1 from *Saccharomyces cerevisiae*. J Mol Biol 272:509-522

Whitby MC, Osman F, Dixon J (2003) Cleavage of model replication forks by fission yeast Mus81-Eme1 and budding yeast Mus81-Mms4. J Biol Chem 278:6928-6935

White MF, Lilley DM (1996) The structure-selectivity and sequence-preference of the junction- resolving enzyme CCE1 of *Saccharomyces cerevisiae*. J Mol Biol 257:330-341

White MF, Lilley DM (1997a) Characterization of a Holliday junction-resolving enzyme from *Schizosaccharomyces pombe*. Mol Cell Biol 17:6465-6471

White MF, Lilley DM (1997b) The resolving enzyme CCE1 of yeast opens the structure of the four-way DNA junction. J Mol Biol 266:122-134

White MF, Lilley DM (1998) Interaction of the resolving enzyme YDC2 with the four-way DNA junction. Nucleic Acids Res 26:5609-5616

Wu L, Hickson ID (2003) The Bloom's syndrome helicase suppresses crossing over during homologous recombination. Nature 426:870-874

Yamada K, Ariyoshi M, Morikawa K (2004) Three-dimensional structural views of branch migration and resolution in DNA homologous recombination. Curr Opin Struct Biol 14:130-137

Yildiz O, Majumder S, Kramer B, Sekelsky JJ (2002) *Drosophila* MUS312 interacts with the nucleotide excision repair endonuclease MEI-9 to generate meiotic crossovers. Mol Cell 10:1503-1509

Zerbib D, Mezard C, George H, West SC (1998) Coordinated actions of RuvABC in Holliday junction processing. J Mol Biol 281:621-630

Whitby, Matthew C.

Department of Biochemistry, University of Oxford, South Parks Road, Oxford OX1 3QU, UK

matthew.whitby@bioch.ox.ac.uk

Replication forks and replication checkpoints in repair

Dana Branzei and Marco Foiani

Abstract

Eukaryotic cells replicate their DNA and coordinate their response to DNA damage and replication blocks by activating appropriate repair processes, regulating recombination, chromatin assembly and chromosome partitioning. Replication forks stall at specific problematic genomic regions, and forks collapse unless protected by replication checkpoint proteins. These events have been associated with recombination and chromosomal rearrangements that lead to genomic instability and cancer development. The replication checkpoints, activated by the checkpoint signals generated by stalled forks, protect the stability of the fork until the replication can resume, regulate recombination pathways, and coordinate the mechanisms that promote replication restart and repair. Domain barriers make easier the topological problems posed by replicating DNA and confine the DNA lesions in manageable units. Here, we focus on the molecular mechanisms that control and promote the stability of replication forks and on the regulation of replication restart, and its coordination with chromatin structure and postreplicative repair.

1 DNA replication, checkpoint proteins, and chromosome integrity

Eukaryotic chromosome replication is a very complex event that must be thoroughly monitored and coordinated with other aspects of chromosome metabolism such as DNA topology, recombination, repair, and sister chromatid cohesion in order to ensure the stability of replication forks, proper resumption of replication, and postreplicative repair of DNA lesions. Checkpoints are cellular surveillance and signaling pathways that coordinate these physiological responses (Elledge 1996; Melo and Toczyski 2002; Branzei and Foiani 2005). Failure of these controls dramatically impairs cell resistance to genotoxic stress or replication blocks and can lead to profound genome instability (Paulovich et al. 1997; Desany et al. 1998; Lopes et al. 2001; Tercero and Diffley 2001; Admire et al. 2006). An increasing number of evidence suggests that both yeast and mammalian checkpoint proteins preserve genome integrity by regulating DNA replication, and in particular, by stabilizing replication forks and fragile sites (Lopes et al. 2001; Tercero and Diffley 2001; Casper et al. 2002; Cha and Kleckner 2002; Sogo et al. 2002;

Lemoine et al. 2005; Admire et al. 2006). Mammalian fragile sites are chromosomal domains prone to breakage when replication is disrupted or in the absence of the replication checkpoint (Casper et al. 2002). This leads to increased recombination, chromosomal rearrangements, activation of oncogenes, and cancer (Glover 2006). In addition, the DNA replication checkpoint plays an important role in the maintenance of chromosomal integrity and repeat stability of trinucleotide repeats (TNRs) sequences (Freudenreich and Lahiri 2004; Lahiri et al. 2004), the expansion of which causes chromosome breakage and several genetic diseases. In line with these observations, many human genetic syndromes that lead to cancer predisposition are caused by mutations in genes that protect the genome integrity during chromosome replication.

This linkage between replication checkpoints and cancer underscores the importance of the regulation of DNA replication to prevent genome abnormalities. In the following sections, we discuss the molecular mechanisms employed by replication checkpoints to stabilize the replication forks and to assist and coordinate different damage-tolerance mechanisms that contribute to repair and chromosome integrity.

2 Stalled versus collapsed replication forks and fork stabilization versus fork restart

Replication fork progression is normally slowed down at certain genomic regions such as those containing specialized protein-mediated replication fork barriers (Dalgaard and Klar 2000; Takeuchi et al. 2003), replication slow zones and fragile sites (see Sect. 1), or tRNA genes (Deshpande and Newlon 1996). Replication fork pausing or stalling is also induced by blocking replication fork progression as is the case of cells experiencing intra-S damage or replicating with limited pools of dNTPs as caused by hydroxyurea (HU) treatment (reviewed in Branzei and Foiani 2005). In most cases, forks stall with a fully assembled replisome. The collapse is prone to occur when the replisome is falling apart as in HU treated replication checkpoint mutants (Sogo et al. 2002), at DNA breaks, telomeres and, likely, at certain replication-risk zones.

The stalled replication forks are thought to be stabilized by the replication checkpoint and to resume replication once the block or the replication impediment has been removed (Fig. 1). Alternatively, when exposed to damage, forks can synthesize primers downstream the lesion, on both leading and lagging strands, and restart replication (Heller and Marians 2006; Lee et al. 2006), generating daughter-strand gaps that can be repaired postreplicatively by recombination or damage tolerance mechanisms (Goldfless et al. 2006; and see Fig. 2). These mechanisms have been mostly described in prokaryotic cells. When forks collapse, the DNA in those regions can be passively replicated by forks converging from adjacent replicons (Branzei and Foiani 2005), or they can restart using alternative pathways such as recombination-mediated mechanisms. The replication checkpoints have

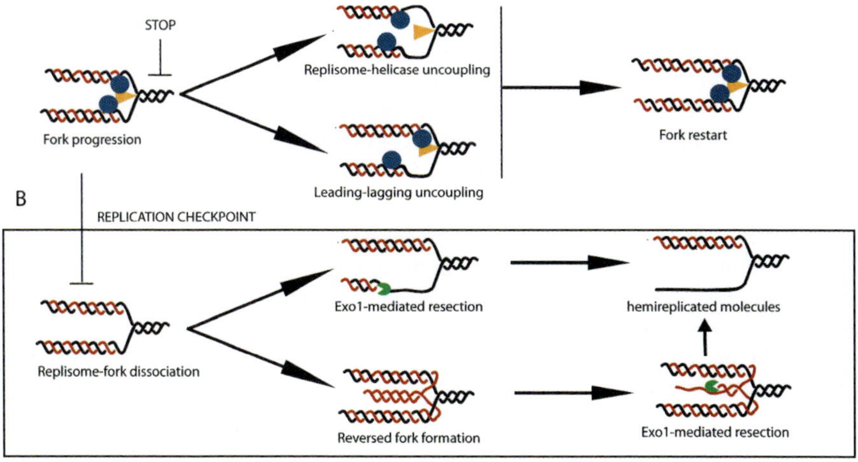

Fig. 1. Physiological and pathological transitions at stalled forks. Replication forks experiencing pausing in wild type cells generate single stranded DNA regions by either uncoupling leading and lagging strand synthesis or the replisome from the helicase. Fork resumption will occur by normal restart. In the absence of a functional checkpoint, the replisome disassembles from stalled forks and forks degenerate by undergoing Exo1-mediated resection of nascent chains or by experiencing fork reversal and further processing by Exo1.

been implicated in fork recovery (Desany et al. 1998; Trenz et al. 2006; and see Sect. 3)

3 Sensing stalled forks and checkpoint mediated stabilization of stalled forks

Large single-stranded DNA (ssDNA) regions are formed at stalled replication forks (Sogo et al. 2002; Zou and Elledge 2003; Byun et al. 2005; Feng et al. 2006), and are coated with RPA to generate checkpoint signals (see Branzei and Foiani 2005). In addition, RNA primers may contribute to checkpoint activation (Michael et al. 2000), together with RPA and ssDNA. DNA synthesis was shown to be required for checkpoint activation in *Xenopus* (Byun et al. 2005) or for efficient DNA unwinding by T7 helicase (Stano et al. 2005). One possibility is that a functional uncoupling between the MCM helicase and polymerase activities at the fork is always required for checkpoint signaling as it has been suggested (Byun et al. 2005). If such is the case, lesions that inhibit the helicase from unwinding are expected not to generate a checkpoint response, and indeed, it was shown that blockage of fork progression in *S. pombe* by DNA-protein complexes at a specific genomic location does not cause checkpoint activation (Lambert et al. 2005). However, extended unwinding in front of the fork upon encountering blocks to DNA synthesis would generate large ssDNA regions on both sides of the same fork that could be detected by electron microscopy (EM) coupled with psoralen crosslinking, but such structures were not detected (Lopes et al. 2006). Thus, it is likely that transient uncoupling between leading and lagging strand synthesis (Sogo et al. 2002) and Exo1 (Cotta-Ramusino et al. 2005) or other protein or recombination factor-mediated processing of structures generated at collapsed forks (Boddy and Russell 2001; Kai et al. 2005) generate most of the ssDNA that becomes signal for the replication checkpoint (see Fig. 1). These ssDNA regions coated by RPA trigger the recruitment of the Mec1 kinase in yeast (or ATR in human cells), which together with other checkpoint factors, mediates Rad53 phosphorylation at multiple sites (Pellicioli and Foiani 2005). It was shown that a certain threshold of ssDNA must be produced in order to activate the checkpoint response (Shimada et al. 2002; Vaze et al. 2002; Byun et al. 2005). Furthermore, in addition to the original checkpoint signal, recent studies suggest that following DNA damage, changes to chromatin structure are involved either in the initial sensing or in the subsequent amplification of the DNA damage response; these aspects will be discussed in Section 5.

One important function of the replication checkpoint in response to replication blocks is to prevent a subset of origins from activation (Santocanale and Diffley 1998; Shirahige et al. 1998) and to maintain the integrity of existing replication forks (Lopes et al. 2001; Tercero and Diffley 2001; Sogo et al. 2002). In replication checkpoint mutants, the replisome dissociates (Cobb et al. 2003; Lucca et al. 2004) and the stalled forks rapidly degenerate accumulating gapped and hemireplicated molecules as well as four branched molecules resembling reversed forks (Sogo et al. 2002; Feng et al. 2006; see Fig. 1). The formation of hemireplicated and gapped molecules is largely due to defects in lagging strand synthesis, perhaps caused by a misregulation of the lagging strand apparatus in replication checkpoint mutants (Pellicioli et al. 1999; Lopes et al. 2001; Sogo et al. 2002), and to the unscheduled processing of nascent chains by Exo1 (Cotta-Ramusino et al. 2005). At least two mechanisms could account for fork reversal in checkpoint mutants: 1) the run off of the sister chromatid junctions (SCJs) resembling hemicatenanes at stalled forks deprived of the replisome (Lopes et al. 2003) and 2) superhelical strain building up at the replication fork (Postow et al. 2001a, 2001b). While the exact mechanism remains to be worked out, the accumulation of reversed forks is counteracted by Exo1 (Cotta-Ramusino et al. 2005)(see Fig. 1). It is generally thought that the abnormal replication intermediates formed during replication or in checkpoint mutants are processed by unscheduled recombination pathways, causing genome instability (Sogo et al. 2002; Branzei and Foiani 2005; Lemoine et al. 2005; Admire et al. 2006). Thus, conceivably, another important task of the replication checkpoint should be to restrain the activity of recombination enzymes at stalled forks (Boddy et al. 2003; Kai et al. 2005), which could efficiently resume replication after the block is removed, or restart through alternative pathways, if not processed by recombination activities (see Sect. 4).

Mutants defective in the replication checkpoint kinase in budding and fission yeast are acutely sensitive to HU or DNA damaging agents and are unable to resume replication even upon removal of the replication block (Lopes et al. 2001; Tercero and Diffley 2001; Tercero et al. 2003). These results suggest either that the replication forks in checkpoint mutants are collapsed and unable to restart, or that the checkpoint mutants are defective in promoting fork restart processes downstream the lesions and that this event leads to extensive fork collapse or breakage. How Rad53 promotes the synthesis of complete chromosomes after replication insults is unclear, but recent evidence suggests a role for the replication checkpoint in facilitating replication restart (Lopes et al. 2006; Trenz et al. 2006). EM and two-dimensional (2D) gel electrophoresis of replicating DNA from UV-irradiated budding yeast cells revealed that besides the degeneration of forks in *rad53* mutants, these cells also contained a high fraction of large gapped forks (Lopes et al. 2006), suggesting that some steps of replication restart after lesion induced DNA synthesis block is defective and that checkpoints might also be needed to assist different pathways contributing to "gap repair" or damage-tolerance mechanisms as it has been proposed by other studies (Paulovich et al. 1998; Kai and Wang 2003). More direct evidence for a role of ATM and ATR in promoting restart of damaged replication forks has come from studies of replicating chromosomes in *Xenopus laevis* egg extracts (Trenz et al. 2006). In this study, recovery of collapsed forks is monitored by transferring replicating forks collapsed by camptothecin or mitomycin C in the absence of active ATM and ATR, into *Xenopus* egg "restarting extracts", lacking damaging agents and in which origin firing is inhibited. In restarting extracts, ATM and ATR get activated by the damaged DNA and promote recovery of the collapsed forks; however, extracts lacking ATM and ATR do not support fork recovery, thus indicating an important role for the replication checkpoints in promoting repair and restart of collapsed forks (Trenz et al. 2006). The main mechanisms implicated in fork restart in eukaryotic organisms are discussed in the following section.

4 Replication fork restart and repair mechanisms

As previously discussed, several lines of evidence suggest that checkpoint proteins are important not only for regulating cell-cycle progression and the integrity of the fork in response to DNA damage or nucleotide depletion events but also for activating replication resumption processes. The effectors of the replication checkpoint relevant for this recovery process are largely unknown, but it is believed that, as in the case of fork stabilization, phosphorylation of targets and their subsequent recruitment to sites of damage is important. In this section, we discuss the present understanding of the mechanisms implicated in restarting and repairing of the replication forks, and their possible association with checkpoint regulated cell cycle events. Once forks in eukaryotic organisms face DNA lesions that block the DNA polymerase, they can restart downstream the lesion and during this process the primase activity is expected to play an important role. This can be followed by bypass of the damage by means of translesion synthesis (TLS), template switch, or by recombination-mediated gap repair (Fig. 2).

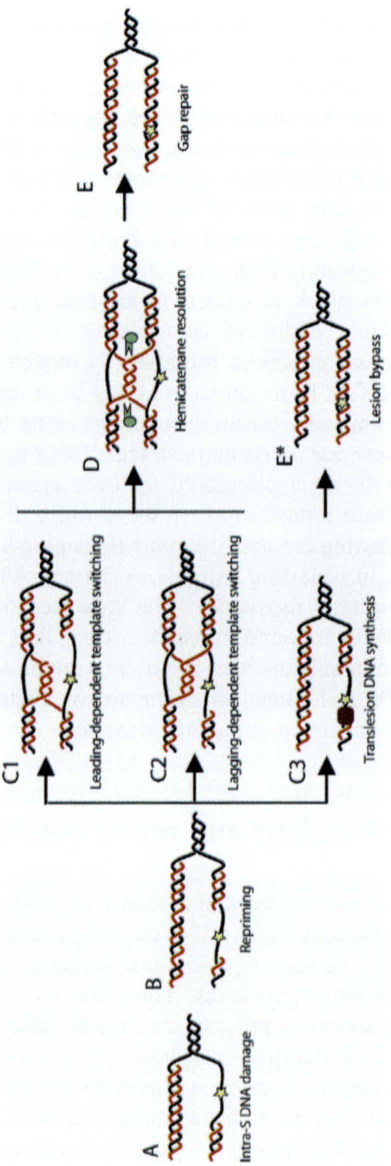

Fig. 2. Alternative mechanisms for repairing intra-S DNA lesions. Replication forks encountering lesions on the template (A) will reinitiate DNA synthesis downstream of the lesion by re-priming (B). Several options are possible at this stage: forks can undergo template switching either at the leading (C1) or at the lagging (C2) strand. This will generate hemicatenane-like structures that will be solved by the coordinated action of Sgs1 and Top3 (D) leading to gap repair (E). Alternatively, translesion DNA polymerases can be recruited at the gap (C3) to synthesize DNA across the lesions leading to gap repair (E*). It should be noted that this can occur both in an error-free or error-prone manner.

4.1 Recombination-mediated fork restart and repair

Although recombination plays an important role in completing replication, especially in certain situations discussed below, it is important to bear in mind that the contribution of recombination to replication completion in eukaryotic cells is different from that in *E. coli*, which have a single origin of replication and therefore rely on recombination mechanisms for fork repair and reconstitution of DNA replication. The ability of recombination to restore forks has been demonstrated *in vivo* for both prokaryotes and eukaryotes and has been termed "recombination-dependent replication", or "break-induced replication", respectively (Kogoma 1997; Kraus et al. 2001).

In eukaryotes, there are two main situations in which recombination mechanisms are thought to be the major pathways assisting completion of replication: 1) when forks collapse in regions where there are no converging forks that could complete replication such as telomeric or subtelomeric regions, and 2) when the DNA lesions or the stalled replication forks are processed to double-strand breaks (DSBs). The assembly of a replication fork by recombination mediated processes appears to require most proteins involved in replication elongation, but not replication initiation factors or MCM (Wang et al. 2004), in line with the view that MCM loading is restricted to G1 and is coupled with regulation of origin firing (Labib et al. 2000) . Interestingly, in *Xenopus* egg extracts, MCM7 is not removed from damaged chromatin in conditions that induce fork collapse (Trenz et al. 2006), and it is known that in vertebrate cell multiple MCM complexes are loaded on the chromatin for every active origin (Edwards et al. 2002). As MCM2 and MCM7 are phosphorylated by ATR (Cortez et al. 2004; Yoo et al. 2004), it is tempting to think that checkpoint mediated phosphorylation of the excess MCM might help to restart collapsed forks, perhaps by promoting the loading of certain replication factors required for replication resumption.

4.2 Checkpoint-mediated regulation of recombination

The DNA damage and replication checkpoint is required for DSB-induced homologous recombination (HR). In response to DSBs, there is a dramatic relocalization of certain damage response or repair proteins to sub-nuclear structures called repair foci (Fernandez-Capetillo et al. 2003; Lisby et al. 2004). One key regulator of repair foci formation, especially in mammalian cells, is the histone H2A variant, H2AX, a component of the nucleosome core structure and comprising over 10% of total H2A in higher organisms (Celeste et al. 2002, 2003). DNA damage induces H2AX phosphorylation by checkpoint kinases ATM, ATR, as well as by DNA-PKcs in human cells or their yeast homologues. The phosphorylated H2AX, referred to as γH2AX, binds specifically to the damaged chromosome region and does not spread throughout the nucleus; the γH2AX is thought to interact with specific repair proteins and to be important for repair of DSBs, perhaps by assisting the efficiency and fidelity of both HR and NHEJ, the main pathways implicated in DSB repair in mammalian cells (reviewed in Stucki and Jackson 2006). In addition, γH2AX together with methylated histone 3 (H3K79) are required for human Rad9 recruitment to DSBs, and in mammalian cells, the γH2AX-MDC1 interaction enhances ATM phosphorylation of H2AX, contributing perhaps to the amplification of the damage response (reviewed in Stucki and Jackson 2006). Thus, checkpoint mediated phosphorylation of H2AX is thought to play an important role in promoting efficient DSB repair.

It is not yet clear which are the polymerases mainly implicated in HR of DSB repair (Kawamoto et al. 2005; McIlwraith et al. 2005; Hirano and Sugimoto 2006). Recent *in vivo* and *in vitro* studies with polη from chicken or human cells suggest that it functions to extend 3' strands exchanged during homologous recombination (Kawamoto et al. 2005; McIlwraith et al. 2005); in addition, Mec1-dependent phosphorylation was shown to promote Polζ-Rev1 association with DSBs (Hirano and Sugimoto 2006). In *S. pombe* the activity and chromatin binding of the translesion synthesis polymerase DinB in response to replication perturbations appears to be regulated by the DNA damage checkpoint (Kai and Wang 2003), and a similar interaction between the damage checkpoint and polζ has been reported also in *S. cerevisiae* (Sabbioneda et al. 2005). Thus, an interesting possibility is that checkpoints might mediate TLS polymerases in HR of DSB repair.

While replication checkpoints seem to promote postreplicative DSB repair, they also function to inhibit inappropriate recombination from occurring during replication at stalled forks. Consistent with this, following intra-S damage, it was shown that in fission yeast recombination foci are rare in S-phase and peak in G2 (Meister et al. 2005), and Rad52 recombination foci are not seen in budding yeast treated with HU, unless the replication checkpoint is not functional (Lisby et al. 2004). The direct effectors and mechanisms of this regulation are just starting to be elucidated. In fission yeast, Mus81 and Rad60 proteins were proposed to promote recombination mediated replication restart, and the replication checkpoint to down-regulate their recombination activities during replication (Boddy et al. 2003; Kai et al. 2005). Mus81 has Holliday junction resolvase activity (Gaillard et al. 2003), and *S. pombe* Rad60 interacts with members of the Smc5-6 complex, implicated in DNA repair and maintenance of chromosome structure and stability, and is required for repair of DSBs (Morikawa et al. 2004). Upon HU treatment, the replication checkpoint Cds1 mediates the phosphorylation of Mus81 and Rad60 and this modification is associated with delocalization of Rad60 from the nucleus and a reduction in the chromatin binding ability of the Mus81-Eme1 endonuclease complex (Boddy et al. 2003; Kai et al. 2005).

4.3 Other fork restart mechanisms: damage tolerance or postreplication repair pathways

In addition to recombination pathways that, as we have already discussed, can "repair", the damage caused by the lesions or the forks collapsed when encountering the lesions, other mechanisms concerned more with the efficient bypass of lesions than with the repair process *per se* have an important role in contributing to cell survival in response to replication stress and they are usually known as damage tolerance processes or postreplication repair mechanisms.

Two processes have been characterized and ascribed to this class: translesion replication or translesion synthesis (TLS), in which specialized DNA polymerases replicate across lesions, often generating mutations, and an error-free damage avoidance pathway, which accounts for the major fraction of damage tolerance, but whose mechanism is still poorly understood.

In budding yeast, the damage tolerance depends on the *RAD6/RAD18* pathway, which has been involved in both TLS and error-free damage avoidance. Rad6/Rad18 mediated monoubiquitination of PCNA promotes translesion synthesis and damage-induced mutagenesis by TLS polymerases (Stelter and Ulrich 2003; Kannouche et al. 2004; Watanabe et al. 2004), which in yeast involve the activity of Polη, encoded by *RAD30*, and Polζ, whose two subunits are encoded by *REV3* and *REV7*, together with Rev1. In contrast, polyubiquitination of PCNA, in which the Rad5/Mms2/Ubc13 complex conjugates ubiquitin moieties to Lys63 of ubiquitin itself (Hoege et al. 2002), promotes the error-free damage avoidance pathway and replication completion in response to replication perturbations induced by replication mutants (Branzei et al. 2004). The high sensitivity of *rad6*, *rad18*, and *rad5* mutants to DNA damaging agents attests to their important contribution in promoting repair or replication fork restart following intra-S damage. The involvement of the Rad6/Rad18/ Rad5/Mms2 branch in the error-free damage bypass mechanisms has been originally known from studies measuring the molecular size of DNA synthesized after UV irradiation in excision repair deficient strains, in different mutant backgrounds (Prakash 1981). In wild type cells, after UV irradiation, the newly synthesized DNA is smaller in size than its counterpart isolated from unirradiated cells, indicative of the presence of gaps; however, after a period of incubation, the DNA reaches the same size as the control one, suggesting that the gaps had been filled. However, the conversion of nascent DNA from small to large size is very much delayed in *rad6/ rad18/rad5/mms2* strains, suggesting an important role for these activities in gap-filling repair. This conversion is not detectably dependent on DNA polymerase ζ, but is partially reduced in *rad52* strains, suggesting a role for this protein in this process. These results are also supported and extended in several important ways by recent studies that measured the contribution of Rad18/Rad5, Rad52, and TLS polymerases in assisting the completion of replication of plasmids with single thymine-thymine pyrimidine (6-4) pyrimidinone photoadducts in each strand at staggered positions 28 base pairs (bp) apart in *rad1* excision defective strains (Zhang and Lawrence 2005). This system permits the unambiguous detection of the two different type of events that can lead to completion of replication, template strand switching or TLS events, on one or the other strands; the results indicate that in more than 90% of the successful events involve the template switch mechanism, with 60-70% of these events depending on the *RAD18/RAD5*, and the remaining events on *RAD52* (Zhang and Lawrence 2005).

Thus, it appears that the majority of the stalled replication forks are restarted by a Rad18/Rad5-mediated template switch mechanism, but a small fraction of them become substrates for recombination. This conclusion is strongly supported by the reports showing that mutants defective in error-free postreplicative repair (*rad18/rad5*), but not in TLS or HR repair, show a high increase in gross chromosomal rearrangements (GCR) rates (Motegi et al. 2006), attesting to their important contribution to gap filling and replication completion (Branzei et al. 2004); furthermore the GCR increase in *rad5/rad18* mutants is dependent on HR activities, suggesting that in the absence of Rad18/Rad5, the ssDNA gaps can be filled in by recombination mechanisms. That some of the Rad18/Rad5/Mms2 template switch substrates at damaged forks could be diverted into the Rad51/Rad52 pathway is clearly indicated also by the phenotype of *srs2* mutations, which can suppress the damage sensitivity of *rad18* and *rad6* mutants, but not their damage-induced mutagenesis defect, into a Rad52-dependent manner (Lawrence and Christensen 1979; Rong et al. 1991). Srs2 is a DNA helicase that disrupts Rad51 nucleoprotein filaments, and perhaps functions to prevent stalled forks from being broken or from generating substrates for recombination repair, while somehow mediating the retaining of the structures generated at stalled or damaged forks in a state compatible with transient template switch recombination events. This presumptive role of Srs2 in stabilizing the stalled forks is somehow reminiscent of that of the replication checkpoint, although by 2D gel no accumulation of pathological structures was so far detected (Liberi et al. 2005). Nevertheless, it is important to mention that Srs2 is a checkpoint target and in response to damage it is engaged in different repair complexes (Chiolo et al. 2005). Whether Srs2 checkpoint mediated phosphorylation is important for its ability to regulate different modes of template switch or repair pathways at the stalled replication forks is an interesting question that remains to be answered.

In addition to Srs2, which is thought to act to prevent the formation of recombinogenic structures in response to intra-S damage, RecQ helicases have also been proposed to regulate recombination events occurring during replication and to promote the stability of replication forks. In budding yeast, the Sgs1-Top3 complex was proposed to act by promoting the resolution of hemicatenane-like structures likely resulting from replication-related SCJs during template switch or during replication termination when replication forks converge (Wang 1991; Liberi et al. 2005). In addition, genetic and *in vitro* studies indicated that the RecQ-Top3 complex from different organisms function to resolve recombination intermediates such as double Holliday junctions (HJs), to lead to noncrossover products (Wu et al. 2002; Ira et al. 2003).

4.4 Damage bypass at the fork versus postreplication repair

The translesion synthesis and template switch mediated processes are largely thought to be damage tolerance pathways occurring at the fork when the DNA polymerase encounters the lesion. However, recent evidence suggests that these processes might involve multiple posttranslational modifications of replication factors and polymerase swap events, which are likely to be slow processes. Recent studies suggest that repriming on both leading and lagging strands might occur more frequently than initially expected (Heller and Marians 2006), and thus, many, if not all, damage bypass events might occur behind the replicating fork, or postreplicatively. In support of this view, EM analysis of replication intermediates from UV-irradiated excision deficient cells showed that internal gaps can be detected on both sides of the fork, and that cells deficient in translesion synthesis accumulate more internal gaps than their wild type counterparts, thus conveying the idea that a large proportion of the translesion synthesis events contribute to restore the integrity of replicated duplexes postreplicatively (Lopes et al. 2006).

Several studies in fission and budding yeast suggest an involvement of the damage checkpoint pathway in promoting the chromatin loading of certain translesion polymerases and in promoting damage-induced mutagenesis in response to replication problems (Paulovich et al. 1998; Kai and Wang 2003; Sabbioneda et al. 2005). It will be of interest to understand whether this function is conserved in other organisms as well, and to determine whether checkpoints have a direct role in promoting template switch, given the important role of this essentially error-free mechanism in promoting fork restart.

5 Coordination between DNA replication, topology, and chromatin structure

Replication and the topological problems posed by replicating DNA are made easier by topological domains, which are regions topologically constrained at their ends in which actively replicating DNA as well as DNA nicks or breaks can be confined, therefore preventing the bulk of the chromosome from precatenation and relaxation. Replication itself causes relaxation, and only fully replicated chromosomes are condensed by supercoiling, an event which promotes decatenation and chromosome disentaglement. The exact nature of the topological domains and their relationship with the moving fork are not known, but these boundaries may serve to concentrate the type-2 topoisomerases to remove positive supercoils in front of the fork or precatenanes behind the fork (Postow et al. 2001a). The control of topology by domains seems to be a widespread mechanism, as eukaryotic cells have also been shown to possess topological domains (Benyajati and Worcel 1976; Kramer and Sinden 1997). Studies in *E. coli* have shown that the topological domains are much smaller than originally thought, of about 10kb on average (Postow et al. 2004); small domains would present the advantage to reduce the amount of DNA that is relaxed by DNA breakage or during DNA replication and to help decatenation. The knowledge regarding the nature of the topological boundaries in eukaryotic cells is not more advanced than that in bacteria, but nevertheless, these topological boundaries are thought to provide structure for chromosomal DNA, to facilitate DNA repair, and to make easier for long DNA molecules to be organized, replicated, and segregated in the cell. The importance of a functional replication checkpoint in setting or maintaining topological domains or their relationship to the moving replication forks is even less understood, although there are intriguing findings that suggest a connection between the replication checkpoints and chromosome disentangling as well as a direct role for the replication checkpoint in regulating spindle dynamics (Krishnan et al. 2004; Bachant et al. 2005).

In addition, or perhaps in cooperation with the topological domain organization of replicating chromosomes, evidence based on genetic and physical interactions suggests that chromatin structure plays a critical role in the initiation and progression through S-phase and for survival of cells exposed to exogenous genotoxic treatments. A coordinated interplay between the replisome and factors involved in reorganizing nucleosomes and promoting establishment of sister chromatid cohesion is required for normal S-phase progression (Zhou and Wang 2004; Kats et al. 2006). In budding yeast, Rad53, but not Mec1 or Tel1, has been implicated in preventing the accumulation of nonnucleosomal histones in the cell (Gunjan and Verreault 2003), thus suggesting that Rad53 can monitor parameters associated with chromatin organization to maintain genome integrity. Several studies indicate that the checkpoint pathway may directly regulate chromatin assembly to promote fork stabilization or restart in response to DNA lesions or replication blocks. Supportive of these models, Rad53 was found to interact with Asf1, a key component of a conserved multisubunit replication-dependent chromatin assembly that acts during S-phase and during DNA repair of DSBs (Le et al. 1997; Tyler et al. 1999; Franco et al. 2005; Linger and Tyler 2005). Rad53 exists in a stable complex with Asf1 in unperturbed conditions, but releases Asf1 in response to both DNA damage and stalled DNA replication (Emili et al. 2001; Hu et al. 2001). In addition, numerous synthetic sick or synthetic lethal interactions have been reported between mutations in DNA replication genes and genes affecting complexes or pathways involved in histone modification, chromatin remodeling, and chromatin assembly (Sharp et al. 2001; Shen et al. 2003; Krogan et al. 2004; Pan et al. 2006) and many mutants affecting chromatin-structure maintenance are sensitive to DNA damaging affects and accumulate checkpoint signals.

In addition to the results discussed above that pinpoint to a role of Rad53 in mediating chromatin structure maintenance, chromatin modifications appear to play a role in checkpoint activation or the amplification of the checkpoint signal.

Methylation of Lys79 (K79) of histone H3 is a genome-wide histone modification, which has been implicated in the activation of the G1 and intra-S phase DNA damage checkpoints. This function is mediated by direct recognition of Dot1-dependent H3 K79 methylation through the Tudor domains of the DNA damage checkpoint proteins Rad9 in budding yeast, and 53BP1 in human (Huyen et al. 2004; Giannattasio et al. 2005; Wysocki et al. 2005). In *S. cerevisiae*, mutations in Dot1, H3 K79, or the Rad9 Tudor domain perturb the G1 and intra-S damage checkpoint, but not the Rad9-dependent G2/M checkpoint (Giannattasio et al. 2005; Wysocki et al. 2005), which could be mediated by recognition of γ-H2A (Nakamura et al. 2004). In addition to Dot1-mediated H3 methylation, histone H2B ubiquitilation by Rad6-Bre1 is also required for activation of Rad53 and cell cycle arrest (Giannattasio et al. 2005). There will be a challenge for future studies to establish the molecular mechanisms through which different type of histone modifications facilitate specific repair pathways, checkpoint activation, and cell cycle arrest.

Acknowledgements

The authors apologize for the many interesting articles that they were not able to discuss or acknowledge due to space limitations. We thank all members of Foiani's laboratory for useful discussions. This work was supported by grants form the Associazione Italiana per la Ricerca sul Cancro, European Union and Telethon-Italy to M. Foiani, and association for International Cancer Research to M. Foiani and D. Branzei. D. Branzei is supported by the European Community Grant LSHC-CT-2005-512113 and by the Buzzati-Traverso foundation.

References

Admire A, Shanks L, Danzl N, Wang M, Weier U, Stevens W, Hunt E, Weinert T (2006) Cycles of chromosome instability are associated with a fragile site and are increased by defects in DNA replication and checkpoint controls in yeast. Genes Dev 20:159-173

Bachant J, Jessen SR, Kavanaugh SE, Fielding CS (2005) The yeast S phase checkpoint enables replicating chromosomes to bi-orient and restrain spindle extension during S phase distress. J Cell Biol 168:999-1012

Benyajati C, Worcel A (1976) Isolation, characterization, and structure of the folded interphase genome of *Drosophila melanogaster*. Cell 9:393-407

Boddy MN, Russell P (2001) DNA replication checkpoint. Curr Biol 11:R953-956

Boddy MN, Shanahan P, McDonald WH, Lopez-Girona A, Noguchi E, Yates IJ, Russell P (2003) Replication checkpoint kinase Cds1 regulates recombinational repair protein Rad60. Mol Cell Biol 23:5939-5946

Branzei D, Foiani M (2005) The DNA damage response during DNA replication. Curr Opin Cell Biol 17:568-575

Branzei D, Seki M, Enomoto T (2004) Rad18/Rad5/Mms2-mediated polyubiquitination of PCNA is implicated in replication completion during replication stress. Genes Cells 9:1031-1042

Byun TS, Pacek M, Yee MC, Walter JC, Cimprich KA (2005) Functional uncoupling of MCM helicase and DNA polymerase activities activates the ATR-dependent checkpoint. Genes Dev 19:1040-1052

Casper AM, Nghiem P, Arlt MF, Glover TW (2002) ATR regulates fragile site stability. Cell 111:779-789

Celeste A, Difilippantonio S, Difilippantonio MJ, Fernandez-Capetillo O, Pilch DR, Sedelnikova OA, Eckhaus M, Ried T, Bonner WM, Nussenzweig A (2003) H2AX haploinsufficiency modifies genomic stability and tumor susceptibility. Cell 114:371-383

Celeste A, Petersen S, Romanienko PJ, Fernandez-Capetillo O, Chen HT, Sedelnikova OA, Reina-San-Martin B, Coppola V, Meffre E, Difilippantonio MJ, Redon C, Pilch DR, Olaru A, Eckhaus M, Camerini-Otero RD, Tessarollo L, Livak F, Manova K, Bonner WM, Nussenzweig MC, Nussenzweig A (2002) Genomic instability in mice lacking histone H2AX. Science 296:922-927

Cha RS, Kleckner N (2002) ATR homolog Mec1 promotes fork progression, thus averting breaks in replication slow zones. Science 297:602-606

Chiolo I, Carotenuto W, Maffioletti G, Petrini JH, Foiani M, Liberi G (2005) Srs2 and Sgs1 DNA helicases associate with Mre11 in different subcomplexes following checkpoint activation and CDK1-mediated Srs2 phosphorylation. Mol Cell Biol 25:5738-5751

Cobb JA, Bjergbaek L, Shimada K, Frei C, Gasser SM (2003) DNA polymerase stabilization at stalled replication forks requires Mec1 and the RecQ helicase Sgs1. Embo J 22:4325-4336

Cortez D, Glick G, Elledge SJ (2004) Minichromosome maintenance proteins are direct targets of the ATM and ATR checkpoint kinases. Proc Natl Acad Sci USA 101:10078-10083

Cotta-Ramusino C, Fachinetti D, Lucca C, Doksani Y, Lopes M, Sogo J, Foiani M (2005) Exo1 processes stalled replication forks and counteracts fork reversal in checkpoint-defective cells. Mol Cell 17:153-159

Dalgaard JZ, Klar AJS (2000) *swi1* and *swi3* perform imprinting, pausing, and termination of DNA replication in *S. pombe*. Cell 102:745-751

Desany BA, Alcasabas AA, Bachant JB, Elledge SJ (1998) Recovery from DNA replicational stress is the essential function of the S-phase checkpoint pathway. Genes Dev 12:2956-2970

Deshpande AM, Newlon CS (1996) DNA replication fork pause sites dependent on transcription. Science 272:1030-1033

Edwards MC, Tutter AV, Cvetic C, Gilbert CH, Prokhorova TA, Walter JC (2002) MCM2-7 complexes bind chromatin in a distributed pattern surrounding the origin recognition complex in *Xenopus* egg extracts. J Biol Chem 277:33049-33057

Elledge SJ (1996) Cell cycle checkpoints: preventing an identity crisis. Science 274:1664-1672

Emili A, Schieltz DM, Yates JR 3rd, Hartwell LH (2001) Dynamic interaction of DNA damage checkpoint protein Rad53 with chromatin assembly factor Asf1. Mol Cell 7:13-20

Feng W, Collingwood D, Boeck ME, Fox LA, Alvino GM, Fangman WL, Raghuraman MK, Brewer BJ (2006) Genomic mapping of single-stranded DNA in hydroxyurea-challenged yeasts identifies origins of replication. Nat Cell Biol 8:148-155

Fernandez-Capetillo O, Celeste A, Nussenzweig A (2003) Focusing on foci: H2AX and the recruitment of DNA-damage response factors. Cell Cycle 2:426-427

Franco AA, Lam WM, Burgers PM, Kaufman PD (2005) Histone deposition protein Asf1 maintains DNA replisome integrity and interacts with replication factor C. Genes Dev 19:1365-1375

Freudenreich CH, Lahiri M (2004) Structure-forming CAG/CTG repeat sequences are sensitive to breakage in the absence of Mrc1 checkpoint function and S-phase checkpoint signaling: implications for trinucleotide repeat expansion diseases. Cell Cycle 3:1370-1374

Gaillard P-HL, Noguchi E, Shanahan P, Russell P (2003) The endogenous Mus81-Eme1 complex resolves Holliday junctions by a nick and counternick mechanism. Mol Cell 12:747-759

Giannattasio M, Lazzaro F, Plevani P, Muzi-Falconi M (2005) The DNA damage checkpoint response requires histone H2B ubiquitination by Rad6-Bre1 and H3 methylation by Dot1. J Biol Chem 280:9879-9886

Glover TW (2006) Common fragile sites. Cancer Lett 232:4-12

Goldfless SJ, Morag AS, Belisle KA, Sutera VA Jr, Lovett ST (2006) DNA repeat rearrangements mediated by DnaK-dependent replication fork repair. Mol Cell 21:595-604

Gunjan A, Verreault A (2003) A Rad53 kinase-dependent surveillance mechanism that regulates histone protein levels in *S. cerevisiae*. Cell 115:537-549

Heller RC, Marians KJ (2006) Replication fork reactivation downstream of a blocked nascent leading strand. Nature 439:557-562

Hirano Y, Sugimoto K (2006) ATR homolog Mec1 controls association of DNA polymerase zeta-Rev1 complex with regions near a double-strand break. Curr Biol 16:586-590

Hoege C, Pfander B, Moldovan GL, Pyrowolakis G, Jentsch S (2002) RAD6-dependent DNA repair is linked to modification of PCNA by ubiquitin and SUMO. Nature 419:135-141

Hu F, Alcasabas AA, Elledge SJ (2001) Asf1 links Rad53 to control of chromatin assembly. Genes Dev 15:1061-1066

Huyen Y, Zgheib O, Ditullio RA Jr, Gorgoulis VG, Zacharatos P, Petty TJ, Sheston EA, Mellert HS, Stavridi ES, Halazonetis TD (2004) Methylated lysine 79 of histone H3 targets 53BP1 to DNA double-strand breaks. Nature 432:406-411

Ira G, Malkova A, Liberi G, Foiani M, Haber JE (2003) Srs2 and Sgs1-Top3 suppress crossovers during double-strand break repair in yeast. Cell 115:401-411

Kai M, Boddy MN, Russell P, Wang TS (2005) Replication checkpoint kinase Cds1 regulates Mus81 to preserve genome integrity during replication stress. Genes Dev 19:919-932

Kai M, Wang TS (2003) Checkpoint activation regulates mutagenic translesion synthesis. Genes Dev 17:64-76

Kannouche PL, Wing J, Lehmann AR (2004) Interaction of human DNA polymerase eta with monoubiquitinated PCNA: a possible mechanism for the polymerase switch in response to DNA damage. Mol Cell 14:491-500

Kats ES, Albuquerque CP, Zhou H, Kolodner RD (2006) Checkpoint functions are required for normal S-phase progression in *Saccharomyces cerevisiae* RCAF- and CAF-I-defective mutants. Proc Natl Acad Sci USA 103:3710-3715

Kawamoto T, Araki K, Sonoda E, Yamashita YM, Harada K, Kikuchi K, Masutani C, Hanaoka F, Nozaki K, Hashimoto N, Takeda S (2005) Dual roles for DNA polymerase eta in homologous DNA recombination and translesion DNA synthesis. Mol Cell 20:793-799

Kogoma T (1997) Stable DNA replication: interplay between DNA replication, homologous recombination, and transcription. Microbiol Mol Biol Rev 61:212-238

Kramer PR, Sinden RR (1997) Measurement of unrestrained negative supercoiling and topological domain size in living human cells. Biochemistry 36:3151-3158

Kraus E, Leung WY, Haber JE (2001) Break-induced replication: a review and an example in budding yeast. Proc Natl Acad Sci USA 98:8255-8262

Krishnan V, Nirantar S, Crasta K, Cheng AY, Surana U (2004) DNA replication checkpoint prevents precocious chromosome segregation by regulating spindle behavior. Mol Cell 16:687-700

Krogan NJ, Baetz K, Keogh MC, Datta N, Sawa C, Kwok TC, Thompson NJ, Davey MG, Pootoolal J, Hughes TR, Emili A, Buratowski S, Hieter P, Greenblatt JF (2004) Regulation of chromosome stability by the histone H2A variant Htz1, the Swr1 chromatin remodeling complex, and the histone acetyltransferase NuA4. Proc Natl Acad Sci USA 101:13513-13518

Labib K, Tercero JA, Diffley JF (2000) Uninterrupted MCM2-7 function required for DNA replication fork progression. Science 288:1643-1647

Lahiri M, Gustafson TL, Majors ER, Freudenreich CH (2004) Expanded CAG repeats activate the DNA damage checkpoint pathway. Mol Cell 15:287-293

Lambert S, Watson A, Sheedy DM, Martin B, Carr AM (2005) Gross chromosomal rearrangements and elevated recombination at an inducible site-specific replication fork barrier. Cell 121:689-702

Lawrence CW, Christensen RB (1979) Metabolic suppressors of trimethoprim and ultraviolet light sensitivities of *Saccharomyces cerevisiae rad6* mutants. J Bacteriol 139:866-876

Le S, Davis C, Konopka JB, Sternglanz R (1997) Two new S-phase-specific genes from *Saccharomyces cerevisiae*. Yeast 13:1029-1042

Lee JB, Hite RK, Hamdan SM, Xie XS, Richardson CC, van Oijen AM (2006) DNA primase acts as a molecular brake in DNA replication. Nature 439:621-624

Lemoine FJ, Degtyareva NP, Lobachev K, Petes TD (2005) Chromosomal translocations in yeast induced by low levels of DNA polymerase a model for chromosome fragile sites. Cell 120:587-598

Liberi G, Maffioletti G, Lucca C, Chiolo I, Baryshnikova A, Cotta-Ramusino C, Lopes M, Pellicioli A, Haber JE, Foiani M (2005) Rad51-dependent DNA structures accumulate at damaged replication forks in sgs1 mutants defective in the yeast ortholog of BLM RecQ helicase. Genes Dev 19:339-350

Linger J, Tyler JK (2005) The yeast histone chaperone chromatin assembly factor 1 protects against double-strand DNA-damaging agents. Genetics 171:1513-1522

Lisby M, Barlow JH, Burgess RC, Rothstein R (2004) Choreography of the DNA damage response: spatiotemporal relationships among checkpoint and repair proteins. Cell 118:699-713

Lopes M, Cotta-Ramusino C, Liberi G, Foiani M (2003) Branch migrating sister chromatid junctions form at replication origins through Rad51/Rad52-independent mechanisms. Mol Cell 12:1499-1510

Lopes M, Cotta-Ramusino C, Pellicioli A, Liberi G, Plevani P, Muzi-Falconi M, Newlon CS, Foiani M (2001) The DNA replication checkpoint response stabilizes stalled replication forks. Nature 412:557-561

Lopes M, Foiani M, Sogo JM (2006) Multiple mechanisms control chromosome integrity after replication fork uncoupling and restart at irreparable UV lesions. Mol Cell 21:15-27

Lucca C, Vanoli F, Cotta-Ramusino C, Pellicioli A, Liberi G, Haber J, Foiani M (2004) Checkpoint-mediated control of replisome-fork association and signalling in response to replication pausing. Oncogene 23:1206-1213

McIlwraith MJ, Vaisman A, Liu Y, Fanning E, Woodgate R, West SC (2005) Human DNA polymerase eta promotes DNA synthesis from strand invasion intermediates of homologous recombination. Mol Cell 20:783-792

Meister P, Taddei A, Vernis L, Poidevin M, Gasser SM, Baldacci G (2005) Temporal separation of replication and recombination requires the intra-S checkpoint. J Cell Biol 168:537-544

Melo J, Toczyski D (2002) A unified view of the DNA-damage checkpoint. Curr Opin Cell Biol 14:237-245

Michael WM, Ott R, Fanning E, Newport J (2000) Activation of the DNA replication checkpoint through RNA synthesis by primase. Science 289:2133-2137

Morikawa H, Morishita T, Kawane S, Iwasaki H, Carr AM, Shinagawa H (2004) Rad62 protein functionally and physically associates with the Smc5/Smc6 protein complex and is required for chromosome integrity and recombination repair in fission yeast. Mol Cell Biol 24:9401-9413

Motegi A, Kuntz K, Majeed A, Smith S, Myung K (2006) Regulation of gross chromosomal rearrangements by ubiquitin and SUMO ligases in *Saccharomyces cerevisiae*. Mol Cell Biol 26:1424-1433

Nakamura TM, Du LL, Redon C, Russell P (2004) Histone H2A phosphorylation controls Crb2 recruitment at DNA breaks, maintains checkpoint arrest, and influences DNA repair in fission yeast. Mol Cell Biol 24:6215-6230

Pan X, Ye P, Yuan DS, Wang X, Bader JS, Boeke JD (2006) A DNA integrity network in the yeast *Saccharomyces cerevisiae*. Cell 124:1069-1081

Paulovich AG, Armour CD, Hartwell LH (1998) The *Saccharomyces cerevisiae RAD9*, *RAD17*, *RAD24* and *MEC3* genes are required for tolerating irreparable, ultraviolet-induced DNA damage. Genetics 150:75-93

Paulovich AG, Margulies RU, Garvik BM, Hartwell LH (1997) *RAD9, RAD17,* and *RAD24* are required for S phase regulation in *Saccharomyces cerevisiae* in response to DNA damage. Genetics 145:45-62

Pellicioli A, Foiani M (2005) Signal transduction: how Rad53 kinase is activated. Curr Biol 15:R769-771

Pellicioli A, Lucca C, Liberi G, Marini F, Lopes M, Plevani P, Romano A, Di Fiore PP, Foiani M (1999) Activation of Rad53 kinase in response to DNA damage and its effect in modulating phosphorylation of the lagging strand DNA polymerase. Embo J 18:6561-6572

Postow L, Crisona NJ, Peter BJ, Hardy CD, Cozzarelli NR (2001a) Topological challenges to DNA replication: conformations at the fork. Proc Natl Acad Sci USA 98:8219-8226

Postow L, Hardy CD, Arsuaga J, Cozzarelli NR (2004) Topological domain structure of the *Escherichia coli* chromosome. Genes Dev 18:1766-1779

Postow L, Ullsperger C, Keller RW, Bustamante C, Vologodskii AV, Cozzarelli NR (2001b) Positive torsional strain causes the formation of a four-way junction at replication forks. J Biol Chem 276:2790-2796

Prakash L (1981) Characterization of postreplication repair in *Saccharomyces cerevisiae* and effects of *rad6*, *rad18*, *rev3* and *rad52* mutations. Mol Gen Genet 184:471-478

Rong L, Palladino F, Aguilera A, Klein HL (1991) The hyper-gene conversion hpr5-1 mutation of *Saccharomyces cerevisiae* is an allele of the SRS2/RADH gene. Genetics 127:75-85

Sabbioneda S, Minesinger BK, Giannattasio M, Plevani P, Muzi-Falconi M, Jinks-Robertson S (2005) The 9-1-1 checkpoint clamp physically interacts with polzeta and is partially required for spontaneous polzeta-dependent mutagenesis in *Saccharomyces cerevisiae*. J Biol Chem 280:38657-38665

Santocanale C, Diffley JFX (1998) A Mec1- and Rad53-dependent checkpoint controls late-firing origins of DNA replication. Nature 395:615-618

Sharp JA, Fouts ET, Krawitz DC, Kaufman PD (2001) Yeast histone deposition protein Asf1p requires Hir proteins and PCNA for heterochromatic silencing. Curr Biol 11:463-473

Shen X, Ranallo R, Choi E, Wu C (2003) Involvement of actin-related proteins in ATP-dependent chromatin remodeling. Mol Cell 12:147-155

Shimada K, Pasero P, Gasser SM (2002) ORC and the intra-S-phase checkpoint: a threshold regulates Rad53p activation in S phase. Genes Dev 16:3236-3252

Shirahige K, Hori Y, Shiraishi K, Yamashita M, Takahashi K, Obuse C, Tsurimoto T, Yoshikawa H (1998) Regulation of DNA-replication origins during cell-cycle progression. Nature 395:618-621

Sogo JM, Lopes M, Foiani M (2002) Fork reversal and ssDNA accumulation at stalled replication forks owing to checkpoint defects. Science 297:599-602

Stano NM, Jeong YJ, Donmez I, Tummalapalli P, Levin MK, Patel SS (2005) DNA synthesis provides the driving force to accelerate DNA unwinding by a helicase. Nature 435:370-373

Stelter P, Ulrich HD (2003) Control of spontaneous and damage-induced mutagenesis by SUMO and ubiquitin conjugation. Nature 425:188-191

Stucki M, Jackson SP (2006) gammaH2AX and MDC1: Anchoring the DNA-damage-response machinery to broken chromosomes. DNA Repair (Amst)

Takeuchi Y, Horiuchi T, Kobayashi T (2003) Transcription-dependent recombination and the role of fork collision in yeast rDNA. Genes Dev 17:1497-1506

Tercero JA, Diffley JF (2001) Regulation of DNA replication fork progression through damaged DNA by the Mec1/Rad53 checkpoint. Nature 412:553-557

Tercero JA, Longhese MP, Diffley JF (2003) A central role for DNA replication forks in checkpoint activation and response. Mol Cell 11:1323-1336

Trenz K, Smith E, Smith S, Costanzo V (2006) ATM and ATR promote Mre11 dependent restart of collapsed replication forks and prevent accumulation of DNA breaks. Embo J 25:1764-1774

Tyler JK, Adams CR, Chen SR, Kobayashi R, Kamakaka RT, Kadonaga JT (1999) The RCAF complex mediates chromatin assembly during DNA replication and repair. Nature 402:555-560

Vaze MB, Pellicioli A, Lee SE, Ira G, Liberi G, Arbel-Eden A, Foiani M, Haber JE (2002) Recovery from checkpoint-mediated arrest after repair of a double-strand break requires Srs2 helicase. Mol Cell 10:373-385

Wang JC (1991) DNA topoisomerases: why so many? J Biol Chem 266:6659-6662

Wang X, Ira G, Tercero JA, Holmes AM, Diffley JF, Haber JE (2004) Role of DNA replication proteins in double-strand break-induced recombination in *Saccharomyces cerevisiae*. Mol Cell Biol 24:6891-6899

Watanabe K, Tateishi S, Kawasuji M, Tsurimoto T, Inoue H, Yamaizumi M (2004) Rad18 guides poleta to replication stalling sites through physical interaction and PCNA monoubiquitination. Embo J 23:3886-3896

Wu L, Davies SL, Levitt NC, Hickson ID (2002) Potential role for the BLM helicase in recombintional repair via a conserved interaction with RAD51. J Biol Chem 276:19375-19381

Wysocki R, Javaheri A, Allard S, Sha F, Cote J, Kron SJ (2005) Role of Dot1-dependent histone H3 methylation in G1 and S phase DNA damage checkpoint functions of Rad9. Mol Cell Biol 25:8430-8443

Yoo HY, Shevchenko A, Shevchenko A, Dunphy WG (2004) Mcm2 is a direct substrate of ATM and ATR during DNA damage and DNA replication checkpoint responses. J Biol Chem 279:53353-53364

Zhang H, Lawrence CW (2005) The error-free component of the RAD6/RAD18 DNA damage tolerance pathway of budding yeast employs sister-strand recombination. Proc Natl Acad Sci USA 102:15954-15959

Zhou Y, Wang TS (2004) A coordinated temporal interplay of nucleosome reorganization factor, sister chromatin cohesion factor, and DNA polymerase alpha facilitates DNA replication. Mol Cell Biol 24:9568-9579

Zou L, Elledge SJ (2003) Sensing DNA damage through ATRIP recognition of RPA-ssDNA complexes. Science 300:1542-1548

Branzei, Dana

FIRC Institute of Molecular Oncology Foundation, Via Adamello 16, 20139 Milan, Italy, and Dipartimento di Scienze Biomolecolari e Biotecnologie, Università degli Studi di Milano, Via Celoria 26, 20133 Milan, Italy
dana.branzei@ifom-ieo-campus.it

Foiani, Marco

FIRC Institute of Molecular Oncology Foundation, Via Adamello 16, 20139 Milan, Italy, and Dipartimento di Scienze Biomolecolari e Biotecnologie, Università degli Studi di Milano, Via Celoria 26, 20133 Milan, Italy
marco.foiani@ifom-ieo-campus.it

Sister chromatid recombination

Felipe Cortés-Ledesma, Félix Prado and Andrés Aguilera

Abstract

Homologous recombination is a DNA repair mechanism that uses the genetic information of a homologous DNA sequence as template for repair of a DNA break. The genetic consequences of recombination depend on the choice of the molecule used as template. While sister-chromatid recombination (SCR), which uses as template for repair the identical and intact sister chromatid, preserves genome integrity, allelic and ectopic recombination can compromise it. SCR is, thus, the most secure mechanism of recombinational repair. This, together with the fact that most DNA breaks may appear spontaneously during replication, makes SCR the major recombination event occurring in mitotic cells from yeast to mammals. Given its physiological relevance, we review here the current knowledge about the mechanism(s) of SCR as well as the genetic and molecular factors controlling it, and how this knowledge open new perspectives to our understanding of genome dynamics.

1 Introduction

DNA repair is essential to prevent genetic instability, a phenomenon associated with most human cancers and some genetic diseases (Lengauer et al. 1998). Among the different mechanisms of DNA repair, homologous recombination (HR), which is believed to account for much of the repair of double-strand breaks (DSBs), is the only one that uses the genetic information of a homologous DNA sequence as a template for repair. As a result of this process, a unidirectional transfer of genetic information from one DNA molecule to another (gene conversion; GC) and/or a reciprocal exchange of genetic information between two DNA molecules (crossover) can occur. Hence, the genetic consequences of DNA repair by HR depend on the choice of the molecule used as template for repair (donor), either the sister chromatid (*s*ister-*c*hromatid *r*ecombination; SCR), the homologous chromosome (allelic recombination) or homologous DNA sequences located elsewhere in the genome (ectopic recombination). While SCR preserves genome integrity, allelic and ectopic recombination can lead to different types of DNA rearrangements (loss of heterozigosity, deletions, inversions and translocations). Therefore, SCR appears as the most secure mechanism for dealing with DNA lesions that need HR for their repair.

The biological relevance of this mechanism is becoming particularly intriguing after the increasing number of data emerging from bacteria and yeast to humans indicating that HR is highly linked to DNA replication (reviewed in Kuzminov 1999; Cox et al. 2000; Rothstein et al. 2000; Michel et al. 2001). These findings, together with observations indicating that SCR is a major mechanism of repair from yeast to mammalian cells (Kadyk and Hartwell 1992; Johnson and Jasin 2000; Gonzalez-Barrera et al. 2003; reviewed in Fasullo 2004) provide a scenario in which SCR, rather than allelic recombination, which is the main event analyzed in mitotic recombination studies, is the major recombination event occurring in mitotic cells. Here we will review our current understanding of the mechanism(s) of SCR as well as the genetic and molecular factors controlling it. The chapter will be mainly devoted to features of SCR that are not shared by other standard types of HR, such as allelic recombination, which is extensively discussed in other chapters of this volume.

2 Homologous recombination: a mechanism with major activity during replication

While the involvement of HR in DNA repair was established soon after the isolation of the first recombination genes (Clark and Chamberlin 1966; Howard-Flanders and Theriot 1966), an overview of its evolutionary conserved relevance in the repair and restart of perturbed replication forks has required the integration of a quantity of evidence from different organisms over the last four decades. These data have led to the proposal of different models by which HR functions during DNA replication. These models can be different depending on whether the replication fork is stalled but competent for resuming replication, or collapsed, which requires replication restart, as it would presumably imply dissociation of the replisome often accompanied by replication fork breakage. While stalled replication forks appear to lead to single-strand DNA (ssDNA) gaps that are repaired by error-prone translesion synthesis (TLS) and error-free mechanisms, collapsed forks render unprotected DNA ends that can be directly processed by HR (Rothstein et al. 2000; Broomfield et al. 2001; Cox 2001; Michel et al. 2001; McGlynn 2004). Before coming to the connection between HR and DNA replication we will review the circumstances that can perturb the advance of the fork and require HR assistance.

2.1 What makes a replication fork stall or collapse?

Different scenarios leading to replication impairment have been envisioned (Fig. 1), though in most of them the structure of the fork – whether stalled or collapsed – still remains unknown. It has been proposed that the advance of the replication fork can convert single-strand breaks (nicks) generated by oxidative damage into DSBs (Lindahl 1993) (Fig. 1, a). Conversion of nicks into DSBs have been shown

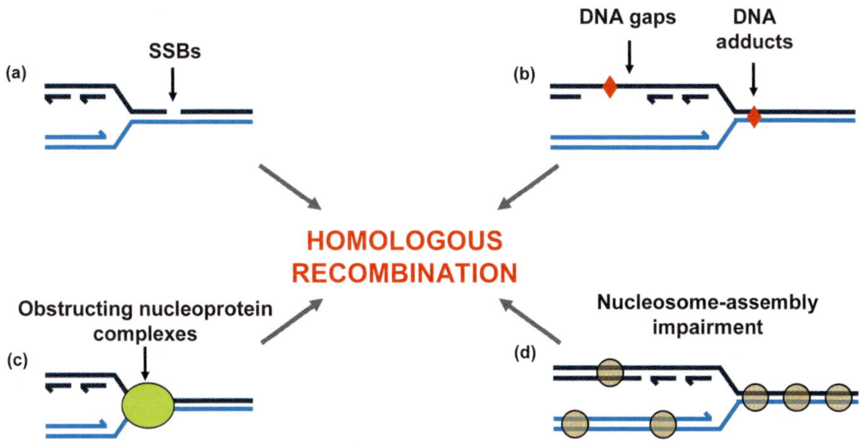

Fig. 1. Different situations related to an impairment of replication-fork progression can lead to HR. **(a)** Replication through a SSB can result in the formation of a DSB, which is repaired by HR. **(b)** DNA adducts can lead to replication-fork stalling and the formation of DNA gaps, which can induce HR. **(c)** Replication is impaired when it encounters nucleoprotein complexes, resulting from factors that strongly bind DNA or metabolic processes such as transcription, leading to an increase in HR. **(d)** Impairment of chromatin assembly stimulates HR possibly by causing replication problems.

upon inhibition of topoisomerase I with camptothecin in mammalian cells (Strumberg et al. 2000), and it has also been proposed for mutants affected in the maturation of the Okazaki fragments in bacteria and yeast, which require for viability the DSB recombinational repair RecBCD and Rad52 pathways, respectively (reviewed in Kuzminov 1995; Aguilera et al. 2000). Molecular evidence for nicks being converted into DSBs by replication was obtained with an *E. coli* system in which the single-strand DNA cleavage site of the M13 phage DNA was inserted into the lambda phage chromosome (Kuzminov 2001). The use of yeast artificial systems has provided additional evidence that in eukaryotes a nick can be converted by replication into a DSB that is processed by HR (Cortés-Ledesma and Aguilera 2006). Therefore, conversion of unrepaired nicks into DSBs by the advance of the replication fork appears as a major source of fork collapse.

The encounter of the replication fork with physical obstacles can be a source of replication stalling and induction of HR (Fig. 1, b). This is the case of DNA adducts left unrepaired like those caused by UV light (Rupp and Howard-Flanders 1968; Rupp et al. 1971; Courcelle et al. 2003). It is also the situation reported for proteins bound to DNA (Fig. 1, c), as is the *E. coli* replication-termination protein Tus, which binds to the Ter termination sites and blocks replication, rendering *E. coli* cells dependent on the RecBCD recombination pathway for viability (Horiuchi and Fujimura 1995). In yeast, the Rrm3 DNA helicase is required for the advance of the replication fork through natural pause sites caused by nucleoprotein complexes, and its absence leads to an increase in replication-fork pausing that could result in fork breakage (Ivessa et al. 2003), which could explain the hyper-recombination phenotype of *rrm3* mutants (Keil and McWilliams 1993).

In eukaryotes, DNA replication has to be accompanied by duplication of nucleosomes and their assembly into daughter chromatids (Adams and Kamakaka 1999; Tyler 2002); failure to properly assemble chromatin may also lead to replication impairment (Fig. 1, d). Although this is a hypothesis yet to be demonstrated, several results would be consistent with this scenario. Thus, it has been shown that the absence of the human chromatin assembly factor CAF1 causes DNA damage and cell cycle arrest (Ye et al. 2003). In *S. cerevisiae* the absence of the chromatin assembly factors Asf1 and CAF1 or a reduction in the pool of histones during replication cause DNA damage, hyper-recombination and a delay during the S and G2/M phases of the cell cycle (Ye et al. 2003; Prado et al. 2004; Ramey et al. 2004; Prado and Aguilera 2005) as well as gross chromosomal rearrangements (Myung et al. 2003). In addition, yeast *asf1* mutants show high levels of spontaneous SCR, suggesting that it is a major mechanism for the repair of the DNA damage associated with defective chromatin assembly (Prado and Aguilera 2005).

Finally, DNA metabolic processes other than replication, and more likely the proteins involved and functioning directly on the DNA, can suppose obstacles for proper replication fork progression that could also demand the assistance of HR (Fig. 1c). A good example is the effect of transcription on the fate of replication forks. In bacteria the high rate and concomitant occurrence of transcription and replication make cells rely on different modulators of RNA polymerase to prevent the accumulation of transcription complexes that would impede the advance of replication forks (McGlynn and Lloyd 2000; Trautinger et al. 2005). In yeast, RNA polIII transcription of transfer RNA (tRNA) genes provided a first demonstration for replication fork impairment by transcription (Deshpande and Newlon 1996; Ivessa et al. 2003). The connection with HR was further established at the ribosomal DNA. At these loci, replication fork barriers impede co-transcriptional advance of replication forks, and disruption of blocking activities causes collisions between RNA pol I-dependent transcription and replication that trigger DNA rearrangements (Takeuchi et al. 2003). An encounter between RNA pol II-dependent transcription and the replication machinery can also cause a replication fork pause and an increase in the frequency of direct-repeat recombination, suggesting that a tight control of both processes along the genome is required to prevent genetic instability (Prado and Aguilera 2005b). In this regard, it has recently been shown that *hpr1* mutants of the THO complex, involved in mRNP biogenesis, show a strong increase in transcription-dependent recombination between direct repeats if they are transcribed during S-phase, the hyper-recombination being associated with replication fork impairment (Wellinger et al. 2006).

2.2 The role of recombination during DNA replication

The connection between HR and replication was first characterized in T4 phage (Dannenberg and Mosig 1981; Luder and Mosig 1982), in which HR provides a mechanism of origin-independent DNA replication (Mosig 1998). However, a role for HR in the assistance to defective replication forks was first suggested in bacteria as early as 1974 by analysis of recombination mutants of the phage lambda (Skalka 1974). Kogoma and collaborators developed the concept of recombination-dependent replication (RDR) in bacteria by showing that HR is required to restart replication forks blocked by DNA damage in an origin-independent manner (Kogoma 1997). The observations that DSBs can activate RDR (Asai et al. 1994) and that one-ended DSBs generated during replication can be processed up to a properly oriented Chi site (Kuzminov et al. 1994) suggested a model in which collapsed forks lead to DSBs that are processed by HR to invade the sister chromatid and restart replication. In eukaryotes, replication impairment can also lead to DNA breaks that are repaired by HR (Saintigny et al. 2001; Ivessa et al. 2003; Lambert et al. 2005). As in bacteria, there is no evidence so far in eukaryotes that collapsed forks are either restarted or held until the arrival of the oncoming fork. Nevertheless, genetic evidence in the yeast *Saccharomyces cerevisiae* suggests that HR can repair DSBs by 3'-ended DNA invasion of a homologous sequence and priming of new DNA synthesis along the rest of the chromosome arm in a process termed *b*reak-*i*nduced *r*eplication (BIR) (Voelkel-Meiman and Roeder 1990; Malkova et al. 1996; Paques and Haber 1999).

To prevent collapse of replication forks and their potential deleterious consequences, cells are endowed with specialized mechanisms that encompass recombination functions to stabilize stalled replication forks and facilitate replication resumption. Early studies with bacteria, yeast and mammalian cells irradiated with UV showed an accumulation of ssDNA gaps that was attributed to unreplicated DNA fragments left opposite to the lesion (Rupp and Howard-Flanders 1968; Lehmann 1972; Prakash 1981). These gaps can result from stalled replication forks with uncoupled leading and lagging strands, or from replication forks that bypass an obstacle leaving a gap behind it (Sogo et al. 2002; Pages and Fuchs 2003; Heller and Marians 2006; Lopes et al. 2006). These gaps can be repaired by error-free mechanisms that involve DNA exchanges between sister chromatids mediated by the strand exchange RecA/RecFOR functions in *E. coli* (Rupp et al. 1971; Horii and Clark 1973) and the recombination protein Rad52 in *S. cerevisiae* (Prakash 1981; Lopes et al. 2006).

A large quantity of genetic and molecular evidence from bacteria and yeast to mammals has supported the idea that a Holliday-Junction (HJ)-like structure forms during the rescue of stalled replication forks (see Branzei and Foiani; Michel et al., this volume). These structures, first reported in mammalian cells after treatment with BrdU, would be formed by pairing of the nascent sister chromatids in a "template switching mechanism" (Higgins et al. 1976). In bacteria, it has been proposed that HJ-like structures form during replication in mutants prone to suffer replication fork stalling (e.g. helicase mutants) (Seigneur et al. 1998) and have been detected after UV irradiation (Courcelle et al. 2003). During replication of UV-damaged DNA, the RecA and RecFOR recombination proteins stabilize this HJ-like structure, facilitating DNA repair and promoting resumption of replication (Courcelle et al. 2003). In yeast, HJ-like structures dependent on the Rad52 recombination function have been observed in checkpoint mutants in the presence of replication inhibitors (Sogo et al. 2002), in mutants defective in the helicase Sgs1 (Liberi et al. 2005) and in wild type cells at the rDNA locus (Zou and Rothstein 1997).

Additional experimental evidence relates HR to replication in eukaryotes. Thus, yeast HR is active almost exclusively during the S and G2 phases (Aylon et al. 2004; Ira et al. 2004), which may allow cells a chance to assist replication with genetic information of the sister chromatid. A preference for HR during S phase relative to G1 has also been shown in chicken, hamster, and human cells (Takata et al. 1998; Rothkamm et al. 2003; Saleh-Gohari and Helleday 2004), and analysis of recombination proteins fused to the green fluorescence protein in yeast has provided evidence that spontaneous recombination foci accumulate during S and G2 (Lisby et al. 2001). These observations are consistent with the facts that DNA damage checkpoint mechanisms become essential during DNA replication, as deduced by the observation that S-phase checkpoint mechanisms are required to prevent genetic instability from yeast to humans (Myung et al. 2001; Kolodner et al. 2002). This may be in part due to the role of checkpoint proteins in keeping the integrity of stalled replication forks (Cobb et al. 2003; Katou et al. 2003; Tercero et al. 2003; Lucca et al. 2004). Consistent with this, checkpoint mutants accumulate DSBs at common fragile sites (regions that exhibit gaps and breaks under conditions of replicative stress) and slow replicating DNA regions in the absence of external DNA damage (Casper et al. 2002; Cha and Kleckner 2002). Therefore, the link of HR with the replication process would guarantee a specific replication-associated repair mechanism as a backup process to prevent genetic rearrangements after replication failures.

3 Methods for the measurement of sister-chromatid recombination

The tight association of HR with replication suggests that SCR must play a major role in co- and post-replicative repair, due to the availability of the sister chromatid that can be used as repair template. Still, the mechanisms of HR and the contribution of sister chromatid versus other templates in repair remains elusive. This is partly due to the fact that current models of HR raised from the analysis of allelic and ectopic recombination systems in which the presence of heterologies allows genetic and physical detection of distinct recombination intermediates and products. The fact that the two newly synthesized sister chromatids are identical has hampered the capability to detect SCR and, therefore, to properly define the involvement of sister chromatids in repair of DNA damages associated with replication. A revision of the different genetic, cytological and molecular approaches existing to detect SCR will help to understand the technical difficulties for the analysis of SCR and its molecular mechanism.

Sister chromatid recombination has been originally detected by cytological means; however, a number of genetic and molecular approaches are emerging that allow a molecular evaluation of the importance of SCR as a DNA repair mechanism, which are summarized and discussed here. Since some methods allow discrimination between crossovers and gene conversion events, these will be referred to from now on as *s*ister-*c*hromatid *e*xchange (SCE) and *s*ister-*c*hromatid *g*ene *c*onversion (SC-GC), respectively.

3.1 5-Bromodeoxyuridine labelling

Developed in mammalian cells, this classical method allows the cytological visualization of crossovers between sister chromatids (SCE) (Fig. 2A) (reviewed in Wolff 1977). Cells are incubated during two replication rounds with the thymidine analog 5-bromodeoxyuridine (BrdU), leading to a differential labelling of sister chromatids that can be followed by staining with DNA dyes such as Giemsa or Hoechst 33259. SCEs are visualized as sister chromatids that are discontinuously labelled according to a reciprocal pattern.

This method is useful for detection of high-level SCE events, which can be seen by treatment of cells with recombinogenic DNA-damaging agents (reviewed in Wolff 1977) or inhibitors of DNA metabolism enzymes such as camptothecin and m-AMSA (Degrassi et al. 1989; Cortes et al. 1993), or in specific mutant cell lines affected in DNA integrity such as PARP-/-, EM9, etc (Bartram et al. 1976; Thompson et al. 1982; Wang et al. 1997). Nevertheless, BrdU labelling is not useful for the analysis of spontaneous SCE events occurring in wild-type cells, because incorporation of BrdU into DNA induces SCE *per se* (Kato 1974), introducing a background noise that makes quantification non-reliable.

3.2 Detection of SCE in circular molecules

This method is based on the fact that one single crossover between two circular sister chromatids leads to a double-sized dimeric molecule (Fig. 2B). Cytological visualization of double-sized dicentric rings in maize somatic cells was the first evidence for SCE (McClintock 1938). Similar dimers were later visualized in human and *Drosophila* cells carrying ring chromosomes (Brewen 1969; Gatti et al. 1979). Incubation with BrdU of Chinese hamster cells with ring chromosomes has also been used to visualize SCEs in dicentric rings (Wolff et al. 1976; Sutou 1997).

Yeast chromosomes are not distinguished cytologically; however, dimers arising from SCE occurring in a circular chromosome III can be detected in *Saccharomyces cerevisiae* by Southern blot of chromosomes separated by pulsed-field gel electrophoresis (Game et al. 1989). Since spontaneous mitotic SCE levels are below detection levels with this assay, in these studies DSBs are induced by allowing cells to synchronously enter meiosis, to allow induction of metiotic DSBs, followed by return-to-growth conditions.

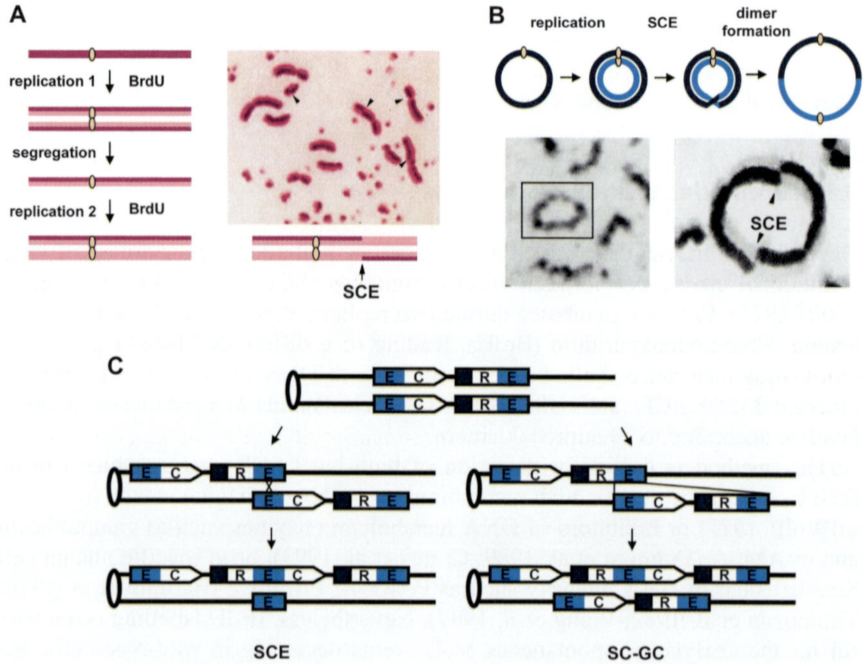

Fig. 2. Cytological and genetic detection of SCR. **A.** 5-Bromodeoxyuridine labelling. Incubation for two replication rounds in the presence of BrdU permits different staining of sister chromatids (left), and, thus, visualization of SCE (bottom right). BrdU substituted DNA strands are shown in light color. Spontaneous SCE in chicken cells is shown (top right). Gene conversion associated with crossover is not represented for simplification. **B.** Detection of SCE in circular molecules. Double-sized dicentric dimers can be formed by a single crossover between two circular sister chromatids (top). A dicentic dimer in *Drosophila* cells (bottom left) and visualization of a SCE in a dicentic ring in Chinese hamster cells (bottom right) is shown. **C.** Genetic assays based on direct repeats. The repeats are displayed in an orientation in which only SCR leads to the formation of a selectable recombinant (REC). Events involving a crossover (SCE, left) or not (SC-GC, right) can be distinguished by the outcome in the other chromatid. Pictures were reprinted from Sonoda et al. 1999 (panel A), Gatti et al. 1979 (panel B left), and Sutou 1997 (panel B right) with permission from American Society for Microbiology, Genetics Society of America and Elsevier, respectively.

3.3 Genetic assays based on direct repeats

Homologous recombination has been traditionally measured genetically by using two mutant heteroalleles, which reconstitute the wild type gene by homologous recombination, so recombinants can be selected phenotypically by prototrophy or drug resistance. In principle, recombination between two identical molecules, like sister chromatids, cannot result in a selectable phenotype, making genetic detection of SCR impossible. This problem can be overcome by using direct repeats, in which case recombination can take place with the same repeat (equal SCR), leading to a genetically undetectable product, but also with the other repeat (unequal SCR) on the sister chromatid, which leads to the formation of a triplication that can result in a detectable recombinant (Fig. 2C). The outcome in the other chromatid can be either the original repeat or a deletion depending on whether the event was a SC-GC or a SCE, respectively (Fig. 2C). The first evidence for unequal SCE resulted from the observation that yeast cells containing a *LEU2* gene inserted at the rDNA locus gave rise to colonies in which half of the colony had lost and the other half had duplicated the *LEU2* gene in both mitosis and meiosis (Petes 1980; Szostak and Wu 1980). In this sense, rDNA repeat expansion has been used as an indication of unequal SCR (Kobayashi et al. 2004). These direct repeat-based substrates have been extensively used in yeast for determination of spontaneous (Jackson and Fink 1981; Fasullo and Davis 1987; Fasullo et al. 2001; Dong and Fasullo 2003) and DNA damage-induced SCR (Fasullo and Davis 1987; Kadyk and Hartwell 1992, 1993; Fasullo et al. 2001; Dong and Fasullo 2003). DSB-induced SCR has also been measured using this type of assay in mammalian cells (Dronkert et al. 2000; Johnson and Jasin 2000; Xie et al. 2004; Puget et al. 2005; Saleh-Gohari et al. 2005), as well as in yeast (Fasullo et al. 2001; Dong and Fasullo 2003), using a site-specific endonuclease. Nevertheless, such endonucleases create an artificial situation in which both chromatids are cleaved, impeding equal SCR. Furthermore, in direct repeat-based systems, repair of the induced DSB can also occur by single-strand annealing (SSA), an efficient mechanism of DSB repair that, although not giving rise to genetically selectable recombinants, could influence the levels of SCR detected. Such genetic assays have been useful to establish the biological relevance of SCR in DSB repair, as well as its genetic requirements in yeast and mammalian cells (Jackson and Fink 1981; Fasullo and Davis 1987; Kadyk and Hartwell 1992, 1993; Dronkert et al. 2000; Johnson and Jasin 2000; Fasullo et al. 2001; Dong and Fasullo 2003).

3.4 Molecular analysis of SCR

The main limitation for the molecular analysis of SCR, as for the cytological analysis, is its difficulty to detect spontaneous events. Therefore, only induced SCR events can be analyzed at the molecular level. In this sense, site-specific endonucleases, such as HO in *S. cerevisiae,* have been extensively used to induce a DSB at a defined location, permitting the molecular analysis of DSB-repair by HR (Paques and Haber 1999). Nevertheless, this approach is not valid for SCR detection, because, as mentioned above, after replication, endonucleases such as HO would cleave both chromatids. To overcome this problem, a molecular assay has recently been developed in yeast that is based on a plasmid harbouring an incomplete 21-bp target for HO (Gonzalez-Barrera et al. 2003) (Fig. 3). This target is cleaved by HO with low efficiency (< 10%) favouring that most DSBs occur at only one chromatid, leaving the other intact and competent as a template for SCE. These events can be detected in time-course experiments by Southern blot as the

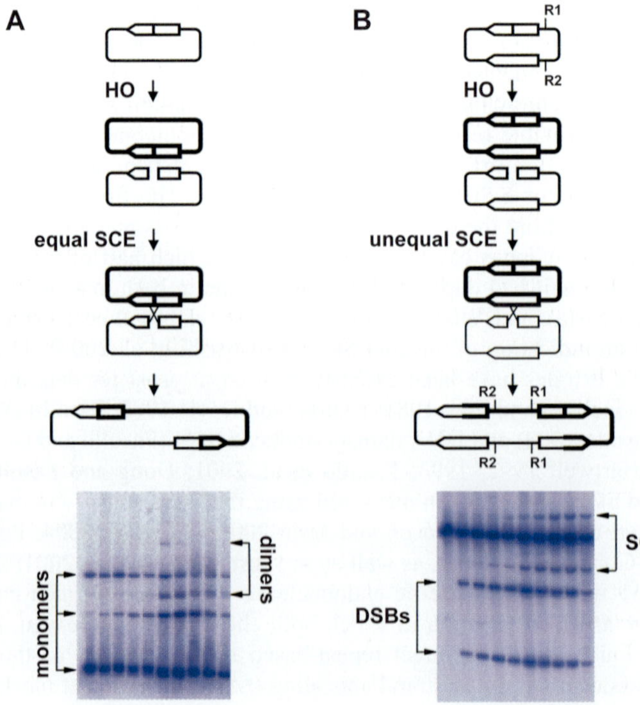

Fig. 3. Molecular analysis of SCE. **A.** A 21-bp HO site with low cutting efficiency in a yeast circular minichromosome permits induction of SCR as mainly only one chromatid is cleaved. SCE can be detected by Southern blot as the appearance of dimers. **B.** The use of inverted repeats allows Southern-blot detection of unequal SCE by the appearance of specific restriction fragments. Alternatively the break can be repaired using the repeat located in the same chromatid.

appearance of double-sized plasmids. In addition, the presence of an inverted copy in the plasmid generates a system in which unequal SCE and intrachromatid recombination (ICR) can compete for DSB repair and the kinetics of both types of events can be followed by the appearance of specific restriction fragments (Gonzalez-Barrera et al. 2003; Cortes-Ledesma and Aguilera 2006) . Notably, the HO endonuclease causes nicks in this incomplete target that are converted into DSBs by the advance of the replication fork, making this system a powerful tool to study DSB repair during DNA replication (Cortes-Ledesma and Aguilera 2006).

4 DNA repair genes required for SCR

Our understanding of HR comes primarily from genetic and molecular studies in *E. coli* and *S. cerevisiae* (reviewed in Kowalczykowski et al. 1994; Kuzminov

1999; Paques and Haber 1999; Symington 2002; Prado et al. 2003; Michel et al., Haber, Lisby and Rothstein, this volume). Since the rest of this chapter will be devoted mainly to SCR in eukaryotic cells, we will summarize the main features of homologous recombination in eukaryotes. In this sense, and, as mentioned above, most of our knowledge of HR mechanisms comes from the study of DSB repair by allelic and ectopic recombination in yeast. In this sense, it is important to notice that spontaneous and induced HR may not present identical genetic requirements. A basic illustration of the relevant molecular steps of HR is shown in Figure 4. Briefly, the 5'-ends of a DSB are resected leaving 3'-OH single-stranded ends. This reaction is regulated by the MRX complex, which also functions in other DNA-end processing events, including NHEJ, DSB signalling and telomere maintenance (reviewed in Haber 1998). Rad52, a protein with DNA annealing activity that is required for all HR events, catalyzes the invasion of the resection-generated 3'-ends on a homologous DNA duplex. The Rad59 protein, which is homologous to Rad52, and also displays DNA annealing activity, can facilitate this step. This invasion is promoted and stabilized by strand exchange catalyzed by the RecA-homolog Rad51, together with its paralogs Rad55 and Rad57, and Rad54, a member of the SNF-SWI family of ATPases. This leads to the formation of a heteroduplex in which the 3'-invading end primes new DNA synthesis using the invaded DNA sequence as a template. These steps are common to the two most widely accepted HR models, *s*ynthesis-*d*ependent *s*trand *a*nnealing (SDSA) (Nassif and Engels 1993) and *d*ouble-*s*trand *b*reak *r*epair (DSBR), (Szostak et al. 1983). In SDSA DNA strand exchange is reverted and newly synthesized DNA molecules re-annealed with each other, leading to gene conversion. In contrast, in the DSBR model, strand exchange and DNA synthesis progress up to reach the resected ends and generate two HJs whose resolution can lead to crossover (Szostak et al. 1983). Besides, DNA synthesis primed by the 3'-invading ends can progress along the invaded molecule without being captured by the other end of the break (*b*reak-*i*nduced *r*eplication; BIR) (Voelkel-Meiman and Roeder 1990; Malkova et al. 1996). Current data suggest that SDSA is the major mechanism of mitotic recombination, consistent with most gene conversions to be unassociated with crossovers, and that DSBR is the major mechanism leading to crossovers.

SCR appears to be mechanistically similar to other mitotic recombination events as deduced by the fact that they share similar genetic requirements (Table 1). It is likely that, therefore, SCR occurs by SDSA, as suggested by the observation that gene conversion unassociated with crossover is the main recombination product of SCR in yeast and mouse cells (Kadyk and Hartwell 1992; Johnson and Jasin 2000) (Fig. 4, e-g). As mentioned above, the importance of SDSA is also supported by the analysis of the genetic requirements for SCR. Genetic approaches in yeast have shown that DNA damage and DSB-induced SCR is abolished in *rad52* (Kadyk and Hartwell 1992, 1993), reduced two to tenfold in *rad51*, *rad54*, *rad55* and *rad57* (Fasullo et al. 2001; Dong and Fasullo 2003) and a defect in *rad59* is only observed for X-ray-induced SCR (Dong and Fasullo 2003). In mammals, damage-induced SCR is reduced twofold in the absence of Rad54 (Dronkert et al. 2000). Nevertheless, SCE, which occurs by crossover and not by SDSA, is also observed in yeast and higher eukaryotes. Molecular analysis of the

Table 1. Protein factors involved in SCR in vertebrate and yeast cells as determined by different methods.

	Vertebrate Cells				*Yeast Cells*		
	BrdU method		Genetic assay		Genetic assay		Molecular assay
	Spontaneous SCR	Induced SCR	Spontaneous SCR	Induced SCR	Spontaneous SCR	Induced SCR	Induced SCR
Rad52	n. d.	n. d.	n. d.	n. d.	+[1-4]	+[3, 4]	+[5]
Rad51	+[6]	+[6]	n. d.	n. d.	-[7]	+[7, 8]	+[5]
Rad54	+[6]	+[6]	+	+[9]	-[7]	+[7]	n. d.
Rad51 paralogs: Rad55-57/ Rad51B-D, Xrcc2-3	+[10, 11]	+[10, 11]	n. d.	n. d.	-[7]	+[7]	n. d.
Rad59	n. d.	n. d.	n. d.	n. d.	-[7]	+/-[7]	+[5]
MRX(N)	n. d.	n. d.	n. d.	n. d.	+[7]	+/-[7, 12]	+[5]
cohesins	n. d.	+[13]	n. d.	n. d.	n. d.	n. d.	+[14]
Smc5-Smc6	+[15]	+[15]	n. d.	+	-[16]	+[16]	+[17]

(+) the protein factor is involved; (-) it is not involved; (+/-) depends on the assay; (n. d.) not determined.

(1) Fasullo and Davis 1987; (2) Jackson and Fink 1981; (3) Kadyk and Hartwell 1992; (4) Kadyk and Hartwell 1993; (5) Gonzalez-Barrera et al. 2003; (6) Sonoda et al. 1999; (7) Dong and Fasullo 2003; (8) Fasullo et al. 2001; (9) Dronkert et al. 2000; (10) Takata et al. 2000; (11) Takata et al. 2001; (12) Bressan et al. 1999; (13) Sonoda et al. 2001; (14) Cortes-Ledesma and Aguilera 2006; (15) Potts et al. 2006; (16) Onoda et al. 2004; (17) De Piccoli et al. 2006

kinetics of DSB repair in yeast has recently demonstrated the importance of Rad52, Rad51, and Rad59 in SCE (Gonzalez-Barrera et al. 2003), and cytological studies in vertebrate cells have revealed that SCE depends on Rad51, Rad54, and the five Rad51 paralogs Rad51B, Rad51C, Rad51D, Xrcc2, and Xrcc3 (Sonoda et al. 1999; Takata et al. 2000, 2001).

The many factors shared by allelic and ectopic recombination with SCR suggest that the mechanisms of HR are similar regardless of the donor of information. This is not in conflict, though, with the idea that the choice of the template may influence some of the required activities. For instance, the existence of a Rad54 homolog in yeast, Rdh54/Tid1, which seems to act particularly in allelic recombination (Klein 1997; Shinohara et al. 1997), has led to the suggestion that Rad54 would act in SCR, leaving other recombination events to Tid1. This could reflect different requirements for the DNA invasion and DNA-strand exchange steps that are determined by the structural features of the donor molecule. Using a meiotic return-to-growth assay it was proposed that Rad54 is required for SCR but not for allelic recombination, and the opposite occurs with Tid1 (Arbel et al. 1999). However, this assay does not demonstrate a specific role of Rad54 in SCR, since it could also be acting in allelic recombination, its function being more difficult to observe due to redundancy with Tid1.

In addition, and in agreement with the tight link of HR with replication discussed previously, spontaneous SCR may reflect the rescue by BIR of collapsed replication forks (reviewed in Helleday 2003; Fasullo 2004) (Fig. 4, b-d). Consistent with this, while spontaneous unequal SCR in yeast depends on Rad52 (Jackson and Fink 1981; Fasullo and Davis 1987; Kadyk and Hartwell 1992; Kadyk and Hartwell 1993), it is independent of Rad51, Rad55, Rad57, and Rad54, indicating that, opposite to DSB-induced SCR, spontaneous SCR can occur in the absence of DNA-strand exchange (Dong and Fasullo 2003). Replication-fork collapse can be mimicked with camptothecin, which produces nicks that are further converted into one-ended DSBs by replication (Strumberg et al. 2000). Recently it has been shown in mammalian cells that the spectrum of spontaneous SCR products resembles the camptothecin-induced and differs from the I-*Sce*I-mediated DSB-induced one (Saleh-Gohari et al. 2005). These results are consistent with early works that reported a low efficiency to induce cytologically-observed SCE of DNA-damaging agents that cause two-ended DSBs (Perry and Evans 1975; Solomon and Bobrow 1975; Morgan et al. 1988), in contrast to the strong SCE increase induced by camptothecin (Degrassi et al. 1989). A possible explanation is that, since crossover is a rare event in mammalian cells (Richardson et al. 1998; Johnson and Jasin 2000), repair of two-ended DSBs may not result in SCE, while repair by BIR of one-ended DSBs can appear as cytological SCE (Helleday 2003).

Also, even though more speculative given the actual experimental evidence, it may be that DNA lesions other than DSBs activate spontaneous SCR (Fabre et al. 2002). In this sense, it is worth mentioning that UV-irradiated yeast cells defective in nucleotide excision repair accumulate DNA gaps that can be repaired by Rad52-dependent HR (Lopes et al. 2006). In addition, the completion of replication through UV-damaged DNA is assisted by Rad52-independent recombination

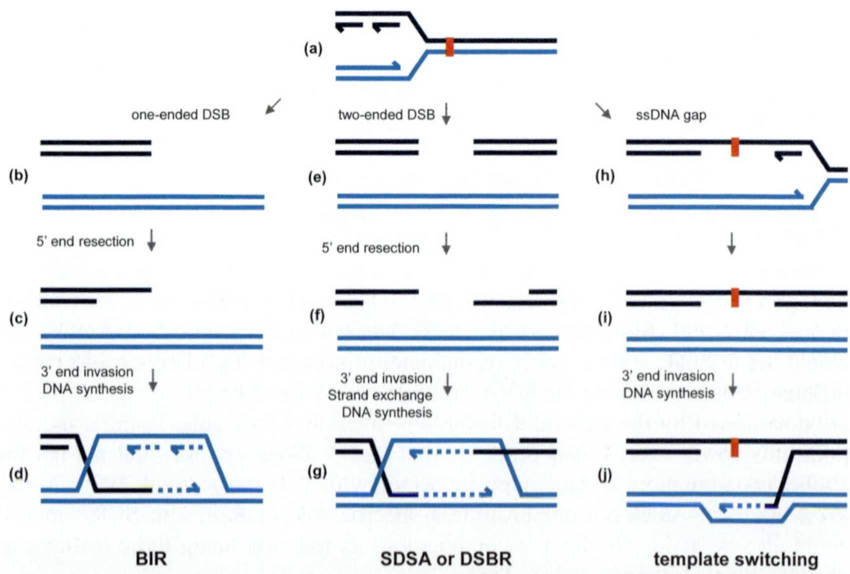

Fig. 4. Different mechanisms can use the sister chromatid to repair DNA damage. Replication through damaged DNA **(a)** can lead to the formation of a two-ended DSB **(e)**, which can be repaired by SDSA or DSBR **(f-g)**. Alternatively, the replication fork can collapse resulting in a one-ended DSB that can restore the replication fork by BIR **(b-d)**. On the other hand, the damage can be bypassed **(h)** leaving a DNA gap **(i)** that could be filled copying the information from the sister chromatid in a "template-switching" mechanism **(j)**.

mechanisms that rely on the *RAD6/RAD18* post-replication repair pathway (Zhang and Lawrence 2005). These recombination-dependent gap-repair events could result from a "template-switching" mechanism (Higgins et al. 1976) (Fig. 4, h-j), which would not necessarily require DNA-strand exchange. Consequently, this mechanism has also been invoked to explain the fact that spontaneous SCR is Rad51 independent (Dong and Fasullo 2003). Trinucleotide repeats have also been shown to induce unequal SCR independent of Rad51, suggesting that a "template switching"-mediated type of SCR could be at least in part responsible for instability of microsatellite DNA (Nag et al. 2004).

Other functions can also influence SCR. Thus, mutants in the *S. cerevisiae RAD9* G2/M checkpoint pathway present a reduction in DNA damage- and DSB-induced SCR (Fasullo et al. 1998; Paulovich et al. 1998) that is accompanied by an increase in HR-dependent translocations (Fasullo et al. 1998). This may be explained because in the absence of G2/M checkpoint, damaged sister chromatids segregate before SCR can occur, and repair can only take place using other homologous sequences that could lead to translocations. Specific SCR suppressor mechanisms could also operate in cells. This idea is favoured by the phenotype of elevated SCE associated with the Bloom Syndrome in humans (Bartram et al. 1976), which can be observed even in UV-irradiated cells (Kurihara et al. 1987). In yeast, an increase in SCE is also observed in the absence of the Sgs1 DNA helicase (Onoda et al. 2000), an ortholog of the Bloom Blm protein. Nevertheless, it cannot be ruled out that the elevated SCE is related to the role of Blm and Sgs1 in crossover suppression (Ira et al. 2003; Wu and Hickson 2003; see Klein, this volume) rather than to a role in SCR suppression.

5 Specific functions required for SCR

SCR is the major mechanism of mitotic recombination in yeast and mammalian cells (Kadyk and Hartwell 1992; Johnson and Jasin 2000; Gonzalez-Barrera et al. 2003) and provides a safe way to repair DNA lesions without affecting genome integrity. Therefore, it is of particular importance to know the elements that determine the choice of the sister as donor of information during the repair by HR. The proximity of sister chromatids may provide an advantage for SCR, and hence, the factors involved in keeping the sisters together may play a role in the regulation of the process. According to this hypothesis, an emerging body of results (see data in Table 1) points to cohesins as required for SCR, and suggests different roles for the MRX(N) complex in HR because of the Rad50 SMC-like structure.

5.1 Cohesins

Cohesins hold sister chromatids together from replication to the onset of anaphase (reviewed in Nasmyth 2002). Cohesins are tetrameric complexes formed by two SMC (*s*tructural *m*aintenance of *c*hromosomes) (Smc1 and Smc3), and two non-SMC proteins (Scc1 and Scc3 in *S. cerevisiae*). Each SMC component is self-folded leaving a hinge, by which both proteins interact, a long coiled-coil and a globular domain. A ring-like structure is formed by binding of Scc1 to both globular domains (Fig. 5A). Cohesins are loaded in G1, tethering sister chromatids by a DNA replication-dependent mechanism and releasing them for chromosome segregation after the Scc1 subunit is cleaved (see Haering and Nasmyth 2003). Besides, association of cohesins with sister chromatids is restrained by DNA topology and not by physical interactions (Ivanov and Nasmyth 2005). These findings have led to a model in which cohesins hold sister chromatids together by embracing them.

A number of results from yeast to humans have shown a role for cohesins in DSB repair that are consistent with a role in SCR. Thus, mutations in *RAD21* (ortholog of *SCC1* in *S. pombe*) cause X-ray, UV light and hydroxyurea sensitivity (reviewed in Lehmann 2005), and haploid, but not diploid, *S. cerevisiae smc1* mutants are severely affected in repair of multiple genomic DNA breaks (Schar et al. 2004). Also, establishment and maintenance of sister chromatid cohesion in *S. cerevisiae* is required for chromosome recovery upon X-ray-induced DNA damage (Sjogren and Nasmyth 2001). Transcriptional repression of *SCC1* in chicken DT40 cells decreases the number of damage-induced cytological SCEs per cell

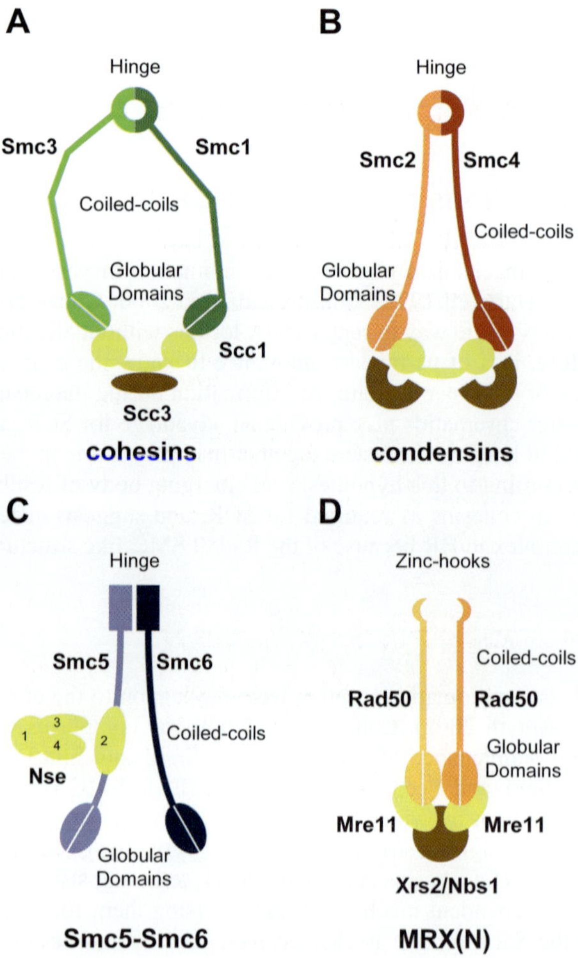

Fig. 5. Structure of SMC and MRX(N) complexes. SMC proteins share some structural characteristics. Each one is self-folded, leaving a globular domain, which contains amino and carboxi termini, and a hinge motif, separated by a long coiled coil. **A.** Cohesins ring-like structure. Smc1 and Smc3 interact by their hinges, and Scc1-Scc3 close the structure by binding to the globular domains. **B.** Condensins V-shaped structure. Three additional factors bind the Smc2-Smc4 core. **C.** The Smc5-Smc6 complex is formed by these two proteins and several *n*on-*s*mc *e*lements (Nse). **D.** Rad50 displays an SMC-like structure in the MRX(N) complex. Two Rad50 molecules interact with each other and with two Mre11 molecules via their globular domains. This constitutes the DNA binding domain of the complex, while a zinc-hook motif, analogous to the hinge domain of SMC proteins, is important for interaction with other MRX(N) complexes. Xrs2 (Nbs1 in mammals) interacts directly with Mre11.

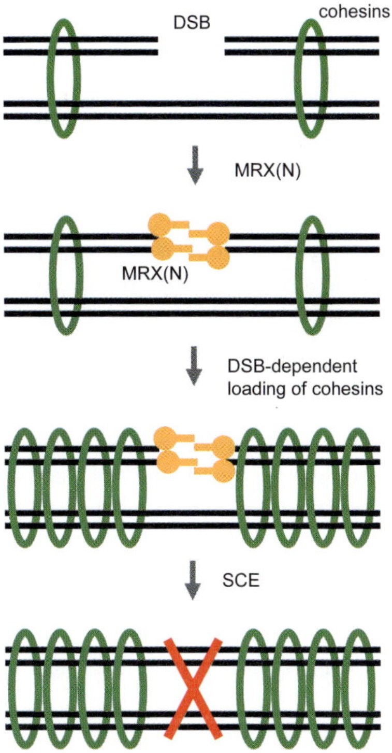

Fig. 6. Model to explain the role of cohesins in SCR. Cohesins (shown as ovals) hold sister chromatids together. In the presence of a DSB additional cohesin is loaded favouring interaction with the sister chromatid and thus SCR. The MRX(N) complex is required for cohesin loading in response to a DSB and tethers DSB-ends together.

and increases at least threefold the frequency of spontaneous and radiation-induced chromosome aberrations (Sonoda et al. 2001). Finally, human Smc1 is phosphorylated after irradiation (Kim et al. 2002b) and accumulates, together with other cohesin components, along laser microbeam-irradiated tracts (Kim et al. 2002a).

Molecular analysis of protein dynamics during DSB-repair in *S. cerevisiae* by Chromatin Immunoprecipitation (ChIP) has revealed that cohesins are loaded along a region expanding several kilobases at both sides of the break. This recruitment is not observed during G1, suggesting that cohesins are required for HR and not for NHEJ. Importantly, newly loaded cohesins upon DNA damage establish cohesion in a replication-independent manner and are required for the recovery of an X-ray-irradiated chromosome but not for ectopic recombination (Strom et al. 2004; Unal et al. 2004). These results strongly suggest a role for cohesins in SCR by holding the broken molecule and the template together (Fig. 6). This role may be mediated by phosphorylation of serine 129 of histone H2A, an early step in DSB formation (Downs et al. 2000), as suggested by the fact that DSB-induced cohesin loading depends on and coincides with the DNA region of H2A phosphorylation in yeast (Strom et al. 2004; Unal et al. 2004). Indeed, H2A phosphorylation has been shown to be required for efficient SCR in mouse cells (Xie et al. 2004). These studies did not address, however, whether the repair deficiency is specific for SCR or represents a general DSB-repair defect.

The involvement of cohesins in SCR has been addressed directly by following the kinetics of DSB repair with a system in which SCE and ICR compete to repair the break (Fig. 3B). While SCE is the major DSB repair mechanism observed in the wild type, it is strongly affected in thermosensitive *smc3* and *scc1* cohesin mutants, as well as in *scc2*, required for cohesin loading (Cortes-Ledesma and Aguilera 2006). Interestingly, ICR is not affected and even increases in these mutants, suggesting a specific role of cohesins in SCR (Cortes-Ledesma and Aguilera 2006). In addition, cohesins at the rDNA have been shown to favour equal rather than unequal events, thus playing a role in rDNA repeat stability (Kobayashi et al. 2004). In this context, it is particularly interesting the recent observation that transcription of a non-coding sequence in the rDNA spacer promotes cohesin dissociation, stimulating changes in repeat-copy number, presumably by unequal SCE (Kobayashi and Ganley 2005). This raises the intriguing possibility that transcription could control the frequency of recombination between DNA repeats by modulating the presence and action of cohesins at particular DNA regions; that is, transcription-associated recombination (TAR) may be modulated by cohesins.

5.2 Other SMC complexes

In addition to cohesins, the other two known eukaryotic SMC complexes, condensins, with a main role in chromosome condensation, and Smc5-Smc6, whose function is not well defined, could also play a role in SCR (Table 1) (Losada and Hirano 2005). In addition to the Smc2-Smc4 core, three other proteins associate to constitute V-shaped condensins (Fig. 5B). While in vertebrate cells two sets of non-SMC components exist, only one is found in yeast. Condensins have been shown to participate in DNA repair in *S. pombe* (reviewed in Lehmann 2005), and to associate with the replication-fork barrier at the rDNA, this being important for the maintenance of repeat copy number in *S. cerevisiae* (Johzuka et al. 2006). The Smc5-Smc6 heterodimer is the centre of an essential complex constituted by at least four additional subunits (Nse1-4) (Fig. 5C). Mutants in components of the Smc5/Smc6 complex are sensitive to various DNA-damaging agents, this sensitivity being genetically related to a defect in HR (reviewed in Lehmann 2005). In addition, *S. cerevisiae* mutants of Smc5-Smc6 are impaired in rDNA segregation (Torres-Rosell et al. 2005), and are affected in MMS-induced ectopic recombination and unequal SCR (Onoda et al. 2004). Furthermore, Smc5-Smc6 has recently been shown to be loaded onto the DNA in response to a DSB, and at collapsed replication forks (de Piccoli et al. 2006; Betts Lindroos et al. 2006; Potts et al. 2006). Importantly, Smc5-Smc6 is required for efficient SCE in yeast, as determined by the molecular HO-based assay (De Piccoli et al. 2006), and in human cells, as determined cytologically and genetically (Potts et al. 2006), but not for other types of homologous recombination. Whether this effect could be mediated, at least in part, by cohesins is an open possibility, as in human cells Smc5-Smc6 is required for DSB-dependent cohesin loading (Potts et al. 2006).

5.3 The MRX(N) complex

Formed by two conserved proteins, Mre11 and Rad50, and a non-conserved one, Xrs2 in yeast and Nbs1 in mammals, the MRX(N) complex binds to DNA ends shortly after DSB formation and participates in its repair by coordinating the activities of non-homologous end joining (NHEJ), HR and DNA damage checkpoint (reviewed in Haber 1998). At least part of this regulation relies on the control of the 5'-ends resection that leads to the generation of ssDNA molecules competent for HR (reviewed in Haber 1998; Symington 2002). In MRX, Rad50 displays an SMC-like structure, but the hinge domain is substituted by a zinc-hook motif (Fig. 5D). Consistent with a general role in DSB repair, *mre11*, *rad50* and *xrs2* null mutants in *S. cerevisiae* are X-ray and MMS sensitive (Game and Mortimer 1974; Ivanov et al. 1992; Ajimura et al. 1993). This sensitivity becomes evident only during S and G2, suggesting that it is a consequence of defective HR (Bressan et al. 1999).

Despite this sensitivity, spontaneous allelic and ectopic recombination is increased five to tenfold in MRX(N) mutants (Malone et al. 1990; Ivanov et al. 1992; Ajimura et al. 1993). This hyper-recombination phenotype has led to the proposal of a specific role for the MRX(N) complex on SCR, based on the idea that the absence of proficient SCR would channel most DNA lesions to allelic and ectopic recombination (reviewed in Haber 1998). Indeed, a role for the MRX(N) complex on SCR has been shown in yeast. Genetic frequencies of spontaneous and DNA damage-induced unequal SCR are decreased around twofold in null MRX mutants (Bressan et al. 1999; Dong and Fasullo 2003), and molecular detection of DSB-induced SCE intermediates is strongly affected in *mre11* mutants (Gonzalez-Barrera et al. 2003).

Nevertheless, radiation-induced allelic recombination is reduced in *mre11Δ* similarly to SCR (Malone et al. 1990; Ivanov et al. 1992; Ajimura et al. 1993; Bressan et al. 1999), suggesting a general role of MRX(N) in HR rather than specifically in SCR. Consistent with this classical view of the MRX(N) complex as a general regulator of HR, null MRX mutants display similar defects in the kinetics of DSB repair by both unequal SCE and ICR (Cortés-Ledesma and Aguilera, unpublished results). The fact that the increase in allelic and ectopic spontaneous recombination in MRX mutants is not observed in damage-induced events suggests that it could be related to an increase in initiation of recombination. On the other hand, hyper-recombination could also be a consequence of the DNA resection defects displayed by MRX mutants, which in turn would lead to an increase in the proportion of gene conversions yielding selectable recombinants (reviewed in Haber 1998; Symington 2002).

A putative structural role of the MRX(N) complex would be consistent with the Rad50 SMC-like structure, holding sister chromatid together and facilitating SCR. *In vitro*, Rad50/Mre11 complexes are able to tether two DNA molecules through interactions between their zinc-hook motifs (Anderson et al. 2001; de Jager et al. 2001; Hopfner et al. 2002). *In vivo* a *rad50* mutant lacking the hook domain causes the same repair defects as the null mutant, despite the hook not being necessary for the structure of the complex. These defects can be suppressed by artificially tethering Rad50 proteins through their coiled-coil domains (Hopfner et al. 2002; Wiltzius et al. 2005). Alternatively, the structural role of the MRX(N) complex in HR could be to hold the two ends of the DSB together in order to facilitate its repair (Fig. 6), as suggested by the observation that the two ends of a DSB remain adjacent in wild type but not in null MRX and hook-less Rad50 mutants (Lobachev et al. 2004; Clerici et al. 2006). Finally, the effect of the MRX(N) complex in SCR could partially be mediated by cohesins, since their DSB-induced loading is impaired in *mre11* mutants (Strom et al. 2004; Unal et al. 2004). Therefore, a possible specific role of MRX in SCR is still possible, but current data favour a general role of the complex in HR, functioning in the processing and/or tethering of DSB ends.

6 Concluding remarks

Different scenarios affecting the normal progression of replication forks can lead to DNA breaks that are repaired by recombination. Despite the fact that most of our knowledge on the different factors and mechanisms responsible for HR come from studies on allelic and ectopic recombination in mitosis and meiosis, the most relevant recombination events are those occurring between sister chromatids. SCR is an essential part of co- and post-replication repair. The technical difficulty to identify the products of recombination between two identical sisters has made studies on SCR scarce, but the availability of new genetic and physical methods to distinguish between parental and recombinant products is promoting new studies on SCR. We have a better knowledge of the relevance of known recombination factors on SCR, and new results indicate that other factors, such as SMC proteins, with a non-detectable or minor effect on allelic recombination, play an important role in SCR. Further research on SCR will contribute to decipher the molecular basis and physiological meaning of the connection between DNA replication and recombinational repair in a cellular context, and to our understanding of genome dynamics.

Acknowledgements

We would like to thank Diane Haun for style supervision. Research in the laboratory of A. A. is funded by grants SAF2003-00204 from the Spanish Ministry of Science and Education and CVI102 from Junta de Andalucía.

References

Adams CR, Kamakaka RT (1999) Chromatin assembly: biochemical identities and genetic redundancy. Curr Opin Genet Dev 9:185-190

Aguilera A, Chavez S, Malagon F (2000) Mitotic recombination in yeast: elements controlling its incidence. Yeast 16:731-754

Ajimura M, Leem SH, Ogawa H (1993) Identification of new genes required for meiotic recombination in *Saccharomyces cerevisiae*. Genetics 133:51-66

Anderson DE, Trujillo KM, Sung P, Erickson HP (2001) Structure of the Rad50 x Mre11 DNA repair complex from *Saccharomyces cerevisiae* by electron microscopy. J Biol Chem 276:37027-37033

Arbel A, Zenvirth D, Simchen G (1999) Sister chromatid-based DNA repair is mediated by *RAD54*, not by *DMC1* or *TID1*. EMBO J 18:2648-2658

Asai T, Bates DB, Kogoma T (1994) DNA replication triggered by double-stranded breaks in *E. coli*: dependence on homologous recombination functions. Cell 78:1051-1061

Aylon Y, Liefshitz B, Kupiec M (2004) The CDK regulates repair of double-strand breaks by homologous recombination during the cell cycle. EMBO J 23:4868-4875

Bartram CR, Koske-Westphal T, Passarge E (1976) Chromatid exchanges in ataxia telangiectasia, Bloom syndrome, Werner syndrome, and xeroderma pigmentosum. Ann Hum Genet 40:79-86

Betts Lindroos H, Strom L, Itoh T, Katou Y, Shirahige K, Sjogren C (2006) Chromosomal association of the Smc5/6 complex reveals that it functions in differently regulated pathways. Mol Cell 22:755-767

Bressan DA, Baxter BK, Petrini JH (1999) The Mre11-Rad50-Xrs2 protein complex facilitates homologous recombination-based double-strand break repair in *Saccharomyces cerevisiae*. Mol Cell Biol 19:7681-7687

Brewen JGP, Peacock WJ(1969) The effect of tritiated thymidineon sister-chromatid exchange in a ring chromosome. Mutat Res 7:433-440

Broomfield S, Hryciw T, Xiao W (2001) DNA postreplication repair and mutagenesis in *Saccharomyces cerevisiae*. Mutat Res 486:167-184

Casper AM, Nghiem P, Arlt MF, Glover TW (2002) ATR regulates fragile site stability. Cell 111:779-789

Cha RS, Kleckner N (2002) ATR homolog Mec1 promotes fork progression, thus averting breaks in replication slow zones. Science 297:602-606

Clark AJ, Chamberlin M (1966) Abnormal metabolic response to ultraviolet light of a recombination deficient mutant of *Escherichia coli* K12. J Mol Biol 19:442-454

Clerici M, Mantiero D, Lucchini G, Longhese MP (2006) The *Saccharomyces cerevisiae* Sae2 protein negatively regulates DNA damage checkpoint signalling. EMBO Rep 7:212-218

Cobb JA, Bjergbaek L, Shimada K, Frei C, Gasser SM (2003) DNA polymerase stabilization at stalled replication forks requires Mec1 and the RecQ helicase Sgs1. EMBO J 22:4325-4336

Cortes F, Pinero J, Palitti F (1993) Cytogenetic effects of inhibition of topoisomerase I or II activities in the CHO mutant EM9 and its parental line AA8. Mutat Res 288:281-289

Cortes-Ledesma F, Aguilera A (2006) Double-strand breaks arising by replication through a nick are repaired by cohesin-dependent sister-chromatid exchange. EMBO Rep 7:919-926

Courcelle J, Donaldson JR, Chow KH, Courcelle CT (2003) DNA damage-induced replication fork regression and processing in *Escherichia coli*. Science 299:1064-1067

Cox MM (2001) Historical overview: searching for replication help in all of the rec places. Proc Natl Acad Sci USA 98:8173-8180

Cox MM, Goodman MF, Kreuzer KN, Sherratt DJ, Sandler SJ, Marians KJ (2000) The importance of repairing stalled replication forks. Nature 404:37-41

Dannenberg R, Mosig G (1981) Semiconservative DNA replication is initiated at a single site in recombination-deficient gene 32 mutants of bacteriophage T4. J Virol 40:890-900

de Jager M, van Noort J, van Gent DC, Dekker C, Kanaar R, Wyman C (2001) Human Rad50/Mre11 is a flexible complex that can tether DNA ends. Mol Cell 8:1129-1135

De Piccoli G, Cortes-Ledesma F, Ira G, Torres-Rosell J, Uhle S, Farmer S, Hwang JY, Machin F, Ceschia A, McAleenan A, Cordon-Preciado V, Clemente-Blanco A, Vilella-Mitjana F, Ullal P, Jarmuz A, Leitao B, Bressan D, Dotiwala F, Papusha A, Zhao X, Myung K, Haber JE, Aguilera A, Aragon L (2006) Smc5-Smc6 mediate DNA double-strand-break repair by promoting sister-chromatid recombination. Nat Cell Biol 8:1032-1034

Degrassi F, De Salvia R, Tanzarella C, Palitti F (1989) Induction of chromosomal aberrations and SCE by camptothecin, an inhibitor of mammalian topoisomerase I. Mutat Res 211:125-130

Deshpande AM, Newlon CS (1996) DNA replication fork pause sites dependent on transcription. Science 272:1030-1033

Dong Z, Fasullo M (2003) Multiple recombination pathways for sister chromatid exchange in *Saccharomyces cerevisiae*: role of *RAD1* and the *RAD52* epistasis group genes. Nucleic Acids Res 31:2576-2585

Downs JA, Lowndes NF, Jackson SP (2000) A role for *Saccharomyces cerevisiae histone* H2A in DNA repair. Nature 408:1001-1004

Dronkert ML, Beverloo HB, Johnson RD, Hoeijmakers JH, Jasin M, Kanaar R (2000) Mouse *RAD54* affects DNA double-strand break repair and sister chromatid exchange. Mol Cell Biol 20:3147-3156

Fabre F, Chan A, Heyer WD, Gangloff S (2002) Alternate pathways involving Sgs1/Top3, Mus81/ Mms4, and Srs2 prevent formation of toxic recombination intermediates from single-stranded gaps created by DNA replication. Proc Natl Acad Sci USA 99:16887-16892

Fasullo M, Bennett T, AhChing P, Koudelik J (1998) The *Saccharomyces cerevisiae RAD9* checkpoint reduces the DNA damage-associated stimulation of directed translocations. Mol Cell Biol 18:1190-1200

Fasullo M, Giallanza P, Dong Z, Cera C, Bennett T (2001) *Saccharomyces cerevisiae rad51* mutants are defective in DNA damage-associated sister chromatid exchanges but exhibit increased rates of homology-directed translocations. Genetics 158:959-972

Fasullo MD, Dong Z (2004) Genetic control of sister chromatid recombination: The role of radiation repair (*RAD*) genes. Curr Genomics 5:123-136

Fasullo MT, Davis RW (1987) Recombinational substrates designed to study recombination between unique and repetitive sequences *in vivo*. Proc Natl Acad Sci USA 84:6215-6219

Game JC, Mortimer RK (1974) A genetic study of x-ray sensitive mutants in yeast. Mutat Res 24:281-292

Game JC, Sitney KC, Cook VE, Mortimer RK (1989) Use of a ring chromosome and pulsed-field gels to study interhomolog recombination, double-strand DNA breaks and sister-chromatid exchange in yeast. Genetics 123:695-713

Gatti M, Santini G, Pimpinelli S, Olivieri G (1979) Lack of spontaneous sister chromatid exchanges in somatic cells of *Drosophila melanogaster*. Genetics 91:255-274

Gonzalez-Barrera S, Cortés-Ledesma F, Wellinger RE, Aguilera A (2003) Equal sister chromatid exchange is a major mechanism of double-strand break repair in yeast. Mol Cell 11:1661-1671

Haber JE (1998) The many interfaces of Mre11. Cell 95:583-586

Haering CH, Nasmyth K (2003) Building and breaking bridges between sister chromatids. Bioessays 25:1178-1191

Helleday T (2003) Pathways for mitotic homologous recombination in mammalian cells. Mutat Res 532:103-115

Heller RC, Marians KJ (2006) Replication fork reactivation downstream of a blocked nascent leading strand. Nature 439:557-562

Higgins NP, Kato K, Strauss B (1976) A model for replication repair in mammalian cells. J Mol Biol 101:417-425

Hopfner KP, Craig L, Moncalian G, Zinkel RA, Usui T, Owen BA, Karcher A, Henderson B, Bodmer JL, McMurray CT, Carney JP, Petrini JH, Tainer JA (2002) The Rad50 zinc-hook is a structure joining Mre11 complexes in DNA recombination and repair. Nature 418:562-566

Horii Z, Clark AJ (1973) Genetic analysis of the *recF* pathway to genetic recombination in *Escherichia coli* K12: isolation and characterization of mutants. J Mol Biol 80:327-344

Horiuchi T, Fujimura Y (1995) Recombinational rescue of the stalled DNA replication fork: a model based on analysis of an *Escherichia coli* strain with a chromosome region difficult to replicate. J Bacteriol 177:783-791

Howard-Flanders P, Theriot L (1966) Mutants of *Escherichia coli* K-12 defective in DNA repair and in genetic recombination. Genetics 53:1137-1150

Ira G, Malkova A, Liberi G, Foiani M, Haber JE (2003) Srs2 and Sgs1-Top3 suppress crossovers during double-strand break repair in yeast. Cell 115:401-411

Ira G, Pellicioli A, Balijja A, Wang X, Fiorani S, Carotenuto W, Liberi G, Bressan D, Wan L, Hollingsworth NM, Haber JE, Foiani M (2004) DNA end resection, homologous recombination and DNA damage checkpoint activation require CDK1. Nature 431:1011-1017

Ivanov D, Nasmyth K (2005) A topological interaction between cohesin rings and a circular minichromosome. Cell 122:849-860

Ivanov EL, Korolev VG, Fabre F (1992) *XRS2*, a DNA repair gene of *Saccharomyces cerevisiae*, is needed for meiotic recombination. Genetics 132:651-664

Ivessa AS, Lenzmeier BA, Bessler JB, Goudsouzian LK, Schnakenberg SL, Zakian VA (2003) The *Saccharomyces cerevisiae* helicase Rrm3p facilitates replication past nonhistone protein-DNA complexes. Mol Cell 12:1525-1536

Jackson JA, Fink GR (1981) Gene conversion between duplicated genetic elements in yeast. Nature 292:306-311

Johnson RD, Jasin M (2000) Sister chromatid gene conversion is a prominent double-strand break repair pathway in mammalian cells. EMBO J 19:3398-3407

Johzuka K, Terasawa M, Ogawa H, Ogawa T, Horiuchi T (2006) Condensin loaded onto the replication fork barrier site in the rRNA gene repeats during S phase in a *FOB1*-dependent fashion to prevent contraction of a long repetitive array in *Saccharomyces cerevisiae*. Mol Cell Biol 26:2226-2236

Kadyk LC, Hartwell LH (1992) Sister chromatids are preferred over homologs as substrates for recombinational repair in *Saccharomyces cerevisiae*. Genetics 132:387-402

Kadyk LC, Hartwell LH (1993) Replication-dependent sister chromatid recombination in *rad1* mutants of *Saccharomyces cerevisiae*. Genetics 133:469-487

Kato H (1974) Spontaneous sister chromatid exchanges detected by a BrdU-labeling method. Nature 251:70-72

Katou Y, Kanoh Y, Bando M, Noguchi H, Tanaka H, Ashikari T, Sugimoto K, Shirahige K (2003) S-phase checkpoint proteins Tof1 and Mrc1 form a stable replication-pausing complex. Nature 424:1078-1083

Keil RL, McWilliams AD (1993) A gene with specific and global effects on recombination of sequences from tandemly repeated genes in *Saccharomyces cerevisiae*. Genetics 135:711-718

Kim JS, Krasieva TB, LaMorte V, Taylor AM, Yokomori K (2002a) Specific recruitment of human cohesin to laser-induced DNA damage. J Biol Chem 277:45149-45153

Kim ST, Xu B, Kastan MB (2002b) Involvement of the cohesin protein, Smc1, in Atm-dependent and independent responses to DNA damage. Genes Dev 16:560-570

Klein HL (1997) *RDH54*, a *RAD54* homologue in *Saccharomyces cerevisiae*, is required for mitotic diploid-specific recombination and repair and for meiosis. Genetics 147:1533-1543

Kobayashi T, Ganley AR (2005) Recombination regulation by transcription-induced cohesin dissociation in rDNA repeats. Science 309:1581-1584

Kobayashi T, Horiuchi T, Tongaonkar P, Vu L, Nomura M (2004) *SIR2* regulates recombination between different rDNA repeats, but not recombination within individual rRNA genes in yeast. Cell 117:441-453

Kogoma T (1997) Stable DNA replication: interplay between DNA replication, homologous recombination, and transcription. Microbiol Mol Biol Rev 61:212-238

Kolodner RD, Putnam CD, Myung K (2002) Maintenance of genome stability in *Saccharomyces cerevisiae*. Science 297:552-557

Kowalczykowski SC, Dixon DA, Eggleston AK, Lauder SD, Rehrauer WM (1994) Biochemistry of homologous recombination in *Escherichia coli*. Microbiol Rev 58:401-465

Kurihara T, Tatsumi K, Takahashi H, Inoue M (1987) Sister-chromatid exchanges induced by ultraviolet light in Bloom's syndrome fibroblasts. Mutat Res 183:197-202

Kuzminov A (1995) Instability of inhibited replication forks in *E. coli*. Bioessays 17:733-741

Kuzminov A (1999) Recombinational repair of DNA damage in *Escherichia coli* and bacteriophage lambda. Microbiol Mol Biol Rev 63:751-813

Kuzminov A (2001) Single-strand interruptions in replicating chromosomes cause double-strand breaks. Proc Natl Acad Sci USA 98:8241-8246

Kuzminov A, Schabtach E, Stahl FW (1994) Chi sites in combination with RecA protein increase the survival of linear DNA in *Escherichia coli* by inactivating exoV activity of RecBCD nuclease. EMBO J 13:2764-2776

Lambert S, Watson A, Sheedy DM, Martin B, Carr AM (2005) Gross chromosomal rearrangements and elevated recombination at an inducible site-specific replication fork barrier. Cell 121:689-702

Lehmann AR (1972) Post-replication repair of DNA in ultraviolet-irradiated mammalian cells. No gaps in DNA synthesized late after ultraviolet irradiation. Eur J Biochem 31:438-445

Lehmann AR (2005) The role of SMC proteins in the responses to DNA damage. DNA Repair 4:309-314

Lengauer C, Kinzler KW, Vogelstein B (1998) Genetic instabilities in human cancers. Nature 396:643-649

Liberi G, Maffioletti G, Lucca C, Chiolo I, Baryshnikova A, Cotta-Ramusino C, Lopes M, Pellicioli A, Haber JE, Foiani M (2005) Rad51-dependent DNA structures accumulate at damaged replication forks in *sgs1* mutants defective in the yeast ortholog of *BLM* RecQ helicase. Genes Dev 19:339-350

Lindahl T (1993) Instability and decay of the primary structure of DNA. Nature 362:709-715

Lisby M, Rothstein R, Mortensen UH (2001) Rad52 forms DNA repair and recombination centers during S phase. Proc Natl Acad Sci USA 98:8276-8282

Lobachev K, Vitriol E, Stemple J, Resnick MA, Bloom K (2004) Chromosome fragmentation after induction of a double-strand break is an active process prevented by the RMX repair complex. Curr Biol 14:2107-2112

Lopes M, Foiani M, Sogo JM (2006) Multiple mechanisms control chromosome integrity after replication fork uncoupling and restart at irreparable UV lesions. Mol Cell 21:15-27

Losada A, Hirano T (2005) Dynamic molecular linkers of the genome: the first decade of SMC proteins. Genes Dev 19:1269-1287

Lucca C, Vanoli F, Cotta-Ramusino C, Pellicioli A, Liberi G, Haber J, Foiani M (2004) Checkpoint-mediated control of replisome-fork association and signalling in response to replication pausing. Oncogene 23:1206-1213

Luder A, Mosig G (1982) Two alternative mechanisms for initiation of DNA replication forks in bacteriophage T4: priming by RNA polymerase and by recombination. Proc Natl Acad Sci USA 79:1101-1105

Malkova A, Ivanov EL, Haber JE (1996) Double-strand break repair in the absence of *RAD51* in yeast: a possible role for break-induced DNA replication. Proc Natl Acad Sci USA 93:7131-7136

Malone RE, Ward T, Lin S, Waring J (1990) The *RAD50* gene, a member of the double strand break repair epistasis group, is not required for spontaneous mitotic recombination in yeast. Curr Genet 18:111-116

McClintock B (1938) The production of homozygous deficient tissues with mutant characteristics by means of the aberrant mitotic behavior of ring-shaped chromosomes. Genetics 23:315-376

McGlynn P (2004) Links between DNA replication and recombination in prokaryotes. Curr Opin Genet Dev 14:107-112

McGlynn P, Lloyd RG (2000) Modulation of RNA polymerase by (p)ppGpp reveals a RecG-dependent mechanism for replication fork progression. Cell 101:35-45

Michel B, Flores MJ, Viguera E, Grompone G, Seigneur M, Bidnenko V (2001) Rescue of arrested replication forks by homologous recombination. Proc Natl Acad Sci USA 98:8181-8188

Morgan WF, Fero ML, Land MC, Winegar RA (1988) Inducible expression and cytogenetic effects of the *EcoRI* restriction endonuclease in Chinese hamster ovary cells. Mol Cell Biol 8:4204-4211

Mosig G (1998) Recombination and recombination-dependent DNA replication in bacteriophage T4. Annu Rev Genet 32:379-413

Myung K, Datta A, Kolodner RD (2001) Suppression of spontaneous chromosomal rearrangements by S phase checkpoint functions in *Saccharomyces cerevisiae*. Cell 104:397-408

Myung K, Pennaneach V, Kats ES, Kolodner RD (2003) *Saccharomyces cerevisiae* chromatin-assembly factors that act during DNA replication function in the maintenance of genome stability. Proc Natl Acad Sci USA 100:6640-6645

Nag DK, Suri M, Stenson EK (2004) Both CAG repeats and inverted DNA repeats stimulate spontaneous unequal sister-chromatid exchange in *Saccharomyces cerevisiae*. Nucleic Acids Res 32:5677-5684

Nasmyth K (2002) Segregating sister genomes: the molecular biology of chromosome separation. Science 297:559-565

Nassif N, Engels W (1993) DNA homology requirements for mitotic gap repair in *Drosophila*. Proc Natl Acad Sci USA 90:1262-1266

Onoda F, Seki M, Miyajima A, Enomoto T (2000) Elevation of sister chromatid exchange in *Saccharomyces cerevisiae sgs1* disruptants and the relevance of the disruptants as a system to evaluate mutations in Bloom's syndrome gene. Mutat Res 459:203-209

Onoda F, Takeda M, Seki M, Maeda D, Tajima J, Ui A, Yagi H, Enomoto T (2004) *SMC6* is required for MMS-induced interchromosomal and sister chromatid recombinations in *Saccharomyces cerevisiae*. DNA Repair 3:429-439

Pages V, Fuchs RP (2003) Uncoupling of leading- and lagging-strand DNA replication during lesion bypass *in vivo*. Science 300:1300-1303

Paques F, Haber JE (1999) Multiple pathways of recombination induced by double-strand breaks in *Saccharomyces cerevisiae*. Microbiol Mol Biol Rev 63:349-404

Paulovich AG, Armour CD, Hartwell LH (1998) The *Saccharomyces cerevisiae RAD9*, *RAD17*, *RAD24* and *MEC3* genes are required for tolerating irreparable, ultraviolet-induced DNA damage. Genetics 150:75-93

Perry P, Evans HJ (1975) Cytological detection of mutagen-carcinogen exposure by sister chromatid exchange. Nature 258:121-125

Petes TD (1980) Unequal meiotic recombination within tandem arrays of yeast ribosomal DNA genes. Cell 19:765-774

Potts PR, Porteus MH, Yu H (2006) Human SMC5/6 complex promotes sister chromatid homologous recombination by recruiting the SMC1/3 cohesin complex to double-strand breaks. EMBO J 25:3377–3388

Prado F, Aguilera A (2005a) Partial depletion of histone H4 increases homologous recombination-mediated genetic instability. Mol Cell Biol 25:1526-1536

Prado F, Aguilera A (2005b) Impairment of replication fork progression mediates RNA pol II transcription-associated recombination. EMBO J 24:1267-1276

Prado F, Cortes-Ledesma F, Aguilera A (2004) The absence of the yeast chromatin assembly factor Asf1 increases genomic instability and sister chromatid exchange. EMBO Rep 5:497-502

Prado F, Cortes-Ledesma F, Huertas P, Aguilera A (2003) Mitotic recombination in *Saccharomyces cerevisiae*. Curr Genet 42:185-198

Prakash L (1981) Characterization of postreplication repair in *Saccharomyces cerevisiae* and effects of *rad6*, *rad18*, *rev3* and *rad52* mutations. Mol Gen Genet 184:471-478

Puget N, Knowlton M, Scully R (2005) Molecular analysis of sister chromatid recombination in mammalian cells. DNA Repair 4:149-161

Ramey CJ, Howar S, Adkins M, Linger J, Spicer J, Tyler JK (2004) Activation of the DNA damage checkpoint in yeast lacking the histone chaperone anti-silencing function 1. Mol Cell Biol 24:10313-10327

Richardson C, Moynahan ME, Jasin M (1998) Double-strand break repair by interchromosomal recombination: suppression of chromosomal translocations. Genes Dev 12:3831-3842

Rothkamm K, Kruger I, Thompson LH, Lobrich M (2003) Pathways of DNA double-strand break repair during the mammalian cell cycle. Mol Cell Biol 23:5706-5715

Rothstein R, Michel B, Gangloff S (2000) Replication fork pausing and recombination or "gimme a break". Genes Dev 14:1-10

Rupp WD, Howard-Flanders P (1968) Discontinuities in the DNA synthesized in an excision-defective strain of *Escherichia coli* following ultraviolet irradiation. J Mol Biol 31:291-304

Rupp WD, Wilde CE 3rd, Reno DL, Howard-Flanders P (1971) Exchanges between DNA strands in ultraviolet-irradiated *Escherichia coli*. J Mol Biol 61:25-44

Saintigny Y, Delacote F, Vares G, Petitot F, Lambert S, Averbeck D, Lopez BS (2001) Characterization of homologous recombination induced by replication inhibition in mammalian cells. EMBO J 20:3861-3870

Saleh-Gohari N, Bryant HE, Schultz N, Parker KM, Cassel TN, Helleday T (2005) Spontaneous homologous recombination is induced by collapsed replication forks that are caused by endogenous DNA single-strand breaks. Mol Cell Biol 25:7158-7169

Saleh-Gohari N, Helleday T (2004) Conservative homologous recombination preferentially repairs DNA double-strand breaks in the S phase of the cell cycle in human cells. Nucleic Acids Res 32:3683-3688

Schar P, Fasi M, Jessberger R (2004) *SMC1* coordinates DNA double-strand break repair pathways. Nucleic Acids Res 32:3921-3929

Seigneur M, Bidnenko V, Ehrlich SD, Michel B (1998) RuvAB acts at arrested replication forks. Cell 95:419-430

Shinohara M, Shita-Yamaguchi E, Buerstedde JM, Shinagawa H, Ogawa H, Shinohara A (1997) Characterization of the roles of the *Saccharomyces cerevisiae RAD54* gene and a homologue of *RAD54*, *RDH54/TID1*, in mitosis and meiosis. Genetics 147:1545-1556

Sjogren C, Nasmyth K (2001) Sister chromatid cohesion is required for postreplicative double-strand break repair in *Saccharomyces cerevisiae*. Curr Biol 11:991-995

Skalka A (1974) A replicator's view of recombination (and repair). In: Grell RR (ed) Mechanisms in recombination. Plenum Press, New York, pp 421-432

Sogo JM, Lopes M, Foiani M (2002) Fork reversal and ssDNA accumulation at stalled replication forks owing to checkpoint defects. Science 297:599-602

Solomon E, Bobrow M (1975) Sister chromatid exchanges: a sensitive assay of agents damaging human chromosomes. Mutat Res 30:273-278

Sonoda E, Matsusaka T, Morrison C, Vagnarelli P, Hoshi O, Ushiki T, Nojima K, Fukagawa T, Waizenegger IC, Peters JM, Earnshaw WC, Takeda S (2001)

Scc1/Rad21/Mcd1 is required for sister chromatid cohesion and kinetochore function in vertebrate cells. Dev Cell 1:759-770

Sonoda E, Sasaki MS, Morrison C, Yamaguchi-Iwai Y, Takata M, Takeda S (1999) Sister chromatid exchanges are mediated by homologous recombination in vertebrate cells. Mol Cell Biol 19:5166-5169

Strom L, Lindroos HB, Shirahige K, Sjogren C (2004) Postreplicative recruitment of cohesin to double-strand breaks is required for DNA repair. Mol Cell 16:1003-1015

Strumberg D, Pilon AA, Smith M, Hickey R, Malkas L, Pommier Y (2000) Conversion of topoisomerase I cleavage complexes on the leading strand of ribosomal DNA into 5'-phosphorylated DNA double-strand breaks by replication runoff. Mol Cell Biol 20:3977-3987

Sutou S (1997) Reversal of DNA polarity as revealed by sister chromatid exchanges in ring chromosomes. Mutat Res 394:69-75

Symington LS (2002) Role of RAD52 epistasis group genes in homologous recombination and double-strand break repair. Microbiol Mol Biol Rev 66:630-670

Szostak JW, Orr-Weaver TL, Rothstein RJ, Stahl FW (1983) The double-strand-break repair model for recombination. Cell 33:25-35

Szostak JW, Wu R (1980) Unequal crossing over in the ribosomal DNA of *Saccharomyces cerevisiae*. Nature 284:426-430

Takata M, Sasaki MS, Sonoda E, Fukushima T, Morrison C, Albala JS, Swagemakers SM, Kanaar R, Thompson LH, Takeda S (2000) The Rad51 paralog Rad51B promotes homologous recombinational repair. Mol Cell Biol 20:6476-6482

Takata M, Sasaki MS, Sonoda E, Morrison C, Hashimoto M, Utsumi H, Yamaguchi-Iwai Y, Shinohara A, Takeda S (1998) Homologous recombination and non-homologous end-joining pathways of DNA double-strand break repair have overlapping roles in the maintenance of chromosomal integrity in vertebrate cells. EMBO J 17:5497-5508

Takata M, Sasaki MS, Tachiiri S, Fukushima T, Sonoda E, Schild D, Thompson LH, Takeda S (2001) Chromosome instability and defective recombinational repair in knockout mutants of the five Rad51 paralogs. Mol Cell Biol 21:2858-2866

Takeuchi Y, Horiuchi T, Kobayashi T (2003) Transcription-dependent recombination and the role of fork collision in yeast rDNA. Genes Dev 17:1497-1506

Tercero JA, Longhese MP, Diffley JF (2003) A central role for DNA replication forks in checkpoint activation and response. Mol Cell 11:1323-1336

Thompson LH, Brookman KW, Dillehay LE, Carrano AV, Mazrimas JA, Mooney CL, Minkler JL (1982) A CHO-cell strain having hypersensitivity to mutagens, a defect in DNA strand-break repair, and an extraordinary baseline frequency of sister-chromatid exchange. Mutat Res 95:427-440

Torres-Ramos CA, Prakash S, Prakash L (2002) Requirement of *RAD5* and *MMS2* for postreplication repair of UV-damaged DNA in *Saccharomyces cerevisiae*. Mol Cell Biol 22:2419-2426

Torres-Rosell J, Machin F, Farmer S, Jarmuz A, Eydmann T, Dalgaard JZ, Aragon L (2005) *SMC5* and *SMC6* genes are required for the segregation of repetitive chromosome regions. Nat Cell Biol 7:412-419

Trautinger BW, Jaktaji RP, Rusakova E, Lloyd RG (2005) RNA polymerase modulators and DNA repair activities resolve conflicts between DNA replication and transcription. Mol Cell 19:247-258

Tyler JK (2002) Chromatin assembly. Cooperation between histone chaperones and ATP-dependent nucleosome remodeling machines. Eur J Biochem 269:2268-2274

Unal E, Arbel-Eden A, Sattler U, Shroff R, Lichten M, Haber JE, Koshland D (2004) DNA damage response pathway uses histone modification to assemble a double-strand break-specific cohesin domain. Mol Cell 16:991-1002

Voelkel-Meiman K, Roeder GS (1990) Gene conversion tracts stimulated by *HOT1*-promoted transcription are long and continuous. Genetics 126:851-867

Wang ZQ, Stingl L, Morrison C, Jantsch M, Los M, Schulze-Osthoff K, Wagner EF (1997) PARP is important for genomic stability but dispensable in apoptosis. Genes Dev 11:2347-2358

Wellinger RE, Prado F, Aguilera A (2006) Replication fork progression is impaired by transcription in hyperrecombinant yeast cells lacking a functional THO complex. Mol Cell Biol 26:3327-3334

Wiltzius JJ, Hohl M, Fleming JC, Petrini JH (2005) The Rad50 hook domain is a critical determinant of Mre11 complex functions. Nat Struct Mol Biol 12:403-407

Wolff S (1977) Sister chromatid exchange. Annu Rev Genet 11:183-201

Wolff S, Lindsley DL, Peacock WJ (1976) Cytological evidence for switches in polarity of chromosomal DNA. Proc Natl Acad Sci USA 73:877-881

Wu L, Hickson ID (2003) The Bloom's syndrome helicase suppresses crossing over during homologous recombination. Nature 426:870-874

Xie A, Puget N, Shim I, Odate S, Jarzyna I, Bassing CH, Alt FW, Scully R (2004) Control of sister chromatid recombination by histone H2AX. Mol Cell 16:1017-1025

Ye X, Franco AA, Santos H, Nelson DM, Kaufman PD, Adams PD (2003) Defective S phase chromatin assembly causes DNA damage, activation of the S phase checkpoint, and S phase arrest. Mol Cell 11:341-351

Zhang H, Lawrence CW (2005) The error-free component of the *RAD6/RAD18* DNA damage tolerance pathway of budding yeast employs sister-strand recombination. Proc Natl Acad Sci USA 102:15954-15959

Zou H, Rothstein R (1997) Holliday junctions accumulate in replication mutants via a RecA homolog- independent mechanism. Cell 90:87-96

Aguilera, Andrés
Departamento de Biología Molecular, CABIMER, CSIC-Universidad de Sevilla, Av. Américo Vespucio s/n, 41092 Sevilla, Spain
aguilo@us.es

Cortés-Ledesma, Felipe
Departamento de Biología Molecular, CABIMER, CSIC-Universidad de Sevilla, Av. Américo Vespucio s/n, 41092 Sevilla, Spain

Prado Félix
Departamento de Biología Molecular, CABIMER, CSIC-Universidad de Sevilla, Av. Américo Vespucio s/n, 41092 Sevilla, Spain

Mating-type switching in *S. pombe*

Benoit Arcangioli, Laura Roseaulin, and Allyson Holmes

Abstract

Mating-type switching in fission yeast consists of replacing genetic information at the transcriptionally active sexual locus, *mat1,* with sequences copied from one of the silent donor loci, *mat2*-P or *mat3*-M. This is a two-step developmental process, involving two consecutive rounds of DNA replication. The first replication introduces a DNA lesion at *mat1* and the second replication collides with the lesion forming a polar one-ended double-strand break. In the wild type strain, the silent donor alleles, embedded into heterochromatin, provide the appropriate intact DNA templates for repair, allowing mating-type switching. In the absence of the donor loci, viability requires the presence of the homologous recombination (HR) machinery, indicating the usage of the sister chromatid for repair. Both pathways differ by their mode of resolution. Therefore this system provides the opportunity to study how the HR enzymes participate in replication fork restart at single or repeated chromosomal sequences.

1 Fission yeast life cycle

The fission yeast, *Schizosaccharomyces pombe* alternates its mating-type (MT) during vegetative growth, forming colonies containing a homogeneously mixed population of P (plus) and M (minus) mating-type cells. The ability of a single haploid cell to produce a cell population of both mating-types demonstrates its switching potential and existence of both mating-type genes within the haploid genome. We will see that mating-type switching reflects programmed recombination events, allowing both mating-types to be alternatively expressed. *S. pombe* cells are not motile since they do not do not contain flagella, fins, wings, or limbs. Their switching ability ensures that a P or M cells will always find a complementary mating partner in close proximity. In growing conditions, the P and M functions are not expressed and the cells are phenotypically sterile. Nutritional deprivation (i.e. lack of nitrogen) slows down the G1-phase and activates the mating-type specific genes, allowing opposite mating-type partners to fuse and form transitory diploids, which directly undergo meiosis and sporulation, thereby producing zygotic asci containing four haploid spores. This dormant phase is the resistant form of *S. pombe* life. When the conditions are favorable, the spores germinate and enter into a new cycle. The homothallic yeast (forming diploids from a single spore) is considered as the wild type strain because it can spontaneously produce

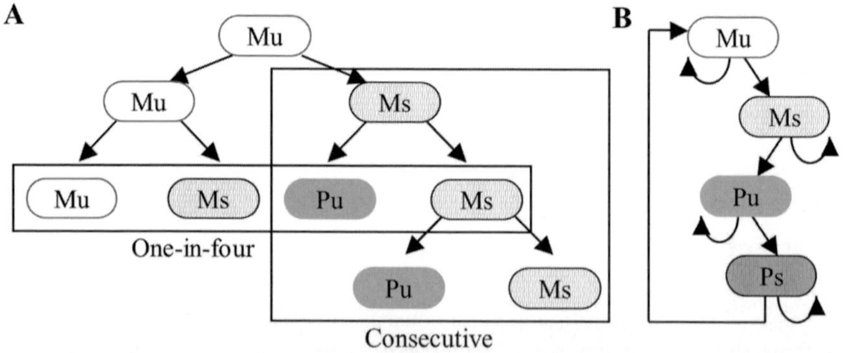

Fig. 1. Mating-type switching pattern in the fission yeast cell lineage. P and M represent the two MT alleles, with "u" and "s" indicating unswitchable and switchable, respectively. The P and M lineage are identical (A) the one-in-four and consecutive switching rules are shown. (B) Summary of MT switching pattern, stressing the analogy with the stem cell lineage.

all of the heterothallic variants generating homogenous and stable populations of P or M cells. A simple and rapid plate assay to determine the switching ability of strains is the classical iodine vapor staining reaction (Leupold 1955). The homothallic colony stained uniformly black is also called h^{90} because the colony contains 90% spores, while colonies formed by sterile or heterothallic (h^+ or h^-) cells appear yellowish. Slow-switching mutants exhibit streaky or mottled iodine staining patterns. The black staining reflects the stain of starch material present in the spores by trapping the iodine vapors, in a reversible manner.

2 The pattern of switching

In growing conditions, the MT alleles are not expressed and the cells are phenotypically sterile; however, MT switching is functional, witnessed by the presence of cells containing the P or M alleles at the *mat1* locus in a clonal growing culture. The first feature of MT switching is that conjugation often occurs between two sister cells (Leupold 1950), demonstrating that only one sister changes its MT. Later Miyata and Miyata (1981) have shown that, among four cousin cells, only two form an ascus. This remarkable feature, termed the one-in-four rule, indicates that two consecutive divisions are required for MT switching. The first division produces two cells with the same MT but with different switching potentials (called "u" for unswitchable and "s" for switchable (see Fig. 1A). During the second division, the "u" cell divides and keeps the same MT and the "s" cell divides and produces one switched cell. Furthermore, the sister (the "s" cell) of a switched cell is able to switch during its next division, forming a chain of recurrent switching, termed the consecutive switching rule (Egel and Eie 1987; Klar 1990). The efficiency of this overall process is above 90%, indicating that the cells are directed to switch primarily to the opposite MT. The observed pattern (Fig.1B) shows analogy to the mammalian stem cell lineage, in which a stem cell produces one daughter cell like itself while the other is different and has moved one step forward in its developmental program.

3 The mating-type region

Genetic studies have mapped the mating-type locus (*mat1*) on the right arm of chromosome II. The mating type of the cell is determined by the allele contained at (*mat1*), in which cells with the P allele (*mat1*-P) exhibit the P mating type and cells with the M allele (*mat1*-M) exhibit the M mating type (Leupold 1950; Egel 1977). Molecular analyses have defined the structure of the mating-type region, covering about 30 kb, in which three cassettes, *mat1*, *mat2*-P and *mat3*-M, are organized as direct repeats (Fig. 2A). The *mat2*-P and *mat3*-M cassettes are silent, not expressed and serve as genetic reservoirs to replace the *mat1* allele. The spacer regions between *mat1*, *mat2*-P, and *mat3*-M are termed the L and K regions, respectively (Beach et al. 1982; Beach 1983). The L region (17 kb) contains six genes, of which at least one is essential. The K region (11 kb) does not contain genes but contains a 4.3 kb *cenH*, a repeat element that shares homology to the centromeric *dg/dh* repeats (Grewal and Klar 1997) and together with *mat2*-P and *mat3*-M forms a heterochromatic-like domain which is transcriptionally silent. The REII and REIII (Fig. 2A) elements located distal to *mat2-P* and *mat3-M*, respectively, cooperate with *cenH* for silencing (Thon et al. 1999; Ayoub et al. 2000). Two 2.1 kb inverted repeats (IR-L and IR-R) delineate the edges of the silent region (Noma et al. 2001; Thon et al. 2002).

The P and M alleles, of about 1 kb each, code for two divergently transcribed genes (Pc, 118 amino acids (aa); Pi, 159 aa; Mc, 181 aa; Mi, 42 aa) (Kelly et al. 1988). Pc and Mc are required for mating and meiosis, Pi and Mi are only required for meiosis, and they code for transcription factors or co-factors, ensuring the expression of mating-type specific genes, that are involved in the synthesis and processing of the two mating-type pheromones and their receptors (reviewed in Nielsen 2004). Interestingly, Mc shares homology with Sry (for sex-determining region on the Y chromosome) and contains a HMG domain (DNA-binding and DNA-bending motif of High-Mobility Group of non-histone proteins) conserved between species. This site is found mutated in almost all clinical forms of XY gonadal dysgenesis (Sinclair et al. 1990).

In addition to their direct repeat organization, a noticeable feature of the three MT cassettes is that they are flanked by short homologous sequences termed, H1 and H2. The H1 sequence, 59 bp long, is located on the centromere-distal (right) side and the H2 sequence, 135 bp long, is located on the centromere-proximal (left) side. A third homologous sequence, the H3 box, 57 bp long, is present on the centromere-proximal (left) of the H2 box, only at the two silent *mat2*-P and *mat3*-M cassettes (Fig. 2A). These homologous sequences are used during the initiation

Fig. 2. (overleaf). MT region. (A) The three MT cassettes are shown, the P allele is indicated by a white box and M allele by a black box. The H1, H2, and H3 box flanking the cassettes are indicated. The imprint is shown by a black arrow; *IR-L* and *IR-R*, inverted region left and right, respectively. Two circles indicate the positions of the RE2 and RE3 silencing elements. The cold region for transcription and recombination is shaded. The L and K regions and *cenH* are shown. (B). *mat1* regulatory elements. The deletions in the distal *mat1* regions *Msmt-0*, *PΔ17* and *Smt-s1*, the mut-elements within the H1 box and the SAS1 and SAS2 *cis*-activating sites are shown. (C) Organization of the MT region in h^{90} and several heterothallic strains described in the text. Each cassette is named, such that the first number indicates the origin of the proximal sequences and the second number indicates the origin of the distal sequences. The K and L regions are shown. The size of the arrow indicates the level of the break. The cooperation between the RE2, RE3, and *cenH cis*-acting elements may provide transient or stable silencing (shaded).

and resolution steps of the gene conversion recombination process required for MT switching and will be discussed in the following sections.

Heterothallic variants that stabilize the M or P cell type are called h^+ or h^- and arise spontaneously from a homothallic culture at a frequency of 10^{-3} to 10^{-6}. Southern blot studies have shown that they contain gross rearrangements of the MT region, including deletions, duplications and mini-circles (Fig. 2C). These rearrangements probably result from aberrant switching events or recombination between direct repeats and provide insight into the mechanism of MT switching. Depending on the structures of the rearranged MT region, variants can exhibit stable or unstable phenotypes (Beach and Klar 1984). A subgroup of heterothallic variants accumulates in some of the *swi* (switching) mutants, discriminating between rearrangements due to abnormal switching events and inappropriate recombination between homologous sequences (Egel et al. 1984).

4 A site- and strand-specific imprint at *mat1*

Southern blot analysis of genomic DNA prepared from a homothallic strain revealed that a constant proportion (20-40%) of cells contain a fragile site at the *mat1* locus, often converted into a double-strand break (DSB) (Beach 1983; Arcangioli 1998). The DSB is localized at the junction between the *mat1* specific allele and the H1-homology box and is persistent throughout cell cycle progression (Nielsen and Egel 1989). By analogy to the mating-type switching process in *Saccharomyces cerevisiae* and by employing mutants with reduced efficiency of switching, it was proposed that a DSB was the initiating event for MT switching. This proposal implies that the program of switching is controlled by the formation of the DSB and that the switching corresponds to the precise replacement (by gene conversion) of the *mat1* allele by the opposite allele present in one of the silent cassettes (Haber JE, this volume).

Intriguingly, when the genomic DNA was gently prepared in low melting agarose plugs, as for pulse field gel analysis, the DSB at *mat1* was not observed

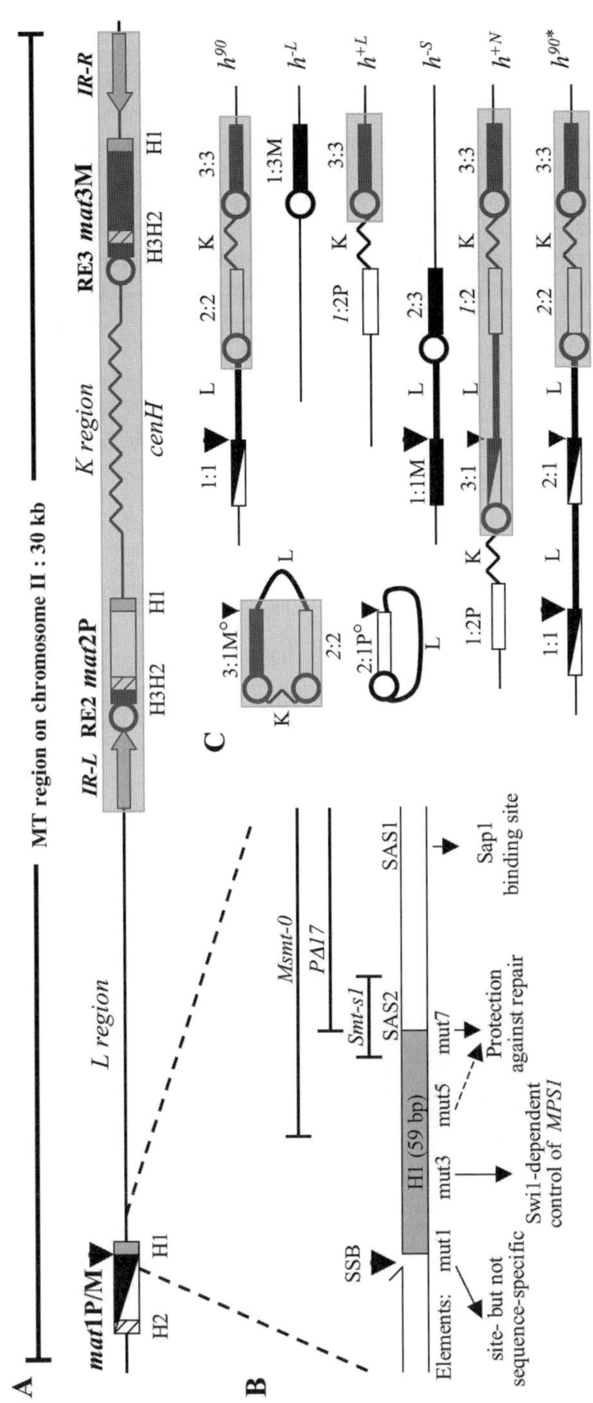
A
MT region on chromosome II : 30 kb
mat1P/M
H2
H1
L region
IR-L
RE2 mat2P
H3H2
H1
K region
cenH
RE3 mat3M
H3H2
H1
IR-R
B
SSB
Elements:
mut1
mut3
mut5
mut7
H1 (59 bp)
SAS2
SAS1
Smt-s1
PΔ17
Msmt-0
site- but not sequence-specific
Swi1-dependent control of MPS1
Protection against repair
Sap1 binding site
C
3:1M°
2:2
K
L
2:1P°
L
1:1
L
2:2
K
3:3
h90
1:3M
h-L
1:2P
K
3:3
h+L
1:1M
L
2:3
h-S
1:2P
K
3:1
L
1:2
K
3:3
h+N
1:1
L
2:1
L
2:2
K
3:3
h90*

(Arcangioli 1998; Dalgaard and Klar 1999). However, the apparently intact *mat1* locus appears physically and enzymatically fragile and easily breaks or shears during standard genomic DNA purification, generating a DSB. Further biochemical experiments revealed that the *mat1* fragile site consists of a single-strand break or alkali-labile modification on the *mat1* upper strand, leaving the lower DNA strand intact (Arcangioli 1998; Dalgaard and Klar 1999). It was proposed that the alkaline-labile modification of the *mat1* upper strand consists of a DNA-RNA-DNA region and that the observed nick is the product of a specific cellular RNAse(s) acting during genomic DNA purification (Dalgaard and Klar 1999). More recently Vengrova and Dalgaard (2004, 2005) state that the lesion consists of one or two ribonuclotides incorporated into the otherwise intact DNA duplex and can be purified intact or cleaved in "some" conditions. We suggest in "all" conditions, since we always observed a nick, with a 3'OH and 5'OH termini and no gap (Kaykov and Arcangioli 2004). A better understanding of the biochemical conditions allowing for the purification of an intact DNA strand at *mat1* is an essential step to confirm and understand this novel mechanism. Thus, in the following sections, we will refer to the lesion as a SSB for single-strand break.

The position of the SSB in the *mat1*-M and *mat1*-P allele is located after the second thymidine in a continuous run of 7 T's in M, and 10 T's in P, shifting the SSB by three bases within M and P with respect to the H1 homology sequences (Nielsen and Egel 1989; Arcangioli 1998). To better understand how the position of the SSB is determined, our laboratory has inserted six *Pst*I sites (CTGCAG) at one nucleotide intervals, overlapping the imprinted site. These substitutions were made in the wild type strain and in a strain in which the donor loci have been removed to avoid potential repair of the mutations by recombination with the H1 homology boxes also present at the silent donor loci. All of the mutant strains analyzed exhibit the same level of cleavage at the wild type position, demonstrating that the position of cleavage is not sequence, but site-specific (mut1 in Fig. 2B) and its location must be controlled from other *cis*-acting elements outside of H1, probably within the M or P sequences (Kaykov and Arcangioli 2004). This result does not support the model proposed earlier, that small imperfect palindromic sequences surrounding the imprint site are necessary and sufficient to direct the position of the SSB (Nielsen and Egel 1989; Vengrova and Dalgaard 2004), since they are removed by the mutations described above.

Finally, the SSB is thought to be the initiating event of switching, so that the switching pattern in a cell lineage is dictated by the regulation of SSB formation and its usage. The stability of the SSB indicates that it is protected against repair during the entire length of the cell cycle. The suggestion is that the SSB acts as a strand-specific imprint, marking one of the two sister chromatids one generation in advance of switching (Arcangioli 1998).

5 Cis-acting elements controlling the imprint

Although, all MT cassettes contain identical sequences at the cut site, only *mat1* is cleaved *in vivo*. This clearly indicates that sequences outside of *mat1* are required for SSB formation. Early work, based on the iodine-staining assay, has isolated two mutations linked to *mat1*. The *smt*-s1 (C13-P11) mutation exhibits a streaky iodine staining pattern (Egel and Gutz 1981) and reduces the SSB level and MT switching (Beach 1983). This mutation is a small deletion of 27 bp between two direct repeats of 7 bp overlapping the distal end of H1 (Klar et al. 1991). Engelke et al. (1987) found a second mutation, called M*smt*-0, since the mutation occurred when *mat1* contained the M allele, which completely abolishes SSB formation and MT switching. The M*smt*-0 mutation consists of a 263 bp deletion between two direct repeats of 9bp, removing about one half of H1 (Styrkarsdottir et al. 1993). To further characterize the *cis*-acting elements important for SSB formation, we have introduced progressive deletions in the distal *mat1* region. This approach revealed the presence of two switch-activating sites (SAS1 and SAS2), localized at 140 bp and 60 bp from the SSB site and it was later shown that SAS1 is a target for the Sap1 protein (described below). Individual deletions exhibit reduced SSB levels and MT switching, whereas double deletions totally abolish the SSB and switching (Arcangioli and Klar 1991). The strain was called *PΔ17*, since the mutation was introduced when *mat1* was containing the P allele. To preserve the distance between already known elements, we systematically introduced 6 bp substitution (*Pst*I site) mutations every 10 bp starting from the SSB site towards SAS1. We found that mutations located at 20, 40, and 60 bp (termed mut3, mut5, and mut7, Fig. 2B) from the SSB site affect its steady-state level and consequently MT switching (note that mut7 located at 60bp from the SSB site overlaps SAS2 element), whereas mutations at 10, 30, 50, 70, 80, 90, and 100 pb have no effects. The specific activity for each site in the imprinting process is further described below. Because two of the three new sites are inside the H1 box, also present next to *mat2*-P and *mat3*-M, then we concluded that these elements are necessary but not sufficient for SSB formation (Kaykov et al. 2004).

6 Trans-acting *swi* (switch) genes

Extensive search for MT switching mutants, using the iodine staining method, allowed the genetic identification of 16 genes which can be grouped into three classes, based on their requirements for particular steps of MT switching (Egel et al. 1984; Gutz and Schmidt 1985). Class Ia includes *swi1, swi3, swi7,* and *sap1*, which encode functions important for SSB formation or maintenance. Class Ib includes *swi2*, *swi5*, and *swi6*, which exhibit wild type levels of SSB but reduced MT switching. Class II, includes *swi4*, *swi8*, *swi9*, *swi10*, and *rad22*, which also exhibit wild type levels of SSB, but produce a high level of heterothallic mutants, containing rearrangements in the MT region (Fig. 2C). Mutants from Class Ia are epistatic to mutants of the other classes, and double or triple mutants exhibit cumulative iodine staining defects. All of the *swi* genes have been cloned and encode for known and conserved gene products, with the exception of *sap1* (Table 1).

6.1 Class Ia

Swi1 belongs to a protein family including Tof1 (Topoisomerase 1 interacting factor) and Tim1 (*Timeless*) in *S. cerevisiae* and metazoans, respectively (Park and Sternglanz 1999; Dalgaard and Klar 2000; Chan et al. 2003). Swi3 is also very well conserved and is homologous to Csm3 and Tipin in *S. cerevisiae* and metazoans, respectively (Gotter 2003; Lee et al. 2004; Mayer et al. 2004). In *S. pombe*, as in other eukaryotes, Swi1/3 form heterodimers *in vivo*, are components of a replication fork protection complex (FPC) and were proposed to coordinates leading- and lagging-strand synthesis, preventing replication fork collapse (Katou et al. 2003; Noguchi et al. 2004). Tof1 interacts with the MCM-Cdc45 replicative DNA helicases and was proposed to retain the helicases from unwinding the DNA when DNA synthesis is inhibited, hence stabilizing the stalled replisome (Katou et al. 2003). It was also shown that Swi1 associates with the Hsk1/Dfp1 kinase complex, essential for DNA replication, homologous to the Cdc7/Dbf4 complex in *S. cerevisiae*. Furthermore, an *hsk1* temperature-sensitive mutant is partially defective in MT switching and requires Swi1 for growth at a semi-permissive temperature (Matsumoto et al. 2005). In *S. pombe*, the *swi1* and *swi3* genes are not essential and their deletions exhibit similar defects for SSB formation and MT switching, as point mutations, indicating that neither proteins are direct components of imprinting activity (Schmidt 1987; Noguchi et al. 2003; Lee et al. 2004). However, Swi1 and Swi3 are required for the activity of replication fork blocks within the rDNA spacer region (Noguchi et al. 2003; Krings and Bastia 2004) and within the *mat1* regions (see below).

The *swi7* gene encodes for the catalytic large subunit of DNA polymerase α (Singh and Klar 1993). Swi7 is part of a larger complex containing the primase enzyme, crucial for replication initiation and lagging-strand DNA synthesis. A single allele of *swi7* was isolated suggesting that this class of MT mutants is far from being saturated. As expected, Swi7 is essential for viability and mapping of the point mutation was not informative about the imprinting mechanism. However, the *swi7* mutant exhibits some growth defects indicating that this mutation affects other general biological functions. Recently, we found that this mutant increases telomere length (B. Xhemalce, unpublished).

Sap1 (for switch-activating protein) was not found among the collection of *swi* gene mutants and was instead isolated by reverse genetics. *sap1* codes for a sequence-specific DNA-binding protein, which interacts with SAS1 (Arcangioli and Klar 1991). Disruption of *sap1* showed that Sap1 played an essential function, independently of MT switching (Arcangioli et al. 1994). Over-expression of the Sap1 C-terminal domain during S phase is associated with chromosomal fragmentation, a cut phenotype and partial loss of sister chromatid cohesion (de Lahondes et al. 2003). Furthermore, these phenotypes are observed only when over-expression occurs in S phase but not in G2 phase. Recently, Sap1 was found to

Table 1. Switching genes in *S. pombe, S. cerevisiae* and *H. sapiens*.

Class	Gene	SSB	Rearrangement	Comment	S cerevisiae	H. sapiens
Ia	*Swi1*	+/-	No	Replication fork protection (FPC)	*TOF1*	*TIM*
	Swi3	+/-	No		*CSM3*	*TIPIN*
	Swi7/PolI	+/-	No	Polymerase α	*POL1*	*POL1*
	Sap1	ND	No	SAS1 DNA binding	Functional homolog?	
Ib	*Swi2*	wt	No	Interact with Swi6 and Rhp51. Rhp51 mediator	*MEI5* related	*AAH20892.1*
	Swi5	wt	No		*SAE3*	*AAH21748.1*
	Swi6	wt	No	heterochromatin	*Not found*	*HP1*
II	*Swi4*	wt	Yes	Mismatch repair (MMR)	*MSH3*	*MSH3*
	Swi8	wt	Yes		*MSH2*	*MSH2*
	Swi9/Rad16	wt	Yes	Nucleotide excision repair (NER)	*RAD1*	*XPF*
	Swi10	wt	Yes		*RAD10*	*ERCC1*
	rad22-67	wt	Yes	HR	*RAD52*	*RAD52*
	Rhp57	wt	Yes	Rhp51 mediator	*RAD57*	*XRCC3*
	Rad50	wt	Yes	MRN/X	*RAD50*	*RAD50*
	Rad32/Mre11	wt	Yes	MRN/X	*MRE11*	*MRE11*
	Pfh1	wt	Yes	DNA helicase	*PIF1/RRM3*	*PIF1*

bind to the strongest polar replication fork barrier (RFB1) in the ribosomal DNA intergenic rDNA spacer regions, which has strong homology with SAS1. Importantly, the Sap1-dependent fork barrier at RFB1 is also under Swi1/3 control (Krings and Bastia 2005; Mejia-Ramirez et al. 2005). This result is consistent with the proposal that Sap1 recruits the Swi1/Swi3 complex to chromatin (E. Noguchi, personal communication).

The double mutants *swi1 swi7* and *swi3 swi7* are dead, independently of MT switching (Schmidt 1987). Biochemical studies have shown that the Swi1/3 complex, Swi7 and Sap1 do not interact with each other, suggesting that they act in different steps of imprinting formation or maintenance (Lee et al. 2004). In addition, an increase in *sap1*$^{+}$ dosage rescues camptothecin sensitivity of a *swi3* mutant (E. Noguchi, personal communication). The discovery that all of the Class Ia gene products are playing direct or indirect roles in the replication process and that the imprint is a site- and strand-specific lesion strongly supports the notion that imprint formation is mechanistically linked to DNA replication.

6.2 Class Ib

The Class Ib mutants are interesting in the sense that they are fully viable, exhibit a wild type level of SSB but have poor MT switching efficiency. This behavior indicates that switching is not the only way to bypass the SSB and that other efficient repair pathway(s) are equally possible. Similar conclusions have been reached with the strains deleted for both donor loci (*mat2*-P and *mat3*-M), which exhibit wild type levels of SSB and are viable (Klar and Miglio 1986). From these data it was proposed that the *swi2*, *swi5* and *swi6* gene products play a role in the accessibility of the silent donors, allowing for efficient MT-switching. Furthermore, the choice of the donor is not random, since switching to the opposite MT occurs 90% of the time. Therefore, these gene products have been proposed to participate in a process referred to as directionality of switching (Thon and Klar 1993).

Swi6 is the ortholog of HP1 (heterochromatin protein 1), an essential component of heterochromatin (Lorentz et al. 1994), and was shown to bind the entire *mat2*-P and *mat3*-M region and participate in the transcriptional silencing and recombination cold spot of this region in mitosis and meiosis (Klar and Bonaduce 1991; Lorentz et al. 1992; Thon and Klar 1992; Nakayama et al. 2000). Swi6 interacts through its chromodomain with histone H3, methylated on lysine 9 by the Clr4 methyltransferase protein (Rea et al. 2000; Nakayama et al. 2001). Sir2 is an NAD^{+}-dependent deacetylase that plays an important role in heterochromatin assembly. It was proposed that deacetylation of H3-K9 is required for Clr4 methylation and subsequent localization of Swi6 (Shankaranarayana et al. 2003). Similar heterochromatic structures are present near centromeres and telomeres, where transcription is also blocked. The establishment of heterochromatin is particularly well described in *S. pombe* and requires small interfering RNA (RNAi) molecules. At the MT silent region, RNAi-dependent heterochromatic nucleation is taking place within *cenH* in the K region (for review see Grewal and Rice 2004). Atf1 and Pcr1, two ATF/CRB family proteins, bind to the RE3 elements and act in a parallel pathways to nucleate constitutive heterochromatin, possibly by recruiting the histone deacetylase Clr6 (Thon et al. 1999; Jia et al. 2004; Hansen et al. 2005). Another set of important functional interactions necessary for heterochromatin formation relies on Clr7 and Clr8, prior the action of Clr4. Importantly, Clr7 was shown to interact with the silent MT region and Clr8, which in turn interacts with the nuclear porin Nup189, suggesting that the heterochromatin is tethered to the nuclear envelope by association with the nuclear-pore complex (Thon et al. 2005). Once assembled, the Swi6-containing heterochromatin promotes a specific chromosomal architecture dedicated to appropriate biological functions. At the mating-type locus, Swi6 serves as a scaffolding protein for different complexes and is also a crucial player for the donor-choice mechanism by promoting an intrachromosomal folding of *mat2* and *mat3* onto *mat1* in a cell type-specific manner (Thon and Klar 1993; Grewal and Klar 1997; Jia et al. 2004). Finally, Rik1 a protein structurally related to the repair protein Ddb1 (Tuzon et al. 2004), required for heterochromatin silencing and chromosomal stability, associates with Clr4 and cullin 4 (Cul4) proteins (Horn et al. 2005; Jia et al. 2005; Petroski and Deshaies 2005).

Swi5 codes for a small protein of 85 aa homologous to Sae3 in *S. cerevisiae* (Akamatsu et al. 2003). It has been implicated in general Rhp51-dependent (ortholog of Rad51) homologous recombination, during mitosis and meiosis (Gutz and Schmidt 1985; Akamatsu et al. 2003; Ellermeier et al. 2004). Recent data from Akamatsu et al. (2003) reveals that Swi5 makes two different complexes, one with Swi2 specialized for MT switching, and one with Sfr1 (homologous to Mei5, in *S. cerevisiae*), required for general recombination. In addition, Swi2 interacts with Rhp51 and Swi6. Sfr1 also interacts with Rhp51 but is lacking the Swi6 interacting domain. Despite the absence of sequence similarity between Sfr1 or Swi5 and Rhp51 paralogs, it was proposed that Swi5/Sfr1 functions as a novel mediator with a biochemical role similar to that of the Rhp55/57 heterodimer, but with some different and important genetic functions. Furthermore, use of an assay for site-specific DSB repair revealed that Rhp51 and Rhp57, but not Swi5/Sfr1, are essential for gene conversion with crossover resolution, during mitotic growth (Y. Akamatsu et al. submitted). It was recently shown that the Swi5/Sfr1 complex stimulates the DNA strand-exchange activity of Rhp51 *in vitro* (H. Iwasaki, personal communication). By analogy with the Swi5/Sfr1 complex it is tempting to propose that the Swi5/Swi2 complex also channels recombination at the MT loci through a "synthesis-dependent strand annealing" (SDSA-like) process, to avoid the formation of crossover products (see below).

6.3 Class II

The common feature of class II mutants is that they produce heterothallic variants at high frequency, indicating that they participate to the MT switching resolution step (Beach and Klar 1984; Egel et al. 1984).

Most of the heterothallic variants arising from class II mutants are of the h^{+N} type, in which the 17kb DNA sequence, including *mat2*-P, the K region and *mat3*-M, are inserted into *mat1* by gene conversion (Fig. 2C). In this configuration, containing four cassettes, the more proximal cassette, *mat1:2*P is expressed but switching deficient, conferring the stable P mating phenotype. The next cassette, *mat3:1P/*M is now localized next to the L region, containing the *mat1 cis*-acting SAS1/2 elements, and thereby exhibits the SSB and is switching proficient, although less efficient than observed in h^{90}, probably due to the lack of the *mat1* proximal DNA sequence (Fig. 2C). From this, it was proposed that during P to M switching, the broken *mat1* DNA invades the H1 sequence of the *mat3*-M cassette and the gene conversion, instead to be resolved in the H2 sequence, continues to the next possible sequence of homology for resolution in the *mat2*-P cassette. During this extensive gene conversion event, the K region, localized between the two silent copies, is also inserted.

Swi4 and Swi8 were identified as homologs of the bacterial mismatch repair proteins MutS or Msh (for MutS homolog) in eukaryotes. Swi4 is related to the Msh3 subfamily (Fleck et al. 1992) and loss of Swi4 does not cause a defect in mismatch repair, but rather causes a reduction of recombinants in intergenic crosses (Tornier et al. 2001). The h^{90} *swi4* mutant generates heterothallic of h^{+N} and homothallic-like variants (Fig. 2C, h^{90*}). The latter variants (called h^{90*} in Fig. 2C) are very unstable and can endure further rounds of duplications, such that strains with up to seven MT cassettes have been found (Fleck et al. 1990). The characterization of these rearrangements relies on the reintroduction of the wild type *swi4* gene to stabilize the rearranged region.

Swi8 is related to Msh2, binds to mismatches and its disruption increases spontaneous mitotic mutation rates and postmeiotic segregation of genetic markers (Fleck et al. 1994; Rudolph et al. 1999). It is known from other eukaryotic systems that Msh2 forms heterodimers with either Msh3 or Msh6, the third protein of the Msh family found in *S. pombe*, while Msh6 is required for mismatch repair and does not seem to be involved in MT switching (Tornier et al. 2001; for review see Marti et al. 2002). h^{90} *swi8* mutants generate iodine-negative variants, the majority being h^{+N}, but contrary to *swi4*, some *swi8* mutants are sterile or h^{-}, together with slow-switching variants. The h^{-} variant contains a fusion between *mat1* and *mat3*-M (*mat1:3*M), removing the intergenic L and K regions on the chromosome (termed h^{-L}). The episomal structure complements the essential function present in the L region, presumably generated by intrachromosomal crossing over between *mat1* and *mat3*-M (Fig. 2C), and the sterile segregants are probably generated by the general mutator effect of *swi8* (Fleck et al. 1994).

Swi9 (also named *rad16*) and *swi10* mutants have pleiotropic effects, such as reduced MT switching and increased UV sensitivities (Egel et al. 1984; Schmidt et al. 1989). Swi9/Swi10 are homologous to the Rad1/Rad10 or ERCC1/XPF structure-specific endonuclease of the nucleotide excision repair pathway (Carr et al. 1994). Rodel et al. (1997) have shown that the human ERCC1 gene complements MT switching and DNA repair of the *swi10* mutants.

By using an intrachromosomal direct repeat system, it was shown that *swi9* and *swi10* mutants exhibit spontaneous hyper-recombinant phenotypes in *S. pombe* (Osman et al. 2000). The initiation of recombination remains uncertain, but it likely reflects repair of spontaneous DNA damage. The nucleotide excision repair (NER), base excision repair (BER) or DNA mismatch repair (MMR) pathways responsible for the repair of endogenous DNA lesions are also recombinogenic (Memisoglu and Samson 2000; Osman et al. 2000; Kunz and Fleck 2001; Marti et al. 2002). In the single-strand annealing (SSA) pathway in *S. cerevisiae*, the structure-specific Rad1/Rad10 endonuclease removes the 3' non-homologous DNA from the 3' invading tail (Fishman-Lobell and Haber 1992; Paques and Haber 1999). Similarly, the ERCC1/XPF complex has been shown to play a major role in recombination-dependent rearrangements in mammalian cells (Sargent et al. 2000). During meiosis in *S. cerevisiae*, the Msh2/Msh3 complex, together with the NER proteins Rad1 and Rad10, are required for the deletion of large loops formed in heteroduplexes containing unpaired DNA (Kearney et al. 2001). Such un-annealed loops or free 3' tails are proposed to be stabilize by the Msh2/Msh3 complex (Saparbaev et al. 1996; Sugawara et al. 1997). The DNA rearrangements in the MT region arising in the h^{90} *swi9* or h^{90} *swi10* mutants are similar to those observed in the previously described *swi4* and *swi8* mutants, and double mutants have similar MT switching defects, indicating that both complexes act together to resolve MT switching recombination intermediates.

Rad22 is the homolog of Rad52 in *S. cerevisiae*, a protein essential for Rhp51-dependent homologous recombination (Ostermann et al. 1993; Osman et al. 2005). The *rad22* mutant was not found in the initial screen of *swi* gene mutants and is not essential for viability in heterothallic strains, but is essential in the homothallic strain. However, a *rad22-67* mutant allele, sensitive to UV and gamma rays, exhibits MT switching defects and produces heterothallic variants (h^{+N} and h^{-L}) (Nasim and Smith 1975; Schmidt et al. 1989). Chromatin immunoprecipitation experiments have shown that Rad22 interacts *in vivo* with *mat1* in h^{90} but not in *Msmt-0* strains, indicating a role for Rad22 in an early step (Y. Yamada and P. Russell, personal communication; Kim et al. 2000). Taken together, these results indicate that the *rad22-67* mutant allele initiates MT switching but fails to correctly resolve the recombination product.

7 The direction of replication model

The specific strand segregation model (Klar 1987) was proposed to explain the switching pattern in fission yeast, where two asymmetric cell divisions are required to produce the observed pattern of switching. Subsequently, it was shown that *swi7* codes for the DNA polymerase α (Singh and Klar 1993). In addition, the SSB is site- and strand-specific and is converted during S phase into a polar one-ended DSB. From these new data, the direction of replication model was proposed and tested (reviewed in Dalgaard and Klar 2001).

By using native two-dimensional gel electrophoresis (2-D gel), the direction of replication at the *mat1* locus was determined (Dalgaard and Klar 1999). The replication termination site 1, *RTS1*, located 700bp to the left side of the H2 box of *mat1*, arrests the replication fork approaching from the centromere-proximal side of *mat1* (Fig. 3a). By doing so, *RTS1* activity constrains the replication of *mat1* from its centromere- distal side. Swi1 and Swi3 are required for *RTS1* termination activity in addition to Rtf1 and Rtf2 (Dalgaard and Klar 2000). A second replication fork control element, the *mat1* pause site, *MPS1*, has been identified on the other side of *mat1* near the H1 box. Swi1 and Swi3 are required for *MPS1* pausing activity, and this site seems to play a major role for imprinting. Deletion of 800bp,

Fig. 3. (overleaf). Recombination-coupled replication model. Starting from a virgin *mat1*-P(u) template; (a) The replication forks arrest at *RTS1* and pause at *MPS1*. (b) Site-specific lagging-strand re-initiation. (c) Upon replication restart *mat1*-P(u) and *mat1*-P(s) are generated. (d left) Replication of *mat1-P(u)*. (d right) During the replication of *mat1*-P(s) template, *RTS1* arrests the fork of replication coming from the left, while the fork coming from the right transforms the SSB into a polar one-ended DSB, recognized by the MRN complex. Mre11 5'-to-3' resection, produces a 5'-phosphate end, and leaves a recombinogenic 3'-end overhang. (e1) In the presence of the donors, the 3' overhang invades the correct donor, by using the Swi5/Swi2 complex, and forms a D-loop, allowing initiation of MT switching. (f) Resolution using Swi9/Swi10 endonuclease (red arrow), the resolution of the old *mat1* strand is not known and two scenarios have been proposed (grey arrow: endonuclease or grey packman: exonuclease). (g) DNA synthesis of the second strand. (h) Site-specific lagging-strand re-initiation. (i) Generation of *mat1*-M(u) and *mat1*-P(s). (e2) Without donors, DNA synthesis and ligation of the sister chromatid is required for invasion by the 3' overhang and D-loop formation. (j) Mus81/Eme1 resolves the D-loop, and resets a replication fork structure without crossover. (l) Site-specific lagging-strand re-initiation. (m) Generation of *mat1*-P(u) and *mat1*-P(s), similar to the products obtained during P(u) replication.

containing three 60bp partially conserved direct repeats, abolishes *RTS1* activity but only slightly reduces the SSB and MT switching efficiency (Dalgaard and Klar 2000; Codlin and Dalgaard 2003). This might be due to the close proximity of an origin of replication localized 2.5kb distal to the *mat1* locus. Altogether, the programmed replication at *mat1*, through the combined replication termination at *RTS1* on one side, and a close origin of replication on the other side, constrains the replication of *mat1* from its distal side to ensure an optimal imprinting efficiency. In turn, this will ensure a particular *mat1* DNA strand to be recurrently replicated by the lagging-strand machinery (Fig. 3a). This is the first-characterized example exploiting the intrinsic asymmetry of the leading- and lagging-strand replication machineries to dictate the formation of two differentiated sister chromosomes (Dalgaard and Klar 2001).

8 Imprinting formation is coupled to DNA replication

The replication program at *mat1* ensures that the fork of replication enters into *mat1* by its centromere-distal side for proper establishment of the imprint. 2D gel experiments revealed that *MPS1* activity depends on Swi1 and Swi3, whereas a *swi7* mutant has no effect on pausing. These results support a model pathway, in which Swi1/Swi3 act upstream of Swi7 in the formation of the SSB (Dalgaard and Klar 2000). Interestingly, in the *swi7* mutant, the imprint appears as multiple faint SSBs, such that this mutation lacks the imprint positioning information or are unstable (Nielsen and Egel 1989). Similarly, the *cis*-acting mutants *Msmt-0* or *PΔ17*, retain full *MPS1* activity but abolish SSB formation. *Msmt-0* deletion removes most of the *cis*-acting elements with the exception of the mut3 element, 20 bp

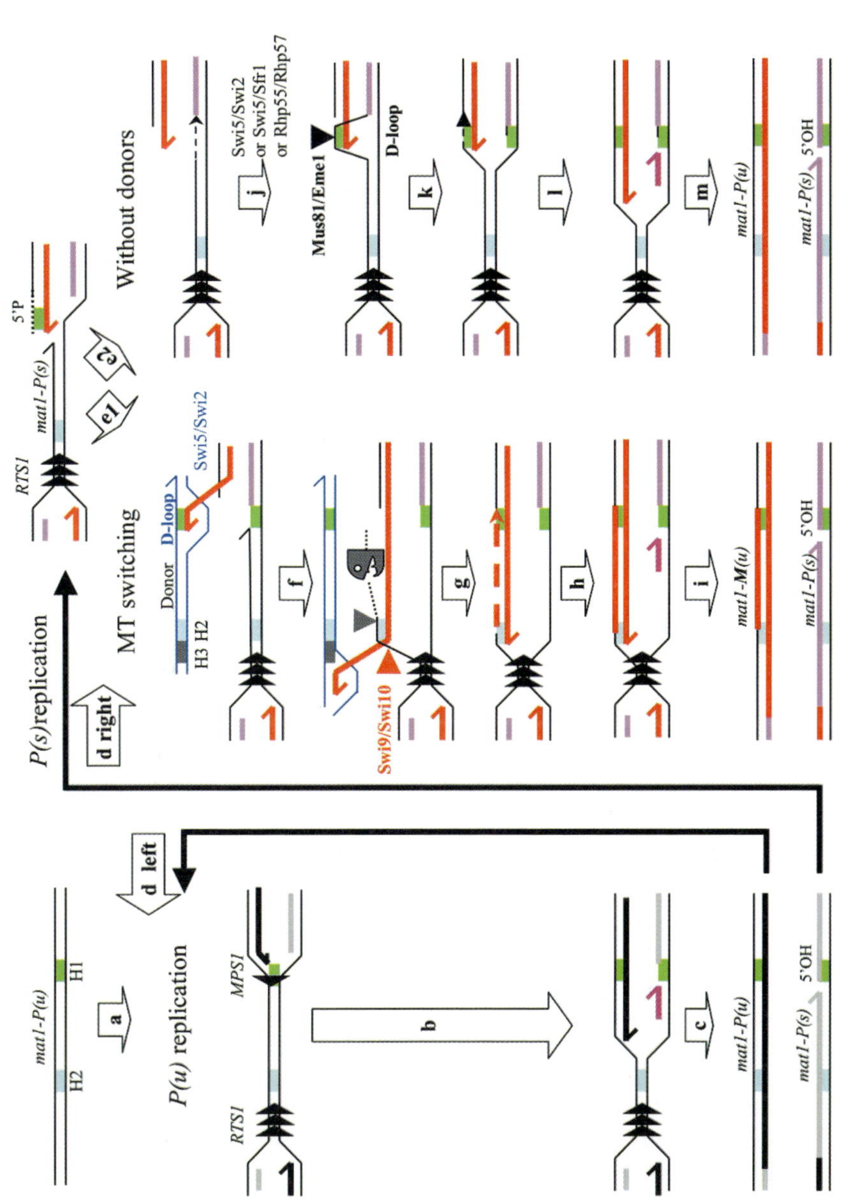

from the position of the nick, suggesting that this site might participate in *swi1*-dependent *MPS1* activity (Fig. 2B). Indeed, mut3 has been shown to be essential for *MPS1* activity and Swi1 *in vivo* interaction (Kaykov et al. 2004). Thus, *MPS1* activity is necessary but not sufficient for formation or maintenance of the SSB, and other functions and interaction with SAS1 and SAS2, are also required.

In 2D gel experiments, the location of *MPS1* can be estimated from the type of restriction DNA fragment analyzed and the position of the accumulating intermediates into the Y-arc. *MPS1* was roughly mapped distal to the *mat1* cassette at the H1 domain (Dalgaard and Klar 1999). Consistent with this, 2D-gel electrophoresis experiments, using a small DNA fragment of 476 bp, overlapping the H1 sequence, show *MPS1* activity. Furthermore, the position of the accumulating intermediates into the Y-arc roughly corresponds to the mut3 position (Holmes et al. 2005). Recently, by using the method for replication initiation point (RIP) mapping, a discrete 5'-end Okazaki lagging-strand fragment was identified 340 bp from the imprint site, within the *mat1*-M locus and was proposed to be part of *MPS1* (Vengrova and Dalgaard 2004). This proposal is not consistent with the position of the pause observed in the short 476 bp DNA fragment, since the 5'end of the Okazaki fragment is localized outside of the restriction DNA fragment. Consequently, if the site-specific Okazaki fragment is not part of the DNA replication pause material at *MPS1*, it is probably synthesized later during the re-initiation of the replication fork and its detection essentially rests on the specificity of the re-initiation site (Fig. 3b, 3h, and 3l). This interpretation is consistent with the observation that the 5'end of the Okazaki fragment was observed in wild type, *swi1*, *swi3* and *swi7* strains in equal amounts, independently of *MPS1* activity (Vengrova and Dalgaard 2004).

The evidence described above supports the notion that the imprint is made during S phase, but was never formally proven. The difficulty originates from the mixture of uncleaved and cleaved molecules in growing cell population. To definitively demonstrate that the SSB is produced during S phase, Holmes et al. (2005) recently developed and validated an inducible system allowing us to follow the kinetics of each molecular event from pausing, imprinting and finally MT switching, in a homogeneous cell populations. By introducing a thiamine repressible promoter, upstream of the *mat1* locus in a neutral position, one can force transcription through the imprinted region, erasing the imprint and inhibiting further MT switching, as long as transcription is maintained. It is not yet clear how the SSB is repaired during the first wave of transcription, but it seems to require a transcription-coupled-repair-like process (A. Holmes and B. Arcangioli, unpublished). This system is fully reversible and when the transcription is switched off, MT switching is restored. Starting from a synchronized and homogeneous Mu-cell population, the SSB is formed when DNA replication intermediates appear at *mat1* during the first S phase. The formation of the SSB is concomitant with the formation of the *MPS1* pause and the beginning of Swi1 accumulation at *mat1* until early G2, and then Swi1 rapidly disappears. The mut3-dependent Swi1 accumulation within the H1 box is consistent with the association of Swi1 to the progressing replication fork accumulating at *MPS1* (Katou et al. 2003), but its persistence until the early G2 phase suggests that it has an additional function.

Two simple models have been proposed to explain the formation of the imprint (for review see Egel 2005), regardless of whether the imprint consists of a 3'OH-5'OH nick or one or two ribonucleotides. In the first mechanism, the modification is introduced in one molecular step, whereas the second mechanism relies on a two-step event. However, at this stage of the research we have still more questions than answers about the important molecular details required for imprint formation.

Since this system is amenable to sophisticated molecular and genetic approaches, answers to the imprinting process should be forthcoming.

9 Imprinting protection

The MT switching pattern infers that the SSB marks one sister chromatid one generation before MT switching. Consequently, following its formation, the SSB has to be actively protected against DNA repair processes, in order to persist for the entire length of the cell cycle. The density and the location of the *cis*-acting elements within the H1 box are suggestive that a protective nucleosomal structure at *mat1* protects the SSB from repair. In this respect, mutation of the element (mut7) located 60 nucleotides from the SSB site exhibits variable levels of SSB during cell cycle progression, which peaks during S phase and then rapidly diminishes, indicating that mut7 participates to its protection/maintenance (Kaykov et al. 2004). Furthermore, the mut3-dependent Swi1 accumulation is observed for almost 60 minutes after the formation of the SSB, which extends beyond the estimated half-life of *MPS1*, suggesting a potential function of Swi1 in the initial protection and/or maintenance of the SSB. Altogether, the mut3, mut5, and mut7 *cis*-acting elements seem to play different but complementary functions in the stability of the SSB (Fig. 2B). Interestingly, the three mutations display temperature sensitive phenotypes, such that at a low temperature (25°C), the level of the SSB and MT switching are nearly wild type, whereas at a higher temperature (33°C or 37°C) the SSB level and MT switching are seriously reduced (A. Kaykov and B. Arcangioli, unpublished). These observations raise interesting questions concerning how single-stranded lesions are maintained and how repair occurs within different chromosomal regions.

10 Mating-type switching

Following the discovery that the imprint is a stable, site- and strand-specific and that MT switching is restricted to only one of the two sister chromatids, it was proposed that MT switching followed a recombination-dependent replication pattern (Arcangioli 1998). The demonstration that MT switching is a two-step developmental process, involving two consecutives rounds of DNA replication, was shown by the inducible system described above. The intimate relationship between replication and recombination was first proposed for the replication of T4 phage (Mosig 1998; Cox et al. 2000). In this section, the MT switching process will be subdivided in three phases: (i) initiation, (ii) choice of the donor, and (iii) resolution.

10.1 Initiation

Genetic and molecular studies have shown that the leading-strand polymerase synthesizes DNA through the H1 box to the last nucleotide of the 5'-end of the SSB, creating a polar blunt-ended DSB (Arcangioli 1998; Kaykov et al. 2004; Vengrova and Dalgaard 2004). Several lines of evidence indicated that MT switching follows a single-end invasion pathway by invading one of the H1 homologous sequences of the silent donor loci. The *Pst*1 substitution mutation overlapping the SSB site does not affect the level and the position of the break but reduces the efficiency of MT switching, without loss of viability. This inhibition is partial and spontaneous revertants appear as half sectors in a colony containing the wild type H1 sequence. This strongly indicates that removal of the heterologous *Pst*1 mutation is an obligatory step for H1-dependent efficient MT switching initiation. The second *Pst*1 mutation, located 10 nucleotides from the SSB also reverts, but less frequently, whereas the *Pst*1 mutations localized at 30 and 50 nucleotides further downstream are stable and MT switching is not affected. This means that only 20 to 30 nucleotides of perfect homology are required for efficient strand invasion. The loss of the *Pst*1 mutation near the SSB site can be due to degradation of the broken end, either before or after DNA invasion of the silent donor loci.

Many proteins are involved in homology search and joint-molecule formation. The *rad22* (*RAD52*^{Sc}) epistasis group includes *Rad50, Rad32*^{ScMre11}*, Nbs1, Rad22, Rhp51, Rhp54, Rhp55, Rhp57, Swi5, Sfr1,* and *Swi2*. To determine the effects on MT switching and viability of these genes, the null alleles were first introduced into the stable non-switching *Msmt-0* strain to be subsequently crossed with an h^{90} strain. None of the mutants is essential for viability in the *Msmt-0* strain, although most of them exhibit variable growth defects, reflecting their role in spontaneous DNA damage repair by homologous recombination (HR). Tetrad dissection of the diploids allowed us to determine the growth and switching phenotypes of the HR mutants in h^{90} switching background. The first observation was that Rad22, Rhp51, and Rhp54 are essential in h^{90}. This result indicates that the one-ended polar DSB requires the HR enzymes for survival and that *MPS1*, still present in *Msmt-0*, is not recombinogenic. Interestingly, the h^{90} *rad50Δ* mutant forms small colonies, with few spores, containing many dead cells, and the h^{90} *exo1Δ* mutant is viable and does not exhibit MT switching defects, whereas the double *rad50Δ exo1Δ* mutant is not viable. These results are consistent with an early recruitment of the Mre11/Rad50/Nbs1 complex at the one-ended DSB. By analogy with MT switching and meiosis in *S. cerevisiae* (Sugawara and Haber 1992; Neale et al. 2005), Rad50 might act through the MRN/X complex, in which the Rad32 nuclease (the ortholog of Mre11) participates in the resection of the 5'OH end-containing strand to generate recombinogenic 3' single-stranded tails (Fig. 3, (d right) *Ps* replication). The synthetic lethality observed in the *rad50Δ exo1Δ* double mutant indicates that the Exo1 nuclease is partially redundant with Mre11 activity (for review see Tran et al. 2004). Interestingly, all checkpoint mutants studied (*csd1Δ, chk1Δ rad3Δ*) do not affect MT switching (L. Roseaulin, unpublished; E. Noguchi et al. 2003; Branzei D and Foiani M, this volume). Since Rad3 (ataxia-telangiectasia Rad3-related (ATR) ortholog in mammals) is thought to recognize single-strand DNA coated with RPA (Zou and Elledge 2003), the resection of the 5'end of the polar DSB is limited (H1=59 bp) during MT switching, which in turn does not trigger the DNA-damage response pathways. The polar one ended DSB appears very transiently in the wild type cells (Kaykov et al. 2004) and accumulates in the *rad50S* mutant strain. Rad50S is a hypomorphic mutation that alters the Rad50 subunit of MRN and blocks 5'-ends resection of DSBs, causing their accumulation (Alani et al. 1990; L. Roseaulin, unpublished).

The absolute requirement for the HR pathway strongly indicates that the replication forks collapse at the SSB, and that the MT switching event not only repairs the polar DSB, but also resets a fork structure appropriate for replication restart. It follows that one-ended DSB repair, as for DSBR pathway, also initiates with the MRN/X complex (Mre11, Rad50, and Nbs1/Xrs2) described as the first DSB end sensor, (de Jager et al. 2002; Petrini and Stracker 2003; Lisby et al. 2004) allowing resection of the 5'end DNA strand (or unwinding) leaving a 3'-end single strand DNA overhang. Concomitantly, the generated ssDNA is bound by replication protein A (RP-A) or directly Rad22 then by Rhp51. Once assembled, the nucleoprotein filament is competent to search, pair, and eventually, together with Swi5/Swi2 and Rhp55/Rhp57 mediators and Rhp54, form a joint molecule also called the D-loop, with one of the intact H1 homologous double stranded DNA elements at the silent loci (Shibata et al. 1979; Sung 1994; Benson et al. 1998).

Interestingly, it was reported that the *swi8Δ* (*msh2*) mutant generates mutations next to the H1 sequence of *mat1*. The sequences of several independent mutants have shown that they contain the same mutation, in which 8 bp of *mat2-P* were transferred to *mat1*, probably by an abnormal gene conversion event initiating outside of H1, and within a 12 bp homologous sequence common between *mat1* and *mat2* (Fleck et al. 1994). This result indicates that Swi8/Swi4 after strand invasion and before DNA repair synthesis is required to stabilize or protect the H1 invading strand against abnormal DNA cleavage by Swi9/Swi10 or other nuclease activities. Such a model was also proposed for several HR events in yeast and mammals (Schiestl and Prakash 1988; Kirkpatrick and Petes 1997; Niedernhofer et al. 2001; Langston and Symington 2005) and indicates that the invading H1 broken end is already under the surveillance of the class II gene products at the initial step of gene conversion, potentially distinguishing homeologous from homologous, or ectopic from sister chromatid, templates for repair (Evans et al. 2000).

10.2 Choice of the donor

We also analyzed the Rad51 paralog, Rhp57 (homologous to Rad57 is *S. cerevisiae*) and the recently described mediator complexes, Swi5/Sfr1 involved in global homologous recombination repair, and Swi5/Swi2 dedicated to mating-type switching. The absence of Rad57 has a mild effect on MT switching, but upon restreaking, the h^{90} *rad57Δ* mutant produces streaky and iodine-negative colonies, indicating that the Rhp57 is also involved for efficient MT switching. The h^{90} *swi5Δ* mutant also produces healthy colonies, but MT switching is drastically reduced. The h^{90} *rad57Δ swi5Δ*double mutant is not viable, indicating that in the absence of Swi5, Rhp57 ensures viability but not MT switching, showing that Rad57 cannot direct the Rhp51 nucleoprotein filament to the correct silent cassette containing the opposite MT allele but instead uses the sister chromatid or the donor containing the same allele.

It was recently shown that the Swi2/Swi5 complex is differentially localized, in a cell type-specific manner, within the silent MT region. In P cells, Swi2 binds to a recombination enhancer element (*SRE*) next to the H1 box of the *mat3* cassette, and in M cells, Swi2 spreads, in a Swi6-dependent manner, across the entire silent region, up to the *mat2* cassette. It is not yet known how Swi2 binds to the *SRE* in P or M cells and spreads along the heterochromatin only in M cells, but its recruitment to the *SRE* is a prerequisite for spreading. It was proposed that Swi2 interacts with AT-rich sequences through two putative AT hook domains (Jia et al. 2004). These results extend the initial observation showing that the location of the silent cassettes in the chromosome, rather than their content, directs the choice of the donor (Thon and Klar 1993). It was proposed that the Swi5/Swi2 complex captures the Rhp51 nucleoprotein filament, to impose the choice of the donor (Jia et al. 2004). Importantly, the interaction between the heterochromatin-bound Swi2 and Rhp51 nucleofilament, in this early step, does not necessarily require homologous DNA sequences for search and recognition. An additional consequence of the Swi2-dependent capture of the Rhp51 nucleofilament is to avoid the usage of the sister chromatid for repair.

10.3 Gene conversion and its resolution

The proposed D-loop structure, joining the H1 sequence of *mat1* with the H1 sequence of the appropriate donor, uses the invading 3'-end as a primer to allow DNA synthesis to proceed through the donor template (Fig. 3e1). Since the donor contains the opposite allele of *mat1*, DNA synthesis has to extend to the other end of the silent cassette and reach the H2 homologous sequence to permit completion of gene conversion. Gene conversion intermediates, longer than 1.4 kb, have been detected, during S phase, using inverted-PCR primers, hybridizing to sequences distal to *mat1* and proximal to *mat2-P* or *mat3-M*. We observed gene conversion intermediates extending DNA synthesis, at least 50 nucleotides beyond the H3 homology box (Arcangioli and de Lahondes 2000). It is not known if the gene conversion intermediates are single- or double-strand or where and how the gene converting DNA polymerase terminates DNA synthesis. The SDSA model predicts that the newly synthesized DNA is unwound from its template as a Rhp51 nucleoprotein filament. The presence of H2 sequences within DNA intermediates will allow recognition and capture of the H2 sequence of the original *mat1* locus, which provides the second end (Fig. 3f). This capture might be facilitated by the degradation of the DNA strand containing the SSB. The MRN complex may play an important role, by maintaining the *mat1* DNA strand in close proximity to the gene conversion machinery via the tethering function of Rad50 and the degradation of the old *mat1* strand via the 3'-to-5' exonuclease activities of Mre11. The proposed function for the MRN complex is also supported by the formation of h^{+N}

arising in the *rad50*Δ strain (L. Roseaulin, unpublished) and may act as a barrier for break-induced replication (BIR) as previously suggested (Aguilera 2001). It was also reported, that a single-stranded H2/H3 can form stem-loop structures, which might help to place the Swi4/Swi8 complex at the resolution site (for review see Rudolph et al. 1999; Egel 2005). We also propose that gene conversion termination is coupled to maturation of the intermediate (Arcangioli and de Lahondes 2000). As suggested above, Swi4 and Swi8 are also involved in the initiation step and may follow the gene conversion machinery in order to be rapidly delivered for resolution at the H2 homology box. The annealing of the two H2 sequences forms a structure with two non-homologous 3' tails (Fig. 3f), the newly synthesized strand can be recognized/stabilized by Swi4/Swi8 and clipped off by Swi9/Swi10, and the old *mat1* strand can be also cleaved by Swi9/Swi10 or degraded by the MRN complex. Maturation of the 3'-end from the old *mat1* DNA strand primes DNA synthesis on the newly synthesized gene conversion DNA intermediate, until it reaches and fuses with the initial 5'-end within H1, completing MT switching (Fig. 3g). The 3'-end in H2, might prime DNA synthesis toward *RTS1* accompanied by the lagging-strand machinery, which synthesizes DNA across *mat1* and forms a new imprint (Fig. 3h). Another possibility will be that this 3'-end continues DNA synthesis by itself until reaching *RTS1*, liberating the arrested leading-strand machinery from the opposite replication fork, which can restart its DNA synthesis. This opens the possibility that the imprinting process could be associated with the leading-strand machinery coming from *MPS1*. These potential scenarios stress another important role of *RTS1*, in blocking the fork of replication coming from the centromere-proximal side of *mat1*, during gene conversion. The overall process generates a switched *mat1* allele without the imprint and an unswitched allele with a new imprint (Fig. 3i).

At least three non-conventional types of DNA synthesis are required to complete MT switching. The first DNA synthesis intermediate has to traverse more than 1.4 kb of heterochromatin, while the second intermediate, using the former one as a template, has to remove the Rhp51 nucleofilament. The last synthesis has to complete the replication of the unswitched template and form a new imprint. It is anticipated that these three events require DNA helicases to unwind the duplex strands, to remove nucleosomes or single-stranded binding proteins (Klein H, this volume). We analyzed the roles of several DNA helicases, including Rqh1, Srs2, Fbh1, Pfh1, and Rhp54. Only Rhp54 is essential for viability and probably for MT switching (L. Roseaulin, unpublished).

Finally, by using isotope density transfer experiments, we have shown that the switched allele contains two newly synthesized DNA strands, whereas the donor is replicated following the conventional semi-conservative mode, as well as the unswitched allele. This result strongly supports the SDSA mechanism for MT switching, since this event is not associated with crossing over, the donor locus is unchanged and the recipient receives two newly synthesized DNA strands, demonstrating that the old DNA strand containing the SSB has been removed. Furthermore, the newly switched *mat1* allele is intact and the unswitched *mat1* allele contains a new SSB on the neo-synthesized strand (Arcangioli 2000). More recently, using the inducible system, we have recreated the MT switching pedigree at the molecular and cellular level (Holmes et al. 2005).

11 Mus81 is the essential nuclease resolving sister chromatid recombination

In the absence of the donor loci, the steady state SSB level is similar to the level observed in wild type cells and a donorless strain is perfectly viable (Klar and Miglio 1986). This result indicates that another process repairs the SSB and allows cell survival without MT switching. All of the single and double mutants described above have been assayed for viability in a donorless strain and exhibit the same viability phenotypes as observed in h^{90}, except for *swi4*Δ, *swi8*Δ, *swi9*Δ, and *swi10*Δ mutant strains, which are not required (L. Roseaulin, unpublished). Since the only homologous sequence available for repair in the donorless strain is the sister chromatid, then it must be used to restart DNA replication (Cortés-Ledesma et al., this volume). In the absence of donors, DNA synthesis primed at the 3'-OH of the SSB, fills the gap remaining on the sister chromatid as a prerequisite for the other free 3' DNA-end to invade its sister (Fig. 3e2), note that the MRN complex might provide a 5'-Phosphate, 5'P). Subsequently, Rhp51-catalysed strand exchange within the H1 sequence of the sister chromatid can proceed, by forming a D-loop intermediate. Formally, invasion of the sister chromatid does not require DNA synthesis from the 3' invading end and the simplest model proposes a resolution by cleavage of the D-loop, restoring the replication fork (Fig. 3j).

Mus81 is a well-conserved DNA structure-specific nuclease, and a member of the XPF family, which forms a heterodimer with $\text{Eme1}^{\text{S.c.Mms4}}$ (Whitby M, this volume). Tetrad analyses from crosses between M*smt*-0 *mus81*Δ and h^{90} or donorless strains, have shown that Mus81 is dispensable for MT switching, but is essential when the sister chromatid is used for repair; conversely, Swi10 is essential for MT-switching but is dispensable in the absence of donors. Genetic and molecular analyses strongly indicate that in the presence or absence of donors, the initial events, forming the polar one-ended DSB, are identical. Therefore, Mus81 is not required in an early step, such as cleavage of the replication fork at *MPS1*, but acts late as the major endonuclease to resolve joint molecules when the sister is used for repair. The *in vivo* requirement of Mus81 at *mat1* in the donorless strain was recently confirmed by ChIP experiments (Y. Yamada and P. Russell, personal communication.). These results are consistent with *in vitro* experiments showing that the Mus81/Eme1 complex exhibits a similar DNA structure specificity among different organisms, and that its favorite substrate is a nicked Holliday Junction (nHJ) or D-loop (Boddy et al. 2001; Chen et al. 2001; Gaillard et al. 2003; Osman et al. 2003). Importantly, this result demonstrates that Mus81/Eme1 is the nuclease dedicated to resolve sister chromatid recombination in *S. pombe*, most likely by cleaving the D-loop to reform a replication fork structure (Fig. 3k).

It was proposed that Mus81/Eme1 and $\text{Rqh1}^{\text{S.c.Sgs1}}$ (the Sgs1 DNA helicase in *S. cerevisiae*) provide alternative activities to process stalled replication forks or recombination intermediates. When a long restriction DNA fragment (around 3 kb) was analyzed by 2D gels a cone-shaped signal was detected at the apex of *MPS1*, only in strains containing the SSB, with or without donors (Kaykov et al. 2004; Vengrova and Dalgaard 2004). The cone-shaped signal, was interpreted as a fork regression, forming a transient "chicken foot or X-shaped" structure (Vengrova and Dalgaard 2004). The signal seems to accumulate in an *rqh1*Δ mutant, consistent with the proposed function of Rqh1 to process X-shaped intermediates. However, the lack of phenotypes in an *rqh1*Δ mutant for SSB formation, MT switching or for viability in the absence of donors, indicates that Rqh1 has only a modest role, if any, at *mat1*. Notably, the Slx1/Slx4 structure-specific endonuclease is required to maintain ribosomal DNA (Mullen et al. 2001; Coulon et al. 2004), but is not required for viability and MT switching (L. Roseaulin and B. Arcangioli, unpublished).

Interestingly, it was previously shown in *S. pombe* and *S. cerevisiae* that *rad52*Δ, *rad51*Δ, *rad54*Δ, *rad50*Δ, and *mus81*Δ mutants are hyper-sensitive (<1 μM) to camptothecin (CPT) drug treatment, an inhibitor of the topoisomerase I, known to induce SSB, and that *rad1*Δ, *sgs1*Δ, and *srs2*Δ (and also *fbh1*Δ in *S. pombe*) mutants are only modestly sensitive to CPT (>1 μM), consistent with the model that topoisomerase I-associated DNA single-strand breaks trapped by CPT are toxic during DNA replication and that HR is the major repair pathway (L. Roseaulin, unpublished; Doe et al. 2002; Pommier et al. 2003). The fact that a *mus81* mutant is only mildly sensitive to CPT in higher eukaryotes has suggested that alternative mechanisms can ensure replication restart (Dendouga et al. 2005). Contrary to *S. pombe*, mammalian cells are diploid and contain less than 50% of unique sequences. Thus, it is tempting to propose that the XPF/ERCC1 complex participates in camptothecin resistance in mammalian *mus81* cells following the SDSA repair pathway, in addition to the MUS81/EME1 sister chromatid HR repair pathway.

12 Outlook and future directions

Increasing evidence from bacteria to mammalian cells indicate that the DNA replication period is the most active phase of the cell cycle for HR repair in nonpathological conditions (Caldecott 2001; Lisby et al. 2001; Vilenchik and Knudson 2003). The MT switching system described above exemplifies how a developmental program directs the replication and recombination processes to generate cellular diversity. In this review, we have postulated that the molecular nature of the imprint is a SSB, although it formally still could be one or two ribonucleotides. The model for MT switching and sister chromatid repair described here, relies on the formation of a polar double strand end appearing during S phase. This implies either that the SSB is converted into a DSB or the DNA is cleaved at the ribonucleotide(s), when the leading strand encounters the lesion. Despite our extensive and recent progress toward the understanding of the molecular events required for MT switching, besides the molecular nature of the lesion, the following questions still need to be answered: What proteins are involved in the establishment and maintenance of the imprint? How much has this process been conserved or adapted through evolution? What determines the cell-type localization of Swi2? How does the heterochromatin control the HR machinery? How is the choice of resolution made with and without donors?

The inducible MT switching system provides a great opportunity to search for genes implicated in imprint formation and maintenance. By transcribing across the imprint region we interfere with SSB formation and preserve the viability of a *rhp51Δ* mutant strain. When transcription is turned off, the SSB is formed and in the absence of Rhp51 the cells are incapable of repairing the polar DSB, and they die. This conditional death assay allows us to directly isolate spontaneous survivors, not able to form (or maintain) the SSB, which are now viable in the absence of a functional HR pathway. The identification of new *swi1* or *swi3* mutant alleles among the spontaneous survivors confirms the tight selection of this approach, and predicts the discovery of new candidates genes required for imprinting formation and maintenance (X. Sun, A. Holmes, and B. Arcangioli, unpublished).

Another interesting question raised above is the accessibility of HR enzymes to DNA sequences embedded in heterochromatin. Indeed, most of the repetitive DNAs, in *S. pombe* and other eukaryotes, are packed in heterochromatic regions. The major functions of heterochromatin are gene silencing, genome defense, and chromosomal structures, which all together play an essential role for chromosome stability. The high density of cohesin complexes (Bernard et al. 2001; Nonaka et al. 2002) and the strong phenotype of the *rad21* mutant (a cohesin subunit) in DSB repair (Birkenbihl and Subramani 1992) strongly indicates that heterochromatin favors the usage of the sister chromatid for gene conversion without crossover resolution, to avoid dramatic chromosomal rearrangements and sister chromatid exchanges. For instance, the *mat2* and *mat3* region lacks meiotic crossovers and has been designated as a cold spot for mitotic and meiotic recombination, is yet accessible for MT switching (Klar and Bonaduce 1991). *mat1, mat2*, and *mat3* loci are replicated during early S phase (Kim et al. 2003) suggesting that heterochromatin may not yet be fully organized allowing strand invasion and subsequent MT switching steps. Several studies have indicated that sumoylation participates in the regulation of recombination and heterochromatin stability in fission yeast (Xhemalce et al. 2004; Andrews et al. 2005; Shin et al. 2005). In this regard, some of the first evidence implicating SUMO in recombination and/or centromere function was obtained by yeast two-hybrid screens with human Rad51 or Rad52 as baits and from screens for high-copy suppressors of temperature-sensitive alleles of the budding yeast MIF2 or of its human homolog, the centromeric CENP-C protein (Meluh and Koshland 1995; Shen et al. 1996).

The MT switching system in *S. pombe* allowed us to distinguish between two different resolution pathways for recombination-dependent replication restart, depending on the DNA sequences used as the template. The mechanism of choice/exclusion of the endonucleases is unknown. Several hypotheses can be considered. The simplest model is that the two joint molecules make different structures. So at the time of resolution the prior history of both D-loops and associated proteins are different and might engage different nucleases for resolution. It is tempting to propose that the decision is made early, during Swi2/Swi5-dependent formation of the joint molecule. Such early decisions to channel the type of resolution have been suggested for meiosis (for review see Bishop and Zickler 2004). In the h^{90} background, the absence of Swi4/Swi8 or Swi9/Swi10 activities triggers the formation of duplications and mini-circles, indicating non-crossover and crossover types of resolution, respectively, suggestive of inaccurate resolution by Mus81/Eme1, thus generating aberrant intermediates. The h^{90} *swi10Δ mus81Δ* double mutant showed a dramatic SSB-dependent growth defect, supporting the idea that Mus81/Eme1 might resolve the recombination intermediates in the absence of Swi9/Swi10, but at the expense of producing aberrant resolutions. It will be interesting in the future to analyze the real-time kinetics of appearance of the aberrant molecular intermediates in the class II mutants, and not only their final products, for which we have little information of their respective stabilities, other than h^{+N} being more stable than mini-circles. Again the inducible MT switching system is a tool of choice to reveal and study the initial structure of the aberrant recombinant molecules.

Acknowledgements

This work was funded by the Association pour la Recherche sur le Cancer (ARC), L. Roseaulin is supported by the Region Martinique and URA2171 du CNRS. Special thanks to R. Egel for helpful comments during preparation of the manuscript.

References

Aguilera A (2001) Double-strand break repair: Are Rad51/RecA--DNA joints barriers to DNA replication? Trends Genet 17:318-321

Akamatsu Y, Dziadkowiec D, Ikeguchi M, Shinagawa H, Iwasaki H (2003) Two different Swi5-containing protein complexes are involved in mating-type switching and recombination repair in fission yeast. Proc Natl Acad Sci USA 100:15770-15775

Andrews EA, Palecek J, Sergeant J, Taylor E, Lehmann AR, Watts FZ (2005) Nse2, a component of the Smc5-6 complex, is a SUMO ligase required for the response to DNA damage. Mol Cell Biol 25:185-196

Arcangioli B (1998) A site- and strand-specific DNA break confers asymmetric switching potential in fission yeast. Embo J 17:4503-4510

Arcangioli B (2000) Fate of mat1 DNA strands during mating-type switching in fission yeast. EMBO Rep 1:145-150

Arcangioli B, Copeland TD, Klar AJ (1994) Sap1, a protein that binds to sequences required for mating-type switching, is essential for viability in *Schizosaccharomyces pombe*. Mol Cell Biol 14:2058-2065

Arcangioli B, de Lahondes R (2000) Fission yeast switches mating type by a replication-recombination coupled process. Embo J 19:1389-1396

Arcangioli B, Klar AJ (1991) A novel switch-activating site (SAS1) and its cognate binding factor (SAP1) required for efficient *mat1* switching in *Schizosaccharomyces pombe*. Embo J 10:3025-3032

Ayoub N, Goldshmidt I, Lyakhovetsky R, Cohen A (2000) A fission yeast repression element cooperates with centromere-like sequences and defines a mat silent domain boundary. Genetics 156:983-994

Beach DH (1983) Cell type switching by DNA transposition in fission yeast. Nature (*London*) 305:682-688

Beach DH, Klar AJ (1984) Rearrangements of the transposable mating-type cassettes of fission yeast. Embo J 3:603-610

Beach DH, Nurse P, Egel R (1982) Molecular rearrangement of mating-type genes in fission yeast. Nature (*London*) 296:682-683

Benson FP, Baumann P, West SC (1998) Synergistic actions of Rad51 and Rad52 in recombination and DNA repair. Nature 391

Bernard P, Maure JF, Partridge JF, Genier S, Javerzat JP, Allshire RC (2001) Requirement of heterochromatin for cohesion at centromeres. Science 294:2539-2542

Birkenbihl RP, Subramani S (1992) Cloning and characterization of *rad21* an essential gene of *Schizosaccharomyces pombe* involved in DNA double-strand-break repair. Nucleic Acids Res 20:6605-6611

Bishop DK, Zickler D (2004) Early decision; meiotic crossover interference prior to stable strand exchange and synapsis. Cell 117:9-15

Boddy MN, Gaillard PH, McDonald WH, Shanahan P, Yates JR 3rd, Russell P (2001) Mus81-Eme1 are essential components of a Holliday junction resolvase. Cell 107:537-548

Caldecott KW (2001) Mammalian DNA single-strand break repair: an X-ra(y)ted affair. Bioessays 23:447-455

Carr AM, Schmidt H, Kirchhoff S, Muriel WJ, Sheldrick KS, Griffiths DJ, Basmacioglu CN, Subramani S, Clegg M, Nasim A, et al. (1994) The *rad16* gene of *Schizosaccharomyces pombe*: a homolog of the *RAD1* gene of *Saccharomyces cerevisiae*. Mol Cell Biol 14:2029-2040

Chan RC, Chan A, Jeon M, Wu TF, Pasqualone D, Rougvie AE, Meyer BJ (2003) Chromosome cohesion is regulated by a clock gene paralogue TIM-1. Nature 423:1002-1009

Chen XB, Melchionna R, Denis CM, Gaillard PH, Blasina A, Van de Weyer I, Boddy MN, Russell P, Vialard J, McGowan CH (2001) Human Mus81-associated endonuclease cleaves Holliday junctions *in vitro*. Mol Cell 8:1117-1127

Codlin S, Dalgaard JZ (2003) Complex mechanism of site-specific DNA replication termination in fission yeast. Embo J 22:3431-3440

Coulon S, Gaillard PH, Chahwan C, McDonald WH, Yates JR 3rd, Russell P (2004) Slx1-Slx4 are subunits of a structure-specific endonuclease that maintains ribosomal DNA in fission yeast. Mol Biol Cell 15:71-80

Cox MM, Goodman MF, Kreuzer KN, Sherratt DJ, Sandler SJ, Marians KJ (2000) The importance of repairing stalled replication forks. Nature 404:37-41

Dalgaard JZ, Klar AJ (1999) Orientation of DNA replication establishes mating-type switching pattern in *S. pombe*. Nature 400:181-184

Dalgaard JZ, Klar AJ (2000) Swi1 and Swi3 perform imprinting, pausing, and termination of DNA replication in *S. pombe*. Cell 102:745-751

Dalgaard JZ, Klar AJ (2001) Does *S. pombe* exploit the intrinsic asymmetry of DNA synthesis to imprint daughter cells for mating-type switching? Trends Genet 17:153-157

de Jager M, Wyman C, van Gent DC, Kanaar R (2002) DNA end-binding specificity of human Rad50/Mre11 is influenced by ATP. Nucleic Acids Res 30:4425-4431

de Lahondes R, Ribes V, Arcangioli B (2003) Fission yeast Sap1 protein is essential for chromosome stability. Eukaryot Cell 2:910-921

Dendouga N, Gao H, Moechars D, Janicot M, Vialard J, McGowan CH (2005) Disruption of murine *Mus81* increases genomic instability and DNA damage sensitivity but does not promote tumorigenesis. Mol Cell Biol 25:7569-7579

Doe CL, Ahn JS, Dixon J, Whitby MC (2002) Mus81-Eme1 and Rqh1 involvement in processing stalled and collapsed replication forks. J Biol Chem 277:32753-32759

Egel R (1977) Frequency of mating-type switching in homothallic fission yeast. Nature 266:172-174

Egel R (2005) Fission yeast mating-type switching: programmed damage and repair. DNA Repair (Amst) 4:525-536

Egel R, Beach DH, Klar AJ (1984) Genes required for initiation and resolution steps of mating-type switching in fission yeast. Proc Natl Acad Sci USA 81:3481-3485

Egel R, Eie B (1987) Cell lineage asymmetry in *Schizosaccharomyces pombe*: Unilateral transmission of a high-frequency state for mating-type switching in diploid pedigrees. Curr Genet 12:429-433

Egel R, Gutz H (1981) Gene activation by copy transposition in mating-type switching of homothallic fission yeast. Curr Genet 3:5-12

Ellermeier C, Schmidt H, Smith GR (2004) Swi5 acts in meiotic DNA joint molecule formation in *Schizosaccharomyces pombe*. Genetics 168:1891-1898

Engelke U, Koster L, Gutz H, Heim L, Schmidt H (1987) Molecular characterization of *h*- mutants of *Schizosaccharomyces pombe*. Curr Genet 12:535-542

Evans E, Sugarawa N, Haber JE, Alani E (2000) The *Saccharomyces cerevisiae* Msh2 mismatch repair protein localizes to recombination intermediates *in vivo*. Mol Cell 5:789-799

Fishman-Lobell J, Haber JE (1992) Removal of nonhomologous DNA ends in double-strand break recombination: the role of the yeast ultraviolet repair gene *RAD1*. Science 258:480-484

Fleck O, Heim L, Gutz H (1990) A mutated *swi4* gene causes duplications in the mating-type region of *Schizosaccharomyces pombe*. Curr Genet 18:501-509

Fleck O, Michael H, Heim L (1992) The *swi4+* gene of *Schizosaccharomyces pombe* encodes a homologue of mismatch repair enzymes. Nucleic Acids Res 20:2271-2278

Fleck O, Rudolph C, Albrecht A, Lorentz A, Schar P, Schmidt H (1994) The mutator gene *swi8* effects specific mutations in the mating-type region of *Schizosaccharomyces pombe*. Genetics 138:621-632

Gaillard PH, Noguchi E, Shanahan P, Russell P (2003) The endogenous Mus81-Eme1 complex resolves Holliday junctions by a nick and counternick mechanism. Mol Cell 12:747-759

Gotter AL (2003) Tipin, a novel timeless-interacting protein, is developmentally co-expressed with timeless and disrupts its self-association. J Mol Biol 331:167-176

Grewal SI, Klar AJ (1997) A recombinationally repressed region between *mat2* and *mat3* loci shares homology to centromeric repeats and regulates directionality of mating-type switching in fission yeast. Genetics 146:1221-1238

Grewal SI, Rice JC (2004) Regulation of heterochromatin by histone methylation and small RNAs. Curr Opin Cell Biol 16:230-238

Gutz H, Schmidt H (1985) Switching genes in *Schizosaccharomyces pombe*. Curr Genet 9:325-331

Hansen KR, Burns G, Mata J, Volpe TA, Martienssen RA, Bahler J, Thon G (2005) Global effects on gene expression in fission yeast by silencing and RNA interference machineries. Mol Cell Biol 25:590-601

Holmes AM, Kaykov A, Arcangioli B (2005) Molecular and cellular dissection of mating-type switching steps in *Schizosaccharomyces pombe*. Mol Cell Biol 25:303-311

Horn PJ, Bastie JN, Peterson CL (2005) A Rik1-associated, cullin-dependent E3 ubiquitin ligase is essential for heterochromatin formation. Genes Dev 19:1705-1714

Jia S, Kobayashi R, Grewal SI (2005) Ubiquitin ligase component Cul4 associates with Clr4 histone methyltransferase to assemble heterochromatin. Nat Cell Biol 7:1007-1013

Jia S, Yamada T, Grewal SI (2004) Heterochromatin regulates cell type-specific long-range chromatin interactions essential for directed recombination. Cell 119:469-480

Katou Y, Kanoh Y, Bando M, Noguchi H, Tanaka H, Ashikari T, Sugimoto K, Shirahige K (2003) S-phase checkpoint proteins Tof1 and Mrc1 form a stable replication-pausing complex. Nature 424:1078-1083

Kaykov A, Arcangioli B (2004) A programmed strand-specific and modified nick in *S. pombe* constitutes a novel type of chromosomal imprint. Curr Biol 14:1924-1928

Kaykov A, Holmes AM, Arcangioli B (2004) Formation, maintenance and consequences of the imprint at the mating-type locus in fission yeast. Embo J 23:930-938

Kearney HM, Kirkpatrick DT, Gerton JL, Petes TD (2001) Meiotic recombination involving heterozygous large insertions in *Saccharomyces cerevisiae*: formation and repair of large, unpaired DNA loops. Genetics 158:1457-1476

Kelly M, Burke J, Smith M, Klar A, Beach D (1988) Four mating-type genes control sexual differentiation in the fission yeast. Embo J 7:1537-1547

Kim SM, Dubey DD, Huberman JA (2003) Early-replicating heterochromatin. Genes Dev 17:330-335

Kim WJ, Lee S, Park MS, Jang YK, Kim JB, Park SD (2000) Rad22 protein, a Rad52 homologue in *Schizosaccharomyces pombe*, binds to DNA double-strand breaks. J Biol Chem 275:35607-35611

Kirkpatrick DT, Petes TD (1997) Repair of DNA loops involves DNA-mismatch and nucleotide-excision repair proteins. Nature 387:929-931

Klar AJ (1987) The mother-daughter mating type switching asymmetry of budding yeast is not conferred by the segregation of parental HO gene DNA strands. Genes Dev 1:1059-1064

Klar AJ (1990) The developmental fate of fission yeast cells is determined by the pattern of inheritance of parental and grandparental DNA strands. Embo J 9:1407-1415

Klar AJ, Bonaduce MJ (1991) *swi6*, a gene required for mating-type switching, prohibits meiotic recombination in the *mat2-mat3* "cold spot" of fission yeast. Genetics 129:1033-1042

Klar AJ, Bonaduce MJ, Cafferkey R (1991) The mechanism of fission yeast mating type interconversion: seal/replicate/cleave model of replication across the double-stranded break site at *mat1*. Genetics 127:489-496

Klar AJ, Miglio LM (1986) Initiation of meiotic recombination by double-strand DNA breaks in *S. pombe*. Cell 46:725-731

Krings G, Bastia D (2004) swi1- and swi3-dependent and independent replication fork arrest at the ribosomal DNA of *Schizosaccharomyces pombe*. Proc Natl Acad Sci USA 101:14085-14090

Krings G, Bastia D (2005) Sap1p binds to Ter1 at the ribosomal DNA of *Schizosaccharomyces pombe* and causes polar replication fork arrest. J Biol Chem 280:39135-39142

Kunz C, Fleck O (2001) Role of the DNA repair nucleases Rad13, Rad2 and Uve1 of *Schizosaccharomyces pombe* in mismatch correction. J Mol Biol 313:241-253

Langston LD, Symington LS (2005) Opposing roles for DNA structure-specific proteins *Rad1*, *Msh2*, *Msh3*, and *Sgs1* in yeast gene targeting. Embo J 24:2214-2223

Lee BS, Grewal SI, Klar AJ (2004) Biochemical interactions between proteins and *mat1* cis-acting sequences required for imprinting in fission yeast. Mol Cell Biol 24:9813-9822

Leupold U (1950) Die Vererbung von Homothallie und Heterothallie bie *Schizosaccharomyces pombe*. CR Trav Lab Carlsberg Ser Physiol 24:381-480

Leupold U (1955) Methodisches zur Genetik von *Schizosaccharomyces pombe*. Schweiz Z Pathol Bakteriol 18:1141-1146

Lisby M, Barlow J, Burgess R, Pothstein R (2004) Choreography of the DNA damage response: spatiotemporal relationships among checkpoint and repair proteins. Cell 118:699-713

Lisby M, Rothstein R, Mortensen UH (2001) Rad52 forms DNA repair and recombination centers during S phase. Proc Natl Acad Sci USA 98:8276-8282

Lorentz A, Heim L, Schmidt H (1992) The switching gene *swi6* affects recombination and gene expression in the mating-type region of *Schizosaccharomyces pombe*. Mol Gen Genet 233:436-442

Lorentz A, Ostermann K, Fleck O, Schmidt H (1994) Switching gene *swi6*, involved in repression of silent mating-type loci in fission yeast, encodes a homologue of chromatin-associated proteins from *Drosophila* and mammals. Gene 143:139-143

Marti TM, Kunz C, Fleck O (2002) DNA mismatch repair and mutation avoidance pathways. J Cell Physiol 191:28-41

Matsumoto S, Ogino K, Noguchi E, Russell P, Masai H (2005) Hsk1-Dfp1/Him1, the Cdc7-Dbf4 kinase in *Schizosaccharomyces pombe*, associates with Swi1, a component of the replication fork protection complex. J Biol Chem 280:42536-42542

Mayer ML, Pot I, Chang M, Xu H, Aneliunas V, Kwok T, Newitt R, Aebersold R, Boone C, Brown GW, Hieter P (2004) Identification of protein complexes required for efficient sister chromatid cohesion. Mol Biol Cell 15:1736-1745

Mejia-Ramirez E, Sanchez-Gorostiaga A, Krimer DB, Schvartzman JB, Hernandez P (2005) The mating type switch-activating protein Sap1 Is required for replication fork arrest at the rRNA genes of fission yeast. Mol Cell Biol 25:8755-8761

Meluh PB, Koshland D (1995) Evidence that the *MIF2* gene of *Saccharomyces cerevisiae* encodes a centromere protein with homology to the mammalian centromere protein CENP-C. Mol Biol Cell 6:793-807

Memisoglu A, Samson L (2000) Contribution of base excision repair, nucleotide excision repair, and DNA recombination to alkylation resistance of the fission yeast *Schizosaccharomyces pombe*. J Bacteriol 182:2104-2112

Miyata H, Miyata M (1981) Mode of conjugaison in homothallic cells of *Schizosaccharomyces pombe*. J Gen Appl Microbiol 27:365-371

Mosig G (1998) Recombination and recombination-dependent DNA replication in bacteriophage T4. Annu Rev Genet 32:379-413

Mullen JR, Kaliraman V, Ibrahim SS, Brill SJ (2001) Requirement for three novel protein complexes in the absence of Sgs1 DNA helicase in *Saccharomyces cerevisiae*. Genetics 157:103-118

Nakayama J, Klar AJ, Grewal SI (2000) A chromodomain protein, Swi6, performs imprinting functions in fission yeast during mitosis and meiosis. Cell 101:307-317

Nakayama J, Rice JC, Strahl BD, Allis CD, Grewal SI (2001) Role of histone H3 lysine 9 methylation in epigenetic control of heterochromatin assembly. Science 292:110-113

Nasim A, Smith BP (1975) Genetic control of radiation sensitivity in *Schizosaccharomyces pombe*. Genetics 79:573-582

Neale MJ, Pan J, Keeney S (2005) Endonucleolytic processing of covalent protein-linked DNA double-strand breaks. Nature 436:1053-1057

Niedernhofer LJ, Essers J, Weeda G, Beverloo B, de Wit J, Muijtjens M, Odijk H, Hoeijmakers JH, Kanaar R (2001) The structure-specific endonuclease Ercc1-Xpf is required for targeted gene replacement in embryonic stem cells. Embo J 20:6540-6549

Nielsen O (2004) Mating-type control and differentiation. In: Egel R (ed) The Molecular Biology of *Schizosaccharomyces pombe*. Springer-Verlag Berlin, Heidelberg, pp 281-296

Nielsen O, Egel R (1989) Mapping the double-strand breaks at the mating-type locus in fission yeast by genomic sequencing. Embo J 8:269-276

Noguchi E, Noguchi C, Du LL, Russell P (2003) Swi1 prevents replication fork collapse and controls checkpoint kinase Cds1. Mol Cell Biol 23:7861-7874

Noguchi E, Noguchi C, McDonald WH, Yates JR 3rd, Russell P (2004) Swi1 and Swi3 are components of a replication fork protection complex in fission yeast. Mol Cell Biol 24:8342-8355

Noma K, Allis CD, Grewal SI (2001) Transitions in distinct histone H3 methylation patterns at the heterochromatin domain boundaries. Science 293:1150-1155

Nonaka N, Kitajima T, Yokobayashi S, Xiao G, Yamamoto M, Grewal SI, Watanabe Y (2002) Recruitment of cohesin to heterochromatic regions by Swi6/HP1 in fission yeast. Nat Cell Biol 4:89-93

Osman F, Adriance M, McCready S (2000) The genetic control of spontaneous and UV-induced mitotic intrachromosomal recombination in the fission yeast *Schizosaccharomyces pombe*. Curr Genet 38:113-125

Osman F, Dixon J, Barr AR, Whitby MC (2005) The F-Box DNA helicase Fbh1 prevents Rhp51-dependent recombination without mediator proteins. Mol Cell Biol 25:8084-8096

Osman F, Dixon J, Doe CL, Whitby MC (2003) Generating crossovers by resolution of nicked Holliday junctions: a role for Mus81-Eme1 in meiosis. Mol Cell 12:761-774

Ostermann K, Lorentz A, Schmidt H (1993) The fission yeast *rad22* gene, having a function in mating-type switching and repair of DNA damages, encodes a protein homolog to Rad52 of *Saccharomyces cerevisiae*. Nucleic Acids Res 21:5940-5944

Paques F, Haber JE (1999) Multiple pathways of recombination induced by double-strand breaks in *Saccharomyces cerevisiae*. Microbiol Mol Biol Rev 63:349-404

Park H, Sternglanz R (1999) Identification and characterization of the genes for two topoisomerase I-interacting proteins from *Saccharomyces cerevisiae*. Yeast 15:35-41

Petrini JH, Stracker TH (2003) The cellular response to DNA double-strand breaks: defining the sensors and mediators. Trends Cell Biol 13:458-462

Petroski MD, Deshaies RJ (2005) Function and regulation of cullin-RING ubiquitin ligases. Nat Rev Mol Cell Biol 6:9-20

Pommier Y, Redon C, Rao VA, Seiler JA, Sordet O, Takemura H, Antony S, Meng L, Liao Z, Kohlhagen G, Zhang H, Kohn KW (2003) Repair of and checkpoint response to topoisomerase I-mediated DNA damage. Mutat Res 532:173-203

Rea S, Eisenhaber F, O'Carroll D, Strahl BD, Sun ZW, Schmid M, Opravil S, Mechtler K, Ponting CP, Allis CD, Jenuwein T (2000) Regulation of chromatin structure by site-specific histone H3 methyltransferases. Nature 406:593-599

Rodel C, Jupitz T, Schmidt H (1997) Complementation of the DNA repair-deficient *swi10* mutant of fission yeast by the human *ERCC1* gene. Nucleic Acids Res 25:2823-2827

Rudolph C, Kunz C, Parisi S, Lehmann E, Hartsuiker E, Fartmann B, Kramer W, Kohli J, Fleck O (1999) The *msh2* gene of *Schizosaccharomyces pombe* is involved in mismatch repair, mating-type switching, and meiotic chromosome organization. Mol Cell Biol 19:241-250

Saparbaev M, Prakash L, Prakash S (1996) Requirement of mismatch repair genes *MSH2* and *MSH3* in the *RAD1-RAD10* pathway of mitotic recombination in *Saccharomyces cerevisiae*. Genetics 142:727-736

Sargent RG, Meservy JL, Perkins BD, Kilburn AE, Intody Z, Adair GM, Nairn RS, Wilson JH (2000) Role of the nucleotide excision repair gene *ERCC1* in formation of recombination-dependent rearrangements in mammalian cells. Nucleic Acids Res 28:3771-3778

Schiestl RH, Prakash S (1988) *RAD1*, an excision repair gene of *Saccharomyces cerevisiae*, is also involved in recombination. Mol Cell Biol 8:3619-3626

Schmidt H (1987) Strains of *Schizosaccharomyces pombe* with a disrupted *swi1* gene still show some mating-type switching. Mol Gen Genet 210:485-489

Schmidt H, Kapitza-Fecke P, Stephen ER, Gutz H (1989) Some of the *swi* genes of *Schizosaccharomyces pombe* also have a function in the repair of radiation damage. Curr Genet 16:89-94

Shankaranarayana GD, Motamedi MR, Moazed D, Grewal SI (2003) Sir2 regulates histone H3 lysine 9 methylation and heterochromatin assembly in fission yeast. Curr Biol 13:1240-1246

Shen Z, Pardington-Purtymun PE, Comeaux JC, Moyzis RK, Chen DJ (1996) UBL1, a human ubiquitin-like protein associating with human RAD51/RAD52 proteins. Genomics 36:271-279

Shibata T, Cunningham RP, DasGupta C, Radding CM (1979) Homologous pairing in genetic recombination: complexes of recA protein and DNA. Proc Natl Acad Sci USA 76:5100-5104

Shin JA, Choi ES, Kim HS, Ho JC, Watts FZ, Park SD, Jang YK (2005) SUMO modification is involved in the maintenance of heterochromatin stability in fission yeast. Mol Cell 19:817-828

Sinclair AH, Berta P, Palmer MS, Hawkins JR, Griffiths BL, Smith MJ, Foster JW, Frischauf AM, Lovell-Badge R, Goodfellow PN (1990) A gene from the human sex-determining region encodes a protein with homology to a conserved DNA-binding motif. Nature 346:240-244

Singh J, Klar AJ (1993) DNA polymerase-alpha is essential for mating-type switching in fission yeast. Nature 361:271-273

Styrkarsdottir U, Egel R, Nielsen O (1993) The *smt-0* mutation, which abolishes mating-type switching in fission yeast is a deletion. Curr Genet 23:184-186

Sugawara N, Haber JE (1992) Characterization of double-strand break-induced recombination: homology requirements and single-stranded DNA formation. Mol Cell Biol 12:563-575

Sugawara N, Paques F, Colaiacovo M, Haber JE (1997) Role of *Saccharomyces cerevisiae* Msh2 and Msh3 repair proteins in double-strand break-induced recombination. Proc Natl Acad Sci USA 94:9214-9219

Sugimoto A, Iino Y, Maeda T, Watanabe Y, Yamamoto M (1991) *Schizosaccharomyces pombe ste11+* encodes a transcription factor with an HMG motif that is a critical regulator of sexual development. Genes Dev 5:1990-1999

Sung P (1994) Catalysis of ATP-dependent homologous DNA pairing and strand exchange by yeast RAD51 protein. Science 265:1241-1243

Thon G, Bjerling KP, Nielsen IS (1999) Localization and properties of a silencing element near the *mat3*-M mating-type cassette of *Schizosaccharomyces pombe*. Genetics 151:945-963

Thon G, Bjerling P, Bunner CM, Verhein-Hansen J (2002) Expression-state boundaries in the mating-type region of fission yeast. Genetics 161:611-622

Thon G, Hansen KR, Altes SP, Sidhu D, Singh G, Verhein-Hansen J, Bonaduce MJ, Klar AJ (2005) The Clr7 and Clr8 directionality factors and the Pcu4 cullin mediate heterochromatin formation in the fission yeast *Schizosaccharomyces pombe*. Genetics 171:1583-1595

Thon G, Klar AJ (1992) The *clr1* locus regulates the expression of the cryptic mating-type loci of fission yeast. Genetics 131:287-296

Thon G, Klar AJ (1993) Directionality of fission yeast mating-type interconversion is controlled by the location of the donor loci. Genetics 134:1045-1054

Tornier C, Bessone S, Varlet I, Rudolph C, Darmon M, Fleck O (2001) Requirement for Msh6, but not for Swi4 (Msh3), in Msh2-dependent repair of base-base mismatches and mononucleotide loops in *Schizosaccharomyces pombe*. Genetics 158:65-75

Tran PT, Erdeniz N, Symington LS, Liskay RM (2004) EXO1-A multi-tasking eukaryotic nuclease. DNA Repair (Amst) 3:1549-1559

Tuzon CT, Borgstrom B, Weilguny D, Egel R, Cooper JP, Nielsen O (2004) The fission yeast heterochromatin protein Rik1 is required for telomere clustering during meiosis. J Cell Biol 165:759-765

Vengrova S, Dalgaard JZ (2004) RNase-sensitive DNA modification(s) initiates *S. pombe* mating-type switching. Genes Dev 18:794-804

Vengrova S, Dalgaard JZ (2005) The *Schizosaccharomyces pombe* imprint--nick or ribonucleotide(s)? Curr Biol 15:R326-R327; author reply R327

Vilenchik MM, Knudson AG (2003) Endogenous DNA double-strand breaks: production, fidelity of repair, and induction of cancer. Proc Natl Acad Sci USA 100:12871-12876

Xhemalce B, Seeler JS, Thon G, Dejean A, Arcangioli B (2004) Role of the fission yeast SUMO E3 ligase Pli1p in centromere and telomere maintenance. Embo J 23:3844-3853

Zou L, Elledge SJ (2003) Sensing DNA damage through ATRIP recognition of RPA-ssDNA complexes. Science 300:1542-1548

Arcangioli, Benoit
Department of Genomes and Genetics, Institut Pasteur, 25-28 rue du Dr Roux, Paris cedex 75724. France
barcan@pasteur.fr

Holmes, Allyson
Department of Genomes and Genetics, Institut Pasteur, 25-28 rue du Dr Roux, Paris cedex 75724. France

Roseaulin, Laura
Department of Genomes and Genetics, Institut Pasteur, 25-28 rue du Dr Roux, Paris cedex 75724. France

Multiple mechanisms of repairing meganuclease-induced double-strand DNA breaks in budding yeast

James E. Haber

Abstract

Double-strand breaks (DSBs) threaten the integrity of chromosomes. Consequently, cells have devised a number of mechanisms to repair broken chromosomes. There are several competing mechanisms of homologous recombination as well as multiple nonhomologous end-joining pathways that can repair chromosome breaks with varying degrees of fidelity. This review summarizes what has been learned about DSB repair in the budding yeast, *Saccharomyces cerevisiae*, where it is possible to create, rapidly and synchronously, a specific DSB using inducible meganucleases, HO or I-SceI. The physical monitoring of DNA undergoing recombination and the binding of various recombination proteins in the vicinity of the DSB provides a picture of the sequence of molecular events during recombination. We first examine an intrachromosomal recombination event, *MAT* gene switching, and then to interchromosomal ectopic events. In addition, repair by single-stand annealing, break-induced replication and nonhomologous end-joining are discussed.

1 Introduction

DSBs arise spontaneously during DNA replication. In vertebrate cells, depletion of the essential Rad51 recombinase protein causes rapid cell death, with cells accumulating perhaps a dozen unrepaired broken chromosomes that presumably would have been repaired by recombination using an intact sister chromatid as a template if Rad51 were active (Sonoda et al. 1998). Budding yeast chromosomes seem to be equally fragile during replication, although with its relatively tiny genome the absence of Rad51 or Rad52 proteins is not lethal. Pedigree analysis of *Saccharomyces* cells lacking Rad52 suggests that about 10% of cells experience at least one broken chromatid each cell cycle (J.E. Haber, unpublished). These breaks may arise at stalled replication forks, by replication over a pre-existing nick, by the excision of DNA transposable elements, by failures of topoisomerases to complete their strand breakings and rejoinings or by mechanical stress (e.g. in dicentric chromosomes). DSBs also arise by exposure to ionizing radiation or other clastogenic agents, or by site-specific cleavage by endonucleases.

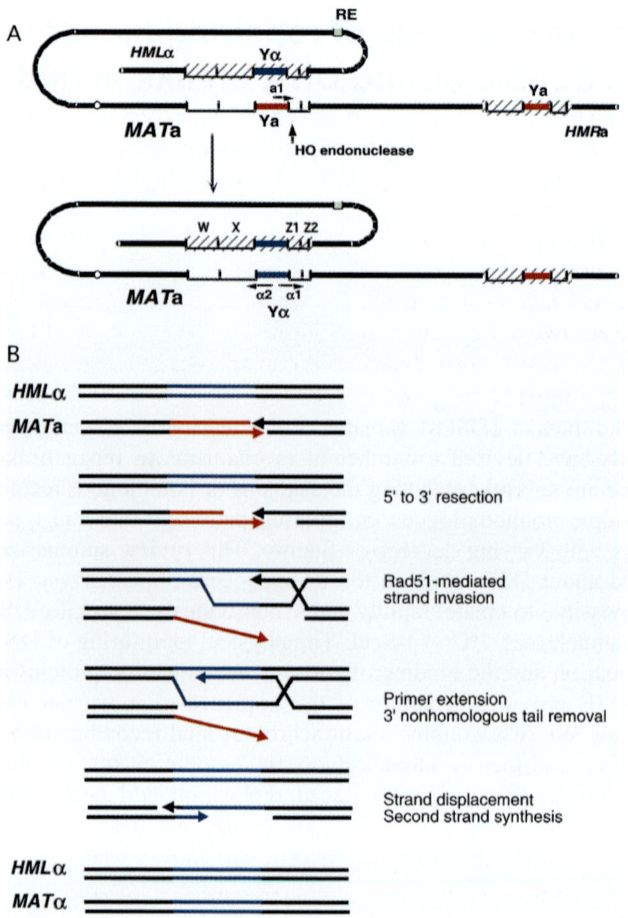

Fig. 1. *MAT* switching in *S. cerevisiae*. A. An HO endonuclease-induced DSB at *MAT***a** is repaired by gene conversion using the heterochromatic and transcriptionally silent *HML*α locus. *MAT***a** cells primarily recombine with *HML* but an equivalent recombination/replacement process occurs between and HO-cut *MAT*α and *HMR***a**. Donor choice is regulated by the Recombination Enhancer (RE). B. Molecular steps in gene conversion, as envisioned by the Synthesis Dependent Strand Annealing (SDSA) mechanism.

2 *MAT* switching in *Saccharomyces*, a paradigm for DSB repair

A detailed understanding of how DSBs are repaired was greatly advanced by the ability to create specific DSBs in a large proportion of cells, so that the fate of the DSB could be monitored by Southern blots where the outcomes produced restriction fragments that were different sizes than the initial strain. Physical monitoring of the timing of meiotic recombination was accomplished in 1984 before there was a demonstration that these events were mediated by DSBs (Borts et al. 1984); but most of our understanding has come from the analysis of mitotic yeast cells undergoing mating-type (*MAT*) gene switching. Homothallic yeasts have developed a mechanism whereby a haploid spore could generate offspring that are of the opposite mating-type, allowing cells to conjugate and form diploids. The key feature of this system is that haploid cells express the HO endonuclease, which creates a DSB at the *MAT* locus, leading to homologous recombination – gene conversion without an associated crossover event – involving an unexpressed donor cassette that acts as a template for repair (Fig. 1). In *Saccharomyces cerevisiae*, there are two such silent donors located at the two ends of the same chromosome carrying *MAT*. *HML*α carries 700 bp of α-specific genes whereas *HMR***a** carries 650 bp of **a**-specific sequences (reviewed in Haber 2002). A cis-acting recombination enhancer (RE) sequence 17 kb from *HML* promotes a biased use of the two donors, so that *MAT***a** recombines with *HML*α and thus switches to *MAT*α 90% of the time, whereas *MAT*α recombines with *HMR***a** (Wu and Haber 1996; and reviewed in Haber 2002). Some strains carry *HML***a** or *HMR*α; the rules of donor preference are not changed in such cells. When HO is expressed in *MAT***a** cells, a DSB is created close to the Y**a**-Z border (see Fig. 1). The DSB is repaired by homologous recombination, specifically a gene conversion event in which the original Y sequences at *MAT* (and some adjacent sequences in W, X, and Z regions) are replaced by a copy of the donor sequences (McGill et al. 1989).

A variety of genetic and molecular biological experiments suggest that *MAT* switching proceeds via a synthesis-dependent strand annealing (SDSA) mechanism (Gloor et al. 1991) in which strand invasion results in new DNA synthesis that is primed from the 3' end of the invading strand (Fig. 1B). The newly synthesized strand is apparently displaced and eventually the DSB end anneals to the newly copied DNA. Nonhomologous regions including the different Y sequences must be cleaved and then the second end can prime copying of the second strand. Consistent with an SDSA model that lacks stable Holliday junctions, crossovers accompanying *MAT* switching are rare; when they occur it is possible that the migrating D-loop during the first-strand synthesis gets trapped, yielding a double Holliday junction that can be cleaved by a resolvase to yield a reciprocal crossover. Alternatively some DSB repair events may proceed through the double-Holliday junction pathway described by Szostak et al. (1983).

2.1 Physical monitoring of MAT switching

The use of a galactose-inducible HO gene (Jensen and Herskowitz 1984) has made it possible to follow the molecular events of *MAT* switching in detail, by using Southern blots and PCR to examine intermediates of DSB repair. Recombination is initiated by creating long 3'-ended ssDNA ends through the action of 5' to 3' exonucleases (White and Haber 1990). In strains lacking donors, the resection process continues at 4 kb/hr, creating very long 3'-ended tails (Fishman-Lobell et al. 1992). The 3'-ended ssDNA strand is remarkably stable and not attacked by endonucleases. When donors are present, resection of at least 700–1000 bp still occurs in a significant fraction of the cells that undergo DSB repair (White and Haber 1990). It is not clear if there is a feedback mechanism to down-regulate resection once recombination is initiated. The Mre11-Rad50-Xrs2 proteins are required for full resection activity (Ivanov et al. 1994; Tsubouchi and Ogawa 1998), but at least in cycling cells, there must be another efficient exonuclease. The nuclease activity of Mre11 itself does not seem to be involved as phosphoesterase mutants that do not disrupt the MRX complex have normal resection (Lee et al. 2002; Moreau et al. 1999); possibly MRX proteins associate with an as-yet-unidentified exonuclease or a helicase-driven endonuclease analogous to the bacterial RecBCD enzyme. Resection is also reduced in mutants of the chromatin remodeling complex, Ino80 (van Attikum et al. 2004). 5' to 3' resection is dependent on Cdk1 activity, which is normally extinguished in G1; hence homologous recombination is repressed in G1 cells and DSBs are repaired primarily by nonhomologous end-joining (NHEJ) (Aylon et al. 2004; Ira et al. 2004). The Cdk1 target controlling exonuclease processing has not yet been identified from several hundred candidates identified by *in vitro* studies (Ubersax et al. 2003).

2.2 Monitoring of recombination protein binding to the DSB

Chromatin immunoprecipitation (ChIP) experiments have made it possible to follow the binding of recombination proteins to specific DNA sequences, in real time *in vivo*. ssDNA ends are first bound by the ssDNA binding protein complex, RPA and then about 10 minutes later by Rad51 recombinase (Wang and Haber 2004). RPA is strongly bound very soon after HO induces a DSB. This suggests that a large fraction of the HO-cut molecules have been resected enough to have sufficient ssDNA to which multiple copies of RPA can bind. This result stands in contrast to the apparent stability of HO cut ends as measured by ligand-mediated PCR, which only detects ends that have not been resected at all (which would be the case in G1 cells) (Frank-Vaillant and Marcand 2002).

Consistent with *in vitro* studies (reviewed in Krogh and Symington 2004), RPA binding to ssDNA is apparently required for efficient Rad51 loading. There is a failure to observe Rad51-GFP foci at DSBs when a temperature-sensitive degron mutant of the large subunit of RPA is used to deplete RPA abundance prior to inducing DSBs (Lisby and Rothstein 2004). Consistent with genetic and *in vitro* biochemical studies, Rad51 loading depends on the mediators, Rad52 and a heterodimer of Rad51 paralogs, Rad55-Rad57 (Sugiyama and Kowalczykowski 2002; Sung 1997a; Sung 1997b); also reviewed by Krogh and Symington 2004). Even when Rad51 is bound, RPA may not be fully displaced (Wang and Haber 2004). It is also possible that *in vivo* the Rad51 filament is not entirely continuous and some patches of RPA remain or that there is a cycle of dissociation and reassociation. Once Rad51 is bound, the nucleoprotein filament promotes a search for a homologous sequence. How this search takes place within the nucleoprotein filament remains poorly understood.

It is possible to visualize the kinetics of this search by using ChIP to examine when synapsis has occurred between the Rad51-bound end of *MAT* and the donor sequence, so that now the donor is also immunoprecipitated by an anti-Rad51 antibody (Sugawara et al. 2003; Wolner et al. 2003). It takes 15-20 minutes from the time of Rad51 binding to HO-cut *MAT* until synapsis is seen. This search is more rapid intrachromosomally than interchromosomally, as seen both by the longer time that is required for ChIP of an ectopic donor on another chromosome (N. Sugawara and J.E. Haber, unpublished results) and by the fact that the latter event promotes activation of the DNA damage checkpoint, which is activated once a threshold length of ssDNA is generated (Vaze et al. 2002).

A RPA mutation encoding Rfa1-L45E (*rfa1-t11*) has revealed a later role for RPA in strand invasion. *MAT* switching and the Rad51-independent process of single-strand annealing (see below) are both strongly reduced in the Rfa1-L45E mutant (Umezu et al. 1998). *In vitro* studies have suggested that this mutant RPA is more difficult to displace from ssDNA (Kantake et al. 2003), yet Rad51 filament formation appears to be normal *in vivo*, as measured by ChIP (Wang and Haber 2004). However strand invasion, i.e. Rad51 synapsis with the donor sequence, is impaired. Possibly RPA is required to stabilize the displaced D-loop resulting from strand invasion, consistent with *in vitro* studies of Rad51-mediated strand invasion (Eggler et al. 2002). Alternatively the association of the RPA with Rad51 (shown at least for human proteins) (Golub et al. 1998) maybe altered in Rfa1-L45E.

2.3 Primer extension

The next step in recombination is primer extension, using the 3' end of the invading strand as the primer for new DNA synthesis. This can be assayed by using a pair of PCR primers, one adjacent to *MAT* and one in the unique Y sequences at *HML* or *HMR*, initially 200 or 100 kb away, but now joined as part of a short, covalent recombination intermediate (White and Haber 1990). This PCR product can be amplified about 30 minutes before the appearance of a completed repair event, which can be seen either on Southern blots or using a second primer pair that determines when the donor Y sequences are joined to the left side of the HO cut.

The ability to capture this strand invasion intermediate by PCR made it possible to demonstrate that mismatch repair, to correct a single base pair difference between the invading *MAT* strand and the template, occurs very rapidly and in a highly biased way so that the invading strand is corrected to the genotype found in the donor (Haber et al. 1993). A *MAT***a**-stk mutation changes a single base pair at position Z11 and greatly reduces HO cutting (Ray et al. 1991). After strand displacement, this correction results in a gene conversion of the original *MAT* mutant by the sequence variant in the donor. In a mismatch repair mutant many colonies arising from a single *MAT***a**-stk cell undergoing switching were sectored, half being *MAT*α and half being *MAT*α-stk retaining the original mutation at position Z11. It is not clear how such sectored colonies could arise if repair proceeds exclusively by a SDSA mechanism in which the second strand would perforce be copied from the first, primer-extended strand, which would either have the original mutation or would have experienced mismatch correction and have the wild type allele.

Because the Y sequences to the left of the HO cut are usually not homologus with those of the donor (i.e. *MAT*α interacts with *HMR***a**), recombination is initiated in the homologous sequences to the right. *MAT* shares 230 bp homology with *HMR* (region Z1) and 320 bp homology with *HML* (regions Z1 and Z2). Strand invasion occurs into a heterochromatic, silenced donor locus with highly positioned nucleosomes (Weiss and Simpson 1998). Surprisingly, the Swi2/Snf2 homologue, Rad54, is not required for strand invasion per se (as judged by synapsis measured by ChIP) (Sugawara et al. 2003). However, the next step in *MAT* switching, the initiation of new DNA synthesis using the 3' invading end as a primer, does not occur without Rad54. Recently, another chromatin remodeler, Snf2-Snf5, has also been implicated in strand invasion, preventing the ChIP of the donor sequences using anti-Rad51 antibody (Chai et al. 2005). Still another chromatin remodeler, RSC, is rapidly recruited to an HO-induced DSB, but its action seems more important for later steps in the completion of repair (Chai et al. 2005; Shim et al. 2005). Recently, we have compared the efficiency of *MAT* switching using a heterochromatic *HML* versus an "open" donor in which adjacent silencer sequences have been deleted (and a non-cleavable inc mutation introduced at the HO cut site). Surprisingly, the open donor was not used more often in competition with the preferred *HMR* donor, but it was used more when the extent of homology shared with *MAT* beyond the Z region was increased (E. Coïc and J.E. Haber, unpublished). This result suggests that at least wild type cells can readily cope with heterochromatic sequences. Whether *snf2/snf5* or *rsc8/rsc30* mutants will have less impact when the donor is more accessible remains to be tested, but Rad54 is required even when the donor is "open" (Keogh et al. 2005).

Primer extension requires the DNA replication clamp PCNA, but the use of temperature-sensitive (ts) mutations in two components of the Mcm helicase complex suggests that this replicative helicase is not required to complete *MAT* switching (Wang et al. 2004). In G2 arrested cells, when there is no other DNA synthesis, neither lagging-strand components Polα nor primase are needed, however arresting replication in S phase with ts primase or Polα mutants traps some other replication component needed for *MAT* switching (Holmes and Haber 1999; Wang et al. 2004). Both Polδ and Polε ts mutations retard but don't prevent the completion of *MAT* switching, suggesting that they may have redundant functions (Wang et al. 2004). As double mutants are inviable, a definitive test of their redundancy has not been performed.

The newly synthesized strand appears to be displaced from its donor sequence, as imagined in SDSA mechanisms, so that the second strand can be copied from this newly-made template. The use of heavy isotope density transfer methods has confirmed that *MAT* switching leaves with donor unaffected, with all the newly synthesized DNA in the repaired *MAT* locus (G. Ira and J.E. Haber, manuscript submitted). Before second-strand synthesis can occur, however, the nonhomologous Y segment must be clipped off the resected ssDNA end. This is done primarily by a novel use of the Rad1-Rad10 (XPF-Ercc1) endonuclease that is required for nucleotide-excision repair (Colaiácovo et al. 1999; Holmes and Haber 1999). However, there is a slower, back-up nuclease (identity unknown) that can also remove this nonhomologous end when Rad1-Rad10 is missing. A pair of PCR primers that show when the new Y sequences are joined to the proximal side of *MAT* signals the completion of the repair event, which takes about 30 minutes after strand invasion is detected. Other than the step of clipping off the nonhomologous tail, we don't know what other slow steps take place, although it seems that the RSC complex acts at this time (Chai et al. 2005).

The choreography of recombination protein recruitment to the DSB has also been monitored visually, both by indirect immunofluorescence and by using GFP- or other fluorescently-tagged fusion proteins (Lisby and Rothstein 2004). By and large the results are all in good agreement with ChIP experiments. Miyazaki et al. (2004) suggest that the Rad52 protein is needed not only to promote Rad51 nuceloprotein filament formation but also at a later step, when Rad51 has largely disappeared; presumably, this late role is in facilitating the annealing of the strands in SDSA. One also can "tag" chromosome segments by binding LacI-GFP or TetR-GFP proteins to LacO or TetO arrays, so that one can watch in real time the dynamics of the search for homology and the synapsis of the donor and *MAT* sequences (Bressan et al. 2004). This approach has also confirmed that, without Rad54, there is a remarkably stable association of *MAT* and *HML* that persists for hours in the absence of being able to repair the DSB (S. Jain, D. Bressan, and J.E. Haber, unpublished).

Many details of the completion of *MAT* switching remain obscure. One mystery is that inhibition of the Cdk1 kinase after DNA ends have already been resected appears to prevent the completion of gene conversion (Ira et al. 2004). It seems that it is the primer extension step after strand invasion that is blocked; but again we do not know the target of repair. What role the RSC chromatin remodeling complex plays late in recombination is also not known. Another question is how the strands are finally ligated together. A strain lacking DNA ligase 4 and carrying a temperature-sensitive allele of the major DNA ligase, Cdc9, still appears to be able to generate a completed *MAT* switching event, so that even on denaturing gels of rather large restriction fragments there is no evidence of unligated ends (Wang et al. 2004). Possibly there is still some residual activity of the ts DNA ligase 1 or else the nicks have been "nick-translated" far from the *MAT* locus, or there is another ligase that has escaped detection.

3 HO and I-SceI-induced ectopic gene conversions and the control of reciprocal crossing-over

The 24-bp minimal HO cleavage site (or larger fragments) can be moved to different locations so that recombination of different sequences, with various arrangements of donor and recipient sequences can be studied. Nickoloff's lab has carried out extensive studies on HO-induced events at the *URA3* locus (Cho et al. 1998; Nickoloff et al. 1999) and Kupiec has done the same at several locations including within the retrotransposon sequences Ty1 (Inbar and Kupiec 1999; Kupiec and Petes 1988). By deleting *HML* and *HMR*, one can also use HO-cut *MAT* as the recipient, with another copy of *MAT*, carrying a mutation that prevents cleavage, as the donor. In all these arrangements, one can examine interchromosomal recombination and loss of heterozygosity in diploids or the formation of reciprocal translocations (gene conversions associated with crossing-over) between dispersed homologous sequences in haploids (Pâques et al. 1998; Ira et al. 2003).

In addition, some studies in yeast – and many more in other organisms – have been carried out using the I-SceI meganuclease, which is present in yeast mitochondria but has been "domesticated" by codon changes and the use of a galactose-inducible promoter to work in the nucleus (Plessis et al. 1992; Wilson 2002). I-SceI has an 18-bp recognition site and cleaves a cloned target in yeast quite efficiently, though with somewhat slower kinetics than HO.

3.1 Most ectopic recombination occurs by SDSA

Ectopic systems have provided very strong evidence that a major DSB repair pathway proceeds by way of SDSA. One set of experiments, based on earlier studies in *Drosophila*, examined recombination in which the two ends of the DSB invade unique sequences but copy across a template that contains repeated sequences (Pâques et al. 1998). When the template contains 8 copies of 375-bp *Drosophila* 5S DNA repeats, 50% of the DSB repair events had either fewer or more than 8 copies in the recipient, whereas with few exceptions, the donor template remained with 8 copies. Similar studies have been done with microsatellite and minisatellite repeats (Pâques et al. 2001; Richard et al. 1999). The interpretation of these results is that each end of the DSB can initiate copying of the template but that dissociation of the partially replicated strands allows annealing within the repeats to yield a variety of sizes (Fig. 2A). Those that carry precisely 8 copies may have arisen by instances in which only one end initiated copying and proceeded all the way across the template until the unique sequences on the opposite side were copied and then annealed to the waiting second, single-stranded end. One cannot entirely rule out that repair replication is inherently prone to replication slippage that could yield more or fewer repeats, but in this case it is hard to understand how 50% of the events would remain with 8 copies in what is a decidedly non-normal distribution of outcomes.

A second demonstration of SDSA was the use of a tripartite recombination system in which the two ends of an HO-induced DSB in a plasmid-borne sequence ("L" and "2") could each recombine with only one of two templates on different chromosomes ("LEU" and "EU2") (Fig. 2B). Reconstitution of an intact *LEU2* sequence on the plasmid requires two strand invasions and primer extensions as well as two dissociations of newly synthesized DNA, followed by an annealing step. These repair events were about 1/40 as efficient as a simple gap repair of "L" and "2" ends by a single intact *LEU2* template (Pâques et al. 1998).

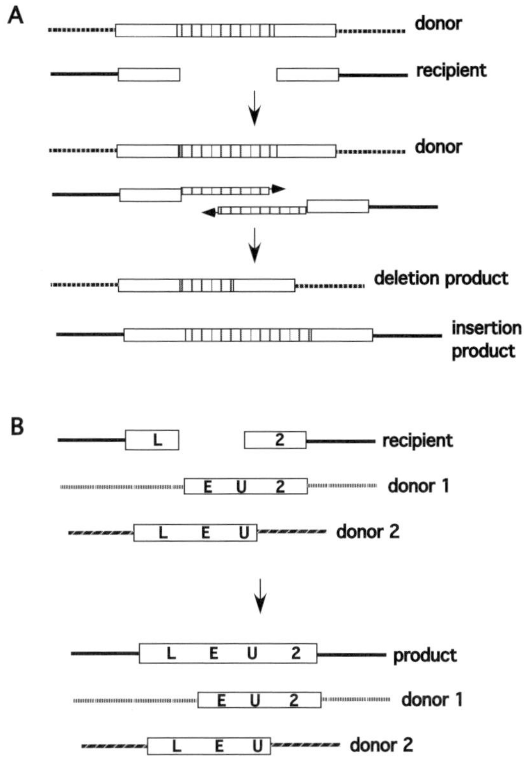

Fig. 2. Evidence supporting SDSA mechanisms. A. Gene conversion in which the ends of a DSB are homologous to sequences surrounding a series of tandemly repeated sequences can lead to repair products containing repeats that are either substantially larger than or smaller than the size of the original repeat array. It is imagined that these changes arise from annealing of two independently-initiated primer extensions from the two ends. B. Tri-parental gene conversion requiring two strand invasions, primer extensions and displacements that will permit the formation of an intact *LEU2* gene from two templates each of which are homologous to only one end of the HO-cleaved recipient.

Although SDSA yielding noncrossovers is clearly the major repair pathway used in mitotic cells, some repair events are associated with crossing-over. Ectopic systems readily provide restriction fragment length polymorphisms so that crossovers associated with gene conversion can be easily identified and quantified on Southern blots (Fig. 3). It is widely believed that crossovers should involve the resolution of Holliday junction intermediates or related branched DNA intermediates. These types of intermediates have to date only been identified in yeast meiotic recombination; in agreement with the canonical DSB repair model of Szostak et al. (1983), they proved to be fully ligated double Holliday junctions (dHJ) (Schwacha and Kleckner 1995). The DSB repair mechanism of Szostak et al. (1983) stands as the canonical alternative to SDSA. Alternative structures, including unligated dHJ, single Holliday junctions, and various nicked intermediates,

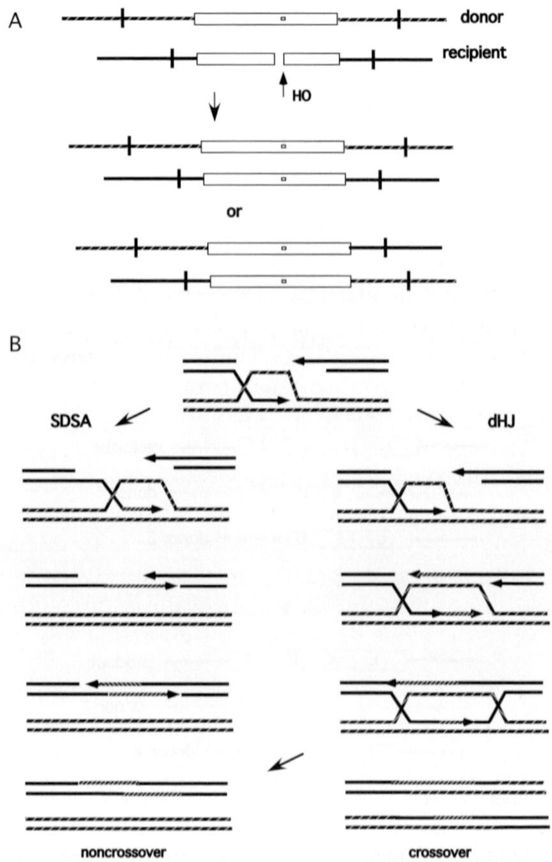

Fig. 3. Ectopic recombination allows the analysis of crossover control. A. Ectopic gene conversion is sometimes accompanied by crossing-over, which is seen on a Southern blot by the appearance of novel pair of restriction fragments. B. Analysis of various mutations suggests that most noncrossovers arise from a SDSA mechanism, but a double Holliday junction intermediate may give rise to crossovers. The double Holliday junction intermediate can also be unwound to produce noncrossovers, as shown *in vitro* using the Sgs1 homologue, BLM helicase, in conjunction with topoisomerase 3 (Wu and Hickson 2003).

have all been suggested to arise under different circumstances (Cromie and Leach 2000; Osman et al. 2003). Hampering our understanding of which of these intermediates is important in mitotic recombination is the lack of identifying the HJ resolvase (or resolvases) that would generate crossovers. One possible resolvase, which seems to act preferentially on nicked branched intermediates is Mus81 and its associated Eme1 protein (Mms4 in budding yeast). In *Schizosaccharomyces pombe* meiosis (Smith et al. 2003), Mus81-Eme1 is apparently the key resolvase, whereas in *S. cerevisiae* meiosis, Mus81-Mms4 plays an important but smaller role (de los Santos et al. 2003).

3.2 Control of crossing-over associated with gene conversion

In general, compared to meiotic recombination, where crossovers are between 30% and 50% of all products, the proportion of gene conversion events between allelic sites on homologous diploid chromosomes is about 10%. In ectopic recombination, the proportion of gene conversions with a reciprocal exchange is strongly influenced by the length of shared homology (Inbar et al. 2000). In the case of an HO-cut *MAT* recombining with another ca. 2-kb *MAT* sequence, only about 4% of gene conversions are crossover-associated (Ira et al. 2003).

A number of mutations significantly change the proportion of recombination events that result in crossovers. In humans, lack of the 3' to 5' BLM helicase promotes a high level of sister-chromatid exchange (SCE). Deletion of its yeast homolog, Sgs1 or its associate Top3 topoisomerase, causes a two to threefold increase in ectopic crossovers (Ira et al. 2003). We suggested, supported by the biochemistry on BLM-Top3α by Hickson's lab (Wu and Hickson 2003) that Sgs1-Top3 unwinds fully ligated dHJ to generate noncrossovers. In *sgs1*Δ, then, crossovers increase because this unwinding pathway is blocked. These results suggest that the BLM phenotype may not be due to an increase in DNA damage leading to more SCE but rather to a higher proportion of intermediates being resolved as crossovers. A reduction in the abundance of Sgs1 also seems to account for the increase in crossovers when HO is induced in G2-arrested yeast cells (Ira et al. 2003).

Another 3' to 5' helicase, Srs2, affects crossovers in an entirely different way. Strains lacking Srs2 show a marked decrease in completing gene conversion and this loss is preferentially among noncrossover products, so that – among cells that can complete recombination – the proportion of crossovers is 3 times higher than in wild type cells. But here the main problem is in completing what appear to be SDSA events (Ira et al. 2003). Srs2 has been shown *in vitro* to be able to remove Rad51 recombinase from ssDNA (Krejci et al. 2003; Veaute et al. 2003), and genetic studies support the idea that the absence of Srs2 might make yeast too avid in carrying out recombination events that have fatal consequences. Yet Rad51-mediated strand invasion is an apparently common step to initiate either SDSA or dHJ formation; but Srs2 seems to be required specifically for SDSA. This result suggests that the main role of Srs2 is to promote a later step, perhaps the displacement of the invaded strand once new DNA synthesis is primed. Interestingly, overexpressing Rad51 in the absence of Srs2 is even more lethal, and again it is the noncrossover outcomes that are selectively eliminated and crossovers among successful recombinants reaches 35% (Ira et al. 2003).

Recently a third helicase, Mph1, moving 3' to 5' (Prakash et al. 2005; Scheller et al. 2000; Schurer et al. 2004), has been shown to play a key role in crossover control. In the absence of Mph1, crossovers increase from 4% to 12%, but in this instance there is no loss of viability (J.E. Haber and G. Ira, unpublished results). The double mutant *mph1*Δ *srs2*Δ is synthetically lethal but can be suppressed by *rad51*Δ, as are many other double mutant combinations of helicases and related proteins. Induction of Rad51 at the same time as HO endonuclease allows us to demonstrate that the defects of *mph1*Δ and *srs2*Δ are independent, as the double mutant combination has nearly 50% crossovers with no less viability than *srs2Δ* alone. The double mutant *mph1Δ sgs1Δ* is viable and again shows additivity compared to the defects of the single mutants. As with sgs1Δ and srs2Δ, the Mph1 deletion mutants shows a two to threefold increase in crossing-over between homologous chromosomes after one of them suffers an HO cut.

There is also an increase in crossovers associated with ectopic gene conversion in strains lacking Rad50. In this case, it is possible that the effect is specifically seen in ectopic recombination. As suggested by Prado and Aguilera (2003), resection of the DSB ends beyond the limits of shared homology between donor and recipient might impair the ability to form Holliday junctions and thus repair would be channeled into SDSA. Hence, if resection is reduced twofold, as it is in a *rad50Δ* mutant, one might expect more of the intermediates to be converted into dHJ and resolved as crossovers. This appears to be the case (G. Ira and J.E. Haber, unpublished). The effect of *rad50Δ* is independent of *sgs1Δ*, as the double mutant has a still higher level of crossovers. Moreover, overexpression of the 5' to 3' exonuclease, *EXO1*, specifically reduces crossovers in the *rad50Δ* but not of the helicase mutants.

Further evidence of the consequences of the limited homology in ectopic recombination has recently been obtained by studying the Mer3 helicase (Mazina et al. 2004). Mer3 is normally expressed only in meiotic cells, where it acts to promote crossing-over between homologous chromosomes. A *MER3* cDNA expressed in mitotic cells also increases crossovers induced by HO between homologous chromosomes, but paradoxically reduces crossing-over with ectopic substrates of 2 kb (G. Ira and J.E. Haber, unpublished). Again we surmise that extensions of the size of strand invasion D-loops might prevent the isomerization of recombination intermediates (requiring that the sequences be homologous) that would be required to get crossing-over.

These results raise another question: Why should isomerization into a double Holliday junction (dHJ) be necessary? Could not the half-crossover side of the D-loop be cleaved and the single HJ be resolved to get the same outcome? Indeed such an outcome is suggested from transformation experiments where the Msh2-Msh3 and Rad1-Rad10 complex is suggested to process the half-Holliday junction (Langston and Symington 2005). One consequence of isomerization of the half-HJ intermediate, to produce a complete dHJ would be to provide the means to stop the otherwise inexorable 5' to 3' exonucleases from degrading the chromosome. Once resection begins, it apparently will continue – at least in mitotic cells – until the repair process sends a more rapidly moving DNA polymerase down the DNA to fill in the ssDNA regions (Vaze et al. 2002). At the "second end" of the DSB, this could only happen either when cells use an SDSA mechanism or when the 1-and-a-half HJ is converted to full dHJ structure (Fig. 3).

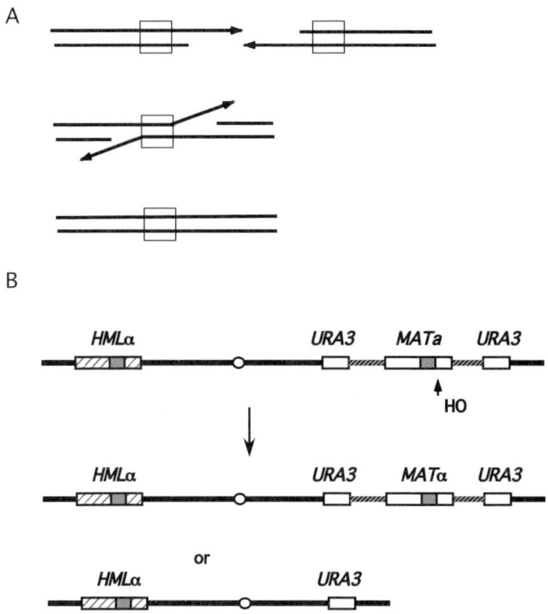

Fig. 4. Single-strand annealing (SSA). A. 5' to 3' resection of DSB ends allows complementary sequences (shown in boxes) to anneal. Removal of the overhanging nonhomologous 3'-ended tails is carried out by the Rad1-Rad10 flap nuclease, with the assistance of the Msh2-Msh3 mismatch repair proteins. B. Competition between SSA and gene conversion. In spite of the fact that *MAT* switching is very efficient, about 30% of the DSBs at *MAT* are repaired by SSA when *MAT* is flanked by homologous sequences.

4 Single-strand annealing (SSA)

Single-strand annealing (Fig. 4A) is the simplest process to repair a DSB by homologous recombination (Fishman-Lobell et al. 1992); reviewed in Pâques and Haber 1999) . Resection of the DSB ends produces long 3'-ended, single-stranded DNA (ssDNA) tails. If the DSB occurs in a region where there are flanking, repeated sequences, the two tails can anneal, creating a deletion of the intervening region. The annealing step is catalyzed by Rad52 and by Rad59, which shares some homology with Rad52, but occurs independently of Rad51, Rad55-Rad57, and Rad54 (Ivanov et al. 1996; McDonald and Rothstein 1994). The requirement for Rad59 depends on the length of the segments that anneal; when homology is about 200 bp, Rad59 is as important as Rad52, but when homology is 1000 bp, deleting Rad59 has a relatively minor effect whereas Rad52 remains essential (Sugawara et al. 2000). When the homology flanking the DSB is exceptionally long, e.g. in the 9-kb repeats of rDNA, SSA appears to be independent even of Rad52 (Ozenberger and Roeder 1991). SSA becomes very inefficient below about

30 nt of homology, but SSA may occur at very low frequencies with very small homology (Sugawara et al. 2000; Wilson 2002).

SSA is also strongly impaired in Rfa1-L45E (*rfa1-t11*) (Umezu et al. 1998). Possibly, RPA is required to prevent self-annealing of a single strand and allow interstrand annealing; but how Rfa1-L45E affects this process is not yet clear.

Completion of SSA requires that the nonhomologous 3' ends be clipped off, so that repair synthesis can occur to fill in the gap. Clipping is carried out by Rad1-Rad10 (Fishman-Lobell and Haber 1992; Ivanov and Haber 1995). Clipping also requires the Msh2-Msh3 mismatch repair complex that apparently recognizes the annealed intermediate and either stabilizes it or promotes Rad1-Rad10 cleavage. Like Rad59, however, the requirement for Msh2-Msh3 is much more pronounced when the annealed region is small (Sugawara et al. 1997). The clipping function of Msh2 can be separated by mutation from its mismatch correction function (Studamire et al. 1999).

Remarkably, SSA can occur between sequences that are very far apart. In one well-studied case, a DSB created by placing an HO cleavage site in the middle of the *LEU2* gene can be repaired by inserting the 3' end of the gene (the "U2" sequences) 25 kb distal on the same chromosome arm, resulting in a 25-kb deletion after SSA is completed (Vaze et al. 2002). The process takes 6 hours, compatible with the observed rate of 5' to 3' resection of 4 kb/hr to render the distant U2 sequence single-stranded. If the homologous sequence is located about 10 kb away, SSA takes 2-3 hours. The first striking conclusion from these experiments is that there is little or no endonucleolytic cleavage of the very long ssDNA strands that will eventually form the deletion; if a single cleavage occurred to the right of the HO-cut *LEU2* region, the homology that would form a deletion with the distant U2 sequence would no longer be attached to the rest of the chromosome. A second surprise arises when SSA takes place in Rad51^{+} cells. ChIP shows that Rad51 promotes synapsis between the two U2 sequences within 2 hr, yet SSA does not appear until 6 hr, when resection would have removed the non-annealed strand (N. Sugawara, S. Jain, and J.E. Haber, unpublished results). Why strand invasion doesn't lead to some sort of new DNA synthesis or break-induced replication (see below) is a mystery. It is possible that a signal to launch new DNA synthesis comes only after the second end of the DSB is also engaged by strand invasion or annealing to an expanding D loop; in the SSA configuration, the second end has no homologous target.

SSA is strongly discouraged if there is a low level of heterology between the flanking sequences. In 205-bp regions, a 3% level of mismatch is sufficient to reduce SSA about six-fold. This reduction is accounted for by an active heteroduplex rejection mechanism that apparently unwinds, rather than degrades, the mismatched paired strands (Sugawara et al. 2004). Heteroduplex rejection requires the Msh2-Msh6 mismatch repair proteins but surprisingly none of the Mlh1, Mlh2, Mlh3, or Pms1 proteins that are needed for mismatch repair. (But indeed without Mlh1 or Pms1, there is clear evidence for unrepaired heteroduplex DNA in those SSA events that are completed). In spontaneous recombination events involving heterologous sequences, Pms1-Mlh1 are needed to suppress homeologous recombination (Chen and Jinks-Robertson 1998); there must be a fundamental difference in how strand invasion of presumably a ssDNA end into a dsDNA donor is treated compared to the two-strand encounter in SSA.

By engineering multiple HO cleavage sites in a situation where SSA could occur either between flanking sequences on the same chromosome or on different chromosomes, we made the surprising discovery that SSA was just as likely to occur between homologous sequences on different chromosomes, producing a pair of reciprocal translocations, as it was to occur between sequences on the same chromosome, producing two deletions (Haber and Leung 1996). This lack of chromosome territoriality is unexpected and may reflect a reorganization of chromosome ends after DNA damage. This result is compatible with the cytological observations by Lisby and Rothstein that regions near DSBs on two different broken chromosomes may associate in a single focus in the absence of homology (Lisby et al. 2003).

It should also be noted that SSA is not confined to situations where there are no efficient alternatives. If the *MAT* locus is deleted and replaced with *URA3* and then a pBR322 plasmid carrying *MAT* and *URA3* is integrated at this site (Fig. 4B), then HO cleavage of *MAT* can be repaired in two ways, either by *MAT* switching which normally occurs with 100% efficiency or by SSA, leading to a deletion of the plasmid sequences and one copy of *URA3* (Wu et al. 1997). Surprisingly, about 30% of the outcomes are the SSA-mediated deletions. This experiment also shows that there is considerable resection of the DSB ends – much more than is required to accomplish *MAT* switching – so that SSA is a competitive process. Whether SSA is evolutionarily advantageous or is an unintended consequence – a "spandrel" (Gould 1997) – of creating ssDNA regions to facilitate gene conversion is an interesting question.

5 Break-induced replication (BIR)

Esposito (Esposito 1978; Esposito et al. 1994) first noted examples of mitotic recombination in which there was a non-reciprocal recombination event that extended hundreds of kb down a chromosome arm. Voekel-Meiman and Roeder (1990) saw similar events promoted by a mitotic "hot spot" and suggested that they could arise by extensive recombination-dependent DNA replication, now termed BIR. The idea that a broken chromosome end could acquire a new telomere by such a recombination process was provided by Dunn et al. (1984), who transformed into yeast a linearized plasmid with one end that lacked a telomere, but had homology to a subtelomeric Y' region that could recombine with another chromosome that had a Y' sequence adjacent to a telomere.

A direct demonstration of the replicative nature of BIR repair was provided by Morrow et al. (1997), who transformed yeast with a DNA fragment containing an origin of replication and a centromere, along with two oppositely oriented identical DNA segments at the ends that shared homology with one region of a yeast chromosome. Transformants contained an entirely new chromosome, in which both ends had to recombine with the same unique homologous sequence located on one chromosome and, both times, replicate all the way to the chromosome end. Bosco and Haber (1998) used HO endonuclease to lop off the end of a chromosome in a diploid, in which the DSB shared homology only centromere-proximal to the DSB. They also showed that BIR could occur, though only in about 5% of cells, in haploids between short (70-bp) homologous sequences located close to the DSB end and on another chromosome arm. This process required Rad52. Other examples of long gene conversion tracts apparently extending to the end of the template chromosome have been seen for HO endonuclease-induced events both in mitotic (Nickoloff et al. 1999) and meiotic cells (Malkova et al. 1996).

BIR appears to be an essential process in survivors of senescent budding yeast cells, lacking functional telomerase (reviewed by McEachern and Haber 2006). Survivors arise by *RAD52*-dependent recombination processes to maintain telomeres (Lundblad and Blackburn 1993). In fact there are two Rad52-dependent pathways that are distinguished both by their genetic requirements and by the structures of the telomeres themselves (Le et al. 1999; Teng et al. 2000). In Type I events essentially all telomere ends acquire subtelomeric Y' sequences by recombination between a deprotected telomere end and similar $TG_{1\text{-}3}$ sequences that are sometimes found between tandem Y' elements or between Y' sequences themselves. It is possible that the template for the proliferation of Y' sequences may be a circular, autonomously replicating form of Y' that also carries telomere sequences (Horowitz and Haber 1985). Evidence for recombination involving telomeres and exogenous circles has been provided by McEachern's lab studying *Kluyveromyces lactis* (Natarajan and McEachern 2002). The appearance of Type I survivors depends on the "canonical" homologous recombination proteins, Rad51, Rad54, Rad55, and Rad57 (Le et al. 1999).

A second Rad52-dependent survivor pathway leads to the substantial elongation of the telomeric sequences themselves (Teng et al. 2000). The template for these telomere elongations could be telomere sequences on a sister chromatid, or at another telomere end. It is also possible that telomere elongations could arise by a form of rolling circle replication if yeast telomeres are capable of forming t-loops (Griffith et al. 1999) and using the 3' end as a primer to extend the telomeres. The type II events depend on another set of "Rad" proteins: the Mre11-Rad50-Xrs2 complex and Rad59 (Chen et al. 2001; Le et al. 1999; Teng et al. 2000; Teng and Zakian 1999; Tsukamoto et al. 2001). As predicted from the individual analyses, nearly all survivors are eliminated in a *rad51Δ rad50Δ* double mutant (Grandin and Charbonneau 2003; Le et al. 1999; Zubko et al. 2004). The Mre11-Rad50-Xrs2 complex and Rad59 both have strand-annealing activity *in vitro*, as does Rad52, but it remains unclear how strand invasion is accomplished without Rad51. One possibility is that the telomere ends are opened up by a helicase; indeed the Sgs1 helicase, related to human BLM and WRN helicases, is required for type II events (Cohen and Sinclair 2001; Huang et al. 2001; Johnson et al. 2001; Watt et al. 1996).

5.1 At least two pathways of BIR can be shown for non-telomere sequences in *S. cerevisiae*

The genetic requirements of BIR have been determined by examining diploids in which there is a single HO-induced DSB in the middle of the right arm of chromosome III. Normally, such a DSB would be repaired by "short patch" gene conversion. A *rad52Δ* diploid shows almost no repair of the broken chromosome; it is simply lost, creating a 2n-1 monosomic derivative. A *rad51Δ* strain eliminates gene conversions but still allows BIR to proceed (Malkova et al. 1996). In colonies derived from single cells suffering a DSB, more than 80% of them give rise to at least a sector of cells that have retained the centromere and left arm of the broken chromosome, while the other cells in the colony had lost the broken chromosome (Fig. 6A & B, below). A similar phenotype is found in *rad54Δ*, *rad55Δ* and *rad57Δ* mutants, all of which eliminate gene conversions but allow BIR (Signon et al. 2001). This *RAD51*-independent BIR pathway is largely dependent on another set of recombination genes: *RAD50*, *MRE11*, *XRS2*, *RAD59* and *TID1* (*RDH54*) (Signon et al. 2001). Double mutants, such a *rad51Δ rad50Δ*, *rad51Δ rad59Δ* or *rad54Δ tid1Δ*, fail to repair the DSB more than 90% of the time, leading to chromosome loss. However, 10% of the cells still give rise to colonies with sectors that appear to derive from BIR events, based on Southern blot and genetic analysis; thus none of these double mutants is as defective as a *rad52Δ* strain. This result is similar to the finding by Bai and Symington (1996) that spontaneous heteroallelic recombination in a *rad52* strain was still threefold more deficient than a *rad51 rad59* double mutant. Perhaps there is still a third *RAD52*-dependent pathway.

The involvement of Rad50 and Rad59 in this Rad51-independent BIR repair in the middle of a chromosome is strongly reminiscent of the Rad51-independent pathway of telomere maintenance without telomerase, discussed above. As noted above, Tid1 is also required for Rad51-independent BIR with a single HO-induced DSB. At telomeres the absence of Tid1 seems to affect both telomere repair pathways (Putnam et al. 2004). There is one distinctive difference; however, Type II telomere recombination needs Sgs1p whereas this helicase had no apparent role in the analogous BIR event measured in the middle of a chromosome (Signon et al. 2001). The need for Sgs1p at telomeres but not for analogous events with other sequences may reflect the presence of G-quartet or other structures formed by G-rich sequences or the inherently mismatched nature of recombination intermediates between variable TG_{1-3} sequences.

The idea that recombination-dependent DNA synthesis to the end of the chromosome can occur without the only known strand exchange protein, Rad51, is a mystery. A clue to how recombination proceeds comes from the finding that the sites where Rad51-independent BIR is initiated are distinctly non-random, as if recombination without Rad51 requires some special DNA sequence environment. Virtually none of the *RAD51*-independent BIR repair events retains a marker on the broken chromosome 10 kb centromere-proximal to the DSB site. There appears to be a cis-acting DNA sequence, named FBI, located 34 kb proximal to the DSB site that is responsible for facilitating the majority of BIR events (Malkova et al. 2001). Recent experiments have revealed that the FBI sequence is in fact a pair of inverted Ty retrotransposon sequences separated by a short unique sequence (A. Malkova, J. Theis, C. Newlon, and J.E. Haber, unpublished results); this sequence is apparently identical to the FSB2 fragile site revealed by being a frequent participant in non-reciprocal translocation events created when the abundance of DNA polymerases is reduced. (Lemoine et al. 2005). Although FSB2 may be a preferential site of chromosome breakage, in the BIR experiments described here the DSB is created by HO 34 kb away and FSB2 is a preferential site for DNA repair, apparently by BIR. Further analysis of Rad51-independent BIR events suggests that these Ty1 elements are used either to promote formation of diploids homozygous for the more distal markers on chromosome III or to create diploids in which the repaired chromosome is a nonreciprocal translocation, apparently from recombination between the FBI/FBS2 Ty sequences and a Ty sequence elsewhere in the genome (A. Malkova, personal communication). Why these Ty sequences would be preferentially used for recombination, while the intervening 34 kb of homologous sequences between the DSB and FBI fail to yield BIR events is unknown. One possibility would be that the Ty sequences somehow prevent further 5' to 3' resection of the DSB ends, thus allowing these sequences to be preferential end-points for repair. However, by examining the resection of the broken chromosome in a *rad52*Δ mutant, where repair cannot occur, it seems that resection is not retarded at the Ty sequences (G. Ira and J.E. Haber, unpublished). It is possible that strand invasion within these sequences somehow facilitates the assembly of a repair replication fork that would then permit BIR to begin. These *RAD51*-independent repair events have not been analyzed in detail, in "real time," by Southern blot or other physical monitoring approaches, so there is yet little insight into the kinetics or intermediates arising during repair.

5.2 RAD51-dependent BIR

Because a DSB in the middle of a chromosome is so efficiently repaired by gene conversion in a *RAD51* cell, one cannot characterize a *RAD51-dependent* BIR process in the same diploid system. To examine such *RAD51*-dependent events, Malkova et al. (2005) created a modified diploid in which the target chromosome is truncated such that there is only a 46-bp segment distal to the DSB that is homologous to the template chromosome. This region is too short to permit efficient repair by gene conversion, although about 10% of the repair events still occur in this way; the remaining events occur by BIR. In this diploid, *RAD51*-mediated BIR is significantly more efficient than what is seen in the absence of *RAD51*. Moreover, the *RAD51*-dependent pathway does not require the distant facilitating sequence that promotes BIR in a *rad51*Δ diploid; a majority of the *RAD51* repair events are initiated within 3 kb of the DSB. BIR also occurs efficiently in haploids in which the centromere-proximal side of the DSB shares ≥ 1 kb of homology with appropriately oriented sequences on another chromosome arm, to produce a nonreciprocal translocation (J. Lydeard and J.E. Haber, unpublished results).

RAD51-dependent BIR is highly efficient; consequently it has been possible to follow repair in real time. Several striking observations have emerged. First, whereas gene conversion events in a diploid with long homology on either side of the DSB occurs with little or no activation of the DNA damage checkpoint, when repair occurs by BIR there is an extended cell cycle delay, even when the sequences proximal to the DSB are completely homologous to the template chromosome (Malkova et al. 2005). This delay is imposed by the Rad9-dependent DNA damage checkpoint that acts through the ATR homolog, Mec1. Second, BIR can take place in G2-arrested cells. Third, DSB repair by BIR, monitored by Southern blots, does not appear until about 6 hours, whereas the 10% of repair occurring by gene conversions appear by 2 hours. Analysis of Rad51-mediated strand invasion, by ChIP, shows that this step occurs within 2 hours, but primer extension, even of as little as 50 nt of new DNA synthesis, is delayed until soon before BIR is completed, at 6 hours (N. Tanguy le Gac, S. Jain, N. Sugawara, and J.E. Haber, unpublished results). This long delay in accomplishing BIR appears to involve a slow step in the formation of a repair replication fork. Understanding how the repair replication fork assembles in G2/M-arrested cells when, for example, the Mcm helicase proteins are believed to be exported from the nucleus (Labib et al. 2001) remains one of the most important goals of current research. Finally, the rate of repair replication itself, once repair is initiated, is comparable to normal replication, as about 100 kb can be replicated in less than 30 minutes (i.e. a rate of about 3 kb/min) (Malkova et al. 2005).

We still lack information about key steps in BIR. We envision that BIR results in a unidirectional replication fork (Fig. 5A), but we do not yet know if it uses the entire replication machinery that loads at origins of replication, or if all three major DNA polymerases are required. Moreover, we don't know if replication is semi-conservative (Fig. 5A) or if both newly synthesized strands remains associated after branch migration accompanying replication. Genetic and chromatin immunoprecipitation experiments should permit a detailed characterization of the process.

5.3 Analysis of BIR using plasmids and transformation assays

A version of BIR has also been studied on a plasmid that contains inverted repeated homologous sequences, one of which is interrupted by an HO endonuclease cleavage site (Ira and Haber 2002). When the homology on both sides of the cleavage site is substantial (several hundred bp), recombination proceeds predominantly by Rad51-mediated gene conversion; however, when homology is reduced to 70-100 bp on either side, Rad51-mediated gene conversions are severely impaired; it appears that to initiate gene conversion, Rad51 requires about 100 bp of homology. With very short homology there is instead a Rad51-independent repair pathway that appears to involve first BIR and then single-strand annealing, as first suggested by Kang and Symington (Kang and Symington 2000). The ability of a *RAD51*-independent, *RAD52*- and *RAD50*-dependent pathway to initiate BIR

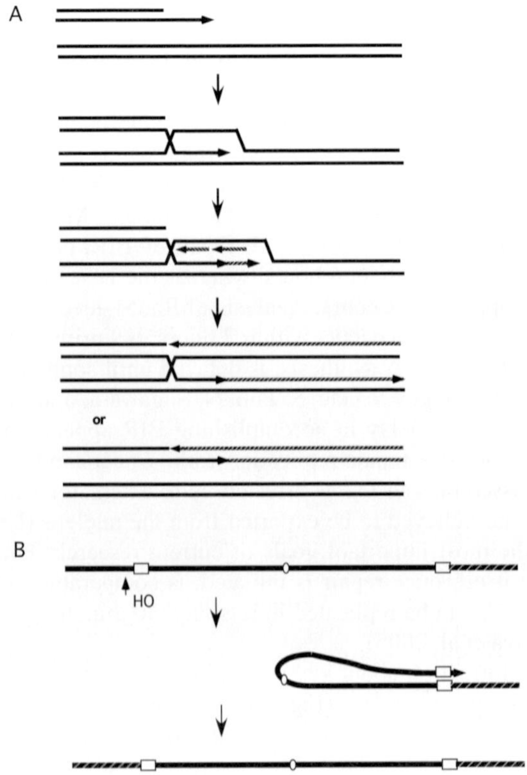

Fig. 5. Break-induced replication (BIR). Strand invasion of a single DSB end can lead to the formation of a repair replication fork that allows both leading and lagging strand synthesis. The products of BIR could either have long semi-conservative regions of replication or – if there is branch migration of the Holliday junction following the replication fork, the repair products could have all the newly synthesized strands annealed to each other (conservative DNA replication).

at short regions of homology is consistent with one of two recombination pathways that can maintain yeast telomeres in the absence of telomerase.

Davis and Symington (2004) have recently used a transformation-based chromosome fragmentation assay similar to that of Morrow et al. (1997) to further analyze BIR. A linearized DNA with telomere sequences on one end, an origin of replication and a centromere is transformed into yeast and the uncapped end is repaired by what appears to be a BIR process. They also demonstrated that there are two different repair mechanisms, one Rad51-dependent and the other Rad51-independent. There are at least two striking differences in the results obtained by this transformation assay and those obtained in analyzing an HO-induced DSB on a chromosome (Malkova et al. 2005). First, in the transformation assay *RAD50* is not required for *RAD51*-independent BIR, whereas it nearly eliminates *RAD51*-independent events for the chromosome break. Moreover, although in the chromosomal assay, deleting *RAD50* has a small negative effect on *RAD51*-dependent BIR, the absence of *RAD50* markedly improved *RAD51*-dependent repair in the transformation assay. Davis and Symington (2004) suggest that eliminating Rad50 will slow down the rate of 5' to 3' degradation of the DNA end, allowing more time for repair before the transforming DNA is degraded; it is possible that in the chromosome assay, where DNA has its normal chromatin structure prior to creating a DSB, such protection is not a rate-limiting step. A second possible difference between the chromosomal Rad51-independent events and those seen by transformation is the apparent absence of a facilitating (FBI) sequence, as unique sequences close to the end of the transformed DNA recombine with their chromosomal template. Again, the events initiated by transformed, "naked" DNA may be quite different from repair of a broken end of a chromatin-assembled chromosome.

6 Nonhomologous end-joining (NHEJ)

Using HO cleavage of a single site that lacks any donor sequences to allow homologous recombination it is possible to examine mechanisms of nonhomologous end-joining for a DSB made in a normal chromatin context. If HO is expressed and then turned off, most ends are re-joined by perfectly rejoining the 4-bp 3' ended overhangs, restoring the HO cleavage site. These ligation events require the yKu70 and yKu80 proteins and the specialized DNA ligase IV, Dnl4 and its associated Xrcc4-homolog, Lifl that are needed in mammalian V(D)J and other NHEJ events (Boulton and Jackson 1996; Milne et al. 1996; Schär et al. 1997; Teo and Jackson 2000; Wilson et al. 1997). Mammals also require DNA PKcs, a protein kinase that is absent in budding yeast. Most NHEJ in budding yeast also requires the Mre11-Rad50-Xrs2 complex (Moore and Haber 1996; Tsukamoto et al. 1996). For a recent review see Daley et al. (2005). In vertebrate cells, whether Mre11-Rad50-Nbs1 is required is uncertain, as null mutants are inviable; however, studies of cells with reduced MRN abundance have not shown a role for this complex (Yamaguchi-Iwai et al. 1999). In fission yeast, MRN is not needed in logarithmically-growing cells (Wilson et al. 1999), but is required for NHEJ in G1-arrested cells (M.G. Ferreira and J.P. Cooper, unpublished results).

In budding yeast, NHEJ is regulated by cell type. In diploid or haploid cells expressing *MAT***a** and *MAT*α, NHEJ is turned off, perhaps to prevent competition between NHEJ and homologous recombination during meiosis where there are about 100 DSBs. Both *MAT***a** or *MAT*α and diploids homozygous for either mating-type (which cannot enter meiosis) are NHEJ-proficient. NHEJ is shut off by the transcriptional repression of Nej1, a protein that interacts with Lifl (Frank-Vaillant and Marcand 2001; Kegel et al. 2001; Ooi and Boeke 2001; Valencia et al. 2001; Wilson 2002). There is disagreement whether Nej1 is involved in the nuclear localization of Lifl and Dnl4 (Frank-Vaillant and Marcand 2001; Kegel et al. 2001; Ooi and Boeke 2001; Valencia et al. 2001; Wilson 2002) but so far Nej1 has not been localized to a DSB end by ChIP, whereas MRX, Lifl, and Ku proteins have been shown to associate with an HO-induced DSB (Lisby et al. 2004; Martin et al. 1999; Shim et al. 2005; Shroff et al. 2004; Teo and Jackson 2000). DNA sequence comparisons show that Nej1 protein diverges rapidly among budding yeast species and it has not been possible to identify a Nej1 homolog in metazoans.

NHEJ has been studied either by transforming in linearized plasmids or by examining repair of HO-cleaved chromosomes when there are no donor sequences that could repair the break by homologous recombination. By and large the results have been quite similar, although the magnitude of reduction of NHEJ by mutants such as *yku70Δ* is nearly 200-fold for the chromosome break compared to about 20-fold for most transformation experiments (Boulton and Jackson 1996; Moore and Haber 1996; Tsukamoto et al. 1996). HO cleavage yields 4-bp 3'-ended overhanging ends. Breaks created by transient induction of HO are most readily rejoined to restore the original cleavage site. In logarithmically growing cells, about 15% of the ends are rejoined in this fashion; apparently resection proceeds quickly enough so that end-joining of long 3'-ended tails compromises the efficiency of end-joining. Indeed model substrates introduced by transformation with increasingly long tails demonstrate that end-joining efficiency diminishes with increased ssDNA lengths (Daley and Wilson 2005). If the Ku proteins preferentially hold DNA ends at the junction between ssDNA and dsDNA, the longer tails might be outside the range where ligation can be accomplished.

If HO expression is continuous, then simple re-ligation of the cut ends is futile, as they will again be cut. Thus survivors have to join ends in some mutagenic fashion that prevents HO cleavage. Unlike mammalian cells where such imprecise end-joinings of I-SceI cut ends appears to occur frequently (Guirouilh-Barbat et al. 2004), in budding yeast, imprecise joining of HO-cut ends arises in only about 2 $x10^{-3}$ cells (Moore and Haber 1996). A very important experiment compared the kinds of end-joinings created by chromosome breaks in a dicentric chromosome with those created by inducing enzymatic cleavage of a chromosome by HO endonuclease. In both cases, most joints appear to arise where 1 or 2 bp can be formed at the junction between the two broken ends, but sites where there are five or more bp homology are not preferentially used (Kramer et al. 1994).

There are in fact several distinct end-joining processes that differ in their genetic and cell cycle requirements. The most common events are 2- and 3-bp insertions created by misaligned joining of the terminal or penultimate T on one strand to the terminal A of the opposite overhang, creating +CA and +ACA insertions (Kramer et al. 1994) (Fig. 6). These events require the participation of DNA polymerase 4 (Wilson and Lieber 1999). Another set of events are –ACA deletions resulting from an A-T base pairing and the removal of 3 nt on each tail. How these cleavage/resections are accomplished is not known. There are also many other deletions, usually formed with one or more nucleotides that can base-pair at the junction. Removal of these tails does not apparently require Rad1-Rad10 or Msh2-Msh3. Interestingly, deletions are just as able to form in G1-arrested cells as in cycling cells, whereas the insertions are nearly absent in G1-arrested cells. Deletions also can arise in the absence of Mre11-Rad50 (Moore and Haber 1996).

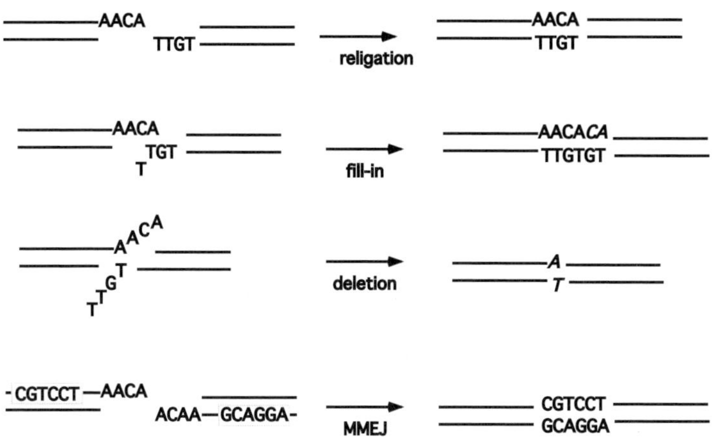

Fig. 6. Nonhomologous end-joining (NHEJ) leads to different types of alterations of the HO cleavage site. In addition to the efficient re-ligation of the 4-bp 3' overhangs of the HO cut, there are both small fill-in insertions and small deletions created by misalignment of the unresected DSB ends, as well as larger deletions of resected ends (not shown). All of these events require the Ku proteins as well as ligase 4 and its associated proteins. In addition, when two HO cuts are made in opposite orientation so that there is no base-pairing of the ends, repair occurs by Ku-independent microhomology-mediated end-joining (MMEJ).

A fourth, Ku-independent NHEJ mechanism was found when the ends of the DSBs lack any base-pairing potential. This was accomplished by creating two HO cut sites in opposite orientation and examining end-joining events that had lost the intervening segment, marked by *URA3*. Surprisingly, these events, which were Ku-independent arose at frequencies seen for Ku-dependent NHEJ when the ends had base-pairing possibilities (Ma et al. 2003). These Ku-independent events are also dependent on Rad1-Rad10 and MRX proteins.

If one creates DSBs on two different chromosomes it is also possible to recover reciprocal translocations. Yu and Gabriel (2004) made the very important finding that even with a single HO cut there are some reciprocal translocations. It appears that in these cases, HO endonuclease was able to cleave a degenerate recognition sequence on another chromosome at a low frequency. This frequency must be low, because cells lacking the HO cut site at *MAT* are viable in *yku70*Δ or *dnl4*Δ (and *rad52*Δ) backgrounds when HO is induced.

One other unusual form of NHEJ has been found: HO and I-SceI DSBs are sometimes repaired by the capture of exogenous DNA fragments, including segments derived from cDNA intermediates of the Ty1 retrotransposon and segments of mitochondrial DNA that apparently enter the nucleus (Moore and Haber 1996; Ricchetti et al. 1999; Teng et al. 1996). This process could explain how segments of mDNA and cDNAs such as Alu or pseudogenes can be inserted into the mammalian genome in the absence of an integrase.

7 Future prospects

The strategy that has proven so effective in studying DSB repair events in yeast is now being applied to metazoans as well as other fungi. HO expression in *Schizosaccharomyces pombe* has been used both to induce recombination and to study DNA damage responses (Prudden et al. 2003). The I-SceI meganuclease has been used to induce recombination in flies (Rong et al. 2002), and I-SceI has become a common reagent in vertebrate cells (Elliott et al. 1998; Johnson and Jasin 2001; Nickoloff and Brenneman 2004; Rodrigue et al. 2006). In addition, the advent of efficient zinc-finger nucleases (ZFN) offers the possibility of creating customized DSBs at many different genomic locations (Bibikova et al. 2001; Porteus 2006; Wright et al. 2005). At present the greatest limitation of these approaches is the absence of an efficient system for rapid induction of the nuclease in plants, mammals or even in fission yeast. For the moment, it is only possible in budding yeast to follow in detail the sequence of molecular events leading to DSB repair. As these other systems become more facile, it will be fascinating to see the degree to which the principles deduced from studying DSB repair in budding yeast will prove to be universal.

Acknowledgements

Work from my lab has been the result of a wonderful collaboration with many students and postdoctoral fellows. Our research has been supported primarily by grants GM20056 and GM61766 from the NIH and by additional support from the NSF and Department of Energy.

References

Aylon Y, Liefshitz B, Kupiec M (2004) The CDK regulates repair of double-strand breaks by homologous recombination during the cell cycle. EMBO J 23:4868-4875

Bai Y, Symington LS (1996) A *RAD52* homolog is required for *RAD51*-independent mitotic recombination in *Saccharomyces cerevisiae*. Genes Dev 10:2025-2037

Bibikova M, Carroll D, Segal DJ, Trautman JK, Smith J, Kim YG, Chandrasegaran S (2001) Stimulation of homologous recombination through targeted cleavage by chimeric nucleases. Mol Cell Biol 21:289-297

Borts RH, Lichten M, Hearn M, Davidow LS, Haber JE (1984) Physical monitoring of meiotic recombination in *Saccharomyces cerevisiae*. Cold Spring Harb Symp Quant Biol 49:67-76

Bosco G, Haber JE (1998) Chromosome break-induced DNA replication leads to nonreciprocal translocations and telomere capture. Genetics 150:1037-1047

Boulton SJ, Jackson SP (1996) *Saccharomyces cerevisiae* Ku70 potentiates illegitimate DNA double- strand break repair and serves as a barrier to error-prone DNA repair pathways. EMBO J 15:5093-5103

Bressan DA, Vazquez J, Haber JE (2004) Mating type-dependent constraints on the mobility of the left arm of yeast chromosome III. J Cell Biol 164:361-371

Callebaut I, Malivert L, Fischer A, Mornon JP, Revy P, de Villartay JP (2006) Cernunnos interacts with the XRCC4 x DNA-ligase IV complex and is homologous to the yeast nonhomologous end-joining factor Nej1. J Biol Chem 281:13857-13860

Chai B, Huang J, Cairns BR, Laurent BC (2005) Distinct roles for the RSC and Swi/Snf ATP-dependent chromatin remodelers in DNA double-strand break repair. Genes Dev 19:1656-1661

Chen Q, IJpma A, Greider CW (2001) Two survivor pathways that allow growth in the absence of telomerase are generated by distinct telomere recombination events. Mol Cell Biol 21:1819-1827

Chen W, Jinks-Robertson S (1998) Mismatch repair proteins regulate heteroduplex formation during mitotic recombination in yeast. Mol Cell Biol 18:6525-6537

Cho JW, Khalsa GJ, Nickoloff JA (1998) Gene-conversion tract directionality is influenced by the chromosome environment. Curr Genet 34:269-279

Cohen H, Sinclair DA (2001) Recombination-mediated lengthening of terminal telomeric repeats requires the Sgs1 DNA helicase. Proc Natl Acad Sci USA 98:3174-3179

Colaiácovo MP, Pâques F, Haber JE (1999) Removal of one nonhomologous DNA end during gene conversion by a *RAD1*- and *MSH2*-independent pathway. Genetics 151:1409-1423

Cromie GA, Leach DR (2000) Control of crossing over. Mol Cell 6:815-826

Daley JM, Palmbos PL, Wu D, Wilson TE (2005) Nonhomologous end joining in yeast. Annu Rev Genet 39:431-451

Daley JM, Wilson TE (2005) Rejoining of DNA double-strand breaks as a function of overhang length. Mol Cell Biol 25:896-906

Davis AP, Symington LS (2004) RAD51-dependent break-induced replication in yeast. Mol Cell Biol 24:2344-2351

de los Santos T, Hunter N, Lee C, Larkin B, Loidl J, Hollingsworth NM (2003) The Mus81/Mms4 endonuclease acts independently of double-Holliday junction resolution to promote a distinct subset of crossovers during meiosis in budding yeast. Genetics 164:81-94

Dunn B, Szauter P, Pardue ML, Szostak JW (1984) Transfer of yeast telomeres to linear plasmids by recombination. Cell 39:191-201

Eggler AL, Inman RB, Cox MM (2002) The Rad51-dependent pairing of long DNA substrates is stabilized by replication protein A. J Biol Chem 277:39280-39288

Elliott B, Richardson C, Winderbaum J, Nickoloff JA, Jasin M (1998) Gene conversion tracts from double-strand break repair in mammalian cells. Mol Cell Biol 18:93-101

Esposito MS (1978) Evidence that spontaneous mitotic recombination occurs at the two-strand stage. Proc Natl Acad Sci USA 75:4436-4440

Esposito MS, Ramirez RM, Bruschi CV (1994) Nonrandomly-associated forward mutation and mitotic recombination yield yeast diploids homozygous for recessive mutations. Curr Genet 26:302-307

Fishman-Lobell J, Haber JE (1992) Removal of nonhomologous DNA ends in double-strand break recombination: the role of the yeast ultraviolet repair gene *RAD1*. Science 258:480-484

Fishman-Lobell J, Rudin N, Haber JE (1992) Two alternative pathways of double-strand break repair that are kinetically separable and independently modulated. Mol Cell Biol 12:1292-1303

Frank-Vaillant M, Marcand S (2001) NHEJ regulation by mating type is exercised through a novel protein, Lif2p, essential to the ligase IV pathway. Genes Dev 15:3005-3012

Frank-Vaillant M, Marcand S (2002) Transient stability of DNA ends allows nonhomologous end joining to precede homologous recombination. Mol Cell 10:1189-1199

Gloor GB, Nassif NA, Johnson-Schlitz DM, Preston CR, Engels WR (1991) Targeted gene replacement in *Drosophila* via P element-induced gap repair. Science 253:1110-1117

Golub EI, Gupta RC, Haaf T, Wold MS, Radding CM (1998) Interaction of human rad51 recombination protein with single-stranded DNA binding protein, RPA. Nucleic Acids Res 26:5388-5393

Gould SJ (1997) The exaptive excellence of spandrels as a term and prototype. Proc Natl Acad Sci USA 94:10750-10755

Grandin N, Charbonneau M (2003) Mitotic cyclins regulate telomeric recombination in telomerase-deficient yeast cells. Mol Cell Biol 23:9162-9177

Griffith JD, Comeau L, Rosenfield S, Stansel RM, Bianchi A, Moss H, de Lange T (1999) Mammalian telomeres end in a large duplex loop [see comments]. Cell 97:503-514

Guirouilh-Barbat J, Huck S, Bertrand P, Pirzio L, Desmaze C, Sabatier L, Lopez BS (2004) Impact of the KU80 pathway on NHEJ-induced genome rearrangements in mammalian cells. Mol Cell 14:611-623

Haber JE, Ray BL, Kolb JM, White CI (1993) Rapid kinetics of mismatch repair of heteroduplex DNA that is formed during recombination in yeast. Proc Natl Acad Sci USA 90:3363-3367

Haber JE, Leung WY (1996) Lack of chromosome territoriality in yeast: promiscuous rejoining of broken chromosome ends. Proc Natl Acad Sci USA 93:13949-13954

Holmes A, Haber JE (1999) Double-strand break repair in yeast requires both leading and lagging strand DNA polymerases. Cell 96:415-424

Horowitz H, Haber JE (1985) Identification of autonomously replicating circular subtelomeric Y' elements in *Saccharomyces cerevisiae.* Mol Cell Biol 5:2369-2380

Huang P, Pryde FE, Lester D, Maddison RL, Borts RH, Hickson ID, Louis EJ (2001) SGS1 is required for telomere elongation in the absence of telomerase. Curr Biol 11:125-129

Inbar O, Kupiec M (1999) Homology search and choice of homologous partner during mitotic recombination. Mol Cell Biol 19:4134-4142

Inbar O, Liefshitz B, Bitan G, Kupiec M (2000) The relationship between homology length and crossing-over during the repair of a broken chromosome. J Biol Chem 275:30833-30838

Ira G, Haber JE (2002) Characterization of RAD51-independent break-induced replication that acts preferentially with short homologous sequences. Mol Cell Biol 22:6384-6392

Ira G, Malkova A, Liberi G, Foiani M, Haber JE (2003) Srs2 and Sgs1-Top3 suppress crossovers during double-strand break repair in yeast. Cell 115:401-411

Ira G, Pellicioli A, Balijja A, Wang X, Fiorani S, Carotenuto W, Liberi G, Bressan D, Wan L, Hollingsworth NM, Haber JE, Foiani M (2004) DNA end resection, homologous recombination and DNA damage checkpoint activation require CDK1. Nature 431:1011-1017

Ivanov EL, Sugawara N, White CI, Fabre F, Haber JE (1994) Mutations in *XRS2* and *RAD50* delay but do not prevent mating-type switching in *Saccharomyces cerevisiae.* Mol Cell Biol 14:3414-3425

Ivanov EL, Haber JE (1995) *RAD1* and *RAD10*, but not other excision repair genes, are required for double-strand break-induced recombination in *Saccharomyces cerevisiae.* Mol Cell Biol 15:2245-2251

Ivanov EL, Sugawara N, Fishman LJ, Haber JE (1996) Genetic requirements for the single-strand annealing pathway of double- strand break repair in *Saccharomyces cerevisiae*. Genetics 142:693-704

Jensen RE, Herskowitz I (1984) Directionality and regulation of cassette substitution in yeast. Cold Spring Harb Symp Quant Biol 49:97-104

Johnson FB, Marciniak RA, McVey M, Stewart SA, Hahn WC, Guarente L (2001) The *Saccharomyces cerevisiae* WRN homolog Sgs1p participates in telomere maintenance in cells lacking telomerase. EMBO J 20:905-913

Johnson RD, Jasin M (2001) Double-strand-break-induced homologous recombination in mammalian cells. Biochem Soc Trans 29:196-201

Kang LE, Symington LS (2000) Aberrant double-strand break repair in rad51 mutants of *Saccharomyces cerevisiae*. Mol Cell Biol 20:9162-9172

Kantake N, Sugiyama T, Kolodner RD, Kowalczykowski SC (2003) The recombination-deficient mutant RPA (rfa1-t11) is displaced slowly from single-stranded DNA by Rad51 protein. J Biol Chem 278:23410-23417

Kegel A, Sjostrand JO, Astrom SU (2001) Nej1p, a cell type-specific regulator of nonhomologous end joining in yeast. Curr Biol 11:1611-1617

Keogh MC, Kim JA, Downey M, Fillingham J, Chowdhury D, Harrison JC, Onishi M, Datta N, Galicia S, Emili A, Lieberman J, Shen X, Buratowski S, Haber JE, Durocher D, Greenblatt JF, Krogan NJ (2005) A phosphatase complex that dephosphorylates gammaH2AX regulates DNA damage checkpoint recovery. Nature 439:497-501

Kramer KM, Brock JA, Bloom K, Moore JK, Haber JE (1994) Two different types of double-strand breaks in *Saccharomyces cerevisiae* are repaired by similar *RAD52*-independent, nonhomologous recombination events. Mol Cell Biol 14:1293-1301

Krejci L, Van Komen S, Li Y, Villemain J, Reddy MS, Klein H, Ellenberger T, Sung P (2003) DNA helicase Srs2 disrupts the Rad51 presynaptic filament. Nature 423:305-309

Krogh BO, Symington LS (2004) Recombination proteins in yeast. Annu Rev Genet 38:233-71

Kupiec M, Petes TD (1988) Allelic and ectopic recombination between Ty elements in yeast. Genetics 119:549-559

Labib K, Kearsey SE, Diffley JF (2001) MCM2-7 proteins are essential components of pre-replicative complexes that accumulate cooperatively in the nucleus during G1-phase and are required to establish, but not maintain, the S-phase checkpoint. Mol Biol Cell 12:3658-3667

Langston LD, Symington LS (2005) Opposing roles for DNA structure-specific proteins Rad1, Msh2, Msh3, and Sgs1 in yeast gene targeting. EMBO J 24:2214-2223

Le S, Moore JK, Haber JE, Greider C (1999) *RAD50* and *RAD51* define two different pathways that collaborate to maintain telomeres in the absence of telomerase. *Genetics* 152:143-152

Lee SE, Bressan DA, Petrini JHJ, Haber JE (2002) Complementation between N-terminal *Saccharomyces cerevisiae mre11* alleles in DNA repair and telomere length maintenance. DNA Repair 1:27-40

Lemoine FJ, Degtyareva NP, Lobachev K, Petes TD (2005) Chromosomal translocations in yeast induced by low levels of DNA polymerase a model for chromosome fragile sites. Cell 120:587-598

Lisby M, Mortensen UH, Rothstein R (2003) Colocalization of multiple DNA double-strand breaks at a single Rad52 repair centre. Nat Cell Biol 5:572-577

Lisby M, Rothstein R (2004) DNA damage checkpoint and repair centers. Curr Opin Cell Biol 16:328-334

Lundblad V, Blackburn EH (1993) An alternative pathway for yeast telomere maintenance rescues *est1*- senescence. Cell 73:347-360

Ma JL, Kim EM, Haber JE, Lee SE (2003) Yeast Mre11 and Rad1 proteins define a Ku-independent mechanism to repair double-strand breaks lacking overlapping end sequences. Mol Cell Biol 23:8820-8828

Malkova A, Ivanov EL, Haber JE (1996) Double-strand break repair in the absence of *RAD51* in yeast: a possible role for break-induced DNA replication. Proc Natl Acad Sci USA 93:7131-7136

Malkova A, Ross L, Dawson D, Hoekstra MF, Haber JE (1996) Meiotic recombination initiated by a double-strand break in *rad50Δ* yeast cells otherwise unable to initiate meiotic recombination. Genetics 143:741-754

Malkova A, Signon L, Schaefer CB, Naylor M, Theis JF, Newlon CS, Haber JE (2001) *RAD51*-independent break-induced replication to repair a broken chromosome depends on a distant enhancer site. Genes Dev 15:1055-1160

Malkova A, Naylor M, Yamaguchi M, Ira G, Haber JE (2005) *RAD51*-dependent break-induced replication differs in kinetics and checkpoint responses from *RAD51*-mediated gene conversion. Mol Cell Biol 25:933-944

Martin S, Laroche T, Suka N, Grunstein M, Gasser SM (1999) Relocalization of telomeric Ku and SIR proteins in response to DBA strand breaks in yeast. Cell 97:621-633

Mazina OM, Mazin AV, Nakagawa T, Kolodner RD, Kowalczykowski SC (2004) *Saccharomyces cerevisiae* Mer3 helicase stimulates 3'-5' heteroduplex extension by Rad51; implications for crossover control in meiotic recombination. Cell 117:47-56

McDonald JP, Rothstein R (1994) Unrepaired heteroduplex DNA in *Saccharomyces cerevisiae* is decreased in *RAD1 RAD52*-independent recombination. Genetics 137:393-405

McEachern MJ, Haber JE (2006) Break-Induced replication and recombinational telomere elongation in yeast. Annu Rev Biochem 75:111-135

McGill C, Shafer B, Strathern J (1989) Coconversion of flanking sequences with homothallic switching. Cell 57:459-467

Milne GT, Jin S, Shannon KB, Weaver DT (1996) Mutations in two Ku homologs define a DNA end-joining repair pathway in *Saccharomyces cerevisiae*. Mol Cell Biol 16:4189-4198

Miyazaki T, Bressan DA, Shinohara M, Haber JE, Shinohara A (2004) *In vivo* assembly and disassembly of Rad51 and Rad52 complexes during double-strand break repair. EMBO J 23:939-949

Moore JK, Haber JE (1996) Capture of retrotransposon DNA at the sites of chromosomal double- strand breaks. Nature 383:644-646

Moore JK, Haber JE (1996) Cell cycle and genetic requirements of two pathways of nonhomologous end-joining repair of double-strand breaks in *Saccharomyces cerevisiae*. Mol Cell Biol 16:2164-2173

Moreau S, Ferguson JR, Symington LS (1999) The nuclease activity of Mre11 is required for meiosis but not for mating type switching, end joining or telomere maintenance. Mol Cell Biol 19:556-566

Morrow DM, Connelly C, Hieter P (1997) "Break copy" duplication: a model for chromosome fragment formation in *Saccharomyces cerevisiae*. Genetics 147:371-382

Natarajan S, McEachern MJ (2002) Recombinational telomere elongation promoted by DNA circles. Mol Cell Biol 22:4512-4521

Nickoloff JA, Sweetser DB, Khalsa GJ, Wheeler SL (1999) Multiple heterologies increase the length of double-strand break-induced allelic gene conversion tracts. Genetics 153:665-679

Nickoloff JA, Brenneman MA (2004) Analysis of recombinational repair of DNA double-strand breaks in mammalian cells with I-SceI nuclease. Methods Mol Biol 262:35-52

Ooi SL, Boeke JD (2001) A DNA microarray-based genetic screen for nonhomologous end-joining mutants in *Saccharomyces cerevisiae*. Science 294:2552-2556

Osman F, Dixon J, Doe CL, Whitby MC (2003) Generating crossovers by resolution of nicked Holliday junctions: a role for Mus81-Eme1 in meiosis. Mol Cell 12:761-774

Ozenberger BA, Roeder GS (1991) A unique pathway of double-strand break repair operates in tandemly repeated genes. Mol Cell Biol 11:1222-1231

Pâques F, Leung WY, Haber JE (1998) Expansions and contractions in a tandem repeat induced by double-strand break repair. Mol Cell Biol 18:2045-2054

Pâques F, Haber JE (1999) Multiple pathways of recombination induced by double-strand breaks in *Saccharomyces cerevisiae*. Microbiol Mol Biol Rev 63:349-404

Pâques F, Richard G-F, Haber JE (2001) Expansions and contractions in 36-bp minisatellite by gene conversion in yeast. Genetics 158:155–166

Plessis A, Perrin A, Haber JE, Dujon B (1992) Site-specific recombination determined by I-SceI, a mitochondrial group I intron-encoded endonuclease expressed in the yeast nucleus. Genetics 130:451-460

Porteus MH (2006) Mammalian gene targeting with designed zinc finger nucleases. Mol Ther 13:438-446

Prado F, Aguilera A (2003) Control of cross-over by single-strand DNA resection. Trends Genet 19:428-431

Prakash R, Krejci L, Van Komen S, Anke Schurer K, Kramer W, Sung P (2005) *Saccharomyces cerevisiae* MPH1 gene, required for homologous recombination-mediated mutation avoidance, encodes a 3' to 5' DNA helicase. J Biol Chem 280:7854-7860

Prudden J, Evans JS, Hussey SP, Deans B, O'Neill P, Thacker J, Humphrey T (2003) Pathway utilization in response to a site-specific DNA double-strand break in fission yeast. EMBO J 22:1419-1430

Putnam CD, Pennaneach V, Kolodner RD (2004) Chromosome healing through terminal deletions generated by de novo telomere additions in *Saccharomyces cerevisiae*. Proc Natl Acad Sci USA 101:13262-13267

Ray BL, White CI, Haber JE (1991) Heteroduplex formation and mismatch repair of the "stuck" mutation during mating-type switching in *Saccharomyces cerevisiae*. Mol Cell Biol 11:5372-5380

Ricchetti M, Fairhead C, Dujon B (1999) Mitochondrial DNA repairs double-strand breaks in yeast chromosomes. Nature 402:96-100

Richard GF, Dujon B, Haber JE (1999) Double-strand break repair can lead to high frequencies of deletions within short CAG/CTG trinucleotide repeats. Mol Gen Genet 261:871-882

Rodrigue A, Lafrance M, Gauthier MC, McDonald D, Hendzel M, West SC, Jasin M, Masson JY (2006) Interplay between human DNA repair proteins at a unique double-strand break *in vivo*. EMBO J 25:222-231

Rong YS, Titen SW, Xie HB, Golic MM, Bastiani M, Bandyopadhyay P, Olivera BM, Brodsky M, Rubin GM, Golic KG (2002) Targeted mutagenesis by homologous recombination in *D. melanogaster*. Genes Dev 16:1568-1581

Schär P, Herrmann G, Daly G, Lindahl T (1997) A newly identified DNA ligase of *Saccharomyces cerevisiae* involved in *RAD52*-independent repair of DNA double-strand breaks. Genes Dev 11:1912-1924

Scheller J, Schurer A, Rudolph C, Hettwer S, Kramer W (2000) MPH1:a yeast gene encoding a DEAH protein, plays a role in protection of the genome from spontaneous and chemically induced damage. Genetics 155:1069-1081

Schurer KA, Rudolph C, Ulrich HD, Kramer W (2004) Yeast MPH1 gene functions in an error-free DNA damage bypass pathway that requires genes from homologous recombination, but not from postreplicative repair. Genetics 166:1673-1686

Schwacha A, Kleckner N (1995) Identification of double Holliday junctions as intermediates in meiotic recombination. Cell 83:783-791

Shim EY, Ma JL, Oum JH, Yanez Y, Lee SE (2005) The yeast chromatin remodeler RSC complex facilitates end joining repair of DNA double-strand breaks. Mol Cell Biol 25:3934-3944

Shroff R, Arbel-Eden A, Pilch D, Ira G, Bonner WM, Petrini JH, Haber JE, Lichten M (2004) Distribution and dynamics of chromatin modification induced by a defined DNA double-strand break. Curr Biol 14:1703-1711

Signon L, Malkova A, Naylor M, Haber JE (2001) Genetic requirements for *RAD51*- and *RAD54*-independent break-induced replication repair of a chromosomal double-strand break. Mol Cell Biol 21:2048-2056

Smith GR, Boddy MN, Shanahan P, Russell P (2003) Fission yeast Mus81.Eme1 Holliday junction resolvase is required for meiotic crossing over but not for gene conversion. Genetics 165:2289-2293

Sonoda E, Sasaki MS, Buerstedde JM, Bezzubova O, Shinohara A, Ogawa H, Takata M, Yamaguchi-Iwai Y, Takeda S (1998) *Rad51*-deficient vertebrate cells accumulate chromosomal breaks prior to cell death. EMBO J 17:598-608

Studamire B, Price G, Sugawara N, Haber JE, Alani E (1999) Separation-of-function mutations in *Saccharomyces cerevisiae* MSH2 that confer mismatch repair defects but do not affect nonhomologous-tail removal during recombination. Mol Cell Biol 19:7558-7567

Sugawara N, Haber JE, unpublished results

Sugawara N, Paques F, Colaiacovo M, Haber JE (1997) Role of *Saccharomyces cerevisiae* Msh2 and Msh3 repair proteins in double-strand break-induced recombination. Proc Natl Acad Sci USA 94:9214-9219

Sugawara N, Ira G, Haber JE (2000) DNA length dependence of the single-strand annealing pathway and the role of *Saccharomyces cerevisiae RAD59* in double-strand break repair. Mol Cell Biol 20:5300-5309

Sugawara N, Wang X, Haber JE (2003) *In vivo* roles of Rad52, Rad54, and Rad55 proteins in Rad51-mediated recombination. Mol Cell 12:209-219

Sugawara N, Goldfarb T, Studamire B, Alani E, Haber JE (2004) Heteroduplex rejection during single-strand annealing requires Sgs1 helicase and mismatch repair proteins Msh2 and Msh6 but not Pms1. Proc Natl Acad Sci USA 101:9315-9320

Sugiyama T, Kowalczykowski SC (2002) Rad52 protein associates with RPA-ssDNA to accelerate Rad51-mediated displacement of RPA and presynaptic complex formation. J Biol Chem 277:31663-31672

Sung P (1997a) Function of yeast Rad52 protein as a mediator between replication protein A and the Rad51 recombinase. J Biol Chem 272:28194-28197

Sung P (1997b) Yeast Rad55 and Rad57 proteins form a heterodimer that functions with replication protein A to promote DNA strand exchange by Rad51 recombinase. Genes Dev 11:1111-1121

Szostak JW, Orr WT, Rothstein RJ, Stahl FW (1983) The double-strand-break repair model for recombination. Cell 33:25-35

Teng S, Chang J, McCowan B, Zakian VA (2000) Telomerase-independent lengthening of yeast telomeres occurs by an abrupt Rad50p-dependent, Rif-inhibited recombinational process. Mol Cell 6:947-952

Teng SC, Kim B, Gabriel A (1996) Retrotransposon reverse-transcriptase-mediated repair of chromosomal breaks. Nature 383:641-644

Teng SC, Zakian VA (1999) Telomere-telomere recombination is an efficient bypass pathway for telomere maintenance in *Saccharomyces cerevisiae*. Mol Cell Biol 19:8083-8093

Teo SH, Jackson SP (2000) Lif1p targets the DNA ligase Lig4p to sites of DNA double-strand breaks. Curr Biol 10:165-168

Tsubouchi H, Ogawa H (1998) A novel *mre11* mutation impairs processing of double-strand breaks of DNA during both mitosis and meiosis. Mol Cell Biol 18:260-268

Tsukamoto Y, Kato J, Ikeda H (1996) Effects of mutations of *RAD50*, *RAD51*, *RAD52*, and related genes on illegitimate recombination in *Saccharomyces cerevisiae*. Genetics 142:383-391

Tsukamoto Y, Taggart AK, Zakian VA (2001) The role of the Mre11-Rad50-Xrs2 complex in telomerase- mediated lengthening of *Saccharomyces cerevisiae* telomeres. Curr Biol 11:1328-1335

Ubersax JA, Woodbury EL, Quang PN, Paraz M, Blethrow JD, Shah K, Shokat KM, Morgan DO (2003) Targets of the cyclin-dependent kinase Cdk1. Nature 425:859-964

Umezu K, Sugawara N, Chen C, Haber JE, Kolodner RD (1998) Genetic analysis of yeast *RPA1* reveals its multiple functions in DNA metabolism. Genetics 148:989-1005

Valencia M, Bentele M, Vaze MB, Herrmann G, Kraus E, Lee SE, Schar P, Haber JE (2001) NEJ1 controls non-homologous end joining in *Saccharomyces cerevisiae*. Nature 414:666-669

van Attikum H, Fritsch O, Hohn B, Gasser SM (2004) Recruitment of the INO80 complex by H2A phosphorylation links ATP-dependent chromatin remodeling with DNA double-strand break repair. Cell 119:777-788

Vaze M, Pellicioli A, Lee S, Ira G, Liberi G, Arbel-Eden A, Foiani M, Haber JE (2002) Recovery from checkpoint-mediated arrest after repair of a double- strand break requires srs2 helicase. Mol Cell 10:373-385

Veaute X, Jeusset J, Soustelle C, Kowalczykowski SC, Le Cam E, Fabre F (2003) The Srs2 helicase prevents recombination by disrupting Rad51 nucleoprotein filaments. Nature 423:309-312

Voelkel-Meiman K, Roeder GS (1990) Gene conversion tracts stimulated by *HOT1*-promoted transcription are long and continuous. Genetics 126:851-867

Wang X, Haber JE (2004) Role of *Saccharomyces* single-stranded DNA-binding protein RPA in the strand invasion step of double-strand break repair. PLoS Biol 2:104-111

Wang X, Ira G, Tercero JA, Holmes AM, Diffley JF, Haber JE (2004) Role of DNA replication proteins in double-strand break-induced recombination in *Saccharomyces cerevisiae*. Mol Cell Biol 24:6891-6899

Watt PM, Hickson ID, Borts RH, Louis EJ (1996) *SGS1*, a homologue of the Bloom's and Werner's syndrome genes, is required for maintenance of genome stability in *Saccharomyces cerevisiae*. Genetics 144:935-945

Weiss K, Simpson RT (1998) High-resolution structural analysis of chromatin at specific loci: *Saccharomyces cerevisiae* silent mating type locus HMLα. Mol Cell Biol 18:5392-5403

White CI, Haber JE (1990) Intermediates of recombination during mating type switching in *Saccharomyces cerevisiae*. EMBO J 9:663-673

Wilson S, Warr N, Taylor DL, Watts FZ (1999) The role of *Schizosaccharomyces pombe* Rad32, the Mre11 homologue, and other DNA damage response proteins in non-homologous end joining and telomere length maintenance. Nucleic Acids Res 27:2655-2661

Wilson TE, Grawunder U, Lieber MR (1997) Yeast DNA ligase IV mediates non-homologous DNA end joining. Nature 388:495-498

Wilson TE, Lieber MR (1999) Efficient processing of DNA ends during yeast nonhomologous end joining. Evidence for a DNA polymerase beta (Pol4)-dependent pathway. J Biol Chem 274:23599-23609

Wilson TE (2002) A genomics-based screen for yeast mutants with an altered recombination/end-joining repair ratio. Genetics 162:677-688

Wolner B, van Komen S, Sung P, Peterson CL (2003) Recruitment of the recombinational repair machinery to a DNA double-strand break in yeast. Mol Cell 12:221-232

Wright DA, Townsend JA, Winfrey RJ Jr, Irwin PA, Rajagopal J, Lonosky PM, Hall BD, Jondle MD, Voytas DF (2005) High-frequency homologous recombination in plants mediated by zinc-finger nucleases. Plant J 44:693-705

Wu L, Hickson ID (2003) The Bloom's syndrome helicase suppresses crossing over during homologous recombination. Nature 426:870-874

Wu X, Haber JE (1996) A 700 bp cis-acting region controls mating-type dependent recombination along the entire left arm of yeast chromosome III. Cell 87:277-285

Wu X, Wu C, Haber JE (1997) Rules of donor preference in *Saccharomyces* mating-type gene switching revealed by a competition assay involving two types of recombination. Genetics 147:399-407

Yamaguchi-Iwai Y, Sonoda E, Sasaki MS, Morrison C, Haraguchi T, Hiraoka Y, Yamashita YM, Yagi T, Takata M, Price C, Kakazu N, Takeda S (1999) Mre11 is essential for the maintenance of chromosomal DNA in vertebrate cells. EMBO J 18:6619-6629

Yu X, Gabriel A (2004) Reciprocal translocations in *Saccharomyces cerevisiae* formed by nonhomologous end joining. Genetics 166:741-751

Zubko MK, Guillard S, Lydall D (2004) Exo1 and Rad24 differentially regulate generation of ssDNA at telomeres of *Saccharomyces cerevisiae* cdc13-1 mutants. Genetics 168:103-115

Haber, James E.

Rosenstiel Center and Department of Biology, Brandeis University, Waltham, MA 02454-9110 USA

haber@brandeis.edu

The cell biology of mitotic recombination in *Saccharomyces cerevisiae*

Michael Lisby and Rodney Rothstein

Abstract

Genetic recombination relies on a number of biochemical activities that must be present at the right time and place in order for two DNA molecules to be recombined properly. Recent advances in real-time fluorescence microscopy provide us with a glimpse of homologous recombination taking place in living cells. These approaches reveal that homologous recombination is highly choreographed *in vivo* with its spatio-temporal organization being dependent on both cell cycle phase and the nature of the initiating DNA lesion. In this chapter, we review the cell biology of homologous recombination in mitotic cells with the main focus on the yeast *Saccharomyces cerevisiae* but also drawing parallels to other eukaryotic organisms.

1 Choreography of DNA double-strand break repair

Homologous recombination (HR) is the primary pathway for repairing DNA double-strand breaks (DSBs) in *S. cerevisiae* during the S and G2 phase, while non-homologous end-joining is the preferred pathway for DSB repair in G1 cells (Karathanasis and Wilson 2002; Wilson TE, this volume). At the DNA level, homology-dependent repair of DSBs has inspired two generalized models for homologous recombination, the double-strand break repair (DSBR) model and the synthesis-dependent strand-annealing (SDSA) model (Fig. 1) (Prado et al. 2003) (Haber JE, this volume). Yet other mechanisms of homologous recombination such as single-strand annealing, break-induced replication and recombinational restart of stalled replication forks are invoked under special circumstances (Krogh and Symington 2004) (Haber J and Foiani M, this volume). In *S. cerevisiae*, efficient homologous recombination requires proteins encoded by genes including *RAD50, RAD51, RAD52, RAD54, RDH54, RAD55, RAD57, RAD59, RFA1, RFA2, RFA3, XRS2*, and *MRE11*. However, the cellular response to DSBs involves a much larger number of proteins responsible for checkpoint signaling, chromatin remodeling and transcriptional regulation (see below).

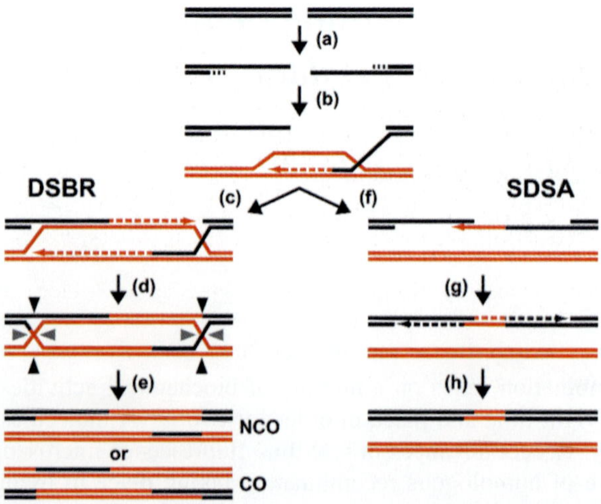

Fig. 1. Double-strand break-repair (DSBR) and synthesis-dependent strand-annealing (SDSA) models for homologous recombination. (a) resection of 5'-ends, (b) strand-invasion and priming of DNA synthesis, (c) second end capture and DNA synthesis, (d) ligation, (e) Holliday-junction resolution to yield non-crossover (NCO, grey arrowheads) or crossover (CO, black arrowheads) products, (f) strand displacement, (g) DNA synthesis, and (h) ligation. Damaged DNA in black and intact homologous sequences in red.

Although DNA damage checkpoint activation and repair are often treated as separate pathways, the two processes are highly intertwined. The cell cycle checkpoint response to DNA damage is initiated by the Mec1 and Tel1 checkpoint kinases, which belong to the phosphatidylinositol 3-kinase-related kinase (PIKK) family including the mammalian ATR, ATM, and DNA-PK. Mec1 and Tel1 launch a signaling cascade that is required for both cell cycle arrest and efficient repair (Melo et al. 2001). Mec1 and Tel1 phosphorylate (S/T)Q motifs in a large number of proteins (Kim et al. 1999), resulting in the activation of downstream effector kinases such as the Rad53, Dun1, and Chk1 effector kinases. Effector kinase activation depends on the Rad9 and Mrc1 adaptor proteins in a lesion dependent manner (Blankley and Lydall 2004; Gilbert et al. 2001; Lee et al. 2004; Osborn and Elledge 2003).

Recombinational repair of a DSB is initiated by the 5' to 3' degradation of one or both DNA ends of the break to yield 3' single-stranded tails. The single-stranded DNA exposed by end-resection is bound by replication protein A (RPA), a heterotrimeric complex found in all organisms. Binding of RPA prevents formation of secondary structures within the single-stranded region but at the same time inhibits binding of the Rad51 recombinase (Alani et al. 1992; Sung 1997a; Heyer W-D, this volume). Rad52 and the Rad55-Rad57 heterodimer can act as co-factors in the displacement of RPA by Rad51 thereby allowing a nucleoprotein filament to form between Rad51 and the single-stranded DNA (Sung 1997a, 1997b). Once bound to the single-stranded DNA, Rad51 can catalyze invasion of the end(s) into

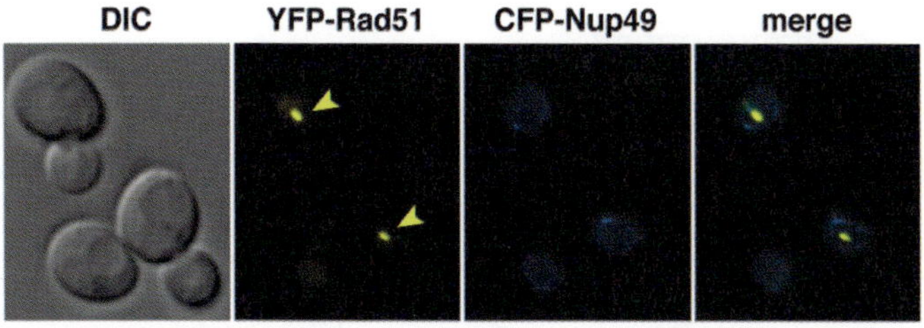

Fig. 2. Rad51 foci induced by bleomycin. Cells expressing YFP-Rad51 and CFP-Nup49 were exposed to 5 μg/ml bleomycin for one hour before imaging (strain ML149-8A). Nup49 is a nuclear pore protein, which serves to mark the nuclear periphery. Arrowheads indicate Rad51 foci formed in S/G2 cells.

a homologous duplex. The 3' invading end can prime DNA synthesis, which ultimately restores genetic information lost at the DSB (Fig. 1).

The relocalization of many checkpoint and HR proteins to a DSB induced by an endonuclease (e.g. *HO* or I-*Sce*I), after exposure to ionizing radiation or by bleomycin treatment has recently been determined by fluorescence microscopy (Lisby et al. 2004) (Fig. 2). This study showed that the Mre11 nuclease, likely as part of the Mre11-Rad50-Xrs2 (MRX) complex, is the first protein detected at a DSB within minutes of DSB formation. The MRX complex is a structure-specific nuclease, which is necessary for both HR and NHEJ (Trujillo and Sung 2001). The binding of Mre11 to a DSB is evidenced by the redistribution of the protein from a diffuse nuclear organization to a distinct focus of high protein concentration that colocalizes with the break, suggesting that Mre11 moves freely around the nucleus until it encounters a binding site(s). Biochemical evidence from experiments with human Mre11 indicates that the Mre11 complex is recruited to DSBs by binding directly to the exposed DNA ends (de Jager et al. 2001). The Tel1 checkpoint kinase appears at a DSB at approximately the same time as Mre11 and its recruitment is dependent on a direct physical association with Xrs2 (Nakada et al. 2003) (Fig. 3). Once resection has begun, the binding site for the Mre11 complex is lost and Mre11 and Tel1 foci disassemble (Lisby et al. 2004). The absence of Mre11 foci at later time-points, when resection is still in progress, supports the notion that Mre11 nuclease activity plays a minor role in processing of DSB ends, possibly only in the trimming of aberrant DNA structures such as hairpins (Llorente and Symington 2004; Lobachev et al. 2002; Moreau et al. 2001), while another thus far unknown nuclease is responsible for the bulk of resection.

Once bound to single-stranded DNA, RPA is necessary for the recruitment of Ddc2, which forms a complex with the Mec1 checkpoint kinase (Lisby et al. 2004; Paciotti et al. 2000). RPA interacts physically with the C terminus of Mec1 via its Rfa1 and Rfa2 subunits as shown by two-hybrid analysis (Nakada et al. 2005). RPA is also required to recruit another checkpoint protein, Rad24, which forms an RFC-like complex with Rfc2-5 (Green et al. 2000; Lisby et al. 2004) (Fig. 3). The recruitment of the Rad24-RFC complex to sites of DNA damage occurs via a direct physical interaction between Rfc4 and Rfa1 (Kim and Brill 2001). The

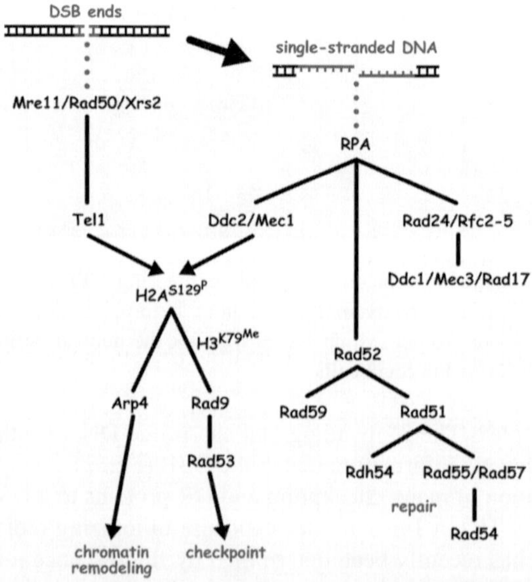

Fig. 3. Order of assembly of proteins at a DNA double-strand break. Two distinct DNA structures, DSB ends and single-stranded DNA, are bound by the Mre11 complex and RPA, respectively. DSB ends and single-stranded DNA can only co-exist if the two ends of a DSB are resected asynchronously.

Rad24-RFC complex loads the Rad17-Mec3-Ddc1 clamp onto DNA in an ATP-dependent manner *in vitro* and is required to recruit Ddc1 to foci *in vivo* (Lisby et al. 2004; Majka and Burgers 2003; Melo et al. 2001).

The Tel1 and Mec1 checkpoint kinases both phosphorylate histone H2A at serine 129 (Redon et al. 2003). Phosphorylation of this residue is an evolutionarily conserved signal for DNA double-strand breaks and is often used as a marker for DSBs using specific antibodies directed towards the epitope (Rogakou et al. 1999). Along with other histone modifications, this phosphorylated histone is important for efficient repair and checkpoint signaling likely because it transforms the chromatin in that region to serve as a docking site for checkpoint proteins, chromatin remodeling proteins and cohesins (Nakamura et al. 2004; Strom et al. 2004; Thiriet and Hayes 2005; Unal et al. 2004). For example, recruitment of the Rad9 checkpoint protein to foci depends on phosphorylation of histone H2A (Toh et al. 2006). A similar requirement for H2AX phosphorylation is found for Crb2 foci, the Rad9 orthologue in *S. pombe*, and for the stability of MDC1 and 53BP1 foci in human cells (Du et al. 2006; Nakamura et al. 2004; Stewart et al. 2003). Furthermore, studies on MDC1 and 53BP1 indicate that proteins of this class interact directly with phosphorylated serine 129 on H2A via their BRCT motifs (Stewart et al. 2003; Ward et al. 2003). Interestingly, Rad9 focus formation in *S. cerevisiae* also requires the methylation of histone H3 at lysine 79 by Dot1 (Ng et al. 2002; Toh et al. 2006). Similar to 53BP1 in human cells, Rad9 associates directly with methylated H3 via its Tudor-related motif (Alpha-Bazin et al. 2005;

Huyen et al. 2004). In contrast, Crb2 focus formation in *S. pombe* requires methylation of histone H4 at lysine 20 but not methylation of histone H3 (Du et al. 2006). In the future, it will be important to determine the relationship between H2A-S129 phosphorylation, H3-K79 methylation and possibly other histone modifications in the recruitment of Rad9 to sites of DNA damage. Activation and recruitment of the Rad53 effector kinase to sites of DNA damage is dependent on Rad9 (Gilbert et al. 2001; Lisby et al. 2004; Sweeney et al. 2005). Moreover, the weak and transient foci formed by Rad53 indicate that it is not retained at the site of DNA damage. Similar findings have been made in mammalian cells, where the ability of Chk2, the human homologue of Rad53, to permeate the nucleus is important for efficient checkpoint signal transduction (Lukas et al. 2003).

In budding yeast, the Rad52 strand annealing protein is the lynchpin required to engage the homologous recombination machinery in DNA repair. Rad52 is recruited to regions of single-stranded DNA by RPA and the subsequent recruitment of the Rad51 recombinase and other recombination proteins is dependent on Rad52 (Hays et al. 1998; Lisby et al. 2004; Sugawara et al. 2003). Both the Rad54 and Rdh54 DNA-dependent ATPases are recruited to DSBs by Rad51 (Clever et al. 1997; Dresser et al. 1997; Jiang et al. 1996; Krejci et al. 2001; Lisby et al. 2004; Petukhova et al. 2000). Furthermore, the Rad51 paralogues Rad55 and Rad57 are dependent on Rad51 for focus formation (Fig. 3) (Lisby et al. 2004; Sung 1997b). The fact that the Rad51 paralogues are required during Rad51 filament formation but not during Rad51 focus formation indicates that Rad51 is recruited to DSBs likely by association with Rad52 prior to and independent of its formation of a nucleoprotein filament (Hays et al. 1995). Moreover, the dependency of Rad54 foci on Rad55 and Rad57 suggest that recruitment of Rad54 to DSBs also requires Rad51 filament formation (Lisby et al. 2004). Interestingly, Rdh54 localizes to kinetochores in a Rad51-independent manner in addition to its Rad51-dependent association with DSBs. In an *rdh54*Δ mutant, Rad54 substitutes for Rdh54 at the kinetochore (unpublished), which is consistent with the genetic interaction between *RAD54* and *RDH54* (Klein 1997). The kinetochore function of these two proteins remains to be established.

A number of DNA helicases, which are important for recombinational DNA repair including Srs2, Sgs1 and Pif1 (Gangloff et al. 1994; Macris and Sung 2005; Wagner et al. 2006; Watt et al. 1996; Klein H, this volume), also relocalize to form foci that colocalize with Rad52 foci (Wagner et al. 2006; unpublished results). Consistent with these observations, it was recently demonstrated that both Srs2 and Sgs1 are required during the synapsis between a DSB and a homologous donor sequence (Houston and Broach 2006).

2 Cell cycle regulation of recombination foci

Foci of Rad52, Rad51, Rad55-Rad57, Rad59, and Rad54 are restricted to S and G2 phase of the cell cycle (Lisby et al. 2003a, 2004, 2001). The molecular mechanism underlying this regulation is unknown, but it possibly involves phosphorylation of Rad52 in S/G2 phase (Antunez de Mayolo et al. 2006; Lisby et al. 2003a). Interestingly, high doses of ionizing radiation (>15 DSBs/cell) will induce Rad52 foci even in G1 cells thereby overriding the cell cycle regulation of focus formation (Lisby et al. 2001). The sigmoidal DNA damage dose response of Rad52 foci suggests that perhaps an inhibitor of Rad52 focus formation in G1 cells can be out-titrated by a large number of DSBs thereby allowing Rad52 to bind to some lesions (Lisby et al. 2001).

Resection of an *HO* endonuclease-induced DSB is restricted to S and G2 phase of the cell cycle and requires cyclin-dependent kinase (CDK1) activity (Ira et al. 2004). As a consequence, recruitment of the single-stranded DNA binding protein, RPA, to an endonuclease-induced DSB only occurs in S/G2 (Ira et al. 2004; Barlow et al. unpublished results). In contrast to endonuclease-induced DSBs, ionizing radiation induces RPA foci at all stages of the cell cycle, indicating that, in addition to CDK1, resection of DSBs is also determined by the molecular properties of the lesion (Barlow et al. unpublished result).

3 The cellular response to stalled and collapsed DNA replication forks

Recombination proteins are recruited to DNA damage structures other than DSBs, consistent with the fact that a variety of DNA lesions including base modifications, DNA single-strand breaks and gaps, abasic sites, and intra- and inter-strand crosslinks are recombinogenic due to stalling or collapsing DNA replication forks (Dronkert and Kanaar 2001; Saffran et al. 1994; Swanson et al. 1999). Some of these lesions, e.g. pyrimidine dimers, can be bypassed in Rad51-mediated template switching (Kadyk and Hartwell 1993) (Fig. 4). Furthermore, initiation of recombination does not require a DNA lesion *per se* as replication blocking proteins, low dNTP pools and DNA secondary structures such as hairpins may also trigger recombination (Galli and Schiestl 1996; Ruskin and Fink 1993; Sommariva et al. 2005).

Recruitment of proteins to foci during recombination at stalled or collapsed replication forks was analyzed by treating wild type, *mec1*Δ, or *rad53*Δ mutant cells with hydroxyurea (HU), which inhibits ribonucleotide reductase thereby depleting dNTP pools (Lopes et al. 2001; Reichard 1988). In wild type cells, replication fork stalling by HU-mediated depletion of dNTP pools results in Rfa1, Ddc1, Ddc2, and Rad53 foci (Lisby et al. 2004). These foci are weak compared to DSB-induced foci, likely reflecting that only small stretches of single-stranded DNA are exposed to binding by RPA at stalled replication forks. Interestingly, the Mre11 and Rad52 proteins do not form foci in response to HU, indicating that the structures recognized by these proteins at DSBs are not exposed at a stalled replication fork. In *mec1*Δ and *rad53*Δ mutant cells, stalled replication forks collapse and the replisome disassembles (Lopes et al. 2001; Lucca et al. 2004; Tercero and Diffley 2001). In contrast to wild type cells, *mec1*Δ and *rad53*Δ mutant cells form Mre11 and Rad52 foci after HU treatment. The preferential binding of Mre11 and Rad52

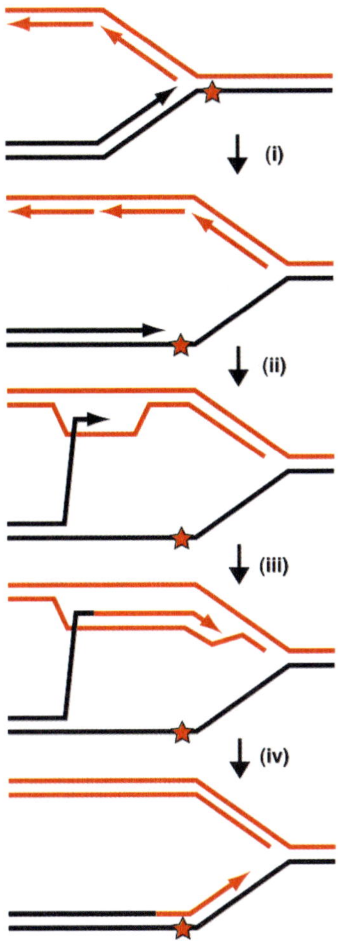

Fig. 4. Recombination at a stalled or collapsed DNA replication fork. Upon replication fork blockage (i), the stalled leading-strand invades the newly synthesized lagging-strand by homologous recombination (ii) to prime DNA synthesis (iii). After copying a segment of the lagging-strand, the extended leading-strand is likely displaced from the lagging strand by branch migration and reanneals to the leading-strand template (iv) to effectively bypass the DNA lesion. Lagging-strand as well as newly synthesized DNA in red. Star, DNA lesion (reproduced with permission from Lisby and Rothstein 2005).

to DNA ends *in vitro* suggests that DNA ends are exposed by fork reversal or some other mechanism during replication fork collapse and that these ends are required for nucleation of Mre11 and Rad52 foci during recombinational restart of the collapsed replication fork (de Jager et al. 2001; Sogo et al. 2002; Van Dyck et al. 1999). In addition, the fluorescence intensity of Rfa1 and Ddc2 foci increases dramatically during fork collapse in *mec1*Δ *sml1*Δ mutant cells, indicating that the amount of single-stranded DNA generated at collapsed replication forks is much greater than at stalled forks (unpublished results).

4 Spontaneous foci

DNA damage checkpoint and recombination proteins form foci at a low frequency even in the absence of exogenous DNA damage. The molecular events that lead to the formation of spontaneous foci are unknown, but the fact that most spontaneous Mre11, Rfa1, Ddc2, Ddc1, and Rad52 foci are observed during S phase suggests that the majority of these foci are triggered by DNA replication (Lisby et al. 2004). For example, the Fob1 replication fork blocking protein in rDNA is a potent inducer of homologous recombination in the absence of exogenous DNA damage (Johzuka and Horiuchi 2002; Kobayashi et al. 1998). Moreover, we find that 20% of the spontaneous Rad52 foci colocalize with Cdc13, suggesting that a telomere end is occasionally recognized as a DSB and targeted for recombination (unpublished result).

5 Dynamics of proteins in foci

The limit for detecting foci of GFP-tagged proteins by fluorescence microscopy varies somewhat with the GFP variant used and the subcellular distribution of the tagged protein. For a YFP-Tel1 fusion protein expressing approximately 25 molecules per cell from the endogenous *TEL1* promoter, we can detect the relocalization of half of the YFP signal into a DNA damage-induced focus, putting the detection limit at ≤10-15 molecules (unpublished result). Thus, the lack of foci formed at DSBs by NHEJ proteins such as Ku80, despite binding of the Ku complex to DSBs as measured by chromatin immunoprecipitation, can be explained by the association of only a few molecules of Ku80 with a DSB (Martin et al. 1999). In contrast, Rad52 foci induced by ionizing radiation contain from 500 to as many as 2100 molecules of the 2500 molecules of Rad52 in the cell. Thus, the local concentration of Rad52 at the DSB is 15-45 μM based on image 3D-reconstruction using Volocity software (Improvision, Coventry, England). This concentration is approximately 50-fold higher than the normal nuclear concentration and much higher than those concentrations regularly assayed in biochemical studies of Rad52. When the amount of Rad52 protein is lowered to 25% of wild type levels *in vivo*, cells exhibit increased sensitivity to gamma-irradiation (Antunez de Mayolo et al. 2006), suggesting that Rad52 protein becomes limiting for repair. Perhaps the cell avoids spurious recombination at undamaged DNA by allowing recombination to take place only at foci where there is a high local concentration of Rad52. This restriction may be particularly important at structures that resemble recombination intermediates such as DNA replication forks, transcription bubbles, regions of DNA supercoiling and telomeres.

Fluorescence loss in photobleaching (FLIP) and fluorescence recovery after photobleaching (FRAP) studies in mammalian cells demonstrate that recombination foci are highly dynamic with focus-associated Rad52 and Rad54 being exchanged with unbound protein at a high rate, while a significant fraction of focus bound Rad51 remains stably associated with the DNA lesion (Essers et al. 2002) (Kanaar R, this volume). At present, we do not have insights into the molecular architecture of foci, but the absence of a preassembled and well-defined repairosome may allow the repair machinery to easily adapt to different kinds of DNA lesions and catalyze an array of different recombination reactions such as direct-repeat recombination, single-strand annealing, break-induced replication, ectopic recombination, and recombinational restart of collapsed replication forks.

6 Centers of recombinational DNA repair

In both yeast and mammalian cells, DSBs exhibit a local mobility of 1-2 μm, which allows for the coalescing of multiple lesions at a single or a few repair foci referred to as centers of recombinational DNA repair (Aten et al. 2004; Lisby et al. 2003b). Since the diameter of the yeast nucleus is approximately 2 μm, haploid cells experiencing up to 80 DSBs after gamma-irradiation form mostly 1 or 2 Rad52 foci (Lisby et al. 2001). Surprisingly, under these conditions diploid cells form 2 to 4 Rad52 foci. This is not due to diploid cells providing twice the amount of Rad52 protein, since overexpression of Rad52 results in fewer and brighter foci rather than more foci (unpublished result). Instead, since the nuclei of haploid and diploid cells are of similar diameter, the difference in the number of foci observed in the two cell types suggests that they exhibit a difference in the mobility of DSBs. Alternatively, the higher number of foci in diploid cells could reflect the recruitment of DSBs to pre-existing Rad52 foci, assuming that the initial nucleation of foci determines the total number of mature Rad52 foci (Lisby et al. 2003b). Thus, for a given gamma-ray source, twice as many repair foci will nucleate before the appearance of mature foci because the diploid genome represents twice the target compared to the haploid genome.

We propose that the propensity for multiple DSBs to colocalize reflects a general mechanism for holding DNA ends together during DSB repair (Kaye et al. 2004; Lisby et al. 2003a; Lobachev et al. 2004). This is likely important for the rejoining of DNA ends, but at the same time the colocalization of multiple DSBs may promote chromosome translocations. The tethering of DNA ends *in vivo* is partially dependent on Rad52 and the Mre11-Rad50-Xrs2 complex but other factors are likely required (Kaye et al. 2004; Lobachev et al. 2004). It remains to be established whether similar requirements exist for the coalescing of multiple DSBs.

7 Nucleolar exclusion of homologous recombination

Three-dimensional mapping of Rad52 foci relative to other nuclear markers has shown that recombination foci are excluded from the nucleolus, where the rDNA is transcribed (manuscript in preparation). The exclusion of Rad52 foci from the nucleolus is mediated by the Smc5-Smc6 complex, which itself is enriched at the rDNA in the nucleolar compartment (Betts Lindroos et al. 2006; Torres-Rosell et al. 2005), and temperature sensitive mutants (*smc6-9* and *nse5-1*) of the complex cause rDNA hyper-recombination. The Smc5-Smc6 complex is recruited to sites of DNA damage in both *S. cerevisiae* and *Candida glabrata* (X. Zhao, personal communication; Betts Lindroos et al. 2006; De Piccoli et al. 2006; Miyazaki et al. 2006). In contrast to Rad52, the Mre11, Rfa1 and Ddc2 proteins are competent in forming foci in the nucleolus, demonstrating that DSBs in the rDNA are recognized by the Mre11-Rad50-Xrs2 complex and resected into single-stranded 3'-tails, which are bound by RPA while inside of the nucleolus. Fluorescence microscopy of an I-*Sce*I endonuclease-induced DSB in the rDNA shows that Rad52 binds to the break during a transient exit from the nucleolus. Apparently, the exclusion of recombination foci from the nucleolus is important for maintaining the stability of the locus by favoring sister-chromatid recombination over unequal sister-chromatid exchange (De Piccoli et al. 2006 and manuscript in preparation).

8 Cohesins

Cohesins are required for efficient DSB repair by homologous recombination (Sjogren and Nasmyth 2001; Strom et al. 2004). The process of chromosome cohesion can be monitored by real-time microscopy using tandem arrays of Lac or Tet repressor binding sites (Michaelis et al. 1997; Straight et al. 1996). In addition, the *de novo* recruitment of cohesin to chromatin surrounding a DSB has been studied by chromatin immunoprecipitation and requires the phosphorylation of histone H2A by the Mec1 and Tel1 checkpoint kinases (Strom et al. 2004; Unal et al. 2004). Moreover, Mre11 is required for cohesin recruitment independently of H2A phosphorylation. Cohesion plays a particularly important role at the repetitive rDNA locus in promoting sister-chromatid recombination and suppressing unequal sister-chromatid exchange. Cohesion at the rDNA locus is dependent on the Sir2 protein which is highly enriched in the nucleolus (Kobayashi et al. 2004; Reid and Rothstein 2004).

9 Molecular switches

Many of the proteins involved in homologous recombination are post-translationally modified by phosphorylation (Rad9, Rad52, Sae2, Srs2, Mre11, RPA, Xrs2, Dun1), ubiquitylation (PCNA) and sumoylation (Rad52 and PCNA)

(M. Sacher 2006 in press; Antunez de Mayolo et al. 2006; Baroni et al. 2004; Chiolo et al. 2005; Hoege et al. 2002; Mallory et al. 2003; Schwartz et al. 2002). For PCNA, its posttranslational modification acts as a molecular switch between different repair pathways at stalled replication forks (reviewed in Watts 2006). Specifically, sumoylated PCNA recruits Srs2 to suppress recombination at stalled replication forks, while mono-ubiquitylation promotes recruitment of low-fidelity DNA polymerases for translesion synthesis and poly-ubiquitylation stimulates Rad5-dependent error-free repair possibly by template switching (Branzei et al. 2004; Hoege et al. 2002). For Rad9, its phosphorylation is required for efficient activation of a Rad53-mediated checkpoint (Schwartz et al. 2002). In contrast, DNA damage-induced phosphorylation of Xrs2, Rfa2, and Dun1 does not appear to be important for DNA repair (Mallory et al. 2003). These latter modifications could be due to a bystander effect causing proteins to be unintentionally modified simply because they are in close proximity with a kinase within a repair focus. The precise role of several posttranslational modifications in recombinational repair remains to be established.

As mentioned, Rad9 focus formation requires histone H2A phosphorylation at serine 129 and H3 methylation at lysine 79 (Toh et al. 2006). However, other chromatin modifications are important for DNA repair proficiency including phosphorylation of histone H2A at serine 122, ubiquitylation of H2B at lysine 123, acetylation of H3 at lysine 56 and acetylation of H4 at lysines 5, 8, 12, and 16 (Bird et al. 2002; Giannattasio et al. 2005; Harvey et al. 2005). The molecular function of these modifications awaits further characterization.

10 Future perspectives

Recombination foci are giga-Dalton structures consisting of thousands of molecules of more than 30 different proteins. The architecture of these structures is largely unknown but is likely to constrain the biochemistry of homologous recombination *in vivo* during such steps as end-resection, homology search, strand-invasion, branch migration, and Holliday junction resolution. The same can be said for chromatin structure for which a wealth of emerging information indicates that chromatin structure is dramatically modified during homologous recombination by post-translational modification of histones and by ATP-dependent processes. The biochemical consequences of these chromatin-related changes are important areas of future research.

It is of considerable interest to understand how the cell integrates molecular signals within a recombination focus and between the focus and the remainder of the cell to yield a productive response from the many inputs. For example, how is the decision made within a recombination focus to terminate resection of DSB ends? Or, how is the choice made between alternative repair mechanisms? It will be important to understand how information about the molecular structure of the DNA lesion is relayed to the many checkpoint and repair proteins and how this information is processed.

References

Alani E, Thresher R, Griffith JD, Kolodner RD (1992) Characterization of DNA-binding and strand-exchange stimulation properties of y-RPA, a yeast single-strand-DNA-binding protein. J Mol Biol 227:54-71

Alpha-Bazin B, Lorphelin A, Nozerand N, Charier G, Marchetti C, Berenguer F, Couprie J, Gilquin B, Zinn-Justin S, Quemeneur E (2005) Boundaries and physical characterization of a new domain shared between mammalian 53BP1 and yeast Rad9 checkpoint proteins. Protein Sci 14:1827-1839

Antunez de Mayolo A, Lisby M, Erdeniz N, Thybo T, Mortensen UH, Rothstein R (2006) Multiple start codons and phosphorylation result in discrete Rad52 protein species. Nucleic Acids Res 34:2587-2597

Aten JA, Stap J, Krawczyk PM, van Oven CH, Hoebe RA, Essers J, Kanaar R (2004) Dynamics of DNA double-strand breaks revealed by clustering of damaged chromosome domains. Science 303:92-95

Baroni E, Viscardi V, Cartagena-Lirola H, Lucchini G, Longhese MP (2004) The functions of budding yeast Sae2 in the DNA damage response require Mec1- and Tel1-dependent phosphorylation. Mol Cell Biol 24:4151-4165

Betts Lindroos H, Strom L, Itoh T, Katou Y, Shirahige K, Sjogren C (2006) Chromosomal association of the Smc5/6 complex reveals that it functions in differently regulated pathways. Mol Cell 22:755-767

Bird AW, Yu DY, Pray-Grant MG, Qiu Q, Harmon KE, Megee PC, Grant PA, Smith MM, Christman MF (2002) Acetylation of histone H4 by Esa1 is required for DNA double-strand break repair. Nature 419:411-415

Blankley RT, Lydall D (2004) A domain of Rad9 specifically required for activation of Chk1 in budding yeast. J Cell Sci 117:601-608

Branzei D, Seki M, Enomoto T (2004) Rad18/Rad5/Mms2-mediated polyubiquitination of PCNA is implicated in replication completion during replication stress. Genes Cells 9:1031-1042

Chiolo I, Carotenuto W, Maffioletti G, Petrini JH, Foiani M, Liberi G (2005) Srs2 and Sgs1 DNA helicases associate with Mre11 in different subcomplexes following checkpoint activation and CDK1-mediated Srs2 phosphorylation. Mol Cell Biol 25:5738-5751

Clever B, Interthal H, Schmuckli-Maurer J, King J, Sigrist M, Heyer WD (1997) Recombinational repair in yeast: functional interactions between Rad51 and Rad54 proteins. EMBO J 16:2535-2544

de Jager M, van Noort J, van Gent DC, Dekker C, Kanaar R, Wyman C (2001) Human Rad50/Mre11 is a flexible complex that can tether DNA ends. Mol Cell 8:1129-1135

De Piccoli G, Cortes-Ledesma F, Ira G, Torres-Rosell J, Uhle S, Farmer S, Hwang JY, Machin F, Ceschia A, McAleenan A, Cordon-Preciado V, Clemente-Blanco A, Vilella-Mitjana F, Ullal P, Jarmuz A, Leitao B, Bressan D, Dotiwala F, Papusha A, Zhao X, Myung K, Haber JE, Aguilera A, Aragon L (2006) Smc5-Smc6 mediate DNA double-strand-break repair by promoting sister-chromatid recombination. Nat Cell Biol 8:1032-1034

Dresser ME, Ewing DJ, Conrad MN, Dominguez AM, Barstead R, Jiang H, Kodadek T (1997) DMC1 functions in a *Saccharomyces cerevisiae* meiotic pathway that is largely independent of the RAD51 pathway. Genetics 147:533-544

Dronkert ML, Kanaar R (2001) Repair of DNA interstrand cross-links. Mutat Res 486:217-247

Du LL, Nakamura TM, Russell P (2006) Histone modification-dependent and -independent pathways for recruitment of checkpoint protein Crb2 to double-strand breaks. Genes Dev 20:1583-1596

Essers J, Houtsmuller AB, van Veelen L, Paulusma C, Nigg AL, Pastink A, Vermeulen W, Hoeijmakers JH, Kanaar R (2002) Nuclear dynamics of *RAD52* group homologous recombination proteins in response to DNA damage. EMBO J 21:2030-2037

Galli A, Schiestl RH (1996) Hydroxyurea induces recombination in dividing but not in G1 or G2 cell cycle arrested yeast cells. Mutat Res 354:69-75

Gangloff S, McDonald JP, Bendixen C, Arthur L, Rothstein R (1994) The yeast type I topoisomerase Top3 interacts with Sgs1, a DNA helicase homolog: a potential eukaryotic reverse gyrase. Mol Cell Biol 14:8391-8398

Giannattasio M, Lazzaro F, Plevani P, Muzi-Falconi M (2005) The DNA damage checkpoint response requires histone H2B ubiquitination by Rad6-Bre1 and H3 methylation by Dot1. J Biol Chem 280:9879-9886

Gilbert CS, Green CM, Lowndes NF (2001) Budding yeast Rad9 is an ATP-dependent Rad53 activating machine. Mol Cell 8:129-136

Green CM, Erdjument-Bromage H, Tempst P, Lowndes NF (2000) A novel Rad24 checkpoint protein complex closely related to replication factor C. Curr Biol 10:39-42

Harvey AC, Jackson SP, Downs JA (2005) *Saccharomyces cerevisiae* histone H2A Ser122 facilitates DNA repair. Genetics 170:543-553

Hays SL, Firmenich AA, Berg P (1995) Complex formation in yeast double-strand break repair: participation of Rad51, Rad52, Rad55, and Rad57 proteins. Proc Natl Acad Sci USA 92:6925-6929

Hays SL, Firmenich AA, Massey P, Banerjee R, Berg P (1998) Studies of the interaction between Rad52 protein and the yeast single-stranded DNA binding protein RPA. Mol Cell Biol 18:4400-4406

Hoege C, Pfander B, Moldovan GL, Pyrowolakis G, Jentsch S (2002) RAD6-dependent DNA repair is linked to modification of PCNA by ubiquitin and SUMO. Nature 419:135-141

Houston PL, Broach JR (2006) The dynamics of homologous pairing during mating type interconversion in budding yeast. PLoS Genet 2:e98

Huyen Y, Zgheib O, Ditullio RA Jr, Gorgoulis VG, Zacharatos P, Petty TJ, Sheston EA, Mellert HS, Stavridi ES, Halazonetis TD (2004) Methylated lysine 79 of histone H3 targets 53BP1 to DNA double-strand breaks. Nature 432:406-411

Ira G, Pellicioli A, Balijja A, Wang X, Fiorani S, Carotenuto W, Liberi G, Bressan D, Wan L, Hollingsworth NM, Haber JE, Foiani M (2004) DNA end resection, homologous recombination and DNA damage checkpoint activation require CDK1. Nature 431:1011-1017

Jiang H, Xie Y, Houston P, Stemke-Hale K, Mortensen UH, Rothstein R, Kodadek T (1996) Direct association between the yeast Rad51 and Rad54 recombination proteins. J Biol Chem 271:33181-33186

Johzuka K, Horiuchi T (2002) Replication fork block protein, Fob1, acts as an rDNA region specific recombinator in *S. cerevisiae*. Genes Cells 7:99-113

Kadyk LC, Hartwell LH (1993) Replication-dependent sister chromatid recombination in *rad1* mutants of *Saccharomyces cerevisiae*. Genetics 133:469-487

Karathanasis E, Wilson TE (2002) Enhancement of *Saccharomyces cerevisiae* end-joining efficiency by cell growth stage but not by impairment of recombination. Genetics 161:1015-1027

Kaye JA, Melo JA, Cheung SK, Vaze MB, Haber JE, Toczyski DP (2004) DNA breaks promote genomic instability by impeding proper chromosome segregation. Curr Biol 14:2096-2106

Kim HS, Brill SJ (2001) Rfc4 interacts with Rpa1 and is required for both DNA replication and DNA damage checkpoints in *Saccharomyces cerevisiae*. Mol Cell Biol 21:3725-3737

Kim ST, Lim DS, Canman CE, Kastan MB (1999) Substrate specificities and identification of putative substrates of ATM kinase family members. J Biol Chem 274:37538-37543

Klein HL (1997) *RDH54*, a *RAD54* homologue in *Saccharomyces cerevisiae*, is required for mitotic diploid-specific recombination and repair and for meiosis. Genetics 147:1533-1543

Kobayashi T, Heck DJ, Nomura M, Horiuchi T (1998) Expansion and contraction of ribosomal DNA repeats in *Saccharomyces cerevisiae*: requirement of replication fork blocking (Fob1) protein and the role of RNA polymerase I. Genes Dev 12:3821-3830

Kobayashi T, Horiuchi T, Tongaonkar P, Vu L, Nomura M (2004) *SIR2* regulates recombination between different rDNA repeats, but not recombination within individual rRNA genes in yeast. Cell 117:441-453

Krejci L, Damborsky J, Thomsen B, Duno M, Bendixen C (2001) Molecular dissection of interactions between Rad51 and members of the recombination-repair group. Mol Cell Biol 21:966-976

Krogh B, Symington L (2004) Recombination proteins in yeast. Annu Rev Genet 38:233-271

Lee SJ, Duong JK, Stern DF (2004) A Ddc2-Rad53 fusion protein can bypass the requirements for *RAD9* and *MRC1* in Rad53 activation. Mol Biol Cell 15:5443-5455

Lisby M, Antúnez de Mayolo A, Mortensen UH, Rothstein R (2003a) Cell cycle-regulated centers of DNA double-strand break repair. Cell Cycle 2:479-483

Lisby M, Barlow JH, Burgess RC, Rothstein R (2004) Choreography of the DNA damage response; spatiotemporal relationships among checkpoint and repair proteins. Cell 118:699-713

Lisby M, Mortensen UH, Rothstein R (2003b) Colocalization of multiple DNA double-strand breaks at a single Rad52 repair centre. Nat Cell Biol 5:572-577

Lisby M, Rothstein R (2005) Localization of checkpoint and repair proteins in eukaryotes. Biochimie 87:579-589

Lisby M, Rothstein R, Mortensen UH (2001) Rad52 forms DNA repair and recombination centers during S phase. Proc Natl Acad Sci USA 98:8276-8282

Llorente B, Symington LS (2004) The Mre11 nuclease is not required for 5' to 3' resection at multiple HO-induced double-strand breaks. Mol Cell Biol 24:9682-9694

Lobachev K, Vitriol E, Stemple J, Resnick MA, Bloom K (2004) Chromosome fragmentation after induction of a double-strand break is an active process prevented by the RMX repair complex. Curr Biol 14:2107-2112

Lobachev KS, Gordenin DA, Resnick MA (2002) The Mre11 complex is required for repair of hairpin-capped double-strand breaks and prevention of chromosome rearrangements. Cell 108:183-193

Lopes M, Cotta-Ramusino C, Pellicioli A, Liberi G, Plevani P, Muzi-Falconi M, Newlon CS, Foiani M (2001) The DNA replication checkpoint response stabilizes stalled replication forks. Nature 412:557-561

Lucca C, Vanoli F, Cotta-Ramusino C, Pellicioli A, Liberi G, Haber J, Foiani M (2004) Checkpoint-mediated control of replisome-fork association and signalling in response to replication pausing. Oncogene 23:1206-1213

Lukas C, Falck J, Bartkova J, Bartek J, Lukas J (2003) Distinct spatiotemporal dynamics of mammalian checkpoint regulators induced by DNA damage. Nat Cell Biol 5:255-260

Macris MA, Sung P (2005) Multifaceted role of the *Saccharomyces cerevisiae* Srs2 helicase in homologous recombination regulation. Biochem Soc Trans 33:1447-1450

Majka J, Burgers PM (2003) Yeast Rad17/Mec3/Ddc1: a sliding clamp for the DNA damage checkpoint. Proc Natl Acad Sci USA 100:2249-2254

Mallory JC, Bashkirov VI, Trujillo KM, Solinger JA, Dominska M, Sung P, Heyer WD, Petes TD (2003) Amino acid changes in Xrs2p, Dun1p, and Rfa2p that remove the preferred targets of the ATM family of protein kinases do not affect DNA repair or telomere length in *Saccharomyces cerevisiae*. DNA Repair (Amst) 2:1041-1064

Martin SG, Laroche T, Suka N, Grunstein M, Gasser SM (1999) Relocalization of telomeric Ku and SIR proteins in response to DNA strand breaks in yeast. Cell 97:621-633

Melo JA, Cohen J, Toczyski DP (2001) Two checkpoint complexes are independently recruited to sites of DNA damage *in vivo*. Genes Dev 15:2809-2821

Michaelis C, Ciosk R, Nasmyth K (1997) Cohesins: chromosomal proteins that prevent premature separation of sister chromatids. Cell 91:35-45

Miyazaki T, Tsai HF, Bennett JE (2006) Kre29p is a novel nuclear protein involved in DNA repair and mitotic fidelity in *Candida glabrata*. Curr Genet 50:11-22

Moreau S, Morgan EA, Symington LS (2001) Overlapping functions of the *Saccharomyces cerevisiae* Mre11, Exo1 and Rad27 nucleases in DNA metabolism. Genetics 159:1423-1433

Nakada D, Hirano Y, Tanaka Y, Sugimoto K (2005) Role of the C terminus of Mec1 checkpoint kinase in its localization to sites of DNA damage. Mol Biol Cell 16:5227-5235

Nakada D, Matsumoto K, Sugimoto K (2003) ATM-related Tel1 associates with double-strand breaks through an Xrs2-dependent mechanism. Genes Dev 17:1957-1962

Nakamura TM, Du LL, Redon C, Russell P (2004) Histone H2A phosphorylation controls Crb2 recruitment at DNA breaks, maintains checkpoint arrest, and influences DNA repair in fission yeast. Mol Cell Biol 24:6215-6230

Ng HH, Feng Q, Wang H, Erdjument-Bromage H, Tempst P, Zhang Y, Struhl K (2002) Lysine methylation within the globular domain of histone H3 by Dot1 is important for telomeric silencing and Sir protein association. Genes Dev 16:1518-1527

Osborn AJ, Elledge SJ (2003) Mrc1 is a replication fork component whose phosphorylation in response to DNA replication stress activates Rad53. Genes Dev 17:1755-1767

Paciotti V, Clerici M, Luchinni G, Longhese MP (2000) The checkpoint protein Ddc2, functionally related to *S. pombe* Rad26, interacts with Mec1 and is regulated by Mec1-dependent phosphorylation in budding yeast. Genes Dev 14:2046-2059

Petukhova G, Sung P, Klein H (2000) Promotion of Rad51-dependent D-loop formation by yeast recombination factor Rdh54/Tid1. Genes Dev 14:2206-2215

Prado F, Cortes-Ledesma F, Huertas P, Aguilera A (2003) Mitotic recombination in *Saccharomyces cerevisiae*. Curr Genet 42:185-198

Redon C, Pilch DR, Rogakou EP, Orr AH, Lowndes NF, Bonner WM (2003) Yeast histone 2A serine 129 is essential for the efficient repair of checkpoint-blind DNA damage. EMBO Rep 4:678-684

Reichard P (1988) Interactions between deoxyribonucleotide and DNA synthesis. Annu Rev Biochem 57:379-374

Reid RJ, Rothstein R (2004) Stay close to your sister. Mol Cell 14:418-420

Rogakou EP, Boon C, Redon C, Bonner WM (1999) Megabase chromatin domains involved in DNA double-strand breaks *in vivo*. J Cell Biol 146:905-916

Ruskin B, Fink GR (1993) Mutations in *POL1* increase the mitotic instability of tandem inverted repeats in *Saccharomyces cerevisiae*. Genetics 134:43-56

Saffran WA, Greenberg RB, Thaler-Scheer MS, Jones MM (1994) Single strand and double strand DNA damage-induced reciprocal recombination in yeast. Dependence on nucleotide excision repair and *RAD1* recombination. Nucleic Acids Research 22:2823-2829

Schwartz MF, Duong JK, Sun Z, Morrow JS, Pradhan D, Stern DF (2002) Rad9 phosphorylation sites couple Rad53 to the *Saccharomyces cerevisiae* DNA damage checkpoint. Mol Cell 9:1055-1065

Sjogren C, Nasmyth K (2001) Sister chromatid cohesion is required for postreplicative double-strand break repair in *Saccharomyces cerevisiae*. Curr Biol 11:991-995

Sogo JM, Lopes M, Foiani M (2002) Fork reversal and ssDNA accumulation at stalled replication forks owing to checkpoint defects. Science 297:599-602

Sommariva E, Pellny TK, Karahan N, Kumar S, Huberman JA, Dalgaard JZ (2005) *Schizosaccharomyces pombe* Swi1, Swi3, and Hsk1 are components of a novel S-phase response pathway to alkylation damage. Mol Cell Biol 25:2770-2784

Stewart GS, Wang B, Bignell CR, Taylor AM, Elledge SJ (2003) MDC1 is a mediator of the mammalian DNA damage checkpoint. Nature 421:961-966

Straight AF, Belmont AS, Robinett CC, Murray AW (1996) GFP tagging of budding yeast chromosomes reveals that protein-protein interactions can mediate sister chromatid cohesion. Curr Biol 6:1599-1608

Strom L, Lindroos HB, Shirahige K, Sjogren C (2004) Postreplicative recruitment of cohesin to double-strand breaks is required for DNA repair. Mol Cell 16:1003-1015

Sugawara N, Wang X, Haber JE (2003) *In vivo* roles of Rad52, Rad54, and Rad55 proteins in Rad51-mediated recombination. Mol Cell 12:209-219

Sung P (1997a) Function of yeast Rad52 protein as a mediator between replication protein A and the Rad51 recombinase. J Biol Chem 272:28194-28197

Sung P (1997b) Yeast Rad55 and Rad57 proteins form a heterodimer that functions with replication protein A to promote DNA strand exchange by Rad51 recombinase. Genes Dev 11:1111-1121

Swanson RL, Morey NJ, Doetsch PW, Jinks-Robertson S (1999) Overlapping specificities of base excision repair, nucleotide excision repair, recombination, and translesion synthesis pathways for DNA base damage in *Saccharomyces cerevisiae*. Mol Cell Biol 19:2929-2935

Sweeney FD, Yang F, Chi A, Shabanowitz J, Hunt DF, Durocher D (2005) *Saccharomyces cerevisiae* Rad9 acts as a Mec1 adaptor to allow Rad53 activation. Curr Biol 15:1364-1375

Tercero JA, Diffley JF (2001) Regulation of DNA replication fork progression through damaged DNA by the Mec1/Rad53 checkpoint. Nature 412:553-557

Thiriet C, Hayes JJ (2005) Chromatin in need of a fix: phosphorylation of H2AX connects chromatin to DNA repair. Mol Cell 18:617-622

Toh GW, O'Shaughnessy A M, Jimeno S, Dobbie IM, Grenon M, Maffini S, O'Rorke A, Lowndes NF (2006) Histone H2A phosphorylation and H3 methylation are required for a novel Rad9 DSB repair function following checkpoint activation. DNA Repair (Amst) 5:693-703

Torres-Rosell J, Machin F, Farmer S, Jarmuz A, Eydmann T, Dalgaard JZ, Aragon L (2005) *SMC5* and *SMC6* genes are required for the segregation of repetitive chromosome regions. Nat Cell Biol 7:412-419

Trujillo KM, Sung P (2001) DNA structure-specific nuclease activities in the *Saccharomyces cerevisiae* Rad50*Mre11 complex. J Biol Chem 276:35458-35464

Unal E, Arbel-Eden A, Sattler U, Shroff R, Lichten M, Haber JE, Koshland D (2004) DNA damage response pathway uses histone modification to assemble a double-strand break-specific cohesin domain. Mol Cell 16:991-1002

Van Dyck E, Stasiak AZ, Stasiak A, West SC (1999) Binding of double-strand breaks in DNA by human Rad52 protein. Nature 398:728-731

Wagner M, Price G, Rothstein R (2006) The absence of Top3 reveals an interaction between the Sgs1 and Pif1 DNA helicases in *Saccharomyces cerevisiae*. Genetics: in press

Ward IM, Minn K, Jorda KG, Chen J (2003) Accumulation of checkpoint protein 53BP1 at DNA breaks involves its binding to phosphorylated histone H2AX. J Biol Chem 278:19579-19582

Watt PM, Hickson ID, Borts RH, Louis EJ (1996) SGS1, a homologue of the Bloom's and Werner's syndrome genes, is required for maintenance of genome stability in *Saccharomyces cerevisiae*. Genetics 144:935-945

Watts FZ (2006) Sumoylation of PCNA: Wrestling with recombination at stalled replication forks. DNA Repair (Amst) 5:399-403

Lisby, Michael

Department of Genetics, Institute of Molecular Biology and Physiology, University of Copenhagen, Øster Farimagsgade 2A, DK-1353 Copenhagen K, Denmark

mlisby@my.molbio.ku.dk

Rothstein, Rodney

Department of Genetics & Development, Columbia University Medical Center, 701 West 168th Street, New York, NY 10032, USA

rothstein@cancercenter.columbia.edu

The cell biology of homologous recombination

Sheba Agarwal, Roland Kanaar, and Jeroen Essers

Abstract

Discontinuities in double-stranded DNA, such as DNA double-strand breaks (DSBs), pose a threat to genome stability. Homologous recombination is a process that not only effectively repairs DSBs, but also promotes preservation of genome integrity by repairing DNA discontinuities arising during DNA replication. Genetic analyses identified many genes involved in DSB repair and placed them in different pathways. Biochemical analyses have aided in placing the protein products in a mechanistic framework for the pathways, while molecular biological approaches, such as chromatin immuno-precipitation, have allowed the monitoring of protein composition near DSBs in populations of fixed cells. Progress in cell biological techniques has now made it possible to analyze proteins in their physiological environment of the living cell. Here, we describe how homologous recombination proteins have been characterized using the methods of cell biology. The current challenge is to integrate insights gained on the spatio-temporal behavior of DSB repair proteins using chromatin immuno-precipitation and live cell imaging in the established genetic and biochemical frameworks for mechanisms of DSB repair.

1 Introduction

Double-strand breaks (DSBs) are detrimental lesions that disrupt the integrity of DNA in the cell. Pathological DSBs can be induced by exogenous factors, for example, ionizing radiation and a wide range of chemical compounds. Certain byproducts of cellular metabolism, such as oxygen free radicals, can also create DSBs in DNA. In contrast, DSBs can also be physiologically relevant intermediates. For example, nuclease-induced DSBs in germ cells trigger meiotic recombination that results in creation of genetic diversity (Hunter, this volume). Another example is the programmed DSB formation during the assembly of active immunoglobulin and T cell receptor genes, as well as in class switch recombination to produce antibodies of different isotypes (Gellert, this volume; Friedberg et al. 2004).

Whether pathological or physiological, the timely and accurate repair of DSBs is critical to the well-being of the cell. Inaccuracies in repair can result in mutations and gross chromosomal rearrangements, which can disrupt the normal working of the cell and might ultimately lead to cancer. If breaks are left unrepaired, the cell can undergo genomic fragmentation, loss of chromosomes and cell death. In order to counteract this, mechanistically diverse methods that differ in their dependence on sequence homology have evolved to rejoin DNA ends: homology-directed repair, including homologous recombination, and non-homologous DNA end joining (Gellert, this volume; Wilson, this volume; Kanaar et al. 1998).

The dissection of the molecular mechanisms of DSB repair has its foundation in genetic experiments, which has revealed several pathways through which DSB can be processed and repaired. The initial studies focused on bacteriophages, bacteria and fungi, including *Saccharomyces cerevisiae* (Shu et al. 1999; Symington 2002; Krogh and Symington 2004; Sanchez et al. this volume; Michel et al. this volume). Since then, it has become apparent that DSB repair pathways are conserved throughout evolution, which together with the advent of reverse genetics, has facilitated their analyses in a variety of other organisms, including mammals such as mice. The genetic approaches have been complemented and extended by biochemical analyses leading to placement of DSB repair proteins at specific steps in the pathways (Heyer, this volume; Cox, this volume). More recently, molecular biological approaches, such as chromatin immuno-precipitation, have allowed the monitoring of protein composition near DSBs in populations of fixed cells. Advances in cell biology have now made it possible to analyze the behavior of DSB repair proteins and their response to DNA damage at the level of the single living cell. In this review, we describe how homologous recombination proteins have been characterized using cell biological techniques. The ultimate goal of these studies is to extend the knowledge of the individual activities of the proteins to their coordinated action within the entire homologous recombination pathway in the context of the living cell.

2 Cell biological analyses of homologous recombination proteins

Tracking of proteins in live cells has become practical due to the discovery of the Green Fluorescent Protein (GFP) family and its subsequent engineering, resulting in a number of different spectral variants (Lippincott-Schwartz and Patterson 2003; Giepmans et al. 2006). Because the GFPs form a chromophore within their core, standard recombinant DNA technology can be used to label proteins of interest in living cells, usually without adverse effects, which can then be monitored over time by epi-fluoresence or confocal microcopy. Besides the simple monitoring of the GFP-fusion protein, the photobleaching property of GFP can be exploited to obtain additional information on the protein of interest. Photobleaching is a phenomenon where a fluorophore loses its fluorescence due to photon-induced chemical damage. While this is a serious drawback of using fluorescent probes for direct observation, the local loss of fluorescence after exposure to excessive excitation light can be used to obtain information on protein mobility by fluorescence recovery after photobleaching (FRAP) experiments (Houtsmuller and Vermeulen 2001).

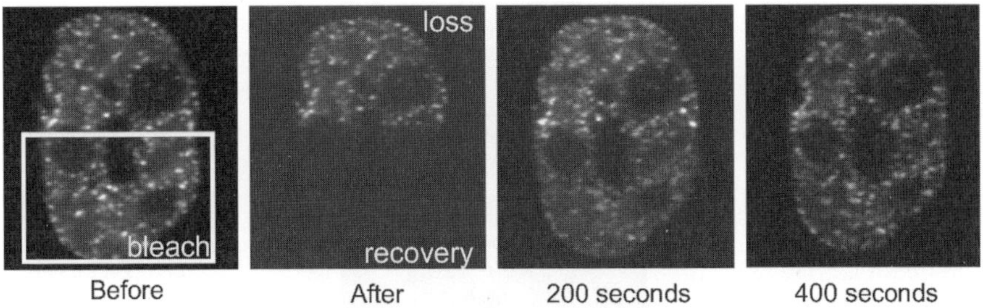

Fig. 1. Example of a photobleaching procedure to determine the mobility of GFP-tagged proteins in living cells. A region, indicated by the rectangle, of a cell containing PCNA-GFP replication foci was photobleached and fluorescence loss and recovery were monitored over time. The cell was imaged at the indicated times after bleaching. Fluorescence loss in photobleaching (FLIP) was measured in foci in the unbleached half of the cell, while fluorescence recovery after photobleaching (FRAP) was measured in foci in the bleached half of the same cell.

During a FRAP experiment (Fig. 1), a localized and very short high-intensity laser pulse is given to quench the fluorescence in a small area within a larger volume containing fluorescent molecules, for example the nucleus. Bleaching of the fluorescence does not cause significant changes in protein functionality and cells retain their viability after long periods of FRAP experiments (Nakata et al. 1998; White and Stelzer 1999). Immediately after the high-intensity laser pulse, the fluorescence over the entire area is monitored. Recovery of fluorescence in the bleached area can be observed if molecules redistribute throughout the cell. Quantitation of fluorescence recovery can yield information on protein diffusion rates and mobile versus immobile fractions, either spontaneously or in response to stimuli such as DNA damage. A number of reviews are available that present in-depth discussions of a large variety of FRAP-based protocols that have been developed for specific purposes (Houtsmuller and Vermeulen 2001; Haraguchi 2002; Carrero et al. 2003; Houtsmuller 2005; Sprague and McNally 2005; Essers et al. 2006).

3 Controlled induction of DNA damage

The study of the cell biology of DSB repair mechanisms involves the documentation of the difference in the behavior of repair proteins in the absence and presence of induced DNA damage. Thus, it is crucial to be able to conveniently and quantitatively induce DNA damage in cells to investigate the response of repair proteins. A number of methods, each with specific advantages and disadvantages, have been developed, which can be classified into various categories, namely global versus local deposition of DNA damage, as well as in the induction of a specific lesion versus a spectrum of different lesions.

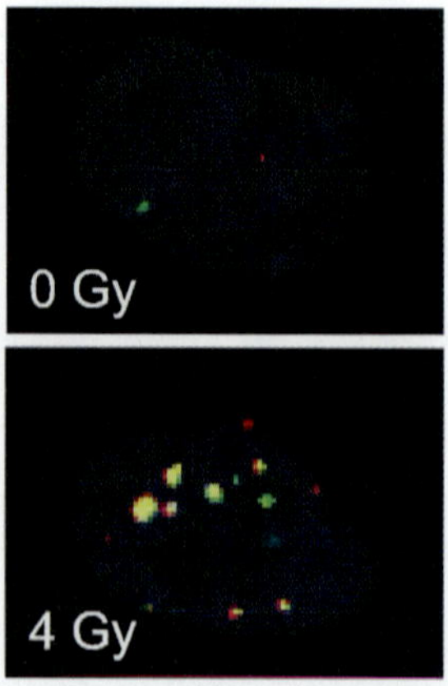

Fig. 2. DNA damage induced local accumulations of DNA damage response proteins into foci. Top panel: Nucleus of a cell before treatment with ionizing radiation. A cell, expressing the DNA repair protein Rad54 as a GFP fusion (in green), was fixed with paraformaldehyde and stained with antibodies against the DNA damage response protein 53BP1 (in red) and DAPI to detect DNA (in blue). Bottom panel: Nucleus of a cell 2 hours after treatment with 4 Gy of ionizing radiation and processed as described above.

In the context of DSB repair, DNA breaks can be introduced in a global manner by irradiating cells with ionizing radiation, for example, by using an X-ray machine or a ^{137}Cs course. These irradiation methods induce DNA damage that is dispersed over the whole nucleus. As discussed below, a number of DSB repair proteins accumulate into so-called foci, which are high local concentrations of the proteins at the sites of DNA damage throughout the nucleus (Fig. 2). However, this method introduces other, although less genotoxic, types of DNA lesions in addition to DSBs. For example, for every DSB made, hundreds of single-strand breaks are introduced. In addition to damage to the phospho-diester backbone of the DNA, base damage also occurs (Friedberg et al. 2006).

The accumulation of repair proteins in response to global damage into nuclear structures that can be visualized by light microscopy is a remarkable phenomenon and can be exploited to characterize the dynamics of repair response. However, such experiments require the creation of breaks in a more controlled fashion, both spatially and temporally. While global DNA damage induction methods introduce damage that is randomly distributed in the nucleus, it is possible to locally induce DNA damage by partially shielding the nucleus from the radiation source. In this way, it can be verified that the accumulation of repair proteins are indeed at sites of DNA damage and the accumulation can be analyzed in time. Also, foci at sites of induced DNA damage can be distinguished from foci that arise 'spontaneously', for example, during S phase (Tashiro et al. 1996). One method of local DSB induction involves the use of synchrotron-generated ultra-soft X-rays that are filtered through a metal grid containing micrometer spaced slits (Nelms et al. 1998). However, the method is not routinely used because facilities to generate ultra soft X-rays are not widely available, the time between irradiation and analysis of the cells is relatively long, and the amount of DSBs introduced is large and difficult to control. An alternative method using α-particle irradiation has been developed recently and has number of advantages (Aten et al. 2004). Exposing cells to α-particles that travel almost horizontally relative to the cells leaves a straight track of DSBs in the nucleus. Proteins accumulating at the breaks can be visualized immediately after irradiation and deviation from the originally linear pattern can yield information on movement of chromosomal domains containing DNA breaks. Similar methods have been developed that use heavy ions instead of α-particles (Jakob et al. 2003; Lukas et al. 2005).

Alternative methods that introduce DNA damage in a spatially controlled manner have been developed with the use of lasers (Cremer et al. 1980). These methods make use of compounds that when bound to or incorporated into DNA and excited by the laser transmit energy to induce DNA lesions. For example, halogenated thymidine analogs, when incorporated in DNA, can induce single-strand breaks and DSBs in living cells when excited by a UV-A laser (Tashiro et al. 2000; Lukas et al. 2003, 2005). Variations of this method use DNA intercalating Hoechst dyes either in the absence or presence of thymidine analogs (Rogakou et al. 1999; Walter et al. 2003; Bradshaw et al. 2005). In addition, there are laser-based micro-irradiation methods available that do not require exogenously added compounds. One such method makes use of the second harmonic of a pulsed neodymium-doped yttrium aluminum garnet laser that will result in DSBs (Kim et al. 2002, 2005). In addition, pulsed multiphoton laser technology can be used to introduce local DNA damage (Meldrum et al. 2003).

A drawback of the methods described above is that the spectrum of the actual induced DNA lesions is not known. Furthermore, for most laser-based methods, the local DNA damage load is unknown and will likely be higher than at sites of DNA damage resulting from global DNA damage induction. Thus, when analyzing DSB repair proteins at the sites of locally induced DNA damage, it should be realized that repair proteins from pathways other than DSB repair might influence the results, as might the artificially high local DNA damage load.

In order to avoid the problem of introducing a large spectrum of lesions by local DNA damage techniques, a site-specific DSB can be created using a rare-cutting endonuclease (Haber, this volume; Haber 1995; Jasin 1996; Porteus and Carroll 2005). An example of a widely used enzyme is I-Sce I, which recognizes an 18-bp nonpalindronic sequence. Cleavage of the site is induced by transfecting cells with an I-Sce I expressing plasmid (Richardson et al. 1999). Expression of the enzyme in mammalian cells appears not to be toxic, presumably because its large recognition site provides sufficient specificity (Rouet et al. 1994a, 1994b). This approach is also followed in the yeast *S. cerevisiae*, but the DSB-inducing enzyme of choice in this system is most often the HO endonuclease, which normally initiates mating switch recombination. Because highly regulated promoters are available in *S. cerevisiae*, events at the induced break can be followed in time (Haber 2000). A disadvantage of the use of these enzymes is that their recognition sequence has to be engineered in the genome and they generate DSBs with complementary single-strand overhangs that can be easily ligated and might therefore not always be processed similarly to ionizing radiation induced DSBs. To overcome some of these limitations, chimeric nucleases are being developed that couple the nuclease domain of the type II restriction enzyme Fok I to Zn-finger DNA binding domains. By combining different Zn-finger DNA binding domains, DSBs at predetermined sites in the genome can be introduced (Durai et al. 2005).

4 Homologous recombination pathways

Homologous recombination is generally an error-free pathway by which DSBs are repaired using the information on an undamaged homologous DNA molecule, usually the sister chromatid (Cortés-Ledesma, this volume). The process is carried out by the proteins of the *RAD52* epistasis group that were originally identified by the genetic analysis of ionizing radiation hypersensitive *S. cerevisiae* mutants (Game and Mortimer 1974; Symington 2002). Many of the *RAD52* group proteins are conserved in mammals. They include the MRN (Rad50/Mre11/NBS1) complex, Rad51, the Rad51 paralogs (Rad51B, Rad51C, Rad51D XRCC2, XRCC3), Rad54 and Rad54B (Dudas and Chovanec 2004). In mammals, homologous recombination is also modulated by the products of the breast cancer susceptibility genes, BRCA1 and BRCA2 (Shivji and Venkitaraman 2004).

The process of DSB repair by homologous recombination can be divided in a number of steps, including DSB detection and processing, joint molecule formation between the broken DNA and the repair template through homologous pairing and strand invasion, and resolution of the recombination partners (Fig. 2). After DSB detection, the DNA ends go through nucleolytic processing resulting in 3' single-stranded DNA tails, which are used for the nucleation of recombination proteins on the DNA. This nucleoprotein complex is capable of pairing with intact homologous duplex DNA resulting in a joint molecule between the two recombining DNA molecules. The joint molecule is used as a template for DNA polymerases such that the information that was lost by processing is restored. The reaction is concluded by ligation of ends and the resolution of the joint molecule to yield two intact DNA copies.

4.1 Detection and processing of DSBs

Once a DSB has occurred in the genome, the global response to its formation starts with the actual detection of the break in the context of the chromosome. A combination of biochemical and cell biological experiments has implicated the highly conserved MRN complex as an initial recognition factor of DSBs (Symington 2002). At the DSB MRN activates the ATM kinase resulting in a signaling cascade leading to cell cycle arrest (Shiloh 2003). The MRN complex is also involved in other cellular functions such as telomere maintenance, cell cycle checkpoint response and nonhomologous DNA end-joining (D'Amours and Jackson 2002). This wide range of MRN complex functions are carried out by a kaleidoscope of activities that exist within the complex, including hydrolysis of ATP hydrolysis, exo- and endo-nuclease, single-strand annealing, DNA end binding, tethering of broken DNA, protein interaction with, among others the damage checkpoint kinase ATM and the signaling mediator protein MDC1 (Maser et al. 1997; Carney et al. 1998; Paull and Gellert 1998, 1999; Stewart et al. 1999; Yamaguchi-Iwai et al. 1999; de Jager et al. 2001a, 2001b, 2002; Hopfner et al. 2002; Kim et al. 2002; Goldberg et al. 2003; Mirzoeva and Petrini 2003; Costanzo et al. 2004; Lukas et al. 2004; Moreno-Herrero et al. 2005; Wiltzius et al. 2005).

The importance of the MRN complex for mammalian cells is underscored by the finding that all three genes that make up the MRN complex are essential for viability (Xiao and Weaver 1997; Luo et al. 1999; Yamaguchi-Iwai et al. 1999; Zhu et al. 2001). The loss of Mre11 in a conditional knockout DT40 cell line results in radiosensitivity, increased levels of chromosome breaks, arrest in G2 and eventual cell death (Yamaguchi-Iwai et al. 1999). In humans, hypomorphic mutations in NBS1 are associated with Nijmegen breakage syndrome (NBS) (Varon et al. 1998) and mutations in Rad50 and Mre11 cause Ataxia telangiectasia-like disease (ATLD) (Stewart et al. 1999; Taylor et al. 2004). NBS and ATLD patients are cancer-prone and their cells are radiosensitive.

The MRN complex relocalizes in response to DNA damage. In primary human fibroblasts, Mre11 and Rad50 are distributed homogeneously throughout the nuclei, but they accumulate in a high local concentration to form colocalizing foci after the global treatment of the cells with ionizing radiation (Maser et al. 1997). The physical association of these proteins is important for their accumulation into foci, since its components do not form DNA damage induced foci in cells from NBS and ATLD patients (Carney et al. 1998; Stewart et al. 1999). In addition, the subnuclear localization of the complex also changes in response to DNA replication; it associates with chromatin in S phase and colocalizes with proliferating cell nuclear antigen (PCNA) throughout S phase (Mirzoeva and Petrini 2003).

A number of local DNA damage induction methods have demonstrated that the DNA damage-induced foci of the MRN complex colocalize with DSBs. These methods include exposure of cells to ultra-soft X-rays through an irradiation mask combined with the labeling of the generated DNA ends (Nelms et al. 1998), dye-dependent laser micro-irradiation (Paull et al. 2000; Lukas et al. 2003), and irradiation with α-particles (Aten et al. 2004). At DSBs, including those created by the local irradiation methods, the histone variant H2AX is modified by phosphorylation (Sedelnikova et al. 2003). This modified version of H2AX, γH2AX, colocalizes with the MRN complex (Paull et al. 2000). Interestingly, chromatin immunoprecipitation studies using *S. cerevisiae* showed that the phosphorylation of H2A in response to a nuclease-induced DSB can be detected up to 50 kb away from the break, but very little of it is found within 1 – 2 kb of the break. On the other hand, almost all Mre11 is concentrated within this region in close vicinity to the break (Shroff et al. 2004). Results such as these emphasize the necessity of different techniques to address one problem, as each provides information at various levels of specificity. For example, while live cell imaging provides the advantage of real time observation of proteins in single cells, it lacks the spatial resolution provided by chromatin immunoprecipitation.

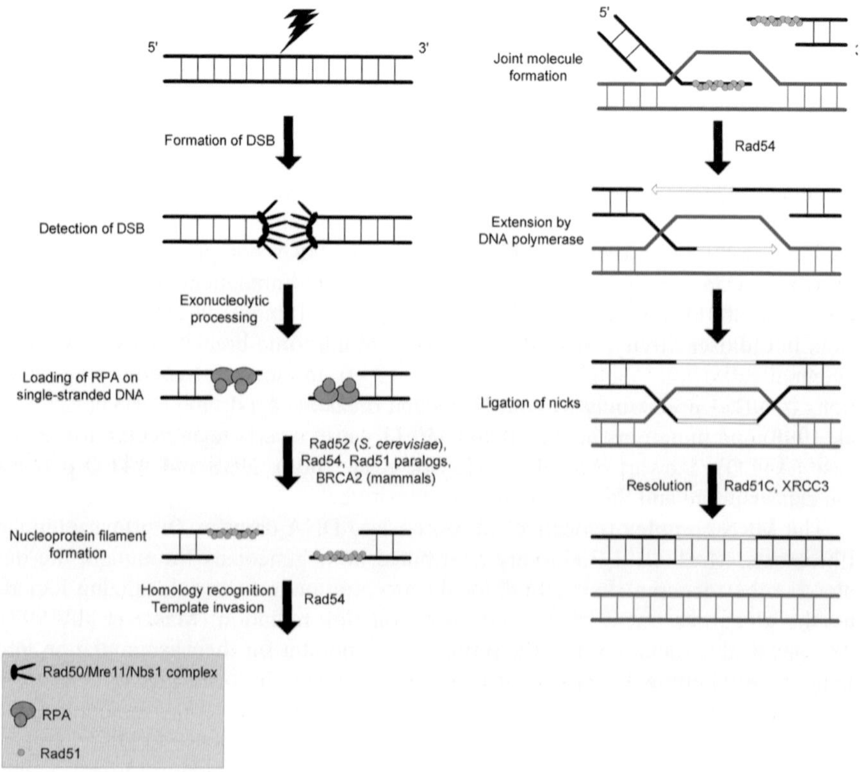

Fig. 3. A model for DSB repair through homologous recombination. The double-stranded DNA, depicted as a black ladder, suffers a DNA damage-induced DSB. Oligomers of the MRN complex tether the broken ends and initiate their processing, resulting in RPA-bound single-stranded DNA overhangs with a 3' polarity. The Rad51 recombinase is loaded on the single-stranded DNA with the assistance of mediators including Rad52 (*S. cerevisiae*), Rad51 paralogs, and BRCA2 (mammalian cells). The gray ladder represents a homologous duplex DNA (sister chromatid). The Rad51 nucleoprotein filament mediates homology recognition, joint molecule formation and strand exchange with the intact homologous duplex repair template. Steps that can be stimulated by the Rad54 protein are indicated. In this model the second DNA end of the DSB is captured by the displaced strand from the D-loop intermediate. DNA polymerization will restore missing nucleotides (indicated in white) and ligation will covalently lock the recombining partner DNA molecules into a structure joined through Holliday junctions. Resolution of the junctions by a resolvase activity that requires the Rad51 paralogs Rad51C and XRCC3 will separate the repaired DNA duplexes.

Once the DSB is detected, it is processed as the next step to its repair through homologous recombination (Fig. 3). In *S. cerevisiae*, the MRX (Mre11/Rad50/Xrs2) complex has been implicated in nucleolytic processing of the DSBs to produce the 3' tailed single-stranded DNA, which is the substrate for Rad51 binding (Lee et al. 1998). The mechanistic details of how the mammalian

MRN complex participates in this reaction are not clear, since its intrinsic exonuclease activity is 3' to 5' (Paull and Gellert 1998). Thus, how the DNA intermediate is handed off from the MRN bound and processed DNA end to Rad51 is still an open question.

4.2 Nucleoprotein filament formation

Rad51 is a critical and central protein in the process of homologous recombination. *S. cerevisiae* cells deleted for *RAD51* display reduced recombination and as a result are sensitive to a range of DNA damaging agents, including ionizing radiation, but they are still viable. In vertebrate cells, Rad51 deficiency has more dire consequences; Rad51 depletion in chicken DT40 cells leads to G2/M cell cycle arrest, genomic fragmentation and cell death (Sonoda et al. 1998), while targeted disruption of *Rad51* in mouse cells results in embryonic lethality (Lim and Hasty 1996; Tsuzuki et al. 1996). Thus, Rad51 is a critical protein for high fidelity DNA damage repair during proliferation of vertebrate cells.

The function and involvement of Rad51 has been characterized extensively by biochemical studies (Fig. 3). After a DSB has been detected and the DNA ends resected to produce 3' single-strand tails, it becomes the substrate for the Rad51 recombinase. Rad51 oligomerizes on the single-stranded DNA giving rise to a nucleoprotein filament, which then recognizes homologous duplex DNA in the genome, mediates joint molecule formation between the broken and intact template DNA, and promotes strand exchange between the recombining partner DNA molecules. Rad51 is aided in its function by accessory proteins, including the single-strand DNA binding protein RPA, Rad52 (in *S. cerevisiae*), the Rad51 paralogs, BRCA2 (in mammals) and Rad54.

At the cellular level, immunofluorescence experiments revealed that Rad51 occasionally forms punctuate nuclear accumulations referred to as foci (Haaf et al. 1995). Such spontaneous Rad51 foci are restricted to S phase (Tashiro et al. 1996), suggesting a role for Rad51 in supporting DNA replication, presumably through its involvement in recombination. In response to various global DNA damaging agents, most cells demonstrate an accumulation of Rad51 into foci, marking presumably the sites of damage. Similar patterns of Rad51 accumulation is also visualized when cells are hit by a number of local DNA damage induction techniques, including UV-A light micro-irradiation on nuclear DNA sensitized by incorporation of halogenated thymidine analogues, irradiation of cells with α-particles or heavy ions (Tashiro et al. 2000; Aten et al. 2004; Hauptner et al. 2004). Although it is attractive to equate Rad51 foci formation with Rad51 nucleoprotein filament formation, there is no evidence that support this assumption. Indeed, other proteins involved in homologous recombination, which show no biochemical evidence of nucleoprotein filament formation, such as BRCA2, ATM, and the MRN complex, have also been shown to form foci. Instead, foci formation might be related to the fact that DNA damage, even when present only locally, induces signals that can spread out into the surrounding chromosomal domains (Shiloh 2003; Fernandez-Capetillo et al. 2004; Lukas et al. 2005).

To form nucleoprotein filaments on single-stranded DNA, Rad51 has to negotiate with RPA. *In vitro*, RPA stimulates Rad51 nucleoprotein filament formation, most likely by removing inhibitory secondary structures in the single-stranded DNA (Alani et al. 1992; Sugiyama et al. 1997; Sugiyama et al. 1998). However, order-of-addition experiments using the *S. cerevisiae* proteins have shown that if RPA is added to single-stranded DNA prior to Rad51, the displacement of RPA does not occur unless the Rad52 mediator protein is also present (Sung 1997; New et al. 1998; Shinohara et al. 1998; Sugiyama et al. 1998; Song and Sung 2000; Sugiyama and Kowalczykowski 2002; Symington 2002). In mammalian cells, colocalization of RPA, Rad51 and single-stranded DNA has been observed in ionizing induced foci (Golub et al. 1998; Raderschall et al. 1999).

RAD52 has been shown to be a very important gene for DSB repair in *S. cerevisiae*: Rad52 mutants display a more severe repair phenotype than *rad51* mutants. This contrasts sharply with the effect of *Rad52* in vertebrate cells, where while *Rad51* is an essential gene, *Rad52* mutants hardly display any phenotypes. The absence of Rad52 does not affect viability or ionizing radiation sensitivity, and efficiency of gene targeting is only reduced by twofold in either mouse or chicken cells (Rijkers et al. 1998; Yamaguchi-Iwai et al. 1998). Furthermore, a systematic cell biological study analyzing foci formation by numerous homologous recombination and checkpoint proteins in living *S. cerevisiae* cells showed that Rad52 is required for Rad51 and Rad54 foci formation (Lisby et al. 2001, 2004; Lisby, this volume). By contrast, in mammalian cells, Rad52 is not required for foci formation of Rad51 and Rad54 (van Veelen et al. 2005b). Thus, while Rad52 in *S. cerevisiae* is placed early in the homologous recombination reaction, its role does not appear central in vertebrate cells. A possible reason for the sharp differences in *rad52* mutant phenotypes in yeast and vertebrates could be functional redundancy of Rad52. It has been shown that Rad52 deficient chicken cells in which the Rad51 paralog XRCC3 has also been deleted through a conditional knockout approach are unable to proliferate (Fujimori et al. 2001). In this respect, it is interesting to note that biochemical experiments support a role for a complex between the Rad51 paralogs Rad51B and Rad51C in Rad51 filament formation analogous to the activity of *S. cerevisiae* Rad52 (Sigurdsson et al. 2001).

In vertebrate cells, a total of five Rad51 paralogs have been identified, namely XRCC2, XRCC3, Rad51B, Rad51C and Rad51D (Thompson and Schild 2001; Thacker 2005). These proteins have 20 to 30% amino acid sequence identity to Rad51, and appear to function in complexes with each other. Two-hybrid and biochemical assay reveal the existence of a number of Rad51 paralog complexes, including complexes containing XRCC3/Rad51C and XRCC2/Rad51B/Rad51C/Rad51D (Schild et al. 2000; Masson et al. 2001a; Liu et al. 2002; Miller et al. 2002; Wiese et al. 2002). They display various biochemical activities, including DNA binding (including synthetic Holliday junctions), ATPase activity, DNA strand exchange activity and Holliday junction processing (Kurumizaka et al. 2002; Lio et al. 2003, 2004; Yamada et al. 2004; Yokoyama et al. 2004). The Rad51 paralogs are required for cell viability because all Rad51 paralog mouse knockouts that have been generated are lethal at the embryonic stage (Shu et al. 1999; Deans et al. 2000; Pittman and Schimenti 2000; Smiraldo et al. 2005; Thacker 2005). In contrast, chicken DT40 cells deleted for the Rad51 paralogs are viable, but this might be due to their additional inactivation of p53 (Takata et al. 2000, 2001). Rad51 paralog deficiency results in reduced homologous recombination efficiency, genome instability and DNA damage sensitivity, including ionizing radiation sensitivity (Johnson et al. 1999; Pierce et al. 1999; Godthelp et al. 2002; Lio et al. 2004), which is partially suppressed by the overexpression of Rad51 (Takata et al. 2001). DNA damage-induced Rad51 foci formation depends on the paralogs (Bishop et al. 1998; Takata et al. 2000, 2001) and the purified Rad51 paralog complex Rad51B/Rad51C stimulated Rad51 mediated strand exchange (Sigurdsson et al. 2001). Thus, the intimate relationship between Rad51 and the Rad51 paralogs is manifested at the genetic, the cell biological and biochemical level.

Another important mediator protein in homologous recombination is the breast cancer susceptibility gene product, BRCA2 (Christ et al. this volume). Carriers of mutations in the *BRCA2* gene are predisposed to breast, ovarian, prostate, and pancreatic cancer (Venkitaraman 2002). Rad51 and BRCA2 have many common features. For example, *BRCA2* is also an essential gene (Gowen et al. 1996; Hakem et al. 1996; Liu et al. 1996; Ludwig et al. 1997; Sharan et al. 1997; Suzuki et al. 1997); both mouse and human hypomorphic mutant cell lines display chromosomal instability and sensitivity to DNA damaging agents (Tutt et al. 1999; Scully et al. 2000; Yu et al. 2000), as well as reduced recombination efficiency (Moynahan et al. 2001). Like the Rad51 paralogs, BRCA2 is required for DNA damage-induced Rad51 foci formation (Yuan et al. 1999, 2000). These similar features suggest that the direct protein-protein interaction between BRCA2 and Rad51 is of functional importance (Scully et al. 1997; Wong et al. 1997; Marmorstein et al. 1998; Davies et al. 2001; Pellegrini et al. 2002; Venkitaraman 2002).

Recently, a mechanistic basis for the mediator function of BRCA2 with respect to Rad51 nucleoprotein filaments formation was suggested. A BRCA2 family member, Brh2, has been discovered in *Ustilago maydis*, which is required for repair and recombination proficiency (Kojic et al. 2002). An ortholog of Dss1, a protein that interacts with BRCA2, has also been found to interact with Brh2, and is important in genome stability and recombination (Kojic et al. 2003, 2005). Brh2 functions to recruit Rad51 to DNA and aids in the nucleation of the Rad51 filament, establishing a function for BRCA2 protein in Rad51 mediated repair of DSBs (Yang et al. 2005). A similar activity has been established for a minimal version of human BRCA2, containing some of its Rad51 interaction domains and the DNA binding domain (San Filippo et al. 2006).

At the cell biological level, BRCA2 also forms DNA damage-induced foci, which colocalize with Rad51 foci (Chen et al. 1998, 1999). In living cells, the interplay between BRCA2 and Rad51 has been investigated using GFP-tagged Rad51. FRAP experiments revealed the existence of two different nuclear pools of Rad51 with respect its mobility; a mobile fraction and a relatively immobile fraction (Essers et al. 2002). The relatively immobile fraction of Rad51 molecules is bound to BRCA2 and this fraction is reduced upon replication arrest with hydroxyurea (Yu et al. 2003). This behavior is lost for Rad51 mutants that no longer interact with BRCA2, suggesting a role for BRCA2 in the Rad51 DNA damage response at the cellular level. Given the behavior of BRCA2 at the biochemical and cellular level and the fact that *S. cerevisiae* does not contain a BRCA2 homolog, BRCA2 is, in addition or in concert with the Rad51 paralogs, also a candidate for the mammalian equivalent of *S. cerevisiase* Rad52 activity.

Once the nucleoprotein filament is formed and has found the target duplex, the next step is the joint molecule formation. This is a critical step in homologous recombination, in which Rad51 is aided by Rad54. Rad54 is a member of the SNF2/SWI2 family of DNA-dependent ATPases, which have been implicated in modulating protein-DNA interactions. Mouse and chicken Rad54 deficient cells show sensitivity to DSB inducing agents, and a reduced level of homologous recombination (Bezzubova et al. 1997; Essers et al. 1997; Dronkert et al. 2000). The absence of Rad54 is compatible with mouse development, in spite of the fact that *Rad54* knockout mice as well as *Rad54* knockout ES cells are sensitive to mitomycin C. By contrast, unlike ES cells, adult *Rad54* knockout mice are not sensitive to ionizing radiation (Essers et al. 2000). However, the contribution of Rad54 to repair of ionizing radiation induced DNA damage in adult mice is clearly evident when the *Rad54* knockout mutation is combined with a defect in the DNA end-joining pathway, either due to mutation of DNA-Pk_{cs}, Ku70, or DNA Ligase IV. The phenotypes related to DNA damage sensitivity and genome instability of the DNA end joining defective mice are dramatically enhanced (Essers et al. 2000; Couedel et al. 2004; Mills et al. 2004).

Biochemical experiments have revealed that the important substrate of Rad54 in recombination is double-stranded template DNA: only double-stranded DNA activates its ATPase activity (Swagemakers et al. 1998; Petukhova et al. 1999). Rad54 has been shown to be a motor protein on DNA, whose translocation can lead to supercoiling of DNA domains thereby lowering the energy required to separate the strands of the double helix (Petukhova et al. 1999; Tan et al. 1999; Mazin et al. 2000; Van Komen et al. 2000; Ristic et al. 2001). This activity is important during the strand invasion step of the Rad51 coated single-stranded DNA into the template duplex. Indeed, the Rad54 protein interacts with Rad51 (Clever et al. 1997; Golub et al. 1997; Tan et al. 1999; Van Komen et al. 2000; Raschle et al. 2004) and this interaction has functional consequences, for example the stimulation of Rad51 mediated joint molecule formation by Rad54 (Petukhova et al. 1998; Mazin et al. 2000). In addition to this early role, biochemical experiments have also suggested a late role for Rad54 in recombination. Rad54 can remove Rad51 filaments from double-stranded DNA (Solinger and Heyer 2001; Solinger et al. 2002). Evidence for the importance of this role of Rad54 at the cellular level is provided by experiments that showed that homologously paired molecules in *S. cerevisiae* cells could not be extended by a DNA polymerase in the absence of Rad54 (Sugawara et al. 2003). In addition, during meiosis in *Rad54* knockout mice, Rad51 protein appears to remain associated with chromatin loops of synapsed chromosomes (Wesoly et al. 2006).

The finding that Rad51 and Rad54 interact closely in biochemical assays is further confirmed by cell biology analysis. Like Rad51, Rad54 forms DNA damaged induced foci and these foci colocalize (Tan et al. 1999). Under conditions in which

Rad51 DNA damage induced foci do not form, such as in the Rad51 paralog mutants, Rad54 also fails to form foci (van Veelen et al. 2005b). In the absence of Rad54, Rad51 foci appear to be destabilized (Tan et al. 1999; van Veelen et al. 2005a). The reduced stability of Rad51 aggregation in cells lacking Rad54 is consistent with the biochemical demonstration that Rad54 can stabilize Rad51 nucleoprotein filaments (Mazin et al. 2003).

The Rad54 protein has also been analyzed in living cells. The first study to analyze DNA damage induced foci in live cells revealed similar aggregations of recombination proteins at sites of DNA damage as seen in fixed cells (Essers et al. 2002). FRAP experiments demonstrated that these foci are highly dynamic; Rad51 and Rad54 proteins actively sample these foci through an equilibrium of association and dissociation, but they display different residence times. Furthermore, even though both recombination proteins have worked together in recombination, they are not present in the cell as a holo-complex in the absence of DNA damage because they diffuse through the cell at different rates. Executing DNA transactions through dynamic multi-protein complexes, rather than stable holo-complexes, allows flexibility. For example, it will facilitate cross-talk between different DNA repair pathways and coupling to other DNA transactions, such as replication.

4.3 Resolution

Once the joint molecule between the nucleoprotein filament and target duplex is formed, the information lost during end processing can be restored by DNA polymerases (Fig. 3). Recently, the translesion DNA polymerase eta has been implicated in this step (Rattray and Strathern 2005). DNA polymerase eta relocalizes into foci upon UV irradiation and those foci colocalize with Rad51 (Kannouche et al. 2001). A chicken B cell derived cell line deficient in DNA polymerase eta displays defects in DSB-induced homologous recombination (Kawamoto et al. 2005). In addition, DNA polymerase eta interacts with Rad51 and can extend DNA synthesis from joint molecule recombination intermediates (McLlwraith et al. 2005). Chromatin immunoprecipitation experiments indicate that the *S. cerevisiae* Rad54 protein is important in promoting the transition from pairing of homologous DNA strands by Rad51 to extension of the invading strand by DNA polymerases (Sugawara et al. 2003). Possibly, Rad54's potential to remove Rad51 nucleoprotein filaments from double-stranded DNA might be important in promoting this step in homologous recombination (Solinger and Heyer 2001; Solinger et al. 2002).

After all sequences are restored, remaining single-strand nicks are sealed by DNA ligase. At this stage the recombined DNA molecules can be physically joined in a structure often referred to as a Holliday junction (Fig. 3). To complete recombination this junction needs to be resolved (Whitby, this volume). In *E. coli*, this reaction is carried out by the RuvABC complex. The RuvA and RuvB proteins promotes ATP-dependent branch migration of the Holliday junction, while RuvC introduces nicks in two of the four DNA strands of the Holliday junction allowing resolution of the junction into recombinant DNA molecules (West 1997). In mammalian cells, less is known about the proteins involved in resolution of Holliday junctions; however, some initial studies have found clues in elucidating this activity (Waldman and Liskay 1988; Hyde et al. 1994; Constantinou et al. 2001, 2002).

As described above, several biochemical and cellular studies have resulted in the suggestion that the Rad51 paralogs have an early function in loading Rad51 onto single-stranded DNA during the assembly of the nucleoprotein filament (Masson et al. 2001b; Sigurdsson et al. 2001; Yonetani et al. 2005). Interestingly, the identification of Rad51C and XRCC3 as components of an activity that promotes Holliday junction resolution suggest that at least some Rad51 paralogs can also have a late role in homologous recombination (Liu et al. 2004). Further evidence for this notion comes from studies showing that Rad51B can bind preferentially to synthetic Holliday junctions (Yokoyama et al. 2003). Support at the cellular level for a late function of Rad51 paralogs associated with resolution of recombination intermediates comes from studies on XRCC3 mutant hamster cells that showed that gene conversion tract lengths are increased in the absence of XRCC3 (Brenneman et al. 2000). Once the Holliday junction has been cleaved by the resolvase, the partner DNA molecules are separated and ligation of the resolvase-induced single strand nicks will produce two completely restored duplex DNA recombinants (Fig. 3).

5 Recombination and replication

Above we focused on homologous recombination in the context of the repair of a pre-existing DSB. Joint molecule formation between the broken DNA and the intact repair duplex catalyzed by Rad51 and accessory factors sets up the substrate for DNA polymerases such that DNA replication can restore information lost by processing of the DSB. Conversely, homologous recombination also plays an important role in supporting DNA replication when the replication fork encounters DNA damage in its template for example thymidine intra-strand dimers induced by UV-light (Michel et al. 2004; Branzei et al. this volume). Depending on the type of DNA damage, processing might or might not result in a DSB. In either case, homologous recombination proteins are involved in helping the replication machinery pass the damage (Cox et al. 2000). The presence of unrepaired DNA damage serves as a block to the passage of the replication machinery. The bypass or repair of these blocks and the subsequent fidelity of DNA replication requires several coordinated processes, including chromatin remodeling, DNA repair and DNA synthesis, which has to occur in an ordered manner to achieve proper cell division. The synthesis of DNA past lesions requires the use of specialized DNA polymerases that bypass them, such as DNA polymerase eta, since the highly stringent replicative DNA polymerases cannot accommodate damaged bases in its active site (Prakash et al. 2005). It is for this reason that DNA synthesis during S phase of cells is blocked in the presence of unrepaired lesions and as a consequence, replication stalls, the replisome dissociates and the forks collapses. The resulting structures that emerge are substrates for homologous recombination.

The central protein in DNA replication and several forms of DNA repair, including nucleotide excision repair (NER) is the DNA polymerase processivity factor proliferating cell nuclear antigen (PCNA), which localizes proteins such as polymerases to DNA (Ellison and Stillman 2003). Recently, the coordination between DNA repair and replication has been studied by determining the behavior of GFP-tagged PCNA in living cells using photobleaching. While PCNA molecules move rapidly through the nucleus during the G1, S, and G2 phases of the cell cycle, they reside for 10-20 minutes in replication foci during S phase (Sporbert et al. 2002; Solomon et al. 2004; Essers et al. 2005; Solovjeva et al. 2005). To simultaneously monitor PCNA action in DNA replication and repair, local irradiation has shown an accumulation of PCNA at sites of UV-light induced DNA damage in brightly fluorescent regions, on top of the typical replication pattern. Photobleaching experiments have revealed that PCNA also binds transiently to these local UV-damaged areas although residence times are considerably longer compared to replication foci (Solomon et al. 2004; Essers et al. 2006). This difference is not found in a PCNA mutant that can no longer be ubiquitinated (PCNA K164R), showing that one function of mono-ubiquitination of PCNA is to modulate the residence time of PCNA at sites of DNA damage (Essers et al. 2006). Similar analysis also revealed the residence time of other replication factors, such as Fen1, DNA ligase I, and RPA, which showed significant faster turnover rates at replication foci compared to PCNA, (Sporbert et al. 2002; Chapados et al. 2004; Solovjeva et al. 2005). This is reminiscent of what has been found for the IR-induced foci of the homologous recombination proteins Rad51 and Rad54 (Essers et al. 2002). Rad51, like PCNA, is a more stable component of the DNA damage-induced foci, while Rad54 reversibly interacts with these structures. The differential mobility of these proteins likely reflects their functional status *in vivo* and can therefore be used as an analytical tool to explore their function.

6 The function of DNA damage induced foci

The formation of foci containing proteins involved in homologous recombination and checkpoint activation at sites of DNA damage is a remarkable phenomenon. Clearly, many molecules of each protein must accumulate at those sites; in order for a focus to be detected by immunofluoresence or by GFP signal, the number of molecules present is in the order of 100. To gain insight into the function of foci, important questions to be answered include: (1) What is the composition of these foci? (2) Why do they contain so many molecules of proteins? (3) How do they form and how are they disassembled?

To determine the identity of all proteins in a focus is not a straightforward problem. Methods that have been successful in the analysis of the proteome of other subnuclear organelles such as nucleoli are not easily adapted to analyze foci (Andersen et al. 2002). Because biochemically isolated nucleoli can be tested for a particular activity, it can be safely assumed that mass spectrometric analysis reveals the proteome of the active subnuclear organelle. Methods to isolate foci containing homologous recombination proteins are yet to be developed. The most promising approach might be to perform *in vivo* crosslinking experiments of complexes near site-specific DSBs and fishing for a specific sequence near the DSB. However, in the absence of an *in vitro* activity assay for foci activity, the interpretation of subsequent proteome analysis will be ambiguous.

Besides its composition, it would be interesting to determine why so many molecules of the homologous recombination proteins accumulate at the sites of DNA damage. It is conceivable that in order to do their job, homologous recombination proteins need to act much more locally at the site of damage, as it is clear from biochemical experiments that the actual stoichiometry of proteins required to take the DNA strands through recombination is much lower than the sheer numbers of proteins that accumulate into foci. In addition, the requirement for high local concentrations of DNA repair proteins at damage sites is not a general prerequisite for repair, as shown by the fact that global UV-light irradiation of cells does not result in the formation of foci of NER proteins, for example (Houtsmuller and Vermeulen 2001). Therefore the necessity for foci in the repair of damage by recombination is still an open question. It is possible that such an accumulation could synchronize the presence and function of the various protein components of recombination both spatially and temporally, since it has been shown that the enzymes of homologous recombination have to work together in a timely and highly coordinated manner.

Foci of homologous recombination proteins near sites of DNA damage might simply form because the proteins could have a higher affinity for damaged compared to undamaged chromosomal domains. This would require a mechanism that distinguishes between damaged and undamaged chromosomal domains. One possible marker for chromosomal domains containing DNA damage is γH2AX. The increased local concentration of γH2AX can be rationalized as a marker of the location of a DSB, since this modification is present in the megabase chromosomal domain that contains the damage (Rogakou et al. 1999). However, the mild phenotype of *H2AX* knockout cells and mice argues that there must be other potential more dominant distinctions.

Thus, once a DSB arises, modification of the chromosomal domain it is contained in might create a site with a slightly increased affinity for the recombination proteins compared to intact chromosomal domains. The difference in affinity ensures that the proteins will be immobilized for a longer time at the damage-containing chromosomal domain, resulting in an accumulation of homologous recombination proteins at the DSB site. Moreover, the homogeneous distribution of freely mobile DNA repair proteins (Essers et al. 2002) ensures that all required factors are always present in the vicinity of DNA lesions wherever they occur, allowing rapid and efficient detection and subsequent repair.

New insight into the mechanism of homologous recombination repair in living cells will come from analyzing the behavior of proteins with biochemically characterized mutations to see how these affect their *in vivo* behavior. It is clear that the technology is in place to sort through the mechanistic possibilities suggested from genetic and biochemical studies of homologous recombination.

References

Alani E, Thresher R, Griffith JD, Kolodner RD (1992) Characterization of DNA-binding and strand-exchange stimulation properties of y-RPA, a yeast single-strand-DNA-binding protein. J Mol Biol 227:54-71

Andersen JS, Lyon CE, Fox AH, Leung AK, Lam YW, Steen H, Mann M, Lamond AI (2002) Directed proteomic analysis of the human nucleolus. Curr Biol 12:1-11

Aten JA, Stap J, Krawczyk PM, van Oven CH, Hoebe RA, Essers J, Kanaar R (2004) Dynamics of DNA double-strand breaks revealed by clustering of damaged chromosome domains. Science 303:92-95

Bezzubova O, Silbergleit A, Yamaguchi-Iwai Y, Takeda S, Buerstedde JM (1997) Reduced X-ray resistance and homologous recombination frequencies in a RAD54-/- mutant of the chicken DT40 cell line. Cell 89:185-193

Bishop DK, Ear U, Bhattacharyya A, Calderone C, Beckett M, Weichselbaum RR, Shinohara A (1998) Xrcc3 is required for assembly of Rad51 complexes *in vivo*. J Biol Chem 273:21482-21488

Bradshaw PS, Stavropoulos DJ, Meyn MS (2005) Human telomeric protein TRF2 associates with genomic double-strand breaks as an early response to DNA damage. Nat Genet 37:193-197

Brenneman MA, Weiss AE, Nickoloff JA, Chen DJ (2000) XRCC3 is required for efficient repair of chromosome breaks by homologous recombination. Mutat Res 459:89-97

Carney JP, Maser RS, Olivares H, Davis EM, Le Beau M, Yates JR 3rd, Hays L, Morgan WF, Petrini JH (1998) The hMre11/hRad50 protein complex and Nijmegen breakage syndrome: linkage of double-strand break repair to the cellular DNA damage response. Cell 93:477-486

Carrero G, McDonald D, Crawford E, de Vries G, Hendzel MJ (2003) Using FRAP and mathematical modeling to determine the *in vivo* kinetics of nuclear proteins. Methods 29:14-28

Chapados BR, Hosfield DJ, Han S, Qiu J, Yelent B, Shen B, Tainer JA (2004) Structural basis for FEN-1 substrate specificity and PCNA-mediated activation in DNA replication and repair. Cell 116:39-50

Chen JJ, Silver D, Cantor S, Livingston DM, Scully R (1999) BRCA1, BRCA2, and Rad51 operate in a common DNA damage response pathway. Cancer Res 59:1752s-1756s

Chen PL, Chen CF, Chen Y, Xiao J, Sharp ZD, Lee WH (1998) The BRC repeats in BRCA2 are critical for RAD51 binding and resistance to methyl methanesulfonate treatment. Proc Natl Acad Sci USA 95:5287-5292

Clever B, Interthal H, Schmuckli-Maurer J, King J, Sigrist M, Heyer WD (1997) Recombinational repair in yeast: functional interactions between Rad51 and Rad54 proteins. Embo J 16:2535-2544

Constantinou A, Chen XB, McGowan CH, West SC (2002) Holliday junction resolution in human cells: two junction endonucleases with distinct substrate specificities. Embo J 21:5577-5585

Constantinou A, Davies AA, West SC (2001) Branch migration and Holliday junction resolution catalyzed by activities from mammalian cells. Cell 104:259-268

Costanzo V, Paull T, Gottesman M, Gautier J (2004) Mre11 assembles linear DNA fragments into DNA damage signaling complexes. PLoS Biol 2:E110

Couedel C, Mills KD, Barchi M, Shen L, Olshen A, Johnson RD, Nussenzweig A, Essers J, Kanaar R, Li GC, Alt FW, Jasin M (2004) Collaboration of homologous recombination and nonhomologous end-joining factors for the survival and integrity of mice and cells. Genes Dev 18:1293-1304

Cox MM, Goodman MF, Kreuzer KN, Sherratt DJ, Sandler SJ, Marians KJ (2000) The importance of repairing stalled replication forks. Nature 404:37-41

Cremer C, Cremer T, Fukuda M, Nakanishi K (1980) Detection of laser--UV microirradiation-induced DNA photolesions by immunofluorescent staining. Hum Genet 54:107-110

D'Amours D, Jackson SP (2002) The Mre11 complex: at the crossroads of dna repair and checkpoint signalling. Nat Rev Mol Cell Biol 3:317-327

Davies AA, Masson JY, McIlwraith MJ, Stasiak AZ, Stasiak A, Venkitaraman AR, West SC (2001) Role of BRCA2 in control of the RAD51 recombination and DNA repair protein. Mol Cell 7:273-282

de Jager M, Dronkert ML, Modesti M, Beerens CE, Kanaar R, van Gent DC (2001a) DNA-binding and strand-annealing activities of human Mre11: implications for its roles in DNA double-strand break repair pathways. Nucleic Acids Res 29:1317-1325

de Jager M, van Noort J, van Gent DC, Dekker C, Kanaar R, Wyman C (2001b) Human Rad50/Mre11 is a flexible complex that can tether DNA ends. Mol Cell 8:1129-1135

de Jager M, Wyman C, van Gent DC, Kanaar R (2002) DNA end-binding specificity of human Rad50/Mre11 is influenced by ATP. Nucleic Acids Res 30:4425-4431

Deans B, Griffin CS, Maconochie M, Thacker J (2000) Xrcc2 is required for genetic stability, embryonic neurogenesis and viability in mice. Embo J 19:6675-6685

Dronkert ML, Beverloo HB, Johnson RD, Hoeijmakers JH, Jasin M, Kanaar R (2000) Mouse RAD54 affects DNA double-strand break repair and sister chromatid exchange. Mol Cell Biol 20:3147-3156

Dudas A, Chovanec M (2004) DNA double-strand break repair by homologous recombination. Mutat Res 566:131-167

Durai S, Mani M, Kandavelou K, Wu J, Porteus MH, Chandrasegaran S (2005) Zinc finger nucleases: custom-designed molecular scissors for genome engineering of plant and mammalian cells. Nucleic Acids Res 33:5978-5990

Ellison V, Stillman B (2003) Biochemical characterization of DNA damage checkpoint complexes: clamp loader and clamp complexes with specificity for 5' recessed DNA. PLoS Biol 1:E33

Essers J, Hendriks RW, Swagemakers SM, Troelstra C, de Wit J, Bootsma D, Hoeijmakers JH, Kanaar R (1997) Disruption of mouse RAD54 reduces ionizing radiation resistance and homologous recombination. Cell 89:195-204

Essers J, Houtsmuller AB, Kanaar R (2006) Analysis of DNA recombination and repair proteins in living cells by photobleaching microscopy. Methods Enzymol 408:463-485

Essers J, Houtsmuller AB, van Veelen L, Paulusma C, Nigg AL, Pastink A, Vermeulen W, Hoeijmakers JH, Kanaar R (2002) Nuclear dynamics of RAD52 group homologous recombination proteins in response to DNA damage. Embo J 21:2030-2037

Essers J, Theil AF, Baldeyron C, van Cappellen WA, Houtsmuller AB, Kanaar R, Vermeulen W (2005) Nuclear dynamics of PCNA in DNA replication and repair. Mol Cell Biol 25:9350-9359

Essers J, van Steeg H, de Wit J, Swagemakers SM, Vermeij M, Hoeijmakers JH, Kanaar R (2000) Homologous and non-homologous recombination differentially affect DNA damage repair in mice. Embo J 19:1703-1710

Fernandez-Capetillo O, Lee A, Nussenzweig M, Nussenzweig A (2004) H2AX: the histone guardian of the genome. DNA Repair (Amst) 3:959-967

Friedberg EC, McDaniel LD, Schultz RA (2004) The role of endogenous and exogenous DNA damage and mutagenesis. Curr Opin Genet Dev 14:5-10

Friedberg EC, Walker GC, Siede W, Wood RD, Schultz RA, Ellenberger T (2006) DNA Repair And Mutagenesis. ASM Press, Washington, D.C.

Fujimori A, Tachiiri S, Sonoda E, Thompson LH, Dhar PK, Hiraoka M, Takeda S, Zhang Y, Reth M, Takata M (2001) Rad52 partially substitutes for the Rad51 paralog XRCC3 in maintaining chromosomal integrity in vertebrate cells. Embo J 20:5513-5520

Game JC, Mortimer RK (1974) A genetic study of x-ray sensitive mutants in yeast. Mutat Res 24:281-292

Giepmans BN, Adams SR, Ellisman MH, Tsien RY (2006) The fluorescent toolbox for assessing protein location and function. Science 312:217-224

Godthelp BC, Wiegant WW, van Duijn-Goedhart A, Scharer OD, van Buul PP, Kanaar R, Zdzienicka MZ (2002) Mammalian Rad51C contributes to DNA cross-link resistance, sister chromatid cohesion and genomic stability. Nucleic Acids Res 30:2172-2182

Goldberg M, Stucki M, Falck J, D'Amours D, Rahman D, Pappin D, Bartek J, Jackson SP (2003) MDC1 is required for the intra-S-phase DNA damage checkpoint. Nature 421:952-956

Golub EI, Gupta RC, Haaf T, Wold MS, Radding CM (1998) Interaction of human rad51 recombination protein with single-stranded DNA binding protein, RPA. Nucleic Acids Res 26:5388-5393

Golub EI, Kovalenko OV, Gupta RC, Ward DC, Radding CM (1997) Interaction of human recombination proteins Rad51 and Rad54. Nucleic Acids Res 25:4106-4110

Gowen LC, Johnson BL, Latour AM, Sulik KK, Koller BH (1996) Brca1 deficiency results in early embryonic lethality characterized by neuroepithelial abnormalities. Nat Genet 12:191-194

Haaf T, Golub EI, Reddy G, Radding CM, Ward DC (1995) Nuclear foci of mammalian Rad51 recombination protein in somatic cells after DNA damage and its localization in synaptonemal complexes. Proc Natl Acad Sci USA 92:2298-2302

Haber JE (1995) *In vivo* biochemistry: physical monitoring of recombination induced by site-specific endonucleases. Bioessays 17:609-620

Haber JE (2000) Lucky breaks: analysis of recombination in *Saccharomyces*. Mutat Res 451:53-69

Hakem R, de la Pompa JL, Sirard C, Mo R, Woo M, Hakem A, Wakeham A, Potter J, Reitmair A, Billia F, Firpo E, Hui CC, Roberts J, Rossant J, Mak TW (1996) The tumor suppressor gene Brca1 is required for embryonic cellular proliferation in the mouse. Cell 85:1009-1023

Haraguchi T (2002) Live cell imaging: approaches for studying protein dynamics in living cells. Cell Struct Funct 27:333-334

Hauptner A, Dietzel S, Drexler GA, Reichart P, Krucken R, Cremer T, Friedl AA, Dollinger G (2004) Microirradiation of cells with energetic heavy ions. Radiat Environ Biophys 42:237-245

Hopfner KP, Craig L, Moncalian G, Zinkel RA, Usui T, Owen BA, Karcher A, Henderson B, Bodmer JL, McMurray CT, Carney JP, Petrini JH, Tainer JA (2002) The Rad50 zinc-hook is a structure joining Mre11 complexes in DNA recombination and repair. Nature 418:562-566

Houtsmuller AB (2005) Fluorescence recovery after photobleaching: application to nuclear proteins. Adv Biochem Eng Biotechnol 95:177-199

Houtsmuller AB, Vermeulen W (2001) Macromolecular dynamics in living cell nuclei revealed by fluorescence redistribution after photobleaching. Histochem Cell Biol 115:13-21

Hyde H, Davies AA, Benson FE, West SC (1994) Resolution of recombination intermediates by a mammalian activity functionally analogous to *Escherichia coli* RuvC resolvase. J Biol Chem 269:5202-5209

Jakob B, Scholz M, Taucher-Scholz G (2003) Biological imaging of heavy charged-particle tracks. Radiat Res 159:676-684

Jasin M (1996) Genetic manipulation of genomes with rare-cutting endonucleases. Trends Genet 12:224-228

Johnson RD, Liu N, Jasin M (1999) Mammalian XRCC2 promotes the repair of DNA double-strand breaks by homologous recombination. Nature 401:397-399

Kanaar R, Hoeijmakers JH, van Gent DC (1998) Molecular mechanisms of DNA double strand break repair. Trends Cell Biol 8:483-489

Kannouche P, Broughton BC, Volker M, Hanaoka F, Mullenders LH, Lehmann AR (2001) Domain structure, localization, and function of DNA polymerase eta, defective in xeroderma pigmentosum variant cells. Genes Dev 15:158-172

Kawamoto T, Araki K, Sonoda E, Yamashita YM, Harada K, Kikuchi K, Masutani C, Hanaoka F, Nozaki K, Hashimoto N, Takeda S (2005) Dual roles for DNA polymerase eta in homologous DNA recombination and translesion DNA synthesis. Mol Cell 20:793-799

Kim JS, Krasieva TB, Kurumizaka H, Chen DJ, Taylor AM, Yokomori K (2005) Independent and sequential recruitment of NHEJ and HR factors to DNA damage sites in mammalian cells. J Cell Biol 170:341-347

Kim JS, Krasieva TB, LaMorte V, Taylor AM, Yokomori K (2002) Specific recruitment of human cohesin to laser-induced DNA damage. J Biol Chem 277:45149-45153

Kojic M, Kostrub CF, Buchman AR, Holloman WK (2002) BRCA2 homolog required for proficiency in DNA repair, recombination, and genome stability in *Ustilago maydis*. Mol Cell 10:683-691

Kojic M, Yang H, Kostrub CF, Pavletich NP, Holloman WK (2003) The BRCA2-interacting protein DSS1 is vital for DNA repair, recombination, and genome stability in *Ustilago maydis*. Mol Cell 12:1043-1049

Kojic M, Zhou Q, Lisby M, Holloman WK (2005) Brh2-Dss1 interplay enables properly controlled recombination in *Ustilago maydis*. Mol Cell Biol 25:2547-2557

Krogh BO, Symington LS (2004) Recombination proteins in yeast. Annu Rev Genet 38:233-271

Kurumizaka H, Ikawa S, Nakada M, Enomoto R, Kagawa W, Kinebuchi T, Yamazoe M, Yokoyama S, Shibata T (2002) Homologous pairing and ring and filament structure

formation activities of the human Xrcc2*Rad51D complex. J Biol Chem 277:14315-14320
Lee SE, Moore JK, Holmes A, Umezu K, Kolodner RD, Haber JE (1998) *Saccharomyces* Ku70, mre11/rad50 and RPA proteins regulate adaptation to G2/M arrest after DNA damage. Cell 94:399-409
Lim DS, Hasty P (1996) A mutation in mouse rad51 results in an early embryonic lethal that is suppressed by a mutation in p53. Mol Cell Biol 16:7133-7143
Lio YC, Mazin AV, Kowalczykowski SC, Chen DJ (2003) Complex formation by the human Rad51B and Rad51C DNA repair proteins and their activities *in vitro*. J Biol Chem 278:2469-2478
Lio YC, Schild D, Brenneman MA, Redpath JL, Chen DJ (2004) Human Rad51C deficiency destabilizes XRCC3, impairs recombination, and radiosensitizes S/G2-phase cells. J Biol Chem 279:42313-42320
Lippincott-Schwartz J, Patterson GH (2003) Development and use of fluorescent protein markers in living cells. Science 300:87-91
Lisby M, Barlow JH, Burgess RC, Rothstein R (2004) Choreography of the DNA damage response: spatiotemporal relationships among checkpoint and repair proteins. Cell 118:699-713
Lisby M, Rothstein R, Mortensen UH (2001) Rad52 forms DNA repair and recombination centers during S phase. Proc Natl Acad Sci USA 98:8276-8282
Liu CY, Flesken-Nikitin A, Li S, Zeng Y, Lee WH (1996) Inactivation of the mouse Brca1 gene leads to failure in the morphogenesis of the egg cylinder in early postimplantation development. Genes Dev 10:1835-1843
Liu N, Schild D, Thelen MP, Thompson LH (2002) Involvement of Rad51C in two distinct protein complexes of Rad51 paralogs in human cells. Nucleic Acids Res 30:1009-1015
Liu Y, Masson JY, Shah R, O'Regan P, West SC (2004) RAD51C is required for Holliday junction processing in mammalian cells. Science 303:243-246
Ludwig T, Chapman DL, Papaioannou VE, Efstratiadis A (1997) Targeted mutations of breast cancer susceptibility gene homologs in mice: lethal phenotypes of Brca1, Brca2, Brca1/Brca2, Brca1/p53, and Brca2/p53 nullizygous embryos. Genes Dev 11:1226-1241
Lukas C, Bartek J, Lukas J (2005) Imaging of protein movement induced by chromosomal breakage: tiny 'local' lesions pose great 'global' challenges. Chromosoma 114:146-154
Lukas C, Falck J, Bartkova J, Bartek J, Lukas J (2003) Distinct spatiotemporal dynamics of mammalian checkpoint regulators induced by DNA damage. Nat Cell Biol 5:255-260
Lukas C, Melander F, Stucki M, Falck J, Bekker-Jensen S, Goldberg M, Lerenthal Y, Jackson SP, Bartek J, Lukas J (2004) Mdc1 couples DNA double-strand break recognition by Nbs1 with its H2AX-dependent chromatin retention. Embo J 23:2674-2683
Luo G, Yao MS, Bender CF, Mills M, Bladl AR, Bradley A, Petrini JH (1999) Disruption of mRad50 causes embryonic stem cell lethality, abnormal embryonic development, and sensitivity to ionizing radiation. Proc Natl Acad Sci USA 96:7376-7381
Marmorstein LY, Ouchi T, Aaronson SA (1998) The BRCA2 gene product functionally interacts with p53 and RAD51. Proc Natl Acad Sci USA 95:13869-13874
Maser RS, Monsen KJ, Nelms BE, Petrini JH (1997) hMre11 and hRad50 nuclear foci are induced during the normal cellular response to DNA double-strand breaks. Mol Cell Biol 17:6087-6096

Masson JY, Stasiak AZ, Stasiak A, Benson FE, West SC (2001a) Complex formation by the human RAD51C and XRCC3 recombination repair proteins. Proc Natl Acad Sci USA 98:8440-8446

Masson JY, Tarsounas MC, Stasiak AZ, Stasiak A, Shah R, McIlwraith MJ, Benson FE, West SC (2001b) Identification and purification of two distinct complexes containing the five RAD51 paralogs. Genes Dev 15:3296-3307

Mazin AV, Alexeev AA, Kowalczykowski SC (2003) A novel function of Rad54 protein. Stabilization of the Rad51 nucleoprotein filament. J Biol Chem 278:14029-14036

Mazin AV, Bornarth CJ, Solinger JA, Heyer WD, Kowalczykowski SC (2000) Rad54 protein is targeted to pairing loci by the Rad51 nucleoprotein filament. Mol Cell 6:583-592

McLlwraith MJ, Vaisman A, Liu Y, Fanning E, Woodgate R, West SC (2005) Human DNA polymerase eta promotes DNA synthesis from strand invasion intermediates of homologous recombination. Mol Cell 20:783-792

Meldrum RA, Botchway SW, Wharton CW, Hirst GJ (2003) Nanoscale spatial induction of ultraviolet photoproducts in cellular DNA by three-photon near-infrared absorption. EMBO Rep 4:1144-1149

Michel B, Grompone G, Flores MJ, Bidnenko V (2004) Multiple pathways process stalled replication forks. Proc Natl Acad Sci USA 101:12783-12788

Miller KA, Yoshikawa DM, McConnell IR, Clark R, Schild D, Albala JS (2002) RAD51C interacts with RAD51B and is central to a larger protein complex *in vivo* exclusive of RAD51. J Biol Chem 277:8406-8411

Mills KD, Ferguson DO, Essers J, Eckersdorff M, Kanaar R, Alt FW (2004) Rad54 and DNA Ligase IV cooperate to maintain mammalian chromatid stability. Genes Dev 18:1283-1292

Mirzoeva OK, Petrini JH (2003) DNA replication-dependent nuclear dynamics of the Mre11 complex. Mol Cancer Res 1:207-218

Moreno-Herrero F, de Jager M, Dekker NH, Kanaar R, Wyman C, Dekker C (2005) Mesoscale conformational changes in the DNA-repair complex Rad50/Mre11/Nbs1 upon binding DNA. Nature 437:440-443

Moynahan ME, Pierce AJ, Jasin M (2001) BRCA2 is required for homology-directed repair of chromosomal breaks. Mol Cell 7:263-272

Nakata T, Terada S, Hirokawa N (1998) Visualization of the dynamics of synaptic vesicle and plasma membrane proteins in living axons. J Cell Biol 140:659-674

Nelms BE, Maser RS, MacKay JF, Lagally MG, Petrini JH (1998) In situ visualization of DNA double-strand break repair in human fibroblasts. Science 280:590-592

New JH, Sugiyama T, Zaitseva E, Kowalczykowski SC (1998) Rad52 protein stimulates DNA strand exchange by Rad51 and replication protein A. Nature 391:407-410

Paull TT, Gellert M (1998) The 3' to 5' exonuclease activity of Mre 11 facilitates repair of DNA double-strand breaks. Mol Cell 1:969-979

Paull TT, Gellert M (1999) Nbs1 potentiates ATP-driven DNA unwinding and endonuclease cleavage by the Mre11/Rad50 complex. Genes Dev 13:1276-1288

Paull TT, Rogakou EP, Yamazaki V, Kirchgessner CU, Gellert M, Bonner WM (2000) A critical role for histone H2AX in recruitment of repair factors to nuclear foci after DNA damage. Curr Biol 10:886-895

Pellegrini L, Yu DS, Lo T, Anand S, Lee M, Blundell TL, Venkitaraman AR (2002) Insights into DNA recombination from the structure of a RAD51-BRCA2 complex. Nature 420:287-293

Petukhova G, Stratton S, Sung P (1998) Catalysis of homologous DNA pairing by yeast Rad51 and Rad54 proteins. Nature 393:91-94

Petukhova G, Van Komen S, Vergano S, Klein H, Sung P (1999) Yeast Rad54 promotes Rad51-dependent homologous DNA pairing via ATP hydrolysis-driven change in DNA double helix conformation. J Biol Chem 274:29453-29462

Pierce AJ, Johnson RD, Thompson LH, Jasin M (1999) XRCC3 promotes homology-directed repair of DNA damage in mammalian cells. Genes Dev 13:2633-2638

Pittman DL, Schimenti JC (2000) Midgestation lethality in mice deficient for the RecA-related gene, Rad51d/Rad51l3. Genesis 26:167-173

Porteus MH, Carroll D (2005) Gene targeting using zinc finger nucleases. Nat Biotechnol 23:967-973

Prakash S, Johnson RE, Prakash L (2005) Eukaryotic translesion synthesis DNA polymerases: specificity of structure and function. Annu Rev Biochem 74:317-353

Raderschall E, Golub EI, Haaf T (1999) Nuclear foci of mammalian recombination proteins are located at single-stranded DNA regions formed after DNA damage. Proc Natl Acad Sci USA 96:1921-1926

Raschle M, Van Komen S, Chi P, Ellenberger T, Sung P (2004) Multiple interactions with the Rad51 recombinase govern the homologous recombination function of Rad54. J Biol Chem 279:51973-51980

Rattray AJ, Strathern JN (2005) Homologous recombination is promoted by translesion polymerase poleta. Mol Cell 20:658-659

Richardson C, Elliott B, Jasin M (1999) Chromosomal double-strand breaks introduced in mammalian cells by expression of I-Sce I endonuclease. Methods Mol Biol 113:453-463

Rijkers T, Van Den Ouweland J, Morolli B, Rolink AG, Baarends WM, Van Sloun PP, Lohman PH, Pastink A (1998) Targeted inactivation of mouse RAD52 reduces homologous recombination but not resistance to ionizing radiation. Mol Cell Biol 18:6423-6429

Ristic D, Wyman C, Paulusma C, Kanaar R (2001) The architecture of the human Rad54-DNA complex provides evidence for protein translocation along DNA. Proc Natl Acad Sci USA 98:8454-8460

Rogakou EP, Boon C, Redon C, Bonner WM (1999) Megabase chromatin domains involved in DNA double-strand breaks *in vivo*. J Cell Biol 146:905-916

Rouet P, Smih F, Jasin M (1994a) Expression of a site-specific endonuclease stimulates homologous recombination in mammalian cells. Proc Natl Acad Sci USA 91:6064-6068

Rouet P, Smih F, Jasin M (1994b) Introduction of double-strand breaks into the genome of mouse cells by expression of a rare-cutting endonuclease. Mol Cell Biol 14:8096-8106

San Filippo J, Chi P, Sehorn MG, Etchin J, Krejci L, Sung P (2006) Recombination mediator and RAD51 targeting activities of a human BRCA2 polypeptide. J Biol Chem 28:11649-657

Schild D, Lio YC, Collins DW, Tsomondo T, Chen DJ (2000) Evidence for simultaneous protein interactions between human Rad51 paralogs. J Biol Chem 275:16443-16449

Scully R, Chen J, Plug A, Xiao Y, Weaver D, Feunteun J, Ashley T, Livingston DM (1997) Association of BRCA1 with Rad51 in mitotic and meiotic cells. Cell 88:265-275

Scully R, Puget N, Vlasakova K (2000) DNA polymerase stalling, sister chromatid recombination and the BRCA genes. Oncogene 19:6176-6183

Sedelnikova OA, Pilch DR, Redon C, Bonner WM (2003) Histone H2AX in DNA damage and repair. Cancer Biol Ther 2:233-235

Sharan SK, Morimatsu M, Albrecht U, Lim DS, Regel E, Dinh C, Sands A, Eichele G, Hasty P, Bradley A (1997) Embryonic lethality and radiation hypersensitivity mediated by Rad51 in mice lacking Brca2. Nature 386:804-810

Shiloh Y (2003) ATM and related protein kinases: safeguarding genome integrity. Nat Rev Cancer 3:155-168

Shinohara A, Shinohara M, Ohta T, Matsuda S, Ogawa T (1998) Rad52 forms ring structures and co-operates with RPA in single-strand DNA annealing. Genes Cells 3:145-156

Shivji MK, Venkitaraman AR (2004) DNA recombination, chromosomal stability and carcinogenesis: insights into the role of BRCA2. DNA Repair (Amst) 3:835-843

Shroff R, Arbel-Eden A, Pilch D, Ira G, Bonner WM, Petrini JH, Haber JE, Lichten M (2004) Distribution and dynamics of chromatin modification induced by a defined DNA double-strand break. Curr Biol 14:1703-1711

Shu Z, Smith S, Wang L, Rice MC, Kmiec EB (1999) Disruption of muREC2/RAD51L1 in mice results in early embryonic lethality which can Be partially rescued in a p53(-/-) background. Mol Cell Biol 19:8686-8693

Sigurdsson S, Van Komen S, Bussen W, Schild D, Albala JS, Sung P (2001) Mediator function of the human Rad51B-Rad51C complex in Rad51/RPA-catalyzed DNA strand exchange. Genes Dev 15:3308-3318

Smiraldo PG, Gruver AM, Osborn JC, Pittman DL (2005) Extensive chromosomal instability in Rad51d-deficient mouse cells. Cancer Res 65:2089-2096

Solinger JA, Heyer WD (2001) Rad54 protein stimulates the postsynaptic phase of Rad51 protein-mediated DNA strand exchange. Proc Natl Acad Sci USA 98:8447-8453

Solinger JA, Kiianitsa K, Heyer WD (2002) Rad54, a Swi2/Snf2-like recombinational repair protein, disassembles Rad51:dsDNA filaments. Mol Cell 10:1175-1188

Solomon DA, Cardoso MC, Knudsen ES (2004) Dynamic targeting of the replication machinery to sites of DNA damage. J Cell Biol 166:455-463

Solovjeva L, Svetlova M, Sasina L, Tanaka K, Saijo M, Nazarov I, Bradbury M, Tomilin N (2005) High mobility of flap endonuclease 1 and DNA polymerase eta associated with replication foci in mammalian S-phase nucleus. Mol Biol Cell 16:2518-2528

Song B, Sung P (2000) Functional interactions among yeast Rad51 recombinase, Rad52 mediator, and replication protein A in DNA strand exchange. J Biol Chem 275:15895-15904

Sonoda E, Sasaki MS, Buerstedde JM, Bezzubova O, Shinohara A, Ogawa H, Takata M, Yamaguchi-Iwai Y, Takeda S (1998) Rad51-deficient vertebrate cells accumulate chromosomal breaks prior to cell death. Embo J 17:598-608

Sporbert A, Gahl A, Ankerhold R, Leonhardt H, Cardoso MC (2002) DNA polymerase clamp shows little turnover at established replication sites but sequential *de novo* assembly at adjacent origin clusters. Mol Cell 10:1355-1365

Sprague BL, McNally JG (2005) FRAP analysis of binding: proper and fitting. Trends Cell Biol 15:84-91

Stewart GS, Maser RS, Stankovic T, Bressan DA, Kaplan MI, Jaspers NG, Raams A, Byrd PJ, Petrini JH, Taylor AM (1999) The DNA double-strand break repair gene hMRE11 is mutated in individuals with an ataxia-telangiectasia-like disorder. Cell 99:577-587

Sugawara N, Wang X, Haber JE (2003) *In vivo* roles of Rad52, Rad54, and Rad55 proteins in Rad51-mediated recombination. Mol Cell 12:209-219

Sugiyama T, Kowalczykowski SC (2002) Rad52 protein associates with replication protein A (RPA)-single-stranded DNA to accelerate Rad51-mediated displacement of RPA and presynaptic complex formation. J Biol Chem 277:31663-31672

Sugiyama T, New JH, Kowalczykowski SC (1998) DNA annealing by RAD52 protein is stimulated by specific interaction with the complex of replication protein A and single-stranded DNA. Proc Natl Acad Sci USA 95:6049-6054

Sugiyama T, Zaitseva EM, Kowalczykowski SC (1997) A single-stranded DNA-binding protein is needed for efficient presynaptic complex formation by the *Saccharomyces cerevisiae* Rad51 protein. J Biol Chem 272:7940-7945

Sung P (1997) Function of yeast Rad52 protein as a mediator between replication protein A and the Rad51 recombinase. J Biol Chem 272:28194-28197

Suzuki A, de la Pompa JL, Hakem R, Elia A, Yoshida R, Mo R, Nishina H, Chuang T, Wakeham A, Itie A, Koo W, Billia P, Ho A, Fukumoto M, Hui CC, Mak TW (1997) Brca2 is required for embryonic cellular proliferation in the mouse. Genes Dev 11:1242-1252

Swagemakers SM, Essers J, de Wit J, Hoeijmakers JH, Kanaar R (1998) The human RAD54 recombinational DNA repair protein is a double-stranded DNA-dependent ATPase. J Biol Chem 273:28292-28297

Symington LS (2002) Role of RAD52 epistasis group genes in homologous recombination and double-strand break repair. Microbiol Mol Biol Rev 66:630-670, table of contents

Takata M, Sasaki MS, Sonoda E, Fukushima T, Morrison C, Albala JS, Swagemakers SM, Kanaar R, Thompson LH, Takeda S (2000) The Rad51 paralog Rad51B promotes homologous recombinational repair. Mol Cell Biol 20:6476-6482

Takata M, Sasaki MS, Tachiiri S, Fukushima T, Sonoda E, Schild D, Thompson LH, Takeda S (2001) Chromosome instability and defective recombinational repair in knockout mutants of the five Rad51 paralogs. Mol Cell Biol 21:2858-2866

Tan TL, Essers J, Citterio E, Swagemakers SM, de Wit J, Benson FE, Hoeijmakers JH, Kanaar R (1999) Mouse Rad54 affects DNA conformation and DNA-damage-induced Rad51 foci formation. Curr Biol 9:325-328

Tashiro S, Kotomura N, Shinohara A, Tanaka K, Ueda K, Kamada N (1996) S phase specific formation of the human Rad51 protein nuclear foci in lymphocytes. Oncogene 12:2165-2170

Tashiro S, Walter J, Shinohara A, Kamada N, Cremer T (2000) Rad51 accumulation at sites of DNA damage and in postreplicative chromatin. J Cell Biol 150:283-291

Taylor AM, Groom A, Byrd PJ (2004) Ataxia-telangiectasia-like disorder (ATLD)-its clinical presentation and molecular basis. DNA Repair (Amst) 3:1219-1225

Thacker J (2005) The RAD51 gene family, genetic instability and cancer. Cancer Lett 219:125-135

Thompson LH, Schild D (2001) Homologous recombinational repair of DNA ensures mammalian chromosome stability. Mutat Res 477:131-153

Tsuzuki T, Fujii Y, Sakumi K, Tominaga Y, Nakao K, Sekiguchi M, Matsushiro A, Yoshimura Y, MoritaT (1996) Targeted disruption of the Rad51 gene leads to lethality in embryonic mice. Proc Natl Acad Sci USA 93:6236-6240

Tutt A, Gabriel A, Bertwistle D, Connor F, Paterson H, Peacock J, Ross G, Ashworth A (1999) Absence of Brca2 causes genome instability by chromosome breakage and loss associated with centrosome amplification. Curr Biol 9:1107-1110

Van Komen S, Petukhova G, Sigurdsson S, Stratton S, Sung P (2000) Superhelicity-driven homologous DNA pairing by yeast recombination factors Rad51 and Rad54. Mol Cell 6:563-572

Van Komen S, Petukhova G, Sigurdsson S, Sung P (2002) Functional cross-talk among Rad51, Rad54, and replication protein A in heteroduplex DNA joint formation. J Biol Chem 277:43578-43587

van Veelen LR, Cervelli T, van de Rakt MW, Theil AF, Essers J, Kanaar R (2005a) Analysis of ionizing radiation-induced foci of DNA damage repair proteins. Mutat Res 574:22-33

van Veelen LR, Essers J, van de Rakt MW, Odijk H, Pastink A, Zdzienicka MZ, Paulusma CC, Kanaar R (2005b) Ionizing radiation-induced foci formation of mammalian Rad51 and Rad54 depends on the Rad51 paralogs, but not on Rad52. Mutat Res 574:34-49

Varon R, Vissinga C, Platzer M, Cerosaletti KM, Chrzanowska KH, Saar K, Beckmann G, Seemanova E, Cooper PR, Nowak NJ, Stumm M, Weemaes CM, Gatti RA, Wilson RK, Digweed M, Rosenthal A, Sperling K, Concannon P, Reis A (1998) Nibrin, a novel DNA double-strand break repair protein, is mutated in Nijmegen breakage syndrome. Cell 93:467-476

Venkitaraman AR (2002) Cancer susceptibility and the functions of BRCA1 and BRCA2. Cell 108:171-182

Waldman AS, Liskay RM (1988) Resolution of synthetic Holliday structures by an extract of human cells. Nucleic Acids Res 16:10249-10266

Walter J, Cremer T, Miyagawa K, Tashiro S (2003) A new system for laser-UVA-microirradiation of living cells. J Microsc 209:71-75

Wesoly J, Agarwal S, Sigurdsson S, Bussen W, Van Komen S, Qin J, van Steeg H, van Benthem J, Wassenaar E, Baarends WM, Ghazvini M, Tafel AA, Heath H, Galjart N, Essers J, Grootegoed JA, Arnheim N, Bezzubova O, Buerstedde JM, Sung P, Kanaar R (2006) Differential contributions of mammalian Rad54 paralogs to recombination, DNA damage repair, and meiosis. Mol Cell Biol 26:976-989

West SC (1997) Processing of recombination intermediates by the RuvABC proteins. Annu Rev Genet 31:213-244

White J, Stelzer E (1999) Photobleaching GFP reveals protein dynamics inside live cells. Trends Cell Biol 9:61-65

Wiese C, Collins DW, Albala JS, Thompson LH, Kronenberg A, Schild D (2002) Interactions involving the Rad51 paralogs Rad51C and XRCC3 in human cells. Nucleic Acids Res 30:1001-1008

Wiltzius JJ, Hohl M, Fleming JC, Petrini JH (2005) The Rad50 hook domain is a critical determinant of Mre11 complex functions. Nat Struct Mol Biol 12:403-407

Wong AK, Pero R, Ormonde PA, Tavtigian SV, Bartel PL (1997) RAD51 interacts with the evolutionarily conserved BRC motifs in the human breast cancer susceptibility gene brca2. J Biol Chem 272:31941-31944

Xiao Y, Weaver DT (1997) Conditional gene targeted deletion by Cre recombinase demonstrates the requirement for the double-strand break repair Mre11 protein in murine embryonic stem cells. Nucleic Acids Res 25:2985-2991

Yamada NA, Hinz JM, Kopf VL, Segalle KD, Thompson LH (2004) XRCC3 ATPase activity is required for normal XRCC3-Rad51C complex dynamics and homologous recombination. J Biol Chem 279:23250-23254

Yamaguchi-Iwai Y, Sonoda E, Buerstedde JM, Bezzubova O, Morrison C, Takata M, Shinohara A, Takeda S (1998) Homologous recombination, but not DNA repair, is reduced in vertebrate cells deficient in RAD52. Mol Cell Biol 18:6430-6435

Yamaguchi-Iwai Y, Sonoda E, Sasaki MS, Morrison C, Haraguchi T, Hiraoka Y, Yamashita YM, Yagi T, Takata M, Price C, Kakazu N, Takeda S (1999) Mre11 is essential for the maintenance of chromosomal DNA in vertebrate cells. Embo J 18:6619-6629

Yang H, Li Q, Fan J, Holloman WK, Pavletich NP (2005) The BRCA2 homologue Brh2 nucleates RAD51 filament formation at a dsDNA-ssDNA junction. Nature 433:653-657

Yokoyama H, Kurumizaka H, Ikawa S, Yokoyama S, Shibata T (2003) Holliday junction binding activity of the human Rad51B protein. J Biol Chem 278:2767-2772

Yokoyama H, Sarai N, Kagawa W, Enomoto R, Shibata T, Kurumizaka H, Yokoyama S (2004) Preferential binding to branched DNA strands and strand-annealing activity of the human Rad51B, Rad51C, Rad51D and Xrcc2 protein complex. Nucleic Acids Res 32:2556-2565

Yonetani Y, Hochegger H, Sonoda E, Shinya S, Yoshikawa H, Takeda S, Yamazoe M (2005) Differential and collaborative actions of Rad51 paralog proteins in cellular response to DNA damage. Nucleic Acids Res 33:4544-4552

Yu DS, Sonoda E, Takeda S, Huang CL, Pellegrini L, Blundell TL, Venkitaraman AR (2003) Dynamic control of Rad51 recombinase by self-association and interaction with BRCA2. Mol Cell 12:1029-1041

Yu VP, Koehler M, Steinlein C, Schmid M, Hanakahi LA, van Gool AJ, West SC, Venkitaraman AR (2000) Gross chromosomal rearrangements and genetic exchange between nonhomologous chromosomes following BRCA2 inactivation. Genes Dev 14:1400-1406

Yuan SS, Lee SY, Chen G, Song M, Tomlinson GE, Lee EY (1999) BRCA2 is required for ionizing radiation-induced assembly of Rad51 complex *in vivo*. Cancer Res 59:3547-3551

Zhu J, Petersen S, Tessarollo L, Nussenzweig A (2001) Targeted disruption of the Nijmegen breakage syndrome gene NBS1 leads to early embryonic lethality in mice. Curr Biol 11:105-109

Agarwal, Sheba

Department of Cell Biology and Genetics, Erasmus MC, PO Box 1738, 3000 DR Rotterdam, The Netherlands

Essers, Jeroen

Department of Cell Biology and Genetics, Erasmus MC, PO Box 1738, 3000 DR Rotterdam, The Netherlands, and Department of Radiation Oncology, Erasmus MC, PO Box 1738, 3000 DR Rotterdam, The Netherlands
j.essers@erasmusmc.nl

Kanaar, Roland
Department of Cell Biology and Genetics, Erasmus MC, PO Box 1738, 3000 DR Rotterdam, The Netherlands, and Department of Radiation Oncology, Erasmus MC, PO Box 1738, 3000 DR Rotterdam, The Netherlands

BRCA2: safeguarding the genome through homologous recombination

Nicole Christ, Mary Ellen Moynahan, Maria Jasin

Abstract

Germline mutations in the tumor suppressor gene *BRCA2* predispose individuals to breast, ovarian, and other cancers. In recent years, the BRCA2 protein has been recognized to have an important function in homologous recombination, a key pathway in mammalian cells for repairing spontaneous and induced DNA lesions and, thus, for maintaining genomic integrity. Loss of BRCA2 leads to embryonic death in mice, but is compatible with cell survival in adult somatic cells and tumor proliferation. This review summarizes recent advances in our understanding of BRCA2 from several perspectives, especially with regard to its broad evolutionary conservation.

1 Introduction

Human chromosomes frequently undergo DNA damage, either as a result of normal cellular processes or DNA damaging agents in the environment or from chemotherapy (Featherstone and Jackson 1999). DNA double-strand breaks (DSBs) pose a particular danger to the integrity of chromosomes because they can result in gross rearrangements, a characteristic of many tumor cells (Lengauer et al. 1998). When a cell encounters DSBs, a complex network of proteins is recruited to ensure accurate and timely repair, and, if that is not possible, cell death (Kastan and Bartek 2004; and reviewed elsewhere in this book by Lisby and Rothstein).

Two major types of DSB repair exist in mammalian cells, homologous recombination (HR), also known as homology-directed repair, and nonhomologous end-joining (NHEJ). As the names imply, the distinction between these repair pathways lies in the requirement (or not) for homology during the repair process, which in turn affects the precision of repair. NHEJ utilizes very little, if any, sequence homology during the repair process, such that the broken ends are ligated back together, often with the addition or deletion of nucleotides at the break site (reviewed elsewhere in this book by Wilson). On the contrary, HR utilizes the identical sister chromatid or, more rarely, the homologous chromosome to template repair, such that HR is usually a precise type of repair (Johnson and Jasin 2000). In addition to DSB repair, HR is likely to be important in mammalian cells for the repair of DNA gaps, as it is in *Escherichia coli* (see e.g. Cromie and Leach 2000). A more specialized type of DSB repair is single-strand annealing, which occurs between DNA repeats, and as its name implies, involves annealing of complementary sequences (Paques and Haber 1999).

Given its crucial role in repairing spontaneous and induced DNA damage, it is not surprising that HR is implicated both in cell and organismal survival and in tumor suppression. A unique biochemical step in HR is strand exchange between the participating homologous DNA molecules, a reaction that is catalyzed by the RAD51 recombinase (Sung et al. 2003). Disruption of RAD51 leads to cell death and early embryonic lethality in the mouse (Lim and Hasty 1996; Tsuzuki et al. 1996; Sonoda et al. 1998). RAD51 interacts with the tumor suppressor proteins BRCA1 and BRCA2; loss of either of these proteins impairs HR in mammalian cells and leads to genomic instability (Scully and Livingston 2000; Jasin 2002; Venkitaraman 2002). This review summarizes recent developments in our understanding of the BRCA2 protein.

2 BRCA2: a tumor suppressor with diverse domain structures in different organisms

2.1 BRCA2 in vertebrates

Human BRCA2 is a large protein of 3418 amino acids (~410 kDa) (Tavtigian et al. 1996). The mouse protein is slightly smaller at 3326 amino acids and it shares 59% identity with the human protein (Connor et al. 1997; McAllister et al. 1997; Sharan and Bradley 1997). This degree of human/mouse conservation is rather low compared with other tumor suppressors, for example, MSH2 and p53 (92% and 77%, respectively), and is extremely low compared to RAD51 (99%). BRCA2 proteins from other vertebrates can exhibit an even lower degree of conservation (e.g. 40% for chicken) (Takata et al. 2002; Warren et al. 2002). Nevertheless, the sequence conservation is sufficient for human *BRCA2* to rescue the embryonic lethality of *Brca2*-null mice (Sharan et al. 2004) and correct the DNA damage sensitivities and other repair phenotypes of *Brca2*-mutant hamster cells (Kraakman-van der Zwet et al. 2002).

The ability of human BRCA2 to complement phenotypes of rodent mutants is almost certainly due to the higher degree of conservation observed in segments of the protein. Two particular regions are notable in this regard: the BRC repeat region and the DNA binding domain (Fig. 1). The BRC repeat region consists of eight repeats of a motif that binds the RAD51 recombinase; the repeats are spaced throughout the central 1643 amino acid region of BRCA2 which is encoded by exon 11 (Bork et al. 1996; Bignell et al. 1997; Wong et al. 1997; Chen et al. 1998). A comparison between mammalian species has revealed a BRC repeat consensus core of ~26 amino acids, although each repeat may span a larger segment of about 70 amino acids (Bork et al. 1996; Bignell et al. 1997). Individual repeats are generally more similar across vertebrate species than to other repeats from the same species, and although the spacing between repeats is variable, it is reasonably well maintained across species. Nevertheless, some variation is noted;

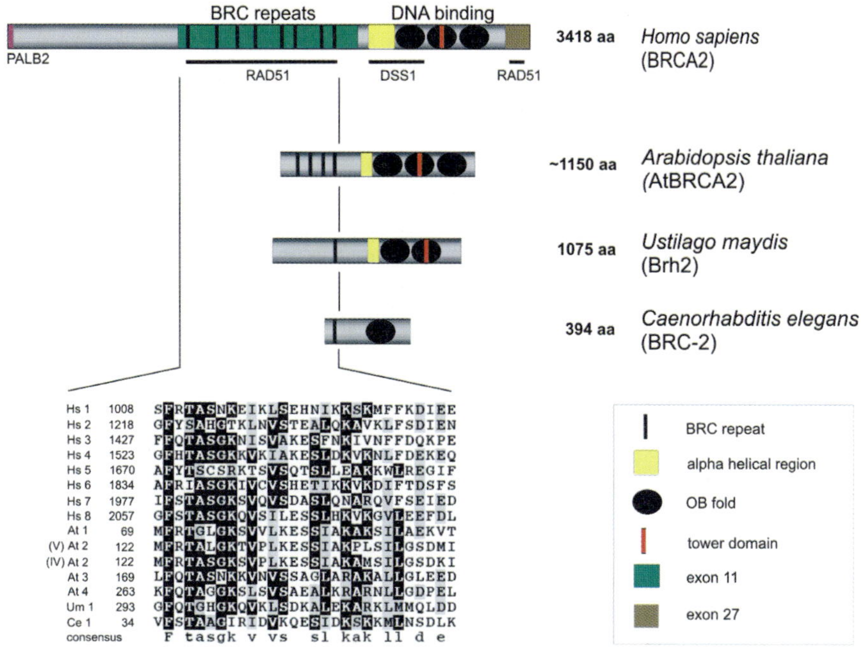

Fig. 1. Domain organization of human BRCA2 and plant, fungus, and worm orthologs. The alignment of the BRC repeat sequences from the various orthologs are also shown.

in particular, the chicken BRC3 is poorly conserved and likely does not bind RAD51 (Takata et al. 2002).

An extended region of higher sequence conservation is found in the C-terminal region of BRCA2, which in human and mouse share 77% identity (Connor et al. 1997; McAllister et al. 1997; Sharan and Bradley 1997). This region binds single-stranded DNA (ssDNA) (Yang et al. 2002), as well as the small protein DSS1 (Marston et al. 1999). Smaller regions of conservation have also been noted, including a segment between BRC repeats 1 and 2 (Sharan and Bradley 1997; Takata et al. 2002), whose function is not yet known, and the extreme N- and C-termini, which bind PALB2 (Xia et al. 2006) and RAD51 (Mizuta et al. 1997; Sharan et al. 1997; Esashi et al. 2005), respectively. Nuclear localization signals (NLSs) have also been described at the C-terminus of the human protein (Spain et al. 1999), although another NLS(s) exists, at least in the mouse protein (Sarkisian et al. 2001). The remaining N-terminal third of BRCA2 is poorly conserved.

2.2 BRCA2 in non-vertebrate species

The identification of functional motifs and well-conserved regions in vertebrate BRCA2 proteins has allowed researchers to identify BRCA2 homologs in other eukaryotic species, including *Arabidopsis thaliana* (Warren et al. 2002; Siaud et al. 2004), *Ustilago maydis* (Kojic et al. 2002), and *Caenorhabditis elegans* (Martin et al. 2005) (Fig. 1). These proteins are strikingly different in size from vertebrate BRCA2: the two *A. thaliana* proteins are ~1150 amino acids, *U. maydis* Brh2 is 1075 amino acids, and *C. elegans* BRC-2 is only 394 amino acids, nearly tenfold smaller than vertebrate BRCA2. The large size range arises from widely variable lengths of the N-terminal sequences, variation in the number of BRC repeats, and domain differences in the DNA binding region. Common to all BRCA2 homologs, however, are at least one BRC repeat capable of binding RAD51 and one domain capable of binding ssDNA. Despite the presence of Brh2 in *U. maydis*, which is a yeast-like fungus, BRCA2 homologs are not found in budding or fission yeast.

3 Binding Partners of BRCA2

3.1 Rad51: the BRC repeats

Of the eight BRC repeats in human BRCA2, only five (BRC1, 3, 4, 7, 8) conform well to the consensus sequence (Fig. 1). Six BRC repeats interact with RAD51 in yeast two-hybrid analysis (Wong et al. 1997; Chen et al. 1998), although the remaining two repeats can interact with RAD51 in pull down experiments *in vitro* (Wong et al. 1997). Given that one BRC repeat is sufficient for activity of the *U. maydis* and *C. elegans* proteins and that the chicken protein may have one fewer repeat, all eight repeats may not be required for function of the mammalian proteins.

The crystal structure of one repeat – BRC4 – has been solved as part of an artificial protein fusion with the nucleotide binding core region of RAD51, revealing that BRC4 structurally mimics the motif within RAD51 required for oligomerization (Pellegrini et al. 2002). This mimicry readily explains the finding that individual BRC repeats or sets of BRC repeats act as dominant negative peptides to inhibit RAD51 filament formation on DNA *in vitro* (Davies et al. 2001), RAD51 focus formation upon DNA damage *in vivo* (Chen et al. 1999; Yuan et al. 1999), and HR (Stark et al. 2002; Saeki et al. 2006). However, all of the repeats may not bind to RAD51 in an identical fashion, since electron micrograph reconstructions suggest that different repeats interact with different regions of RAD51 (Galkin et al. 2005).

3.2 Rad51: exon 27-encoded sequences

A distinct RAD51 interacting domain unrelated to the BRC repeats has been mapped to the C-terminus of mammalian BRCA2 which is encoded by exon 27 (Fig. 1) (Mizuta et al. 1997; Sharan et al. 1997). RAD51 binding to this sequence is regulated by CDK phosphorylation of a conserved serine (S3291 in humans) (Esashi et al. 2005). In S phase or after treatment with ionizing radiation, phosphorylation of this serine is reduced, allowing RAD51 to bind to this sequence on BRCA2. How RAD51 binding to one or more BRC repeats and this exon 27-encoded motif is coordinated is currently unknown.

3.3 DNA

The discovery that BRCA2 is a ssDNA binding protein came about from the crystal structure determination of a ~90 kDa C-terminal fragment of BRCA2 (Fig. 1) (Yang et al. 2002). This region of BRCA2 consists of four globular domains arranged in a linear manner and a fifth domain, which has a tower-like structure. Three globular domains are OB (oligonucleotide/ oligosaccharide-binding) folds, which are found in ssDNA binding proteins such as RPA (Bochkarev and Bochkareva 2004). The ability of BRCA2 to bind ssDNA was confirmed by biochemical studies, and the mode of ssDNA binding was determined by solution of a co-crystal of the BRCA2 fragment with a short oligonucleotide (oligo-dT_9), demonstrating that ssDNA binds to two OB folds in a manner similar to RPA. The unusual tower domain extends from the middle OB fold and consists of two long antiparallel α helices with a three-helix bundle at the apex. The three-helix bundle has a structure similar to double-stranded DNA (dsDNA) binding domains found in the Hin recombinase and some other proteins, and evidence has been provided that BRCA2 binds dsDNA (Yang et al. 2002). Combined ssDNA and dsDNA binding may assist BRCA2 activity at ssDNA-dsDNA junctions (Yang et al. 2005).

3.4 DSS1

The structure determinations of the BRCA2 C-terminal fragment described above were solved in complex with the 70 amino acid DSS1 protein (Yang et al. 2002). DSS1, which was first identified to interact with BRCA2 using two-hybrid analysis (Marston et al. 1999), was found to improve the solubility of the BRCA2 fragments, facilitating the crystallization studies. DSS1 binds BRCA2 in an extended conformation, such that clusters of acidic residues of DSS1 interspersed with hydrophobic residues bind to a surface of BRCA2 that is rich in basic and hydrophobic residues (Yang et al. 2002). DSS1 also interacts with the proteosome (Funakoshi et al. 2004; Krogan et al. 2004), although it is not clear if this interaction is relevant to its interaction with BRCA2.

DSS1 is conserved from yeast to mammalian cells. It was first described to be an important HR component in *U. maydis* (Kojic et al. 2003), and more recently has been proposed to regulate Brh2 activity during HR (Kojic et al. 2005). DSS1 also appears to be important for HR in mammalian cells (Gudmundsdottir et al. 2004). DSS1 may regulate BRCA2 stability *in vivo* as reported by one group (Li et al. 2006), although another group has not observed an effect on BRCA2 levels as a result of DSS1 knockdown (Gudmundsdottir et al. 2004). Sem1, the DSS1 homolog in yeast, has been shown to be recruited to a DSB *in vivo* and to participate in DSB repair by both HR and NHEJ (Krogan et al. 2004). Since yeast does not have a BRCA2 ortholog, these results imply that Seml/DSS1 can have BRCA2-independent functions.

3.5 PALB2 and other proteins

As noted above, the extreme N-terminus of BRCA2 (amino acids 10 – 40) interacts with the recently discovered protein PALB2 (Xia et al. 2006). PALB2 was identified to be in nuclear complexes with BRCA2 and has been implicated in HR. Importantly, BRCA2 missense mutations found in breast cancer patients that disrupt the interaction with PALB2 also interfere with HR.

BRCA2, which itself has been found to be mutated in patients with Fanconi anemia (FA, see Section 5.1), has been reported to bind to two other proteins mutated in FA patients, FANCD2 and FANCG (Hussain et al. 2003, 2004; Wang et al. 2004). These proteins are also involved in the DNA damage response and may have overlapping cellular functions with BRCA2, although loss of these proteins can lead to different consequences than loss of BRCA2. A number of other BRCA2-interacting proteins have been reported which appear to have diverse functions in the cell, including the putative oncoprotein EMSY (Hughes-Davies et al. 2003), the nuclear protein BCCIP (Liu et al. 2001, 2005), the deubiquitinating enzyme USP11 (Schoenfeld et al. 2004), the mitotic Polo-like kinase PLK1 (Lin et al. 2003; Lee et al. 2004), and the androgen receptor (Shin and Verma 2003).

4 BRCA2 and homologous recombination

4.1 Studies *in vitro*

The first biochemical studies reported with a full-length BRCA2 ortholog were with *U. maydis* Brh2 co-purified with Dss1 from insect cells (Yang et al. 2005). These studies found that Brh2 recruits Rad51 to DNA, facilitates Rad51 nucleoprotein filament formation, and, importantly, stimulates Rad51 strand exchange activity in the presence of RPA at substoichiometric concentrations of Brh2. Of note, Brh2 acts preferentially at ssDNA-dsDNA junctions, which are processing intermediates during HR reactions (Paques and Haber 1999). More recently, *C. elegans* BRC-2 has been purified from *E. coli* and shown to promote D-loop formation, in which a ssDNA fragment invades a supercoiled homologous duplex (Petalcorin et al. 2006). D-loop formation occurs at substoichiometric concentrations of BRC-2 and, importantly, DSS1 is not required in the reaction, nor does it promote it. Interestingly, BRC-2 also promotes the annealing of complementary ssDNA. The authors propose that this activity provides a role for BRC-2 in DNA repair events that occur in the absence of RAD-51 (Martin et al. 2005).

Biochemical activities of the human protein have recently been examined using fusions of segments of BRCA2 purified from *E. coli* (San Filippo et al. 2006). In these studies, BRC repeats 3 and 4 were fused to the BRCA2 DNA binding domain, creating the fusion BRC3/4-DBD. This fusion binds both RAD51 and DNA, with a clear preference for ssDNA, and promotes Rad51 filament formation on ssDNA. Moreover, BRC3/4-DBD promotes RAD51 strand exchange activity in the presence of RPA. Interestingly, these experiments were carried out without DSS1. Activities of this fusion likely reflect those of the full-length BRCA2 protein, although presumably the full-length protein has additional complexities.

4.2 Studies *in vivo*

BRCA2 has been shown to be central to HR in several vertebrate cell lines (human, mouse, hamster, chicken) (Moynahan et al. 2001; Tutt et al. 2001; Xia et al. 2001; Hatanaka et al. 2005; Saeki et al. 2006), as well as in plant (Siaud et al. 2004), worm (Martin et al. 2005), and *U. maydis* (Kojic et al. 2002). BRCA1, which does not appear to interact directly with RAD51, also plays an important role in HR (Moynahan et al. 1999). However, the two proteins appear to have distinct functions in DSB repair. BRCA1 promotes annealing between direct repeats in the genome (Moynahan et al. 1999; Stark et al. 2004), while BRCA2 suppresses these events (Tutt et al. 2001; Stark et al. 2004), leading to the proposal that BRCA1 has a role upstream of BRCA2 in DSB repair (Stark et al. 2004). Distinct roles in DSB repair may account for the observation that BRCA1 and BRCA2 mutations do not lead to identical consequences in patients (Moynahan 2002); however, other explanations are also possible given that these proteins interact with distinct sets of cellular proteins.

Recent structure-function analysis of the *U. maydis* protein has indicated that the N-terminal half of Brh2 has residual activity in HR (Kojic et al. 2005). This N-terminal fragment contains the single BRC repeat of Brh2, but the DNA/DSS1 binding domain is completely absent. Of note, this N-terminal half is active in HR in the absence of Dss1, indicating that Dss1 may be required for HR only when the Dss1-interacting domain of Brh2 is present. The ability of the N-terminal fragment to function in HR is consistent with the observation that *Brca2* mouse mutants which maintain at least some of the BRC repeats have a less severe phenotype than those in which all of the BRC repeats are deleted (Moynahan 2002). How the BRC repeats are able to function in HR *in vivo* in the absence of ssDNA binding is currently unknown, but it is interesting to note that BRC repeats can bind (rather than disrupt) RAD51 nucleoprotein filaments under some conditions *in vitro*, suggesting that BRC repeats may have some function(s) independent of the DNA binding domain (Galkin et al. 2005).

Domain swaps have been constructed to further understand the role of different regions of BRCA2. In *U. maydis*, the N-terminal half of Brh2 was fused to RPA70, the large subunit of RPA (Kojic et al. 2005). This Brh2-RPA fusion protein is more active than the N-terminal Brh2 fragment alone, since it is able to fully correct the UV sensitivity of Brh2 mutant cells. However, wild type cells expressing the Brh2-RPA fusion have elevated levels of HR (hyper-recombination). Because the Brh2-RPA fusion does not bind Dss1, the hyper-recombination phenotype is consistent with the proposal that Dss1 is important in controlling the activity of Brh2.

In a related approach in mammalian cells, single or multiple BRC repeats have been fused to RPA70 (Saeki et al. 2006). These BRC-RPA fusions are able to increase HR in *Brca2* mutant hamster cells to nearly normal levels, without causing a hyper-recombination phenotype in wild type cells. Chromosome instability is also suppressed in mutant cells by expression of the BRC-RPA fusion proteins. Because as little as 2% of BRCA2 fused to a ssDNA binding protein is sufficient to suppress cellular defects found in *Brca2* mutant cells, these results are quite intriguing in relation to the observed diversity of BRCA2 domain structures in different organisms. Interestingly, a fusion containing the Rad52 ssDNA binding domain instead of RPA70 also promoted HR; this result supports the speculation that BRCA2 and Rad52, although structurally very distinct, may have related functions in the cell for bringing Rad51 to ssDNA.

5 BRCA2 is essential for development but dispensable for the survival of cancer cells

5.1 *BRCA2* and cancer predisposition in humans

BRCA2 was initially localized on chromosome 13q through linkage analysis of families with early onset breast cancer not attributed to mutations in *BRCA1* (Wooster et al. 1994); the gene was subsequently identified by positional cloning (Wooster et al. 1995; Tavtigian et al. 1996). A large number of disease-causing mutations have been mapped throughout the gene, most of which result in premature truncation of the BRCA2 protein (see e.g. Tavtigian et al. 1996). Notably, as compared to carriers of germline *BRCA1* mutations, families with *BRCA2* mutations were found to have a higher incidence of male breast cancer and a lower predisposition to ovary cancer. In addition to breast and ovary, larger databases of *BRCA2* cancer families have extended the spectrum of cancer predisposition to include other tissues including pancreas, prostate, and gastrointestinal (Moynahan 2002). Tumors from predisposed individuals follow the paradigm of second allele inactivation (Collins et al. 1995), thus supporting *BRCA2* as a tumor suppressor.

Inheritance of bi-allelic *BRCA2* mutations is observed in some patients, resulting in Fanconi anemia (FA), subtype D1 (Howlett et al. 2002). FA is a rare recessive disorder characterized by bone marrow failure, diverse developmental abnormalities, predisposition to solid and hematologic malignancies, and cellular chromosome instability (Kennedy and D'Andrea 2005). Genetically, FA can be separated into at least 12 complementation groups, each with a mutation in a different gene, and thus far 11 genes, termed *FANC* genes, have been identified. The proteins encoded by these genes are involved in a common DNA damage response pathway, although some of the proteins have distinct functions not associated with other FA proteins. An example of this is the severe HR defect associated with loss of BRCA2 as compared to the mild HR defect observed with loss of FA-core complex proteins (Moynahan et al. 2001; Nakanishi et al. 2005). In addition, unlike heterozygous *BRCA2* carriers, other FA subtype heterozygous carriers are not predisposed to cancer. Although the number of cases with bi-allelic mutations in *BRCA2/FANCD1* is small, the clinical course of patients in the FA-D1 subtype is more severe with higher probabilities and earlier age of onset of both hematologic malignancy and solid tumors (Alter et al. 2006). *BRCA2/FANCD1* mutations appear to be partial loss of function (hypomorphic), in keeping with the hypothesis that BRCA2 function is essential for normal development of the mammalian embryo.

5.2 BRCA2 is essential during embryogenesis

In an effort to understand the *in vivo* functions of BRCA2, several mouse models have been established (Moynahan 2002). Embryonic lethality was observed in a number of *Brca2* models at E7.5-E10.5 with mutant embryos exhibiting severe proliferation defects (Ludwig et al. 1997; Sharan et al. 1997; Suzuki et al. 1997). The early embryonic lethality of these mutant mice was delayed by mutation of p53, such that the embryos survived a day or two longer (Ludwig et al. 1997), likely due to abrogation of apoptosis and relief from p53-dependent activation of checkpoints in response to unrepaired damage.

Early embryonic lethality uniformly occurred in mice with *Brca2* mutations that deleted all of the BRC repeats. However, in *Brca2* mutant mice in which some of the BRC repeats were retained, a small percentage of viable mice was obtained (Connor et al. 1997; Friedman et al. 1998). Notably the percentage of viable mice was strain dependent, and these mice nevertheless succumbed early in life to thymic lymphoma and were infertile (Connor et al. 1997; Friedman et al. 1998). Chicken DT40 cells with targeted mutations in BRCA2 also demonstrated a dependence on the BRC repeats, in that *brca2tr* cells which retained BRC1 and BRC2 were viable but markedly deficient for HR (Hatanaka et al. 2005), whereas cells homozygous for a mutation that deleted all of the BRC repeats were reportedly not obtained (Warren et al. 2003).

A markedly less severe phenotype was observed in mice carrying a deletion of the C-terminal exon 27 (McAllister et al. 2002; Donoho et al. 2003). The truncated BRCA2 in these mice retained all of the BRC repeats and the DNA binding domain, but deleted the C-terminal RAD51 binding motif described above which is unrelated to the centrally located BRC repeats. In one study the *Brca2* mutant mice that lack exon 27 have been reported to have a subtle degree of embryonic/peri-natal lethality (McAllister et al. 2002), but otherwise develop normally and are fertile (McAllister et al. 2002; Donoho et al. 2003). A mildly shortened lifespan was noted, as was a modest increase in tumor incidence. Tumors occurred at long latency and in a variety of tissues, with a significant increase in epithelial tumors as compared to control mice.

Unlike in humans, a tumor phenotype has not observed in mice heterozygous for *Brca2* mutations (Bennett et al. 2000), nor was a repair defect observed in these animals (Tutt et al. 2002).

5.3 Tumorigenesis in conditional *Brca2* mutants

To bypass the embryonic lethality seen in *Brca2*-deficient mice, mouse models have been created in which *Brca2* mutation is restricted to specific lineages by Cre-mediated recombination. *Brca2* deletion restricted to mammary epithelium was achieved using the WAP (whey acidic protein) promoter to express Cre-recombinase during late pregnancy and lactation (Ludwig et al. 2001; Cheung et al. 2004). In the mammary epithelium of multiparous female mice, WAP-Cre mediated deletion of *Brca2* sequences occurred in 25-32% of alleles. Normal mammary gland development was observed, and these mice were able to nurse their litters and undergo similar patterns of involution as compared to control mice (Cheung et al. 2004). Significant mammary tumor multifocality was observed, with palpable tumors arising after a long latency of 13–19 months (Ludwig et al. 2001; Cheung et al. 2004). Thus, this approach has been successful to model mammary tumors with *Brca2* loss in mice.

Another conditional mouse model directed Cre-mediated deletion of *Brca2* exon 11 ($Brca2^{F11/F11}$) to stratified epithelium of skin, salivary and mammary glands using the human keratin 14 (K14) gene promoter (Jonkers et al. 2001). If expressed, the predicted BRCA2 product would be deleted for all of the BRC motifs but retain the N- and C-terminal sequences, including the DNA binding domain. In the germline, this allele confers embryonic lethality. In the tissue specific deletion, K14-Cre mediated recombination was estimated to occur in 5-30% of the mammary epithelium in both luminal and myo-epithelial cells, demonstrating overall weak expression of the K14 promoter. A predisposition for tumorigenesis was not observed in $Brca2^{F11/F11}$ mice; however, co-inactivation of p53 in these mice gave rise to mammary, skin, and rare salivary gland tumors.

Conditional loss of *Brca2* in cells of the developing mouse thymus has also been accomplished using an Lck-promoter to express the Cre recombinase (Cheung et al. 2002). These $Brca2^{flox9\text{-}10/flox9\text{-}10}$ mice demonstrated normal T cell development, proliferation, and brisk apoptotic response to various damaging agents as is expected with thymocytes, as well as an unaltered cell cycle as compared to wild type cells. However, there was a notable increase in genetic instability and an increase in spontaneous apoptosis of activated T cells, as well as a small acceleration of thymic lymphoma when p53 was also mutated.

5.4 How do BRCA2-deficient cells escape genome surveillance checkpoints?

The acquisition of mutations was studied in *Brca2*-deficient embryonic tissue and was found to be a rapid, resulting in a striking accumulation of spontaneous and damage-induced mutations, with a preponderance of deletions and rearrangements (Tutt et al. 2002). In differentiated tissues that maintain a small population of tissue renewal stem cells, *Brca2*-deficiency may also be significantly detrimental resulting in increasing apoptosis and depletion of the progenitor pool following DNA damage as demonstrated in the crypts of the small intestine (Hay et al.

2005). However, in many differentiated somatic tissues where tissue renewal is minimal it is possible that cellular BRCA2 deficiency is tolerated as it is primarily required for DNA damage repair during active proliferation (Tutt et al. 2003). DNA damage that remains unrepaired due to BRCA2-deficiency may be repaired by more error prone pathways such as NHEJ during G1 arrest or by Rad51-independent SSA (Tutt et al. 2001; Stark et al. 2004). This increases the likelihood for accumulating mutations that would provide selective growth advantages to cells, which would then allow unstable cells to escape apoptosis and become tumorigenic. Ineffective checkpoints due to mutations in tumor suppressors such as p53 have consistently shown a decrease in tumor latency and an increase in tumor incidence for several *Brca2*-deficient mouse models as noted above as well as for other mouse models mutated for genes involved in DNA repair (Cressman et al. 1999; Brodie et al. 2001; Freie et al. 2003; Houghtaling et al. 2005), highlighting a cooperative role for the DNA damage response in tumor suppression.

Yet not all Brca2-deficient tumors that arise are defective for p53 tumor suppressor functions. The mechanisms for cell survival with persistent genetic damage are not fully known. Recent biochemical and structural studies place BRCA2 directly in the recombination repair pathway, but how this repair defect is managed by the cell may be dependent on cellular checkpoints, the extent of damage and the requirement for continued cellular proliferation. It is worth noting in this regard, that BRCA2 itself may have a role in DNA damage checkpoints (Chen et al. 1999; Kraakman-van der Zwet et al. 2002). If the cell is allowed to propagate fixed genetic damage and the genetic alterations lead to growth advantages then tumorigenesis may occur. The relative rarity of cancer incidence given the extent of genetic aberrations elicited by ineffective repair highlights a proficient but not perfect DNA damage response.

6 Conclusions

In the decade since the BRCA2 gene was cloned, substantial progress has been made in our understanding of the function of this protein, in particular its role in maintaining genomic integrity. Further studies are needed to more precisely define the molecular roles of various domains of BRCA2, as well as to precisely define its role in tumor suppression, in particular the tissue tropism that is observed in adult carriers of BRCA2 mutations.

Acknowledgments

This work was supported by grants from the Human Frontier Science Program (NC), The Susan G. Komen Breast Cancer Foundation BCTR0601221 (MJ), and NIH P01 CA94060 (MEM and MJ).

References

Alter BP, Rosenberg PS, Brody LC (2006) Clinical and molecular features associated with biallelic mutations in FANCD1/BRCA2. J Med Genet

Bennett LM, McAllister KA, Blackshear PE, Malphurs J, Goulding G, Collins NK, Ward T, Bunch DO, Eddy EM, Davis BJ, Wiseman RW (2000) BRCA2-null embryonic survival is prolonged on the BALB/c genetic background. Mol Carcinog 28:174-183

Bignell G, Micklem G, Stratton MR, Ashworth A, Wooster R (1997) The BRC repeats are conserved in mammalian BRCA2 proteins. Hum Mol Genet 6:53-58

Bochkarev A, Bochkareva E (2004) From RPA to BRCA2: lessons from single-stranded DNA binding by the OB-fold. Curr Opin Struct Biol 14:36-42

Bork P, Blomberg N, Nilges M (1996) Internal repeats in the BRCA2 protein sequence. Nat Genet 13:22-23

Brodie SG, Xu X, Qiao W, Li WM, Cao L, Deng CX (2001) Multiple genetic changes are associated with mammary tumorigenesis in Brca1 conditional knockout mice. Oncogene 20:7514-7523

Chen CF, Chen PL, Zhong Q, Sharp ZD, Lee WH (1999) Expression of BRC repeats in breast cancer cells disrupts the BRCA2-Rad51 complex and leads to radiation hypersensitivity and loss of G(2)/M checkpoint control. J Biol Chem 274:32931-32935

Chen PL, Chen CF, Chen Y, Xiao J, Sharp ZD, Lee WH (1998) The BRC repeats in BRCA2 are critical for RAD51 binding and resistance to methyl methanesulfonate treatment. Proc Natl Acad Sci USA 95:5287-5292

Cheung AM, Elia A, Tsao MS, Done S, Wagner KU, Hennighausen L, Hakem R, Mak TW (2004) Brca2 deficiency does not impair mammary epithelium development but promotes mammary adenocarcinoma formation in p53(+/-) mutant mice. Cancer Res 64:1959-1965

Cheung AM, Hande MP, Jalali F, Tsao MS, Skinnider B, Hirao A, McPherson JP, Karaskova J, Suzuki A, Wakeham A, You-Ten A, Elia A, Squire J, Bristow R, Hakem R, Mak TW (2002) Loss of Brca2 and p53 synergistically promotes genomic instability and deregulation of T-cell apoptosis. Cancer Res 62:6194-6204

Collins N, McManus R, Wooster R, Mangion J, Seal S, Lakhani SR, Ormiston W, Daly PA, Ford D, Easton DF, et al. (1995) Consistent loss of the wild type allele in breast cancers from a family linked to the BRCA2 gene on chromosome 13q12-13. Oncogene 10:1673-1675

Connor F, Bertwistle D, Mee PJ, Ross GM, Swift S, Grigorieva E, Tybulewicz VL, Ashworth A (1997) Tumorigenesis and a DNA repair defect in mice with a truncating Brca2 mutation. Nat Genet 17:423-430

Cressman VL, Backlund DC, Hicks EM, Gowen LC, Godfrey V, Koller BH (1999) Mammary tumor formation in p53- and BRCA1-deficient mice. Cell Growth Differ 10:1-10

Cromie GA, Leach DR (2000) Control of crossing over. Mol Cell 6:815-826

Davies AA, Masson JY, McIlwraith MJ, Stasiak AZ, Stasiak A, Venkitaraman AR, West SC (2001) Role of BRCA2 in control of the RAD51 recombination and DNA repair protein. Mol Cell 7:273-282

Donoho G, Brenneman MA, Cui TX, Donoviel D, Vogel H, Goodwin EH, Chen DJ, Hasty P (2003) Deletion of Brca2 exon 27 causes hypersensitivity to DNA crosslinks, chromosomal instability, and reduced life span in mice. Genes Chromosomes Cancer 36:317-331

Esashi F, Christ N, Gannon J, Liu Y, Hunt T, Jasin M, West SC (2005) CDK-dependent phosphorylation of BRCA2 as a regulatory mechanism for recombinational repair. Nature 434:598-604

Featherstone C, Jackson SP (1999) DNA double-strand break repair. Curr Biol 9:R759-761

Freie B, Li X, Ciccone SL, Nawa K, Cooper S, Vogelweid C, Schantz L, Haneline LS, Orazi A, Broxmeyer HE, Lee SH, Clapp DW (2003) Fanconi anemia type C and p53 cooperate in apoptosis and tumorigenesis. Blood 102:4146-4152

Friedman LS, Thistlethwaite FC, Patel KJ, Yu VP, Lee H, Venkitaraman AR, Abel KJ, Carlton MB, Hunter SM, Colledge WH, Evans MJ, Ponder BA (1998) Thymic lymphomas in mice with a truncating mutation in Brca2. Cancer Res 58:1338-1343

Funakoshi M, Li X, Velichutina I, Hochstrasser M, Kobayashi H (2004) Sem1, the yeast ortholog of a human BRCA2-binding protein, is a component of the proteasome regulatory particle that enhances proteasome stability. J Cell Sci 117:6447-6454

Galkin VE, Esashi F, Yu X, Yang S, West SC, Egelman EH (2005) BRCA2 BRC motifs bind RAD51-DNA filaments. Proc Natl Acad Sci USA 102:8537-8542

Gudmundsdottir K, Lord CJ, Witt E, Tutt AN, Ashworth A (2004) DSS1 is required for RAD51 focus formation and genomic stability in mammalian cells. EMBO Rep 5:989-993

Hatanaka A, Yamazoe M, Sale JE, Takata M, Yamamoto K, Kitao H, Sonoda E, Kikuchi K, Yonetani Y, Takeda S (2005) Similar effects of Brca2 truncation and Rad51 paralog deficiency on immunoglobulin V gene diversification in DT40 cells support an early role for Rad51 paralogs in homologous recombination. Mol Cell Biol 25:1124-1134

Hay T, Patrick T, Winton D, Sansom OJ, Clarke AR (2005) Brca2 deficiency in the murine small intestine sensitizes to p53-dependent apoptosis and leads to the spontaneous deletion of stem cells. Oncogene 24:3842-3846

Houghtaling S, Granville L, Akkari Y, Torimaru Y, Olson S, Finegold M, Grompe M (2005) Heterozygosity for p53 (Trp53+/-) accelerates epithelial tumor formation in Fanconi anemia complementation group D2 (Fancd2) knockout mice. Cancer Res 65:85-91

Howlett NG, Taniguchi T, Olson S, Cox B, Waisfisz Q, De Die-Smulders C, Persky N, Grompe M, Joenje H, Pals G, Ikeda H, Fox EA, D'Andrea AD (2002) Biallelic inactivation of BRCA2 in Fanconi anemia. Science 297:606-609

Hughes-Davies L, Huntsman D, Ruas M, Fuks F, Bye J, Chin SF, Milner J, Brown LA, Hsu F, Gilks B, Nielsen T, Schulzer M, Chia S, Ragaz J, Cahn A, Linger L, Ozdag H, Cattaneo E, Jordanova ES, Schuuring E, Yu DS, Venkitaraman A, Ponder B, Doherty A, Aparicio S, Bentley D, Theillet C, Ponting CP, Caldas C, Kouzarides T (2003) EMSY links the BRCA2 pathway to sporadic breast and ovarian cancer. Cell 115:523-535

Hussain S, Wilson JB, Medhurst AL, Hejna J, Witt E, Ananth S, Davies A, Masson JY, Moses R, West SC, de Winter JP, Ashworth A, Jones NJ, Mathew CG (2004) Direct interaction of FANCD2 with BRCA2 in DNA damage response pathways. Hum Mol Genet 13:1241-1248

Hussain S, Witt E, Huber PA, Medhurst AL, Ashworth A, Mathew CG (2003) Direct interaction of the Fanconi anaemia protein FANCG with BRCA2/FANCD1. Hum Mol Genet 12:2503-2510

Jasin M (2002) Homologous repair of DNA damage and tumorigenesis: the BRCA connection. Oncogene 21:8981-8993

Johnson RD, Jasin M (2000) Sister chromatid gene conversion is a prominent double-strand break repair pathway in mammalian cells. EMBO J 19:3398-3407

Jonkers J, Meuwissen R, van der Gulden H, Peterse H, van der Valk M, Berns A (2001) Synergistic tumor suppressor activity of BRCA2 and p53 in a conditional mouse model for breast cancer. Nat Genet 29:418-425

Kastan MB, Bartek J (2004) Cell-cycle checkpoints and cancer. Nature 432:316-323

Kennedy RD, D'Andrea AD (2005) The Fanconi anemia/BRCA pathway: new faces in the crowd. Genes Dev 19:2925-2940

Kojic M, Kostrub CF, Buchman AR, Holloman WK (2002) BRCA2 homolog required for proficiency in DNA repair, recombination, and genome stability in *Ustilago maydis*. Mol Cell 10:683-691

Kojic M, Yang H, Kostrub CF, Pavletich NP, Holloman WK (2003) The BRCA2-interacting protein DSS1 is vital for DNA repair, recombination, and genome stability in *Ustilago maydis*. Mol Cell 12:1043-1049

Kojic M, Zhou Q, Lisby M, Holloman WK (2005) Brh2-Dss1 interplay enables properly controlled recombination in *Ustilago maydis*. Mol Cell Biol 25:2547-2557

Kraakman-van der Zwet M, Overkamp WJ, van Lange RE, Essers J, van Duijn-Goedhart A, Wiggers I, Swaminathan S, van Buul PP, Errami A, Tan RT, Jaspers NG, Sharan SK, Kanaar R, Zdzienicka MZ (2002) Brca2 (XRCC11) deficiency results in radioresistant DNA synthesis and a higher frequency of spontaneous deletions. Mol Cell Biol 22:669-679

Krogan NJ, Lam MH, Fillingham J, Keogh MC, Gebbia M, Li J, Datta N, Cagney G, Buratowski S, Emili A, Greenblatt JF (2004) Proteasome involvement in the repair of DNA double-strand breaks. Mol Cell 16:1027-1034

Lee M, Daniels MJ, Venkitaraman AR (2004) Phosphorylation of BRCA2 by the Polo-like kinase Plk1 is regulated by DNA damage and mitotic progression. Oncogene 23:865-872

Lengauer C, Kinzler KW, Vogelstein B (1998) Genetic instabilities in human cancers. Nature 396:643-649

Li J, Zou C, Bai Y, Wazer DE, Band V, Gao Q (2006) DSS1 is required for the stability of BRCA2. Oncogene 25:1186-1194

Lim DS, Hasty P (1996) A mutation in mouse rad51 results in an early embryonic lethal that is suppressed by a mutation in p53. Mol Cell Biol 16:7133-7143

Lin HR, Ting NS, Qin J, Lee WH (2003) M phase-specific phosphorylation of BRCA2 by Polo-like kinase 1 correlates with the dissociation of the BRCA2-P/CAF complex. J Biol Chem 278:35979-35987

Liu J, Yuan Y, Huan J, Shen Z (2001) Inhibition of breast and brain cancer cell growth by BCCIPalpha, an evolutionarily conserved nuclear protein that interacts with BRCA2. Oncogene 20:336-345

Lu H, Guo X, Meng X, Liu J, Allen C, Wray J, Nickoloff JA, Shen Z (2005) The BRCA2-interacting protein BCCIP functions in RAD51 and BRCA2 focus formation and homologous recombinational repair. Mol Cell Biol 25:1949-1957

Ludwig T, Chapman DL, Papaioannou VE, Efstratiadis A (1997) Targeted mutations of breast cancer susceptibility gene homologs in mice: lethal phenotypes of Brca1, Brca2, Brca1/Brca2, Brca1/p53, and Brca2/p53 nullizygous embryos. Genes Dev 11:1226-1241

Ludwig T, Fisher P, Murty V, Efstratiadis A (2001) Development of mammary adenocarcinomas by tissue-specific knockout of Brca2 in mice. Oncogene 20:3937-3948

Marston NJ, Richards WJ, Hughes D, Bertwistle D, Marshall CJ, Ashworth A (1999) Interaction between the product of the breast cancer susceptibility gene BRCA2 and DSS1, a protein functionally conserved from yeast to mammals. Mol Cell Biol 19:4633-4642

Martin JS, Winkelmann N, Petalcorin MI, McIlwraith MJ, Boulton SJ (2005) RAD-51-dependent and -independent roles of a *Caenorhabditis elegans* BRCA2-related protein during DNA double-strand break repair. Mol Cell Biol 25:3127-3139

McAllister KA, Bennett LM, Houle CD, Ward T, Malphurs J, Collins NK, Cachafeiro C, Haseman J, Goulding EH, Bunch D, Eddy EM, Davis BJ, Wiseman RW (2002) Cancer susceptibility of mice with a homozygous deletion in the COOH-terminal domain of the Brca2 gene. Cancer Res 62:990-994

McAllister KA, Haugen-Strano A, Hagevik S, Brownlee HA, Collins NK, Futreal PA, Bennett LM, Wiseman RW (1997) Characterization of the rat and mouse homologues of the BRCA2 breast cancer susceptibility gene. Cancer Res 57:3121-3125

Mizuta R, LaSalle JM, Cheng HL, Shinohara A, Ogawa H, Copeland N, Jenkins NA, Lalande M, Alt FW (1997) RAB22 and RAB163/mouse BRCA2: proteins that specifically interact with the RAD51 protein. Proc Natl Acad Sci USA 94:6927-6932

Moynahan ME (2002) The cancer connection: BRCA1 and BRCA2 tumor suppression in mice and humans. Oncogene 21:8994-9007

Moynahan ME, Chiu JW, Koller BH, Jasin M (1999) Brca1 controls homology-directed DNA repair. Mol Cell 4:511-518

Moynahan ME, Pierce AJ, Jasin M (2001) BRCA2 is required for homology-directed repair of chromosomal breaks. Mol Cell 7:263-272

Nakanishi K, Yang YG, Pierce AJ, Taniguchi T, Digweed M, D'Andrea AD, Wang ZQ, Jasin M (2005) Human Fanconi anemia monoubiquitination pathway promotes homologous DNA repair. Proc Natl Acad Sci USA 102:1110-1115

Paques F, Haber JE (1999) Multiple pathways of recombination induced by double-strand breaks in *Saccharomyces cerevisiae*. Microbiol Mol Biol Rev 63:349-404

Pellegrini L, Yu DS, Lo T, Anand S, Lee M, Blundell TL, Venkitaraman AR (2002) Insights into DNA recombination from the structure of a RAD51-BRCA2 complex. Nature 420:287-293

Petalcorin MI, Sandall J, Wigley DB, Boulton SJ (2006) CeBRC-2 Stimulates D-loop formation by RAD-51 and promotes DNA single-strand annealing. J Mol Biol 361:231-242

Saeki H, Siaud N, Christ N, Wiegant WW, van Buul PP, Han M, Zdzienicka MZ, Stark JM, Jasin M (2006) Suppression of the DNA repair defects of BRCA2-deficient cells with heterologous protein fusions. Proc Natl Acad Sci USA 103:8768-8773

San Filippo J, Chi P, Sehorn MG, Etchin J, Krejci L, Sung P (2006) Recombination mediator and RAD51 targeting activities of a human BRCA2 polypeptide. J Biol Chem 281:11649-11657

Sarkisian CJ, Master SR, Huber LJ, Ha SI, Chodosh LA (2001) Analysis of murine Brca2 reveals conservation of protein-protein interactions but differences in nuclear localization signals. J Biol Chem 276:37640-37648

Schoenfeld AR, Apgar S, Dolios G, Wang R, Aaronson SA (2004) BRCA2 is ubiquitinated *in vivo* and interacts with USP11, a deubiquitinating enzyme that exhibits prosurvival function in the cellular response to DNA damage. Mol Cell Biol 24:7444-7455

Scully R, Livingston DM (2000) In search of the tumour-suppressor functions of BRCA1 and BRCA2. Nature 408:429-432

Sharan SK, Bradley A (1997) Murine Brca2: sequence, map position, and expression pattern. Genomics 40:234-241

Sharan SK, Morimatsu M, Albrecht U, Lim DS, Regel E, Dinh C, Sands A, Eichele G, Hasty P, Bradley A (1997) Embryonic lethality and radiation hypersensitivity mediated by Rad51 in mice lacking Brca2. Nature 386:804-810

Sharan SK, Pyle A, Coppola V, Babus J, Swaminathan S, Benedict J, Swing D, Martin BK, Tessarollo L, Evans JP, Flaws JA, Handel MA (2004) BRCA2 deficiency in mice leads to meiotic impairment and infertility. Development 131:131-142

Shin S, Verma IM (2003) BRCA2 cooperates with histone acetyltransferases in androgen receptor-mediated transcription. Proc Natl Acad Sci USA 100:7201-7206

Siaud N, Dray E, Gy I, Gerard E, Takvorian N, Doutriaux MP (2004) Brca2 is involved in meiosis in *Arabidopsis thaliana* as suggested by its interaction with Dmc1. EMBO J 23:1392-1401

Sonoda E, Sasaki MS, Buerstedde JM, Bezzubova O, Shinohara A, Ogawa H, Takata M, Yamaguchi-Iwai Y, Takeda S (1998) Rad51-deficient vertebrate cells accumulate chromosomal breaks prior to cell death. EMBO J 17:598-608

Spain BH, Larson CJ, Shihabuddin LS, Gage FH, Verma IM (1999) Truncated BRCA2 is cytoplasmic: implications for cancer-linked mutations. Proc Natl Acad Sci USA 96:13920-13925

Stark JM, Hu P, Pierce AJ, Moynahan ME, Ellis N, Jasin M (2002) ATP hydrolysis by mammalian RAD51 has a key role during homology-directed DNA repair. J Biol Chem 277:20185-20194

Stark JM, Pierce AJ, Oh J, Pastink A, Jasin M (2004) Genetic steps of mammalian homologous repair with distinct mutagenic consequences. Mol Cell Biol 24:9305-9316

Sung P, Krejci L, Van Komen S, Sehorn MG (2003) Rad51 recombinase and recombination mediators. J Biol Chem 278:42729-42732

Suzuki A, de la Pompa JL, Hakem R, Elia A, Yoshida R, Mo R, Nishina H, Chuang T, Wakeham A, Itie A, Koo W, Billia P, Ho A, Fukumoto M, Hui CC, Mak TW (1997) Brca2 is required for embryonic cellular proliferation in the mouse. Genes Dev 11:1242-1252

Takata M, Tachiiri S, Fujimori A, Thompson LH, Miki Y, Hiraoka M, Takeda S, Yamazoe M (2002) Conserved domains in the chicken homologue of BRCA2. Oncogene 21:1130-1134

Tavtigian SV, Simard J, Rommens J, Couch F, Shattuck-Eidens D, Neuhausen S, Merajver S, Thorlacius S, Offit K, Stoppa-Lyonnet D, Belanger C, Bell R, Berry S, Bogden R, Chen Q, Davis T, Dumont M, Frye C, Hattier T, Jammulapati S, Janecki T, Jiang P, Kehrer R, Leblanc JF, Mitchell JT, McArthur-Morrison J, Nguyen K, Peng Y, Samson C, Schroeder M, Snyder SC, Steele L, Stringfellow M, Stroup C, Swedlund B, Swense J, Teng D, Thomas A, Tran T, Tranchant M, Weaver-Feldhaus J, Wong AK, Shizuya H, Eyfjord JE, Cannon-Albright L, Tranchant M, Labrie F, Skolnick MH, Weber B, Kamb A, Goldgar DE (1996) The complete BRCA2 gene and mutations in chromosome 13q-linked kindreds. Nat Genet 12:333-337

Tsuzuki T, Fujii Y, Sakumi K, Tominaga Y, Nakao K, Sekiguchi M, Matsushiro A, Yoshimura Y, MoritaT (1996) Targeted disruption of the Rad51 gene leads to lethality in embryonic mice. Proc Natl Acad Sci USA 93:6236-6240

Tutt A, Bertwistle D, Valentine J, Gabriel A, Swift S, Ross G, Griffin C, Thacker J, Ashworth A (2001) Mutation in Brca2 stimulates error-prone homology-directed repair of DNA double-strand breaks occurring between repeated sequences. EMBO J 20:4704-4716

Tutt A, Connor F, Bertwistle D, Kerr P, Peacock J, Ross G, Ashworth A (2003) Cell cycle and genetic background dependence of the effect of loss of BRCA2 on ionizing radiation sensitivity. Oncogene 22:2926-2931

Tutt AN, van Oostrom CT, Ross GM, van Steeg H, Ashworth A (2002) Disruption of Brca2 increases the spontaneous mutation rate *in vivo*: synergism with ionizing radiation. EMBO Rep 3:255-260

Venkitaraman AR (2002) Cancer susceptibility and the functions of BRCA1 and BRCA2. Cell 108:171-182

Wang X, Andreassen PR, D'Andrea AD (2004) Functional interaction of monoubiquitinated FANCD2 and BRCA2/FANCD1 in chromatin. Mol Cell Biol 24:5850-5862

Warren M, Lord CJ, Masabanda J, Griffin D, Ashworth A (2003) Phenotypic effects of heterozygosity for a BRCA2 mutation. Hum Mol Genet 12:2645-2656

Warren M, Smith A, Partridge N, Masabanda J, Griffin D, Ashworth A (2002) Structural analysis of the chicken BRCA2 gene facilitates identification of functional domains and disease causing mutations. Hum Mol Genet 11:841-851

Wong AK, Pero R, Ormonde PA, Tavtigian SV, Bartel PL (1997) RAD51 interacts with the evolutionarily conserved BRC motifs in the human breast cancer susceptibility gene brca2. J Biol Chem 272:31941-31944

Wooster R, Bignell G, Lancaster J, Swift S, Seal S, Mangion J, Collins N, Gregory S, Gumbs C, Micklem G (1995) Identification of the breast cancer susceptibility gene BRCA2. Nature 378:789-792

Wooster R, Neuhausen SL, Mangion J, Quirk Y, Ford D, Collins N, Nguyen K, Seal S, Tran T, Averill D, et al. (1994) Localization of a breast cancer susceptibility gene, BRCA2, to chromosome 13q12-13. Science 265:2088-2090

Xia B, Sheng Q, Nakanishi K, Ohashi A, Wu J, Christ N, Liu X, Jasin M, Couch FJ, Livingston DM (2006) Control of BRCA2 cellular and clinical functions by a nuclear partner, PALB2. Mol Cell 22:719-729

Xia F, Taghian DG, DeFrank JS, Zeng ZC, Willers H, Iliakis G, Powell SN (2001) Deficiency of human BRCA2 leads to impaired homologous recombination but maintains normal nonhomologous end joining. Proc Natl Acad Sci USA 98:8644-8649

Yang H, Jeffrey PD, Miller J, Kinnucan E, Sun Y, Thoma NH, Zheng N, Chen PL, Lee WH, Pavletich NP (2002) BRCA2 function in DNA binding and recombination from a BRCA2-DSS1-ssDNA structure. Science 297:1837-1848

Yang H, Li Q, Fan J, Holloman WK, Pavletich NP (2005) The BRCA2 homologue Brh2 nucleates RAD51 filament formation at a dsDNA-ssDNA junction. Nature 433:653-657

Yuan SS, Lee SY, Chen G, Song M, Tomlinson GE, Lee EY (1999) BRCA2 is required for ionizing radiation-induced assembly of Rad51 complex *in vivo*. Cancer Res 59:3547-3551

Christ, Nicole

Developmental Biology Program, Memorial Sloan-Kettering Cancer Center, 1275 York Avenue, New York, NY 10021, USA

Jasin, Maria
Developmental Biology Program, Memorial Sloan-Kettering Cancer Center, 1275 York Avenue, New York, NY 10021, USA
m-jasin@ski.mskcc.org

Moynahan, Mary Ellen
Department of Medicine, Memorial Sloan-Kettering Cancer Center, 1275 York Avenue, New York, NY 10021, USA.

Meiotic recombination

Neil Hunter

Abstract

Crossover recombination is essential for homolog segregation during meiosis. In contrast to spontaneous mitotic recombination, meiotic recombination is intrinsic being initiated by the programmed formation of DNA double-strand-breaks. In addition, the tendencies of the core recombination machinery to utilize a sister-chromatid template and to produce a noncrossover outcome are counteracted by meiosis-specific factors, which ultimately ensure the formation of at least one crossover per homolog.

1 Overview

Homologous recombination during meiosis is distinct from recombination in mitotically dividing cells in several ways. First, meiotic recombination is genetically programmed being induced endogenously by Spo11-catalyzed DNA double-strand-breaks (DSBs) (Section 4). Second, recombination does work during meiosis, driving the movement of homologous chromosomes into close juxtaposition though DNA-DNA interactions, and holding homologs together at metaphase through the formation of crossovers (Section 5). Accordingly, and third, meiotic recombination occurs more frequently between homologs than between sister-chromatids (Section 6). Finally, meiotic recombination is uniquely regulated so that, while most initiated events have a noncrossover outcome, at least one event per chromosome is guaranteed to mature as a crossover (Section 7).

This chapter aims to present contemporary ideas regarding the mechanism, regulation, and function of homologous recombination during meiosis. A comprehensive chronological review of the literature is not intended. I will focus on studies done using *Saccharomyces cerevisiae* because this work provides the most detailed and integrated view of meiotic recombination to date. Wherever possible, data from other organisms will be discussed.

2 Meiosis

The alternation of diploid and haploid generations of cells is a fundamental feature of sexually reproducing organisms. To form haploid cells the diploid chromosome

Fig. 1 (overleaf). Meiosis. (A) Meiotic chromosome segregation. i. Diploid cell with a single pair of homologous chromosomes. ii. Chromosomes replicate to give pairs of sister-chromatids (homologs) connected by cohesion. iii. Homologs pair and undergo crossing-over. The resulting chiasma connects the homologs and thereby facilitates stable bipolar attachment to the meiosis-I spindle. iv. Cohesion is lost between chromosome arms and homologs are pulled to opposite poles. v. Maintenance of cohesion between centromeres allows bipolar connection of sister-chromatids pairs to the meiosis-II spindle. vi. The remaining cohesion is lost and sister-chromatids are segregated. Arrows indicate directions of the pulling forces generated by microtubules. Dashed lines indicate the planes of cell division. (B) i. Spread pachytene chromosomes from *Saccharomyces cerevisiae* visualized by EM (inset; bar = 2μm) (from Moens and Pearlman 1988, Copyright 1988, John Wiley and Sons, Inc. Reprinted with permission of Wiley-Liss, Inc. a subsidiary of John Wiley & Sons, Inc). The chromatin loops of sister-chromatids are organized along the two homolog axes, which are intimately connected by the transverse filaments of the synaptonemal complex. ii. and iii. Longitudinal sections of synaptonemal complexes from *S. cerevisiae* showing associated late recombination nodule (indicated by arrows) (from Schmekel 2000, Copyright 2000, Springer-Verlag GmbH. Reprinted with permission Springer Science and Business media). (C) Relative timing of events during meiotic prophase I in *S. cerevisiae.* Top row: stages of meiotic prophase I. Second row: DNA events of meiotic recombination (see Section 3.1 for details). Only the two chromatids involved in the recombination event are shown. Third row: synaptonamal complex formation. Spread meiotic chromosomes immunostained for the synaptonamal complex protein, Zip1 (see Section 7.4.2). Fourth row: Cartoon of chromosome morphogenesis during prophase I. Blue lines, chromatin loops; red lines, homolog axes; striated green lines, transverse filaments of the synaptonemal complex.

complement is precisely halved via the process of meiosis (Hunter 2004). Like mitosis, meiosis begins with replication to produce pairs of sister-chromatids connected by cohesion (Fig. 1A ii.). Then, in contrast to mitosis, a single copy of every chromosome is accurately distributed to four different nuclei (Fig. 1A iii - vi.). Meiosis achieves this in the only logical way: via two successive rounds of nuclear division, first segregating homologs (the maternal and paternal chromatid pairs) and then segregating sister-chromatids, as in mitosis (Petronczki et al. 2003). Homolog segregation is unique to meiosis and in most organisms homologous recombination plays two essential roles in this process. First, homologous recombination drives the two-by-two pairing of parental homologs. Second, paired homologs become connected by chiasmata (cytological manifestations of DNA crossovers), which facilitate their stable bipolar connection to the spindle and thereby promote accurate homolog segregation (Fig. 1A iii.).

2.1 Meiotic chromosome structure and the synaptonemal complex

Meiotic recombination occurs in intimate relationship with a highly organized and dynamically meiotic chromosome structure (Zickler and Kleckner 1999).

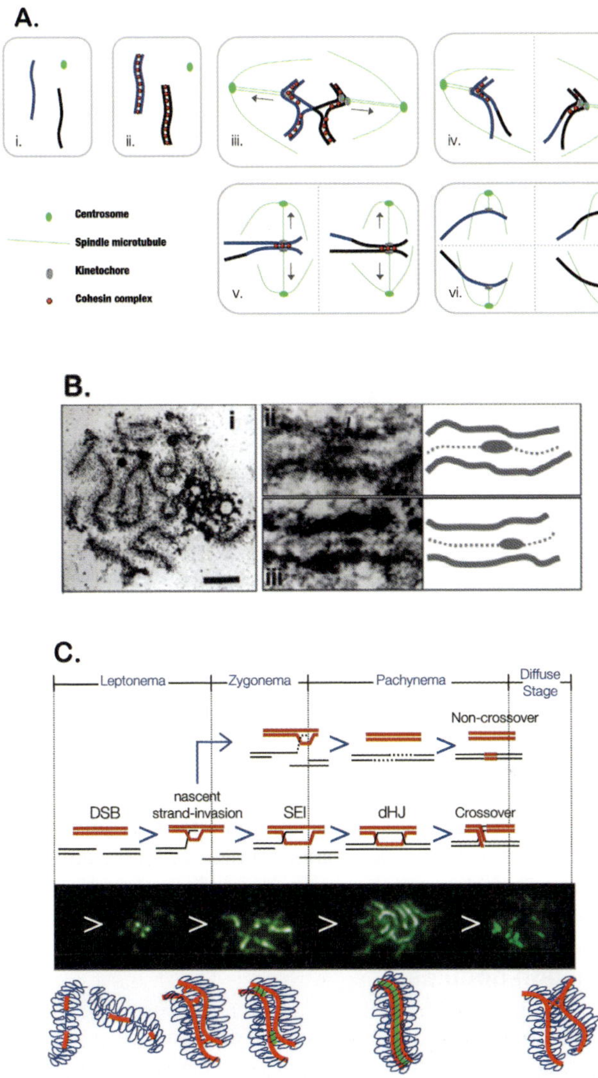

2.1.1 Loops and Axes

Meiotic chromosomes have a well-defined loop-axis structure (Fig. 1B i.) (Moens and Pearlman 1988; Zickler and Kleckner 1999). The loops are loops of chromatin which vary in size from ~20 kb in budding yeast, to ~2500 kb in some insects. The loops of every pair of chromatids are connected at their bases, in a linear arrangement, along a rod-like axis or core, i.e. one axis = two chromatids = one homolog. The basic loop/axis structure is thought to be organized by the cohesin proteins;

this then serves as a platform for the binding of additional proteins which give rise to the distinct axes detected by electron microscopy (Fig. 1B i-iii.).

2.1.2 Synaptonemal complexes

Homolog pairing culminates with formation of synaptonemal complexes (SCs), prominent proteinaceous structures that form between homologs, along their entire lengths (Fig. 1B i.). SCs comprise two lateral elements and a central region, which contains a less dense central element. The lateral elements correspond to the homolog axes, described above. Transverse filaments lie across the central region to create a striated, zipper-like appearance. SCs are thought to assemble by a two step process: nucleation, by installing central region proteins at sites where homolog axes are very closely paired; followed by polymerization between and along the homolog axes. SCs are important for the normal formation of crossovers (see Section 7).

2.2 Stages of meiotic prophase I

Meiotic recombination occurs during the extended prophase that precedes the first meiotic division. This period is divided into six stages, defined by the appearance of the chromosomes and development of the SCs (Fig. 1C) (von Wettstein et al. 1984; Hunter 2004). During *leptonema*, axes develop and homologs begin to pair. SC formation initiates during *zygonema*. When SC has polymerized along the entire lengths of all homologs cells enter *pachynema*. During the *diffuse stage* chromosomes are indistinct and when they reappear at *diplonema*, the homologs are no longer associated along their lengths, but are visibly connected by chiasmata. During *diakinesis*, spindle microtubules develop and attach to the monopolar kinetochores of the homologs. The homolog pairs condense dramatically, becoming progressively shorter and thicker and they move towards the equator of the cell.

2.3 Recombination nodules

During leptonema homologs become coaligned and axes closely associate at multiple sites. These axial associations are the sites where DNA molecules are interacting via homologous recombination and are often associated with densely staining nodular structures that contain recombination proteins such as RecA homologs, Rad51 and Dmc1 (Zickler and Kleckner 1999) (see Sect 5). During zygonema, the polymerization of SCs is nucleated at a subset of these sites. A second type of nodular structure develops during pachynema that is typically larger, denser and rounder, and less numerous than those observed during zygonema. These late recombination nodules mark the sites where chiasmata will appear during diplonema, i.e. the sites of crossovers and contain the Mlh1 and Mlh3 proteins (Fig. 1B ii-iii. and see Section 7.4.3).

3 Overview of meiotic recombination

Molecular models of meiotic recombination, and of DSB-repair in general, owe much to decades of analysis of marker segregation patterns in fungal crosses (Paques and Haber 1999); for contemporary analyses see (Merker et al. 2003; Malkova et al. 2004; Stahl et al. 2004; Hoffmann and Borts 2005; Hoffmann et al. 2005; Jessop et al. 2005). Over the last 16 years, molecular approaches have identified several anticipated intermediates, confirming key features of the canonical DSBR model of Szostak et al. (1983) but also suggesting significant revisions (Fig. 1B).

3.1 The pathway of meiotic recombination

Initiation of meiotic recombination occurs by the formation of programmed DSBs catalyzed by the Spo11 protein (Section 4.1) (Keeney 2001). DSB-ends then undergo nucleolytic resection of the 5'-strands to produce 3'-single-stranded tails of ≥500 nucleotides (Section 4.3) (Sun et al. 1989, 1991; Cao et al. 1990; Bishop et al. 1992; Nag and Petes 1993; Jessop et al. 2005). Successive pairing and strand-exchange at the two DSB-ends yields two joint molecule (JM) intermediates: the Single-End Invasion (SEI) and the double Holliday Junction (dHJ) (Fig. 1B and Section 5) (Schwacha and Kleckner 1995; Hunter and Kleckner 2001). In the SEI, one DSB-end has undergone strand-exchange with a homologous duplex. In a dHJ, the second DSB-end has engaged the same homolog and strand continuity has been restored. Thus, a pair of Holliday junctions and intervening heteroduplexes now connects the homologous duplexes. Significantly, both SEIs and dHJs appear to be crossover-specific intermediates (Section 7) (Allers and Lichten 2001a; Hunter and Kleckner 2001; Clyne et al. 2003; Borner et al. 2004). dHJs are resolved into crossover products. Contrasting the established picture in *S. cerevisiae*, an intriguing study in *S. pombe* indicates that single Holliday junctions are the major precursors of crossovers in this organism (Cromie et al. 2006).

The molecular events leading to non-crossovers are less clear but likely involve a synthesis-dependent strand-annealing (SDSA) mechanism (Allers and Lichten 2001b; Jessop et al. 2005). In this case, one DSB-end invades a homolog, and primes DNA synthesis; the nascent strand is then displaced and anneals to complimentary sequences on the second DSB-end to seal the break (Paques and Haber 1999).

3.2 Monitoring meiotic recombination intermediates

An experimental system for detecting recombination intermediates is shown in Figure 2A (for examples, see Borts et al. 1986; Schwacha and Kleckner 1995; Storlazzi et al. 1995; Allers and Lichten 2001b; Hunter and Kleckner 2001). The basic set-up utilizes a defined DSB "hotspot" flanked by polymorphic restriction

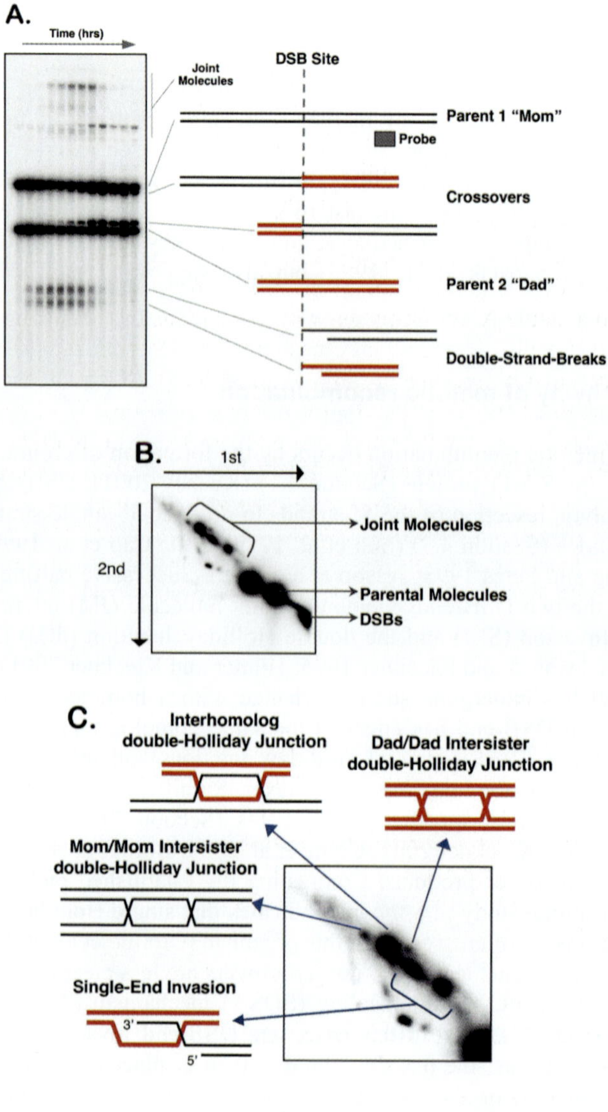

Fig. 2. Molecular assays to monitor meiotic recombination (A) Physical assay system (see Section 3.2). Lines represent polymorphic restriction fragments flanking a defined DSB hotspot. The corresponding bands are detected by Southern analysis of DNA samples prepared from a yeast culture undergoing meiosis. (B) Native/native 2D gel showing the positions of parental, DSB and joint molecule signals. (C) Blowup of the image shown in (B) highlighting the positions of SEIs and the three dHJ species. The prominence of the interhomolog dHJ indicates a ~5-fold interhomolog bias at this site (see Section 6).

sites. Restriction fragments at the assay locus are resolved by gel electrophoresis and detected by Southern hybridization. DSBs produce transient faster-migrating bands. Distinct recombinant-length bands are formed by crossing-over. Detection of noncrossover products requires an additional restriction site polymorphism right at the DSB site (Storlazzi et al. 1995; Allers and Lichten 2001a; Borner et al. 2004; Martini et al. 2006). Detection of strand-exchange or joint molecule (JM) intermediates is complicated by their lability. Standard DNA purification methods promote spontaneous branch migration and dissociation of JMs (Allers and Lichten 2000). Three approaches have been used to stabilize JMs. Psoralen treatment produces inter-strand cross-links that block branch migration under standard DNA purification conditions (Bell and Byers 1983a; Schwacha and Kleckner 1994). Alternatively, junction-migration can be stalled by polyvalent cations, which fold junctions into a stable X conformation with contiguous base-stacking (Allers and Lichten 2000). Finally, preparing chromosomal size DNA in agarose plugs stabilizes JMs, presumably because the immobilized long DNA molecules are topologically constrained (Borde et al. 2000; Cromie et al. 2006). The branched nature of JMs is revealed using native/native two-dimensional gel electrophoresis, which readily resolves JMs from linear molecules of the same molecular weight (Fig. 2B, C) (Bell and Byers 1983a). Finally, the strand composition of JMs is analyzed by first resolving JMs by native electrophoresis, separating strands by alkaline denaturation and then running a perpendicular gel under denaturing conditions (Schwacha and Kleckner 1995; Allers and Lichten 2001b; Hunter and Kleckner 2001).

3.2.1 Mapping DSB sites

Pulsed-field gel electrophoresis can be used to map the positions and intensities of DSB-hotspots along entire yeast chromosomes (Zenvirth et al. 1992). The advent of DNA microarray technology now allows high-resolution mapping of DSB sites on a genome-wide scale. DSB-specific probes are made by purifying the DNA from Spo11-DNA complexes (Gerton et al. 2000) or by purifying the single-stranded DNA that forms specifically at DSB-ends (M. Lichten, personal communication) (see below).

4 Initiation of meiotic recombination

4.1 The Spo11 complex

Meiotic recombination is initiated by Spo11-catalyzed double-strand-breaks (DSBs). In *S. cerevisiae*, Spo11 is the DNA cleaving subunit of a much larger complex comprising at least nine other proteins, all of which are absolutely required for meiotic DSB formation (Keeney 2001).

4.1.1 Spo11

Spo11 is conserved throughout eukaryotes. Its structural relationship to the Top6A subunit of the archeal type-IIB topoisomerase, TopoVI, provides key insights into Spo11 biochemical activity (Bergerat et al. 1997; Nichols et al. 1999; Diaz et al. 2002). All evidence indicates that Spo11 catalyzes DSB formation by a trans-esterification mechanism analogous to type-II topoisomerases (Fig. 3). Cleavage of both strands of a duplex presumably requires the coordinated activities of a dimer of Spo11; structural predictions and the dominant-negative nature of several *spo11* alleles and are consistent with this assumption (Nichols et al. 1999; Diaz et al. 2002). Although demonstration of this activity *in vitro* is thwarted by the recalcitrance of Spo11 to purification (e.g. Wu et al. 2004), detection of covalent Spo11–DNA complexes *in vivo* provides irrefutable evidence that Spo11 is the DNA cleaving enzyme (Keeney et al. 1997; Neale et al. 2005).

Unlike type-II enzymes, Spo11 does not normally function as a topoisomerase during meiosis and is likely incapable of catalyzing the classical two-gate DNA-passage mechanism. Consistently, most organisms lack a homolog of the Top6B subunit that bridges separated Top6A subunits following DNA cleavage (Corbett and Berger 2003b). Top6B uses an ATP binding and hydrolysis cycle both to trap a second DNA duplex and to drive conformational changes that open and close the Top6A gate. *Arabidopsis* and other higher plants do possess a clear homolog of Top6B, BIN3, but this is not required for meiosis. Instead BIN3 appears to function together with one of the three *Arabidopsis* Spo11 homologs (SPO11-3 a.k.a. BIN5) as a *bona fide* Topo6AB topoisomerase, which is required for normal development (Corbett and Berger 2003a). On the other hand, AtSPO11-1, which is required for meiosis, is unable to interact with AtTOP6B/BIN3.

Spo11 has been localized to meiotic chromosomes by immunoflorescence in *S. cerevisiae*, the filamentous ascomycete *Sordaria macrospora* and mouse (Romanienko and Camerini-Otero 2000; Storlazzi et al. 2003; Prieler et al. 2005). Spo11 is bound to chromatin at the time of DSB formation (leptonema) but persists into pachynema, long after DSBs are formed. The role of this late population of Spo11 is unclear. Immunostaining is still observed for a "dead" Spo11, which lacks the catalytic tyrosine (spo11-Y135F), indicating that DSB formation is not required and noncovalently bound Spo11 is being detected. The Spo11 complex proteins Rec102, Rec104 and probably Ski8 are required for Spo11 immunostaining (Storlazzi et al. 2003; Tesse et al. 2003; Arora et al. 2004; Prieler et al. 2005) (Section 4.1.2)

Chromatin immunoprecipitation (ChIP) of Spo11 demonstrates a transient interaction with known DSB sites (Prieler et al. 2005). This interaction is dependent on formaldehyde cross-linking but not Spo11 catalytic activity, again indicating that noncovalent binding is being detected and arguing against the existence of a long-lived covalently bound Spo11 intermediate (but see below and Neale et al. 2005). Consistently, unresected DSBs are never detected as an intermediate in wild-type cells (e.g. Sun et al. 1989; Cao et al. 1990; Bishop et al. 1992). The idea that DSB formation is normally tightly coupled to resection is discussed further below (Section 4.4). Association of Spo11 with DSB sites, as detected by ChIP, requires Rec102, Rec104 and Rec114 but not the other Spo11 complex proteins (Prieler et al. 2005).

4.1.2 Ski8/Rec103

Ski8 is a conserved protein comprising multiple copies of the ~40 amino-acid WD-repeat, a wide-ranging protein-protein interaction motif found in at least 30 functional classes of proteins (Yu et al. 2000). In Ski8, as in other WD-repeat proteins, these motifs fold into a seven-bladed β-propeller structure (Cheng et al. 2004; Madrona and Wilson 2004). This structure is thought to simultaneously interact with multiple proteins to coordinate their interactions. The top and side faces of the Ski8 β-propeller interact directly with Spo11 (Arora et al. 2004; Cheng et al. 2004). 2-hybrid data suggests that Ski8 may also interact with other components of the Spo11 complex, namely Rec104, Rec114, and Mer2 (Arora et al. 2004). An intriguing aspect of Ski8 biology is its dual role in cytoplasmic mRNA decay and in meiotic DSB formation. These roles are distinct as demonstrated by the observations that cytoplasmic partners of Ski8 are not required for meiotic recombination (Arora et al. 2004). Moreover, Ski8 relocalizes to the nucleus specifically during meiosis and this relocalization is strongly Spo11-dependent.

By indirect immunofluorescence, Ski8 is seen to localize to chromatin from leptotene through pachytene. This requires Spo11 but not the other DSB proteins (Arora et al. 2004). Differential extraction and immunoblotting for Spo11 demonstrates that Ski8 stabilizes the association of Spo11 with chromatin suggesting a mutual interdependence between Ski8 and Spo11 for chromatin localization (Arora et al. 2004). This inference is echoed by direct immunoflorescence studies of GFP-tagged Ski8 and Spo11 in *Sordaria macrospora* (Storlazzi et al. 2003; Tesse et al. 2003). At odds with this idea, however, is an immunofluorescence study in which *ski8* mutants showed normal Spo11 immunostaining foci on chromatin (Prieler et al. 2005).

Although Ski8 homologs can be found in most species, a role in meiosis has only been demonstrated in three fungi, *S. cerevisiae, S. pombe* (Rec14) and *Sordaria* (Evans et al. 1997; Gardiner et al. 1997; Tesse et al. 2003; Arora et al. 2004). In fact, *Arabidopsis* Ski8 plays no role in meiosis and sequence analysis suggests that amino-acids in Spo11 required for interaction with Ski8 are not conserved outside of fungi (Jolivet et al. 2006).

4.1.3 Rec102 and Rec104

Rec102 and Rec104 form a putative subcomplex as evidenced by 2-hybrid interaction, coimmunoprecipitation and genetic interactions (Salem et al. 1999; Kee and Keeney 2002; Jiao et al. 2003; Kee et al. 2004). Neither protein appears to be conserved outside of *Saccharomyces* and closely related yeasts. Rec104 is phosphorylated but the function of this phosphorylation and the kinase involved have not

Fig. 3 (overleaf). DSB formation and assembly of strand-exchange complexes. (A) Model for the assembly and activation of the Spo11 complex (see Section 4). The Spo11-Ski8-Rec102/104 and Mer2-Mei4 (±Rec114) complexes assemble independently onto chromatin. High local CDK activity catalyzes the phosphorylation of Mer2 and perhaps Rec104. These modifications promote assembly of the "tight-binding" transition state through Rec114-mediated interactions between Spo11-Ski8-Rec102/104 and Mer2-Mei4, and recruitment of the MRX complex. Concerted Spo11-cleavage and MRX-incision reactions are driven forward by Sae2, Hop1 and conformational changes in Rad50. S-alleles of Rad50 and Mre11, and the absence of Sae2 uncouple these reactions and the tight-binding state decays to form DSBs with covalently attached Spo11. (B) Models of DSB resection, assembly of Dmc1 and Rad51 nucleoprotein filaments and DSB-dependent homolog pairing (see Sections 4.3 and 5.2). DSB-ends are resected for ~500 bp. The Spo11-oligo complex remains associated with one DSB-end. This DSB-end asymmetry establishes asymmetric assembly of Dmc1 and Rad51. Initial pairing interactions are catalyzed by Dmc1, Hop2-Mnd1and Tid1. Hop2-Mnd1 may act locally at DSB sites to promote Dmc1 catalyzed strand-invasion and independently to promote pairing interactions between intact duplexes. The strand-exchange activity of the Rad51 filament is transiently inhibited by Hed1.

been determined. Phosphorylation occurs in early prophase and phospho-Rec104 is the major form after leptotene, the time when Rec104 and Rec102 first associate with chromatin loops and DSBs are made (Kee et al. 2004). Both proteins are mutually interdependent for nuclear localization and chromatin association; Spo11 and Ski8 are also required but other Spo11-complex proteins and DSB-formation are not.

Chromosome-wide ChIP analysis indicates that Rec102–104 binds primarily to chromatin loops but is not preferentially enriched in DSB-hotspot regions and in fact, may be most abundant in regions located a few kb distal to DSB sites (Kee et al. 2004). Chromatin association is precipitously lost during pachynema, which possibly marks a critical regulatory transition that inactivates the Spo11 complex.

The Rec102/104 complex interacts directly with Spo11 and promotes its localization to chromatin and association with DSB sites (Kee and Keeney 2002; Jiao et al. 2003; Arora et al. 2004; Prieler et al. 2005). Rec104 also shows a strong interaction with Rec114 and the Rec102/104 complex likely interacts with the other subcomplexes, Mei4–Mer2–Rec114 and Mre11–Rad50–Xrs2 (Arora et al. 2004) (Section 4.1.4 and 4.1.5).

4.1.4 Mei4, Mer2, and Rec114

The grouping of Mei4, Mer2, and Rec114 is based on 2-hybrid, co-IP and immunolocalization studies, which indicate that these proteins form a dynamic subcomplex (Arora et al. 2004; Li et al. 2006). Identifiable homologs of these proteins are only found in *Saccharomyces* and closely related yeasts, and sequence analysis does not reveal any clear functional motifs. Despite this apparent lack of conservation, analysis of Mer2 function reveals a regulatory mechanism that may prove to be a paradigm for DSB formation in all eukaryotes. Mer2 is a direct target of the cyclin-dependent kinase (CDK) complex Cdk1–Clb5/6, which phosphorylates two

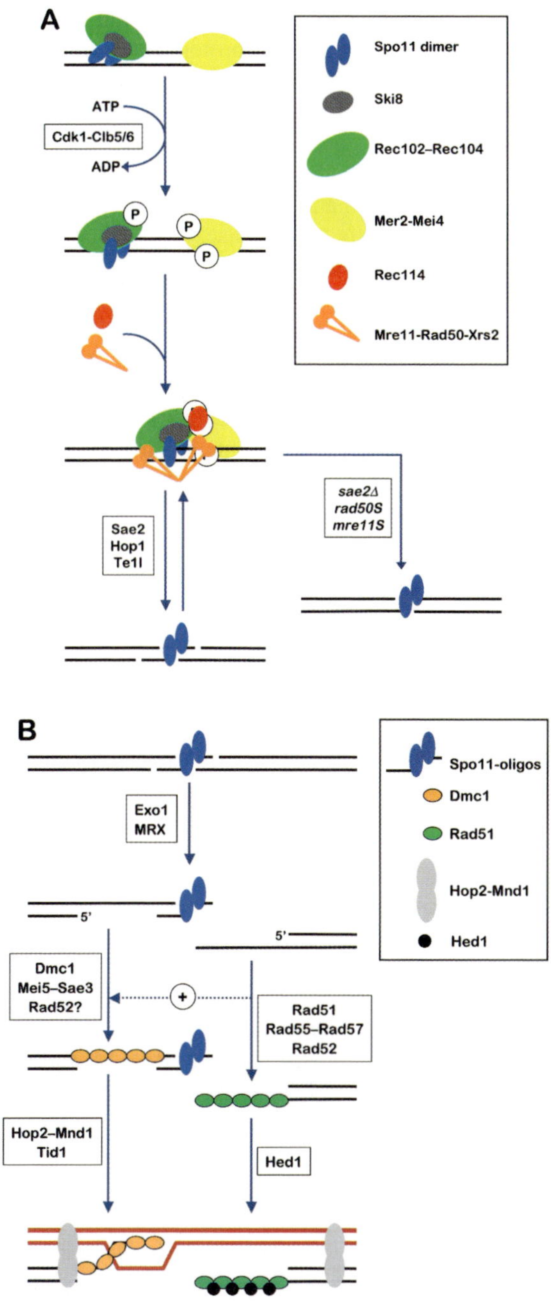

consensus target sites, Ser30 and Ser271 (Henderson et al. 2006) (Fig. 3A). Phospho-Ser30 is essential for DSB formation and mutation of this site causes phenotypes indistinguishable from the *mer2* null mutant. This is because Mer2 phosphorylation is required for various protein-protein interactions within the Spo11 complex. The Mer2-Mer2 self-interaction, and the Mer2-Rec114 and Mer2-Xrs2 interactions are severely compromised by mutation of Ser30. Interaction with Mei4 is also compromised but the effect is much milder (≤2-fold decrease in β-gal units). Ser271 is thought to be a secondary phosphorylation site that also appears to be important for self-interaction and for interaction with Xrs2 (Henderson et al. 2006). The replication-origin activating kinase, Cdc7, is also required for Mer2 phosphorylation and DSB formation (N. Hollingsworth, personal communication; see Section 4.2.3).

The *MER2* transcript is present in non-meiotic cells but efficient splicing requires meiosis-specific factors to process its non-canonical intron (Engebrecht et al. 1991; Nakagawa and Ogawa 1997). Despite this, some Mer2 protein is expressed in non-meiotic cells and chromatin localization can be detected by immunostaining. Chromatin localization becomes abundant very early in meiosis, prior to and independent of DSB formation (Henderson et al. 2006; Li et al. 2006). Moreover, this localization does not require any of the other Spo11 complex proteins nor does it require phosphorylation at Ser30 and Ser271. These observations suggest that Mer2 can bind chromatin autonomously and subsequently is licensed to interact with other Spo11 complex proteins via Cdk1–Clb5/6 dependent phosphorylation (Fig.3A). Phosphorylation of Rec104 could be an analogous event that licenses assembly of the Spo11-Ski8-Rec102/104 subcomplex (above). Paradoxically, Mer2 immunostaining foci do not colocalize with Rec102 or Mre11 (below, Section 4.1.4) foci, suggesting that the Mei4–Mer2–Rec114 complex is distinct and may promote DSB formation indirectly or via ephemeral interactions with the Spo11–Ski8–Rec102/104 and MRX complexes (Li et al. 2006).

Mei4 also localizes to meiotic chromosomes and shows significant colocalization with both Mer2 and Rec114 (Li et al. 2006). Absence of Mei4 prevents the timely dissociation of Spo11 from chromatin as assayed by ChIP (Prieler et al. 2005). Based on known protein-protein interactions, it seems possible that Mei4 could regulate turnover of the Spo11 complex on chromatin by modulating the phospho-Mer2–Rec114 interaction (Arora et al. 2004).

Rec114 interacts strongly with Mei4 and phospho-Mer2 (Arora et al. 2004; Henderson et al. 2006; Li et al. 2006). It also interacts with Rec102 raising the possibility that Rec114 bridges the Mei4–Mer2–Rec114 and Rec102–Rec104–Spo11–Ski8 subassemblies. ChIP analysis suggests that this interaction may mediate the specific association of Spo11 with DSB sites but not its general chromatin localization (Prieler et al. 2005).

Dephosphorylation and degradation of Mer2 requires DSB formation (Li et al. 2006). Possibly, interactions between phospho-Mer2 and other Spo11 complex proteins occur transiently at DSB sites and then Mer2 is dephosporylated, dissociates from chromosomes and is degraded locally. Alternatively, Mei4–Mer2–Rec114 could promote assembly and/or activation of Spo11–Ski8–Rec102/104–MRX supercomplexes, which then dissociate and bind to DSB sites. A third possibility is that Mei4–Mer2–Rec114 promotes DSB formation indirectly, e.g. by inducing favorable chromatin structure or sterically restricting Spo11 binding to DSB hotspots (see Li et al. 2006 for discussion).

4.1.5 The MRX complex

Unlike the Spo11 complex proteins already considered, Mre11, Rad50, and Xrs2 (MRX) are ubiquitously expressed and play central roles in DNA damage signaling and repair. The MRX complex and its analogs have been intensively studied; recent reviews are provided by Krogh & Symmington, and Petrini & colleagues (Krogh and Symington 2004; Stracker et al. 2004). MRX has at least six interrelated functions in DSB-repair: (i) primary sensing and signaling of DNA damage by binding to DSBs and locally activating the PI3K-family kinases, Tel1/Mec1 (ATM and ATR in vertebrates) (Usui et al. 2001); (ii) local recruitment of cohesins and checkpoint proteins via Tel1/Mec1-mediated phosphorylation of histone H2AX (γ-H2AX) (Strom et al. 2004; Unal et al. 2004; Warren et al. 2004; Lisby and Rothstein 2005); (iii) processing of DSB-ends via Mre11-associated nuclease activities (see below); (iv) ATP-dependent DNA unwinding (Paull and Gellert 1999; Chen et al. 2005); (v) intermolecular bridging of DNA-ends (Chen et al. 2001; Kaye et al. 2004; Lobachev et al. 2004; Wiltzius et al. 2005); (vi) remodeling of chromatin at DSB sites in collaboration with the ATP-dependent remodeling complex INO80, and perhaps other chromatin remodeling complexes (Petrini 2005; Tsukuda et al. 2005; van Attikum and Gasser 2005a; van Attikum and Gasser 2005b). MRX is absolutely required for DSB formation in *S. cerevisiae*. In contrast, DSBs can still be detected in the absence of *S. pombe* Rad50 and Rad32 (Mre11) but DSB repair is prevented (Young et al. 2004). Whether DSB formation is normal in *S. pombe rad50* and *rad32* mutants remains unclear; in particular, the fraction of DSBs that form independently of MRX is hard to assess because of pleiotropic effects.

Mre11 has a metallo-phosphoesterase nuclease domain and a DNA binding domain that together constitute a structure-selective nuclease (Furuse et al. 1998; Hopfner et al. 2001). The DNA binding domain is essential for DSB formation (Furuse et al. 1998) but nuclease activity is not. Binding of Mre11 to chromatin, as monitored by ChIP, requires all Spo11 complex proteins with the exception of Rad50 and does not require DSB formation (Borde et al. 2004). It is inferred that MRX joins the Spo11 complex only after the other subcomplexes are assembled, and its interaction is mediated by the Mre11 DNA-binding domain and maybe also Xrs2 dependent interactions (see below).

Rad50 is an SMC-family protein with the signature architecture of a composite ABC-family ATPase domain and an extended coiled-coil. Dimerization occurs upon ATP binding to create a DNA binding domain. Mre11 binds the Rad50 dimer close to this domain to form a $(Rad50)_2$–$(Mre11)_2$ heterotetramer with a composite DNA binding site and two protruding coiled-coils (Hopfner et al. 2000, 2001; Anderson et al. 2001; Moncalian et al. 2004). MRX is able to tether two bound DNA molecules via intermolecular interactions between heterotetramers, mediated by interlocking zinc-hooks found at the apexes of the Rad50 coiled-coils (de Jager et al. 2001; Hopfner et al. 2002; Moreno-Herrero et al. 2005). This intermolecular bridging function is essential for meiotic DSB formation (Wiltzius et al. 2005).

Xrs2, like its human counterpart Nbs1, contains FHA and tandem BRCT domains, which bind phospho-proteins (Becker et al. 2006). FHA domains have a high affinity for phosphothreonine whereas BRCT motifs bind phosphoserine (Becker et al. 2006). By analogy to Nbs1, these motifs may bind phosphoserine on the histone variant H2AX to effect signal amplification during the DNA damage response (Zhao et al. 2002; Kobayashi et al. 2004; Difilippantonio et al. 2005). Xrs2 also contains a well-defined Mre11 binding region and a C-terminal Tel1-binding motif (Nakada et al. 2003; Tsukamoto et al. 2005). An intrinsic DNA binding activity of Xrs2 is thought to help target MRX to DNA ends and other DNA structures (Trujillo et al. 2003).

Surprisingly, an 80 amino-acid peptide (just 1/10th of Xrs2) that includes the Mre11-binding region, has substantial meiotic function (Shima et al. 2005; Tsukamoto et al. 2005). The Tel1 interaction domain is clearly not essential for meiosis but the roles of the FHA and BRCT domains are less clear and merit further investigation (Tsukamoto et al. 2005). It is tempting to think that interaction of Xrs2 with phospho-Mer2 (above) is mediated by FHA/BRCT binding to phosphoserine. In this respect, phospho-Mer2 could mimic phosphorylated histone H2AX (γ-H2AX) and by extension, phospho-Mer2 could help efficiently recruit MRX complexes to DSB sites in anticipation of Spo11-cleavage and, therefore, independently of γ-H2AX.

Remodeling of chromatin by MRX. In meiotic cells, recruitment of MRX to DSB sites occurs in anticipation and independently of DSB formation by Spo11. This correlates with local alteration of chromatin as reported by changes in sensitivity to micrococcal nuclease (MNase) (Ohta et al. 1994, 1998; Murakami et al. 2003). Moreover, hypersensitivity to MNase is reduced in the absence of Mre11 and Mer2 (Ohta et al. 1998). Thus, the post-DSB chromatin-remodeling function described for MRX in mitotic cells (Tsukuda et al. 2005) could function prior to DSB formation during meiosis to facilitate accessibility of DSB sites to the Spo11-complex (Section 4.2.1).

4.2 Other factors that Influence DSB formation

Meiotic DSBs are not randomly distributed, occurring within highly localized (but not sequence specific) hotspots of tens to hundreds of base pairs (reviewed in Lichten and Goldman 1995; Keeney 2001; Petes 2001; Nishant and Rao 2006). DSB hotspots are found throughout the genome and a number of features and factors affecting their activity have been defined. Locally, hotspots usually occur in intergenic regions within relatively GC-rich sequences that are architecturally at the tops of chromatin loops. Regionally, hotspots tend to cluster into GC-rich domains in chromosome arms; AT-rich centromere and telomere domains have fewer and weaker hotspots (Zenvirth et al. 1992; Wu and Lichten 1994; Lichten and Goldman 1995; Klein et al. 1996; Baudat and Nicolas 1997; Blat and Kleckner 1999; Borde et al. 2000; Gerton et al. 2000; Blat et al. 2002; Petes and Merker 2002; reviewed in Keeney 2001; Petes 2001; Kauppi et al. 2004).

4.2.1 Chromatin status

As reported by sensitivity to MNase and DNase nucleases, DSB hotspots have an open chromatin structure and, conversely, positioned nucleosomes inhibit DSB formation (Ohta et al. 1994; Wu and Lichten 1994; Fan and Petes 1996; Keeney and Kleckner 1996; Ben-Aroya et al. 2004). In some cases, hotspots have a constitutively open chromatin structure (Xu and Kleckner 1995; Keeney and Kleckner 1996; Kirkpatrick et al. 1999b) whereas others undergo a transcription-factor dependent opening of chromatin (White et al. 1991, 1993; Fan and Petes 1996; Mizuno et al. 1997; Kirkpatrick et al. 1999a; Fox et al. 2000; Mizuno et al. 2001; Mieczkowski et al. 2006; reviewed in Petes 2001). A cause and effect role for chromatin remodeling is observed at the *ade6-M26* hotspot in *S. pombe* where the Swi2/Snf2A-family ATP-dependent chromatin remodeling factor, Snf22, is required to activate DSB formation (Yamada et al. 2004).

A nuclease-hypersensitive site is not sufficient for hotspot activity and the regional chromatin status appears to dictate the ability of such sites to form DSBs (Wu and Lichten 1995; Fox et al. 1997; Borde et al. 1999). This effect is likely a consequence of histone composition and modifications such as acetylation (Gottlieb and Esposito 1989; San-Segundo and Roeder 1999; Davis et al. 2000; Yamada et al. 2004; Klieger et al. 2005), methylation (Reddy and Villeneuve 2004; Sollier et al. 2004), and ubiquitylation (Yamashita et al. 2004; reviewed in Petes 2001; Maleki and Keeney 2004), all of which have been shown to affect the frequency of DSB formation and/or recombinational activity at some, through typically not at all, hotspots. Thus, the activity of a given DSB hotspot appears to be governed by the combined effects of local chromatin accessibility and regional chromatin-modification status. The latter will alter the ability of chromatin to recruit ATP-dependent remodeling complexes such as INO80, which may act together with the Spo11 complex, via MRX, to actively remodel nucleosomes at sites that are not intrinsically nucleosome-free (see above).

Recent experiments indicate that normal constraints on hotspot activity can be overcome by specific targeting of Spo11. When Spo11 is fused to the DNA binding domain of the Gal4 transcription factor, *de novo* DSB hotspots are created at loci that contain Gal4-binding sites (Pecina et al. 2002). Notably, one of these sites is located in a ~20kb cold domain that is normally devoid of DSBs. This indicates that merely recruiting Spo11 to a specific location is sufficient to overcome natural constraints to hotspot activity. Specific targeting of Spo11 does not, however, overcome the requirement for *all* of the other Spo11 complex proteins (Pecina et al. 2002).

4.2.2 Axis proteins

In *S. cerevisiae*, the meiosis-specific axis-associated proteins, Hop1, Red1, and Mek1, are not absolutely essential for DSB formation but required for normal

DSB levels. These proteins form a complex involved in a number of interrelated functions including (i) DSB formation; (ii) formation of homolog axes; (iii) formation of SCs; (iv) arresting the progression of meiosis in response to unrepaired DSBs; and (v) promoting interhomolog instead of inter-sister recombination (Section 6).

Red1. The coiled-coil protein Red1 is a major structural component of homolog axes and in its absence DSB levels are reduced to ~20-60% of wild type levels (Rockmill and Roeder 1990; Mao-Draayer et al. 1996; Schwacha and Kleckner 1997; Smith and Roeder 1997; Woltering et al. 2000; Hunter and Kleckner 2001; Blat et al. 2002; Pecina et al. 2002). The effect of *red1* appears to vary from locus to locus. In addition, reaching a consensus on the true effect of the *red1* mutation has been confounded by the different backgrounds used to measure DSB levels. In particular, it is now clear that the *rad50S* and *sae2* mutations, which prevent removal of Spo11 from DSB-ends, do not accurately reflect absolute DSB levels and may interact with mutations like *red1* (Xu et al. 1997; Borde et al. 2000; Blat et al. 2002) (M. Lichten, personal communication). I will assume that DSB levels measured in a *dmc1 rad51* background, in which no strand-exchange is possible, most accurately reflect absolute DSB levels in wild type cells. In this case, *red1* mutation reduces DSB levels two to fivefold (Blat et al. 2002). ChIP analysis reveals preferential binding to the GC-rich chromosomal domains where DSBs hotspots tend to cluster but, perhaps counter intuitively, a higher abundance of Red1 is not the reason for higher DSB levels in these regions and rather reflects a post-DSB role of Red1 (Blat et al. 2002). Localization of Spo11 assayed by ChIP indicates that Red1 may restrict Spo11 to hotspot regions (Prieler et al. 2005). In this respect, reduced DSB levels in *red1* mutants could result, at least in part, from mislocalization of Spo11 complexes to non-permissive regions of chromatin. Finally, DSB-independent phosphorylation of Red1 (personal communications from T-F. Wang and N. Hollingsworth) probably promotes its interaction with the Mek1 kinase (Wan et al. 2004).

Mek1(a.k.a. Mre4) is a paralog of the threonine/serine DNA-damage checkpoint-kinase, Rad53 (Rockmill and Roeder 1991; Leem and Ogawa 1992). Like Rad53, Mek1 contains a phospho-protein binding FHA-domain that is required for the interaction between Mek1 and phospho-Red1 (Wan et al. 2004). The effect of *mek1* mutation on DSB levels is similar to *red1* though again, this effect may vary depending on the locus and assay method utilized (Xu et al. 1997; Pecina et al. 2002; Wan et al. 2004).

Hop1. In the absence of Hop1, DSB levels are reduced to ~5-10% of wild type levels (Mao-Draayer et al. 1996; Woltering et al. 2000; Pecina et al. 2002; Niu et al. 2005). In this respect, a *hop1* mutation is epistatic to *red1* but the converse is true with respect to Hop1 chromosomal immunostaining, which requires Red1 (Smith and Roeder 1997). To reconcile these and other observations, it is proposed that a cytologically undetectable and biochemically distinct population of Hop1 acts at DSB sites (Niu et al. 2005) (Section 6). ChIP analysis of Spo11 reveals lower levels of binding but normal hotspot localization in the absence of Hop1 (Prieler et al. 2005). An intimate connection between Hop1 and the Spo11-complex is further suggested by the observation that mutations in *HOP1* can be partially suppressed by overproduction of Rec104 (Hollingsworth and Johnson 1993; Friedman et al. 1994). A genetically defined region that is required for Hop1s' role in DSB formation includes two functional domains: a zinc-finger motif that binds preferentially to GC-rich DNA and a putative protein-protein interaction and/or oligomerization motif called a HORMA domain (Aravind and Koonin 1998; Kironmai et al. 1998; Muniyappa et al. 2000; Anuradha and Muniyappa 2004). A third region, the C-domain, undergoes DSB-dependent phosphorylation, which promotes dimerization of Mek1 and activation of its kinase activity (Section 6) (Niu et al. 2005).

Exactly how Hop1, Red1, and Mek1 promote DSB formation remains unclear. Hop1 is thought to bind to pre-DSB sites where it attracts Red1 and Mek1. One possibility is that this complex promotes or stabilizes interactions between Spo11 complexes, DSB hotspots and the homolog axes (Prieler et al. 2005). Indeed, post-DSB steps of recombination occur in close proximity with the homolog axes and not out in the peripheral chromatin. This arrangement is logical because SC formation requires intimate association of the axes and crossing-over requires coordinated exchange of axes and DNA (Blat et al. 2002). Thus, it makes sense that Spo11-complexes in peripheral chromatin are recruited to the homolog axes at the time of DSB formation. This interaction could favor the Spo11 cleavage reaction. It is also possible that post DSB functions of Hop1 (and Red1 plus Mek1), such as protecting DSB-ends from degradation, could in part contribute to the apparent affect of *hop1* mutations on DSB levels (e.g. Kironmai et al. 1998).

4.2.3 Local Replication

Logically, DSB formation follows meiotic DNA replication. A key observation indicates that replication can influence DSB formation on a region-by-region basis. Delaying the replication of a specific chromosomal region delays DSB formation specifically in that region (Borde et al. 2000; Murakami et al. 2003). DSB formation is not absolutely dependent on replication, however. In the absence of the pre-replicative complex factor, Cdc6, replication does not initiate so sister-chromatids are absent but, regardless, DSB formation and recombination proceed (Hochwagen et al. 2005). An idea that may reconcile these and other observations is that the initiation of replication establishes a local dependence of DSB formation on the completion of replication (Tonami et al. 2005). The molecular nature of this dependence is unclear and could involve at least three components: (i) the inhibition of DSB formation by replication checkpoint proteins (in fission yeast but maybe not in budding yeast) (Tonami et al. 2005); (ii) the dependence of DSB formation on CDK activity via phosphoryaltion of Mer2 (see above) (Henderson et al. 2006); (iii) a novel function of the Cdc7 kinase (Schild and Byers 1978; Ogino et al. 2006; N. Hollingsworth, personal communication). In the mitotic cell cycle, Cdc7 phosphorylates the MCM complex to activate origin unwinding (Masai and Arai 2002). In contrast, Cdc7 is not essential for replication during meiosis but like CDK, it is required for Mer2 phosphorylation and DSB formation (N. Hollingsworth, personal communication). Whether or not Mer2 is a direct target of Cdc7 has not yet been established.

4.3 Resection of DSB-ends

Following DSB formation, covalently bound Spo11 is removed and the 5'-strands of DSB-ends are resected to provide 3'-single-stranded tails as substrates for assembly of Dmc1 and Rad51 nucleoprotein filaments (Sun et al. 1991; Keeney et al. 1997) (see Section 5).

4.3.1 Removal of Spo11-oligos by MRX plus Sae2/Com1

Spo11 is removed from DSB-ends as an oligonucleotide-bound covalent complex (Fig. 3A) (Neale et al. 2005). This reaction requires endonucleolytic incision, which is almost certainly mediated by the MRX complex together with the Sae2 protein. Special separation-of-function alleles of *RAD50* and *MRE11*, and null alleles of *SAE2/COM1* permit DSB formation but block removal of Spo11 (Cao et al. 1990; Keeney and Kleckner 1995; Keeney et al. 1997; McKee and Kleckner 1997a; Nairz and Klein 1997; Prinz et al. 1997; Furuse et al. 1998; Tsubouchi and Ogawa 1998; Moreau et al. 1999). In the case of Mre11, this class of alleles abolishes its nuclease activity. As an endonuclease, Mre11 preferentially cleaves hairpin-ends and ssDNA at single-to-double-stranded transitions (Krogh and Symington 2004). The nature of the substrate that Mre11 cleaves to create Spo11-oligos is not known but it's expected that DNA must be locally unwound, either directly by MRX or via an associated helicase (Krogh and Symington 2004). Sae2 has no obvious functional motifs and no clear homologs outside of fungi. Like MRX, Sae2 functions in both mitotic and meiotic recombination but unlike MRX, Sae2 is not required for DSB formation (McKee and Kleckner 1997a; Prinz et al. 1997; Rattray et al. 2001; Clerici et al. 2005). The observation that high levels of Sae2 can partially suppress the mitotic defects of a *rad50S* mutation is consistent with the idea that Sae2 interacts directly with MRX to regulate its activities at DSB-ends (Clerici et al. 2005, 2006). Moreover, Sae2 undergoes Mec1 and Tel1-dependent phosphorylation, which is stimulated by DSB formation and is required for Sae2s' function in DSB processing. Thus, DNA damage checkpoint activation itself is required for DSB processing (Cartagena-Lirola et al. 2006).

4.3.2 Further processing by Exo1 and MRX

The 5'-ends of DSBs are rapidly resected by several hundred nucleotides (Sun et al. 1989, 1991; Cao et al. 1990; Bishop et al. 1992; Nag and Petes 1993; Jessop et al. 2005). The nucleases that act subsequent to Spo11-oligo formation have not been unambiguously identified but Mre11 and Exo1 are likely candidates. As an exonuclease, Mre11 has a 3'-5' polarity, which is incompatible with the observed 5'-3' processing of DSBs. Thus, for Mre11 to effect 5'-3' resection, it must do so via its endonuclease activity. This would require the unwinding of the DSB-end and cleavage of transient secondary structures (Krogh and Symington 2004). It is possible that Mre11 only acts at the initial step of DSB-end processing to remove Spo11-oligos and provide an entry point for a conventional 5'-3' exonuclease. Exo1 is such a nuclease. Exo1 has multiple functions in DNA repair and recombination and is important for meiotic crossing-over (Khazanehdari and Borts 2000; Kirkpatrick et al. 2000; Tsubouchi and Ogawa 2000; Wei et al. 2003; Hoffmann and Borts 2004; Tran et al. 2004) (Section 7). In *dmc1* null mutants, cells arrest with processed DSBs, which then undergo additional 5'-resection; this "hyper-resection" is dependent on Exo1 but normal resection is not (Tsubouchi and Ogawa 2000). These and other observations raise the possibility that DSB-resection is mainly catalyzed by MRX, as a rapid single step that is tightly coupled to DSB formation (Section 4.4); and that Exo1 may only function to "fine-tune" resection, perhaps in a reaction that is normally coupled to later steps of recombination, such as second-end capture.

Finally, in mitotic cells, the ATP-dependent remodeling complex INO80 promotes DSB-end processing (van Attikum et al. 2004). INO80 and other remodeling complexes may also function in meiotic DSB-resection in conjunction with MRX (Section 4.1.4).

4.4 Assembly of the Spo11 complex and triggering of Spo11 cleavage

The studies summarized above indicate that DSB formation requires the prior assembly and chromatin-association of the ≥10 subunit Spo11 complex. The data are most consistent with a general recruitment of Spo11 to chromatin via a Spo11-Ski8-Rec102-Rec104 subcomplex, followed by assembly of the full Spo11 complex at DSB sites, triggered by CDK-mediated Mer2 (and maybe Rec104) phosphorylation (Fig. 3A). This two-step assembly process may allow assembly of active Spo11 complexes to be coordinated with local replication.

4.4.1 Are Spo11-cleavage and MRX-incision part of a concerted initiation reaction?

Topoisomerase cleavage and ligation reactions are isoenergetic and so cleavage is readily reversible (Wang 2002). Thus, driving the Spo11 transesterification reaction towards an irreversible cleaved product should be a key factor in triggering DSB formation and represents an obvious target for regulatory processes. Ligation of topisomerase-catalyzed DSBs requires that the two post-cleavage monomers remain associated or can readily reassociate; in archeal TopVI, religation is facilitated by the Top6B subunits, which are predicted to tether the two Top6A monomers after cleavage and thereby prevent their dissociation and consequently DSB formation (Corbett and Berger 2003b). It follows that regulation of a Spo11–Spo11 dimer interface may be utilized to license DSB formation and drive the cleavage reaction (de Massy et al. 1995; Keeney 2001). For example, one or more of the Spo11-complex proteins could be a regulatable counterpart of the Top6B subunit.

S-alleles of the *RAD50* and *MRE11* genes, and null mutations of *SAE2* prevent removal of covalently bound Spo11 and block DSB repair (Section 4.3.1). The immediate suggestion is that Spo11-dependent cleavage and MRX-dependent Spo11-oligo formation are sequential steps. However, several features of the *rad50S* phenotype suggest an alternative interpretation in which irreversible Spo11 cleavage and MRX incision are normally tightly coupled reactions. First, unresected DSBs are never detected in wild type cells (e.g. Sun et al. 1989; Cao et al. 1990; Bishop et al. 1992). Second, the *rad50S* mutation actually prevents the formation of a significant subset of DSBs, although DSBs at many other loci form at expected levels. Notably, the absent breaks are those that would normally form in subtelomeric and centromere proximal regions (Borde et al. 2000; M. Lichten, personal communication). Third, DSBs form more slowly in *rad50S* cells than in wild type or *dmc1* cells (e.g. Prieler et al. 2005). Fourth, ChIP analysis reveals that in *rad50S* and *sae2Δ/com1Δ* mutant cells, Spo11 forms an unusually stable complex with hotspot DNA that does not appear to be a covalent Spo11–DNA complex and is salt-labile (Prieler et al. 2005). This "tight-binding" state is detected before DSBs form but still requires the catalytic tyrosine of Spo11. It is proposed that tight-binding of Spo11 represents a normally ephemeral transition state prior to irreversible cleavage.

One interpretation of these and other observations is that irreversible Spo11-cleavage and MRX-incision normally occur as concerted reactions with a single transition state, i.e. tight-binding of Spo11 to substrate DNA. In this transition state, Spo11 cleavage and ligation reactions may be in rapid equilibrium. ATP driven conformational changes in Rad50 may then trigger irreversible Spo11-cleavage and MRX-incision as concerted reactions (Fig. 3A). Under this scenario, the covalent Spo11-DSB complexes detected in *rad50S* and related mutants would not represent a normal intermediate of meiotic recombination but the decay of the tight-binding transition state. In sub-telomeres and centromere-proximal regions, this decay may not be possible.

What event(s) could trigger this reaction? Transient Spo11-cleavage could be sensed by MRX, which would activate Tel1 to rapidly phosphorylate MRX and maybe also Hop1 (Usui et al. 2001; Niu et al. 2005). These phosphorylations could drive conformational changes in Rad50 and consequently promote MRX-incision and irreversible Spo11-claeavage. Thus, any local change that shifts the Spo11 reaction equilibrium towards the cleaved intermediate could be sufficient to trigger this cascade of events. Local changes in DNA/chromatin status, such as supercoiling and compaction, have been proposed as one way to shift the equilibrium towards the cleavage reaction (Keeney 2001; Kleckner et al. 2004).

5 Homolog pairing and formation of joint molecules

In five fungal species, a higher plant (*Arabidopsis*) and a mammal (mouse), mutational analysis has shown directly that Spo11-dependent recombination mediates close homolog juxtaposition and promotes assembly of the SC (Baudat et al. 2000; Celerin et al. 2000; Romanienko and Camerini-Otero 2000; Grelon et al. 2001; Nabeshima et al. 2001; Peoples et al. 2002; Storlazzi et al. 2003; Tesse et al. 2003; Henderson and Keeney 2004; Bowring et al. 2006) (in *S. pombe*, although recombination promotes stable homolog pairing SC is not formed). Two invertebrates (*C. elegans* and *Drosophila*) do not show this dependence and *spo11* mutants can pair and synapse their chromosomes perfectly well (Dernburg et al. 1998; McKim and Hayashi-Hagihara 1998). Intriguingly, only organisms in the first class appear to possess the meiosis-specific RecA homolog, Dmc1 (Bishop et al. 1992; Story et al. 1993), and associated proteins, Mnd1 and Hop2 (Leu et al. 1998; Gerton and DeRisi 2002; Tsubouchi and Roeder 2002), indicating a special role for these proteins in recombination-promoted homolog pairing (Villeneuve and Hillers 2001; Stahl et al. 2004). The biochemistry of Rad51 is reviewed by Wolf Heyer in Chapter 2 of this volume. This section will focus on the function of Dmc1 *vis-a-vis* Rad51 in promoting homolog pairing.

5.1 Dmc1

Dmc1 is expressed exclusively in meiosis and is required for normal homolog pairing and SC formation (Bishop et al. 1992). Substantial pairing and SC formation eventually occurs in *dmc1* mutants but with a severe delay (≥6 hrs) relative to wild type cells (Weiner and Kleckner 1994; Rockmill et al. 1995; Xu et al. 1997). Synapsis is frequently incomplete in *dmc1* cells, however, and small chromosomes in particular often fail to pair (Weiner and Kleckner 1994; Rockmill et al. 1995; Tsubouchi and Roeder 2003). By all criteria, Dmc1 is a *bona-fide* RecA homolog although strand-exchange activity *in vitro* is limited by a strong tendency to assemble into inactive ring structures (Passy et al. 1999; Hong et al. 2001; Kinebuchi et al. 2004; Chang et al. 2005). Under conditions that promote assembly into the archetypal right-handed nucleoprotein filament, Dmc1 has robust strand-exchange activity (Sehorn et al. 2004; Bugreev et al. 2005; Lee et al. 2005).

5.2 Assembly of the strand-exchange complex

The requirements for assembly of Dmc1 and Rad51 nucleofilaments have been inferred by monitoring their assembly into chromosomal immunostaining foci. Beyond a common requirement for DSB formation, Dmc1 and Rad51 assembly show distinct requirements (Fig. 3B).

5.2.1 Dmc1 assembly

Assembly of Dmc1 is dependent on the meiosis-specific Mei5–Sae3 complex (McKee and Kleckner 1997b; Hayase et al. 2004; Tsubouchi and Roeder 2004; Okada and Keeney 2005). Absence of Dmc1, Mei5, or Sae3 causes identical phenotypes, the three proteins show mutual interdependence for localization at DSB sites, and they appear to interact to form a complex. Mei5 and Sae3 are conserved but contain no obvious functional motifs beyond predicted coiled-coils. Clues to their function, however, come from the homologies between Mei5 and *S. pombe* Sfr1 and Swi2 proteins, and between Sae3 and *S. pombe* Swi5 (Akamatsu et al. 2003; Hayase et al. 2004). In contrast to Mei5 and Sae3, the functions of Sfr1,

Swi2 and Swi5 functions are not confined to meiosis. These three proteins appear to form two complexes, which physically interact with the Rad51 homolog, Rhp51, and define one of three distinct Rhp51-dependent recombination pathways, whose specific utilization may reflect the type of initiating lesion. Swi5-Swi2 functions in mating-type switching whereas Swi5–Sfr1 has a role in DNA repair that functions in parallel to the pathway defined by the Rad51 paralogs, Rhp55 and Rhp57 (Akamatsu et al. 2003). Swi5 (and probably Sfr1) is also important for meiotic recombination in *S. pombe* (Ellermeier et al. 2004; Okada and Keeney 2005). Thus, Mei5/Sae3-related protein complexes may function as substrate-specific mediators for eukaryotic RecA proteins.

Only very faint Dmc1 foci are assembled in the absence of Rad51 (Shinohara et al. 1997a) but these assemblies are capable of strand-exchange (Schwacha and Kleckner 1997). Predictably, Mei5 and Sae3 foci are also largely dependent on Rad51 (Hayase et al. 2004). Bright Dmc1 foci can form without Rad52, even though Rad51 foci are completely absent (Gasior et al. 2001; Shinohara and Shinohara 2004). This apparent paradox can be reconciled by the proposal that the efficient assembly and/or stabilization of Dmc1 nucleoprotein filaments requires a lower-order oligomer of Rad51 but does not require extensive Rad51 filaments (Gasior et al. 2001). The efficiency of Dmc1 focus formation in *rad52* mutant cells is, however, reduced to ~30% of wild type levels suggesting that Rad52 is a mediator of Dmc1 assembly at many DSBs (A. Shinohara & M. Shinohara, personal communication). Interaction between Rad52 and Dmc1 has not been demonstrated but it is tempting to draw an analogy between Rad52 and the mediator Brca2, which has been shown to interact with both Dmc1 and Rad51 in *Arabidopsis* (Siaud et al. 2004; Dray et al. 2006).

Tid1/Rdh54 and Rad54 are Swi2/Snf2-family DNA translocases that promote Dmc1- and Rad51-promoted strand-exchange reactions (Section 5.5.1). One function of Tid1/Rdh54 and Rad54 is to compensate for the fact that both Dmc1, like Rad51, shows little preference for single- versus double-stranded DNA (W.D. Heyer, this volume) (Holzen et al. 2006; Symington and Heyer 2006). In the absence of DSBs, Dmc1 tends to randomly assemble onto chromosomes. Formation of these non-productive assemblies is suppressed by Tid1/Rdh54, which presumably acts as an ATP-dependent translocase to kick Dmc1 off double-stranded DNA. In the presence of DSBs, Tid1/Rdh54 and Rad54 appear to act in the same way to bias the assembly of Dmc1-filaments onto single-stranded DSB-tails (Holzen et al. 2006).

5.2.2 Rad51 assembly

Assembly of Rad51 requires both the Rad51-paralogs, Rad55–Rad57, and Rad52 but not Dmc1, Mei5, or Sae3 (Bishop 1994; Shinohara et al. 1997a; Gasior et al. 2001). A requirement for the recently inferred Rad51-paralog complex, Shu1-Shu2-Psy3-Csm2, has not yet been tested but expression of these proteins appears to be increased during meiosis (Shor et al. 2005; Martin et al. 2006) (http://germonline.yeastgenome.org/).

5.2.3 Does Dmc1 assemble onto one DSB-end and Rad51 onto the other?

Although Dmc1 and Rad51 foci normally colocalize, in the absence of Tid1/Rdh54, Rad24 or Rad17, and in a subset of wild type nuclei, dual side-by-side foci are observed (Shinohara et al. 2000). This observation has led to the idea that Dmc1 and Rad51 assemble onto opposite DSB-ends, and that different properties and regulation of these two complexes are responsible for the asymmetric behavior of DSB-ends observed *in vivo* (Figs. 1B, 3B, and 5) (Shinohara et al. 2000; Hunter and Kleckner 2001). Although other scenarios may be equally possible, this idea sits well with the distinct properties and behaviors of the two proteins (Dresser et al. 1997; Shinohara et al. 1997a).

How could asymmetric loading of Dmc1 and Rad51 be imposed? An intriguing possibility is suggested by the analysis of the Spo11-oligonucleotide complexes formed during DSB processing (Neale et al. 2005) (Section 4.3.1). Two distinct Spo11-oligos are detected in equal amounts, differing in the length of covalently bound DNA (≤12 nucleotides versus ~21-37 nucleotides). Unexpectedly, the kinetics of appearance and disappearance of Spo11-oligos mirrors that of DSBs. The favored interpretation is that DSBs are asymmetrically incised by the MRX complex and the resulting Spo11-oligos remain associated with the DSB until strand-exchange occurs (Neale et al. 2005). Specifically, the Spo11-oligo pair is proposed to remain base-paired, via the long oligo, to only one DSB-end. Thus, following 5'-resection, structurally distinct DNA substrates are created: one canonical DSB-end with a free 3'-hydroxyl; and a short double-stranded-end blocked by Spo11 proteins (Fig. 3B). Distinct targeting activities of the mediator complexes could result in the differential loading of Dmc1 and Rad51 onto these substrates; for example, Mei5-Sae3 could target Dmc1 to the 3'-single-/double-strand junction on the Spo11-associated end. Potential advantages of this theoretical arrangement are discussed below.

5.3 The Hop2–Mnd1 complex

The conserved Hop2 and Mnd1 proteins are essential for homolog pairing and strand-exchange (Leu et al. 1998; Nabeshima et al. 2001; Gerton and DeRisi 2002; Tsubouchi and Roeder 2002; Petukhova et al. 2003; Schommer et al. 2003; Tsubouchi and Roeder 2003; Saito et al. 2004; Zierhut et al. 2004; Domenichini et al. 2006; Kerzendorfer et al. 2006; Lui et al. 2006; Panoli et al. 2006). Budding yeast *hop2* and *mnd1* mutants initiate recombination and assemble Rad51 and Dmc1 complexes but pairing and strand-exchange completely fail. In fact, chromosomes show extensive non-homologous pairing, leading to the proposal that Dmc1/Rad51 engage in ectopic interactions in the absence of Hop2–Mnd1 (Leu et al. 1998; Tsubouchi and Roeder 2003). Hop2–Mnd1 are distinct from other recombination proteins in that their localization to meiotic chromosomes does not require DSB formation (Leu et al. 1998; Zierhut et al. 2004). Moreover, imunostaining foci of Mnd1 and Rad51 do not colocalize and ChIP analysis indicates that Mnd1 is not preferentially bound to DSB sites. These data suggest that Hop2–Mnd1 binds generally along chromosomes and facilitates recombination indirectly (Zierhut et al. 2004; Henry et al. 2006). In fact, the *S. pombe* Hop2 homolog, Meu13, appears to promote homolog pairing independently of DSBs (Nabeshima et al. 2001).

A direct role for Hop2-Mnd1 in recombination has also been inferred, however. Studies of yeast and mammalian proteins show that Hop2 and Mnd1 form a 1:1 dimer that binds DNA and stimulates *in vitro* strand-exchange reactions promoted by Dmc1 (Chen et al. 2004; Enomoto et al. 2004; Petukhova et al. 2005; Pezza et al. 2006). Stimulation may occur via direct interaction with Dmc1 and/or by binding the branch point created during nascent D-loop formation (Enomoto et al. 2004; Petukhova et al. 2005). Mouse Hop2–Mnd1 also stimulates Rad51 catalyzed strand-exchange but the two *in vitro* reactions show an intriguing difference. Hop2–Mnd1 overcomes the requirement to preassemble Dmc1 onto ss-DNA whereas Rad51 still requires preassembly (Petukhova et al. 2005). Perhaps Hop2–Mnd1 stimulates turnover of Dmc1 from ds-DNA. Mouse Hop2–Mnd1 binds directly to Dmc1 *in vitro*, and to a lesser extent to Rad51; and the efficiency of binding correlates with the relative stimulation of Dmc1 (35-fold) and Rad51 (10-fold) strand-exchange reactions (Petukhova et al. 2005). Interaction between yeast Hop2-Mnd1 and Dmc1/Rad51 have not been established and evidence suggests that such interactions are at best weak and/or transient (Chen et al. 2004; Henry et al. 2006).

In the absence of Mnd1, dimers and/or tetramers of mouse Hop2 efficiently catalyze ATP-independent D-loop formation via an ill-defined mechanism (Petukhova et al. 2005; Pezza et al. 2006). This novel activity of Hop2 is attenuated by its interaction with Mnd1. Hop2-Mnd1 interaction also promotes interaction between Hop2 and Dmc1. Thus, Hop2 could potentially act both autonomously and in conjunction with Mnd1 to promote homolog-pairing and strand-exchange. This idea predicts that *mnd1* mutants might be less defective than *hop2* mutants with respect to homolog pairing. The phenotypes of *hop2* and *mnd1* mutants appear to be indistinguishable, however, arguing that Hop2 and Mnd1 always work together (Tsubouchi and Roeder 2002).

Perhaps Hop2–Mnd1 can promote general recognition and alignment of homologs via unstable DSB-independent interactions such as paranemic DNA-DNA or DNA-RNA joints. Such interactions could act in parallel to Dmc1/Rad51 promoted DSB-dependent interactions, which would promote more stable association of homolog axes via plectonemic joints. In addition, Hop2-Mnd1 may act directly at DSB sites to promote Dmc1/Rad51-mediated strand-exchange (Fig. 3B). If as seems likely, the latter function is catalytic, Hop2–Mnd1 may not be detectable at recombination sites by immunostaining, as is observed in yeast (Zierhut et al. 2004).

5.4 How do strand-exchange proteins promote homolog pairing?

When a mitotic cell suffers a DSB, the sole aim of recombination is to effect its repair. In contrast, meiotic cells use DSBs and recombination to do the work of homolog pairing (Storlazzi et al. 2003; Tsubouchi and Roeder 2003). Logically, this must require additional constraints and modifications to the basic recombination process. Specifically, "free-running" DSB-repair will be inefficient for homolog pairing and it can be predicted that steps downstream of initial pairing interactions must be blocked until SC formation has locked homologs together. A logical point to transiently block recombination is prior to strand-extension. This could occur by preventing the disassembly of Dmc1 and/or Rad51 filaments and thus preventing access to the terminal 3'-hydroxyl, which would have the additional advantage of stabilizing the nascent heteroduplex intermediate. Alternatively, the 3'-hydroxyl could be sequestered away from the replication machinery by displacing it from the initial D-loop. A variation on this idea is provided by the scenario in which one DSB-end has a double-stranded terminus (above, Section 5.2.2), which would be incapable of priming DNA synthesis until the short terminal strand is removed.

Stable strand-exchange products, in the form of SEIs, are first detected in late zygotene/early pachytene stages, i.e. after homologs are fully paired (Hunter and Kleckner 2001). This implies that Dmc1/Rad51-mediated homolog-pairing interactions are not detected by molecular assays (see Section 3.2). Possible structures include ephemeral or short plectonemic joints, and paranemic joints. The advantage of pairing via such unstable intermediates is that they can be readily disassembled in cases where inappropriate interactions have occurred, e.g. between dispersed repeats. The disadvantage is that individually such unstable interactions will provide only a weak force for association of homolog axes. Numerous, unstable interactions could, however, create a relatively strong "Velcro" effect. This could be the reason why meiotic recombination events are so numerous: an estimated ~250 DSBs per nucleus, or ~6-30 DSBs per homolog pair, depending on its size.

5.4.1 Are pairing interactions normally catalyzed by Dmc1 or Rad51 or both?

The pleiotropic effects of *dmc1* and *rad51* single mutants, such as the regulatory arrest of *dmc1* cells and defective assembly of Dmc1 filaments in *rad51* mutants, makes it hard to tell whether pairing interactions are normally catalyzed by Dmc1, Rad51 or both. Several lines of evidence favor the idea that Dmc1-filaments directly catalyze pairing interactions. First, *DMC1* genes are not found in *Drosophila* and *C. elegans*, two organisms that pair their chromosomes through recombination-independent mechanisms (Section 5). Second, the pairing defect of a *dmc1* mutant is made no worse by deletion of *RAD51* (Weiner and Kleckner 1994; Rockmill et al. 1995; Tsubouchi and Roeder 2003; Chen et al. 2004). Third, strand-exchange is blocked if Dmc1 is not incorporated into the recombination complex, i.e. in *dmc1, mei5* and *sae3* mutants (Schwacha and Kleckner 1997; Xu et al. 1997; Hayase et al. 2004; Tsubouchi and Roeder 2004). Fourth, incorporation of Dmc1 into the recombination complex makes homolog pairing and strand-exchange completely dependent on the Hop2–Mnd1 complex. Moreover, *HOP2* and *MND1* genes are also absent from the genomes of *Drosophila* and *C. elegans* which, together with other evidence, implies a Dmc1-associated homolog pairing function for this complex (see above, Section 5.3).

On the other hand, a direct role for Rad51-filaments in homolog pairing has been inferred from the observation that *dmc1* mutant phenotypes can be more or less overcome by activating Rad51-mediated recombination. Normally, the meiosis-specific Hed1 protein inhibits Rad51 strand-exchange activity when Dmc1 is absent (Tsubouchi and Roeder 2006). Deletion of *HED1* allows *dmc1* mutants to progress through meiosis and repair DSBs, although crossover levels only reach ~40% of wild type levels. Hed1-inhibition of Rad51 in *dmc1* cells can be overcome by massive overproduction of Rad54 or Rad51 (~40 to 100 or more gene copies per cell); remarkably, in these cases, crossing-over approaches wild type levels (Shinohara et al. 2003b; Tsubouchi and Roeder 2003). These observations argue for a model in which Rad51 contributes to pairing following incorporation of Dmc1 into the recombination complex and elimination of Hed1 inhibition (Tsubouchi and Roeder 2003). Moreover, Dmc1 and the inferred Dmc1/Rad51 asymmetry are not absolutely required to achieve crossing-over. It should be emphasized, however, that physiological levels of Rad51 do not efficiently substitute for Dmc1 (above).

A working model of recombination-promoted pairing interactions is presented in Figure 3B. In this model, Rad51 promotes the assembly of the Dmc1-Mei5-Sae3 complex but is, itself, inhibited by Hed1. Thus, stable homolog pairing is promoted by the combined activities of Dmc1-Mei5-Sae3 and Hop2–Mnd1. It is proposed that Rad51 inhibition by Hed1 continues throughout the pairing period and is only lifted when homologs are associated. Subsequently, Rad51 plays an important role in facilitating DSB-repair.

5.5 Strand-exchange and joint molecule formation

5.5.1 Tid1/Rdh54 and Rad54

Stable strand-exchange and the progression of recombination also require the translocases, Tid1/Rdh54 and Rad54, which are thought to stimulate Dmc1 and Rad51 filaments and facilitate accessibility to chromatin templates (W.D. Heyer, this volume) (Dresser et al. 1997; Klein 1997; Shinohara et al. 1997b; Arbel et al. 1999; Schmuckli-Maurer and Heyer 2000; Shinohara et al. 2000, 2003b; for review see Tan et al. 2003; Heyer et al. 2006; Symington and Heyer 2006). Several activities of Tid1/Rdh54 and Rad54 may contribute to strand-exchange and post-exchange steps of recombination, e.g. local unwinding of DNA (Petukhova et al. 1999, 2000); displacement of nucleosomes on target templates (Alexeev et al. 2003; Jaskelioff et al. 2003); and displacement of Rad51 and Dmc1 from duplex

DNA (see Section 5.2.1) (Solinger et al. 2002; Holzen et al. 2006; Wesoly et al. 2006; Symington and Heyer 2006).

In meiosis, the two proteins appear to be partially redundant although evidence suggests that Rad54 plays a major role in intersister strand-exchange whereas Tid1/Rdh54 promotes interhomolog interactions (Shinohara et al. 1997b; Arbel et al. 1999; Schmuckli-Maurer and Heyer 2000). This template specialization and partial redundancy between Rad54 and Tid1/Rdh54 appears to be conserved in *S. pombe* (Catlett and Forsburg 2003). A logical possibility is that Tid1/Rdh54 mostly promotes strand-exchange catalyzed by Dmc1, whereas Rad54 works best with Rad51.

5.5.2 Post-invasion steps

Events following pairing and initial strand-invasion are relatively poorly characterized. Along the crossover pathway, efficient formation and stabilization of SEIs and dHJs requires modulation of the core recombination machinery by a number of meiosis-specific factors, which are discussed in Section 7.4. More generally, post-invasion steps require extension of an invading 3'-end by a DNA polymerase, and coordination of the two DSB-ends. In the case of crossovers, the primary invasion must be stabilized while the second end is "captured." Additional DNA synthesis and ligation is required to complete dHJ formation. In contrast, synthesis-dependent strand-annealing to form noncrossovers requires displacement of the invading strand from the template and annealing of the two DSB-ends (see Figs. 1B and 5, and Section 7.5).

Recombination-associated DNA synthesis. The nature of recombination-associated DNA synthesis during meiosis, and exactly which replication factors are involved is unclear. Recombination-associated DNA synthesis is expected to be distinct from chromosome replication in several ways (e.g. Wang et al. 2004). First, chromosome replication is initiated at specific origins and requires *de novo* primer synthesis by the Polα/primase complex. In contrast, recombination-associated DNA synthesis is initiated by the free 3'-hydroxyl of the invading DSB-end. Second, chromosome replication initiates bi-directional replication forks whereas recombination-associated DNA synthesis is expected to proceed in only one direction. Third, replication forks synthesize leading and lagging strands simultaneously. Recombination-associated DNA synthesis is predicted to require only leading-strand synthesis. Fourth, DNA synthesis during chromosome replication is highly processive whereas recombination associated DNA synthesis need only proceed for a few hundred nucleotides.

Several replication factors have been implicated in meiotic recombination. *S. cerevisiae*, *pol3-ct* is an allele of the major replicative polymerase **Pol δ**. In *pol3-ct* cells, gene conversion tracts are shorter and crossing-over is reduced by ~35% (Maloisel et al. 2004). *Drosophila rec* mutants have an analogous phenotype but the reduction in crossing-over is more severe (~95%) (Grell 1984; Matsubayashi and Yamamoto 2003; Blanton et al. 2005). REC encodes a widely conserved MCM2-6 family protein called **MCM8** (Matsubayashi and Yamamoto 2003; Blanton et al. 2005). *Xenopus* MCM8 has autonomous DNA helicase activity, which appears to stimulate DNA replication (Maiorano et al. 2005, 2006). *MCM8* genes are notably absent from the genomes of *C. elegans* and fungi, including *S. cerevisiae*. The crossover defects of *pol3-ct* and *rec* mutants have led to the idea crossing-over is favored by more processive DNA synthesis. Specifically, that longer extension of invading 3'-ends will facilitate stable interaction of the second DSB-end, e.g. by annealing (Fig. 5). Absence of the family-X DNA polymerase, **Pol IV**, is reported to increase intragenic recombination and steady-state DSB levels, although crossing-over and spore viability are not affected (Leem et al. 1994) (our unpublished data). In mouse, expression of the Pol IV homologs, **Pol λ** and **Pol β**, is greatly increased in meiotic cells (Plug et al. 1997; Garcia-Diaz et al. 2000). Moreover, mouse Pol β localizes as immunostaining foci to zygotene and pachytene stage chromosomes (Plug et al. 1997). Rice Pol λ is also highly expressed in meiotic tissues (Uchiyama et al. 2004). Pol IV and Pol λ have been implicated in DSB-repair by non-homologous end-joining (Wilson and Lieber 1999; Garcia-Diaz et al. 2005) but their function in meiotic recombination remains uncertain and is clearly not essential (Leem et al. 1994; Kobayashi et al. 2002). Finally, in *Drosophila*, conditional alleles of **PCNA** are defective in crossover control (see Section 7) and homolog segregation (Henderson et al. 2000).

6 Interhomolog bias

In mitotic cells, the core recombinational machinery is intrinsically biased towards use of a sister-chromatid template (Fabre et al. 1984; Kadyk and Hartwell 1992). This poses a problem for meiotic cells because inter-sister recombination does not contribute to homolog pairing and chiasma formation. In actuality, meiotic recombination occurs with an estimated ~5-fold bias towards inter-homolog events in *S. cerevisiae* (see Fig. 2C) (Haber et al. 1984; Jackson and Fink 1985; Schwacha and Kleckner 1994, 1997; Hunter and Kleckner 2001). Surprisingly, interhomolog bias is not observed in *S. pombe* (Cromie et al. 2006). Interhomolog bias in budding yeast appears to comprise two main elements, suppression of intersister recombination and active promotion of interhomolog interactions.

6.1 Suppression of intersister recombination

A number of observations imply that a block to intersister recombination is actively imposed during meiosis. For example, in *dmc1* mutants recombination is completely blocked but if cells are returned to vegetative growth conditions, DSBs are repaired efficiently via inter-sister recombination (Schwacha and Kleckner 1997; Zenvirth et al. 1997). Also, in haploid cells tricked into undergoing meiosis, DSBs are formed but their repair (which must occur between sisters) is delayed, presumably until constraints that prevent use of the sister-template are removed (De Massy et al. 1994).

6.1.1 Establishment of the block

A fundamental step in establishing the block to intersister recombination appears to be activation of the Mek1 kinase at DSB sites (Fig. 4) (Wan et al. 2004; Niu et al. 2005). This occurs through assembly and activation of a Hop1–Red1–Mek1 complex (See Section 4.2.2) (Niu et al. 2005). Assembly is mediated by interactions between the FHA domain of Mek1 and phospho-Red1; and between Hop1 and Red1 (Bailis and Roeder 1998; de los Santos and Hollingsworth 1999). In response to DSB formation, the C-domain of Hop1 becomes phosphorylated by an unknown kinase (Niu et al. 2005). Tel1/Mec1 are candidates for the initial phosphorylation because of their roles in sensing nascent DSBs via the MRX complex (Section 4.1.4) (Usui et al. 2001). Phosphorylation of the Hop1 C-domain is thought to activate Mek1 by promoting its dimerization and trans-autophosphorylation (Niu et al. 2005). In strong support of this idea a GST-Mek1 fusion, which can self-dimerize, completely bypasses mutations in the Hop1 C-domain (Niu et al. 2005). In this respect, Red1–Hop1 acts like a meiotic analog of the DNA-damage signal-transducer, Rad9. In response to DNA damage, Rad9 undergoes Mec1/Tel1-mediated phosphorylation. Phospho-threonines on Rad9 then recruit the Mek1 analog Rad53 via its FHA domain, and promote its trans-autophosphorylation and/or trans-phosphorylation by Mec1/Tel1 (Pellicioli and Foiani 2005). An important difference is that Red1 undergoes DSB-*independent* phosphorylation, which means that Mek1 will be recruited to DSB sites in anticipation of Spo11 cleavage (T-F. Wang and N. Hollingsworth, personal communications). This feature will ensure rapid local implementation of the block to intersister recombination at nascent DSBs.

6.1.2 The nature of the block to intersister recombination

A general block to intersister recombination would also constitute a general block to interhomolog recombination. Thus, any block to intersister recombination must be established locally, on a DSB-by-DSB basis. What is the nature of this block? One possibility is suggested by the spatial relationships between sister-chromatid loops, DSB sites and recombination complexes (Fig. 4) (Blat et al. 2002; Sheridan and Bishop 2006). DSB sites localize to sequences that are located in the tops of chromatin loops but which become tethered to the homolog axes once breaks are formed (Blat et al. 2002). If sister loops dissociate as the DSB-ends become associated with the axis, and if homology search then takes place from the axis, the separation of DSB-ends from homologous sequences on the sister-chromatid will impede intersister recombination (Fig. 4).

tid1 mutants show a severe block at meiotic anaphase I due to unresolved sister-chromatid cohesion (Kateneva et al. 2005; Kateneva and Dresser 2006). A cohesin complex based on the meiosis-specific Kleisin component, Rec8, mediates most meiotic cohesion, although the mitotic Kleisin Scc1/Mcd1 is present at low levels (Klein et al. 1999). Surprisingly, in *tid1* mutants, it is a separase-resistant population of Scc1/Mcd1 that causes the block to homolog segregation. Moreover,

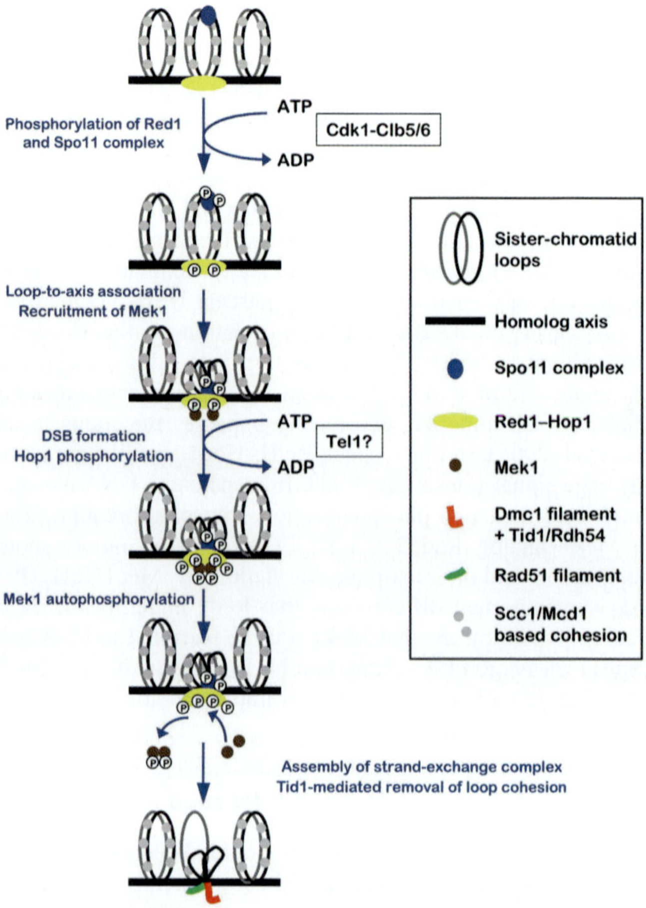

Fig. 4. Model for the establishment of interhomolog bias. Spo11 complexes are bound to DSB sites at the tops of chromatin loops, whereas the Red1-Hop1 complex binds the loop bases in association with the homolog axis (Section 6.1.2). Phosphorylation of the two complexes by CDK promotes their interaction to establish loop-to-axis association. Phospho-Red1 binds Mek1, which is activated following DSB formation and phosporylation of Hop1 (Section 6.1.1). DSBs form in a region of Scc1/Mcd1-based cohesion; alternatively, Scc1/Mcd1-based loop cohesion is established *de novo* via DSB-dependent recruitment of Scc1/Mcd1-based cohesin complexes, as shown for mitotic DSBs (Unal et al. 2004 Strom et al. 2004; Strom and Sjogren 2005). Loop cohesion is removed by the Tid1/Rdh54 translocase following axis-associated assembly of Dmc1 and Rad51 strand-exchange complexes (Section 6.1.2). The resulting spatial constraints disfavor intersister interactions.

this block is overcome by a *dmc1* mutation (Kateneva and Dresser 2006). It is proposed that a separase-resistant domain of Scc1/Mcd1-based cohesion is established specifically at DSB sites, via Dmc1, and can only be resolved following recruitment of Tid1, presumably acting as a translocase to kick Scc1/Mcd1-based cohesin complexes off the DNA. This series of events may be required for the controlled separation of sister-chromatid loops and association of DSB-ends with the homolog axis (Kateneva and Dresser 2006) (Fig.4).

An alternative but not exclusive possibility is that access to the sister-chromatid isn't blocked *per se* but interhomolog bias is implemented by preferentially stabilizing interhomolog interactions. This would require that: (1) initial strand-pairing interactions are rapidly reversed in the absence of active stabilization, e.g. via helicase action; (2) progression of recombination beyond initial strand-pairing interactions requires homolog pairing; and (3) interhomolog interactions are preferentially stabilized, e.g. by bridging the two homolog axes.

To understand the nature of the block, targets of activated Mek1 kinase must be identified. It's important to note that previous reports that Red1 is phosphorylated by Mek1 appear to have been erroneous (Bailis and Roeder 1998; de los Santos and Hollingsworth 1999; Wan et al. 2004). With respect to interhomolog bias, Mek1 targets have yet to be confirmed but candidates include: (i) proteins important for intersister recombination, such as Rad51 and Rad54 (Wan et al. 2004); (ii) proteins that inhibit Rad51, i.e. the Hed1 protein (Sect. 5.4.1) (Sheridan and Bishop 2006; Tsubouchi and Roeder 2006); (iii) cohesins and other axis components (Bailis and Roeder 1998; Sjogren and Nasmyth 2001; Zierhut et al. 2004; Kateneva and Dresser 2006); (iv) negative regulators of recombination such as Sgs1 and Srs2 (Ira et al. 2003; Krejci et al. 2003; Veaute et al. 2003); and (v) factors required for interhomolog interactions such as Mnd1–Hop2 and Dmc1–Mei5–Sae3 (Zierhut et al. 2004). Mek1 can phosphorylate Hop1 *in vitro* but whether this activity is relevant *in vivo* is unclear. Regardless, Mek1-kinase activity is required for the timely dissociation of Hop1 from chromosomes, an event that could mark elimination of the block to intersister recombination (Bailis and Roeder 1998).

6.1.3 Regulating the block to intersister recombination

Following DSB formation, continued activation of Mek1 may involve signaling by Rad24 and Rad17, and subsequent trans-phosphorylation via activated Mec1 (Bailis and Roeder 2000). Rad24 and Rad17 are also required for the normal assembly and disassembly of Rad51 filaments (Shinohara et al. 2003a) suggesting a coordinated role for these components in maintaining the block to intersister recombination. Consistent with this idea, interhomolog bias is lost in *rad51* mutants (Schwacha and Kleckner 1997).

Formally, the block to intersister recombination could be maintained throughout meiotic prophase. It's apparent, however, that access to sister templates is essential in the majority of meioses. For example, in the absence of Rad54 (Section 5.5), which appears to be specifically required for intersister recombination (Arbel et al. 1999), interhomolog exchange is unaffected but spore viability is reduced by ~50% (Schmuckli-Maurer and Heyer 2000). This defect is exacerbated in hybrid strains where homologs are diverged and DSB-repair is presumably more dependent on the presence of a sister-chromatid. These and other observations support the idea that the block to sister-chromatid recombination is lifted after selected recombination events have irreversibly committed to crossing-over with a homolog

(Sheridan and Bishop 2006). Maintaining the barrier beyond this point would seem futile given that access to the sister may greatly improve the efficiency of DSB-repair.

How attenuation of Mek1 activity and elimination of the block occur is uncertain. The loss of Hop1 from synapsing homologs is potentially a marker for this transition, however (Smith and Roeder 1997). Hop1 chromosomal staining is normally lost from regions where homologs have synapsed and is expected to prevent further activation of Mek1, thereby lifting the block to intersister recombination on a region-by-region basis as homologs synapse. Loss of Hop1 could have the dual affects of attenuating Mek1 activation and releasing tethered recombination complexes from the axes, thereby easing architectural constraints on the use of the sister template (above). Dephosphorylation of Red1 by protein phosphatase type 1, Glc7, could be involved in this transition (Bailis and Roeder 2000).

6.2 Interhomolog only functions

The idea of a differentiated, interhomolog pathway derives from two basic observations (Schwacha and Kleckner 1997; Zierhut et al. 2004). First, *red1* mutation specifically reduces interhomolog-dHJs but not IS-dHJs. The interpretation is that Red1-dependent DSBs are already committed to an interhomolog fate (but see Niu et al. 2005). Second, even in the absence of Red1, interhomolog dHJs and crossovers are higher when Dmc1 and Hop2–Mnd1 are present than when strand-exchange is promoted solely by Rad51. This indicates an intrinsic tendency for the Dmc1/Hop2-Mnd1 pathway to utilize the homolog.

7 Crossover control

The raison d'être of meiotic recombination is to create interhomolog connections in the form of crossovers. Two aspects of the meiotic crossover distribution reveal that crossing-over is highly regulated to produce a set of events that is optimized to facilitate homolog segregation.

7.1 Crossover assurance

Every pair of homologs acquires at least one crossover—the obligatory event that ensures normal segregation—and this is achieved irrespective of chromosome length and despite a low average number of crossovers per chromosome (Jones 1984). The observation that short chromosomes have higher rates of crossing-over per unit length is thought to be a manifestation of crossover assurance (Kaback et al. 1989, 1992, 1999; Kaback 1996). Kaback (1992) showed that this chromosome-size effect is not coincidental but the consequence of an active process: when chromosomes were artificially shortened the recombination rate increased and conversely, when chromosomes were lengthened recombination rate decreased (but note that in a similar experiment, Turney et al. (2004) observed no affect of chromosome size on recombination rate).

The nature and mechanism of crossover assurance and the chromosome-size effect are unclear. Theoretically, this process could affect recombination between short chromosomes in two ways: (i) by increasing the fraction of DSBs that mature as crossovers or (ii) by increasing the rate of DSB formation. Kaback et al. (1999) favored the first possibility because they found that crossover interference (see below) was weaker for short chromosomes. Recent analysis of Spo11 hypomorphs has established that the crossover/noncrossover ratio can be modulated in order to maintain crossover levels (Martini et al. 2006). Specifically, when genome-wide DSB levels were decreased by hypomorphic *spo11* alleles, crossovers were maintained at near wild-type levels while noncrossovers were decreased. Consistent with the possibility that the rate of DSB formation is increased along short chromosomes, however, is a genome-wide analysis of yeast DSB hotspots showing that there are more DSBs along short chromosomes (Gerton et al. 2000). This inference could, in part, reflect effects of the *rad50S* mutation on DSB formation in some chromosomal regions and, as such, should be treated cautiously (Section 4.4.1).

7.2 Crossover interference

When two or more crossovers occur between the same pair of homologs, adjacent events are located further apart than expected from a random distribution, i.e. crossovers exhibit mutual interference (Muller 1916; Mather 1933; Hillers 2004). Interference is complete at short distances and decreases in intensity with increasing distance along a chromosome. Interference can extend over huge physical distances: ~0.1 Mb in *S. cerevisiae* and ≥100 Mb in mammals. The effects of interference on the final distribution of crossovers are two fold: first, the minimum interference distance sets a limit on the absolute number of crossovers that may be realized per chromosome; and second, when a chromosome is significantly longer than the minimum interference distance, crossovers along that chromosome will be widely spaced.

The mechanism of interference is not known but interference is weakened by disturbing a variety of meiotic processes, e.g. chromosome structure (Nabeshima et al. 2004), the rate of chromosome pairing (Chua and Roeder 1997; Carlton et al. 2006) and DNA strand-exchange (Shinohara et al. 2003b). Common to all models of interference is the idea that an inhibitory zone is established around sites that have become committed to crossing-over. DSBs within this zone of inhibition are prevented from becoming crossovers and mature instead as non-crossovers (e.g. Kleckner et al. 2004).

The biological relevance of interference is unclear but theories fall into three non-exclusive classes (reviewed in Hillers 2004): (i) interference is a manifestation of the crossover assurance process (Kleckner et al. 2004; Jones and Franklin 2006; Martini et al. 2006); (ii) interference increases the evolutionary benefit of crossing-over by optimizing the efficiency with which new allele combinations are formed (Gorlov and Gorlova 2001); (iii) interference safeguards chiasmata because closely-spaced crossovers may leave insufficient intervening cohesion to hold homologs together (Nilsson and Sall 1995; van Veen and Hawley 2003).

7.3 Crossover and noncrossover pathways

At the molecular level, crossover and non-crossover pathways diverge at an early stage, prior to the formation of extensive strand-exchange intermediates (Allers and Lichten 2001a; Hunter and Kleckner 2001; Bishop and Zickler 2004; Borner et al. 2004). Along the crossover pathway successive strand-exchange at the two DSB-ends forms first an SEI and then a dHJ (Fig. 1B). In contrast, non-crossovers can form with normal timing and levels in mutants that form very few SEIs or dHJs or that fail to resolve dHJs (Allers and Lichten 2001a; Borner et al. 2004). Thus, the decision to make a crossover is synonymous with the decision to form SEI and dHJ intermediates. Noncrossovers are generally understood to be specifically interhomolog noncrossovers (or "gene conversions" without associated crossing-over). To encompass the fact that some DSBs are repaired via inter-sister recombination, I will define noncrossovers as any DSB-repair event that does not yield an interhomolog crossover. Noncrossovers may simply be the default outcome of DSB-repair, similar to repair in mitotic cells except that the homolog is the usual template (Borner et al. 2004). It seems unlikely that DSBs not assigned a crossover function would play any role beyond homolog pairing and as such, the repair of these breaks becomes paramount, be it via homolog or sister templates. Current evidence favors a synthesis-dependent strand-annealing pathway of non-crossover formation (e.g. Allers and Lichten 2001b; Jessop et al. 2005) which is expected to utilize the core DSB-repair machinery. In contrast, ensuring a crossover outcome at designated sites involves substantial modulation of the basic recombination machinery (below).

7.4 Pro-crossover factors

While the nature of the crossover decision remains mysterious, we are beginning to understand how crossing-over is implemented. In *S. cerevisiae*, mutations in more than a dozen known genes lead to a specific deficit of crossovers but not of non-crossovers (e.g. Borner et al. 2004). A subset of these mutations also cause coordinate defects in SC formation, which defines the meiosis-specific "*ZMM*" class of genes. The *ZMMs* encode proteins of diverse function, i.e. a DNA helicase, Mer3 (Nakagawa and Ogawa 1999; Nakagawa and Kolodner 2002); two homologs of the MutS DNA mismatch-repair proteins, Msh4 and Msh5 (Ross-Macdonald and Roeder 1994; Hollingsworth et al. 1995; Pochart et al. 1997), the SC transverse filament protein, Zip1 (Sym et al. 1993; Sym and Roeder 1994); a large WD-like repeat protein, Zip2 (Chua and Roeder 1998; Perry et al. 2005); a SUMO E3 ligase, Zip3 (Agarwal and Roeder 2000; Cheng et al. 2006); a TPR-repeat protein, Zip4/Spo22 (Perry et al. 2005; Tsubouchi et al. 2006); and an orphan protein with no obvious functional motifs, Spo16 (A. Shinohara, personal communication). *In vivo, zmm* mutants show a common (though somewhat variable) recombination phenotype, being defective in the formation of crossover-specific joint molecules, SEIs and dHJs (Borner et al. 2004).

Mutations in five additional genes reduce crossing-over but unlike the *zmm* mutants have no defects in SC formation. These genes include two homologs of the MutL DNA mismatch-repair proteins, Mlh1 and Mlh3 (Baker et al. 1996; Edelmann et al. 1996; Hunter and Borts 1997; Wang et al. 1999; Woods et al. 1999; Lipkin et al. 2002; N. Hunter, A. Jambhekar, J.P. Lao, S.D. Oh, N. Kleckner, and V.B. Borner submitted); the 5'-3' exonuclease, Exo1 (Khazanehdari and Borts 2000; Kirkpatrick et al. 2000; Tsubouchi and Ogawa 2000; Wei et al. 2003) (Section 4.3.2); and the structure-selective nuclease complex, Mus81–Mms4 (Boddy et al. 2001; de los Santos et al. 2001; De Los Santos et al. 2003; Osman et al. 2003; Heyer 2004; Hollingsworth and Brill 2004). Notably, expression of these genes is not confined to meiosis.

7.4.1 Mer3 and Msh4/5

Mer3 is an ATP-dependent DNA helicase that can extend nascent heteroduplexes formed *in vitro* by Rad51 (Nakagawa and Kolodner 2002; Mazina et al. 2004). This reaction shows a strong 3'-to-5' polarity bias (with respect to the displaced strand) and therefore will only extend heteroduplexes initiated from a 3'-end, i.e. the polarity of strand-invasion predicted *in vivo* (e.g. Hunter and Kleckner 2001). In fact, Mer3 will prevent extension of heteroduplex initiated from a 5'-end. Thus, consistent with the *in vivo* recombination defect of *mer3* mutants, Mer3 likely acts to stabilize nascent strand-exchange intermediates and thereby promote the formation of SEIs and dHJs (Borner et al. 2004; Mazina et al. 2004). Whether Mer3 works with Rad51 and/or Dmc1 *in vivo* is not clear.

Msh4 and Msh5 are meiosis-specific homologs of the highly conserved MutS protein family (Ross-Macdonald and Roeder 1994; Hollingsworth et al. 1995; Pochart et al. 1997; Hoffmann and Borts 2004). MutS proteins function during DNA mismatch-correction in the specific recognition of mismatched bases and coordination of downstream repair events (Iyer et al. 2006). Msh4 and Msh5 are not involved in mismatch recognition but function specifically in meiotic recombination. *In vitro*, the heterodimeric Msh4–Msh5 complex binds specifically to D-loop and Holliday-Junction structures (Snowden et al. 2004). This binding provokes ADP$\rightarrow$ATP exchange by Msh4-Msh5, inducing a conformational change that converts the enzyme into a sliding clamp, which encircles the two duplexes adjacent to the exchange junction. Through this mechanism, like Mer3, Msh4-Msh5 will also stabilize nascent strand-exchange intermediates and promote SEI and dHJ formation.

7.4.2 Zip1

Zip1 is a member of a family of SC transverse filament (TF) proteins that polymerize between homolog axes to effect synapsis (Sym et al. 1993; Zickler and Kleckner 1999; de Boer and Heyting 2006). TFs are poorly conserved at the sequence level but share a broadly similar secondary structure; a long coiled-coil flanked by globular C- and N-termini. Coiled-coil lengths of ~70-90 nm fit the prediction that these proteins span the 100 nm wide SC in a head-to-head configuration (Schmekel et al. 1996; Dong and Roeder 2000). Zip1 polymerization nucleates at sites where homolog axes have become closely associated via recombination.

How does the SC promote crossing-over? Lack of sequence conservation between TFs and the absence of any clear catalytic motifs suggests that they are not intrinsic components of the recombination machinery and are unlikely to act directly to promote crossing-over. More likely, TFs have been adopted in meiosis for their ability to lock paired homologs together, effectively superceding recombinational interactions. This ability will allow recombination to progress while maintaining intimate homolog juxtaposition. Moreover, Zip1 still promotes crossing-over even when it does not efficiently polymerize along the chromosomes (in a *red1* mutant) indicating that even local stabilization of interhomolog association can facilitate recombination (Storlazzi et al. 1996). In addition, since SCs nucleate at sites of homology, SC polymerization will also tend to dissociate ectopic interactions and minimize the risk of chromosome translocations (e.g. Higgins et al. 2005).

7.4.3 Zip2, Zip3, and Zip4

Zip2, Zip3 and Zip4 (together with Mer3 and Msh4-Msh5) are thought to function together as a synapsis initiation complex (SIC) that nucleates Zip1 polymerization at sites of recombination (Chua and Roeder 1998; Agarwal and Roeder 2000; Cheng et al. 2006; Tsubouchi et al. 2006). Sites that have been assigned a crossovers fate are favored sites of SC nucleation (Zickler and Kleckner 1999; Fung et al. 2004; Henderson and Keeney 2005).

Zip2 is a poorly conserved WD-like repeat protein with identifiable homologs in *Saccharomyces* species and related yeasts (Perry et al. 2005). Meticulous sequence analysis indicates that Zip2 homologs may share identical structures, however, comprising as many as 14 β-propeller blade structures (Perry et al. 2005) (Section 4.1.2). In this respect, Zip2 could be structurally related to Aip1 (a protein involved in actin dynamics), which folds into a clamshell-shaped molecule comprising two interconnected, seven-bladed β-propellers that present multiple protein-interaction surfaces with specific orientations (Voegtli et al. 2003). Thus, Zip2 could simultaneously bind multiple substrates or particularly large substrates (e.g. Zip1 dimers). By 2-hybrid analysis, Zip2 interacts with Zip3 and Cdc53, the cullin component of SCF-type ubiquitin E3 ligases (Willems et al. 2004); these and other considerations have led to the proposal that Zip2, Zip3 and Zip4 comprise a multisubunit ubiquitin E3 ligase, perhaps related to the APC (Perry et al.

2005) (but see Zip3, below); however, co-chaperone or transport functions are equally likely possibilities for Zip2 and Zip4.

Zip4 (a.k.a. Spo22) encodes a large TPR-repeat protein with identifiable homologs in fungi, plants and animals (Perry et al. 2005) (N. Jackson and N. Hunter, unpublished). TPR repeats form antiparallel α-helices and repeat arrays assemble into an extended right-handed superhelical structure with an amphipathic groove, which is thought to mediate protein-protein interactions (Blatch and Lassle 1999). TPRs are found in proteins with diverse biological functions, often mediating the assembly of multiprotein complexes. TPR and WD repeat proteins partner in a number of protein complexes and, consistently, Zip4 appears to function interdependently with Zip2, although direct interaction has not been demonstrated (Tsubouchi et al. 2006). *zip4* mutants (and presumably *zip2*) show a novel recombination phenotype in which residual crossovers show Zip1-dependent clustering. It is proposed that the numerous short patches of SC formed in *zip4* cells are responsible for this clustering. In other *zmm* mutants, residual crossovers are randomly distributed, not clustered, and extensive SC often forms between some chromosomes, whereas others can be completely unsynapsed (Novak et al. 2001; Borner et al. 2004; Tsubouchi et al. 2006) (N. Hunter, A. Jambhekar, J.P. Lao, S.D. Oh, N. Kleckner, and V.B. Borner, submitted).

Zip3 is a conserved RING-finger E3 ligase (Jantsch et al. 2004; Perry et al. 2005; Cheng et al. 2006) (N. Jackson and N. Hunter, unpublished). Cheng *et al.* recently provided evidence that *S. cerevisiae* Zip3 is an E3 ligase for SUMO (Small Ubiquitin-related MOdifier) (Cheng et al. 2006). SUMO is one of a family of small ubiquitin-like post-translational modifications and is conjugated to target substrates through a mechanism analogous to that of ubiquitlyation but which involves a dedicated set of proteins. SUMOylation appears to modify protein localization, activity and protein-protein interactions (Johnson 2004). Unlike ubiquitin, however, SUMOylation does not target proteins for degradation and in some cases may directly antagonize ubiquitin by competing for identical lysine target residues. A role for SUMOylation during meiosis has been generally inferred. Most notably, the sole SUMO E2 conjugase, Ubc9, assembles into immunostaining foci along mouse meiotic chromosomes, and interacts both with the recombination enzyme Rad51 (Kovalenko et al. 1996) and synaptonemal complex proteins (hamster Cor1 and Syn1) (Tarsounas et al. 1997). Yeast studies demonstrate a critical role for SUMOylation in the formation of SCs (Cheng et al. 2006; Hooker and Roeder 2006). SUMOylation of homolog axes is proposed to promote SC polymerization by attracting the SC building block Zip1, which can bind noncovalently to SUMO (Cheng et al. 2006). 2-hybrid and co-IP experiments show that Zip3 interacts with a number of SC and recombination proteins, including Zip1, Zip2, Msh5, Mre11 and Rad57 (Agarwal and Roeder 2000). Thus, Zip3 may modify multiple target proteins which would be consistent with the dramatic changes in the global pattern of SUMO conjugates observed in *zip3* mutants (Cheng et al. 2006) (E. Mortensen and N. Hunter, unpublished). By extension, Zip3 via SUMO could help couple recombination enzymes to SC formation (Agarwal and Roeder 2000). Paradoxically, *C. elegans zhp-3* (*zip3*) mutants have no defect in SC formation, questioning a conserved role for Zip3 proteins in SC formation (Jantsch et al.

2004). Crossing-over in *zhp-3* mutants is abolished, however. A proposal that can reconcile budding yeast and *C. elegans* data, is that the conserved function of Zip3 proteins, like other ZMMs, is to stabilize recombination intermediates. In yeast, this would facilitate both SC formation and crossing-over; whereas in *C. elegans*, which does not require DSBs for pairing and SC formation, stabilization would only be required for crossing-over. SUMOylation and thus stabilization of recombination proteins at designated crossover sites could achieve this.

7.4.4 Mlh1 and Mlh3

The MutL homologs, Mlh1 and Mlh3, have a widely conserved role in crossing-over (Baker et al. 1996; Edelmann et al. 1996; Hunter and Borts 1997; Wang et al. 1999; Woods et al. 1999; Lipkin et al. 2002). MutL-family proteins are GHKL-type ATPases, which function as transducers between mismatch recognition by MutS proteins and proteins involved in processing of mismatches, such as helicases and nucleases (Dutta and Inouye 2000; Yang 2000; Iyer et al. 2006). ATP binding and hydrolysis induces conformational changes in MutL that regulate protein-protein interactions and DNA binding. Like MutS, MutL proteins function as dimers. Mlh1 is the common component of three eukaryotic MutL heterodimers, Mlh1-Pms1, Mlh1-Mlh2, and Mlh1-Mlh3; only the latter is involved in crossing-over, although it also plays a minor role in mismatch correction (Flores-Rozas and Kolodner 1998; Wang et al. 1999; Hoffmann and Borts 2004). In mammalian cells, immunostaining foci of MLH1 and MLH3 assemble onto chromosomes at early-to-mid pachynema, around the time that crossovers form (Baker et al. 1996; Guillon et al. 2005). Moreover, MLH1/3 foci specifically mark the sites where chiasmata will appear at diplonema, i.e. crossover sites (Marcon and Moens 2003). Consistent with the idea that Mlh1-Mlh3 coordinates events that occur downstream of junction binding by Msh4-Msh5, mammalian MLH1-MLH3 and MSH4-MSH5 complexes interact, (Santucci-Darmanin et al. 2000, 2002). Mutant phenotypes also indicate that Mlh1-Mlh3 acts after Msh4-Msh5 and other ZMMs to promote crossing-over (e.g. Hunter and Borts 1997; N. Hunter, A. Jambhekar, J.P. Lao, S.D. Oh, N. Kleckner, and V.B. Borner, submitted). Intriguingly, the human MLH1–PMS2 complex (Mlh1-Pms1 in budding yeast) has recently been shown to have a latent nuclease activity (Kadyrov et al. 2006). A candidate nuclease motif identified in PMS2 is also present in Mlh3 proteins (but not in Mlh1 or Mlh2). This raises the possibility that Mlh1-Mlh3 directly catalyzes meiotic dHJ resolution.

7.4.5 Exo1

The Exo1 nuclease is discussed in Section 4.3.2. Exo1 acts along the major crossover pathway, as defined by its epistasis with Msh4, and its apparent role in DSB resection suggests an early role in recombination (Khazanehdari and Borts 2000; Tsubouchi and Ogawa 2000). The phenotypes of *Exo1*$^{-/-}$ *vis-a-vis Msh4*$^{-/-}$ mutant mice are distinct, however, and suggest a later role for Exo1. In *Exo1*$^{-/-}$ mutants, pairing and synapsis are normal but crossing-over is severely reduced. In this respect, the *Exo1*$^{-/-}$ phenotype is indistinguishable from those of *Mlh1*$^{-/-}$ and *Mlh3*$^{-/-}$ mutants (Edelmann et al. 1996; Lipkin et al. 2002; Wei et al. 2003). In contrast, in *Msh4*$^{-/-}$ mice pairing and synapsis fails (Kneitz et al. 2000). Several models have been proposed for the function of Exo1 in meiotic recombination (Hoffmann and Borts 2004). Longer single-stranded tails created by Exo1-catalyzed resection could promote more extensive strand-exchange and, consequently more efficient SEI and dHJ formation. Alternatively, Exo1 might function at later steps of meiotic recombination to facilitate the ligation of DNA strands during dHJ formation. Following interaction of the second DSB-end, Exo1, acting either as an exonuclease or a 5'-flap endonuclease, could facilitate formation of the ligatable intermediate required to complete dHJ formation. Consistently, Exo1 has a 5'-flap endonucleolytic activity (Lee and Wilson 1999). This model would reflect the late defects in meiotic recombination demonstrated in mice.

7.4.6 Mus81–Mms4(Eme1)

Mus81 and Mms4 interact to form a structure-selective endonuclease of the XPF-family, which functions in Rad52-dependent replication-fork restart as well as meiotic recombination (Heyer et al. 2003). The requirement for XPF-type nucleases for meiotic crossing-over varies with organism: Mus81–Eme1 and MEI9–MUS312 are required for most crossovers in *S. pombe* and *Drosophila*, respectively (Boddy et al. 2001; Yildiz et al. 2002); whereas, in *S. cerevisiae*, Mus81-Mms4 represents a minor pathway of crossing-over that parallels the major ZMM-dependent pathway (above) (de Los Santos et al. 2003; Tsubouchi et al. 2006). However, Mus81-Mms4 is essential for the repair of at least a subset of DSBs in *S. cerevisiae*. Mus81-Mms4 interacts with Rad54, which may recruit it to sites of recombination (Interthal and Heyer 2000).

A debate has raged as to whether Mus81-Mms4 and Mus81-Eme1 act as *bona fide* Holliday Junction resolvases or instead catalyze crossing-over through a non-HJ mechanism, via sequential cleavage of D-loops formed by strand-exchange at the two DSB-ends (de los Santos et al. 2003; Smith et al. 2003; Heyer 2004; Hollingsworth and Brill 2004; Whitby 2005). Budding yeast phenotypes are not obviously indicative of a defect HJ resolution and suggest that Mus81–Mms4 may instead promote SEI and dHJ formation (de los Santos et al. 2003). On the other hand, the phenotypes of fission yeast *mus81* mutants provide a persuasive case for a role in Holliday junction resolution (Boddy et al. 2001; Yildiz et al. 2002).

7.5 A molecular model of crossover and noncrossover pathways

Implicit in canonical models of DSB-repair is the idea that only one DSB-end invades a template duplex while the second end interacts by annealing (e.g. Szostak et al. 1983; Paques and Haber 1999). What then, could be the advantage of assembling RecA proteins at both DSB-ends, as suggested for meiotic recombination

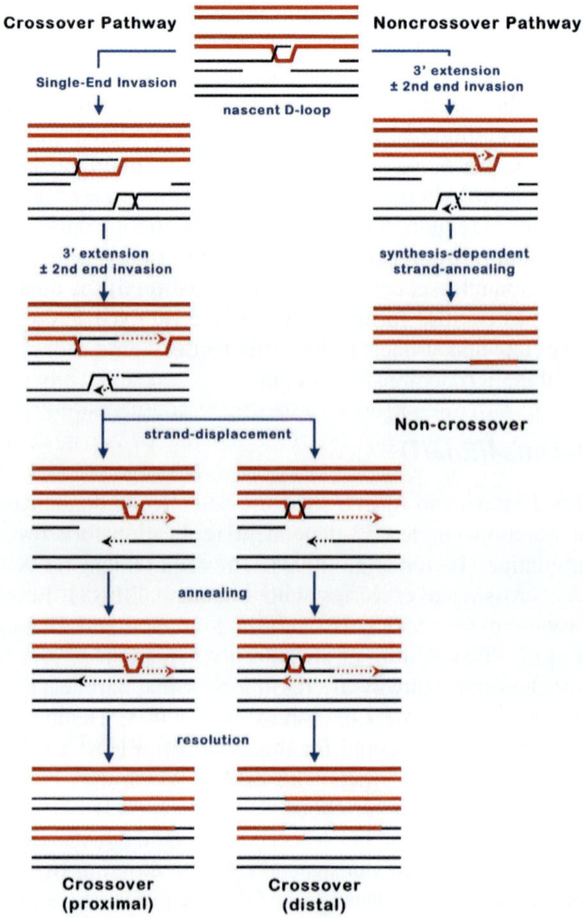

Fig. 5. Molecular model of crossover and noncrossover pathways. Nascent interhomolog D-loop formation is catalyzed by Dmc1. The crossover pathway is differentiated by Zmm-mediated stabilization of the nascent D-loop to form a SEI. Second-end invasion catalyzed by Rad51 follows homolog pairing and SC formation, and removal of inhibition by Hed1. The second DSB-end may interact with a homolog template or, as shown here, the sister-chromatid. Following DNA synthesis from the two DSB-ends, distinct strand-displacement reactions are proposed. Along the noncrossover pathway, both DSB-ends are completely dissociated and anneal to seal the break. Along the crossover pathway, the first DSB-end is exposed by D-loop displacement but the homologs remain connected (Allers and Lichten 2001b). The second DSB-end is completely dissociated and anneals to the displaced D-loop to form a dHJ. The extent of initial D-loop displacement will dictate the position of the dHJ and ultimately the position of the crossover relative to the DSB site and associated heteroduplex.

(Section 5)? Two non-exclusive ideas can be suggested: (i) that the Rad51-associated DSB-end serves as a back-up to repair DSBs that have failed to engage a homolog (Sheridan and Bishop 2006); and (ii) that efficient meiotic recombination frequently involves strand-invasion by *both* DSB-ends. The latter possibility is illustrated in Figure 5. In this model, the Dmc1-associated end promotes homolog pairing and SC formation via nascent D-loop formation. Then, following removal of Hed1 inhibition, the Rad51-associated end becomes capable of strand-exchange, invades *either* a homolog or a sister template and primes DNA synthesis. In this respect, the Rad51-end is proposed to function as it would during mitotic DSB-repair. It is assumed that this end is ultimately displaced from the template and then undergoes one of two annealing reactions: either it anneals to the SEI formed by invasion of the first end; or alternatively, it anneals to the first DSB-end after it has been displaced from the homolog. Thus, consistent with the defects of *zmm* mutants, control over the crossover or noncrossover outcome is determined by whether the initial D-loop is stabilized and converted into a SEI (crossover) or is disassembled (noncrossover). Strand-invasion by the Rad51-end may not be essential but will improve the efficiency of the subsequent annealing step. This will be particularly important if DNA synthesis from the Dmc1-associated end is limited, e.g. by topological constraints in the homolog template.

No single model, including the one described here, seems capable of reconciling all of the features of meiotic recombination observed by genetic and molecular analyses (see Merker et al. 2003). The model presented in Figure 5 and variations on its basic themes can accommodate the following features, however:

(1) Heteroduplex is generally formed on only one side of the DSB site (e.g. Merker et al. 2003; Jessop et al. 2005).

(2) The extent of heteroduplex is dictated by the extent of DNA synthesis that occurs prior to annealing and not by the length of initial DSB resection (Merker et al. 2003; Maloisel et al. 2004; Blanton et al. 2005; Jessop et al. 2005).

(3) The inter-junction distance measured for dHJs visualized by electron microscopy averages ~260 bp (Bell and Byers 1983b; Cromie et al. 2006). This is significantly less than the total length of DSB-resection, which is estimated at ~1000 bp (~500 bp on both sides of the DSB). The fact that heteroduplex tracts regularly extend much further than the inter-junction distance implies that a significant fraction of heteroduplex forms adjacent to the dHJ.

(4) A large fraction of crossovers points are located distal to the DSB site but do not form detectable heteroduplex at an intervening marker (Allers and Lichten 2001b; Jessop et al. 2005). This observation is readily accommodated by the idea that the second DSB-end may engage a sister-chromatid template.

8 Closing remarks

Studies of meiotic recombination are providing unique insights, not only into the general mechanism of DSB-repair by homologous recombination but particularly into ways in which the core recombination machinery can be modulated to control template choice, strand-exchange activity, recombination outcome, and the coordination of DNA transitions with the chromosomal and cellular events of meiosis.

Conspicuous gaps in our understanding are highlighted throughout this chapter. Particularly enigmatic are the mechanism and control of Spo11 cleavage; the relationships between Dmc1 and Rad51, and their respective roles in homolog-pairing and joint molecule formation; the post-invasion steps of recombination; the nature of the block to intersister recombination; and the crossover/noncrossover decision. More generally, our understanding of the molecular mechanisms of meiotic recombination is still in its infancy. In this respect, the wealth of molecular and genetic tools available in budding yeast promise to keep it at the forefront of the field.

Acknowledgements

I am grateful to all those who contributed the data and ideas reviewed in this chapter. Special thanks go to Scott Keeney, Nancy Hollingsworth, Nancy Kleckner, Michael Lichten, Malgosia Bzymek and members of my laboratory. Work in my laboratory is generously supported by grants from the NIGMS (GM074223-01), the Damon Runyon Cancer Research Foundation (Scholar Award 40-04), and the Concern Foundation.

References

Agarwal S, Roeder GS (2000) Zip3 provides a link between recombination enzymes and synaptonemal complex proteins. Cell 102:245-255

Akamatsu Y, Dziadkowiec D, Ikeguchi M, Shinagawa H, Iwasaki H (2003) Two different Swi5-containing protein complexes are involved in mating-type switching and recombination repair in fission yeast. Proc Natl Acad Sci USA 100:15770-15775

Alexeev A, Mazin A, Kowalczykowski SC (2003) Rad54 protein possesses chromatin-remodeling activity stimulated by the Rad51-ssDNA nucleoprotein filament. Nat Struct Biol 10:182-186

Allers T, Lichten M (2000) A method for preparing genomic DNA that restrains branch migration of Holliday junctions. Nucleic Acids Res 28:e6

Allers T, Lichten M (2001a) Differential timing and control of noncrossover and crossover recombination during meiosis. Cell 106:47-57

Allers T, Lichten M (2001b) Intermediates of yeast meiotic recombination contain heteroduplex DNA. Mol Cell 8:225-231

Anderson DE, Trujillo KM, Sung P, Erickson HP (2001) Structure of the Rad50.Mre11 DNA repair complex from *Saccharomyces cerevisiae* by electron microscopy. J Biol Chem 276:37027-37033

Anuradha S, Muniyappa K (2004) *Saccharomyces cerevisiae* Hop1 zinc finger motif is the minimal region required for its function *in vitro*. J Biol Chem 279:28961-28969

Aravind L, Koonin EV (1998) The HORMA domain: a common structural denominator in mitotic checkpoints, chromosome synapsis and DNA repair. Trends Biochem Sci 23:284-286

Arbel A, Zenvirth D, Simchen G (1999) Sister chromatid-based DNA repair is mediated by *RAD54*, not by *DMC1* or *TID1*. EMBO J 18:2648-2658

Arora C, Kee K, Maleki S, Keeney S (2004) Antiviral protein Ski8 is a direct partner of Spo11 in meiotic DNA break formation, independent of its cytoplasmic role in RNA metabolism. Mol Cell 13:549-559

Bailis JM, Roeder GS (1998) Synaptonemal complex morphogenesis and sister-chromatid cohesion require Mek1-dependent phosphorylation of a meiotic chromosomal protein. Genes Dev 12:3551-3563

Bailis JM, Roeder GS (2000) Pachytene exit controlled by reversal of Mek1-dependent phosphorylation. Cell 101:211-221

Baker SM, Plug AW, Prolla TA, Bronner CE, Harris AC, Yao X, Christie DM, Monell C, Arnheim N, Bradley A, Ashley T, Liskay RM (1996) Involvement of mouse *Mlh1* in DNA mismatch repair and meiotic crossing over. Nat Genet 13:336-342

Baudat F, Manova K, Yuen JP, Jasin M, Keeney S (2000) Chromosome synapsis defects and sexually dimorphic meiotic progression in mice lacking *Spo11*. Mol Cell 6:989-998

Baudat F, Nicolas A (1997) Clustering of meiotic double-strand breaks on yeast chromosome III. Proc Natl Acad Sci USA 94:5213-5218

Becker E, Meyer V, Madaoui H, Guerois R (2006) Detection of a tandem BRCT in Nbs1 and Xrs2 with functional implications in the DNA damage response. Bioinformatics 22:1289-1292

Bell L, Byers B (1983a) Separation of branched from linear DNA by two-dimensional gel electrophoresis. Analyt Biochem 130:527-535

Bell LR, Byers B (1983b) Homologous association of chromosomal DNA during yeast meiosis. Cold Spring Harb Symp Quant Biol 47 Pt 2:829-840

Ben-Aroya S, Mieczkowski PA, Petes TD, Kupiec M (2004) The compact chromatin structure of a Ty repeated sequence suppresses recombination hotspot activity in *Saccharomyces cerevisiae*. Mol Cell 15:221-231

Bergerat A, de Massy B, Gadelle D, Varoutas PC, Nicolas A, Forterre P (1997) An atypical topoisomerase II from *Archaea* with implications for meiotic recombination. Nature 386:414-417

Bishop DK (1994) RecA homologs Dmc1 and Rad51 interact to form multiple nuclear complexes prior to meiotic chromosome synapsis. Cell 79:1081-1092

Bishop DK, Park D, Xu L, Kleckner N (1992) *DMC1*: a meiosis-specific yeast homolog of *E. coli* recA required for recombination, synaptonemal complex formation, and cell cycle progression. Cell 69:439-456

Bishop DK, Zickler D (2004) Early decision; meiotic crossover interference prior to stable strand exchange and synapsis. Cell 117:9-15

Blanton HL, Radford SJ, McMahan S, Kearney HM, Ibrahim JG, Sekelsky J (2005) REC, *Drosophila* MCM8, drives formation of meiotic crossovers. PLoS Genet 1:e40

Blat Y, Kleckner N (1999) Cohesins bind to preferential sites along yeast chromosome III, with differential regulation along arms versus the centric region. Cell 98:249-259

Blat Y, Protacio RU, Hunter N, Kleckner N (2002) Physical and functional interactions among basic chromosome organizational features govern early steps of meiotic chiasma formation. Cell 111:791-802

Blatch GL, Lassle M (1999) The tetratricopeptide repeat: a structural motif mediating protein-protein interactions. Bioessays 21:932-939

Boddy MN, Gaillard PH, McDonald WH, Shanahan P, Yates JR 3rd, Russell P (2001) Mus81-Eme1 are essential components of a Holliday junction resolvase. Cell 107:537-548

Borde V, Goldman AS, Lichten M (2000) Direct coupling between meiotic DNA replication and recombination initiation. Science 290:806-809

Borde V, Lin W, Novikov E, Petrini JH, Lichten M, Nicolas A (2004) Association of Mre11p with double-strand break sites during yeast meiosis. Mol Cell 13:389-401

Borde V, Wu TC, Lichten M (1999) Use of a recombination reporter insert to define meiotic recombination domains on chromosome III of *Saccharomyces cerevisiae*. Mol Cell Biol 19:4832-4842

Borner GV, Kleckner N, Hunter N (2004) Crossover/noncrossover differentiation, synaptonemal complex formation, and regulatory surveillance at the leptotene/zygotene transition of meiosis. Cell 117:29-45

Borts RH, Lichten M, Haber JE (1986) Analysis of meiosis-defective mutations in yeast by physical monitoring of recombination. Genetics 113:551-567

Bowring FJ, Yeadon PJ, Stainer RG, Catcheside DE (2006) Chromosome pairing and meiotic recombination in *Neurospora crassa spo11* mutants. Curr Genet 50:115-123

Bugreev DV, Golub EI, Stasiak AZ, Stasiak A, Mazin AV (2005) Activation of human meiosis-specific recombinase Dmc1 by Ca2+. J Biol Chem 280:26886-26895

Cao L, Alani E, Kleckner N (1990) A pathway for generation and processing of double-strand breaks during meiotic recombination in *S. cerevisiae*. Cell 61:1089-1101

Carlton PM, Farruggio AP, Dernburg AF (2006) A link between meiotic prophase progression and crossover control. PLoS Genet 2:e12

Cartagena-Lirola H, Guerini I, Viscardi V, Lucchini G, Longhese MP (2006) Budding Yeast Sae2 is an *in vivo* target of the Mec1 and Tel1 checkpoint kinases during meiosis. Cell Cycle 5:1549-1559

Catlett MG, Forsburg SL (2003) *Schizosaccharomyces pombe* Rdh54 (TID1) acts with Rhp54 (RAD54) to repair meiotic double-strand breaks. Mol Biol Cell 14:4707-4720

Celerin M, Merino ST, Stone JE, Menzie AM, Zolan ME (2000) Multiple roles of Spo11 in meiotic chromosome behavior. EMBO J 19:2739-2750

Chang YC, Lo YH, Lee MH, Leng CH, Hu SM, Chang CS, Wang TF (2005) Molecular visualization of the yeast Dmc1 protein ring and Dmc1-ssDNA nucleoprotein complex. Biochemistry 44:6052-6058

Chen L, Trujillo K, Ramos W, Sung P, Tomkinson AE (2001) Promotion of Dnl4-catalyzed DNA end-joining by the Rad50/Mre11/Xrs2 and Hdf1/Hdf2 complexes. Mol Cell 8:1105-1115

Chen L, Trujillo KM, Van Komen S, Roh DH, Krejci L, Lewis LK, Resnick MA, Sung P, Tomkinson AE (2005) Effect of amino acid substitutions in the Rad50 ATP binding domain on DNA double strand break repair in yeast. J Biol Chem 280:2620-2627

Chen YK, Leng CH, Olivares H, Lee MH, Chang YC, Kung WM, Ti SC, Lo YH, Wang AH, Chang CS, Bishop DK, Hsueh YP, Wang TF (2004) Heterodimeric complexes of Hop2 and Mnd1 function with Dmc1 to promote meiotic homolog juxtaposition and strand assimilation. Proc Natl Acad Sci USA 101:10572-10577

Cheng CH, Lo YH, Liang SS, Ti SC, Lin FM, Yeh CH, Huang HY, Wang TF (2006) SUMO modifications control assembly of synaptonemal complex and polycomplex in meiosis of *Saccharomyces cerevisiae*. Genes Dev 20:2067-2081

Cheng Z, Liu Y, Wang C, Parker R, Song H (2004) Crystal structure of Ski8p, a WD-repeat protein with dual roles in mRNA metabolism and meiotic recombination. Protein Sci 13:2673-2684

Chua PR, Roeder GS (1997) Tam1, a telomere-associated meiotic protein, functions in chromosome synapsis and crossover interference. Genes Dev 11:1786-1800

Chua PR, Roeder GS (1998) Zip2, a meiosis-specific protein required for the initiation of chromosome synapsis. Cell 93:349-359

Clerici M, Mantiero D, Lucchini G, Longhese MP (2005) The *Saccharomyces cerevisiae* Sae2 protein promotes resection and bridging of double strand break ends. J Biol Chem 280:38631-38638

Clerici M, Mantiero D, Lucchini G, Longhese MP (2006) The *Saccharomyces cerevisiae* Sae2 protein negatively regulates DNA damage checkpoint signalling. EMBO Rep 7:212-218

Clyne RK, Katis VL, Jessop L, Benjamin KR, Herskowitz I, Lichten M, Nasmyth K (2003) Polo-like kinase Cdc5 promotes chiasmata formation and cosegregation of sister centromeres at meiosis I. Nat Cell Biol 5:480-485

Corbett KD, Berger JM (2003a) Emerging roles for plant topoisomerase VI. Chem Biol 10:107-111

Corbett KD, Berger JM (2003b) Structure of the topoisomerase VI-B subunit: implications for type II topoisomerase mechanism and evolution. EMBO J 22:151-163

Cromie GA, Hyppa RW, Taylor AF, Zakharyevich K, Hunter N, Smith GR (2006) Single Holliday junctions are intermediates of meiotic recombination. Cell in press

Davis ES, Shafer BK, Strathern JN (2000) The *Saccharomyces cerevisiae RDN1* locus is sequestered from interchromosomal meiotic ectopic recombination in a *SIR2*-dependent manner. Genetics 155:1019-1032

de Boer E, Heyting C (2006) The diverse roles of transverse filaments of synaptonemal complexes in meiosis. Chromosoma 115:220-234

de Jager M, van Noort J, van Gent DC, Dekker C, Kanaar R, Wyman C (2001) Human Rad50/Mre11 is a flexible complex that can tether DNA ends. Mol Cell 8:1129-1135

de los Santos T, Hollingsworth NM (1999) Red1p, a *MEK1*-dependent phosphoprotein that physically interacts with Hop1p during meiosis in yeast. J Biol Chem 274:1783-1790

de los Santos T, Hunter N, Lee C, Larkin B, Loidl J, Hollingsworth NM (2003) The Mus81/Mms4 endonuclease acts independently of double-holliday junction resolution to promote a distinct subset of crossovers during meiosis in budding yeast. Genetics 164:81-94

de los Santos T, Loidl J, Larkin B, Hollingsworth NM (2001) A role for *MMS4* in the processing of recombination intermediates during meiosis in *Saccharomyces cerevisiae*. Genetics 159:1511-1525

De Massy B, Baudat F, Nicolas A (1994) Initiation of recombination in *Saccharomyces cerevisiae* haploid meiosis. Proc Natl Acad Sci USA 91:11929-11933

de Massy B, Rocco V, Nicolas A (1995) The nucleotide mapping of DNA double-strand breaks at the *CYS3* initiation site of meiotic recombination in *Saccharomyces cerevisiae*. EMBO J 14:4589-4598

Dernburg AF, McDonald K, Moulder G, Barstead R, Dresser M, Villeneuve AM (1998) Meiotic recombination in *C. elegans* initiates by a conserved mechanism and is dispensable for homologous chromosome synapsis. Cell 94:387-398

Diaz RL, Alcid AD, Berger JM, Keeney S (2002) Identification of residues in yeast Spo11p critical for meiotic DNA double-strand break formation. Mol Cell Biol 22:1106-1115

Difilippantonio S, Celeste A, Fernandez-Capetillo O, Chen HT, Reina San Martin B, Van Laethem F, Yang YP, Petukhova GV, Eckhaus M, Feigenbaum L, Manova K, Kruhlak M, Camerini-Otero RD, Sharan S, Nussenzweig M, Nussenzweig A (2005) Role of Nbs1 in the activation of the Atm kinase revealed in humanized mouse models. Nat Cell Biol 7:675-685

Domenichini S, Raynaud C, Ni DA, Henry Y, Bergounioux C (2006) *Atmndl-Δ1* is sensitive to gamma-irradiation and defective in meiotic DNA repair. DNA Repair (Amst) 5:455-464

Dong H, Roeder GS (2000) Organization of the yeast Zip1 protein within the central region of the synaptonemal complex. J Cell Biol 148:417-426

Dray E, Siaud N, Dubois E, Doutriaux MP (2006) Interaction between *Arabidopsis* Brca2 and its partners Rad51, Dmc1, and Dss1. Plant Physiol 140:1059-1069

Dresser ME, Ewing DJ, Conrad MN, Dominguez AM, Barstead R, Jiang H, Kodadek T (1997) *DMC1* functions in a *Saccharomyces cerevisiae* meiotic pathway that is largely independent of the *RAD51* pathway. Genetics 147:533-544

Dutta R, Inouye M (2000) GHKL, an emergent ATPase/kinase superfamily. Trends Biochem Sci 25:24-28

Edelmann W, Cohen PE, Kane M, Lau K, Morrow B, Bennett S, Umar A, Kunkel T, Cattoretti G, Chaganti R, Pollard JW, Kolodner RD, Kucherlapati R (1996) Meiotic pachytene arrest in MLH1-deficient mice. Cell 85:1125-1134

Ellermeier C, Schmidt H, Smith GR (2004) Swi5 acts in meiotic DNA joint molecule formation in *Schizosaccharomyces pombe*. Genetics 168:1891-1898

Engebrecht JA, Voelkel-Meiman K, Roeder GS (1991) Meiosis-specific RNA splicing in yeast. Cell 66:1257-1268

Enomoto R, Kinebuchi T, Sato M, Yagi H, Shibata T, Kurumizaka H, Yokoyama S (2004) Positive role of the mammalian TBPIP/HOP2 protein in DMC1-mediated homologous pairing. J Biol Chem 279:35263-35272

Evans DH, Li YF, Fox ME, Smith GR (1997) A WD repeat protein, Rec14, essential for meiotic recombination in *Schizosaccharomyces pombe*. Genetics 146:1253-1264

Fabre F, Boulet A, Roman H (1984) Gene conversion at different points in the mitotic cycle of *Saccharomyces cerevisiae*. Mol Gen Genet 195:139-143

Fan QQ, Petes TD (1996) Relationship between nuclease-hypersensitive sites and meiotic recombination hot spot activity at the *HIS4* locus of *Saccharomyces cerevisiae*. Mol Cell Biol 16:2037-2043

Flores-Rozas H, Kolodner RD (1998) The *Saccharomyces cerevisiae MLH3* gene functions in *MSH3*-dependent suppression of frameshift mutations. Proc Natl Acad Sci USA 95:12404-12409

Fox ME, Virgin JB, Metzger J, Smith GR (1997) Position- and orientation-independent activity of the *Schizosaccharomyces pombe* meiotic recombination hot spot *M26*. Proc Natl Acad Sci USA 94:7446-7451

Fox ME, Yamada T, Ohta K, Smith GR (2000) A family of cAMP-response-element-related DNA sequences with meiotic recombination hotspot activity in *Schizosaccharomyces pombe*. Genetics 156:59-68

Friedman DB, Hollingsworth NM, Byers B (1994) Insertional mutations in the yeast *HOP1* gene: evidence for multimeric assembly in meiosis. Genetics 136:449-464

Fung JC, Rockmill B, Odell M, Roeder GS (2004) Imposition of crossover interference through the nonrandom distribution of synapsis initiation complexes. Cell 116:795-802

Furuse M, Nagase Y, Tsubouchi H, Murakami-Murofushi K, Shibata T, Ohta K (1998) Distinct roles of two separable *in vitro* activities of yeast Mre11 in mitotic and meiotic recombination. EMBO J 17:6412-6425

Garcia-Diaz M, Bebenek K, Gao G, Pedersen LC, London RE, Kunkel TA (2005) Structure-function studies of DNA polymerase lambda. DNA Repair (Amst) 4:1358-1367

Garcia-Diaz M, Dominguez O, Lopez-Fernandez LA, de Lera LT, Saniger ML, Ruiz JF, Parraga M, Garcia-Ortiz MJ, Kirchhoff T, del Mazo J, Bernad A, Blanco L (2000) DNA polymerase lambda (Pol lambda), a novel eukaryotic DNA polymerase with a potential role in meiosis. J Mol Biol 301:851-867

Gardiner JM, Bullard SA, Chrome C, Malone RE (1997) Molecular and genetic analysis of *REC103*, an early meiotic recombination gene in yeast. Genetics 146:1265-1274

Gasior SL, Olivares H, Ear U, Hari DM, Weichselbaum R, Bishop DK (2001) Assembly of RecA-like recombinases: distinct roles for mediator proteins in mitosis and meiosis. Proc Natl Acad Sci USA 98:8411-8418

Gerton JL, DeRisi J, Shroff R, Lichten M, Brown PO, Petes TD (2000) Inaugural article: global mapping of meiotic recombination hotspots and coldspots in the yeast *Saccharomyces cerevisiae*. Proc Natl Acad Sci USA 97:11383-11390

Gerton JL, DeRisi JL (2002) Mnd1p: an evolutionarily conserved protein required for meiotic recombination. Proc Natl Acad Sci USA 99:6895-6900

Gorlov IP, Gorlova OY (2001) Cost-benefit analysis of recombination and its application for understanding of chiasma interference. J Theor Biol 213:1-8

Gottlieb S, Esposito RE (1989) A new role for a yeast transcriptional silencer gene, *SIR2*, in regulation of recombination in ribosomal DNA. Cell 56:771-776

Grell RF (1984) Time of recombination in the *Drosophila melanogaster* oocyte.

III. Selection and characterization of temperature-sensitive and -insensitive recombination-deficient alleles in *Drosophila*. Genetics 108:425-443

Grelon M, Vezon D, Gendrot G, Pelletier G (2001) AtSPO11-1 is necessary for efficient meiotic recombination in plants. EMBO J 20:589-600

Guillon H, Baudat F, Grey C, Liskay RM, de Massy B (2005) Crossover and noncrossover pathways in mouse meiosis. Mol Cell 20:563-573

Haber JE, Thorburn PC, Rogers D (1984) Meiotic and mitotic behavior of dicentric chromosomes in *Saccharomyces cerevisiae*. Genetics 106:185-205

Hayase A, Takagi M, Miyazaki T, Oshiumi H, Shinohara M, Shinohara A (2004) A protein complex containing Mei5 and Sae3 promotes the assembly of the meiosis-specific RecA homolog Dmc1. Cell 119:927-940

Henderson DS, Wiegand UK, Norman DG, Glover DM (2000) Mutual correction of faulty PCNA subunits in temperature-sensitive lethal *mus209* mutants of *Drosophila melanogaster*. Genetics 154:1721-1733

Henderson KA, Kee K, Maleki S, Santini PA, Keeney S (2006) Cyclin-dependent kinase directly regulates initiation of meiotic recombination. Cell 125:1321-1332

Henderson KA, Keeney S (2004) Tying synaptonemal complex initiation to the formation and programmed repair of DNA double-strand breaks. Proc Natl Acad Sci USA 101:4519-4524

Henderson KA, Keeney S (2005) Synaptonemal complex formation: where does it start? Bioessays 27:995-998

Henry JM, Camahort R, Rice DA, Florens L, Swanson SK, Washburn MP, Gerton JL (2006) Mnd1/Hop2 facilitates Dmc1-dependent interhomolog crossover formation in meiosis of budding yeast. Mol Cell Biol 26:2913-2923

Heyer WD (2004) Recombination: Holliday junction resolution and crossover formation. Curr Biol 14:R56-58

Heyer WD, Ehmsen KT, Solinger JA (2003) Holliday junctions in the eukaryotic nucleus: resolution in sight? Trends Biochem Sci 28:548-557

Heyer WD, Li X, Rolfsmeier M, Zhang XP (2006) Rad54: the Swiss Army knife of homologous recombination? Nucleic Acids Res 4115-4125

Higgins JD, Sanchez-Moran E, Armstrong SJ, Jones GH, Franklin FC (2005) The *Arabidopsis* synaptonemal complex protein ZYP1 is required for chromosome synapsis and normal fidelity of crossing over. Genes Dev 19:2488-2500

Hillers KJ (2004) Crossover interference. Curr Biol 14:R1036-1037

Hochwagen A, Tham WH, Brar GA, Amon A (2005) The FK506 binding protein Fpr3 counteracts protein phosphatase 1 to maintain meiotic recombination checkpoint activity. Cell 122:861-873

Hoffmann ER, Borts RH (2004) Meiotic recombination intermediates and mismatch repair proteins. Cytogenet Genome Res 107:232-248

Hoffmann ER, Borts RH (2005) Trans events associated with crossovers are revealed in the absence of mismatch repair genes in *Saccharomyces cerevisiae*. Genetics 169:1305-1310

Hoffmann ER, Eriksson E, Herbert BJ, Borts RH (2005) *MLH1* and *MSH2* promote the symmetry of double-strand break repair events at the *HIS4* hotspot in *Saccharomyces cerevisiae*. Genetics 169:1291-1303

Hollingsworth NM, Brill SJ (2004) The Mus81 solution to resolution: generating meiotic crossovers without Holliday junctions. Genes Dev 18:117-125

Hollingsworth NM, Johnson AD (1993) A conditional allele of the *Saccharomyces cerevisiae HOP1* gene is suppressed by overexpression of two other meiosis-specific genes: *RED1* and *REC104*. Genetics 133:785-797

Hollingsworth NM, Ponte L, Halsey C (1995) *MSH5*, a novel MutS homolog, facilitates meiotic reciprocal recombination between homologs in *Saccharomyces cerevisiae* but not mismatch repair. Genes Dev 9:1728-1739

Holzen TM, Shah PP, Olivares HA, Bishop DK (2006) Tid1/Rdh54 promotes dissociation of Dmc1 from nonrecombinogenic sites in meiotic chromatin. Genes Dev 20: 2593-604

Hong EL, Shinohara A, Bishop DK (2001) *Saccharomyces cerevisiae* Dmc1 protein promotes renaturation of single-strand DNA (ssDNA) and assimilation of ssDNA into homologous super-coiled duplex DNA. J Biol Chem 276:41906-41912

Hooker GW, Roeder GS (2006) A Role for SUMO in meiotic chromosome synapsis. Curr Biol 16:1238-1243

Hopfner KP, Craig L, Moncalian G, Zinkel RA, Usui T, Owen BA, Karcher A, Henderson B, Bodmer JL, McMurray CT, Carney JP, Petrini JH, Tainer JA (2002) The Rad50 zinc-hook is a structure joining Mre11 complexes in DNA recombination and repair. Nature 418:562-566

Hopfner KP, Karcher A, Craig L, Woo TT, Carney JP, Tainer JA (2001) Structural biochemistry and interaction architecture of the DNA double-strand break repair Mre11 nuclease and Rad50-ATPase. Cell 105:473-485

Hopfner KP, Karcher A, Shin DS, Craig L, Arthur LM, Carney JP, Tainer JA (2000) Structural biology of Rad50 ATPase: ATP-driven conformational control in DNA double-strand break repair and the ABC-ATPase superfamily. Cell 101:789-800

Hunter N (2004) Meiosis. The Encyclopedia of Biological Chemistry 2: 610-616

Hunter N, Borts RH (1997) Mlh1 is unique among mismatch repair proteins in its ability to promote crossing-over during meiosis. Genes Dev 11:1573-1582

Hunter N, Kleckner N (2001) The single-end invasion: an asymmetric intermediate at the double-strand break to double-holliday junction transition of meiotic recombination. Cell 106:59-70

Interthal H, Heyer WD (2000) *MUS81* encodes a novel helix-hairpin-helix protein involved in the response to UV- and methylation-induced DNA damage in *Saccharomyces cerevisiae*. Mol Gen Genet 263:812-827

Ira G, Malkova A, Liberi G, Foiani M, Haber JE (2003) Srs2 and Sgs1-Top3 suppress crossovers during double-strand break repair in yeast. Cell 115:401-411

Iyer RR, Pluciennik A, Burdett V, Modrich PL (2006) DNA mismatch repair: functions and mechanisms. Chem Rev 106:302-323

Jackson JA, Fink GR (1985) Meiotic recombination between duplicated genetic elements in *Saccharomyces cerevisiae*. Genetics 109:303-332

Jantsch V, Pasierbek P, Mueller MM, Schweizer D, Jantsch M, Loidl J (2004) Targeted gene knockout reveals a role in meiotic recombination for ZHP-3, a Zip3-related protein in *Caenorhabditis elegans*. Mol Cell Biol 24:7998-8006

Jaskelioff M, Van Komen S, Krebs JE, Sung P, Peterson CL (2003) Rad54p is a chromatin remodeling enzyme required for heteroduplex DNA joint formation with chromatin. J Biol Chem 278:9212-9218

Jessop L, Allers T, Lichten M (2005) Infrequent co-conversion of markers flanking a meiotic recombination initiation site in *Saccharomyces cerevisiae*. Genetics 169:1353-1367

Jiao K, Salem L, Malone R (2003) Support for a meiotic recombination initiation complex: interactions among Rec102p, Rec104p, and Spo11p. Mol Cell Biol 23:5928-5938

Johnson ES (2004) Protein modification by SUMO. Annu Rev Biochem 73:355-382

Jolivet S, Vezon D, Froger N, Mercier R (2006) Non conservation of the meiotic function of the Ski8/Rec103 homolog in *Arabidopsis*. Genes Cells 11:615-622

Jones GH (1984) The control of chiasma distribution. Symp Soc Exp Biol 38:293-320

Jones GH, Franklin FC (2006) Meiotic crossing-over: obligation and interference. Cell 126:246-248

Kaback DB (1996) Chromosome-size dependent control of meiotic recombination in humans. Nat Genet 13:20-21

Kaback DB, Barber D, Mahon J, Lamb J, You J (1999) Chromosome size-dependent control of meiotic reciprocal recombination in *Saccharomyces cerevisiae*: the role of crossover interference. Genetics 152:1475-1486

Kaback DB, Guacci V, Barber D, Mahon JW (1992) Chromosome size-dependent control of meiotic recombination. Science 256:228-232

Kaback DB, Steensma HY, de Jonge P (1989) Enhanced meiotic recombination on the smallest chromosome of *Saccharomyces cerevisiae*. Proc Natl Acad Sci USA 86:3694-3698

Kadyk LC, Hartwell LH (1992) Sister chromatids are preferred over homologs as substrates for recombinational repair in *Saccharomyces cerevisiae*. Genetics 132:387-402

Kadyrov FA, Dzantiev L, Constantin N, Modrich P (2006) Endonucleolytic function of MutLalpha in human mismatch repair. Cell 126:297-308

Kateneva AV, Dresser ME (2006) Sister chromatid cohesion remodeling and meiotic recombination. Cell Cycle 5:467-471

Kateneva AV, Konovchenko AA, Guacci V, Dresser ME (2005) Recombination protein Tid1p controls resolution of cohesin-dependent linkages in meiosis in *Saccharomyces cerevisiae*. J Cell Biol 171:241-253

Kauppi L, Jeffreys AJ, Keeney S (2004) Where the crossovers are: recombination distributions in mammals. Nat Rev Genet 5:413-424

Kaye JA, Melo JA, Cheung SK, Vaze MB, Haber JE, Toczyski DP (2004) DNA breaks promote genomic instability by impeding proper chromosome segregation. Curr Biol 14:2096-2106

Kee K, Keeney S (2002) Functional interactions between *SPO11* and *REC102* during initiation of meiotic recombination in *Saccharomyces cerevisiae*. Genetics 160:111-122

Kee K, Protacio RU, Arora C, Keeney S (2004) Spatial organization and dynamics of the association of Rec102 and Rec104 with meiotic chromosomes. EMBO J 23:1815-1824

Keeney S (2001) Mechanism and control of meiotic recombination initiation. Curr Top Dev Biol 52:1-53

Keeney S, Giroux CN, Kleckner N (1997) Meiosis-specific DNA double-strand breaks are catalyzed by Spo11, a member of a widely conserved protein family. Cell 88:375-384

Keeney S, Kleckner N (1995) Covalent protein-DNA complexes at the 5' strand termini of meiosis-specific double-strand breaks in yeast. Proc Natl Acad Sci USA 92:11274-11278

Keeney S, Kleckner N (1996) Communication between homologous chromosomes: genetic alterations at a nuclease-hypersensitive site can alter mitotic chromatin structure at that site both *in cis* and *in trans*. Genes Cells 1:475-489

Kerzendorfer C, Vignard J, Pedrosa-Harand A, Siwiec T, Akimcheva S, Jolivet S, Sablowski R, Armstrong S, Schweizer D, Mercier R, Schlogelhofer P (2006) The *Arabidopsis* thaliana *MND1* homologue plays a key role in meiotic homologous pairing, synapsis and recombination. J Cell Sci 119:2486-2496

Khazanehdari KA, Borts RH (2000) *EXO1* and *MSH4* differentially affect crossing-over and segregation. Chromosoma 109:94-102

Kinebuchi T, Kagawa W, Enomoto R, Tanaka K, Miyagawa K, Shibata T, Kurumizaka H, Yokoyama S (2004) Structural basis for octameric ring formation and DNA interaction of the human homologous-pairing protein Dmc1. Mol Cell 14:363-374

Kirkpatrick DT, Fan Q, Petes TD (1999a) Maximal stimulation of meiotic recombination by a yeast transcription factor requires the transcription activation domain and a DNA-binding domain. Genetics 152:101-115

Kirkpatrick DT, Ferguson JR, Petes TD, Symington LS (2000) Decreased meiotic intergenic recombination and increased meiosis I nondisjunction in *exo1* mutants of *Saccharomyces cerevisiae*. Genetics 156:1549-1557

Kirkpatrick DT, Wang YH, Dominska M, Griffith JD, Petes TD (1999b) Control of meiotic recombination and gene expression in yeast by a simple repetitive DNA sequence that excludes nucleosomes. Mol Cell Biol 19:7661-7671

Kironmai KM, Muniyappa K, Friedman DB, Hollingsworth NM, Byers B (1998) DNA-binding activities of Hop1 protein, a synaptonemal complex component from *Saccharomyces cerevisiae*. Mol Cell Biol 18:1424-1435

Kleckner N, Zickler D, Jones GH, Dekker J, Padmore R, Henle J, Hutchinson J (2004) A mechanical basis for chromosome function. Proc Natl Acad Sci USA 101:12592-12597

Klein F, Mahr P, Galova M, Buonomo SB, Michaelis C, Nairz K, Nasmyth K (1999) A central role for cohesins in sister chromatid cohesion, formation of axial elements, and recombination during yeast meiosis. Cell 98:91-103

Klein HL (1997) *RDH54*, a *RAD54* homologue in *Saccharomyces cerevisiae*, is required for mitotic diploid-specific recombination and repair and for meiosis. Genetics 147:1533-1543

Klein S, Zenvirth D, Dror V, Barton AB, Kaback DB, Simchen G (1996) Patterns of meiotic double-strand breakage on native and artificial yeast chromosomes. Chromosoma 105:276-284

Klieger Y, Yizhar O, Zenvirth D, Shtepel-Milman N, Snoek M, Simchen G (2005) Involvement of Sir2/4 in silencing of DNA breakage and recombination on mouse YACs during yeast meiosis. Mol Biol Cell 16:1449-1455

Kneitz B, Cohen PE, Avdievich E, Zhu L, Kane MF, Hou H Jr, Kolodner RD, Kucherlapati R, Pollard JW, Edelmann W (2000) MutS homolog 4 localization to meiotic chromosomes is required for chromosome pairing during meiosis in male and female mice. Genes Dev 14:1085-1097

Kobayashi J, Antoccia A, Tauchi H, Matsuura S, Komatsu K (2004) NBS1 and its functional role in the DNA damage response. DNA Repair (Amst) 3:855-861

Kobayashi Y, Watanabe M, Okada Y, Sawa H, Takai H, Nakanishi M, Kawase Y, Suzuki H, Nagashima K, Ikeda K, Motoyama N (2002) Hydrocephalus, situs inversus, chronic sinusitis, and male infertility in DNA polymerase lambda-deficient mice: possible implication for the pathogenesis of immotile cilia syndrome. Mol Cell Biol 22:2769-2776

Kovalenko OV, Plug AW, Haaf T, Gonda DK, Ashley T, Ward DC, Radding CM, Golub EI (1996) Mammalian ubiquitin-conjugating enzyme Ubc9 interacts with Rad51 recombination protein and localizes in synaptonemal complexes. Proc Natl Acad Sci USA 93:2958-2963

Krejci L, Van Komen S, Li Y, Villemain J, Reddy MS, Klein H, Ellenberger T, Sung P (2003) DNA helicase Srs2 disrupts the Rad51 presynaptic filament. Nature 423:305-309

Krogh BO, Symington LS (2004) Recombination proteins in yeast. Annu Rev Genet 38:233-271

Lee BI, Wilson DM 3rd (1999) The RAD2 domain of human exonuclease 1 exhibits 5' to 3' exonuclease and flap structure-specific endonuclease activities. J Biol Chem 274:37763-37769

Lee MH, Chang YC, Hong EL, Grubb J, Chang CS, Bishop DK, Wang TF (2005) Calcium ion promotes yeast Dmc1 activity via formation of long and fine helical filaments with single-stranded DNA. J Biol Chem 280:40980-40984

Leem SH, Ogawa H (1992) The *MRE4* gene encodes a novel protein kinase homologue required for meiotic recombination in *Saccharomyces cerevisiae*. Nucleic Acids Res 20:449-457

Leem SH, Ropp PA, Sugino A (1994) The yeast *Saccharomyces cerevisiae* DNA polymerase IV: possible involvement in double strand break DNA repair. Nucleic Acids Res 22:3011-3017

Leu JY, Chua PR, Roeder GS (1998) The meiosis-specific Hop2 protein of *S. cerevisiae* ensures synapsis between homologous chromosomes. Cell 94:375-386

Li J, Hooker GW, Roeder GS (2006) *Saccharomyces cerevisiae* Mer2, Mei4 and Rec114 form a complex required for meiotic double-strand break formation. Genetics 173:1969-1981

Lichten M, Goldman AS (1995) Meiotic recombination hotspots. Annu Rev Genet 29:423-444

Lipkin SM, Moens PB, Wang V, Lenzi M, Shanmugarajah D, Gilgeous A, Thomas J, Cheng J, Touchman JW, Green ED, Schwartzberg P, Collins FS, Cohen PE (2002) Meiotic arrest and aneuploidy in MLH3-deficient mice. Nat Genet 31:385-390

Lisby M, Rothstein R (2005) Localization of checkpoint and repair proteins in eukaryotes. Biochimie 87:579-589

Lobachev K, Vitriol E, Stemple J, Resnick MA, Bloom K (2004) Chromosome fragmentation after induction of a double-strand break is an active process prevented by the RMX repair complex. Curr Biol 14:2107-2112

Lui DY, Peoples-Holst TL, Mell JC, Wu HY, Dean EW, Burgess SM (2006) Analysis of close stable homolog juxtaposition during meiosis in mutants of *Saccharomyces cerevisiae*. Genetics 173:1207-1222

Madrona AY, Wilson DK (2004) The structure of Ski8p, a protein regulating mRNA degradation: Implications for WD protein structure. Protein Sci 13:1557-1565

Maiorano D, Cuvier O, Danis E, Mechali M (2005) MCM8 is an MCM2-7-related protein that functions as a DNA helicase during replication elongation and not initiation. Cell 120:315-328

Maiorano D, Lutzmann M, Mechali M (2006) MCM proteins and DNA replication. Curr Opin Cell Biol 18:130-136

Maleki S, Keeney S (2004) Modifying histones and initiating meiotic recombination; new answers to an old question. Cell 118:404-406

Malkova A, Swanson J, German M, McCusker JH, Housworth EA, Stahl FW, Haber JE (2004) Gene conversion and crossing over along the 405-kb left arm of *Saccharomyces cerevisiae* chromosome VII. Genetics 168:49-63

Maloisel L, Bhargava J, Roeder GS (2004) A role for DNA polymerase delta in gene conversion and crossing over during meiosis in *Saccharomyces cerevisiae*. Genetics 167:1133-1142

Mao-Draayer Y, Galbraith AM, Pittman DL, Cool M, Malone RE (1996) Analysis of meiotic recombination pathways in the yeast *Saccharomyces cerevisiae*. Genetics 144:71-86

Marcon E, Moens P (2003) MLH1p and MLH3p localize to precociously induced chiasmata of okadaic-acid-treated mouse spermatocytes. Genetics 165:2283-2287

Martin V, Chahwan C, Gao H, Blais V, Wohlschlegel J, Yates JR 3rd, McGowan CH, Russell P (2006) Sws1 is a conserved regulator of homologous recombination in eukaryotic cells. EMBO J 25:2564-2574

Martini E, Diaz RL, Hunter N, Keeney S (2006) Crossover homeostasis in yeast meiosis. Cell 126: 285-95

Masai H, Arai K (2002) Cdc7 kinase complex: a key regulator in the initiation of DNA replication. J Cell Physiol 190:287-296

Mather K (1933) The relations between chiasmata and crossing-over in diploid and triploid *Drosophila melanogaster*. J Genet 27:243-259

Matsubayashi H, Yamamoto MT (2003) REC, a new member of the MCM-related protein family, is required for meiotic recombination in *Drosophila*. Genes Genet Syst 78:363-371

Mazina OM, Mazin AV, Nakagawa T, Kolodner RD, Kowalczykowski SC (2004) *Saccharomyces cerevisiae* Mer3 helicase stimulates 3'-5' heteroduplex extension by Rad51; implications for crossover control in meiotic recombination. Cell 117:47-56

McKee AH, Kleckner N (1997a) A general method for identifying recessive diploid-specific mutations in *Saccharomyces cerevisiae*, its application to the isolation of mutants blocked at intermediate stages of meiotic prophase and characterization of a new gene *SAE2*. Genetics 146:797-816

McKee AH, Kleckner N (1997b) Mutations in *Saccharomyces cerevisiae* that block meiotic prophase chromosome metabolism and confer cell cycle arrest at pachytene identify two new meiosis-specific genes *SAE1* and *SAE3*. Genetics 146:817-834

McKim KS, Hayashi-Hagihara A (1998) *mei-W68* in *Drosophila melanogaster* encodes a Spo11 homolog: evidence that the mechanism for initiating meiotic recombination is conserved. Genes Dev 12:2932-2942

Merker JD, Dominska M, Petes TD (2003) Patterns of heteroduplex formation associated with the initiation of meiotic recombination in the yeast *Saccharomyces cerevisiae*. Genetics 165:47-63

Mieczkowski PA, Dominska M, Buck MJ, Gerton JL, Lieb JD, Petes TD (2006) Global analysis of the relationship between the binding of the Bas1p transcription factor and meiosis-specific double-strand DNA breaks in *Saccharomyces cerevisiae*. Mol Cell Biol 26:1014-1027

Mizuno K, Emura Y, Baur M, Kohli J, Ohta K, Shibata T (1997) The meiotic recombination hot spot created by the single-base substitution *ade6-M26* results in remodeling of chromatin structure in fission yeast. Genes Dev 11:876-886

Mizuno K, Hasemi T, Ubukata T, Yamada T, Lehmann E, Kohli J, Watanabe Y, Iino Y, Yamamoto M, Fox ME, Smith GR, Murofushi H, Shibata T, Ohta K (2001) Counteracting regulation of chromatin remodeling at a fission yeast cAMP response element-related recombination hotspot by stress-activated protein kinase, cAMP-dependent kinase and meiosis regulators. Genetics 159:1467-1478

Moens PB, Pearlman RE (1988) Chromatin organization at meiosis. Bioessays 9:151-153

Moncalian G, Lengsfeld B, Bhaskara V, Hopfner KP, Karcher A, Alden E, Tainer JA, Paull TT (2004) The Rad50 signature motif: essential to ATP binding and biological function. J Mol Biol 335:937-951

Moreau S, Ferguson JR, Symington LS (1999) The nuclease activity of Mre11 is required for meiosis but not for mating type switching, end joining, or telomere maintenance. Mol Cell Biol 19:556-566

Moreno-Herrero F, de Jager M, Dekker NH, Kanaar R, Wyman C, Dekker C (2005) Mesoscale conformational changes in the DNA-repair complex Rad50/Mre11/Nbs1 upon binding DNA. Nature 437:440-443

Muller HJ (1916) The mechanism of crossing over. Am Nat 50:193-221

Muniyappa K, Anuradha S, Byers B (2000) Yeast meiosis-specific protein Hop1 binds to G4 DNA and promotes its formation. Mol Cell Biol 20:1361-1369

Murakami H, Borde V, Shibata T, Lichten M, Ohta K (2003) Correlation between premeiotic DNA replication and chromatin transition at yeast recombination initiation sites. Nucleic Acids Res 31:4085-4090

Nabeshima K, Kakihara Y, Hiraoka Y, Nojima H (2001) A novel meiosis-specific protein of fission yeast, Meu13p, promotes homologous pairing independently of homologous recombination. EMBO J 20:3871-3881

Nabeshima K, Villeneuve AM, Hillers KJ (2004) Chromosome-wide regulation of meiotic crossover formation in *Caenorhabditis elegans* requires properly assembled chromosome axes. Genetics 168:1275-1292

Nag DK, Petes TD (1993) Physical detection of heteroduplexes during meiotic recombination in the yeast *Saccharomyces cerevisiae*. Mol Cell Biol 13:2324-2331

Nairz K, Klein F (1997) *mre11S*--a yeast mutation that blocks double-strand-break processing and permits nonhomologous synapsis in meiosis. Genes Dev 11:2272-2290

Nakada D, Matsumoto K, Sugimoto K (2003) ATM-related Tel1 associates with double-strand breaks through an Xrs2-dependent mechanism. Genes Dev 17:1957-1962

Nakagawa T, Kolodner RD (2002) *Saccharomyces cerevisiae* Mer3 is a DNA helicase involved in meiotic crossing over. Mol Cell Biol 22:3281-3291

Nakagawa T, Ogawa H (1997) Involvement of the *MRE2* gene of yeast in formation of meiosis-specific double-strand breaks and crossover recombination through RNA splicing. Genes Cells 2:65-79

Nakagawa T, Ogawa H (1999) The *Saccharomyces cerevisiae MER3* gene, encoding a novel helicase-like protein, is required for crossover control in meiosis. EMBO J 18:5714-5723

Neale MJ, Pan J, Keeney S (2005) Endonucleolytic processing of covalent protein-linked DNA double-strand breaks. Nature 436:1053-1057

Nichols MD, DeAngelis K, Keck JL, Berger JM (1999) Structure and function of an archaeal topoisomerase VI subunit with homology to the meiotic recombination factor Spo11. EMBO J 18:6177-6188

Nilsson NO, Sall T (1995) A model of chiasma reduction of closely formed crossovers. J Theor Biol 173:93-98

Nishant KT, Rao MR (2006) Molecular features of meiotic recombination hot spots. Bioessays 28:45-56

Niu H, Wan L, Baumgartner B, Schaefer D, Loidl J, Hollingsworth NM (2005) Partner choice during meiosis is regulated by Hop1-promoted dimerization of Mek1. Mol Biol Cell 16:5804-5818

Novak JE, Ross-Macdonald PB, Roeder GS (2001) The budding yeast Msh4 protein functions in chromosome synapsis and the regulation of crossover distribution. Genetics 158:1013-1025

Ogino K, Hirota K, Matsumoto S, Takeda T, Ohta K, Arai K, Masai H (2006) Hsk1 kinase is required for induction of meiotic dsDNA breaks without involving checkpoint kinases in fission yeast. Proc Natl Acad Sci USA 103:8131-8136

Ohta K, Nicolas A, Furuse M, Nabetani A, Ogawa H, Shibata T (1998) Mutations in the *MRE11, RAD50, XRS2*, and *MRE2* genes alter chromatin configuration at meiotic DNA double-stranded break sites in premeiotic and meiotic cells. Proc Natl Acad Sci USA 95:646-651

Ohta K, Shibata T, Nicolas A (1994) Changes in chromatin structure at recombination initiation sites during yeast meiosis. EMBO J 13:5754-5763

Okada T, Keeney S (2005) Homologous recombination: needing to have my say. Curr Biol 15:R200-202

Osman F, Dixon J, Doe CL, Whitby MC (2003) Generating crossovers by resolution of nicked Holliday junctions: a role for Mus81-Eme1 in meiosis. Mol Cell 12:761-774

Panoli AP, Ravi M, Sebastian J, Nishal B, Reddy TV, Marimuthu MP, Subbiah V, Vijaybhaskar V, Siddiqi I (2006) *AtMND1* is required for homologous pairing during meiosis in *Arabidopsis*. BMC Mol Biol 7:24

Paques F, Haber JE (1999) Multiple pathways of recombination induced by double-strand breaks in *Saccharomyces cerevisiae*. Microbiol Mol Biol Rev 63:349-404

Passy SI, Yu X, Li Z, Radding CM, Masson JY, West SC, Egelman EH (1999) Human Dmc1 protein binds DNA as an octameric ring. Proc Natl Acad Sci USA 96:10684-10688

Paull TT, Gellert M (1999) Nbs1 potentiates ATP-driven DNA unwinding and endonuclease cleavage by the Mre11/Rad50 complex. Genes Dev 13:1276-1288

Pecina A, Smith KN, Mezard C, Murakami H, Ohta K, Nicolas A (2002) Targeted stimulation of meiotic recombination. Cell 111:173-184

Pellicioli A, Foiani M (2005) Signal transduction: how Rad53 kinase is activated. Curr Biol 15:R769-771

Peoples TL, Dean E, Gonzalez O, Lambourne L, Burgess SM (2002) Close, stable homolog juxtaposition during meiosis in budding yeast is dependent on meiotic recombination, occurs independently of synapsis, and is distinct from DSB-independent pairing contacts. Genes Dev 16:1682-1695

Perry J, Kleckner N, Borner GV (2005) Bioinformatic analyses implicate the collaborating meiotic crossover/chiasma proteins Zip2, Zip3, and Spo22/Zip4 in ubiquitin labeling. Proc Natl Acad Sci USA 102:17594-17599

Petes TD (2001) Meiotic recombination hot spots and cold spots. Nat Rev Genet 2:360-369

Petes TD, Merker JD (2002) Context dependence of meiotic recombination hotspots in yeast: the relationship between recombination activity of a reporter construct and base composition. Genetics 162:2049-2052

Petrini JH (2005) At the end, remodeling leads to eviction. Nat Struct Mol Biol 12:1028-1029

Petronczki M, Siomos MF, Nasmyth K (2003) Un Ménage à Quatre: the molecular biology of chromosome segregation in meiosis. Cell 112:423-440

Petukhova G, Sung P, Klein H (2000) Promotion of Rad51-dependent D-loop formation by yeast recombination factor Rdh54/Tid1. Genes Dev 14:2206-2215

Petukhova G, Van Komen S, Vergano S, Klein H, Sung P (1999) Yeast Rad54 promotes Rad51-dependent homologous DNA pairing via ATP hydrolysis-driven change in DNA double helix conformation. J Biol Chem 274:29453-29462

Petukhova GV, Pezza RJ, Vanevski F, Ploquin M, Masson JY, Camerini-Otero RD (2005) The Hop2 and Mnd1 proteins act in concert with Rad51 and Dmc1 in meiotic recombination. Nat Struct Mol Biol 12:449-453

Petukhova GV, Romanienko PJ, Camerini-Otero RD (2003) The Hop2 protein has a direct role in promoting interhomolog interactions during mouse meiosis. Dev Cell 5:927-936

Pezza RJ, Petukhova GV, Ghirlando R, Camerini-Otero RD (2006) Molecular activities of meiosis-specific proteins Hop2, Mnd1, and the Hop2-Mnd1 complex. J Biol Chem 281:18426-18434

Plug AW, Clairmont CA, Sapi E, Ashley T, Sweasy JB (1997) Evidence for a role for DNA polymerase beta in mammalian meiosis. Proc Natl Acad Sci USA 94:1327-1331

Pochart P, Woltering D, Hollingsworth NM (1997) Conserved properties between functionally distinct MutS homologs in yeast. J Biol Chem 272:30345-30349

Prieler S, Penkner A, Borde V, Klein F (2005) The control of Spo11's interaction with meiotic recombination hotspots. Genes Dev 19:255-269

Prinz S, Amon A, Klein F (1997) Isolation of *COM1*, a new gene required to complete meiotic double-strand break-induced recombination in *Saccharomyces cerevisiae*. Genetics 146:781-795

Rattray AJ, McGill CB, Shafer BK, Strathern JN (2001) Fidelity of mitotic double-strand-break repair in *Saccharomyces cerevisiae*: a role for *SAE2/COM1*. Genetics 158:109-122

Reddy KC, Villeneuve AM (2004) *C. elegans* HIM-17 links chromatin modification and competence for initiation of meiotic recombination. Cell 118:439-452

Rockmill B, Roeder GS (1990) Meiosis in asynaptic yeast. Genetics 126:563-574

Rockmill B, Roeder GS (1991) A meiosis-specific protein kinase homolog required for chromosome synapsis and recombination. Genes Dev 5:2392-2404

Rockmill B, Sym M, Scherthan H, Roeder GS (1995) Roles for two RecA homologs in promoting meiotic chromosome synapsis. Genes Dev 9:2684-2695

Romanienko PJ, Camerini-Otero RD (2000) The mouse Spo11 gene is required for meiotic chromosome synapsis. Mol Cell 6:975-987

Ross-Macdonald P, Roeder GS (1994) Mutation of a meiosis-specific MutS homolog decreases crossing over but not mismatch correction. Cell 79:1069-1080

Saito TT, Tougan T, Kasama T, Okuzaki D, Nojima H (2004) Mcp7, a meiosis-specific coiled-coil protein of fission yeast, associates with Meu13 and is required for meiotic recombination. Nucleic Acids Res 32:3325-3339

Salem L, Walter N, Malone R (1999) Suppressor analysis of the *Saccharomyces cerevisiae* gene *REC104* reveals a genetic interaction with *REC102*. Genetics 151:1261-1272

San-Segundo PA, Roeder GS (1999) Pch2 links chromatin silencing to meiotic checkpoint control. Cell 97:313-324

Santucci-Darmanin S, Neyton S, Lespinasse F, Saunieres A, Gaudray P, Paquis-Flucklinger V (2002) The DNA mismatch-repair MLH3 protein interacts with MSH4 in meiotic cells, supporting a role for this MutL homolog in mammalian meiotic recombination. Hum Mol Genet 11:1697-1706

Santucci-Darmanin S, Walpita D, Lespinasse F, Desnuelle C, Ashley T, Paquis-Flucklinger V (2000) MSH4 acts in conjunction with MLH1 during mammalian meiosis. FASEB J 14:1539-1547

Schild D, Byers B (1978) Meiotic effects of DNA-defective cell division cycle mutations of *Saccharomyces cerevisiae*. Chromosoma 70:109-130

Schmekel K (2000) Methods for immunoelectron microscopic and fine structural analysis of synaptonemal complexes and nodules in yeast. Chromosoma 109:110-116

Schmekel K, Meuwissen RL, Dietrich AJ, Vink AC, van Marle J, van Veen H, Heyting C (1996) Organization of SCP1 protein molecules within synaptonemal complexes of the rat. Exp Cell Res 226:20-30

Schmuckli-Maurer J, Heyer WD (2000) Meiotic recombination in RAD54 mutants of *Saccharomyces cerevisiae*. Chromosoma 109:86-93

Schommer C, Beven A, Lawrenson T, Shaw P, Sablowski R (2003) *AHP2* is required for bivalent formation and for segregation of homologous chromosomes in *Arabidopsis* meiosis. Plant J 36:1-11

Schwacha A, Kleckner N (1994) Identification of joint molecules that form frequently between homologs but rarely between sister chromatids during yeast meiosis. Cell 76:51-63

Schwacha A, Kleckner N (1995) Identification of double Holliday junctions as intermediates in meiotic recombination. Cell 83:783-791

Schwacha A, Kleckner N (1997) Interhomolog bias during meiotic recombination: meiotic functions promote a highly differentiated interhomolog-only pathway. Cell 90:1123-1135

Sehorn MG, Sigurdsson S, Bussen W, Unger VM, Sung P (2004) Human meiotic recombinase Dmc1 promotes ATP-dependent homologous DNA strand exchange. Nature 429:433-437

Sheridan S, Bishop DK (2006) Red-Hed regulation: recombinase Rad51, though capable of playing the leading role, may be relegated to supporting Dmc1 in budding yeast meiosis. Genes Dev 20:1685-1691

Shima H, Suzuki M, Shinohara M (2005) Isolation and characterization of novel *xrs2* mutations in *Saccharomyces cerevisiae*. Genetics 170:71-85

Shinohara A, Gasior S, Ogawa T, Kleckner N, Bishop DK (1997a) *Saccharomyces cerevisiae* recA homologues *RAD51* and *DMC1* have both distinct and overlapping roles in meiotic recombination. Genes Cells 2:615-629

Shinohara A, Shinohara M (2004) Roles of RecA homologues Rad51 and Dmc1 during meiotic recombination. Cytogenet Genome Res 107:201-207

Shinohara M, Gasior SL, Bishop DK, Shinohara A (2000) Tid1/Rdh54 promotes colocalization of rad51 and dmc1 during meiotic recombination. Proc Natl Acad Sci USA 97:10814-10819

Shinohara M, Sakai K, Ogawa T, Shinohara A (2003a) The mitotic DNA damage checkpoint proteins Rad17 and Rad24 are required for repair of double-strand breaks during meiosis in yeast. Genetics 164:855-865

Shinohara M, Sakai K, Shinohara A, Bishop DK (2003b) Crossover interference in *Saccharomyces cerevisiae* requires a *TID1/RDH54*- and *DMC1*-dependent pathway. Genetics 163:1273-1286

Shinohara M, Shita-Yamaguchi E, Buerstedde JM, Shinagawa H, Ogawa H, Shinohara A (1997b) Characterization of the roles of the *Saccharomyces cerevisiae RAD54* gene and a homologue of *RAD54, RDH54/TID1*, in mitosis and meiosis. Genetics 147:1545-1556

Shor E, Weinstein J, Rothstein R (2005) A genetic screen for *top3* suppressors in *Saccharomyces cerevisiae* identifies *SHU1, SHU2, PSY3* and *CSM2*: four genes involved in error-free DNA repair. Genetics 169:1275-1289

Siaud N, Dray E, Gy I, Gerard E, Takvorian N, Doutriaux MP (2004) Brca2 is involved in meiosis in *Arabidopsis thaliana* as suggested by its interaction with Dmc1. EMBO J 23:1392-1401

Sjogren C, Nasmyth K (2001) Sister chromatid cohesion is required for postreplicative double-strand break repair in *Saccharomyces cerevisiae*. Curr Biol 11:991-995

Smith AV, Roeder GS (1997) The yeast Red1 protein localizes to the cores of meiotic chromosomes. J Cell Biol 136:957-967

Smith GR, Boddy MN, Shanahan P, Russell P (2003) Fission yeast Mus81.Eme1 Holliday junction resolvase is required for meiotic crossing over but not for gene conversion. Genetics 165:2289-2293

Snowden T, Acharya S, Butz C, Berardini M, Fishel R (2004) hMSH4-hMSH5 recognizes Holliday junctions and forms a meiosis-specific sliding clamp that embraces homologous chromosomes. Mol Cell 15:437-451

Solinger JA, Kiianitsa K, Heyer WD (2002) Rad54, a Swi2/Snf2-like recombinational repair protein, disassembles Rad51:dsDNA filaments. Mol Cell 10:1175-1188

Sollier J, Lin W, Soustelle C, Suhre K, Nicolas A, Geli V, de La Roche Saint-Andre C (2004) Set1 is required for meiotic S-phase onset, double-strand break formation and middle gene expression. EMBO J 23:1957-1967

Stahl FW, Foss HM, Young LS, Borts RH, Abdullah MF, Copenhaver GP (2004) Does crossover interference count in *Saccharomyces cerevisiae*? Genetics 168:35-48

Storlazzi A, Tesse S, Gargano S, James F, Kleckner N, Zickler D (2003) Meiotic double-strand breaks at the interface of chromosome movement, chromosome remodeling, and reductional division. Genes Dev 17:2675-2687

Storlazzi A, Xu L, Cao L, Kleckner N (1995) Crossover and noncrossover recombination during meiosis: timing and pathway relationships. Proc Natl Acad Sci USA 92:8512-8516

Storlazzi A, Xu L, Schwacha A, Kleckner N (1996) Synaptonemal complex (SC) component Zip1 plays a role in meiotic recombination independent of SC polymerization along the chromosomes. Proc Natl Acad Sci USA 93:9043-9048

Story RM, Bishop DK, Kleckner N, Steitz TA (1993) Structural relationship of bacterial RecA proteins to recombination proteins from bacteriophage T4 and yeast. Science 259:1892-1896

Stracker TH, Theunissen JW, Morales M, Petrini JH (2004) The Mre11 complex and the metabolism of chromosome breaks: the importance of communicating and holding things together. DNA Repair (Amst) 3:845-854

Strom L, Lindroos HB, Shirahige K, Sjogren C (2004) Postreplicative recruitment of cohesin to double-strand breaks is required for DNA repair. Mol Cell 16:1003-1015

Strom L, Sjogren C (2005) DNA damage-induced cohesion. Cell Cycle 4:536-539

Sun H, Treco D, Schultes NP, Szostak JW (1989) Double-strand breaks at an initiation site for meiotic gene conversion. Nature 338:87-90

Sun H, Treco D, Szostak JW (1991) Extensive 3'-overhanging, single-stranded DNA associated with the meiosis-specific double-strand breaks at the *ARG4* recombination initiation site. Cell 64:1155-1161

Sym M, Engebrecht JA, Roeder GS (1993) Zip1 is a synaptonemal complex protein required for meiotic chromosome synapsis. Cell 72:365-378

Sym M, Roeder GS (1994) Crossover interference is abolished in the absence of a synaptonemal complex protein. Cell 79:283-292

Symington LS, Heyer WD (2006) Some disassembly required: role of DNA translocases in the disruption of recombination intermediates and dead-end complexes. Genes Dev 20:2479-2486

Szostak JW, Orr-Weaver TL, Rothstein RJ, Stahl FW (1983) The double-strand-break repair model for recombination. Cell 33:25-35

Tan TL, Kanaar R, Wyman C (2003) Rad54, a Jack of all trades in homologous recombination. DNA Repair (Amst) 2:787-794

Tarsounas M, Pearlman RE, Gasser PJ, Park MS, Moens PB (1997) Protein-protein interactions in the synaptonemal complex. Mol Biol Cell 8:1405-1414

Tesse S, Storlazzi A, Kleckner N, Gargano S, Zickler D (2003) Localization and roles of Ski8p protein in *Sordaria* meiosis and delineation of three mechanistically distinct steps of meiotic homolog juxtaposition. Proc Natl Acad Sci USA 100:12865-12870

Tonami Y, Murakami H, Shirahige K, Nakanishi M (2005) A checkpoint control linking meiotic S phase and recombination initiation in fission yeast. Proc Natl Acad Sci USA 102:5797-5801

Tran PT, Erdeniz N, Symington LS, Liskay RM (2004) EXO1-A multi-tasking eukaryotic nuclease. DNA Repair (Amst) 3:1549-1559

Trujillo KM, Roh DH, Chen L, Van Komen S, Tomkinson A, Sung P (2003) Yeast Xrs2 binds DNA and helps target Rad50 and Mre11 to DNA ends. J Biol Chem 278:48957-48964

Tsubouchi H, Ogawa H (1998) A novel *mre11* mutation impairs processing of double-strand breaks of DNA during both mitosis and meiosis. Mol Cell Biol 18:260-268

Tsubouchi H, Ogawa H (2000) Exo1 roles for repair of DNA double-strand breaks and meiotic crossing over in *Saccharomyces cerevisiae*. Mol Biol Cell 11:2221-2233

Tsubouchi H, Roeder GS (2002) The Mnd1 protein forms a complex with Hop2 to promote homologous chromosome pairing and meiotic double-strand break repair. Mol Cell Biol 22:3078-3088

Tsubouchi H, Roeder GS (2003) The importance of genetic recombination for fidelity of chromosome pairing in meiosis. Dev Cell 5:915-925

Tsubouchi H, Roeder GS (2004) The budding yeast Mei5 and Sae3 proteins act together with Dmc1 during meiotic recombination. Genetics 168:1219-1230

Tsubouchi H, Roeder GS (2006) Budding yeast Hed1 down-regulates the mitotic recombination machinery when meiotic recombination is impaired. Genes Dev 20:1766-1775

Tsubouchi T, Zhao H, Roeder GS (2006) The meiosis-specific Zip4 protein regulates crossover distribution by promoting synaptonemal complex formation together with Zip2. Dev Cell 10:809-819

Tsukamoto Y, Mitsuoka C, Terasawa M, Ogawa H, Ogawa T (2005) Xrs2p regulates Mre11p translocation to the nucleus and plays a role in telomere elongation and meiotic recombination. Mol Biol Cell 16:597-608

Tsukuda T, Fleming AB, Nickoloff JA, Osley MA (2005) Chromatin remodelling at a DNA double-strand break site in *Saccharomyces cerevisiae*. Nature 438:379-383

Turney D, de Los Santos T, Hollingsworth NM (2004) Does chromosome size affect map distance and genetic interference in budding yeast? Genetics 168:2421-2424

Uchiyama Y, Kimura S, Yamamoto T, Ishibashi T, Sakaguchi K (2004) Plant DNA polymerase lambda, a DNA repair enzyme that functions in plant meristematic and meiotic tissues. Eur J Biochem 271:2799-2807

Unal E, Arbel-Eden A, Sattler U, Shroff R, Lichten M, Haber JE, Koshland D (2004) DNA damage response pathway uses histone modification to assemble a double-strand break-specific cohesin domain. Mol Cell 16:991-1002

Usui T, Ogawa H, Petrini JH (2001) A DNA damage response pathway controlled by Tel1 and the Mre11 complex. Mol Cell 7:1255-1266

van Attikum H, Fritsch O, Hohn B, Gasser SM (2004) Recruitment of the INO80 complex by H2A phosphorylation links ATP-dependent chromatin remodeling with DNA double-strand break repair. Cell 119:777-788

van Attikum H, Gasser SM (2005a) ATP-dependent chromatin remodeling and DNA double-strand break repair. Cell Cycle 4:1011-1014

van Attikum H, Gasser SM (2005b) The histone code at DNA breaks: a guide to repair? Nat Rev Mol Cell Biol 6:757-765

van Veen JE, Hawley RS (2003) Meiosis: when even two is a crowd. Curr Biol 13:R831-833

Veaute X, Jeusset J, Soustelle C, Kowalczykowski SC, Le Cam E, Fabre F (2003) The Srs2 helicase prevents recombination by disrupting Rad51 nucleoprotein filaments. Nature 423:309-312

Villeneuve AM, Hillers KJ (2001) Whence meiosis? Cell 106:647-650

Voegtli WC, Madrona AY, Wilson DK (2003) The structure of Aip1p, a WD repeat protein that regulates Cofilin-mediated actin depolymerization. J Biol Chem 278:34373-34379
von Wettstein D, Rasmussen SW, Holm PB (1984) The synaptonemal complex in genetic segregation. Annu Rev Genet 18:331-413
Wan L, de los Santos T, Zhang C, Shokat K, Hollingsworth NM (2004) Mek1 kinase activity functions downstream of *RED1* in the regulation of meiotic double strand break repair in budding yeast. Mol Biol Cell 15:11-23
Wang JC (2002) Cellular roles of DNA topoisomerases: a molecular perspective. Nat Rev Mol Cell Biol 3:430-440
Wang TF, Kleckner N, Hunter N (1999) Functional specificity of MutL homologs in yeast: evidence for three Mlh1-based heterocomplexes with distinct roles during meiosis in recombination and mismatch correction. Proc Natl Acad Sci USA 96:13914-13919
Wang X, Ira G, Tercero JA, Holmes AM, Diffley JF, Haber JE (2004) Role of DNA replication proteins in double-strand break-induced recombination in *Saccharomyces cerevisiae*. Mol Cell Biol 24:6891-6899
Warren CD, Eckley DM, Lee MS, Hanna JS, Hughes A, Peyser B, Jie C, Irizarry R, Spencer FA (2004) S-phase checkpoint genes safeguard high-fidelity sister chromatid cohesion. Mol Biol Cell 15:1724-1735
Wei K, Clark AB, Wong E, Kane MF, Mazur DJ, Parris T, Kolas NK, Russell R, Hou H Jr, Kneitz B, Yang G, Kunkel TA, Kolodner RD, Cohen PE, Edelmann W (2003) Inactivation of Exonuclease 1 in mice results in DNA mismatch repair defects, increased cancer susceptibility, and male and female sterility. Genes Dev 17:603-614
Weiner BM, Kleckner N (1994) Chromosome pairing via multiple interstitial interactions before and during meiosis in yeast. Cell 77:977-991
Wesoly J, Agarwal S, Sigurdsson S, Bussen W, Van Komen S, Qin J, van Steeg H, van Benthem J, Wassenaar E, Baarends WM, Ghazvini M, Tafel AA, Heath H, Galjart N, Essers J, Grootegoed JA, Arnheim N, Bezzubova O, Buerstedde JM, Sung P, Kanaar R (2006) Differential contributions of mammalian Rad54 paralogs to recombination, DNA damage repair, and meiosis. Mol Cell Biol 26:976-989
Whitby MC (2005) Making crossovers during meiosis. Biochem Soc Trans 33:1451-1455
White MA, Dominska M, Petes TD (1993) Transcription factors are required for the meiotic recombination hotspot at the *HIS4* locus in *Saccharomyces cerevisiae*. Proc Natl Acad Sci USA 90:6621-6625
White MA, Wierdl M, Detloff P, Petes TD (1991) DNA-binding protein Rap1 stimulates meiotic recombination at the *HIS4* locus in yeast. Proc Natl Acad Sci USA 88:9755-9759
Willems AR, Schwab M, Tyers M (2004) A hitchhiker's guide to the cullin ubiquitin ligases: SCF and its kin. Biochim Biophys Acta 1695:133-170
Wilson TE, Lieber MR (1999) Efficient processing of DNA ends during yeast nonhomologous end joining. Evidence for a DNA polymerase beta (Pol4)-dependent pathway. J Biol Chem 274:23599-23609
Wiltzius JJ, Hohl M, Fleming JC, Petrini JH (2005) The Rad50 hook domain is a critical determinant of Mre11 complex functions. Nat Struct Mol Biol 12:403-407
Woltering D, Baumgartner B, Bagchi S, Larkin B, Loidl J, de los Santos T, Hollingsworth NM (2000) Meiotic segregation, synapsis, and recombination checkpoint functions require physical interaction between the chromosomal proteins Red1p and Hop1p. Mol Cell Biol 20:6646-6658

Woods LM, Hodges CA, Baart E, Baker SM, Liskay M, Hunt PA (1999) Chromosomal influence on meiotic spindle assembly: abnormal meiosis I in female Mlh1 mutant mice. J Cell Biol 145:1395-1406

Wu H, Gao J, Sharif WD, Davidson MK, Wahls WP (2004) Purification, folding, and characterization of Rec12 (Spo11) meiotic recombinase of fission yeast. Protein Expr Purif 38:136-144

Wu TC, Lichten M (1994) Meiosis-induced double-strand break sites determined by yeast chromatin structure. Science 263:515-518

Wu TC, Lichten M (1995) Factors that affect the location and frequency of meiosis-induced double-strand breaks in *Saccharomyces cerevisiae*. Genetics 140:55-66

Xu L, Kleckner N (1995) Sequence non-specific double-strand breaks and interhomolog interactions prior to double-strand break formation at a meiotic recombination hot spot in yeast. EMBO J 14:5115-5128

Xu L, Weiner BM, Kleckner N (1997) Meiotic cells monitor the status of the interhomolog recombination complex. Genes Dev 11:106-118

Yamada T, Mizuno K, Hirota K, Kon N, Wahls WP, Hartsuiker E, Murofushi H, Shibata T, Ohta K (2004) Roles of histone acetylation and chromatin remodeling factor in a meiotic recombination hotspot. EMBO J 23:1792-1803

Yamashita K, Shinohara M, Shinohara A (2004) Rad6-Bre1-mediated histone H2B ubiquitylation modulates the formation of double-strand breaks during meiosis. Proc Natl Acad Sci USA 101:11380-11385

Yang W (2000) Structure and function of mismatch repair proteins. Mutat Res 460:245-256

Yildiz O, Majumder S, Kramer B, Sekelsky JJ (2002) *Drosophila* MUS312 interacts with the nucleotide excision repair endonuclease MEI-9 to generate meiotic crossovers. Mol Cell 10:1503-1509

Young JA, Hyppa RW, Smith GR (2004) Conserved and nonconserved proteins for meiotic DNA breakage and repair in yeasts. Genetics 167:593-605

Yu L, Gaitatzes C, Neer E, Smith TF (2000) Thirty-plus functional families from a single motif. Protein Sci 9:2470-2476

Zenvirth D, Arbel T, Sherman A, Goldway M, Klein S, Simchen G (1992) Multiple sites for double-strand breaks in whole meiotic chromosomes of *Saccharomyces cerevisiae*. EMBO J 11:3441-3447

Zenvirth D, Loidl J, Klein S, Arbel A, Shemesh R, Simchen G (1997) Switching yeast from meiosis to mitosis: double-strand break repair, recombination and synaptonemal complex. Genes Cells 2:487-498

Zhao S, Renthal W, Lee EY (2002) Functional analysis of FHA and BRCT domains of NBS1 in chromatin association and DNA damage responses. Nucleic Acids Res 30:4815-4822

Zickler D, Kleckner N (1999) Meiotic chromosomes: integrating structure and function. Annu Rev Genet 33:603-754

Zierhut C, Berlinger M, Rupp C, Shinohara A, Klein F (2004) Mnd1 is required for meiotic interhomolog repair. Curr Biol 14:752-762

Hunter, Neil
Sections of Microbiology and Molecular & Cellular Biology, University of California Davis, Davis, CA 95616, USA
nhunter@ucdavis.edu

Site-specific recombination

Ian Grainge and David J. Sherratt

Abstract

Site-specific recombination is a reaction in which a pair of genetically defined sites undergoes reciprocal exchange ("crossing-over") via a recombinase-mediated DNA breakage and joining process. Such reactions have a wide range of biological outcomes, from integration and excision of virus genomes into and out of host chromosomes, to acquisition of novel genes and drug resistance, and even facilitating bacterial chromosome segregation. Two distinct families of recombinases exist, designated by their active site residues. In both these families recombination is carried out by a core of four recombinase monomers acting at two synapsed DNA sites. In many cases additional recombinase monomers and/or accessory proteins act at adjacent DNA sites to facilitate synapsis and often play a critical role in determining reaction topology. Here, the mechanism of site-specific recombination reactions is examined for both site-specific recombinase families, as well as for related proteins that mediate variant reactions, such as integrons and the integrases of conjugative transposons.

1 Introduction

Site-specific recombination is used by a wide variety of biological systems in all three domains of life to produce a localised recombination event. The minimal requirement of each recombination system is a relatively small core DNA site at which a pair of recombinase proteins binds. When two such DNA sites are brought together with four bound recombinase monomers then the reaction can proceed. The proteins that catalyse recombination are conserved, as are their reaction mechanisms, and yet, despite all these similarities in the reactions between these relatively small pairs of DNA sites, site-specific recombination has been utilised by these diverse systems to produce an amazing range of biological outcomes.

The recombination reaction has three distinct physical outcomes depending upon the relative orientation and disposition of the two target sites for recombination (Fig. 1). Recombination between two sites on circular DNA molecules can lead to their fusion, with one inserted into the other, and conversely, the reverse of this reaction can produce two separate molecules from one progenitor. Site-specific recombination between a site on a circular DNA and a site on a linear

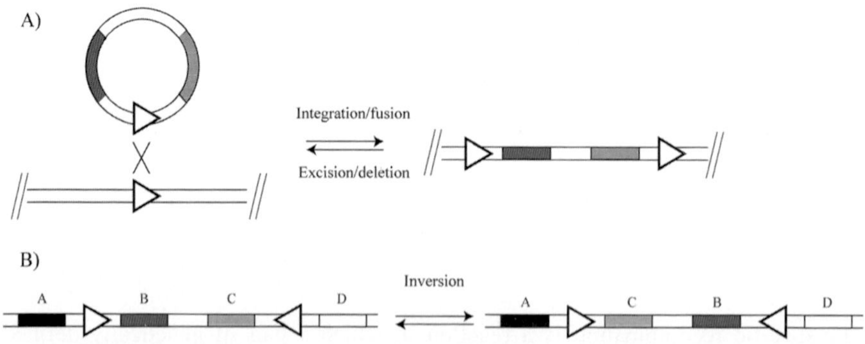

Fig. 1. Cartoon representation of reactions catalysed by site-specific recombinases. A) Recombination of a site on a circular DNA with a second site, be it on a linear or circular template, leads to integration. The reverse reaction, recombination between two directly repeated sites, yields a circular excision product. B) Inversion reaction.

piece of DNA also results in the fusion of the two molecules, whereas recombination between two linear DNAs will simply result in a crossover event. Although some recombinase proteins can catalyse recombination between two linear DNAs *in vitro*, no system so far examined utilises this as the natural substrate *in vivo*.

The third alternative is to invert a segment of DNA with respect to the surrounding region (Fig. 1B). Inversion can change the orientation of genes with respect to a flanking promoter resulting in differential gene expression. Alternatively, inversion that occurs during replication can reverse the direction of a replication fork with respect to the surrounding DNA, as proposed for the copy number amplification of the 2μm plasmid of yeast (Futcher 1986).

These basic mechanisms, based on site orientation and disposition, are used *in vivo* by diverse systems to produce the wide variety of different outcomes observed, allowing chromosome or plasmid segregation by resolution of dimeric forms (e.g. XerC/D (Sherratt et al. 1995), Cre (Hoess et al. 1982; Abremski and Hoess 1984), phage integration (e.g. λ Int, (Azaro and Landy 2002)), regulation of gene expression leading to antigenic variation in pathogens (e.g. Hin, Gin (van de Putte and Goosen 1992)), resolution of transposition intermediates (e.g. resolvase (Stark et al. 1989)), acquisition of novel drug resistances (e.g. integrons (Hall and Collis 1995)), and a myriad of other variations.

Each core site, typically ~30bp, consists of two recombinase binding sites, inverted with respect to each other, and separated by a few bases. The reaction specificity and outcome is determined by controlling the way these sites are brought together, which is either determined by sequence asymmetry within the core site or by the use of accessory proteins and sequences. The use of accessory sequences and proteins provides a topological specificity to the reaction, and the huge variety of biological outcomes stems from the range of different genetic contexts that site-specific recombination occurs within.

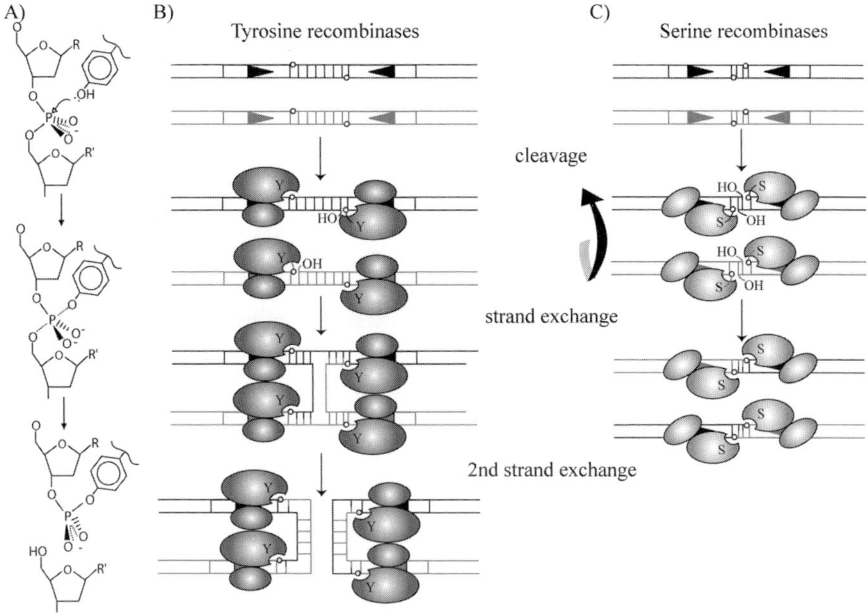

Fig. 2. A) A schematic diagram of the S_N2 reaction mechanism of cleavage by tyrosine recombinases. In-line attack of the tyrosine hydroxyl produces a transient trigonal bipyramidal intermediate, from which the 5'OH of DNA leaves. Rejoining is simply the reverse of this reaction. For simplicity the other catalytic site residues have been omitted. Cartoon comparison of the mechanism of tyrosine and serine family recombinases. DNA is shown as parallel horizontal lines with each recombinase binding site as a box with an arrow to show its direction. The scissile phosphates are open circles and the base pairs between the cleavage sites are shown as vertical lines. Note that the tyrosine recombinases produce and then resolve a Holliday junction intermediate, whereas the serine recombinases produce two transient double strand breaks and strand exchange requires a 180 degree rotation of two half-sites before rejoining.

2 The two families of recombinases: tyrosine and serine

The proteins that catalyse recombination can be classified by the active site residue used to provide the nucleophile for DNA cleavage, either a tyrosine or a serine. The two families have some common themes in their reaction mechanism: in both families a recombinase monomer binds to each of the two inverted binding sites within the core site. Each recombinase monomer cleaves one DNA phosphodiester during the reaction using an S_N2 in-line attack of the side-chain nucleophile, and rejoins the DNA using a second S_N2 reaction using the DNA hydroxyl as the nucleophile (shown for tyrosine mediated cleavage in Fig. 2A). The production, and subsequent resolution, of a covalent protein-DNA intermediate conserves the overall bond energy making the break-joining reaction iso-energetic.

However, there also exist a number of differences between the mechanisms of recombination of the two families (Fig. 2B). Tyrosine recombinases normally catalyse two consecutive pairs of strand exchanges, the first pair produces a Holliday junction intermediate and the second pair of exchanges resolves this. Conversely, the serine recombinases catalyse both pairs of strand exchanges simultaneously, effectively producing two transient double strand breaks. The stereochemistry of strand cleavage is also different between the two classes; cleavage produces a 3' phosphotyrosine and a 5' hydroxyl in one family, and a 5' phosphoserine and a 3' hydroxyl in the other. The relative position of the cleavage sites on the two strands of a DNA duplex are also different between these two classes: tyrosine recombinase cleavage sites are separated by 6-8 bp and the cleavage points would give a 5' overhang if cut simultaneously. Conversely, serine family recombinases have cleavage sites that are usually two bases apart and produce a 3' overhang when cleaved.

Most site-specific recombination systems use a homo-tetramer of the recombinase to catalyse recombination. However, there are notable exceptions to this. Xer-mediated recombination requires two related recombinases, XerC and XerD, both of which mediate the exchange of one pair of strands adjacent to their binding site during a full recombination reaction. The Xer families are conserved in a very wide range of bacteria, but evidence from several archaeal genomes suggests that they contain only a single Xer homologue. Perhaps then the heterodimeric bacterial Xer system evolved after the split between these prokaryotic domains of life. Phase variation of fimbriae in *E. coli* is catalysed by two proteins of the tyrosine recombinase family, FimB and FimE (Smith and Dorman 1999; Burns et al. 2000). However, some studies suggest that these two proteins can both work independently upon the same inversion sites and both can bind and carry out inversion as homotetramers (Gally et al. 1996). Whether or not there is some co-operation between the two *in vivo* remains to be seen.

Despite all the differences, the outcome of recombination is the same for the all these systems, namely, a precise breaking and joining reaction proceeding through a protein-DNA covalent complex that requires neither a high energy co-factor such as ATP, nor DNA digestion or synthesis, resulting in a precise DNA crossover. This sets site-specific recombination apart from homologous recombination.

3 The tyrosine recombinase family

3.1 Topoisomerases and tyrosine recombinase active sites

Members of the tyrosine recombinase family show very little conservation at the amino acid level, outside the catalytic site residues (Esposito and Scocca 1997), and yet crystal structures of several recombinases have revealed similar folds and domain conservation (Guo et al. 1997; Hickman et al. 1997; Kwon et al. 1997; Subramanya et al. 1997; Chen et al. 2000). Tyrosine recombinases generally consist of two domains, an N-terminal DNA binding domain that contacts the inner

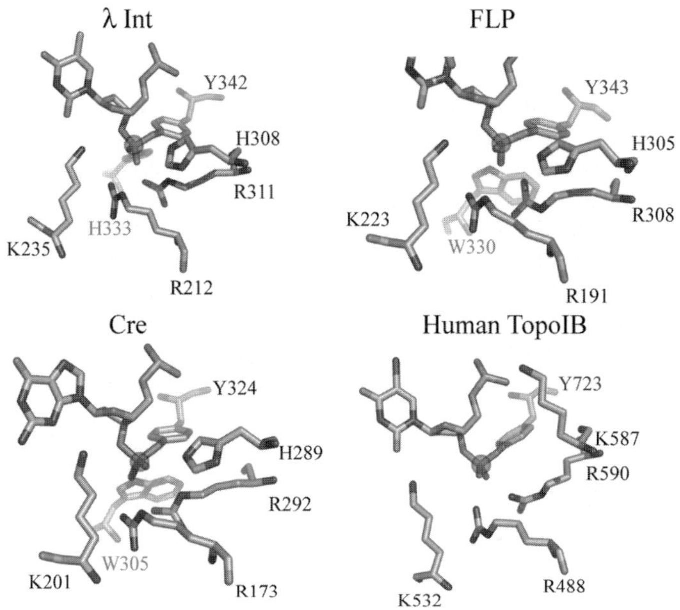

Fig. 3. The active site residues of cleaved intermediates from crystal structures are shown to emphasize their similarity and demonstrate the disposition of the catalytic pentad around the cleavage site. Active site residues are identified with the amino acid number beside them. The nucleotide to which the tyrosine is attached is also shown, with the phosphate picked with a sphere superimposed on top of it.

part of the core site, and a C-terminal domain that both binds DNA and provides all the catalytic residues; together these two domains form a C-clamp around DNA. Therefore, over evolution, it appears that the protein sequences have diverged substantially without altering the overall fold of the protein to a great extent.

The active site of the tyrosine recombinase family is comprised by six conserved residues (Fig. 3). Five of these, sometimes called the catalytic pentad, surround the scissile phosphate at the site where the recombinase is bound: two arginines and two histidines (one of which is a tryptophan in some family members) and a lysine. The sixth is the nucleophilic tyrosine. The catalytic pentad could potentially act to align the scissile phosphate for cleavage, activate it for the incoming nucleophile, stabilise the pentavalent cleavage intermediate or act as a general acid/base catalyst (reviewed in Grainge and Jayaram 1999; Chen and Rice 2003). The nucleophile for cleavage of the DNA is provided by the tyrosine hydroxyl, but there exists some variability in the way it is provided. Almost all characterised prokaryotic and phage encoded recombinases use the tyrosine from the same monomer that provides the rest of the catalytic pocket to carry out cleavage (termed cleavage *in cis*). However, the recombinase FLP of the yeast 2 micron plasmid, and other related yeast recombinases, uses a tyrosine donated *in trans* from another monomer bound across the spacer (Lee et al. 1994, 1999). Recent evidence suggests that the recombinase SSV1 from an archaeal virus also cleaves *in trans* (Letzelter et al. 2004).

Mechanistically, DNA cleavage and joining reactions of tyrosine recombinases are very closely related to type IB topoisomerases. Type IB topoisomerases are found in eukaryotic nuclei where they play a role in DNA replication and transcription, as well as in cytoplasmic poxvirus encoded versions, as well as being present in *Deinococcus radiodurans* and the hyperthermophile *Methanopyrus kandleri*. However, type IB topoisomerases are not found in other archaea (to date) or eubacteria, where type IA topisomerases predominate.

The main difference from recombinases is that type IB topoisomerases function as a monomer and use the liberated 5'OH to rejoin directly back to its original phosphate, thereby, restoring the integrity of the DNA. During recombination the 5'OH is used to rejoin across synapsed partners to produce and resolve a Holliday junction. A series of crystal structures has revealed that the active sites of these proteins are very similar, with human topoisomerase having two arginines and a histidine in a similar disposition as seen in Cre structures (Fig 3). One exception is that the first conserved histidine from the recombinases is a lysine in the topoisomerases (Redinbo et al. 1997; Cheng et al. 1998). This high degree of conservation strongly suggests an evolutionary relationship between tyrosine recombinases and topoisomerases.

3.2 Control of the recombination reaction

With the exception of Xer-mediated recombination, which uses the two recombinases XerC and XerD, the tyrosine recombinases employ a homotetramer at a pair of sites, each of which has dyad symmetrical binding sites. How then does the reaction achieve directionality? To form and then resolve the Holliday junction intermediate the DNA must be first cut and exchanged at one end of the core site, and then at the other. Within each naturally occurring site there is asymmetry, usually provided by the central sequence between the two cleavage sites, but sometimes also found within the binding sites themselves. The asymmetry of the spacer DNA sequence lends the site an intrinsic direction (hence a core site is often represented as an arrow, see Fig. 4A); exchange of a pair of strands only occurs if both sites are cut at the same end of the spacer DNA. Both biochemical and crystallographic evidence suggests that three bases are exchanged at a time following cleavage, and that the base pairing of the exchanged strand with the complementary DNA is necessary to align the 5'OH for the rejoining reaction (Senecoff and Cox 1986; Lee and Jayaram 1995; Arciszewska et al. 1997; Nunes-Duby et al. 1995; Guo et al. 1997).

Asymmetry in the recombination core site also plays a role in determining which pair of strands are exchanged first during a reaction. Which end of the spacer is cleaved first depends upon the asymmetric interactions between monomers described below, and represented in Figure 4A as a simple ball-and-socket interaction between monomers. Upon initial binding of the recombinase proteins

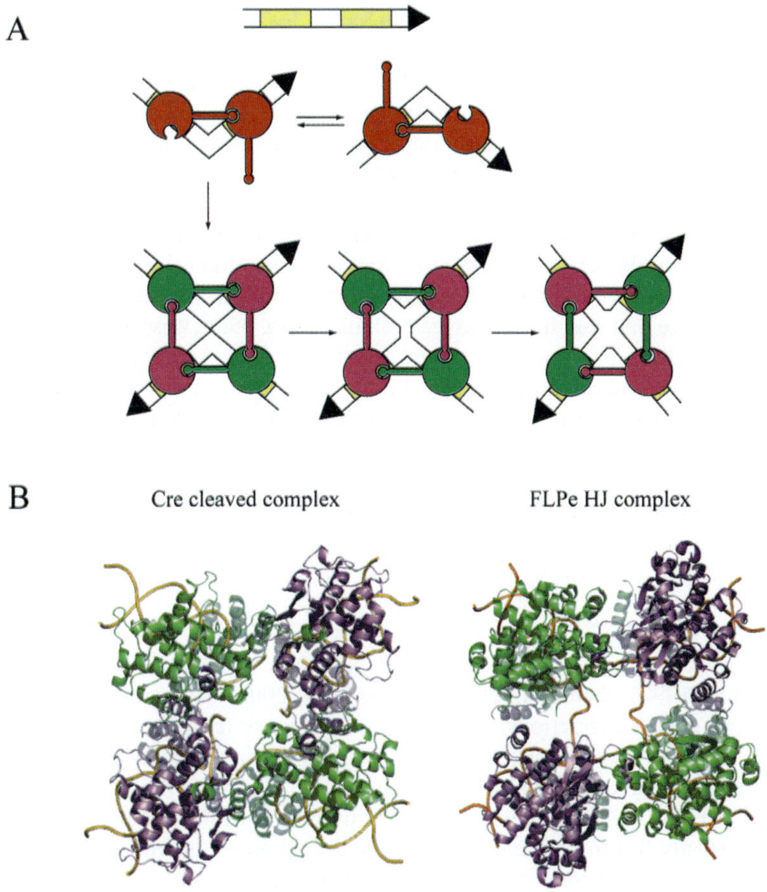

Fig. 4. A) A recombination site is represented with yellow boxes as binding sites and the black arrowhead shows the overall site directionality given by the asymmetric spacer DNA. Each site is bound by two recombinase monomers, initially shown in red, which bend the DNA two interact with each other as two non-equivalent dimeric forms. A bending preference will favour one form over the other. Within a synaptic complex of two sites aligned antiparallel with the same bend direction, two monomers are active and two inactive. Using the convention from the Cre crystal structure, active monomers are green and inactive magenta. Strand exchange forms a Holliday junction and subsequent isomerisation of the junction accompanied by allosteric protein changes switches the active and inactive monomers. Note the twofold symmetry in interactions throughout the reaction. B) Crystal structures of a Cre-*lox* cleaved complex and the cleaved FLPe- Holliday junction complex viewed from the C-terminal face. Monomers bound to the site where cleavage has occurred are green. In the Cre tetramer donation of the C-terminal N helix occurs cyclically in a clockwise manner, whereas for FLPe donation of the M helix with the catalytic tyrosine is anticlockwise. In the FLPe structure only the M helices from the two purple monomers are seen, with those from the green monomers being disordered. This reflects the correct and stable donation of the M helix and the catalytic tyrosine to the "active" green monomers.

to DNA, the site can be bent in either direction, leading to mutually exclusive activation and inactivation of each monomer (Fig. 4A). Thus, the direction of the bend could be seen to determine which monomer is active, and sites with a strong bending preference will yield recombination reactions with a strong preference to start at one end of the site. The requirement for a bend in the DNA in order to achieve the necessary protein-protein interactions to activate one monomer also leads to the outcome that the other partner monomer is held inactive. It is impossible to achieve mutual exchange of activating interactions between a dimer on a single DNA duplex. Thus, cleavage is limited to one end of the spacer at a time, and double strand breaks are avoided. It is not simply the intrinsic bend preference of the core site that can determine reaction direction, but as described in Section 5, accessory factors can also influence the reaction.

Potentially, a strong bending preference could also prevent direct reversal of the reaction following one round of strand exchanges; the tetramer that has just carried out recombination could reverse the reaction in a stepwise fashion to recreate the starting substrate. However, if the product of recombination is bent in the less favourable direction then the complex will be more likely to disassemble before recombination occurs.

Biochemical investigations suggested that synapsis of recombination sites occurs by aligning the two recombining sites in an antiparallel configuration (Arciszewska et al. 1997; Azaro and Landy 1997). This view was confirmed by the crystal structures of Cre, FLP and λ Int bound to their cognate DNA elements (Guo et al. 1997; Chen et al. 2000; Biswas et al. 2005). However, bending of the DNA and cyclic interactions around the recombinase tetramer produced a structure, which was close to planar, with only minor deviation from fourfold symmetry when looking at the flanking DNA emerging from the synaptic nucleoprotein complex (Fig. 4B). The implication of this pseudo fourfold symmetric complex is that there only has to be a modest movement to achieve the allosteric control between the first and second strand exchanges (Fig. 4).

What, then, is the nature of the allosteric control between monomers pairs to control cleavage? It was clear from both cleaved and uncleaved DNA-protein complexes that recombinase binding positions the catalytic pentad around the scissile phosphodiester, activating it for cleavage (Guo et al. 1997; Gopaul et al. 1998; Rice et al. 2000; Biswas et al. 2005), yet cleavage does not automatically occur, but is tightly regulated. Control of cleavage, thus, depends upon altering the availability of the tyrosine nucleophile.

The positioning of the tyrosine nucleophile is determined by the cyclic interactions around the tetramer most specifically at the very C-terminus of the protein, whether it be in the donation of the C-terminal N helix as in Cre (Guo et al. 1997), or donation of the catalytic tyrosine on the M helix as in FLP (Chen et al. 2000) or insertion of the C-terminal β-sheet in λ Int (Biswas et al. 2005). There is twofold symmetry in these interactions ensuring that one pair of monomers is active and the other pair inactive at any given time (Fig. 4). Indeed, the cyclic nature of the interactions and the need to alternate between active and inactive monomers around the tetramer implies that the sites have to be brought together in an antiparallel configuration for recombination to occur. This was shown to be the case using symmetrised DNA core sites (Grainge and Jayaram 2000). Even when there is no cue from the DNA sequence as to how to align the sites, the recombination reaction proceeds as normal as though the sites were antiparallel.

4 Serine family recombinases

4.1 Domain organisation and active site of serine family recombinases

The serine recombinases that are best characterised are of the resolvase/invertase class, typified by γ/δ and Tn3 resolvase and the invertases Hin and Gin, and these will be the major focus here. The resolvases are found on transposons and they are necessary to recombine the co-integrate transposition intermediates. Invertases play a role in variation of gene expression; by inverting a segment of DNA, genes can be switched into and out of an expressed context. In the case of Hin this produces flagellar phase variation in *Salmonella typhimurium*, whereas Gin is encoded by Mu and is responsible for a variation in host range. However, it has become apparent that the family is more diverse than once thought, both in structure and function (Smith and Thorpe 2002), and the other family members such as "large serine" recombinases will be dealt with separately from the better characterised resolvase/invertases (see Section 7.1).

In contrast to tyrosine recombinases, most serine recombinases have their active site residues at the N-terminus of the protein, with the C-terminus being solely involved in DNA binding, also through a HTH motif. The exception to this is a group of enzymes typified by the recombinase encoded by IS607, which has an N-terminal HTH motif (Kersulyte et al. 2000). Cleavage by all the serine recombinases characterised occurs *in cis*: all the residues required for the activation and cleavage of the DNA backbone are provided by the monomer bound at the site proximal to the cleavage. Indeed, as discussed later, the strand exchange mechanism employed by serine recombinases would seem to preclude *trans* cleavage.

The structure of γ/δ resolvase shows that the scissile phosphate is surrounded by three residues which make a hydrogen bonding network with each other and with the phosphate: two arginines and an aspartate (Fig. 5; Li et al. 2005). A further arginine is seen to interact with the 3'OH that is formed upon strand cleavage. These residues have been shown biochemically to be vital for catalysis (Boocock et al. 1995). Comparison of cleaved and uncleaved resolvase/DNA co-crystal structures reveals that there is a large movement necessary to bring the active site serine into close proximity with the scissile phosphate (Li et al. 2005). Thus it appears that for serine recombinases too, control of cleavage occurs primarily by limiting the availability of the active site nucleophile. Mutational analyses have shown that protein-protein interactions with accessory site proteins are necessary to form this active complex within which cleavage can occur (Arnold et al. 1999; Merickel et al. 1998).

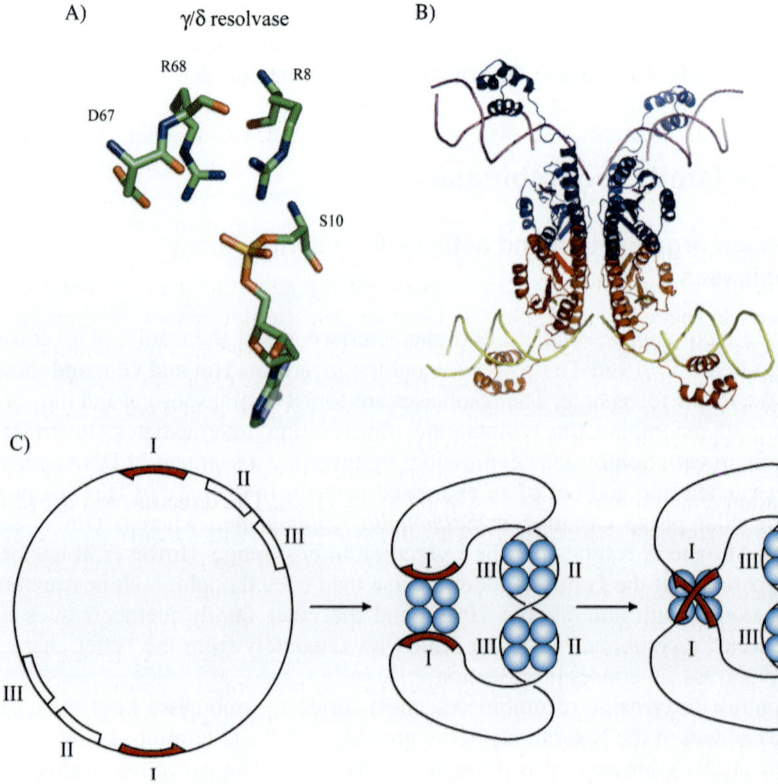

Fig. 5. A) Catalytic site of the cleaved γ/δ resolvase–*res* complex. Amino acids are numbered. B) Structure of the cleaved complex showing the interaction interface between pairs of resolvase monomers and the DNA on the outside of the protein core. Monomers are coloured to represent a blue dimer of resolvase bound to a magenta *res* site recombining with a red dimer bound to a yellow site. Note the flat interaction surface between the left pair and the right pair of resolvase monomers. C) Topological path of resolvase recombination. The complete *res* site contains three sites (numbered I-III) each with an inverted pair of recombinase binding sites. A dimer of resolvase, represented as blue spheres, binds to each site and the interwrapping of sites II and III juxtaposes the two site Is where recombination occurs. For simplicity, any interactions between recombinase molecules at sites I and sites II and III have been omitted.

4.2 Mechanism of recombination by serine family recombinases

The cleaved complex of γ/δ resolvase has the presumptive 3'OH and 5' phosphoserine that should be rejoined to form the recombinant product separated by ~50 Å (Fig. 5B; Li et al. 2005). Furthermore, this intervening space is occupied by protein. The crystal structure is consistent with data on a complex of Tn3 resolvase in solution suggesting a similar arrangement with the DNA essentially on the outside of a protein core (Nöllman et al. 2004). These two factors taken together suggest that in order for strand exchange to occur there has to be a major rearrangement of protein and DNA. The subunit rotation model, originally proposed on the basis of biochemical results (Stark et al. 1991), explains this movement; following cleavage two subunits rotate by 180° with respect to the other two in order to now juxtapose the 3'OH from one DNA partner with the 5'phosphoserine on the other DNA substrate. Such a large movement of both DNA and protein explains the necessity for all four phosphodiesters to be cleaved simultaneously, effectively producing a double strand break, which allows for the rearrangement.

The nature of the subunit interface between the necessary pair of monomers which have to rotate is an unusually flat surface, consisting of mainly hydrophobic interactions and is thus ideal to accommodate rotation of the subunits without dissociation of the complex (Li et al. 2005) This interface is clear in the representation shown in Figure 5B, where the complex is shown side on, and there is an obvious gap between the left and right halves dimer pairs. Following one round of rotation the DNA can then be re-ligated to restore its continuity. Therefore, the recombination reaction introduces a crossing in the DNA. The direction that the protein/DNA rotates, and hence the sign of the crossing introduced, seems to be determined solely by the supercoiling state of the DNA. Therefore, in a negatively supercoiled substrate, the supercoiling energy drives rotation to give a positive crossing (loss of negative supercoiling). Similarly to the strand rejoining reaction of tyrosine recombinases, base pairing following subunit rotation is necessary to correctly align the 3'OH with the phosphoserine. If the central 2 bp of the recombining sites are different then the absence of base pairing upon subunit rotation leads to a second 180° rotation, presumably by the same mechanism and without rejoining having occurred (McIlwraith et al. 1997). Thus, rejoining after 360° rotation occurs because of the restoration of base pairing and a non-recombinant product with altered topology is produced.

There are no cleaved complex structures available for an invertase, but evidence for this sub-family is also consistent with the subunit rotation model (Dhar et al. 2005).

5 Directing recombination outcome

5.1 Accessory proteins, sequences, and topological selectivity

Some site-specific recombinases appear to act at very simple sites, consisting of just a recombination core site, without the need for any accessory sites or proteins, for example Cre from phage P1, and the 2μ plasmid encoded FLP. In a simple recombination system like Cre, recombination of the target *loxP* sites yields mainly unlinked deletion products, and presumably this preference is largely a consequence of entropy (Abremski and Hoess 1984). In the case of FLP, the 2 micron plasmid exists in an equilibrium between the two inversion products with each being roughly equally represented.

However, many site-specific recombination systems appear to function directionally. For example, two directly repeated recombination sites on a circular DNA element may recombine to delete the intervening DNA and yet not catalyse the reverse reaction to re-integrate the DNA back into the circle. This reaction directionality is achieved through the use of accessory DNA sequences and proteins. These work to synapse the two recombining sites only under certain conditions. For example, the Tn3 and γ/δ resolvase sites, *res*, have three sites at each of which a resolvase dimer binds, and are among the most extensively studied for understanding the relationship between the accessory sites and recombination. Site I contains the scissile phosphodiesters that undergo exchange, whereas sites II and III are bound by resolvase dimers that remain catalytically inactive throughout recombination (Fig. 5C). When two *res* sites come together the interactions between the dimers at sites II and III of the two sites leads to interwrapping of these sites to form a specific structure wherein three DNA crossings are trapped (Fig. 5C). As a result of this arrangement, the two site Is are brought into close proximity and can recombine. In the absence of site II/III interwrapping, recombination does not occur. The specific topological structure formed in synapsis is interwound in such a way that its formation is energetically unfavourable unless the two sites are present on the same DNA molecule and in direct repeat. In this way, the topology of synapsis is used as an energetic filter to allow only recombination that will lead to a deletion product. This kind of 3 noded topological filter is similarly used by Xer recombinases to resolve plasmid dimers in *E. coli*. In the case of the ColE1 plasmid *cer* site the accessory proteins are ArgR and PepA, and for the *psi* site of pSC101 PepA and ArcA form the three noded structure.

By altering the type of synapse that is selected, the outcome of recombination can also be driven along a certain topological path. Invertases, for example, use various DNA bending proteins, such as Hu and Fis, and secondary DNA binding sites in a so called enhancer element, to select for inversion between two sites contained on the same molecule, either side of the enhancer. Two Fis dimers bind to the enhancer region and stabilise the bent conformation necessary to bring the sites together. On top of this protein-protein interactions between Fis and the invertase are required to activate cleavage (Merickel et al. 1998).

5.2 Recombination between asymmetric accessory sites can give reaction directionality

The accessory sites and proteins that facilitate recombination do not have to be symmetric as in the examples above. One of the most complex sets of interactions necessary for formation of a productive complex for recombination is found in the λ integrase *attP* site (see Azaro and Landy 2002; Radman-Livaja 2005). The target site for integration of phage λ in the *E. coli* chromosome, *attB*, is a simple core site with no accessory sequences necessary. Conversely, *attP* contains the recombination core site flanked by binding sites for IHF, Fis and high affinity "arm" binding sites for lambda integrase itself (Fig. 6). Biochemical evidence indicated that integrase monomers bind simultaneously to the high affinity arm sites and to

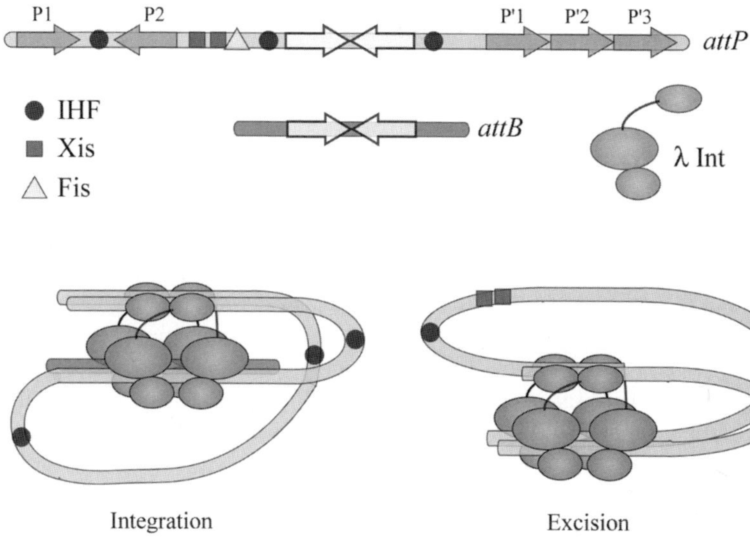

Fig. 6. Schematic representation of the *attP* and *attB* sites. The core recombination sites are represented as inverted open arrows. Arm binding sites are shown as grey arrows and are numbered above each site. IHF, Xis and Fis binding sites are represented by symbols in the key below the *attP* site. B) Integration and excision synapses are shown in cartoon form to demonstrate the different paths of the DNA involved and hence the way that reaction directionality is achieved.

the lower affinity core sites where catalysis occurs (Azaro and Landy 2002). This view was backed up by recent crystal structures of full-length λ Int bound simultaneously to arm and core sites (Biswas et al. 2005). The binary DNA binding sites on each monomer causes a wrapping of the two phage arms and synapsis with the chromosomal site, a structure stabilised by the binding of the DNA bending proteins Fis and IHF. When this complex is assembled *attP* can then be recombined with *attB* (Fig. 6). The resulting two recombination sites flanking the λ genome are termed *attL* and *attR*. These two sites are therefore not identical to either of the two starting sites and thus the directionality of integration or excision can be controlled. Modelling has shown that in order to achieve the same arrangement of λ Int monomers bound to core and arm sites the arm DNA must be bent into different structures during integration and excision reactions. Therefore, excision is not merely a reversal of the integration reaction. For excision to occur efficiently another protein, Xis, is required to bind to the accessory sequences, which stabilises specifically the arm wrapping needed for synapsis and subsequent catalysis to occur to excise phage λ. By using non-identical recombination sites, the phage is thus able to control reaction directionality but still catalyse both integration and excision reactions. Therefore, the excision and integration reactions can both be thought of as directional reactions, which are irreversible under their respective conditions.

6 Applications of site-specific recombination

There are many applications of site-specific recombination in modern genetics, and this section gives a flavour of the varied potential uses. Site-specific recombinases have been used for some years in the generation of stable transgenic DNA in eukaryotes. Due to the simplicity of the systems, the ability to integrate and delete DNA without accessory factors, FLP and Cre are the most widely used recombinases for this purpose (Sauer 1998; Bode et al. 2000). Due to its inherently greater stability and activity at 37°C, Cre has been more widely adopted. However, a temperature stable variant of FLP, FLPe, has been engineered (Bucholz et al. 1998). Similarly, Cre and FLP are used to generate large insertions to create BACs in bacteria. Variations on this theme have allowed for tissue specific expression of genes to be examined. By placing the Cre or FLP gene under the regulation of a tissue specific promoter it can be turned on in only a subset of cells. Its action can then delete or invert a gene, to up or downregulate gene expression in a tissue specific manner. This allows a much finer control of gene expression and circumvents lethality of mutants early in development.

FLP is also widely used in bacterial genetics as a method of producing targeted deletions. Directly repeated *FRT* sites are placed either side of a selectable marker, and the gene is replaced by this cassette using homologous recombination or λ Red recombination using flanking homologies. Recombination by FLP can then be used to delete the selectable marker to leave an in frame deletion that minimises effects on flanking genes or in operons. Using this method a complete library of in frame gene deletions has been constructed for *E. coli* (Baba et al. 2006).

A hybrid resolvase protein has also been developed which can function in eukaryotes (Akopian et al. 2003). By fusing a zinc-finger DNA binding domain to the catalytic domain of resolvase containing a mutation that allows function in the absence of accessory sites, a recombinase with novel specificity was created (Akopian et al. 2003). Similar chimaeric proteins should allow a much larger range of sites to be used for DNA manipulations in the future, and the possibility of selecting specific novel integration sites with "custom built" recombinases.

Site-specific recombination frequency can also be used as a readout of how close two sections of DNA are to each other *in vivo*, and whether two sections of DNA are contained within the same topological domain, by exploiting the idiosyncrasies of individual recombination systems. This approach has been used to map domain organisation in *E. coli* using γ–δ resolvase (Staczek and Higgins 1998), determine relative DNA concentrations using Cre-*loxP* (Hildebrandt and Cozzarelli 1995), and to assess relative disposition of chromosome segments in *Drosophila* using FLP (Golic and Golic 1996) and in yeast using Cre-*loxP* (Burgess and Kleckner 1999.)

7 Related proteins

It has been already noted that the active sites of tyrosine recombinases and those of type IB topoisomerases are very similar. There are, however, several other proteins, which look like site-specific recombinases of the tyrosine family and yet do not catalyse the types of recombination reaction that is canonical for this type of enzymes. Similarly the serine recombinases contain several members, which catalyse very different reactions from the resolvases/invertases already described. These related proteins will be dealt with in this section.

7.1 Large serine recombinases

The majority of the serine recombinases which do not fit into the resolvase/invertase paradigm come from phage and transposons of Gram$^+$ bacteria (Smith and Thorpe 2002), and their name derives from the larger molecular weight of these proteins compared to the rest of the serine family recombinases. Primary amino acid sequence suggests that the large serine recombinases have their catalytic residues at the N-terminus of the protein, and mutation of the serine predicted to correspond to the active site serine of resolvases abolished recombination activity (Crellin and Rood 1997; Thorpe and Smith 1998; Wang and Mullany 2000). Furthermore, it was shown that the reactions catalysed by these proteins resulted in the integration or excision of DNA, reactions previously only known to be catalysed by tyrosine family recombinases. In the case of ϕC31, the integration/excision reactions are both directional and specific, reminiscent of phage λ, with excision requiring Xis protein, but the extent of the similarity remains to be investigated (Smith and Thorpe 2002).

ϕC31 Int has also been shown to be active *in vivo* in eukaryotic cells. The need for Xis to efficiently excise a DNA fragment means that the integration reaction is essentially unidirectional, an advantage over the more promiscuous Cre/*loxP* system.

7.2 Integrons

The term integron refers to an assembly on DNA of several factors: a gene *IntI*, which encodes a protein of the tyrosine recombinase family, a linked integration site, *attI*, a divergent promoter, and a gene cassette. The divergent promoter drives expression of the *IntI* gene in one direction and in the other direction expresses whatever gene is put into the integration site, the integron cassette. Integrons have been identified as major factors in the capture and spread of antibiotic resistance genes among bacteria (Hall and Collis 1998). Integrons can also be found clustered, sometimes in the hundreds, each expressing a protein involved in an environmental adaptive response. These chromosomal clusters, called superintegrons, have been identified in several bacterial species (Rowe-Magnus et al. 2001).

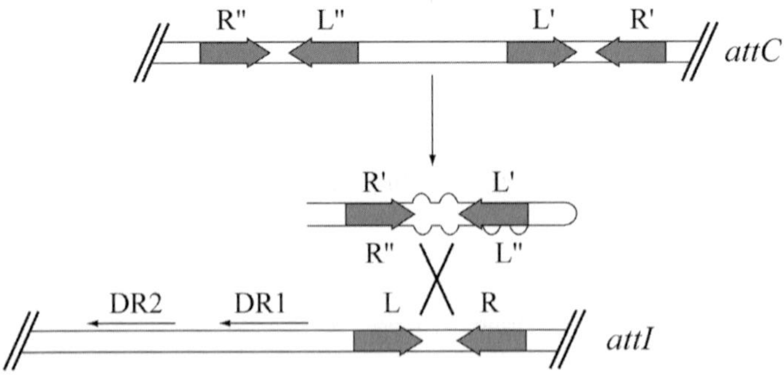

Fig. 7. Representation of integron *att* sites. *AttC* is palindromic and so can form a structure where each single strand fold back on itself. One of these folded strands is preferentially bound by the IntI protein and used as a substrate to recombine with the *attI* site. Each core site is represented by the inverted arrows. The two direct repeats of the *attI* binding site are shown (DR1 and 2), but their contribution to the recombination reaction is not well understood. Different *AttC* sites can have various mismatches when the single stranded form folds upon itself, both within the binding sites and in the spacer DNA.

Although the *IntI* gene resembles the tyrosine family recombinases, evidence is accumulating that the recombination mechanism that shuffles the integron cassettes is quite different from that described for other tyrosine recombinases. IntI can catalyse recombination between the integron *attI* site and the *attC* sites that flank the cassette, as well as between two *attC* sites. The structure of the *att* sites is very different from those seen for the canonical tyrosine recombinases (Fig. 7). *attI* contains four binding sites for IntI, an inverted pair, separated by five base pairs that resembles a core site, flanked by two direct repeats. All four sites can be bound *in vitro* (Collis et al. 1998; Gravel et al. 1998), despite the fact that one of the inverted repeat sites has a degenerate sequence compared to the others. *attC* sites are more complex, consisting of two pairs of inverted sites, each pair resembling a core site, separated by a variable number of base pairs, of variable sequence. Recombination occurs at one of these sites and yet the other is essential for this recombination to occur. The explanation for this came from more *in vitro* DNA binding studies that showed that IntI does not bind to a double stranded form of the *attC* site but binds specifically to one of the single strands of the site (Francia et al. 1999; Bouvier et al. 2005). The structure of the *attC* site allows for extensive cruciform formation, the result of which is that the single stranded form can fold back on itself to recreate a core site, albeit with mismatches (Fig. 7). The recombination reaction appears to be between the double stranded *attI* site and a single strand of the *attC* site with extensive secondary structure. One pair of strand exchange yields a Holliday junction intermediate in which the entire single strand is integrated into the *attC* site on one strand (Fig. 8). Replication through this region then yields one daughter chromosome with the parental *attC* site and the other daughter with the inserted cassette. This differentiates the reaction from other recombinases, because only one pair of strand exchanges occurs, and the reaction requires replication in order to complete the double strand integration. This

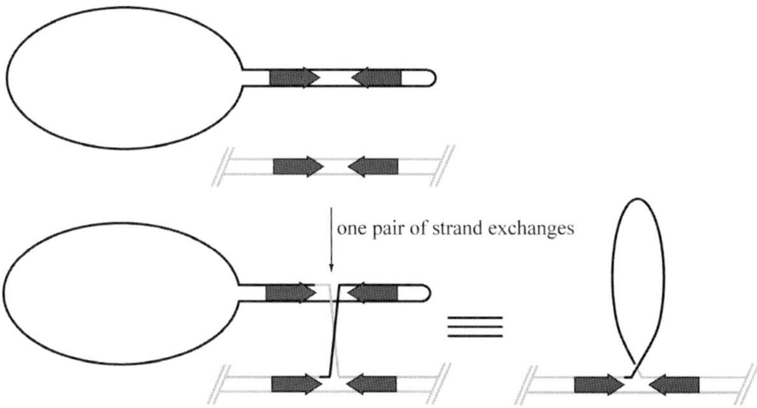

Fig. 8. Single stranded insertion by recombinase proteins. A DNA substrate is represented as a black line, which has the ability to fold a single stranded region back upon itself to form a recombinase core site, represented by inverted grey arrows. The remainder of the molecule can either be double stranded DNA or the whole molecule can be single stranded as shown here. This transient core site can then exchange one pair of strands with a second core site, here represented on a light grey DNA backbone. The second pair of exchanges can be prevented by mismatches between the spacers of the core sites involved. Whether the substrate was initially single or double stranded the outcome is the same- insertion of a single strand into a double stranded partner.

mechanism may have evolved to allow a greater promiscuity in target site recombination than seen for recombinases, which need to exchange both pairs of strands.

7.3 Conjugative transposons

"Conjugative transposons", such as Tn916, were first identified in Gram$^+$ bacteria, but are also present in Gram$^-$, and indeed can be transferred between the two. Conjugative transposons are thus a vehicle for the spread of antibiotic resistance within and between species. The transposons are normally integrated into the host genome but can be excised to give a circular intermediate. This intermediate can then either be used to integrate back into the chromosome in the same or a different position (transposition), or can be used for conjugative transfer. The protein that carries out the excision and integration reaction is related to site-specific recombinases. In some cases, it can be a member of the tyrosine class, as for Tn916, and in others it is a serine family recombinase, as in Tn5397 and Tn4451 (see Smith and Thorpe 2002). In the case of those members that use a tyrosine recombinase an accessory factor, Xis, is also required for efficient and directional recombination, whereas the serine recombinase class of transposons lack this directionality. The target sites for integration can be quite degenerate and the mechanism by which the transposition overcomes this non-identity in the core sites it recombines remains to be seen. It is an intriguing possibility that the single stranded form transferred to a recipient cell may itself be the form competent for integration in a reaction analogous to that proposed for integron single stranded recombination.

The Tn916 Int protein, like its counterpart in phage λ, contains domains for binding two distinct DNA sequences. This allows it to simultaneously bind to the imperfect repeats of each core site at which it cleaves DNA and to accessory sites, termed DR2 sites. The N-terminus of the Tn916 protein is responsible for the arm site binding and the catalytic residues are contained in the C-terminal domain. Binding of the N-terminal domain is unusual in that it employs a three stranded beta sheet rather than the HTH motif often seen in recombinases (Wojciak et al. 1999) Another similarity to lambda is that excision is stimulated by an accessory protein excisionase (Xis). At the sequence level though, neither Int nor Xis of Tn916 show any similarity to the corresponding proteins of λ and so give no hints as to the likely mechanism of excision. The structure of Tn916 Xis shows that it binds DNA through a winged helix motif but lacks the C-terminal tail of λ Xis that interacts with Int (Abbani et al. 2005) explaining its modest stimulation of the excision reaction.

7.4 telomeres of linear prokaryotic chromosomes

The genome of *Borrelia* spirochetes, pathogens responsible for diseases such as Lymes disease, is organised into linear chromosomes, covalently closed at each end. Telomere maintenance and hairpin formation is carried out by the ResT protein, which has homology to tyrosine recombinases (Kobryn and Chaconas 2001 2002). A difference is the addition at the C-terminus of a hairpin binding motif, similar to that seen in transposases such as those from Tn5 and Tn10, which proceed through a hairpin intermediate. Each telomere contains a single binding site for ResT, but upon replication of the closed circular chromosome an inverted repeat of this site is produced at each end, separated by 6 bp. Cleavage by the active site tyrosine at each end produces the necessary 5'OH to rejoin the ends into hairpins (Kobryn and Chaconas 2002). Although these reactions do not have to be coordinated, both cleavages have to be present at the same time for formation of the two hairpin ends (Kobryn et al. 2005). Therefore, the mechanism controlling activity of the monomers appears quite different for ResT than for other tyrosine recombinases. With ResT both monomers on a duplex can be active simultaneously, which is never observed for canonical tyrosine recombinases, and the result is effectively a transient double strand break.

Covalently closed hairpin telomeres are also found in poxviruses and *E. coli* phage N15. The N15 phage has also been shown to have a telomere resolution system similar to that of *Borrelia,* involving a tyrosine mediated cleavage-rejoining reaction (Huang et al. 2004) by a recombinase-like protein. The question of what happens at poxvirus telomeres now awaits resolution.

7.5 Xer recombination: a multifunctional recombination system

The Xer recombination system is remarkable for its adaptability, and unique in absolutely requiring two distinct recombinases for catalysis. It was first discovered in *E. coli* for its role in plasmid stability (Summers and Sherratt 1984), and subsequent studies revealed that it acted at the site *cer* of ColE1 when present in a plasmid dimer: accessory sequences bound by the DNA binding proteins ArgR and PepA provide a topological filter to ensure activity only in a dimeric molecule. However, only one pair of strands is exchanged at *cer,* by XerC, and resolution requires subsequent processing (McCulloch et al. 1994; Colloms et al. 1996). Another plasmid encoded recombination site, *psi* of pSC101, is also acted upon by Xer recombination (Cornet et al. 1994). Again, a topological filter ensures that only dimeric forms are recombined, provided by sequences bound by PepA and ArcA. However, at the *psi* site both XerC and XerD exchange a pair of strands, in that order, and thus complete the recombination reaction alone, yielding a four-noded catenated product (Colloms et al. 1996). XerC/D also acts in the resolution of chromosomal dimers by recombination at a site, *dif,* in the terminus region of the chromosome (Kuempel et al. 1991; Blakely et al. 1991). Unlike the plasmid encoded sites, *dif* has no apparent accessory sequences immediately adjacent, and during recombination it is XerD that exchanges the first pair of strands and XerC that resolves the Holliday junction intermediate. Some phages also integrate into the chromosome of bacteria using Xer-mediated recombination at the *dif* site in a non-canonical reaction, which is described in more detail in the following section (7.5.2) (Huber and Waldor 2002; Val et al. 2005). Thus the range of natural sites and the different ways in which they are acted upon by the pair of recombinases, demonstrates the diversity of reaction control mechanisms available to Xer recombination.

7.5.1 FtsK- an accessory protein for co-ordinating chromosome monomerisation and segregation?

One of the most intriguing recombination sites identified to date is the *dif* site. All prokaryotes examined to date appear to encode an Xer recombination system, most of them having two related recombinases, XerC and XerD, which act at *dif.* Recombination at *dif* is responsible for the resolution of chromosome dimers that arise by an odd number of homologous recombination events that produce a crossover, events which occur roughly every six generations in *E. coli* (Barre and Sherratt 2005). Resolution of dimeric chromosomes is a prerequisite for segregation. What makes *dif* so different from many other sites is that it appears to consist of a core recombination site without any accessory DNA sequences. However, recombination does require an accessory protein, FtsK (Aussel et al. 2002). FtsK is a membrane associated hexameric DNA translocase, related to the AAA^+ family of ATPases. The N-terminal portion is predicted to contain 4 transmembrane helices, and is required for viability due to interactions with proteins involved in cell division. The C-terminus contains the AAA^+ motifs, hexamerises and is an active translocase in the absence of the N-terminus. These two portions are connected by a long linker rich in proline and glycine, predicted to be very flexible. Various *in vitro* and *in vivo* analyses show that FtsK translocates along DNA in a directional manner (Pease et al. 2005), with the directionality imposed by polarised DNA sequences (KOPS) (Levy et al. 2005; Bigot et al. 2005). KOPS sequences are overrepresented in the terminus region and are polarised with the direction switching at *dif* itself. FtsK interacts directly with the recombinase XerD and this interaction stimulates recombination to proceed. FtsK has also been shown to interact with TopoIV, a type II topoisomerase with a preference for resolving catenation or knot nodes (Espeli et al. 2003). It is therefore possible that FtsK hexamers will translocate along DNA towards *dif* from both directions, possibly even along both sister chromosomes and this may push any chromosomal interlinks ahead and localise them to the *dif* region where they can be efficiently removed by TopoIV, then Xer recombination can be activated if needed. Recombination occurring after the removal of the interlinks between sister chromosomes would yield unlinked monomeric chromosomes which can be segregated to daughter cells. This proposal would provide a link between recombination and unlinking of chromosomes, through the accessory protein FtsK. It also would suggest a mechanism whereby a seemingly simple pair of recombination sites could be provided with reaction directionality and topological specificity. Another intriguing addition to this model comes from the finding that FtsK-XerC/D-*dif* can act to remove catenation nodes by multiple rounds of site-specific recombination, yielding unlinked products (Ip et al. 2003). Therefore, it seems that XerC/D can also contribute more actively in the unlinking process.

7.5.2 Phage CTX integration

The cholera toxin is encoded on a phage, CTX, and this phage is inserted into the chromosome at the *dif* site in *Vibrio cholerae* dependent upon XerC/D recombination (Huber and Waldor 2002). However, the reaction of phage insertion is not a canonical recombination reaction: it is the single stranded form of phage CTX that is used as the substrate with for Xer-mediated recombination (Val et al. 2005). The single stranded form of the phage can fold into a secondary structure to create a transiently double stranded recombination core site. A single pair of strand exchanges, catalysed by XerC inserts the phage into the chromosomal *dif* site, where it can then be converted to the double stranded form by replication, analogous to integron recombination (see Fig. 8).

8 Concluding remarks

Site-specific recombination plays a crucial role in the biology of microbes, and their mobile genetic elements, by precisely integrating, excising, and inverting segments of DNA. Reducing the fidelity of these reactions to allow more illegitimate crossovers to occur has enabled some systems to widen their target site selection or broaden their host range. It is intriguing that this amazing array of functions may have evolved from a humble monomeric topoisomerase precursor protein.

References

Abbani M, Iwahara M, Clubb RT (2005) The structure of the excisionase (Xis) protein from conjugative transposon Tn916 provides insights into the regulation of heterobivalent tyrosine recombinases. J Mol Biol 347:11-25

Abremski K, Hoess R (1984) Bacteriophage P1 site-specific recombination. Purification and properties of the Cre recombinase protein. J Biol Chem 259:1509-1514

Akopian A, He J, Boocock MR, Stark WM (2003) Chimeric recombinases with designed DNA sequence recognition. Proc Natl Acad Sci USA 100:8688-8691

Arciszewska LK, Grainge I, Sherratt DJ (1997) Action of site-specific recombinases XerC and XerD on tethered Holliday junctions. EMBO J 16:731-743

Arnold PH, Blake DG, Grindley NDF, Boocock MR, Stark WM (1999) Mutants of Tn*3* resolvase, which do not require accessory binding sites for recombination activity. EMBO J 18:1407–1414

Azaro MA, Landy A (1997) The isomeric preference of Holliday junctions influences resolution bias by lambda integrase. EMBO J 16:3744-3755

Azaro MA, Landy A (2002) The integration/excision cycle of lambda and other bacteriophages. In, NL Craig, R Craigie, M Gellert, A Lambowitz (eds.), Mobile DNA II. ASM Press, Washington, DC, pp. 117-148

Baba T , Ara T , Okumura Y , Hasegawa M , Takai Y , Okumura Y , Baba M , Datsenko KA , Tomita M , Wanner BL , Mori H (2006) Construction of *Escherichia coli* K-12 in-frame, single-gene knockout mutants—the Keio collection. *Mol Syst Biol* doi:10.1038/msb4100050

Burgess SM, Kleckner N (1999) Collisions between yeast chromosomal loci *in vivo* are governed by three layers of organization. Genes Dev 13:1871-1883

Bigot S, Saleh OA, Lesterlin C, Pages C, El Karoui M, Dennis C, Grigoriev M, Allemand JF, Barre FX, Cornet F (2005) KOPS: DNA motifs that control *E. coli* chromosome segregation by orienting the FtsK translocase. EMBO J 24:3770-3780

Biswas T, Aihara H, Radman-Livaja M, Filman D, Landy A, Ellenberger, T (2005) A structural basis for allosteric control of DNA recombination by lambda integrase. Nature 435:1059–1066

Blakely G, Colloms S, May G, Burke M, Sherratt D (1991) *Escherichia coli* XerC recombinase is required for chromosomal segregation at cell division. New Biol 3:789-798

Bode J, Schlake T, Iber M, Schubeler D, Seibler J, Snezhkov E, Nikolaev L (2000) The transgeneticist's toolbox: novel methods for the targeted modification of eukaryotic genomes. Biol Chem 381:801–813

Boocock MR, Zhu X, Grindley ND (1995) Catalytic residues of gamma delta resolvase act in cis. EMBO J 14:5129-5140

Buchholz F, Angrand PO, Stewart AF (1998) Improved properties of FLP recombinase evolved by cycling mutagenesis. Nat Biotechnol 16:657-662

Burns LS, Smith SG, Dorman CJ (2000) Interaction of the FimB integrase with the fimS invertible DNA element in *Escherichia coli in vivo* and *in vitro*. J Bacteriol 82:2953-2959

Chen Y, Narendra U, Iype L E, Cox MM, Rice PA (2000) Crystal structure of a Flp recombinase-Holliday junction complex: assembly of an active oligomer by helix swapping. Mol Cell 6:885–897

Chen Y, Rice P (2003) New insight into site-specific recombination from Flp recombinase-DNA structures. Annu Rev Biophys Biomol Struct 32:135–159

Cheng C, Kussie P, Pavletich N, Shuman S (1998) Conservation of structure and mechanism between eukaryotic topoisomerase I and site-specific recombinases. Cell 92:841-850

Collis CM, Kim MJ, Stokes HW, Hall RM (1998) Binding of the purified integron DNA integrase IntI1 to integron- and cassette associated recombination sites. Mol Microbiol 29:477–490

Cornet F, Mortier I, Patte J, Louarn JM (1994) Plasmid pSC101 harbors a recombination site, *psi*, which is able to resolve plasmid multimers and to substitute for the analogous chromosomal *Escherichia coli* site dif. J Bacteriol 176:3188-195

Crellin PK, Rood JI (1997) The resolvase/invertase domain of the site-specific recombinase TnpX is functional and recognizes a target sequence that resembles the junction of the circular form of the *Clostridium perfringens* transposon Tn*4451*. J Bacteriol 179:5148–5156

Dhar G, Sanders ER, Johnson RC (2004) Architecture of the Hin synaptic complex during recombination: the recombinase subunits translocate with the DNA strands. Cell 119:33-45

Esposito D, Scocca JJ (1997) The integrase family of tyrosine recombinases: evolution of a conserved active site domain. Nucleic Acids Res 25:3605–3614

Espeli O, Lee C, Marians KJ (2003) A physical and functional interaction between Escherichia coli FtsK and topoisomerase IV. J Biol Chem 278:44639-44644

Francia MV, Zabala JC, de la Cruz F, Garcia-Lobo JM (1999) The IntI1 integron integrase preferentially binds single-stranded DNA of the attC site. J Bacteriol 181:6844–6849

Futcher AB (1986) Copy number amplification of the 2 micron circle plasmid of *Saccharomyces cerevisiae*. J Theor Biol 119:197-204

Gally DL, Leathart J, Blomfield IC (1996) Interaction of FimB and FimE with the fim switch that controls the phase variation of type 1 fimbriae in *Escherichia coli* K-12. Mol Microbiol 21:725-738

Golic, MM, Golic KG (1996) A quantitative measure of the mitotic pairing of alleles in *Drosophila melanogaster* and the influence of structural heterozygosity. Genetics 143:385-400

Gopaul DN, Guo F, Van Duyne GD (1998) Structure of the Holliday junction intermediate in Cre-loxP site-specific recombination. EMBO J 17:4175–4187

Guo F, Gopaul DN, van Duyne GD (1997) Structure of Cre recombinase complexed with DNA in a site-specific recombination synapse. Nature 389:40–46

Gravel A, Fournier B, Roy PH (1998) DNA complexes obtained with the integron integrase IntI1 at the attI1 site. Nucleic Acids Res 26:4347–4355

Hall RM, Collis CM (1995) Mobile gene cassettes and integrons: capture and spread of genes by site-specific recombination. Mol Microbiol 15:593-600

Hickman AB, Waninger S, Scocca JJ, Dyda F (1997) Molecular organization in site-specific recombination: the catalytic domain of bacteriophage HP1 integrase at 2.7 A resolution. Cell 89:227-237

Hildebrandt ER, Cozzarelli NR (1995) Comparison of recombination *in vitro* and in *E. coli* cells: Measure of the effective concentration of DNA *in vivo.* Cell 81:331-340

Hoess R, Ziese M, Sternberg N (1982) P1 Site-specific recombination: nucleotide sequence of the recombining sites. Proc Natl Acad Sci USA 79:3398-3402

Huang WM, Joss L, Hsieh T, Casjens S (2004) Protelomerase uses a topoisomerase IB/Y-recombinase type mechanism to generate DNA hairpin ends. J Mol Biol 337:77-92

Huber KE, Waldor MK (2002) Filamentous phage integration requires the host recombinases XerC and XerD. Nature 417:656-659

Ip SC, Bregu M, Barre FX, Sherratt DJ (2003) Decatenation of DNA circles by FtsK-dependent Xer site-specific recombination. EMBO J 22:6399-6407

Kersulyte D, Mukhopadhyay AK, Shirai M, Nakazawa T, Berg, DE (2000) Functional organization and insertion specificity of IS607, a chimeric element of *Helicobacter pylori*. J Bacteriol 182:5300–5305

Kobryn K, Chaconas G (2001) The circle is broken: telomere resolution in linear replicons. Curr Opin Microbiol 4:558–564

Kobryn K, Chaconas G (2002) ResT, a telomere resolvase encoded by the Lyme disease spirochete Mol Cell 9:195-201

Kobryn K, Burgin AB, Chaconas G (2005) Uncoupling the chemical steps of telomere resolution by ResT. J Biol Chem 280:26788–26795

Kuempel PL, Henson JM, Dircks L, Tecklenburg M, Lim DF (1991) *dif*, a *recA*-independent recombination site in the terminus region of the chromosome of *Escherichia coli*. New Biol 3:799-811

Kwon HJ, Tirumalai R, Landy A, Ellenberger T (1997) Flexibility in DNA recombination: structure of the lambda integrase catalytic core. Science 276:126-131

Lee J, Whang I, Lee J, Jayaram M (1994) Directed protein replacement in recombination full sites reveals trans-horizontal DNA cleavage by Flp recombinase. EMBO J 13:5346-5354

Lee J, Jayaram M (1995) Role of partner homology in DNA recombination. Complementary base pairing orients the 5'-hydroxyl for strand joining during Flp site-specific recombination. J Biol Chem 270:4042–4052

Lee J, Jayaram M, Grainge I (1999) Wild-type Flp recombinase cleaves DNA in trans. EMBO J 18:784-791

Levy O, Ptacin JL, Pease PJ, Gore J, Eisen MB, Bustamante C, Cozzarelli NR (2005) Identification of oligonucleotide sequences that direct the movement of the *Escherichia coli* FtsK translocase. Proc Natl Acad Sci USA 102:17618-17623

Letzelter C, Duguet M, Serre M (2004) Mutational analysis of the archaeal tyrosine recombinase SSV1 untegrase suggests a mechanism of DNA cleavage in trans. J Biol Chem 279:28936-28944

Li W, Kamtekar S, Xiong Y, Sarkis GJ, Grindley ND, Steitz TA (2005) Structure of a synaptic γδ resolvase tetramer covalently linked to two cleaved DNAs. Science 309:1210-1215

McCulloch R, Coggins LW, Colloms SD, Sherratt DJ (1994) Xer-mediated site-specific recombination at cer generates Holliday junctions *in vivo*. EMBO J 13:1844-1855

McIlwraith MJ, Boocock MR, Stark WM (1997) Tn3 resolvase catalyses multiple recombination events without intermediate rejoining of DNA ends. J Mol Biol 266:108-121

Merickel SK, Haykinson MJ, Johnson RC (1998) Communication between Hin recombinase and Fis regulatory subunits during coordinate activation of Hin-catalyzed site-specific DNA inversion. Genes Dev 12:2803-2816

Nöllman M, He J, Byron O, Stark WM (2004) Solution structure of the Tn3 resolvase-crossover site synaptic complex. Mol Cell 16:127–137

Nünes-Duby SE, Azaro MA, Landy A (1995) Swapping DNA strands and sensing homology without branch migration in lambda site-specific recombination. Curr Biol 5:139-148

Pease PJ, Levy O, Cost GJ, Gore J, Ptacin JL, Sherratt D, Bustamante C, Cozzarelli NR (2005) Sequence-directed DNA translocation by purified FtsK. Science 307:586-590

Radman-Livaja M, Biswas T, Ellenberger T, Landy A, Aihara H (2006) DNA arms do the legwork to ensure the directionality of lambda site-specific recombination. Curr Opin Struct Biol 16: in press

Redinbo MR, Stewart L, Kuhn P, Champoux JJ, Hol WG (1998) Crystal structures of human topoisomerase I in covalent and noncovalent complexes with DNA. Science 279:1504-1513

Rowe-Magnus DA, Guerout A-M, Ploncard P, Dychinco B, Davies J, Mazel D (2001) The evolutionary history of chromosomal super-integrons provides an ancestry for multiresitant integrons. Proc Natl Acad Sci USA 98:652-657

Sauer B (1998) Inducible gene targeting in mice using the Cre/*lox* system. Methods Enzymol 14:381–392

Senecoff JF, Cox MM (1986) Directionality in FLP protein-promoted site-specific recombination is mediated by DNA-DNA pairing. J Biol Chem 261:7380–7386

Sherratt DJ, Arciszewska LK, Blakely G, Colloms S, Grant K, Leslie N, McCulloch R (1995) Site-specific recombination and circular chromosome segregation. Philos Trans R Soc Lond B Biol Sci 347:37-42

Smith MCM, Thorpe HM (2002) Diversity in the serine recombinases. Mol Micro 44:299–307

Smith SG, Dorman CJ (1999) Functional analysis of the FimE integrase of *Escherichia coli* K-12: isolation of mutant derivatives with altered DNA inversion preferences. Mol Microbiol 34:965-979

Stark WM, Boocock MR, Sherratt DJ (1989) Site-specific recombination by Tn3 resolvase. Trends Genet 5:304-309

Stark M, Grindley ND, Hatfull GF, Boocock MR (1991) Resolvase-catalysed reactions between res sites differing in the central dinucleotide of subsite I. EMBO J 10:3541–3548

Staczek P, Higgins NP (1998) Gyrase and Topo IV modulate chromosome domain size *in vivo.* Mol Microbiol 29:1435-1448

Subramanya HS, Arciszewska LK, Baker RA, Bird LE, Sherratt DJ, Wigley DB (1997) Crystal structure of the site-specific recombinase, XerD. EMBO J 16:5178-187

Summers DK, Sherratt DJ (1984) Multimerization of high copy number plasmids causes instability: ColE1 encodes a determinant essential for plasmid monomerization and stability. Cell 36:1097-1103

Thorpe HM, Smith MCM (1998) *In vitro* site-specific integration of bacteriophage DNA catalyzed by a recombinase of the resolvase/invertase family. Proc Natl Acad Sci USA 95:5505–5510

Val ME, Bouvier M, Campos J, Sherratt D, Cornet F, Mazel D, Barre FX (2005) The single-stranded genome of phage CTX is the form used for integration into the genome of *Vibrio cholerae.* Mol Cell 19:559–566

van de Putte P, Goosen N (1992) DNA inversions in phages and bacteria. Trends Genet 8:457-462

Wang H, Mullany P (2000) The large resolvase TndX is required and sufficient for integration and excision of derivatives of the novel conjugative transposon Tn*5397*. J Bacteriol 182:6577–6583

Wojciak JM, Connolly KM, Clubb RT (1999) NMR structure of the Tn916 integrase-DNA complex. Nat Struct Biol 6:366-373

Grainge, Ian
Department of Biochemistry, University of Oxford, South Parks Road, Oxford. OX1 3QU, United Kingdom

Sherratt, David J
Department of Biochemistry, University of Oxford, South Parks Road, Oxford. OX1 3QU, United Kingdom
sherratt@bioch.ox.ac.uk

V(D)J recombination: mechanism and consequences

Martin Gellert

Abstract

V(D)J recombination is responsible for assembling the functional immunoglobulin and T cell receptor genes in cells of the immune system. This specialized DNA rearrangement is initiated by the RAG1 and RAG2 proteins, lymphoid-specific factors which collaborate to make double-strand breaks at specific sites that flank segments of coding sequence. During breakage, the ends of the coding DNA are converted to DNA hairpins. Joining of the broken ends is then carried out by the non-homologous end-joining pathway, after the hairpin ends are cut open by the Artemis protein. The RAG1/2 complex can also carry out transpositional recombination, a reaction that helps to explain how the RAG proteins work, and supports previous suggestions that V(D)J recombination evolved from a mobile DNA.

1 Introduction

How does the body cope with all the various infectious agents that may attack it? This is mainly the job of the adaptive immune system, which relies on its ability to produce such a large variety of antigen-binding proteins (immunoglobulins and T cell receptors) that any infection will meet an immune response. The diverse repertoire is present before the immune challenge, on B and T cells that each express only one antigen receptor species. The adaptive immune system is called "anticipatory", because the specific receptors are made before they are needed.

In many animal species including humans and mice, the diversity of the receptors is due to the assembly of the functional genes from gene fragments by the process called V(D)J recombination. In the germ-line chromosomal arrangement, the genes for immunoglobulin and T cell receptor chains are linear arrays of fragments separated into variable (V), diversity (D), and joining (J) regions, all linked to a constant (C) region. While lymphoid cells are developing, V(D)J recombination joins the segments into functional coding sequences. In mammals, there are seven antigen receptor loci: the immunoglobulin (Ig) H, κ and λ loci, and the T cell receptor (TCR) α, β, γ, and δ loci. All the loci contain sets of V and J segments, and the IgH and TCR β and δ loci also have D segments between the V's and J's. A typical locus has multiple segments of each kind, up to fifty or more of one category, and the combinatorial choice of segments, combined with sequence changes during recombination, leads to a large number of possible Ig and TCR chains. Joining of VJ or VDJ segments produces a variable region exon, which is coupled by RNA splicing to the constant region that makes up the rest of the antigen receptor.

Immunoglobulins and TCRs are heterodimeric proteins: IgH chains combine with either a κ or λ light chain, and the TCR proteins are αβ or γδ dimers. The pairing of two chains, each already diversified, increases the diversity of the antigen receptors even more. The number of possible Ig or TCR molecules must be above 10^7. By comparison, there are only 20, 000-30, 000 genes in the mammalian genome. Because each B or T cell expresses only one antigen receptor, the initial diversity of the cells produced by the immune system is also large.

This article presents a survey of work on V(D)J recombination; other reviews go into greater detail (Lewis 1994; Fugmann et al. 2000a; Gellert 2002; Jung et al. 2006) and discuss the different diversity-generating mechanisms used in species such as chickens or sheep. Regulation of V(D)J recombination, which is not a topic of this article, has also been reviewed (see e.g. (Jung et al. 2006)).

2 General properties of V(D)J recombination

2.1 Recombination sites

V(D)J recombination takes place at "recombination signal sequences" (RSSs) next to each V, D, and J segment. An RSS contains fairly well conserved heptamer and nonamer sequences (Fig. 1A) separated by 12 or 23 bp (±1 bp) of non-conserved spacer DNA. Spacer length is important because efficient recombination occurs only between RSSs with 12 and 23-bp spacers (Tonegawa 1983). At each antigen receptor locus, the RSSs are partitioned so that all elements of the same type have the same spacer length. For example, the Igκ locus has all its V segments attached to 12-spacer RSSs and all J segments to 23-spacer RSSs, so that V-to-J joining is greatly favored over V-to-V or J-to-J joining, which would be futile (Fig. 1B). Similar rules hold at the IgH and TCRβ and δ loci, with the complication that D segments must join to V's on one side and J's on the other, so they are flanked by RSSs of suitable spacer lengths on each side (see Fig. 1B).

DNA breakage occurs between the heptamer and its neighboring coding segment, and recombination then joins the coding regions and in parallel joins the pair of RSSs.

V(D)J rearrangement differs from many other types of site-specific recombination, in that the essential sequence is almost exclusively on one side of the junction; the flanking sequence can be varied almost at will. (A few flanking sequences, such a run of T's reading 5' to 3' into the heptamer, do not work well). This one-sided recognition is similar to that of many transposons (Sakano et al. 1979), a theme that will be one focus of this article.

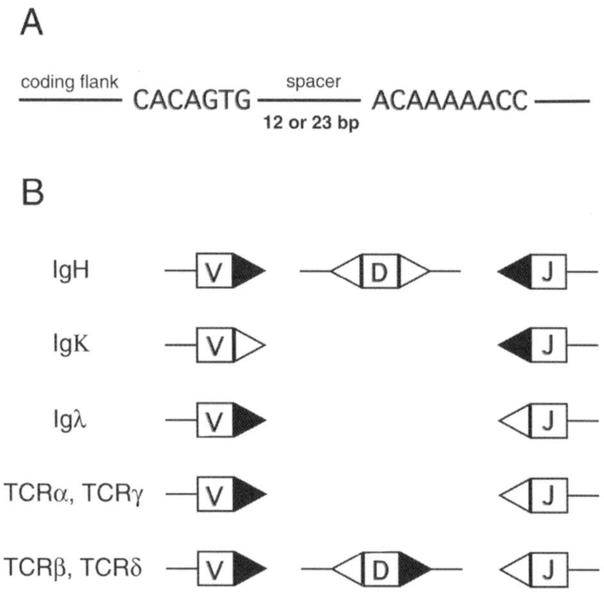

Fig. 1. (A) The consensus heptamer and nonamer of a recombination signal sequence (RSS), with the two possible spacer lengths indicated. Cleavage takes place at the left edge of the heptamer. (B) RSS arrangements at immunoglobulin and T cell receptor loci. An open triangle is a 12-spacer RSS, a black triangle is a 23-spacer RSS. At each locus, all elements of one type (V, D, or J) are flanked by the same type of RSS.

RSS sequences are not tightly conserved, as shown by variations at the antigen receptor loci and by tests of mutated substrates. The three heptamer nucleotides next to the recombination site are the most important, but single mutations at the other heptamer positions still allow recombination. Single changes in the nonamer are even more permissive, but positions 5, 6, and 7 are more conserved than the others (Hesse et al. 1989). RSS variations in the antigen receptor loci may affect the usage of gene segments. For example, the RSSs of mouse Igκ are closer to the consensus sequence than those of the alternative Igλ locus. This may explain why Igκ light chains are used more often than Igλ in mouse immunoglobulins. The preference can be repeated in synthetic substrates, in which a κ RSS pair is used at least 100-fold more frequently than a typical λ RSS pair (Ramsden and Wu 1991; Feeney et al. 2000).

Even with satisfactory RSSs and correct 12/23 pairing, some RSS pairs recombine much more efficiently than others. This "beyond 12/23" effect is thought to be important in targeting recombination, for example by preventing direct V to J joining (skipping the D regions) at TCRβ (Bassing et al. 2000).

In the Ig and TCR loci, RSSs are usually arranged so that the joined coding segments remain in the chromosome and the joined RSSs are excised on a circular DNA which is later lost from the cells. However, some segments are inverted so that both the coding joint and signal joint stay in the chromosome. One example is the human Igκ locus, where roughly half the V segments are inverted relative to J and C.

In the picture of V(D)J recombination that has developed during the last several years, the process has two distinct stages. In the first stage, the RAG1 and RAG2 proteins cooperate to recognize the RSSs and to ensure their correct 12/23 pairing, and to break the DNA between each heptamer and the neighboring coding sequence. In the later stage, the factors that are also used in other types of "non-homologous end joining" (NHEJ) act to process and link the ends into coding joints and signal joints. The biochemistry of the first stage is by now fairly clear, and many factors required in the second stage are known. The two parts of the process will be discussed separately.

3 The RAG genes and proteins

The RAG1 and RAG2 proteins carry out the enzymatic first step of the V(D)J reaction. As the only lymphoid-specific factors needed for V(D)J recombination, they normally restrict this reaction to precursors of B and T cells. However, co-expression of RAG1 and RAG2 leads to recombination of test substrates in non-lymphoid mammalian cells, where it would not normally occur ((Oettinger et al. 1990)), implying that all other required factors must be generally available. Conversely, mice with disruptions of either the RAG1 or RAG2 gene completely lack V(D)J recombination (Mombaerts et al. 1992; Shinkai et al. 1992), and therefore contain no mature B or T cells, but have no other defects.

Recombination requires the cooperation of RAG1 and RAG2, but large parts of both RAG genes can be deleted while still retaining recombination activity. The mouse RAG1 protein, whose full length is 1040 amino acids, can have its N-terminal 383 and C-terminal 32 residues deleted. A large section of the mouse RAG2 C-terminus can also be deleted without destroying activity. Activity requires only the first 383 amino acids, out of the full-length sequence of 527, even though the dispensable region is highly conserved. It has been noted that the truncated proteins can initiate recombination efficiently, but are not so good at completing it (Steen et al. 1999). A necessary post-cleavage complex (see below) may be less stable with the truncated proteins. The dispensable parts of both proteins may also help discriminate between receptor loci, as mentioned later.

3.1 DNA cleavage by the RAG proteins

The enzymatic function of the RAG1/2 complex is to cut the DNA between the RSS heptamer and the flanking sequence. The purified RAG1 and RAG2 proteins by themselves are sufficient for cleavage (McBlane et al. 1995). The truncated forms of the proteins described above were used in all earlier studies, because of solubility problems with the full-length proteins. These shorter proteins (mouse RAG1 amino acids 384-1008 and RAG2 amino acids 1-383 or 1-387) have been purified from various systems: insect cells infected with baculovirus vectors, or HeLa cells infected with recombinant vaccinia virus, or mouse cells transfected with an expression vector. The protein is usually linked to a fusion partner (maltose binding protein or glutathione S-transferase) and/or a polyhistidine tail, for ease of purification. More recently, active full-length RAG2 has been purified (Elkin et al. 2003; Tsai and Schatz 2003; Swanson et al. 2004); so far, full-length preparations of RAG1 have very low activity.

In the presence of Mn^{2+}, RAG1/2 efficiently cuts an RSS in a DNA fragment, to yield blunt 5'-phosphorylated signal ends and hairpin coding ends that retain the full coding sequence (McBlane et al. 1995). These are the same cleavage products found in vivo (Roth et al., 1992a, Roth et al., 1992b). Cleavage by RAG1/2 is a two-step process. A nick is made at the 5' end of the signal heptamer, leaving a 5'-phosphoryl group on the RSS and a 3'-hydroxyl on the coding end (this is shown for coupled cleavage at a pair of RSSs in Fig. 2). The second step joins this 3'-hydroxyl to the phosphoryl group at the same nucleotide position on the opposite strand, resulting in a DNA hairpin coding end and blunt signal end. Both steps require the RAG1/2 complex and the specific sequence of the RSS.

When RAG1/2 acts alone, it cleaves a 12-RSS more efficiently than a 23-RSS. Cleavage is increased, particularly at a 23-RSS, by adding one of the chromosomal high-mobility-group proteins, HMGB1 or HMGB2 (Sawchuk et al. 1997; van Gent et al. 1997). These are non-specific DNA binding and bending proteins; they may act to deform a 23-RSS, allowing better RAG1/2 binding. It is possible the HMGB proteins are significant cofactors of V(D)J recombination in vivo. However, no good test has yet been done.

3.2 Coupled cleavage

Coupled cleavage obeying the 12/23 rule has been shown in vivo (Steen et al., 1996), and coupled cleavage of a 12/23 RSS pair by RAG1/2 can likewise be demonstrated in vitro. Although the RAG proteins cut a single RSS in Mn^{2+}, cleavage in Mg^{2+} requires a pair of RSSs. Coupled cleavage is produced with the purified RAG proteins (van Gent et al. 1996), or with crude extracts (Eastman et al. 1996). Cleavage is best with a 12/23 pair, and depending on reaction conditions the preference (over a 12/12 pair, for example) can approximate the 50-fold factor found in V(D)J recombination in vivo (Kim and Oettinger 1998).Thus the 12/23 rule of V(D)J recombination is determined at the level of RAG interactions with the RSSs.

Even in Mg^{2+}, the RAG proteins nick DNA at a single RSS. The 12/23 pair is required only for hairpinning, reflecting the use of a synaptic complex in this second step. Nicking at RSSs in vivo is similarly independent of synaptic complex formation (Yu and Lieber 2000).

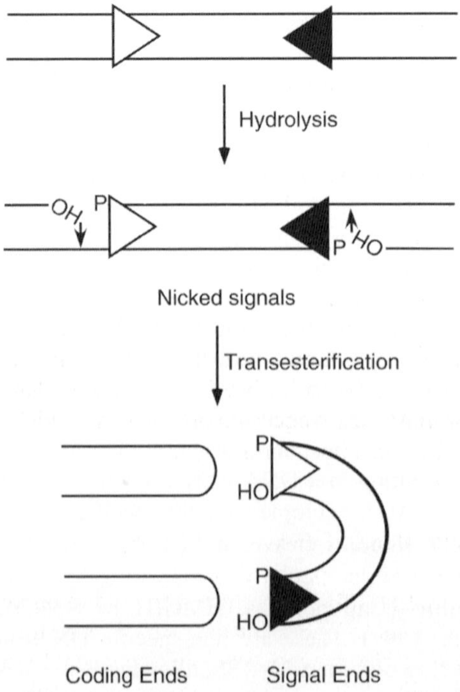

Fig. 2. DNA cleavage by RAG1/2. Coupled cleavage at a pair of RSSs is shown. The first step nicks the 5' ends of the RSS heptamers, leaving the coding flanks with 3'-OH groups which then attack the opposite strands to generate hairpin coding ends and blunt signal ends.

3.3 RSS recognition

Sequence recognition for cleavage by RAG1/2 is similar to that for V(D)J recombination (Cuomo et al. 1996; Ramsden et al. 1996). Once again, the three bases of the heptamer nearest the cutting site are most important, but specifically for the hairpinning step; nicking is less sensitive. If all bases of the heptamer are mutated, no hairpins are made, but there is still some imprecise nicking near the position where the heptamer border would be (i.e. 19 bp from the nonamer in a 12-spacer RSS). The length of the spacer is also significant for RAG1/2 cleavage. The normal 12 and 23 bp spacer lengths differ by almost exactly one turn of DNA; proteins bound to the heptamer and nonamer would be in the same rotational phase on a 12-RSS or a 23-RSS. Changes of the spacer length conform to this view. If the spacer length is changed by half a turn, to 18 or 29 bp, cleavage is inhibited, but it is partly restored at the next integral number of turns, 33 or 34 bp.

Thus the heptamer and nonamer are to some extent recognized independently. An isolated heptamer directs some cleavage to the normal site, and a nonamer by itself induces the RAG proteins to nick where the heptamer border would be.

When the two motifs are in the right helical phase, they increase cleavage synergistically, but in the wrong phase they conflict.

Distortion of the DNA structure may be important for hairpinning. It is known that the CACA/GTGT sequence that is found in the heptamer has a very distorted structure, both in solution and in crystals. And if the RSS in an oligonucleotide is made single-stranded, with only the coding flank remaining double-stranded, hairpinning is still efficient (Cuomo et al. 1996; Ramsden et al. 1996) , suggesting that the heptamer may become partly unpaired in normal cleavage. The nonamer has no effect in this substrate; only the heptamer is recognized. Thus cleavage at the heptamer may involve both specific sequence recognition and DNA unpairing. In fact, the RAG proteins may also help unwind the RSS for hairpinning, as is true for the Tn5 transposase, where a base near the center of the hairpin is flipped out of the double helix and stabilized by binding to a tryptophan. Recent studies of abasic substrates for RAG cleavage support such a model (G.J. Grundy, J.E. Hesse, and M. Gellert, manuscript submitted).

3.4 RAG protein binding to DNA

Although the core RAG1 and RAG2 proteins can bind to duplex DNA separately, the RAG1/2 complex binds to RSS sequences much more tightly and specifically (Hiom and Gellert 1997).The synaptic complex between 12 and 23-spacer RSSs and RAG1/2 (usually with HMG1 added) is even more stable, being resistant to a 100-fold excess of non-specific DNA (Hiom and Gellert 1998). A 12/23 signal pair is strongly preferred over 12/12 or 23/23 combinations. Thus the specificity of RAG cleavage and of V(D)J recombination appears to be largely set by the time the synaptic complex is formed, although possible added specificity at the cleavage step has also been suggested (West and Lieber 1998).

The synaptic complex assembles stepwise in vitro. A complete set of RAG1 and RAG2 protomers (two RAG2's and at least two RAG1's (Bailin et al. 1999; Mundy et al. 2002; Swanson, 2002) binds one RSS, and the other RSS then enters as bare DNA (Jones and Gellert 2002; Mundy et al. 2002). This is an obligate pathway; binding of RAG1/2 to both RSSs before they are combined prevents synaptic complex formation. The synaptic complex is most specific for a 12/23 RSS pair when RAG1/2 is first bound to the 12 RSS.

Recent results show that the same assembly pathway is used in vivo. When cells engaged in V(D)J recombination are tested for nicks at RSSs (nicking can occur before the synaptic complex is formed), the nicks are found only at 12 RSSs, not at their 23 RSS partners (Curry et al. 2005). It seems that once the 23 RSS is also engaged, recombination proceeds to completion too quickly for a population of nicked 23 RSSs to accumulate. Quite remarkably, the same preference for12 RSSs is found, whether they are at V segments in Igκ, at J segments in TCRα, or at D segments in IgH. These observations open the possibility of a previously unsuspected level of regulation of V(D)J recombination.

After in vitro cleavage, the RAG proteins remain bound to DNA. A very tight complex of RAG1/2 with the signal ends can be isolated (Agrawal and Schatz

1997; Jones and Gellert 2001). It is possible that the same complex in vivo may largely account for the long lifetime of broken signal ends before they are joined. A less stable complex of RAG1/2 containing both the signal ends and coding ends has also been described (Hiom and Gellert 1998). Such a four-ended "post-cleavage complex" plays a part in the later joining process in a cell-free system that carries out complete V(D)J recombination, which requires the continued presence of the RAG proteins after cleavage (Ramsden et al. 1997). Studies of mutations described below also support a role for the RAG proteins in vivo in V(D)J joining.

3.5 DNA transposition by RAG1/2

A better understanding of the RAG1/2 complex has come from its similarities to transpositional recombination systems. The most striking result is that RAG1/2 can transpose RSS ends into a target DNA in vitro (Agrawal et al. 1998; Hiom et al. 1998). Transposition requires a 12/23 pair of RSSs, and can produce either a coupled insertion of both ends or a single-ended attack. Transposition can be coupled to cleavage at the RSSs, or can use a pre-cut pair of RSS ends, in either case joining the RSS to target DNA exactly at the heptamer end. The sequence of the DNA target site is not specific, but GC-rich regions are somewhat preferred. When both RSSs attack the target DNA, they insert into opposite strands at positions staggered by 3-5 bp. An attack with such a defined stagger is typical of transposases.

Transpositional attack by purified RAG1/2 can be quite efficient; up to 5% of the RSS ends may insert into another DNA. Transposition inside cells is enormously less efficient. A few naturally occurring transposition events have been identified in human cells (Messier et al. 2003), and more recently a low level of transposition has been detected upon RAG expression in human (Chatterji et al. 2006) or mouse (Reddy et al., 2006) cell lines. This last cited paper is the most successful effort so far, resulting in roughly one transposition per 50,000 V(D)J recombination events.

A low level of RAG-mediated transposition can also be observed in S. cerevisiae (Clatworthy et al. 2003); here the level is about the same as that of signal joint formation, but both require a very sensitive assay for detection. Another reaction that resembles transposase activity is the production of hybrid joints or open-and-shut joints by RAG1/2 in vitro (Melek et al. 1998). An open-and-shut junction rejoins a signal end to the coding end from which it was cleaved (detectable when the joining is imprecise). A hybrid joint links a signal end to the coding end of the partner RSS after coupled cleavage. These reactions are related to "disintegration" catalyzed by HIV integrase and other transposases, in which a transpositional strand transfer is reversed by cutting the attacking DNA end from a target DNA and in the same reaction re-sealing the target. Hybrid and open-and-shut joints are also found as rare byproducts of V(D)J recombination in cells, but it is an unsettled issue whether they can be produced by RAG1/2 alone, or also require the action of non-homologous end-joining factors after RAG cleavage (Han et al. 1997; Sekiguchi et al. 2001; Raghavan et al. 2006).

The RAG proteins evidently have a preference for hairpin DNA structures, because such self-complementary DNA sequences are selectively targeted for a form of transposition in vitro (Lee et al. 2002). Mechanistically, this process has much in common with hybrid joint formation, because the opposing hairpins on the two strands are attacked roughly on center, so the normal 4-5 base pair stagger of transposition is not present.

Now that it is known that DNA hairpins are intermediates in several other transposition reactions (Tn5/Tn10 and the hAT group of transposons (Kennedy et al. 1998; Bhasin et al. 1999; Zhou et al. 2004)), the hairpins made in RAG cleavage can be seen as a natural outcome of the RAG system's relation to this family. RAG-initiated V(D)J recombination deviates from normal transposition by usually joining the signal ends to each other instead of using them to insert into other DNA. Such "transposon circles" are minor side products of some other transposition processes, but represent the vast majority of outcomes in the RAG system.

3.6 implications of RAG1/2 transposition for the evolution of the immune system and for chromosomal translocations

All jawed vertebrates contain a diversified immune system, and have similar RAG1 and RAG2 genes. The level of conservation of both protein sequences ranges from 50 and 90% between sharks, fishes, amphibians, birds and mammals. Below the evolutionary level of the sharks, there is a discontinuity; in the lower eukaryotes there is no V(D)J recombination.

RAG-mediated transposition has led to interesting speculation about the evolutionary origin of V(D)J recombination. Even earlier, the normal inverted repeat arrangement of RSSs in the Ig and TCR loci, similar to that at transposon ends. had led to the proposal that V(D)J recombination might derive from a mobile genetic element (Sakano et al. 1979). The unusual compactness of the RAG locus also pointed to a transpositional origin (Oettinger et al. 1990). In all species tested, the RAG1 and RAG2 genes are nearest neighbors, convergently transcribed, and in most genomes, such as Xenopus, chicken, mouse, and human, lack introns in either structural gene. Only the RAG1 genes of some fish have introns.

Until recently, all evidence fit the suggestion that RAG1 and RAG2 arrived together at the level of early jawed vertebrates, because the gene cluster and its expression through V(D)J recombination had been found in sharks and higher vertebrates, but not in more primitive animals. However, more recent sequence analyses have identified genes with some homology to RAG1 (Transib transposase genes) in insects, sea urchin, and even in some plants (Kapitonov and Jurka 2005). In most cases, no RAG2 homolog has been found, but in sea urchin a gene with weak homology to RAG2 lies adjacent to the previously identified Transib (Fugmann et al. 2006). It is interesting to ask whether primitive RAG1 acquired a functional RAG2 partner later in evolution, whether the "modern" RAG1 and RAG2 arrived together in jawed vertebrates by a separate jump, or whether ancestral versions of RAG1 and RAG2 were present together but acquired their immune system function only later.

Lymphoid tumors commonly contain chromosomal translocations. Some of them must arise from aberrant V(D)J recombination, because one break point is at a V(D)J site in an Ig or TCR gene cluster. It has been pointed out that some of these translocations could result from the attempted transposition by RAG1/2, either single-ended or double-ended, and hypothetical reaction schemes have been proposed (Hiom et al. 1998). Some of these events would link an RSS-ended chromosome fragment to a non-RSS site on the partner chromosome, and a few such translocations have been identified (discussed in (Melek and Gellert 2000)).

3.7 Sequence motifs and mutational studies of the RAG proteins

So far there is no direct structural information on the RAG1 or RAG2 core domains, but sequence analysis has produced useful insights, helped by comparisons among the ~900 species whose RAG genes have been sequenced. The RAG2 core (aa 1-383) is proposed to contain a six-fold repeat of 50 residues each of a so-called kelch motif (named after a *Drosophila* regulatory protein), (Callebaut and Mornon 1998; Aravind and Koonin 1999). Each repeat has a four-stranded twisted antiparallel beta sheet, with the whole pattern circling back on itself like a six-bladed propeller. Similar structures are known to be involved in protein-protein interactions, and it is suggested that this portion of RAG2 is involved in binding RAG1, and possibly also DNA. The beta propeller region by itself (aa 1-351) is sufficient to complement RAG1 for V(D)J cleavage (G. Grundy and M. Gellert, unpublished results).

Beyond this domain, RAG2 has a highly acidic region (residues 352-410), reported to be involved in histone binding (West et al. 2005). Further downstream, there is a Cys-His rich "PHD" motif spanning residues 420-480 whose structure has been determined (Elkin et al. 2005). This domain, related to homeodomain zinc fingers, binds phosphoinositides and is also likely to be involved in binding to chromatin.

Sequence analysis of the RAG1 core in comparison to the Hermes transposase (Zhou et al. 2004) has suggested an RNaseH-like fold containing the three conserved acidic residues D600, D708, and E962 (see below). This fold is often found in transposases. There is also a zinc-finger motif starting at position 720.

The crystal structure of a segment (aa 265-380) of the N-terminal non-core region of RAG1 has been shown to contain a RING finger motif as well as a zinc finger of the C_2H_2 type (Bellon et al. 1997). These have been suggested to be dimerization motifs, but could also be involved in protein-DNA interactions. The RING finger domain has been shown to have a ubiquitin ligase activity (see below).

Intensive mutagenesis has been used to identify the functionality of each protein. We have already pointed out that the N-terminus of RAG1 and the C-terminus of RAG2 can be deleted, and the proteins retain basal recombination activity. More systematic studies have identified catalytically essential residues presumably involved in the metal-binding site. A common motif in transposases is an acidic amino acid triad, often DDE or DDD, that coordinates the catalytic divalent metal ion. In RAG1, mutation of any one of the three residues D600, D708, and E962 abolishes cleavage in vitro and recombination in vivo (Kim et al. 1999; Landree et al. 1999; Fugmann et al. 2000b).

A major result of these investigations is that the metal-binding component of the active site is located in RAG1. No essential acidic residues have turned up in RAG2 (Landree et al. 1999). But some other mutations in RAG2 interfere with activity (Qiu et al. 2001) , so it is possible that the full active site contains elements from both proteins.

Other mutated forms of both RAG1 and RAG2 carry out RSS cleavage and transposition but do not complete joining in vivo, supporting the idea that the RAG proteins have a necessary function after cleavage (Qiu et al. 2001; Yarnell Schultz et al. 2001). Some cases of human severe combined immune deficiency (scid) or the related condition known as Omenn's syndrome have been shown to be due to mutations of RAG1 or RAG2 (Schwarz et al. 1996; Villa et al. 1998; Villa et al. 2001). Mutations such as frameshifts that would destroy RAG activity lead to complete immunodeficiency, whereas hypomorphic mutations cause a partial scid phenotype or the more complex properties of Omenn's syndrome (Villa et al. 2001).

3.8 Other functions of the RAG proteins

An unusual RAG reaction that re-cuts signal joints both in vitro and in vivo by nicking each strand at the junction of the two RSSs, without the formation of a hairpin, has been reported. This "nick-nick" reversal of signal joining has been proposed to be at least partly responsible for the prolonged lifetime of signal ends in cells (Neiditch et al. 2002).

It has also been reported that RAG1/2, in addition to forming hairpins, can cut them open in an in vitro reaction (Besmer et al. 1998; Shockett and Schatz 1999; Qiu et al. 2001), and also has a more general endonuclease activity that cuts single-strand DNA and flaps.The biological significance of these activities is unclear because they overlap those of the Artemis protein, and Artemis-deficient cells are ineffective at least in opening hairpins (Moshous et al. 2001; Ma et al. 2002; Ma et al. 2005).

The RING finger domain in the N-terminal part of RAG1 has a ubiquitin ligase activity (Jones and Gellert 2003; Yurchenko et al. 2003) that acts at least on a highly conserved lysine (K233) in the same general region. RAG1 is also ubiquitylated in vivo, but the biological effects of this modification have not yet been worked out.

The RAG proteins may have a role in regulating which loci are recombined. The N-terminally truncated RAG1 and C-terminally truncated RAG2 proteins used in cell-free experiments support recombination of test substrates in vivo, although their efficiency is lower than the full-length versions. Thus, a naturally occurring RAG1 N-terminal truncation supports rearrangement of TCR but not Ig genes (Noordzij et al. 2000). The function of the RAG2 C-terminus has been studied in more detail. In pre-B cell lines or in mouse primary cells, the truncated form of RAG2 allows Igκ and IgH D to J rearrangement, but IgH V to DJ recombination is greatly reduced (Kirch et al. 1998; Akamatsu et al. 2003). The histone binding properties of the acidic and PHD portions of the RAG2 C-terminal domain may well be involved (see section 3.7).

4 End processing and joining in V(D)J recombination

The later steps of V(D)J recombination use the non-homologous end-joining pathway (NHEJ) that is fully discussed in the article by T. Wilson in this volume. This section will concentrate on those aspects that are special to V(D)J recombination. Because the NHEJ factors are globally expressed, ectopic expression of RAG1/2 allows V(D)J recombination to be studied in diverse cell types, conveniently including the standard cell lines with mutations in NHEJ factors.

In outline, the pathway of V(D)J joining works as follows. After cleavage, the RAG1/2 complex holds on to the broken ends and, in some way not yet clarified, helps to transfer the ends to the NHEJ machinery. (With RAG mutants that do not form a stable post-cleavage complex, the ends can be diverted into homologous recombination (Lee et al. 2004)). The one step that is special to V(D)J joining and not involved in most NHEJ repair is the cutting of the hairpin coding ends in order to make them available for joining. This step is performed by the Artemis protein in conjunction with DNA-PK_{CS}, as shown by studies both in vivo and in vitro. Patients with a defect in Artemis are immunodeficient because of impaired V(D)J recombination (Moshous et al. 2001), as are mice with the gene knocked out (Rooney et al. 2002), and cells defective in either DNA-PK_{CS} or Artemis display reduced hairpin opening (Roth et al. 1992; Rooney et al. 2003). Finally, in the presence of DNA-PK_{CS}, Artemis is able to cut DNA hairpins in vitro. The kinase activity of DNA-PK_{CS} is required for this reaction, but only to autophosphorylate DNA-PK_{CS}. Phosphorylation of Artemis is not necessary for its endonuclease activity, or for V(D)J recombination in cells (Goodarzi et al., 2006).

Because signal joints and coding joints are both made by a similar NHEJ process, one might expect them to have similar structures, but this is not true. Signal joints are usually precise end-to-end fusions of two heptamer sequences (though they occasionally have nucleotides added between the heptamers). Coding joints are much more variable, often with a few nucleotides deleted from one or both ends, and also with nucleotides added in the junction (Tonegawa 1983).

These sequence changes are highly significant for the immune system. In both Ig and TCR genes, the junctional sequence is part of the antigen binding site, so local additions or deletions make the receptors much more diverse than simple combinatorial joining of the gene segments. But this variability also leads to a lot of wastage, because the length of DNA added or lost is essentially random, so two-thirds of coding joints change the reading frame and cause premature termination of the protein chain. If the rearrangement is unsuccessful, a second attempt is possible on the other allele. Or, in loci with a V-J array, a second recombination on the same allele is possible, by the use of a V region upstream and a J region downstream of the erroneous junction.

Nucleotide insertions in coding joints can be templated or non-templated. Templated "P nucleotide" insertions (P for palindromic) (Lafaille et al. 1989), resulting from off-center nicking of the hairpin DNA intermediates, add a few nucleotides complementary to the terminal bases of the coding end nearest the RSS. Not all coding junctions have P nucleotide insertions; if the hairpin is nicked exactly at its center, there is no self-complementary overhang, or an overhang may be resected before the ends are joined.

Non-templated "N region" insertions up to 15 nucleotides in length are added by the enzyme terminal deoxynucleotidyl transferase (TdT). TdT is normally expressed only in early lymphoid cells where V(D)J recombination is active, so these insertions are relatively specific to this type of recombination. TdT adds deoxynucleotides to the ends of DNA chains without the need for a template, but with a preference for G residues that results in N regions being generally GC-rich.

The reasons for nucleotide loss in coding junctions are not so well understood. Removal of a few nucleotides occurs in all cell types that perform V(D)J recombination, including non-lymphoid cells that ectopically express RAG1 and RAG2. It has been suggested that the exonuclease and/or endonuclease activity of the Artemis protein may be responsible (Ma et al. 2005), and an attractive possibility is that end-trimming may follow immediately after hairpin opening by Artemis. As mentioned above, nuclease activities of RAG1/2 itself could also be involved, but no in vivo evidence supporting this idea has yet been found.

The role of the Mre11/Rad50/Nbs1 complex (MRN) in mammalian NHEJ is unclear (see the article by Wilson in this volume), but some of its properties might suggest a possible role in V(D)J joining. For one thing, MRN can nick DNA hairpins in vitro, and so could serve as a backup activity for Artemis (Artemis knockouts are not fully defective in hairpin opening (Rooney et al. 2002)). MRN can also bridge DNA ends, similarly to Ku and DNA-PK_{CS}. Such bridging has been shown for Mre11 alone (Paull and Gellert 2000)and for an Mre11-Rad50 complex (de Jager et al. 2001). Because knockouts of either Mre11, Rad50, or Nbs1 are lethal, the role of MRN in V(D)J joining, or more generally in mammalian NHEJ, is not yet fully worked out.

V(D)J recombination is largely confined to the G1 phase of the cell cycle. This restriction is at least partly determined by cell cycle control of the amount of RAG2. At the G1-S transition, RAG2 is phosphorylated on T490, and targeted for export to the cytoplasm, followed by ubiquitylation and degradation (Li et al. 1996; Mizuta et al. 2002). Limiting the V(D)J process to G1 will direct joining into the NHEJ track, instead of allowing an undesired homolog search in G2, when the chromosomes have been duplicated. This restriction will add to the channeling of joining into the NHEJ pathway by properly functioning RAG1, noted above.

As compared to the fairly detailed picture of RAG1/2 action, the understanding of later stages of V(D)J recombination is still incomplete, and indeed there may be more NHEJ factors still to be discovered. However, the rapid progress being made in dissecting NHEJ repair both in vivo and in vitro should soon allow a more complete picture to be built up.

References

Agrawal A, Eastman QM, Schatz DG (1998) Transposition mediated by RAG1 and RAG2 and its implications for the evolution of the immune system. Nature 394:744-751

Agrawal A, Schatz DG (1997) RAG1 and RAG2 form a stable postcleavage synaptic complex with DNA containing signal ends in V(D)J recombination. Cell 89:43-53

Akamatsu Y, Monroe R, Dudley DD, Elkin SK, Gartner F, Talukder SR, Takahama Y, Alt FW, Bassing CH, Oettinger MA (2003) Deletion of the RAG2 C terminus leads to impaired lymphoid development in mice. Proc Natl Acad Sci USA 100:1209-1214

Aravind L, Koonin EV (1999) Gleaning non-trivial structural, functional and evolutionary information about proteins by iterative database searches. J Mol Biol 287:1023-1040

Bailin T, Mo X, Sadofsky MJ (1999) A RAG1 and RAG2 tetramer complex is active in cleavage in V(D)J recombination. Mol Cell Biol 19:4664-4671

Bassing CH, Alt FW, Hughes MM, D'Auteuil M, Wehrly TD, Woodman BB, Gartner F, White JM, Davidson L, Sleckman BP (2000) Recombination signal sequences restrict chromosomal V(D)J recombination beyond the 12/23 rule. Nature 405:583-586

Bellon SF, Rodgers KK, Schatz DG, Coleman JE, Steitz TA (1997) Crystal structure of the RAG1 dimerization domain reveals multiple zinc-binding motifs including a novel zinc binuclear cluster. Nat Struct Biol 4:586-591

Besmer E, Mansilla-Soto J, Cassard S, Sawchuk DJ, Brown G, Sadofsky M, Lewis SM, Nussenzweig MC, Cortes P (1998) Hairpin coding end opening is mediated by RAG1 and RAG2 proteins. Mol Cell 2:817-828

Bhasin A, Goryshin IY, Reznikoff WS (1999) Hairpin formation in Tn5 transposition. J Biol Chem 274:37021-37029

Callebaut I, Mornon JP (1998) The V(D)J recombination activating protein RAG2 consists of a six- bladed propeller and a PHD fingerlike domain, as revealed by sequence analysis. Cell Mol Life Sci 54:880-891

Chatterji M, Tsai CL, Schatz DG (2006) Mobilization of RAG-generated signal ends by transposition and insertion *in vivo*. Mol Cell Biol 26:1558-1568

Clatworthy AE, Valencia MA, Haber JE, Oettinger MA (2003) V(D)J recombination and RAG-mediated transposition in yeast. Mol Cell 12:489-499

Cuomo CA, Mundy CL, Oettinger MA (1996) DNA sequence and structure requirements for cleavage of V(D)J recombination signal sequences. Mol Cell Biol 16:5683-5690

Curry JD, Geier JK, Schlissel MS (2005) Single-strand recombination signal sequence nicks *in vivo*: evidence for a capture model of synapsis. Nat Immunol 6:1272-1279

de Jager M, van Noort J, van Gent DC, Dekker C, Kanaar R, Wyman C (2001) Human Rad50/Mre11 is a flexible complex that can tether DNA ends. Mol Cell 8:1129-1135

Eastman QM, Leu TM, Schatz DG (1996) Initiation of V(D)J recombination *in vitro* obeying the 12/23 rule. Nature 380:85-88

Elkin SK, Ivanov D, Ewalt M, Ferguson CG, Hyberts SG, Sun ZY, Prestwich GD, Yuan J, Wagner G, Oettinger MA, Gozani OP (2005) A PHD finger motif in the C terminus of RAG2 modulates recombination activity. J Biol Chem 280:28701-28710

Elkin SK, Matthews AG, Oettinger MA (2003) The C-terminal portion of RAG2 protects against transposition *in vitro*. EMBO J 22:1931-1938

Feeney AJ, Tang A, Ogwaro KM (2000) B-cell repertoire formation: role of the recombination signal sequence in non-random V segment utilization. Immunol Rev 175:59-69

Fugmann SD, Lee AI, Shockett PE, Villey IJ, Schatz DG (2000a) The RAG proteins and V(D)J recombination: complexes, ends, and transposition. Annu Rev Immunol 18:495-527

Fugmann SD, Messier C, Novack LA, Cameron RA, Rast JP (2006) An ancient evolutionary origin of the Rag1/2 gene locus. Proc Natl Acad Sci USA 103:3728-3733

Fugmann SD, Villey IJ, Ptaszek LM, Schatz DG (2000b) Identification of two catalytic residues in RAG1 that define a single active site within the RAG1/RAG2 protein complex. Mol Cell 5:97-107

Gellert M (2002) V(D)J recombination:RAG proteins, repair factors, and regulation. Annu Rev Biochem 71:101-132

Goodarzi A, Yu Y, Riballo E, Douglas P, Walker SA, Harer C, Ye R, Marchetti C, Morrice N, Jeggo PA, Lees-Miller SP (2006) DNA-PK autophosphorylation facilitates Artemis endonuclease activity. EMBO J: in press

Han JO, Steen SB, Roth DB (1997) Ku86 is not required for protection of signal ends or for formation of nonstandard V(D)J recombination products. Mol Cell Biol 17:2226-2234

Hesse JE, Lieber MR, Mizuuchi K, Gellert M (1989) V(D)J recombination: a functional definition of the joining signals. Genes Dev 3:1053-1061

Hiom K, Gellert M (1997) A stable RAG1-RAG2-DNA complex that is active in V(D)J cleavage. Cell 88:65-72

Hiom K, Gellert M (1998) Assembly of a 12/23 paired signal complex: A critical control point in V(D)J recombination. Mol Cell 1:1011-1019

Hiom K, Melek M, Gellert M (1998) DNA transposition by the RAG1 and RAG2 proteins: A possible source of oncogenic translocations. Cell 94:463-470

Jones JM, Gellert M (2001) Intermediates in V(D)J recombination: A stable RAG1/2 complex sequesters cleaved RSS ends. Proc Natl Acad Sci USA 98:12926-12931

Jones JM, Gellert M (2002) Ordered assembly of the V(D)J synaptic complex ensures accurate recombination. EMBO J 21:4162-4171

Jones JM, Gellert M (2003) Autoubiquitylation of the V(D)J recombinase protein RAG1. Proc Natl Acad Sci USA 100:15446-15451

Jung D, Giallourakis C, Mostoslavsky R, Alt FW (2006) Mechanism and control of V(D)J recombination at the immunoglobulin heavy chain locus. Annu Rev Immunol 24:541-570

Kapitonov VV, Jurka J (2005) RAG1 core and V(D)J recombination signal sequences were derived from Transib transposons. PLoS Biol 3:e181

Kennedy AK, Guhathakurta A, Kleckner N, Haniford DB (1998) Tn10 transposition via a DNA hairpin intermediate. Cell 95:125-134

Kim DR, Dai Y, Mundy CL, Yang W, Oettinger MA (1999) Mutations of acidic residues in RAG1 define the active site of the V(D)J recombinase. Genes Dev 13:3070-3080

Kim DR, Oettinger MA (1998) Functional analysis of coordinated cleavage in V(D)J recombination. Mol Cell Biol 18:4679-4688

Kirch SA, Rathbun GA, Oettinger MA (1998) Dual role of RAG2 in V(D)J recombination: catalysis and regulation of ordered Ig gene assembly. EMBO J 17:4881-4886

Lafaille JJ, DeCloux A, Bonneville M, Takagaki Y, Tonegawa S (1989) Junctional sequences of T cell receptor gamma delta genes: implications for gamma delta T cell lineages and for a novel intermediate of V-(D)-J joining. Cell 59:859-870

Landree MA, Wibbenmeyer JA, Roth DB (1999) Mutational analysis of RAG1 and RAG2 identifies three catalytic amino acids in RAG1 critical for both cleavage steps of V(D)J recombination. Genes Dev 13:3059-3069

Lee GS, Neiditch MB, Salus SS, Roth DB (2004) RAG proteins shepherd double-strand breaks to a specific pathway, suppressing error-prone repair, but RAG nicking initiates homologous recombination. Cell 117:171-184

Lee GS, Neiditch MB, Sinden RR, Roth DB (2002) Targeted transposition by the V(D)J recombinase. Mol Cell Biol 22:2068-2077

Lewis SM (1994) The mechanism of V(D)J joining: Lessons from molecular, immunological, and comparative analyses. Adv Immunol 56:27-150

Li Z, Dordai DI, Lee J, Desiderio S (1996) A conserved degradation signal regulates RAG-2 accumulation during cell division and links V(D)J recombination to the cell cycle. Immunity 5:575-589

Ma Y, Pannicke U, Schwarz K, Lieber MR (2002) Hairpin opening and overhang processing by an Artemis/DNA-dependent protein kinase complex in nonhomologous end joining and V(D)J recombination. Cell 108:781-794

Ma Y, Schwarz K, Lieber MR (2005) The Artemis:DNA-PKcs endonuclease cleaves DNA loops, flaps, and gaps. DNA Repair (Amst) 4:845-851

McBlane JF, van Gent DC, Ramsden DA, Romeo C, Cuomo CA, Gellert M, Oettinger MA (1995) Cleavage at a V(D)J recombination signal requires only RAG1 and RAG2 proteins and occurs in two steps. Cell 83:387-395

Melek M, Gellert M (2000) RAG1/2-mediated resolution of transposition intermediates: Two pathways and possible consequences. Cell 101:625-633

Melek M, Gellert M, van Gent DC (1998) Rejoining of DNA by the RAG1 and RAG2 proteins. Science 280:301-303

Messier TL, O'Neill JP, Hou SM, Nicklas JA, Finette BA (2003) *In vivo* transposition mediated by V(D)J recombinase in human T lymphocytes. EMBO J 22:1381-1388

Mizuta R, Mizuta M, Araki S, Kitamura D (2002) RAG2 is down-regulated by cytoplasmic sequestration and ubiquitin-dependent degradation. J Biol Chem 277:41423-41427

Mombaerts P, Iacomini J, Johnson RS, Herrup K, Tonegawa S, Papaioannou VE (1992) RAG-1-deficient mice have no mature B and T lymphocytes. Cell 68:869-877

Moshous D, Callebaut I, de Chasseval R, Corneo B, Cavazzana-Calvo M, Le Deist F, Tezcan I, Sanal O, Bertrand Y, Philippe N, Fischer A, de Villartay J-P (2001) Artemis, a novel DNA double-strand break repair/V(D)J recombination protein, is mutated in human severe combined immune deficiency. Cell 105:177-186

Mundy CL, Patenge N, Matthews AG, Oettinger MA (2002) Assembly of the RAG1/RAG2 synaptic complex. Mol Cell Biol 22:69-77

Neiditch MB, Lee GS, Huye LE, Brandt VL, Roth DB (2002) The V(D)J recombinase efficiently cleaves and transposes signal joints. Mol Cell 9:871-878

Noordzij JG, Verkaik NS, Hartwig NG, de Groot R, van Gent DC, van Dongen JJ (2000) N-terminal truncated human RAG1 proteins can direct T-cell receptor but not immunoglobulin gene rearrangements. Blood 96:203-209

Oettinger MA, Schatz DG, Gorka C, Baltimore D (1990) RAG-1 and RAG-2, adjacent genes that synergistically activate V(D)J recombination. Science 248:1517-1523

Paull TT, Gellert M (2000) A mechanistic basis for Mre11-directed DNA joining at microhomologies. Proc Natl Acad Sci USA 97:6409-6414

Qiu JX, Kale SB, Yarnell Schultz H, Roth DB (2001) Separation-of-function mutants reveal critical roles for RAG2 in both the cleavage and joining steps of V(D)J recombination. Mol Cell 7:77-87

Raghavan SC, Tong J, Lieber MR (2006) Hybrid joint formation in human V(D)J recombination requires nonhomologous DNA end joining. DNA Repair (Amst) 5:278-285

Ramsden DA, McBlane JF, van Gent DC, Gellert M (1996) Distinct DNA sequence and structure requirements for the two steps of V(D)J recombination signal cleavage. EMBO J 15:3197-3206

Ramsden DA, Paull TT, Gellert M (1997) Cell-free V(D)J recombination. Nature 388:488-491

Ramsden DA, Wu GE (1991) Mouse κ light-chain recombination signal sequences mediate recombination more frequently than do those of λ light chain. Proc Natl Acad Sci USA 88:10721-10725

Reddy YV, Perkins EJ, Ramsden DA (2006) Genomic instability due to V(D)J recombination-associated transposition. Genes Dev 20:1575-1582

Rooney S, Alt FW, Lombard D, Whitlow S, Eckersdorff M, Fleming J, Fugmann S, Ferguson DO, Schatz DG, Sekiguchi J (2003) Defective DNA repair and increased genomic instability in Artemis-deficient murine cells. J Exp Med 197:553-565

Rooney S, Sekiguchi J, Zhu C, Cheng HL, Manis J, Whitlow S, DeVido J, Foy D, Chaudhuri J, Lombard D, Alt FW (2002) Leaky Scid phenotype associated with defective V(D)J coding end processing in Artemis-deficient mice. Mol Cell 10:1379-1390

Roth DB, Nakajima PB, Menetski JP, Bosma MJ, Gellert M (1992) V(D)J recombination in mouse thymocytes: double-strand breaks near T cell receptor δ rearrangement signals. Cell 69:41-53

Roth DB, Menetski JP, Nakajima PB, Bosma MJ, Gellert M (1992) V(D)J recombination: broken DNA molecules with covalently sealed (hairpin) coding ends in scid mouse thymocytes. Cell 70:983-991

Sakano H, Huppi K, Heinrich G, Tonegawa S (1979) Sequences at the somatic recombination sites of immunoglobulin light-chain genes. Nature 280:288-294

Sawchuk DJ, Weis-Garcia F, Malik S, Besmer E, Bustin M, Nussenzweig MC, Cortes P (1997) V(D)J recombination: modulation of RAG1 and RAG2 cleavage activity on 12/23 substrates by whole cell extract and DNA-bending proteins. J Exp Med 185:2025-2032

Schwarz K, Gauss GH, Ludwig L, Pannicke U, Li Z, Lindner D, Friedrich W, Seger RA, Hansen-Hagge TE, Desiderio S, Lieber MR, Bartram CR (1996) RAG mutations in human B cell-negative SCID. Science 274:97-99

Sekiguchi JA, Whitlow S, Alt FW (2001) Increased accumulation of hybrid V(D)J joins in cells expressing truncated versus full-length RAGs. Mol Cell 8:1383-1390

Shinkai Y, Rathbun G, Lam K-P, Oltz EM, Stewart V, Mendelsohn M, Charron J, Datta M, Young F, Stall AM, Alt FW (1992) RAG-2-deficient mice lack mature lymphocytes owing to inability to initiate V(D)J rearrangement. Cell 68:855-867

Shockett PE, Schatz DG (1999) DNA hairpin opening mediated by the RAG1 and RAG2 proteins. Mol Cell Biol 19:4159-4166

Steen SB, Gomelsky L, Roth DB (1996) The 12/23 rule is enforced at the cleavage step of V(D)J recombination *in vivo*. Genes Cells 1:543-553

Steen SB, Han JO, Mundy C, Oettinger MA, Roth DB (1999) Roles of the "dispensable" portions of RAG-1 and RAG-2 in V(D)J recombination. Mol Cell Biol 19:3010-3017

Swanson PC (2002) A RAG-1/RAG-2 tetramer supports 12/23-regulated synapsis, cleavage, and transposition of V(D)J recombination signals. Mol Cell Biol 22:7790-7801Swanson PC, Volkmer D, Wang L (2004) Full-length RAG-2, and not full-length RAG-1, specifically suppresses RAG-mediated transposition but not hybrid joint formation or disintegration. J Biol Chem 279:4034-4044

Tonegawa S (1983) Somatic generation of antibody diversity. Nature 302:575-581

Tsai CL, Schatz DG (2003) Regulation of RAG1/RAG2-mediated transposition by GTP and the C-terminal region of RAG2. EMBO J 22:1922-1930

van Gent DC, Hiom K, Paull TT, Gellert M (1997) Stimulation of V(D)J cleavage by high mobility group proteins. EMBO J 16:2265-2670

van Gent DC, Ramsden DA, Gellert M (1996) The RAG1 and RAG2 proteins establish the 12/23 rule in V(D)J recombination. Cell 85:107-113

Villa A, Santagata S, Bozzi F, Giliani S, Frattini A, Imberti L, Gatta LB, Ochs HD, Schwarz K, Notarangelo LD, Vezzoni P, Spanopoulou E (1998) Partial V(D)J recombination activity leads to Omenn syndrome. Cell 93:885-896

Villa A, Sobacchi C, Notarangelo LD, Bozzi F, Abinun M, Abrahamsen TG, Arkwright PD, Baniyash M, Brooks EG, Conley ME, Cortes P, Duse M, Fasth A, Filipovich AM, Infante AJ, Jones A, Mazzolari E, Muller SM, Pasic S, Rechavi G, Sacco MG, Santagata S, Schroeder ML, Seger R, Strina D, Ugazio A, Valiaho J, Vihinen M, Vogler LB, Ochs H, Vezzoni P, Friedrich W, Schwarz K (2001) V(D)J recombination defects in lymphocytes due to RAG mutations: severe immunodeficiency with a spectrum of clinical presentations. Blood 97:81-88

West KL, Singha NC, De Ioannes P, Lacomis L, Erdjument-Bromage H, Tempst P, Cortes P (2005) A direct interaction between the RAG2 C terminus and the core histones is required for efficient V(D)J recombination. Immunity 23:203-212

West RB, Lieber MR (1998) The RAG-HMG1 complex enforces the 12/23 rule of V(D)J recombination specifically at the double-hairpin formation step. Mol Cell Biol 18:6408-6415

Yarnell Schultz H, Landree MA, Qiu JX, Kale SB, Roth DB (2001) Joining-deficient RAG1 mutants block V(D)J recombination *in vivo* and hairpin opening *in vitro*. Mol Cell 7:65-75

Yu K, Lieber MR (2000) The nicking step in V(D)J recombination is independent of synapsis: implications for the immune repertoire. Mol Cell Biol 20:7914-7921

Yurchenko V, Xue Z, Sadofsky M (2003) The RAG1 N-terminal domain is an E3 ubiquitin ligase. Genes Dev 17:581-585

Zhou L, Mitra R, Atkinson PW, Hickman AB, Dyda F, Craig NL (2004) Transposition of hAT elements links transposable elements and V(D)J recombination. Nature 432:995-1001

Gellert, Martin

Laboratory of Molecular Biology, National Institute of Diabetes and Digestive and Kidney Diseases, National Institutes of Health, Building 5, Room 241, 9000 Rockville Pike, Bethesda, MD 20892-0540, USA
gellert@helix.nih.gov

Nonhomologous end-joining: mechanisms, conservation and relationship to illegitimate recombination

Thomas E. Wilson

Abstract

Illegitimate recombination (IR) is the exchange of genetic information by formation of junctions between nonhomologous chromosome segments. IR is intimately associated with the process of nonhomologous end-joining (NHEJ) that is dependent on Ku, DNA ligase IV, and associated proteins. Other processes also appear to be at play at in IR which have varying degrees of overlap with NHEJ, homologous recombination and single-strand break repair, and which depend on structural features of the initiating double-strand breaks (DSBs). This review takes a broad view of these issues by considering the conservation of Ku-dependent NHEJ and related processes from bacteria to man. Both extensive conservation and marked variability are evident, providing a rich future framework for correlating life cycles, DSB rejoining mechanisms, repair accuracy, and other selectable evolutionary benefits of NHEJ.

1 Introduction

Recombination is the exchange of genetic information by formation of junctions between existing but previously separate chromosomal regions. The majority of this volume addresses the mechanisms and consequences of homologous recombination (HR), whose hallmark is the presence of common sequences at the point of information exchange. This chapter addresses illegitimate recombination (IR). IR is defined by its contrast to HR, in that long homologous sequences are not apparent at the junction of new molecule formation. Clearly, distinct mechanisms must be at play. Foremost among these is the pathway of double-strand break (DSB) repair that has come to be called nonhomologous end-joining (NHEJ). Unfortunately, this term is often loosely applied and variably used to refer both to the general process of IR joining as well as a specific enzymatic pathway. In fact, the joining of DNA ends without homology encompasses distinct mechanisms that overlap with each other and with HR. Beyond mechanism, I also consider the conservation of the main Ku-dependent NHEJ pathway across phylogeny, as well as the outcomes of NHEJ and its deficiency. Just as the repair of DSBs by HR is beneficial even without true recombination, many outcomes of NHEJ are beneficial, explaining its substantial conservation despite an associated increased frequency of mutagenic IR.

2 DNA mechanisms of nonhomologous end-joining

2.1 Double strand breaks

NHEJ begins with a DSB; at present there are no known mechanisms for transposase-independent nonhomologous joining via nicks. A DSB is the least stable and most threatening of all DNA lesions because the linear integrity of the chromosome has been lost. But a DSB is neither a single nor a simple entity, and its precise structure will have profound impacts on its repair (Fig. 1). Every DSB is by definition a compound lesion in which at least two DNA damage events have occurred near each other and on opposite strands. Moreover, DSB termini will vary with the agents that create them. Restriction endonucleases typically create simple religatible DSBs. Of greater physiological importance are clustered chemical lesions created by oxidizing agents, ionizing radiation, and radiomimetic chemicals. Here, termini may be associated with base loss or terminal blocking lesions (i.e. fragments of the sugar-phosphate backbone) created either primarily or when repair enzymes attempt to excise a damaged nucleotide (Friedberg et al. 2005). Other sources of DSBs include replication fork failure (Haber 1999) or simple mechanical stress on a chromosome, such as occurs during mitosis.

2.2 Overhang-to-overhang joining

The one constant feature of a DSB is that the enthalpy provided by base-pairing in any overhangs is insufficient to overcome the entropy gained by separation of the ends. More accurately, the free energy relationship dictates an association/dissociation equilibrium that will change in a predictable way with overhang length or GC content (Daley and Wilson 2005; Sandoval and Labhart 2004). It is much more difficult to predict whether association and dissociation are kinetically equivalent for a single break *in vivo*, as compared to a population of randomly mixed ends *in vitro*. DSBs may appear with time if dissociation is in fact irreversible in the absence of protein catalysis due to chromosomal tension forces that might "pull apart" ends even when the equilibrium favors association. Overcoming both entropic gain and tension forces through the input of protein binding energy is thus critical for any joining mechanism.

The limiting case of NHEJ is when DSB overhangs are fully compatible and unblocked, so that repair can be achieved by simple religation (Fig. 1D). Importantly, any overhangs at a DSB comprised of only two strand lesions will be complementary by definition as they represent the two halves of the original duplex (Fig. 1A). However, such overhangs need not be directly ligatible if base loss or terminal blocking lesions are present. Thus, additional processing by trimming and

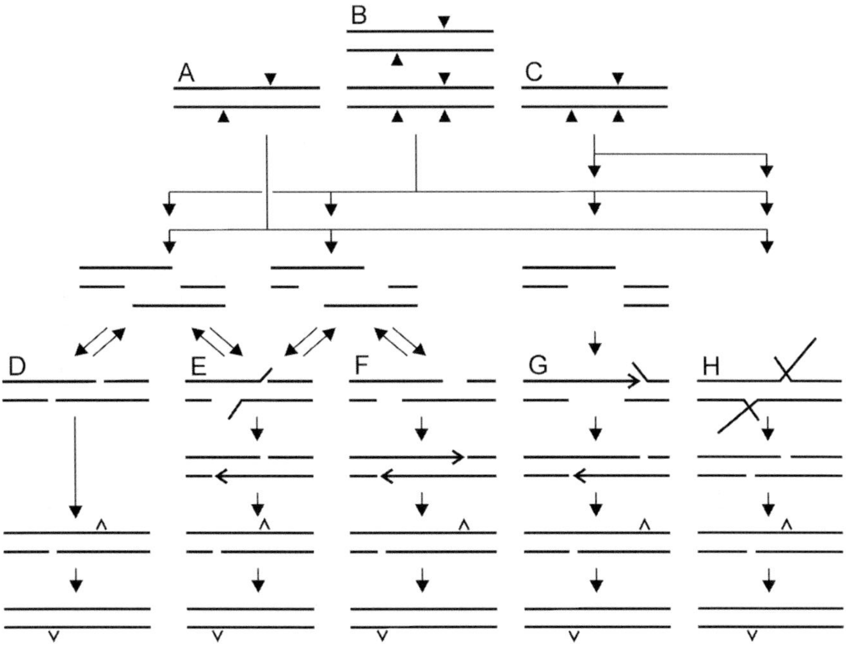

Fig. 1. Disposition of DNA strands during NHEJ and IR. Three distinct classes of DSBs are illustrated (top): a simple 2-lesion DSB (A), two simultaneous DSBs at distant chromosomal sites (B), and a complex 3-lesion DSB (C). Each can give rise to various DNA joining mechanisms, including simple religation (D), imprecise rejoining via misalignment of overhangs (E), processing-dependent but precise rejoining of overhangs (F), polymerization across a break (G), and MMEJ (H). Note that the different outcomes are initially in competition and in equilibrium, and that the accuracy of each mechanism varies. Strand lesions are indicated with a filled triangle, polymerization by an arrowhead, and ligation by a carat.

filling enzymes can be important even when overhangs are complementary (Fig. 1F). The challenges for the repair machinery are first to find what might be very limited complementarity, as little as one base pair. Further, it may be necessary to choose between competing fortuitous overhang pairings (Fig. 1E), only one of which will lead to accurate repair. Finally, NHEJ repair enzymes that are dependent on a duplex, including polymerases, ligases and perhaps nucleases, must be uniquely suited to dealing with the limiting substrate presented by annealed overhangs.

Overhangs that are not fully complementary can only occur when three or more strand lesions gave rise to the DSB ends being paired. At a single DSB this means that at least one end was multiply damaged. Accurate NHEJ of such DSBs may be impossible if part of the original DNA duplex was irretrievably lost. Incompatible overhangs might also arise when attempting to join ends from different DSBs (Fig. 1B). This is obviously mutagenic and will lead to IR, but possible if even one base-pair can be found. There is the additional challenge of removing any unpaired nucleotides, but this is similar to the processing of damaged compatible ends, making rejection of IR joints difficult.

2.3 Blunt end joining and polymerization across the break

Not all DSBs will have useful overhangs. The limiting case is when ends are unblocked and blunt, again allowing simple religation. More general are complex or mixed DSB ends where ligation is not possible, such as opposed 5' and 3' overhangs, fully non-complementary overhangs, or an overhang and a blunt end (Fig. 1C). Such ends might be made blunt by trimming or filling. Alternatively, specialized polymerases might synthesize across the DSB breakpoint, extending the 3' terminus of one end using a strand from the other end as template, but with no initiating primer template base pair(s) (Fig. 1G) (Pfeiffer et al. 1994). Such polymerization might still lead to accurate repair if the intervening duplex was not irretrievably lost.

2.4 Use of internal microhomologies

A fundamentally different mode of DSB joining resects or unwinds into the adjacent DNA duplex to expose hidden base-pairing potential (Fig. 1H). Bases paired in this way were obviously not paired in the parent duplex so the outcome is always mutagenic. They are inferred in the final joint as nucleotides that could have arisen from either of the parent DSB ends. Such nucleotides are known as "microhomologies" because they represent a common junction sequence that is too short (typically 1-10 base pairs) to be accounted for by HR. The term microhomology-mediated end-joining (MMEJ) has thus been applied to this joining process (Ma et al. 2003). Importantly, microhomologies observed at IR junctions might have arisen from either MMEJ or overhang pairing; the mechanism cannot be inferred from joint sequence alone unless the structures of the initiating DSBs are known.

2.5 The balance between joining modes

What determines whether overhang pairing, cross-break polymerization or resection is used? Initial insight can be gained by understanding that nearly all of the above mechanisms are driven by base pairing. Thus, outcomes will be strongly influenced by the relative energetics of that pairing, and by the kinetics of the different mechanisms that create and use it. When an overhang pairing of significant length is present, this would represent an energetically and kinetically favorable mechanism that would be expected to predominate. By a similar logic, correct overhang alignments would be expected to predominate over misalignments. In contrast, limited pairing at complex overhangs may be energetically unfavorable or slow enough that the balance may tip toward cross-break polymerization or resection.

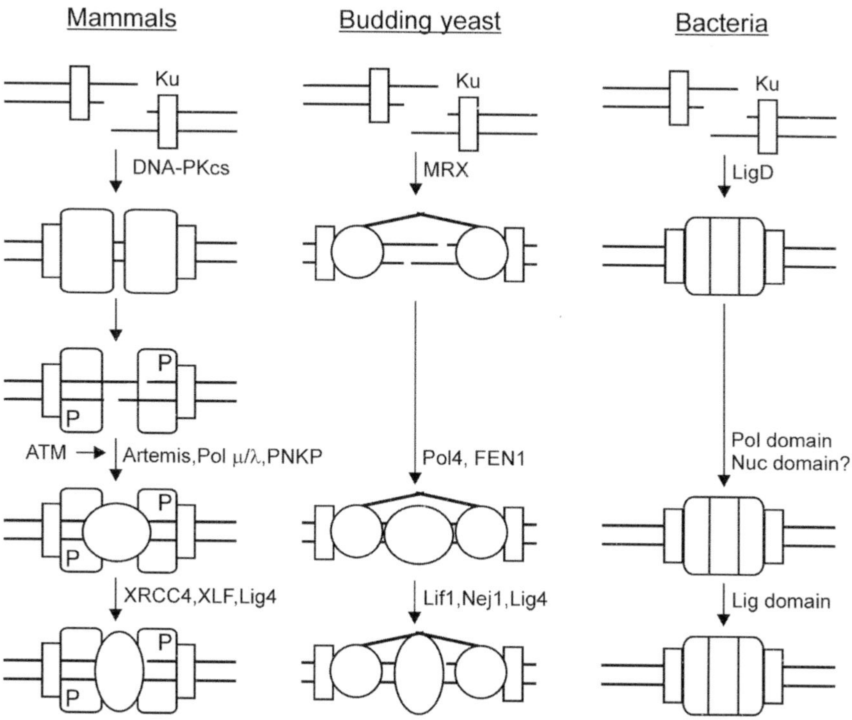

Fig. 2. Schematic of Ku-dependent NHEJ mechanisms. The three best characterized NHEJ systems are illustrated, from mammals, budding yeast, and bacteria. Other organisms are similar to these, but with notable differences in the proteins available (Table 1). In all organisms, end binding by Ku is an early and perhaps first step. Further binding and bridging of ends by distinct proteins is followed by end processing when needed, and finally ligation. The drawing is highly schematized because more precise interactions and mechanisms are not known. See text for further discussion.

3 Protein pathways for nonhomologous end-joining

3.1 Ku- and Lig4-dependent NHEJ

The first breakthrough into a mechanism for IR/NHEJ (Fig. 2) came in 1986 when it was discovered that the nuclear autoantigen Ku is a protein that binds strongly and specifically to linear but not circular DNA (Mimori and Hardin 1986). This property can now be understood based on crystallographic evidence that Ku, a heterodimer of ~70 and ~80 kDa subunits, forms an integral closed ring through which a DSB end is passed (Walker et al. 2001). Ku80 was found to be mutated in a radiosensitive Chinese hamster ovary (CHO) cell line (xrs-6,), all but establishing its role in the early steps of a DSB repair pathway (Taccioli et al. 1994). The second breakthrough was the recognition that Ku is the primary DNA-binding component of the DNA-dependent protein kinase (DNA-PK) (Gottlieb and Jackson 1993), whose catalytic subunit (DNA-PK_{cs}) is stimulated by free DSB ends and again shown to be deficient in a radiosensitive CHO line (V-3) (Peterson et al. 1995). DNA-PK_{cs} is a member of the polyinositol kinase family of enzymes, but clearly acts as a protein rather than a lipid kinase. Perhaps the most important substrate of DNA-PK_{cs} is DNA-PK_{cs} itself, with autophosphorylation controlling both its kinase activity and the accessibility of the bound DNA ends (Reddy et al. 2004). DNA-PK_{cs} itself binds DNA independently of Ku, probably by winding the DNA strands into a complex system of grooves and channels in this very large 469 kDa protein (Rivera-Calzada et al. 2005). Biochemical and structural evidence points to DNA-bound DNA-PK_{cs} as a main contributor to bridging of the two DSB ends (DeFazio et al. 2002; Spagnolo et al. 2006).

Insight into downstream NHEJ steps came with the identification that the XRCC4 protein, deficient in yet another radiosensitive CHO cell line (XR-1) (Li et al. 1995), interacted with DNA ligase IV (Lig4) via association of a dimeric coiled-coil region in XRCC4 and an inter-BRCT domain linker in Lig4 (Grawunder et al. 1997; Sibanda et al. 2001). It is consistently seen that Lig4 is the only ligase capable of coordinating with Ku and DNA-PK_{cs} during NHEJ (e.g. Ma et al. 2004), and in fact Lig4 has no other known function. XRCC4 both stabilizes and recruits Lig4 (Bryans et al. 1999; Teo and Jackson 2000). XRCC4 can also bind DNA, but nothing is known of the contribution of this binding to NHEJ, or of higher order XRCC4 multimers observed *in vitro* (Modesti et al. 1999; 2003). XRCC4 is a target of phosphorylation by DNA-PK_{cs}, but this in fact appears unnecessary for NHEJ (Yu et al. 2003). Recently, yet another component of this ligase complex was identified that interacts directly with XRCC4. It was predicted to fold similarly to XRCC4, with a globular head and extend alpha-helical segment, leading to the name XRCC4-like factor (XLF, also called Cernunnos) (Ahnesorg et al. 2006; Buck et al. 2006). It will be of considerable interest to establish the stoichiometry of XLF in the DNA ligase IV complex, whether it forms true heterodimers with XRCC4, regulates Lig4 binding, and other important issues.

The above proteins can account for simple religation NHEJ. For complex joints, two polymerases, Pol μ and Pol λ, have been directly implicated in mammalian NHEJ (Ma et al. 2004; Nick McElhinny et al. 2005; Capp et al. 2006). These enzymes are related to each other, and to the larger Pol X family of polymerases typified by Pol β. Unlike Pol β, Pol μ and Pol λ each contain amino-terminal BRCT domains that partially account for their NHEJ function by interaction with Ku and DNA ligase IV (Ma et al. 2004; Mahajan et al. 2002). Moreover, the catalytic domains of Pol μ and Pol λ display unusual properties such as a high error rate with a propensity toward frameshifts and an ability to bypass abasic sites (Bebenek et al. 2003; Covo et al. 2004). Combined with extensive crystallographic analyses (Garcia-Diaz et al. 2006), these properties indicate that Pol μ and Pol λ have the expected reduced dependence on a stable primer-template pair (2.2). Pol μ may in fact catalyze cross-break polymerization (2.3) by placing an active site loop in the position ordinarily occupied by the proximal template strand (Nick

McElhinny et al. 2005). Such differences may allow Pol μ and Pol λ to be optimally efficient for different joint classes, explaining why individual knockouts have only subtle DSB repair phenotypes (Nick McElhinny et al. 2005).

The mammalian nuclease known to function in NHEJ is Artemis/Snm1C, discovered as the gene defective in certain individuals with immunodeficiency combined with radiosensitivity (Moshous et al. 2001). Artemis is a member of the Snm family of β-CASP domain DNA hydrolases (Callebaut et al. 2002), although, unlike Snm1A and Snm1B, Artemis is not implicated in repair of DNA interstrand crosslinks. Instead, Artemis on its own displays single-strand-specific 5' to 3' exonuclease activity. Further, Artemis acts in close conjunction with DNA-PK_{cs} which confers upon it regulated endonucleolytic activities (Ma et al. 2002), phosphorylates it, and recruits it to DSB sites (Drouet et al. 2006). Curiously, this regulation appears to depend more on accessibility of the substrate ends than on Artemis phosphorylation (Goodarzi et al. 2006). Artemis can act in Ku-mediated NHEJ *in vitro* (Ma et al. 2004). It is not known whether other nucleases may also act in NHEJ, but the *in vivo* phenotype conferred by Artemis deficiency establishes that at least certain end trimming events are unique to it.

A less severe means of dealing with blocking lesions is to reverse them without removing entire nucleotides. One such NHEJ enzyme is polynucleotide kinase/3' phosphatase (PNKP). The polynucleotide kinase activity of PNKP was selectively utilized to repair 5' hydroxyl termini during NHEJ *in vitro* (Chappell et al. 2002). NHEJ repair of irradiation damage in cells also depended on PNKP, and more specifically on interaction of its forkhead-associated (FHA) domain with the constitutively phosphorylated C-terminus of XRCC4 (Koch et al. 2004). It has not been shown that the 3' phosphatase activity of PNKP participates in NHEJ, but this seems likely. Other enzymes may also participate in reversal of blocking lesions. For example, the lyase activity of Pol λ is a prime candidate for removal of 5' deoxyribose phosphates, although this may be more relevant to base excision repair (Garcia-Diaz et al. 2001).

Putting it together (Fig. 2), one can envision an NHEJ reaction in which Ku binds to the DSB ends, recruiting DNA-PK_{cs}, which in conjunction with inward translocation of Ku (Kysela et al. 2003), binds and protects the DSB ends and drives their association. Autophosphorylation of DNA-PK_{cs} at least partially releases the ends, exposing them to processing enzymes and ultimately Lig4. Very little is known about the coordination of these downstream enzymatic steps of NHEJ. Strand ligation can be sequential *in vitro* (Fig. 1) (Ma et al. 2004), consistent with the fact that ligases encompass the DNA duplex in a manner that would preclude simultaneous ligation of both strands (Pascal et al. 2004). It is not necessary that the second strand be ligated by Lig4 once linear integrity of the chromosome is restored, but this seems likely.

3.2 MMEJ

MMEJ (2.4) is quite different than NHEJ. However, is widely stated that Ku-dependent NHEJ is "error prone" and tends to promiscuously create deletions even when presented with compatible overhangs. This remains a difficult question. Models of strand-unwinding by DNA-PK_{cs} could account for the exposure of internal microhomologies (Hammarsten et al. 2000). However, this does not account for the fact that internal microhomologies require more extensive processing than overhangs (Fig. 1). Indeed, *in vitro* studies have suggested that Ku-dependent end-joining is very accurate when presented with overhangs (Budman and Chu 2005). In cell studies, DSBs are frequently joined with deletions suggestive of MMEJ, but mutation of NHEJ factors does not always correlate with reduced MMEJ-type repair. Instead, the overall efficiency of repair remains similar with a shift toward lower fidelity and increased microhomology use (Kabotyanski et al. 1998; Verkaik et al. 2002). One study using blunt ends provided a clear distinction, where the Ku-dependent process could be specifically associated with accurate religation (van Heemst et al. 2004). Thus, although mammalian Ku-dependent NHEJ may be capable of catalyzing MMEJ-type events, there appears to be a distinct MMEJ mechanism.

A useful comparison is to the budding yeast *Saccharomyces cerevisiae*, in which MMEJ has been given some biochemical definition. Although many questions remain, yeast MMEJ is nearly or entirely Ku-independent, and in fact Ku inhibits MMEJ consistent with an end-protective function (Boulton and Jackson 1996; Daley et al. 2005; Ma et al. 2003). Curiously, there is an apparent partial dependence of yeast MMEJ on DNA ligase IV (Ma et al. 2003), but not its Pol X polymerase Pol4 (Daley et al. 2005). Yeast MMEJ is partially dependent on the nuclease Rad1-Rad10 (Ma et al. 2003), which has yet to be implicated in NHEJ but functions in 3' tail removal in HR (Ivanov and Haber 1995). Yeast MMEJ is finally known to require the Rad50-Mre11 complex (4.3) that functions in both HR and NHEJ (Ma et al. 2003). In total, yeast MMEJ appears to be a hybrid pathway, but that is nonetheless distinct from NHEJ.

3.3 SSA and related mechanisms

Fully understanding MMEJ requires comparison to single-strand annealing (SSA). SSA describes DSB joining that occurs by base-pairing of strands in larger blocks of homology present in direct repeats on either side of a DSB (Ivanov and Haber 1995). SSA is strongly dependent on resection, and, in yeast, on the HR protein Rad52, and so is a clear subset of HR (Haber JE, this volume). However, there is a relationship between SSA and MMEJ in that each require resection to expose base-pairing potential between ends, so that each can be considered one point on a spectrum in which base-pairing length and distance from the ends are critical parameters (Fig. 3) (Karathanasis and Wilson 2002; Sugawara et al. 2000). The impact of these parameters probably lies in the requirements for facilitated annealing by Rad52. However, there will necessarily be a transition over which more than one annealing mechanism might contribute to joining, blurring the distinction between SSA and MMEJ. Importantly, this also applies to overhangs (Fig. 3). In yeast, Ku-dependent NHEJ is only required at overhangs of less than ~6 bases (Daley and Wilson 2005). At longer overhangs, joining becomes Ku-independent

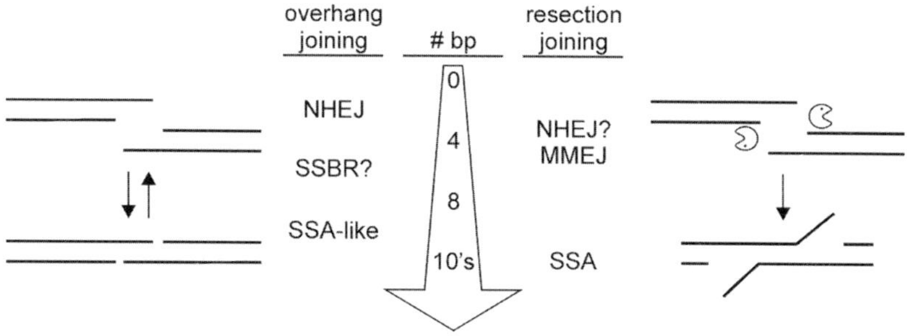

Fig. 3. Relationship of base-pairing length to joining mechanism. Joining of DSB ends can occur via overhangs or via microhomologies exposed by resection into the adjacent duplex. For each, increasing the number of bases in the overhang or microhomology impacts the possible joining mechanisms, with probable overlaps in the range. The relative contribution of these various mechanisms to IR is not well established.

and considerably more efficient, as predicted by base-pairing energetics (2.2). An active joining process is still required, however, in the form of Rad52, presumably to again facilitate annealing. Important implications are that a form of HR can indeed contribute to joining of microhomologies as short as 8 bases in special circumstances (Daley and Wilson 2005; Karathanasis and Wilson 2002). When these microhomologies are the DSB overhangs, repair is both efficient and accurate (Daley and Wilson 2005).

3.4 SSBR applied to DSBs

There was a striking gray zone of overhang length (4 to 8 bases) over which DSBs could be repaired in yeast in both a Ku- and Rad52-independent manner (Daley and Wilson 2005). A reasonable model for such joining is that overhang annealing occurred spontaneously followed by repair of the DNA nicks by single-strand break repair (SSBR) (Fig. 3). This is not a trivial solution because the lesions are still DSBs with an equilibrium that favors dissociation. Evidence supporting this notion has been building in mammalian systems. A "backup NHEJ" pathway has been observed *in vivo* as kinetically slower and Ku-independent DSB repair, and *in vitro* as a DSB joining capacity in extracts depleted for Ku and/or DNA-PK$_{cs}$ (Perrault et al. 2004). The *in vitro* activity appears to be attributable to DNA ligase III (Lig3), whose main function is to catalyze mammalian SSBR (Wang et al. 2005). Inhibition of poly(ADP-ribose) polymerase 1 (PARP-1), again ordinarily associated with SSBR, antagonized DSB repair independently of DNA-PK$_{cs}$ (Audebert et al. 2004). Reconstitution *in vitro* verified that PARP-1, Lig3, XRCC1 and PNKP could indeed cooperate in the bridging and joining of ends (Audebert et al. 2006). Many questions remain, especially how DSBs are committed to these different pathways, their interactions, and their relative fidelity given that backup NHEJ was described as error-prone.

Table 1. Conservation of NHEJ proteins among eukaryotes.

		Ku	Lig4	XRCC4	XLF/ Nej1	PolX	DNA-PK_{cs}	Artemis
Metazoa								
human	*Homo sapiens*	+[a]	+	+	+	+	+	+
mouse	*Mus musculus*	+	+	+	+	+	+	+
chicken	*Gallus gallus*	+	+	+	+	+	+	+
frog	*Xenopus laevis*	+	+	+	+		+	
zebrafish	*Danio rerio*	+	+	+	+	+	+	+
sea squirt	*Ciona intestinalis*	+	+	+	+		+	+
sea urchin	*Strongylocentrotus purpuratus*	+	+	+		+	+	+
mosquito	*Anopheles gambiae*	+	+	–[b]	+	–	+	+
honeybee	*Apis mellifera*	+	+	+	–[b]	–	+	+
fruit fly	*Drosophila melanogaster*	+	+	+	+	–	–	–
worm	*Caenorhabditis elegans*	+	+	–	–	–	–	–
Fungi								
Microsporidia	*Encephalitozoon cuniculi*	–	–	–	–	–	–	–
Basidiomycota	*Cryptococcus neoformans*	+	+	–	–	+	–	–
Schizosaccharomycetes	*Schizosaccharomyces pombe*	+	+	–	+	+	–	–
Pezizomycotina	*Aspergillus fumigatus*	+	+	+	+	+	–	?[c]
	Neurospora crassa	+	+	+	+	+		?
Saccharomycetales	*Yarrowia lipolytica*	+	+	+	+	+	–	?
	Candida albicans	+	+	–	+	–	–	?
pre-duplication[d]	*Ashbya gossypii*	+	+	–	+	+	–	–
	Kluyveromyces lactis	+	+	+	+	+	–	–
post-duplication[d]	*Candida glabrata*	+	+	+	+	+	–	–
	Saccharomyces cerevisiae	+	+	+	+	+	–	–
Protozoa								
	Cyanophora paradoxa	–	–	–	–	–	–	–
	Cryptosporidium parvum	–	–	–	–	–	–	–
	Plasmodium falciparum	–	–	–	–	–	–	–
	Theileria parva	–	–	–	–	–	–	–
	Paramecium tetraurelia	+	+				+	
	Tetrahymena thermophila	+	+				+	
slime mold	*Dictyostelium discoideum*	+	+	+	+	–	+	–[e]
	Leishmania major	+	–	–	–	–	–	–
	Trypanosoma brucei	+	–	–	–	–	–	–
Plants								
thale cress	*Arabidopsis thaliana*	+	+	+	–	+	–	–
rice	*Oryza sativa*	+	+	+	–	+	–	–

[a] "+" indicates presence and "-" indicates absence of a gene from a complete genome. No entry indicates absence of a gene from an incomplete genome where no conclusion can be drawn. In all cases, NHEJ gene orthologues were distinguished from related genes of different inferred function. For example, ATP-dependent DNA ligases were included only if they were more related to DNA ligase IV than to ligase I or III. All entries are based on current best data and are subject to change with new information.

[b] Although too diverged to call based on sequence alone, very tentative mosquito XRCC4 and honeybee XLF assignments are possible.

[c] "?" indicates the presence of an Artemis-like gene in these fungi whose assignment within the Snm family is unclear.
[d] "Pre-duplication" and "post-duplication" indicate whether these yeast species were derived before or after the genome duplication event that occurred in the *Saccharomyces* lineage (Kellis et al. 2004).
[e] Block and Lees-Miller (2005) previously reported the presence of an Artemis homologue in *D. discoideum*, but this sole Snm family member (DictyBase DDB0169391) is significantly more related to Snm1A/Pso2 than to Artemis.

4 Species conservation of Ku-dependent NHEJ

4.1 Vertebrates and related

Table 1 depicts the conservation of the core human NHEJ proteins among various eukaryotes in many taxa, with an emphasis on completely sequenced genomes; a similar analysis of prokaryotes was previously provided (Wilson et al. 2003). All mammals, and indeed all vertebrates appear to have a mechanism for NHEJ that utilizes all of the protein components discussed above, notably including DNA-PK_{cs}. In a recent surprise, this system was also discovered in *Ciona*, or sea squirts, one of the more rudimentary forms of chordate life (Block and Lees-Miller 2005). Still further, the sea urchin *Strongylocentrotus purpuratus* shows that mammalian-type NHEJ can be found in non-chordate deuterostomes. There is insufficient data to determine the full breadth of conservation, but it seems clear that DNA-PK-associated NHEJ was established early in the eukaryotic lineage. In addition to mammals (see other sections), NHEJ has been studied in the tractable chicken DT40 cell system, which was instrumental in establishing the cell cycle dependence of DSB repair in vertebrates (5.3) (Takata et al. 1998).

4.2 Insects and worms

Variations in NHEJ components begin to appear at the greater levels of divergence from humans represented by *Drosophila melanogaster (*fruitfly) and *Caenorhabditis elegans* (worm) (Table 1). Although each have Ku and Lig4, DNA-PK_{cs}, Artemis, and Pol X family polymerases are absent. *C. elegans* even lacks apparent XRCC4 and XLF homologues, although these are substantially less conserved overall and might be missed by current algorithms. Biochemical confirmation of these proteins has not been provided, but functional studies of *C. elegans* have revealed that NHEJ is selectively utilized in non-cycling somatic cells, while HR is used in cycling and germ cells (Clejan et al. 2006). Genetic data support a role of the *D. melanogaster* proteins in DSB repair in response to radiation and P-element excision (McVey et al. 2004; Min et al. 2004; Romeijn et al. 2005). Importantly, these same studies also support the existence of efficient Ku-independent MMEJ-type pathways. A revealing observation in the insect lineage is the presence of DNA-PK_{cs} and Artemis homologues in mosquito and honeybee (Dore et al. 2004), suggesting that fruitfly is unusual in lacking these and that DNA-PK$_{cs}$ is probably present in most metazoans. This pattern also suggests an evolutionary association between DNA-PK$_{cs}$ and Artemis that reflects their functional interaction (Ma et al. 2002).

4.3 *S. cerevisiae*

Aside from mammals, NHEJ has been best characterized in *S. cerevisiae* (Fig. 2). Unlike mammals, early mutant screens for DSB-sensitive yeast consistently failed to identify NHEJ genes. It was thus striking when Ku homologues were observed in the yeast genome that proved critical for NHEJ as assessed by transformation of linear plasmids (Boulton and Jackson 1996), in chromosomal assays (Karathanasis and Wilson 2002), and in biochemical reconstitution (Chen et al. 2001). The presumption is still that HR predominates in growing yeast. Indeed, recent screens of all viable yeast mutants that targeted NHEJ by precluding HR have revealed nine genes, including Ku, that participate directly in NHEJ (Ooi et al. 2001; Wilson 2002) (Table 1). The commonality with mammalian NHEJ is established by dependence on homologues of Lig4 and XRCC4 (Lif1 in yeast) (Teo and Jackson 2000; Wilson et al. 1997). Interestingly, the yeast Nej1 protein acts as a third member of the DNA ligase IV complex via interaction with Lif1 (Frank-Vaillant and Marcand 2001; Wilson 2002), in direct parallel with mammalian XLF. There is a low degree of conservation between Nej1 and XLF which, combined with their common function and size, suggests that these proteins are orthologues (Callebaut et al. 2006); it will be important to determine if they act similarly in support of Lig4. Finally, yeast NHEJ also uses a BRCT-containing and low fidelity Pol X polymerase, Pol4 (Wilson and Lieber 1999). Recent studies of Pol4 established that dual-strand gaps and 3' overhangs are critical parameters of Pol X dependence in NHEJ (Daley et al. 2005). The requirement for 3' overhangs underscores the ability of Pol X polymerases to extend limiting primer-template pairs. Evidence from telomere fusions where base-pairing is not possible further suggests that Pol4 catalyzes cross-break polymerization (2.3) (Pardo et al. 2006).

Important differences between yeast and mammalian NHEJ begin with the marked absence of DNA-PK$_{cs}$ from yeast. This established Ku as having DNA-PK-independent NHEJ functions, and that DNA-PK$_{cs}$ is not essential for NHEJ. But what mediates end-bridging in yeast? Part of the answer lies in the final complex essential for yeast NHEJ, Mre11-Rad50-Xrs2 (MRX). Rad50 is a structural maintenance of chromosomes-like protein with DNA-binding globular heads connected via long (50 nm) coiled coils that dimerize at their tails. This structure and much supporting evidence has led to the model that MRX tethers DNA molecules (Connelly and Leach 2002). MRX also directly contacts DNA ligase IV and probably Ku (Palmbos et al. 2005), and in total appears to replace at least some functions of DNA-PK$_{cs}$. This raises the question whether the vertebrate Mre11-Rad50-Nbs1 (MRN) complex might also participate in NHEJ. This has been difficult to address genetically since MRN is required for viability, but Mre11-depeleted *Xenopus* cell extracts did not show an NHEJ defect (Di Virgilio and

Gautier 2005). A secondary yeast-like NHEJ pathway is nonetheless possible, and indeed MRN can stimulate DNA-PKcs-independent NHEJ in vitro (Huang and Dynan 2002). This is important given that yeast NHEJ is generally seen to be more accurate (Boulton and Jackson 1996; Karathanasis and Wilson 2002).

Final differences lie in end processing. Yeast lack a 5' kinase correspondent to PNKP, and its homologous 3' phosphatase, Tpp1, appears dispensable for NHEJ (Daley and Wilson 2005). Indeed, yeast NHEJ overall appears to preferentially utilize nucleases for terminus resolution. These nucleases are not yet clearly identified, but importantly yeast lack Artemis; Pso2, the only β-CASP protein in yeast, is not required for typical NHEJ. Instead, both *in vivo* and *in vitro* evidence indicate that the 5' flap endonuclease FEN-1/Rad27 might participate (Tseng and Tomkinson 2004; Wu et al. 1999). Its precise contribution at naturally occurring DSBs is unknown, nor can it account for all required nucleolytic activities.

4.4 Other fungi

Ku and Lig4 can be found in nearly all fungi (Table 1). Intriguingly, though, the unusual intracellular fungus *Encephalitozoon cuniculi* lacks these proteins, providing a first indication that NHEJ is not essential for eukaryotic life. Among NHEJ-positive fungi, the fission yeast *Schizosaccharomyces pombe* is the next best studied. Like *S. cerevisiae*, *S. pombe* executes NHEJ that is dependent on Ku and Lig4 but not DNA-PK$_{cs}$ or Artemis (Manolis et al. 2001; Wilson et al. 1999). A Pol X polymerase is also present, but its cellular function is not established (Gonzalez-Barrera et al. 2005). Unlike *S. cerevisiae*, the *S. pombe* Mre11-Rad50 complex is very largely dispensable for NHEJ, although some intermolecular events may require it (Decottignies 2005). Like mammalian cells, fission yeast NHEJ has often been observed to be disproportionately inaccurate in plasmid assays (Manolis et al. 2001; Wilson et al. 1999), although this may not be true for chromosomal DSBs (Hope et al. 2006). Unfortunately, studies of *S. pombe* have proceeded more slowly due to the less tractable systems for monitoring NHEJ, but recent studies of cells fixed in the usually short G1 stage hold great promise (Ferreira and Cooper 2004). Another importance of *S. pombe* is that it establishes that Ku- and Lig4-dependent NHEJ can proceed in the apparent absence of XRCC4, although *S. pombe* does possess a probable XLF. This same situation is seen in other yeasts including *Ashbya gossypii*, where synteny relationships make it certain that XRCC4/Lif1 is in fact absent (Kellis et al. 2004). In such species XLF/Nej1 might perhaps serve functions separated into XLF/Nej1 and XRCC4/Lif1 in other species.

Recent efforts have produced Ku disruptions in several other fungi. The main phenotypic manifestation has been an increase in HR, at the expense of IR, with practical importance to gene targeting (Kooistra et al. 2004; Krappmann et al. 2006). These mutants establish that NHEJ can predominate in fungi, and in the case of *Kluyveromyces lactis* that IR can be uniquely dependent on Ku- and Lig4-dependent NHEJ (Kegel et al. 2006). Finally, it is noteworthy that some fungi possess an Artemis-like gene in addition to a separate Snm1A/Pso2 homologue (Bonatto et al. 2005). Assignment of this gene as Artemis is uncertain in the absence of supporting data, but it could represent the first instance of Artemis without DNA-PK_{cs} and outside of metazoans.

4.5 Protozoa

Many protozoa lack NHEJ outright. Surprisingly, then, others have NHEJ pathways with striking conservation to mammals (Table 1). In particular, the slime mold *Dictyostelium discoideum* has a DNA-PK_{cs} homologue, in addition to Ku and Lig4, that is required for normal DSB repair in G1 (Hudson et al. 2005). Artemis is apparently absent, however. One odd pattern is the presence of only Ku homologues in the trypanosomes *Trypanosoma brucei* and *Leishmania major*. Perhaps Ku functions with another ligase in these organisms, but it might also be preserved for NHEJ-independent Ku functions.

4.6 Plants

Arabidopsis thaliana (thale cress) and *Oryza sativa* (rice) each have Ku- and Lig4-dependent NHEJ systems, establishing the presence of NHEJ in all major branches of eukaryotic life (Table 1). However, plants present yet another pattern with XRCC4 but no apparent XLF, and a putative Pol X polymerase but no DNA-PK_{cs} or Artemis. Genetic studies in *A. thaliana* establish that its Ku and Lig4 are required for normal resistance to radiation and handling of T DNA insertions (Friesner and Britt 2003; Gallego et al. 2003; West et al. 2002). In parallel with insects and other organisms, however, high levels of end joining of uncertain mechanism continue to be observed in NHEJ mutant plants.

4.7 Bacteria

The greatest surprise came when genome projects revealed a Ku homologue in a subset of bacteria and archaea (Fig. 2) (Aravind and Koonin 2001; Doherty et al. 2001). Prokaryotic Ku is much smaller than eukaryotic Ku, corresponding only to the presumptive central β-barrel ring. Also, it is encoded in a single polypeptide which homo- rather than hetero-dimerizes (Weller et al. 2002). Prokaryotic Ku nonetheless displays the same pattern of DNA binding as eukaryotic Ku, and so appears to correspond to an ancestral scaffold from which the related Ku70 and Ku80 genes diverged (Gell and Jackson 1999). The first clue that prokaryotic Ku catalyzes NHEJ was its presence in an operon with an ATP-dependent DNA ligase (LigD in mycobacteria) distinct from the NAD-dependent ligase used for replication (Aravind and Koonin 2001; Weller and Doherty 2001). Outside of being a typical ATP-dependent ligase (Akey et al. 2006), this gene bears no obvious homology to Lig4, and so, not surprisingly, bacteria lack XRCC4 or XLF. Ku is thus the protein that most universally defines NHEJ. Indeed, the Ku and ligase from several bacteria can catalyze NHEJ both *in vitro* and *in vivo*, and even when reconstituted in yeast (Della et al. 2004; Gong et al. 2005; Weller et al. 2002).

Still further, the Ku operons in most bacteria also contain a polymerase and nuclease/phosphoesterase domain, which in some cases are fused in frame to the ligase (Aravind and Koonin 2001; Weller and Doherty 2001), suggesting that only two polypeptides of less than 1000 residues can remarkably even contain all required processing activities. These domains have highly unusual properties. *Mycobacterium tuberculosis* LigD displays DNA polymerase, RNA polymerase, terminal transferase and even RNA primase activities in the same active site (Della et al. 2004), and mispair extension and abasic site bypass reminiscent of eukaryotic NHEJ polymerases (Yakovleva and Shuman 2006). Structurally, this polymerase family has a fold similar to archaeal DNA primase (Zhu et al. 2006). The *Pseudomonas aeruginosa* LigD nuclease is even more odd in that it preferentially removes ribonucleotides from the end of a DNA molecule, leaving one terminal ribonucleotide (Zhu and Shuman 2005). The polymerase could have added these ribonucleotides, but whether this use of RNA in DNA helps coordinate NHEJ is unknown; certainly current results do not preclude non-NHEJ functions of these proteins. A critical question is how these processing functions contribute to the (in)accuracy of bacterial NHEJ. When *M. tuberculosis* proteins were used to reconstitute NHEJ in yeast, joining was found to be highly accurate (Della et al. 2004), and polymerase function was required for precise NHEJ of DSBs with gaps meant to mimic terminal damage (our unpublished data). In contrast, when linear plasmids were transformed into *M. smegmatis*, repair was found to be only 45% accurate (Gong et al. 2005). Polymerase mutations led to a two-fold increase in fidelity with only a mild 30% decrease in efficiency (Zhu et al. 2006). LigD-dependent processing may thus contribute to both precise and mutagenic NHEJ.

While NHEJ presence is the rule in eukaryotes, it is the exception in prokaryotes. So why do any bacteria possess NHEJ? The operonic nature of bacterial NHEJ almost certainly facilitated its transfer horizontally rather than vertically, so that not all bacteria likely had equal opportunity to acquire NHEJ. Even so, the maintenance of NHEJ indicates a special role in a subset of bacteria. It has been noted that NHEJ-positive bacteria tend to show prolonged stationary phases of growth, such as spores, biofilms or host granulomas (Wilson et al. 2003). Such cell states have potentially vulnerable single-copy genomes, and yet are highly resistant to many stresses that can create DSBs (Fig. 4). NHEJ may be especially critical in preserving genome stability or mediating adaptive mutagenesis in such states (6.2). Similar arguments have been made for yeast (Karathanasis and Wilson 2002).

4.8 Viruses

Strikingly, the influence of NHEJ extends even beyond cellular organisms. Certain bacteriophage encode a Ku homologue, which has recently been shown to mediate genome circularization after cell entry in conjunction with the host NHEJ ligase

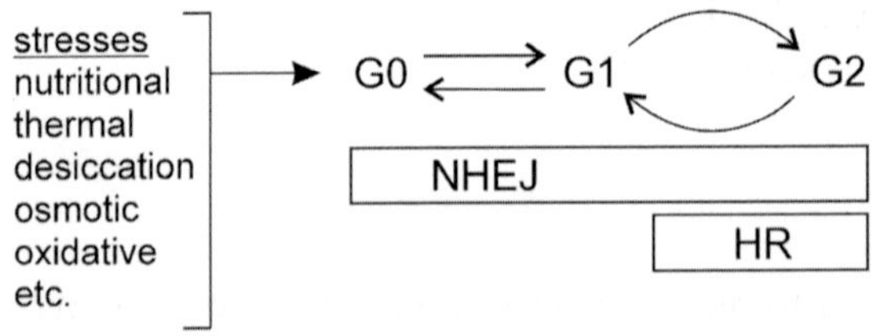

Fig. 4. Relationship between cell cycle and NHEJ. NHEJ is best understood as being active in all cell cycles stages, but in S/G2 is in competition with HR, more specifically with 5' resection that blocks NHEJ. NHEJ is probably especially important in G0, i.e. stationary, phase cells to help mediate resistance to various environmental stresses.

(Pitcher et al. 2006). In contrast to spontaneous circularization of typical long cos overhangs, NHEJ-dependent phage have 4-base overhangs that reinforce the intimate link between NHEJ and unstable DSB substrates. In eukaryotes, genome circularization is also believed to be necessary for lytic infectivity of episomal dsDNA viruses such as HSV (Strang and Stow 2005), although the mechanism is not established. Both DNA-PK and DNA ligase IV are required for full infectivity of HIV in cultured human cells (Daniel et al. 2004). This could reflect a role in completing retroviral genome integration (Skalka and Katz 2005), but this is difficult to reconcile given that the key phosphoryl transfer is mediated by integrase with no DSB intermediate. Alternatively, the effect could be mediated indirectly through interactions with the HIV genome prior to integration, and indeed Ku binds to linear dsDNA forms of both HIV and yeast Ty1 retrotransposable elements (Downs and Jackson 1999; Li et al. 2001). This interaction may help to promote an alternative pathway of HIV genome circularization (Li et al. 2001).

5 NHEJ interplay with host cell processes

5.1 Chromatin

Eukaryotic chromatin has the potential to impede DNA transactions. Much recent literature, mostly from budding yeast, has demonstrated that the same types of histone modifications and ATP-dependent chromatin remodeling enzymes that expose DNA during transcription also facilitate NHEJ (reviewed in Daley et al. 2005; van Attikum and Gasser 2005). In fact, it seems the rule that disrupting chromatin disturbs NHEJ. This fact, combined with the potential for indirect effects, results in a currently poor understanding of the precise mechanistic relationship between chromatin and NHEJ. For example, nucleosome displacement around a DSB occurs in a wide window relevant for HR (Tsukuda et al. 2005), but the local displacement of two, one, or even no nucleosomes might be sufficient for NHEJ reactions at the terminus. The alternative is that chromatin modifications affect NHEJ more globally, via influences on protein recruitment, bridging of chromosome ends, or recovery after repair.

5.2 Checkpoints

Checkpoints describe phenomenological delays in the cell cycle after DNA damage, suggesting that they provide time for DNA repair. Initial reports in yeast suggested that checkpoint function was required for plasmid NHEJ (de la Torre-Ruiz and Lowndes 2000) although this was not observed in a chromosomal assay (our unpublished data), and even in plasmid studies extrinsically imposed cell-cycle delays did not restore NHEJ. Repair time is thus likely only one component of checkpoint function. Insight is provided by mammalian cells deficient in the ATM checkpoint kinase. ATM-dependent radio-resistance correlates with a minor portion (10%) of DSBs that are repaired more slowly over 24 hours (Riballo et al. 2004). Strikingly, only this slow repair depends on Artemis, DNA-PK, ATM, ATM checkpoint partners such as MRN and 53BP1, and phosphorylation of H2AX. The implication is that DSBs that require extensive processing are particularly dependent on checkpoint chromatin modifications and other global functions.

5.3 Cell cycle

NHEJ has a higher intrinsic error rate than HR, but HR is entirely dependent on the ability to find an appropriate homologous donor. NHEJ is thus optimal and potentially more accurate in G1/G0 when the HR donor search is impeded, but less desirable in late S/G2 when the sister chromatid is available for HR (Fig. 4; Cortés-Ledesma et al., this volume). Consequently, budding yeast NHEJ is regulated indirectly by licensing of 5' resection, the committed step to HR, at and after the G1-S boundary by the Cdc28 cyclin-dependent kinase (Ira et al. 2004). Importantly, this does not preclude NHEJ in S/G2 if it can occur more rapidly than the initiation of resection. Evidence for a similar regulation has also been provided in *S. pombe* (Ferreira and Cooper 2004). It remains to be seen how reproducible this regulatory mechanism is in higher eukaryotes, but it is clear that HR does have an important role in S/G2 (Takata et al. 1998) which may help explain observations that HR is preferentially utilized in early developmental lineages (Clejan et al. 2006; Orii et al 2006). Evidence supporting a direct regulation of NHEJ in a cell cycle dependent fashion is scant, but this is distinctly possible. For example, PARP-1, perhaps by direct ribosylation of Ku, appears to inhibit negative effects of NHEJ that antagonize HR (Hochegger et al. 2006).

6 Outcomes of NHEJ and its deficiency

6.1 Accurate repair and maintenance of genome integrity

The conservation of many genes that appear dedicated exclusively to Ku-dependent NHEJ leads inevitably to the conclusion that NHEJ provides a strong fitness gain. It is thus puzzling that NHEJ is often seen to be mutagenic. One resolution to the balance of benefit and mutagenesis would be if Ku-dependent NHEJ is in fact predominantly accurate at physiological DSBs, and there is a paucity of data on this point. A more general resolution is that NHEJ mutagenesis, which nonetheless stabilizes chromosomes, may be more tolerable than the alternatives. A persistent DSB could lead to gross genomic instability through extensive degradation of the ends, loss of a chromosome fragment during mitosis, or complex genomic rearrangements and breakage-fusion-bridge cycles precipitated by even less accurate DSB repair pathways. NHEJ mouse knockout models are indeed associated with excessive p53-mediated cell death, presumably activated by persistent DSBs (Frank et al. 2000). Interestingly, the extent of this phenotype differs between NHEJ genes, being strongest for Lig4 mutants, prominent for Ku, and absent for DNA-PK$_{cs}$ and Artemis. When p53 is additionally mutated, NHEJ-deficient mice develop lymphoid tumors characterized by "complicon" rearrangements interpreted to arise from aberrant handling of persistent DSBs (Zhu et al. 2002). Similar phenomena occur in humans with Lig4 and XLF mutations, in the form of growth retardation and microcephaly, and perhaps an increased incidence of cancer (Buck et al. 2006; O'Driscoll et al. 2004).

6.2 Adaptive and targeted mutagenesis

A non-exclusive resolution to the balance of benefit and mutagenesis is that NHEJ mutagenesis may itself be beneficial. One clear instance is the usurpation of NHEJ during generation of antigen receptor diversity in the vertebrate immune system, most notably during V(D)J recombination (Jung et al. 2006; Gellert M, this volume). No examples of NHEJ-mediated site-directed IR outside of the immune system are known, but cells under stress may benefit from general NHEJ mutations as a "last ditch" effort to adapt by genome alteration. Frameshift mutations in particular have been associated with yeast NHEJ in the stressed stationary phase, where NHEJ promotes reversion of mutant alleles under selection (Heidenreich and Eisler 2004). The benefit of such adaptive mutagenesis must be understood on a population or species level because any individual cell has a low likelihood of improving its lot in this way.

7 Concluding remarks

NHEJ has come a long way since its earliest descriptions in mammals. The basic Ku-dependent joining reaction is now documented or inferred in nearly all branches of life. There is a tremendous variability in NHEJ implementation, however, including NHEJ absence from many organisms. Important continuing questions are the degree to which this variability and the interplay of Ku-dependent NHEJ with less well described alternative rejoining pathways correlate with the frequency of accurate repair, local mutations and IR. Answers will require a detailed description of the NHEJ reaction mechanism(s) on different DSB substrates, and of their interaction with cell cycle, chromatin, and other regulatory process. Correlation of these parameters with organism life cycles across phylogeny should continue to provide powerful tools for this endeavor.

References

Ahnesorg P, Smith P, Jackson SP (2006) XLF interacts with the XRCC4-DNA ligase IV complex to promote DNA nonhomologous end-joining. Cell 124:301-313

Akey D, Martins A, Aniukwu J, Glickman MS, Shuman S, Berger JM (2006) Crystal structure and nonhomologous end joining function of the ligase component of Mycobacterium DNA ligase D. J Biol Chem 281:13412-13423

Aravind L, Koonin EV (2001) Prokaryotic homologs of the eukaryotic DNA-end-binding protein Ku, novel domains in the Ku protein and prediction of a prokaryotic double-strand break repair system. Genome Res 11:1365-1374

Audebert M, Salles B, Calsou P (2004) Involvement of poly(ADP-ribose) polymerase-1 and XRCC1/DNA ligase III in an alternative route for DNA double-strand breaks rejoining. J Biol Chem 279:55117-55126

Audebert M, Salles B, Weinfeld M, Calsou P (2006) Involvement of polynucleotide kinase in a poly(ADP-ribose) polymerase-1-dependent DNA double-strand breaks rejoining pathway. J Mol Biol 356:257-265

Bebenek K, Garcia-Diaz M, Blanco L, Kunkel TA (2003) The frameshift infidelity of human DNA polymerase lambda. Implications for function. J Biol Chem 278:34685-34690

Block WD, Lees-Miller SP (2005) Putative homologues of the DNA-dependent protein kinase catalytic subunit (DNA-PK_{cs}) and other components of the non-homologous end joining machinery in *Dictyostelium discoideum*. DNA Repair 4:1061-1065

Bonatto D, Brendel M, Henriques JA (2005) *In silico* identification and analysis of new Artemis/Artemis-like sequences from fungal and metazoan species. Protein J 24:399-411

Boulton SJ, Jackson SP (1996) *Saccharomyces cerevisiae* Ku70 potentiates illegitimate DNA double- strand break repair and serves as a barrier to error-prone DNA repair pathways. EMBO J 15:5093-5103

Bryans M, Valenzano MC, Stamato TD (1999) Absence of DNA ligase IV protein in XR-1 cells: evidence for stabilization by XRCC4. Mutat Res 433:53-58

Buck D, Malivert L, de Chasseval R, Barraud A, Fondaneche MC, Sanal O, Plebani A, Stephan JL, Hufnagel M, le Deist F, Fischer A, Durandy A, de Villartay JP, Revy P

(2006) Cernunnos, a novel nonhomologous end-joining factor, is mutated in human immunodeficiency with microcephaly. Cell 124:287-299

Budman J, Chu G (2005) Processing of DNA for nonhomologous end-joining by cell-free extract. EMBO J 24:849-860

Callebaut I, Moshous D, Mornon JP, de Villartay JP (2002) Metallo-beta-lactamase fold within nucleic acids processing enzymes: the beta-CASP family. Nucleic Acids Res 30:3592-3601

Callebaut I, Malivert L, Fischer A, Mornon JP, Revy P, de Villartay JP (2006) Cernunnos interacts with the XRCC4 x DNA-ligase IV complex and is homologous to the yeast nonhomologous end-joining factor Nej1. J Biol Chem 281:13857-13860

Capp JP, Boudsocq F, Bertrand P, Laroche-Clary A, Pourquier P, Lopez BS, Cazaux C, Hoffmann JS, Canitrot Y (2006) The DNA polymerase lambda is required for the repair of non-compatible DNA double strand breaks by NHEJ in mammalian cells. Nucleic Acids Res 34:2998-3007

Chappell C, Hanakahi LA, Karimi-Busheri F, Weinfeld M, West SC (2002) Involvement of human polynucleotide kinase in double-strand break repair by non-homologous end joining. EMBO J 21:2827-2832

Chen L, Trujillo K, Ramos W, Sung P, Tomkinson AE (2001) Promotion of Dnl4-catalyzed DNA end-joining by the Rad50/Mre11/Xrs2 and Hdf1/Hdf2 complexes. Mol Cell 8:1105-1115

Clejan I, Boerckel J, Ahmed S (2006) Developmental modulation of nonhomologous end joining in *Caenorhabditis elegans*. Genetics 173:1301-1317

Connelly JC, Leach DR (2002) Tethering on the brink: the evolutionarily conserved Mre11-Rad50 complex. Trends Biochem Sci 27:410-418

Covo S, Blanco L, Livneh Z (2004) Lesion bypass by human DNA polymerase mu reveals a template-dependent, sequence-independent nucleotidyl transferase activity. J Biol Chem 279:859-865

Daley JM, Laan RL, Suresh A, Wilson TE (2005) DNA joint dependence of pol X family polymerase action in nonhomologous end joining. J Biol Chem 280:29030-29037

Daley JM, Palmbos PL, Wu D, Wilson TE (2005) Nonhomologous end joining in yeast. Annu Rev Genet 39:431-451

Daley JM, Wilson TE (2005) Rejoining of DNA double-strand breaks as a function of overhang length. Mol Cell Biol 25:896-906

Daniel R, Greger JG, Katz RA, Taganov KD, Wu X, Kappes JC, Skalka AM (2004) Evidence that stable retroviral transduction and cell survival following DNA integration depend on components of the nonhomologous end joining repair pathway. J Virol 78:8573-8581

de la Torre-Ruiz M, Lowndes NF (2000) The *Saccharomyces cerevisiae* DNA damage checkpoint is required for efficient repair of double strand breaks by non-homologous end joining. FEBS Lett 467:311-315

Decottignies A (2005) Capture of extranuclear DNA at fission yeast double-strand breaks. Genetics 171:1535-1548

DeFazio LG, Stansel RM, Griffith JD, Chu G (2002) Synapsis of DNA ends by DNA-dependent protein kinase. EMBO J 21:3192-3200

Della M, Palmbos PL, Tseng HM, Tonkin LM, Daley JM, Topper LM, Pitcher RS, Tomkinson AE, Wilson TE, Doherty AJ (2004) Mycobacterial Ku and ligase proteins constitute a two-component NHEJ repair machine. Science 306:683-685

Di Virgilio M, Gautier J (2005) Repair of double-strand breaks by nonhomologous end joining in the absence of Mre11. J Cell Biol 171:765-771

Doherty AJ, Jackson SP, Weller GR (2001) Identification of bacterial homologues of the Ku DNA repair proteins. FEBS Lett 500:186-188

Dore AS, Drake AC, Brewerton SC, Blundell TL (2004) Identification of DNA-PK in the arthropods. Evidence for the ancient ancestry of vertebrate non-homologous end-joining. DNA Repair (Amst) 3:33-41

Downs JA, Jackson SP (1999) Involvement of DNA end-binding protein Ku in Ty element retrotransposition. Mol Cell Biol 19:6260-6268

Drouet J, Frit P, Delteil C, de Villartay JP, Salles B, Calsou P (2006) Interplay between Ku, artemis and DNA-PKcs at DNA ends. J Biol Chem: in press

Ferreira MG, Cooper JP (2004) Two modes of DNA double-strand break repair are reciprocally regulated through the fission yeast cell cycle. Genes Dev 18:2249-2254

Frank KM, Sharpless NE, Gao Y, Sekiguchi JM, Ferguson DO, Zhu C, Manis JP, Horner J, DePinho RA, Alt FW (2000) DNA ligase IV deficiency in mice leads to defective neurogenesis and embryonic lethality via the p53 pathway. Mol Cell 5:993-1002

Frank-Vaillant M, Marcand S (2001) NHEJ regulation by mating type is exercised through a novel protein, Lif2p, essential to the ligase IV pathway. Genes Dev 15:3005-3012

Friedberg EC, Walker GC, Siede W, Wood RD, Schultz RA, Ellenberger T (2005) DNA repair and mutagenesis. ASM Press, Washington DC

Friesner J, Britt AB (2003) Ku80- and DNA ligase IV-deficient plants are sensitive to ionizing radiation and defective in T-DNA integration. Plant J 34:427-440

Gallego ME, Bleuyard JY, Daoudal-Cotterell S, Jallut N, White CI (2003) Ku80 plays a role in non-homologous recombination but is not required for T-DNA integration in *Arabidopsis*. Plant J 35:557-565

Garcia-Diaz M, Bebenek K, Krahn JM, Pedersen LC, Kunkel TA (2006) Structural analysis of strand misalignment during DNA synthesis by a human DNA polymerase. Cell 124:331-342

Garcia-Diaz M, Bebenek K, Kunkel TA, Blanco L (2001) Identification of an intrinsic 5'-deoxyribose-5-phosphate lyase activity in human DNA polymerase lambda: a possible role in base excision repair. J Biol Chem 276:34659-34663

Gell D, Jackson SP (1999) Mapping of protein-protein interactions within the DNA-dependent protein kinase complex. Nucleic Acids Res 27:3494-3502

Gong C, Bongiorno P, Martins A, Stephanou NC, Zhu H, Shuman S, Glickman MS (2005) Mechanism of nonhomologous end-joining in mycobacteria: a low-fidelity repair system driven by Ku, ligase D and ligase C. Nat Struct Mol Biol 12:304-312

Gonzalez-Barrera S, Sanchez A, Ruiz JF, Juarez R, Picher AJ, Terrados G, Andrade P, Blanco L (2005) Characterization of SpPol4, a unique X-family DNA polymerase in *Schizosaccharomyces pombe*. Nucleic Acids Res 33:4762-4774

Goodarzi AA, Yu Y, Riballo E, Douglas P, Walker SA, Ye R, Harer C, Marchetti C, Morrice N, Jeggo PA, Lees-Miller SP (2006) DNA-PK autophosphorylation facilitates Artemis endonuclease activity. EMBO J: in press

Gottlieb TM, Jackson SP (1993) The DNA-dependent protein kinase: requirement for DNA ends and association with Ku antigen. Cell 72:131-142

Grawunder U, Wilm M, Wu X, Kulesza P, Wilson TE, Mann M, Lieber MR (1997) Activity of DNA ligase IV stimulated by complex formation with XRCC4 protein in mammalian cells. Nature 388:492-495

Haber JE (1999) DNA recombination: the replication connection. Trends Biochem Sci 24:271-275

Hammarsten O, DeFazio LG, Chu G (2000) Activation of DNA-dependent protein kinase by single-stranded DNA ends. J Biol Chem 275:1541-1550

Kellis M, Birren BW, Lander ES (2004) Proof and evolutionary analysis of ancient genome duplication in the yeast *Saccharomyces cerevisiae*. Nature 428:617-624

Heidenreich E, Eisler H (2004) Non-homologous end joining dependency of gamma-irradiation-induced adaptive frameshift mutation formation in cell cycle-arrested yeast cells. Mutat Res 556:201-208

Hochegger H, Dejsuphong D, Fukushima T, Morrison C, Sonoda E, Schreiber V, Zhao GY, Saberi A, Masutani M, Adachi N, Koyama H, de Murcia G, Takeda S (2006) Parp-1 protects homologous recombination from interference by Ku and ligase IV in vertebrate cells. EMBO J 25:1305-1314

Hope JC, Mense SM, Jalakas M, Mitsumoto J, Freyer GA (2006) Rqh1 blocks recombination between sister chromatids during double strand break repair, independent of its helicase activity. Proc Natl Acad Sci USA 103:5875-5880

Huang J, Dynan WS (2002) Reconstitution of the mammalian DNA double-strand break end-joining reaction reveals a requirement for an Mre11/Rad50/NBS1-containing fraction. Nucleic Acids Res 30:667-674

Hudson JJ, Hsu DW, Guo K, Zhukovskaya N, Liu PH, Williams JG, Pears CJ, Lakin ND (2005) DNA-PK_{cs}-Dependent Signaling of DNA Damage in *Dictyostelium discoideum*. Curr Biol 15:1880-1885

Ira G, Pellicioli A, Balijja A, Wang X, Fiorani S, Carotenuto W, Liberi G, Bressan D, Wan L, Hollingsworth NM, Haber JE, Foiani M (2004) DNA end resection, homologous recombination and DNA damage checkpoint activation require CDK1. Nature 431:1011-1017

Ivanov EL, Haber JE (1995) *RAD1* and *RAD10*, but not other excision repair genes, are required for double-strand break-induced recombination in *Saccharomyces cerevisiae*. Mol Cell Biol 15:2245-2251

Jung D, Giallourakis C, Mostoslavsky R, Alt FW (2006) Mechanism and control of V(D)J recombination at the immunoglobulin heavy chain locus. Annu Rev Immunol 24:541-570

Kabotyanski EB, Gomelsky L, Han JO, Stamato TD, Roth DB (1998) Double-strand break repair in Ku86- and XRCC4-deficient cells. Nucleic Acids Res 26:5333-5342

Karathanasis E, Wilson TE (2002) Enhancement of *Saccharomyces cerevisiae* end-joining efficiency by cell growth stage but not by impairment of recombination. Genetics 161:1015-1027

Kegel A, Martinez P, Carter SD, Astrom SU (2006) Genome wide distribution of illegitimate recombination events in *Kluyveromyces lactis*. Nucleic Acids Res 34:1633-1645

Koch CA, Agyei R, Galicia S, Metalnikov P, O'Donnell P, Starostine A, Weinfeld M, Durocher D (2004) XRCC4 physically links DNA end processing by polynucleotide kinase to DNA ligation by DNA ligase IV. EMBO J 23:3874-3885

Kooistra R, Hooykaas PJ, Steensma HY (2004) Efficient gene targeting in *Kluyveromyces lactis*. Yeast 21:781-792

Krappmann S, Sasse C, Braus GH (2006) Gene targeting in *Aspergillus fumigatus* by homologous recombination is facilitated in a nonhomologous end- joining-deficient genetic background. Eukaryot Cell 5:212-215

Kysela B, Doherty AJ, Chovanec M, Stiff T, Ameer-Beg SM, Vojnovic B, Girard PM, Jeggo PA (2003) Ku stimulation of DNA ligase IV-dependent ligation requires inward movement along the DNA molecule. J Biol Chem 278:22466-22474

Li L, Olvera JM, Yoder KE, Mitchell RS, Butler SL, Lieber M, Martin SL, Bushman FD (2001) Role of the non-homologous DNA end joining pathway in the early steps of retroviral infection. EMBO J 20:3272-3281

Li Z, Otevrel T, Gao Y, Cheng HL, Seed B, Stamato TD, Taccioli GE, Alt FW (1995) The XRCC4 gene encodes a novel protein involved in DNA double-strand break repair and V(D)J recombination. Cell 83:1079-1089

Ma JL, Kim EM, Haber JE, Lee SE (2003) Yeast Mre11 and Rad1 proteins define a Ku-independent mechanism to repair double-strand breaks lacking overlapping end sequences. Mol Cell Biol 23:8820-8828

Ma Y, Lu H, Tippin B, Goodman MF, Shimazaki N, Koiwai O, Hsieh CL, Schwarz K, Lieber MR (2004) A biochemically defined system for mammalian nonhomologous DNA end joining Mol Cell 16:701-713

Ma Y, Pannicke U, Schwarz K, Lieber MR (2002) Hairpin opening and overhang processing by an Artemis/DNA-dependent protein kinase complex in nonhomologous end joining and V(D)J recombination. Cell 108:781-794

Mahajan KN, Nick McElhinny SA, Mitchell BS, Ramsden DA (2002) Association of DNA polymerase μ with Ku and ligase IV: Role for pol μ in rnd-joining double-strand break repair Mol Cell Biol 22:5194-5202

Manolis KG, Nimmo ER, Hartsuiker E, Carr AM, Jeggo PA, Allshire RC (2001) Novel functional requirements for non-homologous DNA end joining in *Schizosaccharomyces pombe*. EMBO J 20:210-221

McVey M, Radut D, Sekelsky JJ (2004) End-joining repair of double-strand breaks in *Drosophila melanogaster* is largely DNA ligase IV independent. Genetics 168:2067-2076

Mimori T, Hardin JA (1986) Mechanism of interaction between Ku protein and DNA. J Biol Chem 261:10375-10379

Min B, Weinert BT, Rio DC (2004) Interplay between *Drosophila* Bloom's syndrome helicase and Ku autoantigen during nonhomologous end joining repair of P element-induced DNA breaks. Proc Natl Acad Sci USA 101:8906-8911

Modesti M, Hesse JE, Gellert M (1999) DNA binding of XRCC4 protein is associated with V(D)J recombination but not with stimulation of DNA ligase IV activity. EMBO J 18:2008-2018

Modesti M, Junop MS, Ghirlando R, van de Rakt M, Gellert M, Yang W, Kanaar R (2003) Tetramerization and DNA ligase IV interaction of the DNA double-strand break repair protein XRCC4 are mutually exclusive. J Mol Biol 334:215-228

Moshous D, Callebaut I, de Chasseval R, Corneo B, Cavazzana-Calvo M, Le Deist F, Tezcan I, Sanal O, Bertrand Y, Philippe N, Fischer A, de Villartay J (2001) Artemis, a novel DNA double-strand break repair/V(D)J recombination protein, is mutated in human severe combined immune deficiency. Cell 105:177-186

Nick McElhinny SA, Havener JM, Garcia-Diaz M, Juarez R, Bebenek K, Kee BL, Blanco L, Kunkel TA, Ramsden DA (2005) A gradient of template dependence defines distinct biological roles for family X polymerases in nonhomologous end joining. Mol Cell 19:357-366

O'Driscoll M, Gennery AR, Seidel J, Concannon P, Jeggo PA (2004) An overview of three new disorders associated with genetic instability: LIG4 syndrome, RS-SCID and ATR-Seckel syndrome. DNA Repair 3:1227-1235

Ooi SL, Shoemaker DD, Boeke JD (2001) A DNA microarray-based genetic screen for nonhomologous end-joining mutants in *Saccharomyces cerevisiae*. Science 294:2552-2556

Orii KE, Lee Y, Kondo N, McKinnon PJ (2006) Selective utilization of nonhomologous end-joining and homologous recombination DNA repair pathways during nervous system development. Proc Natl Acad Sci USA 103:10017-10022

Palmbos PL, Daley JM, Wilson TE (2005) Mutations of the Yku80 C terminus and Xrs2 FHA domain specifically block yeast nonhomologous end joining. Mol Cell Biol 25:10782-10790

Pardo B, Ma E, Marcand S (2006) Mismatch tolerance by DNA polymerase Pol4 in the course of nonhomologous end joining in *Saccharomyces cerevisiae.* Genetics 172:2689-2694

Pascal JM, O'Brien PJ, Tomkinson AE, Ellenberger T (2004) Human DNA ligase I completely encircles and partially unwinds nicked DNA. Nature 432:473-478

Perrault R, Wang H, Wang M, Rosidi B, Iliakis G (2004) Backup pathways of NHEJ are suppressed by DNA-PK. J Cell Biochem 92:781-794

Peterson SR, Kurimasa A, Oshimura M, Dynan WS, Bradbury EM, Chen DJ (1995) Loss of the catalytic subunit of the DNA-dependent protein kinase in DNA double-strand-break-repair mutant mammalian cells. Proc Natl Acad Sci USA 92:3171-3174

Pfeiffer P, Thode S, Hancke J, Vielmetter W (1994) Mechanisms of overlap formation in nonhomologous DNA end joining. Mol Cell Biol 14:888-895

Pitcher RS, Tonkin LM, Daley JM, Palmbos PL, Green AJ, Velting TL, Brzostek A, Korycka-Machala M, Cresawn SC, Dziadek J, Hatfull GF, Wilson TE, Doherty AJ (2006) Mycobacteriophage exploit prokaryotic NHEJ to facilitate genome circularization. Mol Cell: in press

Reddy YV, Ding Q, Lees-Miller SP, Meek K, Ramsden DA (2004) Non-homologous end joining requires that the DNA-PK complex undergo an autophosphorylation-dependent rearrangement at DNA ends J Biol Chem 279:39408-39413

Riballo E, Kuhne M, Rief N, Doherty A, Smith GC, Recio MJ, Reis C, Dahm K, Fricke A, Krempler A, Parker AR, Jackson SP, Gennery A, Jeggo PA, Lobrich M (2004) A pathway of double-strand break rejoining dependent upon ATM, Artemis, and proteins locating to gamma-H2AX foci. Mol Cell 16:715-724

Rivera-Calzada A, Maman JP, Spagnolo L, Pearl LH, Llorca O (2005) Three-dimensional structure and regulation of the DNA-dependent protein kinase catalytic subunit (DNA-PK_{cs}). Structure (Camb) 13:243-255

Romeijn RJ, Gorski MM, van Schie MA, Noordermeer JN, Mullenders LH, Ferro W, Pastink A (2005) Lig4 and Rad54 are required for repair of DNA double-strand breaks induced by p-element excision in *Drosophila.* Genetics 169:795-806

Sandoval A, Labhart P (2004) High G/C content of cohesive overhangs renders DNA end joining Ku-independent. DNA Repair 3:13-21

Sibanda BL, Critchlow SE, Begun J, Pei XY, Jackson SP, Blundell TL, Pellegrini L (2001) Crystal structure of an XRCC4-DNA ligase IV complex. Nat Struct Biol 8:1015-1019

Skalka AM, Katz RA (2005) Retroviral DNA integration and the DNA damage response. Cell Death Differ 12:971-978

Strang BL, Stow ND (2005) Circularization of the herpes simplex virus type 1 genome upon lytic infection. J Virol 79:12487-12494

Spagnolo L, Rivera-Caldeza A, Pearl LH, Llorca O (2006) Three dimensional structure of the human DNA-PK/Ku70/Ku80 complex assembled on DNA and its implications for DNA DSB repair. Mol Cell 22:511-519

Sugawara N, Ira G, Haber JE (2000) DNA length dependence of the single-strand annealing pathway and the role of *Saccharomyces cerevisiae RAD59* in double-strand break repair. Mol Cell Biol 20:5300-5309

Taccioli GE, Gottlieb TM, Blunt T, Priestley A, Demengeot J, Mizuta R, Lehmann AR, Alt FW, Jackson SP, Jeggo PA (1994) Ku80: product of the XRCC5 gene and its role in DNA repair and V(D)J recombination. Science 265:1442-1445

Takata M, Sasaki MS, Sonoda E, Morrison C, Hashimoto M, Utsumi H, Yamaguchi-Iwai Y, Shinohara A, Takeda S (1998) Homologous recombination and non-homologous end-joining pathways of DNA double-strand break repair have overlapping roles in the maintenance of chromosomal integrity in vertebrate cells. EMBO J 17:5497-5508

Teo SH, Jackson SP (2000) Lif1p targets the DNA ligase Lig4p to sites of DNA double-strand breaks. Curr Biol 10:165-168

Tseng HM, Tomkinson AE (2004) Processing and joining of DNA ends coordinated by interactions among Dnl4/Lif1, Pol4, and FEN-1. J Biol Chem 279:47580-47588

Tsukuda T, Fleming AB, Nickoloff JA, Osley MA (2005) Chromatin remodelling at a DNA double-strand break site in *Saccharomyces cerevisiae*. Nature 438:379-383

van Attikum H, Gasser SM (2005) ATP-dependent chromatin remodeling and DNA double-strand break repair. Cell Cycle 4:1011-1014

van Heemst D, Brugmans L, Verkaik NS, van Gent DC (2004) End-joining of blunt DNA double-strand breaks in mammalian fibroblasts is precise and requires DNA-PK and XRCC4. DNA Repair 3:43-50

Verkaik NS, Esveldt-van Lange RE, van Heemst D, Bruggenwirth HT, Hoeijmakers JH, Zdzienicka MZ, van Gent DC (2002) Different types of V(D)J recombination and end-joining defects in DNA double-strand break repair mutant mammalian cells. Eur J Immunol 32:701-709

Walker JR, Corpina RA, Goldberg J (2001) Structure of the Ku heterodimer bound to DNA and its implications for double-strand break repair. Nature 412:607-614

Wang H, Rosidi B, Perrault R, Wang M, Zhang L, Windhofer F, Iliakis G (2005) DNA ligase III as a candidate component of backup pathways of nonhomologous end joining. Cancer Res 65:4020-4030

Weller GR, Doherty AJ (2001) A family of DNA repair ligases in bacteria? FEBS Lett 505:340-342

Weller GR, Kysela B, Roy R, Tonkin LM, Scanlan E, Della M, Devine SK, Day JP, Wilkinson A, di Fagagna F, Devine KM, Bowater RP, Jeggo PA, Jackson SP, Doherty AJ (2002) Identification of a DNA nonhomologous end-joining complex in bacteria. Science 297:1686-1689

West CE, Waterworth WM, Story GW, Sunderland PA, Jiang Q, Bray CM (2002) Disruption of the *Arabidopsis* AtKu80 gene demonstrates an essential role for AtKu80 protein in efficient repair of DNA double-strand breaks *in vivo*. Plant J 31:517-528

Wilson S, Warr N, Taylor DL, Watts FZ (1999) The role of *Schizosaccharomyces pombe* Rad32, the Mre11 homologue, and other DNA damage response proteins in non-homologous end joining and telomere length maintenance. Nucleic Acids Res 27:2655-2661

Wilson TE (2002) A genomics-based screen for yeast mutants with an altered recombination/end-joining repair ratio. Genetics 162:677-688

Wilson TE, Grawunder U, Lieber MR (1997) Yeast DNA ligase IV mediates non-homologous DNA end joining. Nature 388:495-498

Wilson TE, Lieber MR (1999) Efficient processing of DNA ends during yeast nonhomologous end joining. Evidence for a DNA polymerase β (Pol4)-dependent pathway. J Biol Chem 274:23599-23609

Wilson TE, Topper LM, Palmbos PL (2003) Non-homologous end-joining: bacteria join the chromosome breakdance. Trends Biochem Sci 28:62-66

Wu X, Wilson TE, Lieber MR (1999) A role for FEN-1 in nonhomologous DNA end joining: the order of strand annealing and nucleolytic processing events. Proc Natl Acad Sci USA 96:1303-1308

Yakovleva L, Shuman S (2006) Nucleotide misincorporation, 3'-mismatch extension, and responses to abasic sites and DNA adducts by the polymerase component of bacterial DNA ligase D. J Biol Chem 281:25026-25040

Yu Y, Wang W, Ding Q, Ye R, Chen D, Merkle D, Schriemer D, Meek K, Lees-Miller SP (2003) DNA-PK phosphorylation sites in XRCC4 are not required for survival after radiation or for V(D)J recombination. DNA Rep 2:1239-1252

Zhu C, Mills KD, Ferguson DO, Lee C, Manis J, Fleming J, Gao Y, Morton CC, Alt FW (2002) Unrepaired DNA breaks in p53-deficient cells lead to oncogenic gene amplification subsequent to translocations. Cell 109:811-821

Zhu H, Nandakumar J, Aniukwu J, Wang LK, Glickman MS, Lima CD, Shuman S (2006) Atomic structure and nonhomologous end-joining function of the polymerase component of bacterial DNA ligase D. Proc Natl Acad Sci USA 103:1711-1716

Zhu H, Shuman S (2005) Novel 3'-ribonuclease and 3'-phosphatase activities of the bacterial non-homologous end-joining protein, DNA ligase D. J Biol Chem 280:25973-25981

Wilson, Thomas E.

Department of Pathology, University of Michigan Medical School, Ann Arbor MI, Rm. 2065 BSRB, 109 Zina Pitcher Place, Ann Arbor, MI 48109-2200, USA

wilsonte@umich.edu

Abbreviations

IR: illegitimate recombination
NHEJ: nonhomologous end joining
HR: homologous recombination
DSB: double-strand break
MMEJ: microhomology-mediated end joining
CHO: Chinese hamster ovary
DNA-PK: DNA-dependent protein kinase
FHA: forkhead associated
SSA: single-strand annealing

SSBR: single-strand break repair
MRX: Mre11-Rad50-Xrs2
MRN: Mre11-Rad50-Nbs1

Index

Printing: Krips bv, Meppel
Binding: Stürtz, Würzburg